NIELS BOHR
COLLECTED WORKS
VOLUME 4

NIELS BOHR: COLLECTED WORKS, VOLUME 3

ERRATA

	Instead of	Read
p. [9] line 5 from bottom	p. [167]	p. [8]
p. [13] first footnote	p. [682]	p. [686]
p. [30] line 12 from bottom	p. [23]	p. [24]
p. [34] line 12 from bottom	. . ."	. . ."†
p. [40] line 16	Wolfkehl	Wolfskehl
p. [42] first footnote	and [674]	, [674] and [661]
pp. [211] and [212]	$\dfrac{c^3h^3K^3(M+m)}{\pi^2N^2e^4Mm}$	$\dfrac{2c^3h^3K^3(M+m)}{\pi^2N^2e^4Mm}$
p. [212] eq. (8)	$\dfrac{2\pi^2e^4NmM}{h(M+m)}$	$\dfrac{2\pi^2e^4N^2mM}{ch^3(M+m)}$
p. [289] line 13	af at	af
p. [292] line 6	Brøndsted	Brønsted
p. [300] line 15 and footnote	Brøndsted	Brønsted
p. [331] line 18 from bottom	failute	failure
p. [366] line 8	E_3	E_1
p. [377] lines 1 and 2 of second footnote	paragraph	section
p. [503] line 7 from bottom	be andeln	behandeln
p. [526] line 18	xei	bei
p. [532] first footnote	σ_0	ω_0
p. [533] eq. (5)	$\dfrac{2\pi^2N^2e^4M\mu}{M+\mu}\dfrac{1}{\mathcal{J}_0^2}$	$-\dfrac{2\pi^2N^2e^4M\mu}{M+\mu}\dfrac{1}{\mathcal{J}_0^2}$

NIELS BOHR
COLLECTED WORKS

VOLUME 4

THE PERIODIC SYSTEM (1920 - 1923)

EDITED BY

J. RUD NIELSEN

GEORGE LYNN CROSS RESEARCH PROFESSOR EMERITUS

UNIVERSITY OF OKLAHOMA

1977

NORTH-HOLLAND PUBLISHING COMPANY

AMSTERDAM · NEW YORK · OXFORD

PUBLISHERS:

NORTH-HOLLAND PUBLISHING COMPANY
AMSTERDAM . NEW YORK . OXFORD

SOLE DISTRIBUTORS FOR THE U.S.A. AND CANADA:

ELSEVIER NORTH-HOLLAND, INC.
52 VANDERBILT AVENUE NEW YORK, N.Y. 10017

Library of Congress Catalog Card Number: 70-126498
ISBN North-Holland, Collected Works: 0 7204 1800 3
ISBN North-Holland, Volume 4: 0 7204 1804 6

Printed in The Netherlands

FOREWORD

As stated in the foreword to Volume 3, Niels Bohr's researches in the period from 1918 to the end of 1923 had a double aim: to develop a consistent and adequate quantum theory and to determine the structure of atoms and thus explain the physical and chemical properties of the elements. In Volume 3, the papers and manuscripts dealing primarily with Bohr's elaboration of the quantum theory were presented. The present volume contains his papers and manuscripts treating the problem of atomic structure. These are closely related to the papers included in Volume 3; for, as Bohr stressed repeatedly, theory is necessary to understand phenomena, while, on the other hand, a close study of the phenomena is required to suggest the theory.

Although the shortcomings of the quantum theory became increasingly apparent to Bohr in the period from 1920 to 1923, his explanation of the structure of atoms reached a peak in this period, and exerted a profound influence upon the development of physics and chemistry.

Since some of the most important papers included in the present volume were first presented as lectures, and have retained the lecture form, mathematical computations (mostly rather simple), supplementing the painstaking analysis of experimental evidence, are generally not included in these papers.

As in the case of Volume 3, the present volume consists of two parts. Part I contains the published papers and some unpublished manuscripts, preceded by an introduction. Part II gives the complete original texts of the letters quoted in the introduction. The few Danish letters are followed by translations. While the parts of these letters that are not quoted may have little or no bearing on Bohr's work on atomic structure, they bear witness to his participation in the discussion of most of the problems challenging physicists in the early nineteen twenties.

In the preparation of this volume, we have been helped in various ways by the expert advice of Professor Duane H. D. Roller, E. Rüdinger and K. Stolzenburg. Mrs. Lise Madsen and the typing service of the University of Oklahoma have provided expert typing. The work of Mrs. S. Hellmann in making available and preparing the material from the Niels Bohr Archive has been invaluable as have her editorial advice and extensive help with the reading of the proofs. The printing has again been masterly supervised by North-Holland Publishing Company.

J. R. N.

The Institute wishes to acknowledge the continued generous support of the Carlsberg Foundation, which has been of decisive importance for the preparation of the present as well as the previous volumes of these Collected Works.

The Niels Bohr Institute

CONTENTS

PART II: SELECTED CORRESPONDENCE 1920–1924

PART I

ATOMIC STRUCTURE

INTRODUCTION

by

J. RUD NIELSEN

In his Introduction to *On the Quantum Theory of Line Spectra*, the first two parts of which were published in 1918*, Niels Bohr announced that it would contain two additional parts, a third part dealing with the theory of the spectra of elements with several electrons and a fourth part treating the structure of the atoms of such elements and of molecules. Although Bohr spent much effort on revising his early draft of Part III, the development of the quantum theory of series spectra was so rapid in the years following 1918 that he was unable to complete this task, and Part III was published in its original form in 1922**. However, a large part of Vol. 3 of Bohr's Collected Works deals with the envisioned subject matter. Part IV of *On the Quantum Theory of Line Spectra* was never completed***. In a sense, the present volume may be said to contain part of the material that was more or less dimly envisioned when the subject matter of Part IV was announced in 1918.

Bohr's progress was slow to begin with, but toward the end of 1920 significant advances commenced to occur at a rapid rate. Before describing these, we shall attempt to illustrate, by means of a few excerpts from Bohr's correspondence, what his views were just before this development began to take place.

* See Vol. 3, p. [65].

** See Vol. 3, p. [167].

*** For fragmentary drafts of it, see Vol. 3, p. [185].

On June 18, 1920, Rudolf Ladenburg wrote to Bohr and told him that, at a Bunsen Congress in Halle, he had given a report on atomic structure and the periodic system,

> "and in the discussion the question was raised about the difference between your attempt (from the year 1913) to determine the arrangement of electrons in the elements and the arrangement proposed by Kossel in 1916. May I now ask what you think of Kossel's idea that the halogens F, Cl, Br, I in their simple compounds (say with the alkalis) by capture of a single electron, just as the alkalis by losing an electron, assume the particularly stable configuration of the noble gases with 8 electrons in the outer shell ... indeed, Born's calculation of the electron affinity of the halogens and sulphur appears to me to be a strong support for this conception ... Or do you still prefer your old view of two electrons in the outer shell of the noble gases? ..."

Bohr answered Ladenburg on July 16, 1920. Since his reply expresses well the views he held around the middle of 1920, it is given here, in English translation:

Letter from Niels Bohr to R. Ladenburg (16 July 20) German text p. [711]

Dear Professor Ladenburg:

In reply to your kind letter, I must begin by saying that I regret very much that, compelled by circumstances, I only now have occasion to answer it. With regard to your question as to my views about the constitution of the atoms of the elements, I must confess that I do not consider any conception sufficiently assured as yet to make it possible to take a definite standpoint. In any case, the considerations in my first papers, as far as this point is concerned, should only be regarded as a tentative orientation. The difficulties of the problem lie particularly in the rational utilization of the different proposed electron configurations to explain the chemical properties of the elements in question. In fact, this depends not only on the geometrical character of the configuration, but above all on its stability properties. Besides the difficulties, emphasized from many sides, which the assumption of electron rings has in explaining the properties of crystals, band spectra, ionization potentials, etc., it also seems that this assumption must be abandoned simply because of insufficient stability, and that we are forced to reckon with far more complicated motions of the electrons in an atom. However, until some definite case or other of such a motion has been thoroughly investigated, especially as to its stability properties, it seems difficult to express a final opinion as to the utilization of such motions to the problem of the chemical properties.

As regards your special question about the constitution of the atoms of the noble gases, I believe that considerations such as those of Kossel may not present such weighty arguments as one is inclined to believe at a first glance, and it seems to me that this question can be settled only after a much more thorough investigation, which so far is still lacking.

Apologizing for the sketchy form of these remarks which, I hope, nevertheless express the trend of my views, I remain

with kind regards
sincerely yours,
N. Bohr

On August 27, 1920, Ladenburg thanked Bohr for his letter and wrote:

"... I quite agree with you that stability considerations are decisive. But, as long as one cannot say anything certain about the details of the electronic motions, I believe that, from the enormous experimental material on the chemical and physical properties of the elements, and with the aid of the periodic system, one can nevertheless draw some provisional conclusions and state some surmises, i.e., working hypotheses ... For the present, I find the chemical reasons for the stability of the noble gas configurations convincing ..."*

On September 29, 1920, Kramers informed Ladenburg that Bohr had for some time had a draft lying for a letter to him but, since he had not found time to revise this draft and had now left on a brief vacation, he had asked Kramers to send the draft. In this draft Bohr wrote, in part:

"In reply to your kind letter, I should like to remark first that ... although I am of the opinion that ... we do not yet know enough about the conditions of motion and stability to reach a decision on the question of electron configurations, I agree with your view, of course, that it is extremely useful to try to work with the analogies offered by the chemical phenomena, and in this respect I value the work of Kossel and others very highly; however, I feel myself strongly cautioned by the conviction that ultimately the whole question of explaining the chemical stability cannot be regarded as primarily a geometrical problem but must rather be conceived as a dynamical one ..."**

* For the complete text, see p. [712].

** For the text of this draft, see p. [713].

As mentioned in Vols. 2 and 3*, in the autumn of 1916 Bohr and Kramers had carried out extensive calculations on the structure and mechanics of the helium atom. In particular, they had found two different classes of stationary states which, they believed, could explain the existence of the two observed non-combining term systems of this element. On his visit to Leiden in April, 1919, Bohr had given a brief account of this work, but its publication had been delayed.

In 1919, A. Landé published a paper on the structure of helium (Phys. Zeitschr. **20** (1919) 228) which, at least superficially, resembled the work of Bohr and Kramers. This led to a renewed correspondence between Bohr and Landé. Early in 1920, Landé sent Bohr the proof of an excerpt of his "Habilitationsschrift", and Bohr thanked him in a letter of February 24, in which he discussed briefly the difference between Landé's work on helium and that of Kramers and himself. He wrote:

> "... Our publication has unfortunately been delayed longer than I thought then**, because Kramers' recovery has taken longer than I had hoped, and because in the meantime I have had to be busy with other work. Now I am again working on these matters and I hope that ... the joint work of Kramers and myself on the spectrum of helium, which has been lying unprinted so long, will soon appear ..."***

In April, 1920, Bohr and Landé met in Berlin where Bohr delivered a lecture before the German Physical Society†. On July 25 Landé sent Bohr a letter in which he enclosed a typed sheet headed "On Cubical Atoms ("Würfelatome"), Periodic System and Molecular Structure" which, he wrote,

> "... summarizes my investigations of the dynamics of spatial atomic structures and draws a series of consequences therefrom ... If you are interested I would be very pleased to report on it to you personally in Copenhagen ..."††

On July 29 Bohr invited Landé to give a lecture before the Physical Society of Copenhagen. Landé came in October and lectured on atoms with spatial symmetry, and he discussed the problem of helium with Bohr and Kramers. Although Bohr could not accept Landé's calculations, the discussions with Landé undoubtedly provided an important impetus to Bohr's work on the structure of atoms with several electrons.

* See Vol. 2, Part III, Sect. 6 and Vol. 3, p. [3].

** [Bohr referred to a long letter to Landé, dated June 26, 1919.]

*** For the text of this letter, see p. [719].

† See Vol. 3, p. [241].

†† For the text of this letter, see p. [720].

1. BOHR'S LECTURE BEFORE THE PHYSICAL SOCIETY OF COPENHAGEN (15 DECEMBER 1920)

At about the time of Landé's visit, Bohr mentioned to some of his colleagues in Copenhagen that he had found an explanation of the metastable state of helium. and on December 15, 1920, he gave a lecture before the Physical Society. This lecture represented a landmark in Bohr's efforts to explain the structure of atoms with several electrons. It was never published; however, in the Niels Bohr Archive there exists a handwritten manuscript, apparently dictated to O. Klein after the delivery of the lecture. This manuscript is fairly rough; part of it is written in pencil, two pages are missing, and the manuscript has evidently not been corrected by Bohr. In addition, there exist two fragments marked "Fysisk Forening 1920". One of these, consisting of three typed and one handwritten pages with the heading "Some Considerations of Atomic Structure" ("Nogle Betragtninger over Atomernes Bygning"), is believed to have been intended as an introduction to the lecture and may have been written before the lecture was held. The other fragment "On our Knowledge of Atoms" ("Om vort Kendskab til Atomerne"), consisting of $2\frac{1}{2}$ typed pages, may have had the same purpose and was, most likely, written after the lecture. Translations of the two fragments and the manuscript are given on pp. [44], [46], and [48].

In the first half of his lecture, Bohr discussed previous work on atomic structure and the interpretation of optical and X-ray spectra, and he mentioned some open questions and difficulties. He recognized the need for atomic models of spatial symmetry:

> "... it is of great significance that, as pointed out by Born and Landé, there exist possible electronic motions with cubical symmetry which are consistent with the assumptions of the quantum theory and which appear to allow a determination of stationary states from quantum-theoretical principles. Born and Landé were led to the study of such possible motions not from a consideration of the chemical properties of the elements but from their investigation of crystal structure and, especially, of the compressibility of crystals."

He acknowledged his indebtedness to Landé:

> "In a paper recently published by Landé, he points out that agreement in order of magnitude can be attained in many cases by introducing the assumption, similar to that used by Sommerfeld to explain the origin of the L-radiation, that the motion of the outer electrons in the normal state corresponds to a stationary state of the point atom characterized by $n = 2$ ... After getting

acquainted with this work ⟨on the structure of diamond⟩ through a lecture given by Landé before the Physical Society of Copenhagen in October this year, it has occurred to me that it might be possible from simple points of view to give a rational explanation of this at a first glance strange assumption."

Bohr then proceeded to the second part of his lecture which began with some simple calculations showing that certain states might not be mechanically possible. This explained the metastable state of orthohelium and led to the assumption that the valence electrons in the lithium and sodium atoms are bound in orbits for which $n = 2, k = 1$. He pointed out that this meant that the quantum numbers of the stationary states of sodium given by him in a diagram in his lecture in Berlin in April (Z. Phys. **2** (1920) 423) must be revised. He also pointed out that electrons in elliptical orbits for which $n > 1$ may penetrate into the region of the inner electrons and thus become more firmly bound than electrons having the same quantum numbers in hydrogen. He spoke about a coupling between outer and inner electrons and stressed the importance of this new and somewhat vague concept. He surmised that the carbon atom contains two electrons in crossed 1-quantic orbits, as in parhelium, and four electrons in 2-quantic orbits, probably having a tetrahedral configuration. He assumed that these four electrons pass their perihelions at regular intervals and not simultaneously, as in the models of Born and Landé:

"... First of all, it must be emphasized that these authors have given no thought to the influence of the coupling between different electron layers in the atom and, hence, have based their arguments for possible configurations solely on symmetry properties ... However, ... we see in the most striking manner that this coupling plays a decisive role for the properties of the elements ... When, as in Landé's and Born's papers, no special regard is paid to the presence of the inner electrons, it is impossible to give any reason for excluding a state analogous to the normal state of hydrogen ..."

At the end of the lecture, Bohr considered very briefly the structures of nitrogen, oxygen, and fluorine. He assumed that the first two electrons will move in equivalent one-quantic orbits with different orientations in space and surmised that the remaining electrons will move in equivalent two-quantic orbits, but he realized that five, six, or seven such orbits cannot be distributed in as symmetrical a manner as is the case for two, three, or four such electrons.

"... However, it is difficult to decide what significance should be attributed to this fact; for, as already mentioned, we consider the existence of a coupling

> between the inner and the outer electrons to be of primary importance and the symmetry of the outer electrons to play a secondary role ..."

It is interesting that he did not then mention anything about the structure of neon, the last element in the second period.

The principal key to the new conceptions in Bohr's lecture of December 15, 1920, was the explanation he had found for the exclusion of certain states and for the presence in the normal state of atoms of electron orbits with principal quantum number larger than one. As shown by a letter to Landé of February 14, 1921*, he had already discussed this explanation with Landé during the latter's visit to Copenhagen in October 1920. Among the new conceptions, perhaps the most important was the recognition that electrons in elliptic orbits for which $n > 1$ may penetrate into the region of the inner electrons and be coupled to these.

It was Bohr's intention to publish a comprehensive paper on atomic structure based on the new insights expressed in this lecture. Thus, in a letter to Irving Langmuir, dated December 3, 1920, he wrote:

> "... In this letter I allow me to enclose a reprint of a lecture, which I gave in Berlin last April. It contains a general discussion of the problem of series spectra, but do⟨es⟩ only touch very lightly on the internal structure of the atom, which however will be dealt with in greater details in the new paper, of which I hope to send you a reprint very shortly ..."**

In his letter of Feb. 14, 1921, to Landé, mentioned above, he wrote:

> "... I believe now, as I told you then, that one cannot merely achieve a natural explanation of multiple quantic orbits in the normal state of an atom, but an elaboration of the same argument has led me in a natural way to the assumption of a quite definite arrangement of these orbits in the atom. It is true that this arrangement exhibits a pronounced spatial symmetry, but it does not belong to the type of motion proposed by you; rather it can perhaps be best characterized by the statement that each electron in the atom in a certain sense moves independently of the others. I have developed this viewpoint to a theory of atomic structure ... and I intend to report these results, which are of a rather detailed character, in a comprehensive publication very soon ..."

In a letter to Landé of February 28, 1921, Kramers stated that it was Bohr's intention to publish the paper in German***.

* For the text of this letter, see p. [721].

** For the text of this letter, see p. [725].

*** For the text of this letter, see p. [723].

2. ATOMIC STRUCTURE. LETTER TO "NATURE" (24 MARCH 1921)

At this time, Bohr was very busy with the arrangement of the new Institute for Theoretical Physics in Copenhagen. He was overworked and had trouble with his health during most of 1921. Hence, he did not succeed in completing the intended publication. However, on February 14, 1921, he wrote a rather long letter to "Nature", which appeared on March 24 (Nature **107** (1921) 104), in which he gave a brief account of the main conclusions of his lecture of December 15, 1920, and of results obtained in the two months since that time. The choice of "Nature" as a medium of publication was prompted by a discussion of atomic structure carried out there at the time. On November 25, 1920, Norman Campbell had pointed out apparent inconsistencies between conceptions of atomic structure based on the quantum-theoretical interpretation of series spectra and those developed to account for the physical and chemical properties of the elements. He suggested that this apparent inconsistency might not be real, but might arise from the formal character of the principles of the quantum theory, in particular of the correspondence principle. Bohr, at the beginning of his letter, took issue with this view

> "and in this letter I should like briefly to indicate how it seems possible by an extended use of this principle to overcome certain fundamental difficulties hitherto involved in the attempts to develop a general theory of atomic constitution based on the application of the quantum theory to the nucleus atom."

While he had not in his lecture discussed the arrangement of electron orbits in any element beyond fluorine, he now suggested the following electron arrangements in all of the inert gases:

Helium (2_1)	Krypton $(2_1\,8_2\,18_3\,8_2)$
Neon $(2_1\,8_2)$	Xenon $(2_1\,8_2\,18_3\,18_3\,8_2)$
Argon $(2_1\,8_2\,8_2)$	Niton $(2_1\,8_2\,18_3\,32_4\,18_3\,8_2)$,

> "where the large figures denote the number of electrons in the groups starting from the innermost one, and the small figures the total number of quanta characterising the orbits of electrons within each group."

He assumed that in the family of the rare earths

> "we may be ... witnessing the successive formation of an inner group of thirty-two electrons at that place in the atom where formerly the corresponding group possessed only eighteen electrons."

In the case of the iron, palladium and platinum families, stages of the formation of groups of eighteen electrons occur, but conditions are more complicated because the group lies closer to the surface of the atom. At the end of the note Bohr stated that a fuller account would be published soon, and he indicated that this would treat other properties of the elements, in particular the magnetic properties.

This letter to "Nature" is reproduced on p. [71].

The publication of this letter to "Nature", on March 24, 1921, gave rise to a considerable amount of correspondence. Landé, to whom Bohr had sent a manuscript of the letter, wrote on February 21:

> "Many thanks for sending me your manuscript which will be of the very greatest significance for the future atomic theory. It seems to me that, until the appearance of your detailed account, it makes no sense at all to work theoretically on the atomic theory. Therefore, I should like to propose that you let me have your letter to "Nature" for a while, in order that I may translate it into German and let it reach Zeitschrift für Physik as soon as possible, so that also the German physicists can be informed as quickly as possible about your new advances ..."*

A few days later (on February 28) Kramers replied to Landé's proposal on Bohr's behalf:

> "... With regard to your kind offer to translate the letter to "Nature" into German and, for example, publish it in Zeitschrift für Physik, he believes that this would not be appropriate. In fact, the letter should not be considered as more than a contribution to the discussion of atomic structure carried out at present in "Nature", and it would, if it were to be published separately elsewhere, appear far too pretentious, especially since the arguments for the new conceptions contained in it have practically not been developed at all. Besides, Professor Bohr hopes within a few weeks to complete a detailed German article about the subject in question ... "**

On February 22, 1921, Bohr sent Sommerfeld a copy of his letter to "Nature". Sommerfeld acknowledged its receipt on a postcard on March 7:

> "Many thanks for your letter to "Nature". It evidently represents the greatest advance in atomic structure since 1913. As you did in 1916, I have drawn the

* For the text of this letter, see p. [722].

** For the text of this letter, see p. [723].

consequence from it that I have withdrawn part of a manuscript that I had sent to Zeitschr. Phys. I had expressed some thoughts there about how the quantum numbers, which increase systematically in the interior of the atom, might decrease at its surface. But you have now substantiated that firmly ..."*

It is an interesting fact that Bohr shortly afterwards reversed his views on this point.

On April 25, Sommerfeld wrote a long letter to Bohr. It dealt in part with Bohr's failing health which had forced him to give up a trip to Göttingen where Sommerfeld had hoped to get more details about Bohr's new points of view, in order that he might incorporate them in a new edition of his book *Atombau und Spektrallinien*. In this letter he wrote:

"... To be sure, I am fully convinced that your road is the right one; if, as it appears, you can reconstruct mathematically the numbers 2, 8, 18, ... of the elements in the periods, that is, in fact, the fulfilment of the boldest hopes of atomic physics. However, I am enough of a heretic to believe that some day this will be possible in a less formal, more uniform manner ..."**

On April 18, 1921, A. Rubinowicz after thanking Bohr for reprints, especially of the Letter to "Nature", wrote:

"... Even if your letter to "Nature" does not let us see the finer details of the wonder-world which you now again have opened up to us, but only foreshadows them, I believe nevertheless to have gathered so much from it that by your work you once more lead us to a summit from which we now again can view a wide field as a clear and beautiful harmonic unity ..."***

On February 7, 1921, E. Schrödinger, who had then just succeeded Cl. Schaefer as professor of theoretical physics at Breslau, wrote to Bohr (in English) and asked for reprints of the letter to "Nature" and others of Bohr's papers. He enclosed a reprint of his own (Z. Phys. **4** (1921) 347). Toward the end of his letter he wrote:

"... I keep a college ⟨i.e., a seminar⟩ on the atom model just now, and so you may imagine how thankful I would be to possess your important papers, especially because Sommerfeld's beautiful work (which is of course our standard

* For the text of this postcard, see p. [740].

** For the text of this letter, see p. [740].

*** For the text of this letter, see p. [733].

work at present in Germany in all questions of the atom) differs, as you will know, in many respects widely from your own most fascinating mode of viewing the problems ..."*

Bohr answered this letter on June 15. Referring to Schrödinger's paper, he wrote:

"... Besides, I have myself employed exactly the same considerations a considerable time ago; for example, in a lecture before the Copenhagen Physical Society in December last year, in which I presented my results on the explanation of the periodic system, indicated in the note to "Nature", I reported as illustration the value which the radius of such a spherical distribution would have to have in order to accurately reproduce the terms of the sharp subordinate series of lithium. Because of overwork, I have unfortunately not yet been able to publish a comprehensive account of the contents of this lecture. However, I am now trying to begin the work again and hope to complete such a publication in the next few weeks ..."**

It is natural that chemists should be particularly interested in Bohr's note in "Nature". On June 25, 1921, K. Fajans wrote to Bohr:

"... The announcement in your letter to "Nature" of the solution of the riddle of the periodic system has filled us chemists with particularly great expectations. When we once get oriented about the electron configurations in a large number of the elements, I believe that many questions of chemistry will appear in quite a new light. In fact, just the simple assumption that all the ions of the alkali halides (except Li^+) have the same structure which, through your assertion of an 8-shell in all of the noble gases, may now be regarded as demonstrated, has led to many surprising clarifications ..."***

On July 19, 1921, Bohr thanked W. Kossel for a reprint and wrote:

Letter from Niels Bohr to W. Kossel (19 July 21) German text p. [707]

"... Perhaps you have heard from Sommerfeld that I have not felt well because of overwork. However, I have now resumed my work, and I hope in the next few weeks to complete a report of my views of atomic structure which I reported briefly in my note in "Nature". I need not tell you how much I rely on your ideas in this work. If I have long had reservations about your valency theory,

* For the text of this letter, see p. [737].

** For the text of this letter, see p. [738].

*** For the text of this letter, see p. [691].

it was because I believed I saw great difficulties in reconciling it in details with the picture of atomic structure. However, I now believe that I can see how it is possible to present the valency theory in the nicest harmony with the principles that underlie the applications of the quantum theory, and I am looking forward to soon being able to send you my detailed account ..."

At the end of his letter Bohr reminded Kossel of their meeting in Munich in the summer of 1914, and he expressed his pleasure that Kossel had recently moved to Kiel, not far from Copenhagen.

Kossel replied to Bohr's letter on August 15.

"... Now we are all extremely anxious to learn about the remarkable advances that you have announced in your letter to "Nature". I need not say how particularly important they are to me; I have studied your letter very carefully and have prepared a review of it for the "Physikalische Berichte". Hence, I am especially happy that we may now expect your complete exposition of this subject very soon, and it is a great pleasure to me that you find the conclusions that I tried to draw from chemical and other phenomena useful from your general point of view ..."*

3. OUR PRESENT KNOWLEDGE OF ATOMS (1921)

Either in late 1920 or early 1921 Bohr wrote an article for general readers with the title "Unsere heutige Kenntnis vom Atom"; it was published in "Die Umschau" on April 30, 1921 (Die Umschau **25** (1921) 229). A translation of it is given on p. [83].

4. SECOND HALF OF THE SOLVAY REPORT (1921)

On March 3, 1921, the Institute for Theoretical Physics of the University of Copenhagen was dedicated (see Vol. 3, p. [283]). Bohr had undertaken to attend the Solvay Congress in Brussels in April, 1921, and to give a report "On the Application of the Quantum Theory to Atomic Problems". However, because of overwork and ill health, he was unable to go to Brussels (and also had to cancel trips to Göttingen and Cambridge). Bohr sent a manuscript of the first part of his report to P. Ehrenfest who presented it to the Congress (see Vol. 3, p. [364]).

* For the text of this letter, see p. [708].

He promised in several letters to submit a second part of this report to H. A. Lorentz, but, as he worked on it, it grew to an impractical length, and on September 1, 1921, he informed Lorentz that he would not be able to complete this part and suggested that only the first part be published in the Proceedings of the Congress, and this was done. (See Vol. 3, p. [364].) On October 10, he informed Ehrenfest that he was now working on a large new paper on atomic structure (see Vol. 3, p. [630]).

In the Niels Bohr Archive, there are several manuscripts representing parts of Part B of the Solvay report, or of the large new paper mentioned in the letter to Ehrenfest. Although these manuscripts are not dated, their contents indicate that they were written during the summer and autumn of 1921. One of them, consisting of six typed pages, is headed "Brief extract of contents of second half of Professor Bohr's report". This heading, and a parenthesis at the bottom of the last page, indicate that it has been written by, or perhaps dictated to, one of Bohr's collaborators, probably Kramers. It represents a plan for the report in question. On the last page are two references to Bohr's letter to "Nature" of March 24, and the same arrangements of electron orbits in the inert gases are given as in that letter. There seems to be an inconsistency between statements on two of the pages. On p. 3 it is assumed that the series electron in the normal state of the alkali metals describes a two-quantic orbit ($n = 2$, $k = 1$). This is repeated on the last page where it is stated explicitly that in sodium

> "... the valence electron will be bound in a two-quanta orbit again ($n = 2, k = 1$) just as in lithium."

On the other hand, on the previous page it is stated that in atoms of large nuclear charge the eleventh electron will move in a 3-quantic orbit. This manuscript is reproduced on p. [91].

Another manuscript is headed "B. Constitution of Atoms. § 1. Interpretation of the series spectra of the elements. General considerations of the process of binding of electrons by atoms." This manuscript, consisting of 32 typed pages, begins as follows:

> "Having ⟨discussed⟩ the considerations of the general principle of the quantum theory, we shall now return to the problem, mentioned in the beginning of this report, of the arrangement and motions of the particles in the actual atoms ..."

This manuscript is clearly intended as the first section of the second part of the Solvay report. Thus, on the second page reference is made to "the first part of this report", and on a couple of later pages this is referred to as "Part A". Furthermore, reference is made by number to equations in the first part of the Solvay report.

The present manuscript covers such topics as: the origin of X-ray spectra, the theory of series spectra, central motions, penetrating orbits, and an attempt to explain the doublet structure by the assumption of a precession of elliptical orbits. The quantum numbers of the valence electron in the normal state of the alkali atoms are assumed to be $n = 2$, $k = 1$. The manuscript appears to be an early version of the two papers *On the Application of the Quantum Theory to Atomic Structure. I and II*, written some two years later (see Vol. 3, p. [455] and p. [501]). It closes with the following sentence:

> "As we shall see in the following chapters, a consideration of the formation of atoms by successive binding of the electrons leads us to a view as regards the constitution and stability of the normal state of the atom which, at the same time as it seems to offer a suitable basis for the interpretation of the periodic law of the chemical properties of the elements, actually fulfils this claim ⟨i.e., the simultaneous presence of orbits of various shapes⟩."

This manuscript is reproduced on p. [100].

A third manuscript, consisting of 22 typed pages, is marked "§2. Helium". The greater part of this manuscript is devoted to an account of the early work on the structure of the helium atom by Bohr and Kramers. The results of this work are given with a minimum of mathematics. Formulas are given without derivation. The metastable state of orthohelium is explained, and the discovery of the ultraviolet helium line at 585 Å by T. Lyman and H. Fricke is mentioned. (This result was published in Science **56** (1922) 167, but Bohr had received the information before its publication, probably from Fricke.) Attention is then given to a new model of the normal state of the helium atom, assumed to contain two equivalent, approximately circular, one-quantum orbits with planes making an angle of 60° with each other. It is mentioned that Kramers is investigating this model, and a preliminary value is given for its energy.

As indicated by a number of typographical errors, this manuscript has not been corrected. References to formulas in this and previous sections of the Solvay report have either been omitted or are indicated by empty parentheses. On p. 11 of the manuscript, a formula given as $\sigma = \omega \frac{9}{32} h/P^5$ is wrong dimensionally. It has been replaced by $\sigma = \omega \frac{9}{16} h^3/P^3$ (see eq. (73) of the fourth lecture held in Göttingen in June 1922, p. [383]), and a couple of the following formulas have been corrected accordingly. Typographical errors have been corrected, and an attempt has been made to number the formulas consistently. This manuscript is given on p. [122].

A fourth manuscript, consisting of 11 typed and some 25 handwritten pages, is headed: "On the elements within the second* period of the periodic system." It seems to be older than the other manuscripts in this group. About three-quarters of the manuscript deals with lithium. The spectrum of this element is discussed in some detail. It is pointed out that all Li terms, except those of the S-sequence, deviate very little from the hydrogen terms. It is concluded that this arises from the circumstance that the third electron to be bound in the lithium atom cannot be accommodated in a one-quantum orbit but

> "... in the normal state of lithium the orbit of the last bound electron will correspond to an intermediate stage of the binding of an electron in one of these atoms ⟨i.e., hydrogen and helium⟩."

No definite value of the principal quantum number is suggested for this orbit, which is assumed to consist of a number of equal loops, each of which is nearly a Kepler ellipse.

On the last two pages of the typed manuscript the binding of the fourth, fifth and sixth electrons is briefly discussed. Since no spectra of beryllium, boron and carbon were known at the time, only very tentative statements are made to the effect that each of these electrons is bound in a

> "deformed central orbit of the same type as that performed by the third electron in the absence of the fourth".

No mention is made of N, O, F or Ne. The handwritten pages are largely illegible. However, N, O, F and Ne are discussed here, as are such concepts as "penetrating orbits" and "coupling". The typed pages are given on p. [140].

The next manuscript, consisting of five typed pages, is headed "§4. The correspondence principle". The first four pages give a statement of this principle as applied to multiply periodic systems. On the last page it is stated that it will be shown in §5 that, although the motions in atoms containing several electrons are far more intricate, the correspondence principle can nevertheless serve as a guide to the examination of the intra-atomic processes:

> "As we shall see, this leads again to a picture of atomic constitution which ... presents aspects suitable for the interpretation of the specific properties in accordance with the results embodied in the periodic table."

* [The manuscript has "first", but this is not in conformity with the usual terminology used by Bohr and others.]

This manuscript is reproduced on p. [148]; some typographical errors have been corrected.

The last manuscript in this group consists of three pages marked "Chapter V.1", "V_2 2a" and "V Chapter 5", respectively, plus 40 pages marked "V_2 1–40"*. All of the manuscript is handwritten, part of it in pencil. Most of it is in English, but about one-quarter is in Danish. A few remarks or corrections in Bohr's own hand have apparently been made during the dictation, rather than afterwards. In the margin, or on part of the pages, are several rough sketches and numerical estimates of orbital dimensions or firmness of binding. The manuscript has been transcribed, and the Danish parts have been translated. Punctuation, very scant in the manuscript, has been added. Obvious writing errors have been corrected, and a few prepositions have been altered. In half a dozen places a few words have been added in pointed brackets, in attempts to make the meaning clearer. Illegible words or passages are indicated by question marks or footnotes. The numbering of the periods in the periodic system has been changed, so as to agree with common usage. The manuscript is given on p. [152]

It begins with a discussion of the binding of the eleventh electron. The various term sequences of the sodium spectrum are discussed, and it is concluded that the valence electron in the normal state of sodium moves in a penetrating orbit with quantum numbers $n = 3$, $k = 1$ (and not $n = 2$, $k = 1$, as assumed in earlier manuscripts of this group). It is assumed that the doubling of the P and D terms is caused by different orientations of the outer orbit relative to the axis of the inner electron system. The binding of the next seven electrons is discussed with references to the spectra of the elements in the third period of the periodic system. It is of interest that the electron configuration proposed here for argon differs from that suggested in the March letter to "Nature" in that the eight outermost electrons are assumed to move in three-quantic rather than in two-quantic orbits. In potassium and calcium it is assumed that the outermost electrons are bound in orbits with $n = 4$, $k = 1$. However, in the next elements of the fourth period, three-quantic circular orbits may be more firmly bound, and in the iron family a transformation of the three-quantic groups is assumed to occur, giving rise to the magnetic properties of these elements, and resulting in the presence of 18 elements in this period.

A similar, although somewhat less detailed, interpretation is given of the properties of the elements in the fifth period. The valence electron in rubidium is assumed to move in an orbit with $n = 5$, $k = 1$, and the two outermost electron

* The fact that the manuscript uses the term "chapter" rather than section (§) indicates that it may have been dictated after September 1, 1921, when the plan of submitting a second part of the Solvay report was given up.

groups in xenon are assumed to be in four- and five-quantic orbits, rather than in three- and two-quantic orbits, as suggested in the March letter to "Nature". In the sixth period, the occurrence of the rare earths is explained by a gradual transformation in the inner part of the atom of the group of four-quantic electrons from a group of 18 electrons (in three subgroups) into a group of 32 electrons (distributed in four subgroups). The electron configuration of niton (radon) is given as 86 ($2_1\ 8_2\ 18_3\ 32_4\ 18_5\ 8_6$) in the notation used in the letter to "Nature". This differs from the configuration suggested in this letter in the quantum numbers assigned to the two outermost electron groups.

5. ATOMIC STRUCTURE. LETTER TO "NATURE" (13 OCTOBER 1921)

As shown by the manuscript "On the general interpretation of the periodic table", by the autumn of 1921 Bohr's views of atomic structure had been greatly extended and somewhat modified. While attempting to complete a comprehensive paper for publication by the Royal Danish Academy of Sciences and Letters, he sent another letter to "Nature". In this letter, which was dated September 16 and appeared on October 13, 1921 (Nature **108** (1921) 208), Bohr pointed out the main changes in characterizing electron orbits by quantum numbers made since the publication of his previous letter to "Nature" (March 24, 1921). In that letter he had assumed that the outermost electrons in heavy elements move in orbits characterized by quantum numbers equal to or even smaller than those of some of the electrons that are more strongly bound and move closer to the nucleus. That assumption was based on a consideration of

"the larger part of the orbit which lies outside the region of the inner groups".

However, when considering the mechanical properties of the whole orbit, including the penetrating inner loops, he was led to assign larger principal quantum numbers to the outer electrons and, in fact, was led to the assumption that the principal quantum number always increases by one unit as one goes from one electron group to the next.

"Thus, the orbits in the outermost "shell" in the niton atom must be characterised as six-quanta orbits instead of as two-quanta orbits",

as done in the previous letter to "Nature". The second letter to "Nature" is reproduced on p. [175].

Even before its publication, Bohr sent copies of this letter to a few of his fellow physicists. Thus, on September 16, 1921, he sent copies to Sommerfeld and James Franck. To Sommerfeld he wrote:

> "... In the last few months I have been feeling much better, and I have been busy with the completion of a comprehensive paper on my views of atomic structure ... During the work, my views have become much clearer, and much seems even simpler than I had dared hope ... I enclose a copy of a letter which I have just sent to "Nature" and which, as you will see, represents a correction and an improvement on an essential point ...*

He wrote to Franck on the same day much the same as to Sommerfeld, and added that he intended to let the paper on atomic structure replace the third and fourth parts of *On the Quantum Theory of Line Spectra*. He mentioned that Vieweg had undertaken to publish a German translation of the two first parts of this work and asked Franck to help him find a young German physicist who could translate the new paper which would be published in English by the Danish Academy, so that it might appear almost simultaneously in the two languages**.

6. THE STRUCTURE OF THE ATOM AND THE PHYSICAL AND CHEMICAL PROPERTIES OF THE ELEMENTS. LECTURE DELIVERED ON 18 OCTOBER 1921 BEFORE A JOINT MEETING OF THE PHYSICAL AND CHEMICAL SOCIETIES OF COPENHAGEN

It is not known just when Bohr abandoned the plan of publishing a comprehensive treatise on atomic structure as a memoir of the Royal Danish Academy. However, on October 18, 1921, he gave a lecture on this subject before the Physical Society of Copenhagen, at a joint meeting with the Copenhagen Chemical Society. The text of this lecture must have been prepared either before the lecture was delivered or immediately afterwards, for it was published in Fysisk Tidsskrift (**19** (1921) 153) just before the end of the year. A reprint of this lecture was published as a separate booklet by Jul. Gjellerups Forlag, Copenhagen, in January 1922 (Bohr's preface is dated December 1921). The lecture was quickly translated into German and was received by Zeitschrift für Physik on January 3, 1922, and published shortly afterwards (Z. Phys. **9** (1922) 1). It was next included as the third part of *Drei Aufsätze über Spektren und Atombau* (Friedr. Vieweg & Sohn, Braunschweig 1922). Bohr's preface to this book was dated April 1922. An English translation was made by A. D. Udden and published as the third essay

* For the text of this letter, see p. [741].

** For the text of this letter, see p. [693].

in *The Theory of Spectra and Atomic Constitution* (Cambridge University Press 1922), Bohr's preface being dated May 1922. A year later a French version appeared in *Les spectres et la structure de l'atome* (J. Herman, Paris 1923). A second edition of *The Theory of Spectra and Atomic Constitution* appeared in 1924.

As regards contents, this lecture differs from the unpublished manuscripts given on pp. [99]–[174] mainly in that greater attention is paid to the elements of the second group of the periodic system and in the inclusion of a brief discussion of X-ray spectra.

However, the style of the exposition is different. After a fairly long introduction, the problem of atomic constitution is attacked by asking the question: "How may an atom be formed by the successive capture of the electrons one by one in the field of force surrounding the nucleus?" The answers are found primarily from an analysis of the observed spectra, with guidance from the chemical properties and the correspondence principle. More or less tacit use is also made of assumptions which later became articulated as the postulate of the invariance and permanence of quantum numbers*. Mathematical calculations are omitted altogether. This undoubtedly puzzled some readers. However, it is obvious that detailed calculations could not possibly have been made for the large number of atoms considered. The calculations omitted were rough estimates of binding energies and dimensions of simple models of central motion character.

The lecture contains only one figure, a diagram of the sodium atom, already obsolete, taken from Z. Phys. **2** (1920) 423. Three notes, referring to recent papers, were added after the lecture was delivered.

The version published by Gjellerup is reproduced on p. [183]. It is identical with that published in Fysisk Tidsskrift, except for rows of asterisks separating it into four sections and spacings indicating pauses in the lecture and changes of subject matter. Also, the pagination has been changed, and a preface has been added.

The German version which appeared in Zeitschrift für Physik is a very careful, almost verbatim, translation, rendering the Danish original quite accurately. A diagram of the sodium spectrum with revised quantum numbers has been added together with a few lines of explanation, as well as a footnote on the first page giving a brief explanation of the origin and arrangement of the paper. In one place a clause has been added referring to recent work. Emphasis has been added in a number of places by the use of italics.

The version of the lecture published in *Drei Aufsätze über Spektren und Atombau* differs in several respects from that which appeared in Z. Phys. Four new figures have been added: a diagram of the arc spectra of sodium and potassium and of

* See p. [632] and Vol. 3, p. [473].

the spark spectra of magnesium and calcium, diagrams of the X-ray levels of krypton, xenon and niton (radon), and an arrangement of the periodic system devised in 1895 by the Danish chemist Julius Thomsen. There are also four additions to the text dealing with recent work, especially on X-ray spectra. In one of these, it is mentioned that the unknown element with atomic number 72 should be homologous to zirconium and thorium. However, the book has an appendix ("Nachschrift") in which Bohr refers to a recent letter to "Nature" by Rutherford ("Nature" **109** (1922) 781) dealing with an investigation by A. Dauvillier of the X-ray spectra of a mixture of rare earths which seems to show that element 72 is a rare earth and is identical with an element celtium, the existence of which had previously been surmised by G. Urbain from optical spectroscopy. In the appendix, Bohr attempts to reconcile this apparent fact with his views of atomic structure.

The English version of "The Structure of the Atom and the Physical and Chemical Properties of the Elements" is essentially equivalent with the German version in *Drei Aufsätze*. However, it differs from it in several respects. The translation is not nearly as close a rendering of the Danish original as the German version. This is partly due to the greater difficulty of translating Danish into English than into German, but it is also deliberate, as indicated by Bohr in the preface. Certain matters are omitted, and some remarks are added. Headings are used, not only for the four main parts but also for subsections. It is stated again that element 72 must be expected to be homologous with zirconium and thorium. The work of A. Dauvillier, according to which this element should be a rare earth, is mentioned but, instead of trying to reconcile this conclusion with his theory, Bohr now questions its validity.

A second edition of *The Theory of Spectra and Atomic Constitution* was published in 1924. In his preface, dated May 1924, Bohr states that, although work published since the appearance of the first edition had both given further support to the theory of atomic structure and also revealed difficulties at various points, he had decided to retain the original text and to confine himself to giving a brief account of the later developments in an appendix.

This appendix contains a table giving the electron configurations in nearly half of the elements in the periodic system. The element 72 is included outside the family of the rare earths, and with the symbol Hf (hafnium). In addition, the appendix contains five figures giving diagrams of nearly a score of selected spectra and one figure showing the variation of the various X-ray levels with atomic number. A recently published spark spectrum of carbon indicates that the electron configuration previously assumed for that element must be revised slightly. The discovery of hafnium by Dirk Coster and George de Hevesy is briefly reported. The English version of "The Structure of the Atom and the Physical and Chemical Properties of the Elements", 2nd ed. 1924, is reproduced on p. [257].

7. SEVEN LECTURES ON THE THEORY OF ATOMIC STRUCTURE HELD IN GÖTTINGEN (12–22 JUNE 1922)

Bohr had undertaken to deliver a series of lectures, sponsored by the Wolfskehl Foundation, in Göttingen in the summer of 1921. However, because of overwork and ill health, he had to postpone them until the following year.

A. Rubinowicz, who had been in Copenhagen in the spring of 1921, was contemplating spending some time there also in the spring of 1922. On January 3, Bohr wrote to him:

> "I was very glad to learn that you believe that it will be possible for you to come here this spring, and it would be a great help to me if you actually could consider helping with the preparation of my Göttingen lectures ... As I had told Hevesy, in March I must give the lectures in Cambridge which I had to cancel last year because of my illness, I shall therefore only have April and May in which to prepare the lectures that I must give in Göttingen in June ..."*

In the Niels Bohr Archive are typed manuscripts of the seven lectures, prepared by Rudolph Minkowski and another person**. Three typed pages seem to be a discarded introduction to the lectures. There are half a dozen pages of computation of some of the numerical data given in the lectures. Also, there is a list (handwritten in Danish) of 20 slides prepared for the lectures. These slides are no longer extant, but three sheets contain prints of 10 of them, and of three slides not on the list. Also, there is copy from which six of the missing slides may have been made.

The three first lectures dealt with the fundamental principles of the quantum theory of line spectra, as applied to hydrogen and to the influence of external fields upon the hydrogen spectrum. The fourth lecture treated the general features of the spectra of elements with several electrons. In the second half of this lecture the spectrum of helium was discussed. The fifth lecture dealt with the structure of the elements in the second period of the periodic system, and with the structure of the sodium atom. The sixth lecture dealt with the structure of all the remaining elements, in particular with the explanation of the family of rare earths and the platinum family. It was stated that the unknown element with atomic number 72 should not be a rare earth but should have chemical properties similar to those of zirconium. The last lecture was devoted to a discussion of X-ray spectra, and level diagrams of krypton, xenon and niton (radon), taken from recent work of Coster, were shown.

An English translation of these lectures is given on p. [341].

* For the text of this letter, see p. [734].

** [See *Sources for History of Quantum Physics*, The American Philosophical Society, Philadelphia, 1967, p. 108.]

Letter from Niels Bohr to J. Franck (15 July 22) German text p. [695]

After his return from Göttingen, on July 15, 1922, Bohr wrote a long letter to Franck:

"My entire stay in Göttingen was a wonderful and instructive experience for me, and I cannot say how happy I was for all the friendship shown me by everybody. In particular, I can hardly describe how closely I felt myself attached to you, and how grateful I am to you for all your help and sympathy in both small and large matters ... I am writing at the same time to Courant; however, please extend my greetings and thanks to the other colleagues and friends ... I shall write to Born later when I know a little more about my plans for the publication of the lectures ..."

After a first postscript, Bohr continued:

"P.S.2. The only thing I know for certain as yet about my lectures in Göttingen is that several of the results I reported there are already wrong. A first point is the constitution of the element 72 which, contrary to my expectation, has after all been shown by Urbain and Dauvillier to be a rare earth. I have referred to this, as well as to the paper by Götze, in an Appendix to a booklet which has just appeared at Vieweg. Another point is the mercury spectrum, and the spectra of similar type, for which the quantum numbers on the slides I showed cannot all be maintained. On the other hand, I now believe that I have reached a rather certain interpretation of these spectra."

Excerpt of letter from J. Franck to Niels Bohr (29 July 22) German text p. [697]

Franck replied on July 29, 1922:

"Many thanks for your kind and cordial letter which, as well as that sent to Courant, has pleased us very much. If I had not been hindered by various circumstances, I should certainly have replied before now and attempted to express what the time you granted us meant to all of us. Among other things, I unfortunately had to go to Berlin together with Pohl to attend the obsequies for our deceased old chief Geheimrat Rubens. It is sad that he should die so young (57 years). He suffered from pernicious leukemia. However, let me now say that we considered every minute of your stay here as a great gift, and we have set our hopes on your coming here soon again, and then bringing your dear wife with you ..."

After a long discussion pertaining to the first postscript in Bohr's letter, Franck mentioned that after he had broached an idea to Einstein, Einstein had told him that he had held the same view for over a year. Franck then continued:

"... With such an idea it is always like this: Either it is wrong, or Bohr and Einstein have known it for a long time ..."

In Göttingen Bohr met a considerable number of the German physicists; in particular, he met several young physicists, among them W. Pauli and W. Heisenberg. These physicists referred to the lecture series as the "Bohr Festspiele".

8. ON THE EXPLANATION OF THE PERIODIC SYSTEM. ABSTRACT OF A LECTURE DELIVERED BEFORE THE SECOND NORDIC PHYSICS CONFERENCE HELD AT UPPSALA, SWEDEN, 24–26 AUGUST 1922

This abstract "Om Forklaringen af det periodiske System" was prepared by Bohr himself. It was printed in Sweden by Edv. Berlings Boktryckeri A.-B in 1922 and appeared in Denmark in Fysisk Tidsskrift **20** (1921–22) 112. An English translation is given on p. [421].

On August 30, 1922, Bohr wrote to Oseen:

Excerpt of letter from Niels Bohr to C. W. Oseen (30 Aug. 22) Danish text p. [731]

> "I should like to thank you and your wife for all your kindness and hospitality during the interesting and successful physics conference in Uppsala. I cannot say how great a joy it is for me every time we meet to feel how close we are to each other in our general view of the fundamental problems of physics.
>
> I take this opportunity to send you a little calculation carried out by Kramers on the question treated by Wereide ..."

The Norwegian physicist Th. Wereide* had given a paper on relativity and atomic physics at Uppsala.

9. THE NOBEL PRIZE (1922)

In 1922 the Swedish Academy awarded the 1921 Nobel Prize to Einstein and the 1922 prize to Bohr for his work on atomic structure and radiation. On December 11 Bohr delivered his Nobel Lecture "On the Structure of Atoms" ("Om Atomernes Bygning") in Stockholm. This lecture, in its original Danish version, was printed at almost the same time in Stockholm (Les Prix Nobel en 1921–22, P. A. Norstedt & Söner, 1923) and Copenhagen (Jul. Gjellerups Forlag,

* Thorstein Wereide (1882–1969) was docent at the University of Oslo from 1919 to 1952. He worked on statistical mechanics, diffusion in liquids, and the theory of relativity.

1923; Fysisk Tidsskrift **21** (1923) 6). German and English translations appeared shortly afterwards (Über den Bau der Atome, Naturwissenschaften **11** (1923) 606, Springer, Berlin 1924; The structure of the atom, Nature **112** (1923) 29). The original Danish version is reproduced on p. [427] and the English translation on p. [467].

The Nobel Lecture was of a rather broad and general nature. As Bohr said at the outset:

> "... I have thought that I would act in accordance with the traditions of the Nobel Foundation if I shaped this account as a survey of the development which has taken place in recent years within the field of physics to which the work in question belongs ..."

At the end of the lecture, Bohr was able to announce that Coster and Hevesy had just succeeded in establishing the presence of element 72 in various minerals containing zirconium, thus proving that this element is not a rare earth. In a footnote to the printed lecture, Bohr mentioned that Coster and Hevesy had been able to obtain the new element, for which they proposed the name hafnium (after the Latin name of Copenhagen, Hafnia), in such a pure state that a preliminary determination of its atomic weight could be carried out.

After receiving the Nobel Prize at the banquet in Stockholm, Bohr made the following speech (given here in translation):

Speech by Niels Bohr at the banquet in Stockholm after receiving the Nobel Prize (December 1922)

> "As I rise to try to express my deeply felt gratitude for the great honor which the Royal Swedish Academy [Kungl. Svenska Vetenskapsakademien] has shown me by awarding me the Nobel Prize for physics this year, it is especially natural for me to call to mind the emphasis on the international character of science on which Alfred Nobel's great foundation is based. This is so much more natural to me as the small contribution that I may have had the good fortune to make to the development of physical science has consisted in joining together contributions to our knowledge of nature which we owe to investigators of various nations, who have built on widely different scientific traditions.
>
> When the great experimental discoveries around the turn of the century, in which investigators from many countries took such a prominent part, gave us an unsuspected insight into the constitution of atoms, we owe this above all to the great researchers of the English school, Sir Joseph Thomson and Sir Ernest Rutherford, who have inscribed their names in the history of

science as shining examples of how imagination and acute insight are capable of looking through the multiplicity of experiences and laying bare to our eyes the simplicity of nature. On the other hand, abstract thought, which has always been one of mankind's most powerful aids in lifting the veil that conceals the laws of nature to the immediate observer, has been of decisive significance for applying the insight gained into atomic structure to explain the properties of the elements directly accessible to our senses. Also in this work, men of many nations have made important contributions; but it was the great German scientists, Planck and Einstein, who, through their abstract and systematic studies, first taught us that the laws holding for the motions of the atomic particles, which govern the properties of the elements, are of an essentially different nature than the laws by means of which science hitherto had attempted to order our observations of the phenomena of nature.

That it has been my undeserved good fortune to be a connecting link at a stage in this development is only one piece of evidence among many of the fruitfulness, in the world of science, of the closest possible intercommunication of research work developing under different human conditions.

However, when a Danish scientist is here in Stockholm, on an occasion such as this, he must not only think of the international character of science but fully as much of the intellectual fellowship among the Nordic countries which we all feel, not least in the realm of science. It would be tempting to try to describe the great debt which science, and, hence, also Danish research work, owes to Swedish scientists in old and recent times. However, that would take us too far, even if I should limit myself to describing only the most important contributions of the eminent representatives of Swedish natural science present here tonight, whose work in various ways has been of fundamental significance also for atomic research. Hence, I shall call to mind only a single Swedish physicist, the late Professor Rydberg of Lund, whose ingenious work on unravelling the spectral laws has been of such great importance for deepening our knowledge of atoms and, not least, for the contribution that it fell to my lot to make.

Wishing again to express my deep gratitude for the honor the Academy has shown me and Danish science by awarding me the Nobel Prize, I should like at this banquet to be permitted to propose a toast to the vigorous growth of the international work on the advancement of science, which is one of the bright points of human existence in these, in so many respects, sorrowful times and, particularly, to the scientific fellowship among the Nordic nations which, in spite of characteristic differences, feel themselves so closely joined together by the bond of kinship."

Excerpt of letter from Niels Bohr to A. Einstein (11 Nov. 22) German text p. [685]

On November 11, 1922, Bohr wrote to Einstein:

"... To me it was the greatest honor and joy ... that I should be considered at the awarding of the prizes at the same time as you. I know how little I have deserved it, but I should like to say that I have felt it as the greatest good fortune that—quite apart from your great contribution to the world of human thought—the fundamental contribution that you have made to the more special field in which I am working, just as the contributions of Rutherford and Planck, should be recognized, also quite publicly, before I was considered for such an honor ..."

Excerpt of letter from A. Einstein to Niels Bohr (10 Jan. 23) German text p. [686]

Bohr's letter reached Einstein in Japan, and he wrote from the SS "Maruna Maru" on January 10, 1923:

"Your cordial letter reached me shortly before my departure from Japan. I can say without exaggeration that it pleased me as much as the Nobel Prize. I find especially charming your fear that you might have got the prize before me—that is truly "bohrisch". Your new investigations of the atom have accompanied me on the trip, and they have made my fondness of your mind even greater ..."

Bohr received letters of congratulations from a large number of physicists. Excerpts from a few of them will be given here. A. Rubinowicz wrote on November 10:

"... Of course, this was no surprise for us; everyone who knows your work, if only from hearsay, would gladly have given you the Nobel prize a long time ago. Even though, as mentioned, it is one of the highest honors, it is after all only a feeble expression of the great gratitude that the scientific world owes you for the flood of advances you have initiated and pursued in a unique manner. How proud we now can be that we have had the privilege of knowing you personally ..."*

R. Ladenburg wrote on November 12:

"... I would like to use this occasion to tell you what incomparable significance your investigations of atomic structure exert on the entire physical thought and perception of our scientific circles; for many years, hardly a day goes by

* For the text of this letter, see p. [735].

in which we, your colleagues, do not discuss questions raised by your work ..."*

H. A. Lorentz wrote in English on December 10:

Letter from H. A. Lorentz to Niels Bohr (10 Dec. 22)

Dear Professor Bohr,

Allow me to congratulate you most heartily on the high distinction that is conferred upon you today. Surely, no physicist could better deserve it, and I am very glad to hear that by awarding you the Nobel prize the Swedish Academy of Science has given expression to the universal admiration of your work, by which our understanding of Nature has been so wonderfully deepened.

With kind greetings
Very sincerely yours,
H. A. Lorentz.

P. Ehrenfest wrote to Mrs. Bohr in November 1922:

"It was a great pure joy when I learned that Einstein and Niels had received the Nobel prizes for 1921 and 1922.

I wish you from the bottom of my heart happiness and above all good health for both of you, and for your children and all those close to you.—It is so nice to think of your *happy* home ..."**

Ehrenfest wrote a long letter to Bohr himself on February 13, 1923.

P. S. Epstein wrote on December 23, 1922:

"... The closer colleagues who were privileged to collaborate on your proud edifice of atomistics, even though on a modest scale, have long recognized the grandiose significance of your work, but it is gratifying that still wider circles can no longer ignore its importance ..."***

On December 23, 1922, Franck wrote to Bohr in a long letter:

"It was long my intention to write to you at length, apart from the pitiful telegraphic congratulation, in order to tell you how heartily we have rejoiced

* For the text of this letter, see p. [714].

** For the text of this letter, see p. [683].

*** For the text of this letter, see p. [688].

here that you finally got your well-deserved Nobel Prize. There was a general great enthusiasm here that you and Einstein now both have it. One sees that a committee can at times actually make sensible decisions ..."*

10. THE DISCOVERY OF HAFNIUM (1922)

In April 1922 Bohr added a note to the version of "The Structure of the Atom and the Physical and Chemical Properties of the Elements", which was to appear in *Drei Aufsätze über Spektren und Atombau*, in which he stated that the unknown element with atomic number 72 should not be a rare earth but should resemble zirconium in its chemical properties. In May, A. Dauvillier and G. Urbain published notes (C.R. **174** (1922) 1347, 1349) in which they claimed to have detected element 72 by its X-ray spectrum in a mixture of two rare earths, lutecium and ytterbium, and to have identified it with an element celtium, the existence of which had previously been surmised by Urbain. In the same month Bohr added a note to the English version of "The Structure of the Atom and the Physical and Chemical Properties of the Elements" in which he referred to the notes of Dauvillier and Urbain, but he now questioned the validity of their result and pointed out that its acceptance would lead to serious difficulties and would violate

> "the otherwise general rule that the valency never increases by more than one when passing from one element to the next in the periodic table".

However, on June 17, 1922, Rutherford published a note in "Nature" (**109** (1922) 781) "Identification of a Missing Element" in which he translated Dauvillier and Urbain's notes and apparently accepted their finding. This led Bohr to add an appendix to *Drei Aufsätze* in which he pointed out the consequence of the claim of Dauvillier and Urbain. In his sixth lecture in Göttingen, delivered before he could have seen Rutherford's article, Bohr had simply stated that element 72 must have chemical properties similar to those of zirconium "if our conceptions of atomic structure are correct".

The most direct way of detecting element 72 and settling the question of its chemical properties was by X-ray spectroscopy, and, quite apart from this problem, X-ray spectroscopy was an important field for testing Bohr's theory of atomic structure.

Bohr had been in touch with the young Dutch X-ray spectroscopist Dirk Coster

* For the text of this letter, see p. [698].

as early as the summer of 1920, and shortly afterwards he had recommended him to Manne Siegbahn* in Lund, Sweden. Coster spent the better part of two years in Siegbahn's laboratory. In the meantime, Bohr's institute had acquired suitable X-ray spectroscopic equipment, and Coster moved to Copenhagen in the autumn of 1922. Even before he came to Copenhagen, Coster had sent his papers to Bohr for his approval or correction before their publication. It was natural, therefore, that, when Bohr became aware of Rutherford's note in "Nature", he should ask Coster for his opinion. He wrote on July 3, 1922 (in Danish, since Coster, after his long stay in Sweden and frequent visits to Copenhagen, was now able to understand this language):

Letter from Niels Bohr to D. Coster (3 July 22) Danish text p. [675]

> "... On my return from Göttingen, I found a note by Rutherford in "Nature" of July** 17, which you probably have seen; it mentions some notes by Dauvillier and Urbain concerning the identification of the element with atomic number 72 as celtium. I should be extremely grateful to you if you would write to me about your opinion of the reliability of Dauvillier's identification of the X-ray lines in question. The problem is of the greatest interest, since, as you know, the conceptions of atomic structure seem to demand essentially different chemical properties of an element with atomic number 72 than those exhibited by the rare earths. The question is apparently rather clear, but one must, of course, always be prepared for complications. They may arise from the circumstance that we have to do with a simultaneous development of two inner electron groups."

Coster replied in detail from Utrecht on July 15 that he was not convinced by Dauvillier's claim. The following day he met Siegbahn in the Hague, and he wrote immediately from there and told Bohr that Siegbahn, who had just visited Dauvillier in Paris, had authorized him to say that

> "if Dauvillier claims that there was 0.01 % (a hundredth of a percent) of the element 72 present in his sample, then one must grant him that the plates he took do not contradict this—but they confirm it just as little ..."***

Shortly afterwards Coster sent a manuscript to Bohr and asked him to submit it to the Philosophical Magazine as soon as possible. He enclosed a separate note about Dauvillier's work. On August 5 Bohr wrote to Coster:

* Siegbahn, Manne (1886–); Swedish physicist. Became professor at the University of Lund in 1923. Received the Nobel Prize in 1924 for his work on X-ray spectroscopy; was director of the Nobel Institute from 1937 to 1964.

** [It should be June 17.]

*** For the text of this letter, see p. [677].

"I have sent your paper on to Francis ... I deleted the reference to the letter to "Nature" from your letter ... like you, I am convinced that one can hardly ascribe much significance to Dauvillier's result. However, I think that it is very difficult for you to express doubt directly and publicly about Dauvillier's result based on the consideration of his other experimental investigations, without having yourself been in a position to examine Dauvillier's plates or Urbain's sample.

By the same mail I have written to Rutherford and quite privately informed him of your's and Siegbahn's doubt about the reliability of Dauvillier's result ..."*

Before Coster came to Copenhagen, Bohr had obtained samples of the rare earths, and promises of more such samples, from Auer von Welsbach. In Copenhagen, Coster had the advice and help of the distinguished chemist, Bohr's old friend, George de Hevesy.

As mentioned above, Bohr was able to announce on December 11, 1922, at the end of his Nobel Lecture, that Coster and Hevesy had succeeded in detecting the presence of element 72 in various zirconium minerals, thus proving that it is not a rare earth. Coster and Hevesy announced their discovery in several notes (Nature **111** (1923) 79, 182, 252, 462). In the first of these, which appeared on January 20, 1923, they wrote:

"For the new element we propose the name Hafnium (Hafniae = Copenhagen)."

Hevesy published several papers on the preparation and chemical properties of the new element. When the element had been prepared in sufficient purity, H. M. Hansen and Sven Werner, working in Bohr's institute, determined its optical spectrum. Their first paper, covering the ultraviolet part of the spectrum, was published on March 10, 1923 (Nature **111** (1923) 322). None of Urbain's lines were found to be present in the sample.

11. X-RAY SPECTRA AND THE PERIODIC SYSTEM OF THE ELEMENTS (WITH D. COSTER, 1923)

Shortly after Coster had arrived in Copenhagen, he and Bohr wrote a joint paper on X-ray spectra and the periodic system. The paper was completed by

* For the text and translation of this letter, see p. [678].

the end of October 1922, was received by Zeitschrift für Physik on November 2, and appeared towards the end of January 1923 (Röntgenspektren und periodisches System der Elemente, Z. Phys. **12** (1923) 342). The paper is reproduced on p. [483] and a translation is given on p. [519].

The first section of this paper is devoted to a brief account of Bohr's conceptions about atomic structure, and a table is given of the assumed types of electronic orbits in some 40 elements. Element 72 is included, but it is still indicated as unknown. In the second section of the paper a classification is given of the known X-ray spectra of 48 elements, from magnesium to uranium. One table gives the term values divided by Rydberg's constant, and another table gives the square root of this ratio. To each energy level is assigned a symbol $n\ (k_1, k_2)$. Lines corresponding to combinations between the levels occur for $\Delta k_1 = \pm 1$, $\Delta k_2 = 0, \pm 1$. The determination of these energy values from the experimental data is discussed in § 3. In § 4 an attempt is made to estimate theoretically the energy levels in terms of "inner" and "outer" screening constants and a relativity correction. In the next section the empirical X-ray levels are compared with the theory of atomic structure. Levels for which $k_1 = k_2$ are called "normal" levels, and those for which $k_1 \neq k_2$ "abnormal" levels. Each of the latter levels is assumed to be formally related to two normal levels.

> "Thus, each of these levels appears in the atom together with a normal level for which k_1 has the same value, and with which it also has the screening constant in common. On the other hand, it is related to a normal level having the same value of k_2, in that they have the same relativity term, provided the difference in the effective nuclear charge occurring in this term be ignored. This state of affairs leads one to seek the origin of these abnormal levels in a close interplay between the electronic motions in two neighboring subgroups of the excited atom, . . ."

However, the authors emphasize that a complete explanation of the complex structure of the levels cannot be given at the present state of the theory. § 6 deals with the dependence of the energy levels upon the atomic number. This dependence is illustrated in two figures. The various Moseley curves show sharp changes in slope at elements where an electron group is assumed to be completed, and new levels occur at the atomic numbers where new types of orbits are predicted by the theory.

On December 23, 1922, Bohr wrote to Ladenburg who had asked his consent to publish in a book the tables of the distribution of electron orbits in the elements given by Bohr in June, 1922, in Göttingen:

"... my only concern is that they be correct ... I should like now to express myself with greater caution on certain points; a table which I consider more defensible is included in a joint paper with Coster on X-ray spectra and the periodic system which will appear in Z. f. Phys. in a few weeks. There you will also find an opinion concerning the problem of the number of X-ray levels, a question that cannot yet be regarded as settled ..."*

In a letter dated March 14, 1923, Bohr wrote to J. C. McLennan:

"... In the paper on X-ray spectra and periodic system by Dr. Coster and myself you will find on p. 344 a table giving a survey of the electronic configurations as far as I believe that the present state of the theory offers a valid foundation. Already in the table some uncertainties are marked for elements, where the spectroscopic evidence at hand is not sufficient to settle the number of electrons in the groups of highest quantum numbers. In view of the rapid progress in spectroscopy in the last years, one may hope, however, in short time not only to settle these points, but also to reach definite conclusions regarding the elements omitted in the table ..."**

12. LINE SPECTRA AND ATOMIC STRUCTURE (1923)

On December 16, 1922, H. Konen wrote to Bohr and asked if he would be prepared to write an article, even a very brief one, for the "Gratulationsheft" of Annalen der Physik which the editor and publisher intended to issue in honor of the seventieth birthday of H. Kayser on March 16, 1923. P. Zeeman, W. Julius, M. Siegbahn, C. Runge, F. Paschen, and others, had already made commitments to contribute. It would be soon enough if Bohr could have his article ready by the end of January. Konen continued:

"... I take this opportunity to express my pleasure and satisfaction that the Nobel Prize for this year has gone to a man whose valuable contributions to spectroscopy, if I may use my own words from 1913, can only be compared to Darwin's contribution to biology ..."***

Bohr accepted Konen's request on December 23, 1922. He was not able to

* For the text of this letter, see p. [716].

** For the text of this letter, see p. [729].

*** For the text of this letter, see p. [702].

meet the original deadline, and on March 13, 1923, he wrote to Konen:

> "I am exceedingly sorry that I was not able to send you the manuscript of my contribution to the Kayser issue at the promised time. The reason for the delay is not that I have been very burdened with other work but above all that, during the work with the article itself, I hit upon scientific difficulties, to which I had to devote much time in order to try to clarify my ideas ..."

He enclosed the first four sections of the article and promised to send the last two (which were now being typed) in a few days. The work on the paper had given him much pleasure

> "since it gave me the occasion to reconsider once more many old and new matters and to elucidate them from a uniform point of view. This is especially true of the last two sections where the problems in many respects are in a ferment ..."*

Bohr also enclosed a letter of congratulation to Kayser in which he expressed regrets for the delay of his article. He did not wish to send this letter directly to Kayser, since he did not know whether Kayser was aware of the plans for the special issue of Annalen der Physik. Bohr submitted the rest of the manuscript on April 4. In his letter to Konen he wrote:

> "... As mentioned in my last letter, the delay was caused by the work with the subject matter which gradually resulted in a complete change of the originally planned short article. That is to say, I realized how difficult it was to give a satisfactory formal exposition of the theory of the series spectra without a discussion of the deeper problems of atomic structure. As the matter now stands, I am very grateful for the incentive to prepare this paper which your request gave me ..."**

Two days later, on April 6, 1923, Bohr wrote to Konen that he had returned the proof to the printer, and added:

> "... My entire behavior in this matter will undoubtedly appear strange, but, without attempting to excuse myself, I can assure you that only now, at the completion of the article, have I become clear about how little it may be suited

* For the text of this letter, see p. [703].
** For the text of this letter, see p. [704].

to its purpose. Under the impression of the riddles which always arise when one attempts to clarify one's ideas about the problems treated, the form of the exposition gradually changed and expanded, and in the last month I have had to postpone all other work and duties in order to finish the article at all."*

This paper, "Linienspektren und Atombau", Ann. d. Phys. **71** (1923) 228, is closely related to several of the papers included in Vol. 3, in particular to "Über die Anwendung der Quantentheorie auf den Atombau. I" (Z. Phys. **13** (1923) 135), which appeared only a few weeks earlier, and in which the formal postulate of the invariance and permanence of quantum numbers, essential to the present paper, was first set forth. This postulate, which must have been conceived by Bohr during his attempts to interpret the observed spectra, entails in particular that the addition of an electron to a group or subgroup does not in general alter the quantum numbers of the electrons already present in this group.

The last section of the present paper, dealing with complex structure, may be regarded as an early version of the treatment given of this subject in the unpublished manuscript "Über die Anwendung der Quantentheorie auf den Atombau. II", also included in Vol. 3. On the other hand, as Bohr indicated in his letter of April 4, 1923, to Konen, use has been made of the detailed views of the structure of the chemical elements developed by Bohr during 1921 and 1922. This is the reason for placing the paper in Vol. 4, rather than in Vol. 3. On p. 344 of the present paper, a table of the types of electron orbits in some forty elements is given, which is identical with the table given in the paper by Bohr and Coster (Z. Phys. **12** (1923) 342), except that the symbol Hf (hafnium) has been inserted for the element with atomic number 72, and that 71 Lu (lutecium) has been changed to 71 Cp (cassiopeium) and 86 Nt (niton) has been changed to 86 Em (emanation). The paper contains 8 diagrams of term values for some two dozen selected spectra. The quantum numbers *nkj*, as well as the effective quantum number n^*, are indicated in the diagrams.

This paper is reproduced on p. [549]; a translation is given on p. [611].

13. THE PROBLEM OF HELIUM (1916–1923)

In an attempt to determine the structure of the helium atom, and to explain the presence of two entirely separate spectra, Bohr and Kramers made extensive calculations in the autumn of 1916. They considered the mutual perturbations of the two electrons and found two different classes of orbits that might represent ortho- and parhelium, respectively. This work was not prepared for publication.

* For the text of this letter, see p. [705].

In the Niels Bohr Archive are some 200 handwritten pages of calculations. Although several of these are dated, it would be difficult to untangle them, and they are not included in the Collected Works. However, the results of these calculations are mentioned in several of Bohr's published and unpublished papers, and in many of his letters.

In 1919, A. Landé suggested a possible structure of the normal state of the helium atom, and during 1919 and 1920 some correspondence on the problem took place between Bohr and Landé. In his lecture of December 15, 1920, to the Physical Society of Copenhagen, Bohr was able to explain the metastable state of ortho-helium and to explain, on the basis of the correspondence principle, that the old ring model of helium had to be abandoned. In his letter to "Nature" of March 24, 1921, he indicated in a table that in the normal state of helium the two electrons are both in 1_1-orbits, but he gave no details. In the unpublished manuscript headed "§2. Helium", written sometime in 1921 and apparently intended originally as belonging to the second part of his Solvay report, Bohr gave an outline of his and Kramers' early work (see p. [122]). He mentioned that T. Lyman and H. Fricke had determined the value of the lowest term of the parhelium spectrum which corresponds to the normal state of helium. He mentioned that in this state the planes of the two 1_1-orbits presumably make an angle of 60° with each other and that the model has axial symmetry, and he stated that Kramers had recently undertaken to calculate the energy of this model. In his lecture of October 18, 1921, on "The Structure of the Atom and the Physical and Chemical Properties of the Elements" (see p. [183] and p. [257]), Bohr again referred briefly to the 1916 work with Kramers and to Kramers' calculations of the new model (which had subsequently been proposed by E. C. Kemble). He assumed here, and in papers published in 1923, that the second electron to be bound in all the elements is also bound in a 1_1-state. An outline of the early calculations, somewhat more mathematical than that in "§2. Helium", was given by Bohr in the fourth lecture delivered in June 1922 in Göttingen.

In a letter of December 23, 1922, P. S. Epstein wrote to Bohr:

> "... I myself am at present diverted from the quantum theory by other investigations. The only thing that may perhaps interest you is that some time ago I calculated the ionization potential of the symmetrical helium model. Unfortunately, my result is too low: 21 volts. I had certain doubts; for my calculation turned out to be surprisingly short. However, a couple of weeks ago I had occasion to talk with Prof. Brown about it, and he expressed the opinion that when the perturbations of the first and second orders are almost the same (as in my case), the probability of getting the correct result is considerable. Later, the paper by Van Vleck appeared ⟨Phil. Mag. **44** (1922) 842⟩; he

finds the same result by a quite different method ... It would interest me very much to learn if the results of Messrs. Kramers and Pauli agree with mine. For the time being I do not consider publishing ..."*

Letter from M. Born to Niels Bohr (4 May 23) German text p. [669]

On March 4, 1923, M. Born wrote to Bohr:

"A short time ago Heisenberg wrote to Pauli ⟨then in Copenhagen⟩ that, together with me, he had carried out a calculation of the terms of the excited helium atom. Since Pauli has not yet replied, I should like to ask you yourself what you think about it. Our result is quite negative. When one seeks out *all* possible types of orbits of excited He by means of perturbation theory, then one finds no energy level (except the p-term) that agrees with the observed terms. Hence, even here, where one of the electrons is far removed, there must occur deviations from the laws of mechanics, or from the quantum rules. Now, I shall not pester you by discussing all the ways in which one might avoid this catastrophe; but I should like to learn from you whether you, or Kramers or Pauli, have made similar calculations and, if that is not the case, if you consider it in order that we publish the result briefly. In fact, the calculation is a direct continuation of considerations that you and Pauli (together with me) made about ortho-He . . ."

Letter from Niels Bohr to M. Born (9 Apr. 23) German text p. [671]

Bohr answered this letter on April 9:

"... I ... can only recommend that you publish your calculations without any regard to us. In fact, the problem of the helium spectrum is one which, after all, is still far from being solved. In the survey article mentioned ⟨Linienspektren und Atombau, Ann. d. Phys. **71** (1923) 228⟩, I have indicated some general views, and there I have also announced the publication of Kramers' and my calculations which I hope to send off next week ... I have emphasized the deep-going difficulties which we meet everywhere, as soon as we have anything to do with systems with several electrons. I plan to give a more detailed exposition of the indications given there in the second part of my paper on the quantum theory of atomic structure in Zs. f. Phys. ⟨"Über die Anwendung der Quantentheorie auf den Atombau. II", see Vol. 3, p. [501]⟩."

Letter from M. Born to Niels Bohr (17 Apr. 23) German text p. [672]

On April 17, Born answered this letter:

"... Although you write that we should let our paper on helium be printed without regard to you and your collaborators, Heisenberg and I have never-

* For the text of this letter, see p. [688].

theless decided to show it to you first. One reason for this is that Pauli several times has expressed the opinion in letters to Heisenberg that we must have calculated wrongly; we regard this as rather improbable, but we should after all like to win Pauli's approval. I am sure that you will be good enough to show him the enclosed manuscript; then he can see to it that it is returned (to me) very soon. Perhaps you yourself find time to have a look at the paper. I should not like to publish anything in your closest field of work which you could not approve of.

I am anxiously awaiting your contemplated publications. I also ponder much over why it is that mechanics seems to fail in cases of several electrons ..."

On May 2, 1923, Bohr again wrote to Born:

Letter from Niels Bohr to M. Born (2 May 23) German text p. [673]

"... In fact, the result is very important as evidence of the inadequacy of the present basis of the quantum theory, as far as systems with several electrons are concerned. However, I do not see this difficulty in quite the same light as you indicate; for I believe, as I have already pointed out in my last paper in Zeitschrift für Physik, that it is possible to give a uniform interpretation of the quantum theory in which the failure of mechanics for the stationary states fits naturally. The manner in which the conditions for fixing the stationary states are to be formulated on this view is probably the principal problem at the present state of the theory ...

With regard to the old work on the helium problem by Kramers and myself, we have been ashamed that it was not carried to greater completion; for also the investigation of the non-coplanar orbits could in principle just as well have been carried out by our method ... As the matter now stands, we hardly think at present of publishing our calculations."

Bohr's and Kramers' old calculations of the helium atom were never published. The calculations by H. A. Kramers on the new Bohr–Kemble model of the helium atom were published in Z. Phys. **13** (1923) 312. He found an energy value 3.9 eV higher than the experimental value and concluded, not that the model was wrong, but that the laws of mechanics fail even for the stationary states of this simple atom. The paper by Max Born and Werner Heisenberg appeared in Z. Phys. **16** (1923) 229. Epstein's calculations were apparently never published.

14. REVISION OF THE DISTRIBUTION OF ELECTRONS IN ATOMS SUGGESTED BY MAIN SMITH AND STONER (1924)

On March 28, 1924, there appeared a paper by J. D. Main Smith in *Chemistry and Industry* (**43** (1924) 325) in which he suggested a distribution of electrons into subgroups that differed considerably from that proposed by Bohr. From the fact that all but one X-ray level in each main group are double, he divided each *n*-quantic group into a larger number of subgroups than had Bohr. Moreover, on the basis of a large amount of chemical evidence (valencies and coordination numbers), he concluded that the number of electrons in each of these completed subgroups was equal to twice the value of the quantum number k. (He used the old symbols for X-ray levels shown on the left side in Fig. 2 of Bohr and Coster's paper, see p. [491] and p. [524].) Thus, the electrons in the completed 2-quantic group were distributed in three subgroups with 2, 2 and 4 electrons, rather than in two subgroups with 4 electrons each. The distribution of the 32 electrons in the completed 4-quantic group he assumed to be given by the numbers 2244668 rather than by the numbers 8888 as proposed by Bohr.

Because it was published in a chemical journal, Main Smith's work did not immediately become known to physicists. However, in October 1924 E. C. Stoner published a paper (Phil. Mag. **48** (1924) 719) in which he independently proposed the same distribution of electrons. Stoner used three quantum numbers nkj, as proposed by Landé (Z. Phys. **16** (1923) 391). Admittedly arbitrarily, he identified the "inner" quantum number j with the symbol k_2 used by Bohr and Coster. He noticed that twice the sum of all the inner quantum numbers in a completed group characterized by a given value of n was equal to the number of electrons in this group assumed by Bohr. This led him to surmise that

> "the number of electrons in each subgroup separately is also equal to double the inner quantum number".

This led Stoner to the same scheme as that proposed by Main Smith (the former's j being equal to the latter's k). Stoner was able to marshall some experimental evidence, admittedly not conclusive, in favor of the proposed electron distribution. The most direct physical evidence would seem to be X-ray intensities, but unfortunately the measurements of these are not very accurate. One advantage of the Main Smith–Stoner distribution is that it does not require any re-organization of electron subgroups, such as occurs, e.g., for the 3-quantic group between argon and krypton according to Bohr's theory.

After having read Stoner's paper, Coster, who had just been appointed to a professorship at Groningen, wrote to Bohr on December 7, 1924:

"... I regret very much that I have no occasion at the moment to discuss a matter with you or Kramers which interests me very much, namely Stoner's paper on the distribution of electrons in groups. I believe that one should consider the results of this work by themselves and disregard the actually rather doubtful justification. Kramers writes me that you take a negative attitude towards this paper, and I am sorry that I do not know your arguments ... It seems to me that it is actually possible to build up an electron group in such a way that one gets Stoner's subgroups ... The presently available experimental material does not permit a decision with certainty. However, I hope soon to try to obtain more detailed information here ..."

Letter from D. Coster to Niels Bohr (7 Dec. 24) German text p. [679]

Bohr replied on December 10, 1924:

Letter from Niels Bohr to D. Coster (10 Dec. 24) Danish text p. [681]

"... I was very interested in what you wrote about Stoner's paper; I have from the first understood the formal beauty and simplicity of his classification of the levels; however, in the discussions here I have merely emphasized that, from quantum-theoretical points of view, it cannot mean a final solution of the problem, since we do not yet possess any possibility of connecting the classification of levels in a rational manner with a quantum-theoretical analysis of electron orbits. Quite apart from this, it is, of course, of the greatest significance to get cleared up the point emphasized by you about the first appearance of the levels. In fact, from that one would get important new information about the building-up of the groups, even if one could hardly use it to decide directly whether a consideration such as Stoner's is preferable to the view of the complexity of the levels indicated in our joint paper in Zeitschrift für Physik. Above all, in my opinion one can hardly emphasize strongly enough the ambiguous character of any comparison between experiments and theory, as long as the latter does not permit a rational explanation of the main experimental results concerning the character of the doublets. I am attempting at present to write a little note about the general nature of the difficulties we meet at the moment within the atomic theory and, as soon as I have finished it, I shall send you a copy of the manuscript."

It is not known when Bohr learned about the work of Main Smith. However, in the Niels Bohr Archive are undated proofs of a letter to "Nature" by Main Smith and an answer to it by Bohr. Since these notes were evidently never published, and since Bohr's answer is similar to his letter of December 10, 1924, to Coster, they will not be reproduced here. It may be of interest, however, to quote the last sentence of Bohr's note:

"It may be added that the difficulties encountered in the interpretation of the chemical evidence, as far as it depends on the finer features of the coupling of the electronic orbits, are perhaps even greater than in the case of the spectral evidence, since the orbits of electrons in molecules may often differ so largely from central orbits that quantum symbols, essentially different from those used for the classification of the electronic orbits in the isolated atom, may be required for their description."

Stoner's paper undoubtedly furnished an important stimulation to Pauli, who a year or so later formulated the so-called exclusion principle which finally settled the problem of the number of electrons in the various subgroups.

In an article *Atom* in the *Encyclopædia Britannica* (**13** (1926) 262), Bohr has given a table of electron arrangements in a large number of elements. This table is in general agreement with those given by Main Smith and Stoner, but is less detailed in that only the two quantum numbers n and k are used in the classification. Bohr writes:

"... The question how many electrons there are in the various groups and subgroups has been subject to much discussion in the last few years. Table II is the temporary result of this discussion and seems to give an adequate description of the spectral as well as the chemical evidence ..."

The article is reproduced on p. [657].

I. SOME CONSIDERATIONS OF ATOMIC STRUCTURE

LECTURE DELIVERED BEFORE THE PHYSICAL SOCIETY OF COPENHAGEN, 15 DECEMBER 1920

(TRANSLATION*)

* [See Introduction, sect. 1.]

OUR KNOWLEDGE OF ATOMS

Introduction to the lecture

Physical science has long been occupied with the problem of the existence and the structure of atoms. While an increasing amount of material giving evidence of the existence of atoms has accumulated in the course of time, our knowledge about the structure of atoms has begun in very recent years. The first possibility of obtaining any closer knowledge of atomic structure was reached with the discovery of the so-called electrons, i.e., negatively charged small particles, more than a thousand times lighter than the lightest atoms. The discovery of these particles grew out of a close study of phenomena connected with electric discharges in rarefied gases, and, since these particles have the same mass and electric charge independently of the gas used or the material of which the electrodes were made, it became clear that in the electrons we are dealing with common building stones of the atoms of all chemical elements. This result was confirmed shortly afterwards in an exceedingly interesting manner by Zeeman's discovery of the peculiar change suffered by the spectral lines of an element when its luminous atoms are placed in a magnetic field; for, as shown by Lorentz, this effect is just of the kind to be expected if the emission of the spectrum is due to the motion of the electrons within the atom. At the turn of the century, these results strongly revived the question of atomic structure, and in this period the work of J. J. Thomson stands out; for he gave many hints of the direction in which advances in atomic physics were to be sought. However, no decisive results were obtained until Rutherford's discovery of the atomic nucleus about ten years ago. In fact, until then one was restricted to pure fancy with regard to the question of the forces holding the electrons in the atom. Since the entire atom is neutral, it was necessary to assume that, besides the electrons, it also contained positive electricity, but nothing was known about the distribution of this charge until Rutherford's work. However, from experiments on the rays emitted by radioactive elements, Rutherford concluded that the positive electricity in an atom is concentrated in a nucleus which is very small compared to the size of the entire atom but nevertheless contains almost the whole mass of the atom. Rutherford has pursued this result in a series of extremely important papers and has recently been able to show that, from the passage of rays from radioactive elements through matter, it has not only been possible to obtain knowledge of the existence of the atomic nucleus, but also that, in certain collisions with one kind of these rays, the nucleus splits into two parts, each one of which forms the nucleus of a new atom. However, I shall not enter further into the great importance of these experiments, in which for the first time a lasting change in an atom of an element, brought about by external means, has been observed; rather, I shall discuss the prospects and results in the field of atomic research to which the discovery of the building stones of atoms, the nucleus and the electrons, has given rise. We shall assume, then, that an atom consists of a heavy, positively charged, nucleus and a number of lighter electrons, and the careful study of a large number of phenomena has made it clear that the number of electrons in a neutral atom is just equal to the number which marks the place of the element in question in the periodic system. This result, which was first stated by van den Broek, is especially indicative of the formal simplicity presented by the problem of atomic structure. In the closer study of the motion of the particles in an atom we meet, at first sight, a problem quite similar to that met with when we consider the motion of the heavenly bodies; since the electrons, as well as the nucleus, are extremely small compared to their distances apart in the atom, the forces between the particles will depend only on their charges and not on their internal structure, just as, in computing the force exerted on a planet by the sun or another planet, we need only consider the total mass of each body and not its internal structure. In spite of this similarity, there is, nevertheless, an extremely great difference between the problems we meet in astronomy and those we meet in atomic research; for example, we must assume that the motions in the solar system are determined not only by the masses of the sun and the planets but also by the circumstances existing when the solar system was formed. However, the matter is quite different with atoms; since an element always has the same properties, e.g., always emits the same spectrum consisting of the same lines, it is clear that the motions in the atom, at least in what may be called the normal state, i.e., the state to which it returns when it is not subjected to external influence, depend solely upon the nature of the atomic system as characterized by the electric charges and masses of the particles it contains. This mechanically inexplicable property of atoms is perhaps most strikingly exhibited in experiments in which we observe a change in the motions of an atom caused by an external influence.

SOME CONSIDERATIONS ABOUT THE CONSTITUTION OF ATOMS

Introduction to and outline of the lecture

The study of atomic structure is at the moment characterized by the fact that we have come to know the building stones of atoms. It seems hardly possible to doubt that an atom consists of a positive nucleus, which contains almost the entire mass of the atom, and a number of much lighter negative electrons, a number which in the neutral atom is simply equal to the number giving the place of the element in the periodic system, the so-called atomic number. Since the dimensions of the particles of which an atom consists are very small compared to their mutual distances, i.e., to the size of the atom, the mechanical problem we meet in atomic structure is, at a first glance, very similar to the mechanical problem we meet in astronomy, i.e., in our solar system. However, a closer consideration shows, nevertheless, that there is an essential difference between the laws holding for the motion of the particles in an atom and those holding for the mechanical systems we consider in astronomy. In fact, for the latter we must assume that the motions may be varied within wide limits, so that their determination requires the knowledge of certain initial conditions; for example, we must assume that the motion of the planets in our solar system is determined not merely by the masses of the sun and the planets but also depends to an essential degree on the conditions that existed when the solar system was formed, and which we do not know in detail. On the other hand, in order to account for the definite physical and chemical properties of the elements, it is necessary to assume that each atom possesses a state, the so-called normal state, in which the motion is determined solely by the masses and charges of the particles it contains. This peculiar stability of the atom reveals itself perhaps most clearly in experiments in which, by external influence, e.g., absorption of radition or bombardment by free electrons, we observe a change in the state of the atom. For in such experiments we observe how the atom seeks to return to its normal state after this change, and, at the same time, we learn something about the peculiar nature of this process of reorganization. In fact, it turns out that the atom, during this reorganization, emits radiation which forms a part at least of the characteristic spectrum of the element in question, and the very fact that these spectral lines are independent of the influence that brought about the original change in the atom makes it clear that it is impossible to describe the process of reorganization in detail in terms of our usual mechanical and electrodynamical conceptions. However, as you know, by taking recourse to the quantum theory, we have achieved results of such a kind that it is impossible to doubt that our approach is fruitful. These applications of the quantum theory rest on the assumption that an atom possesses a series of states, which we shall call the stationary states, which are distinguished from the other mechanically conceivable states in that the atom can exist in them without emitting radiation. On the other hand, every emission of radiation occurs during a transition between two such stationary states, and we assume that the radiation emitted is monochromatic and has a frequency ν that is determined in a simple way by the difference of energy in these states, being given by the relation $h\nu = 1-2$, where h is Planck's constant and 1 and 2 are the values of the energy in the two states. Of these stationary states, the one of lowest energy will be the normal state of the atom, since, when in one of the other stationary states, the atom has the possibility of passing, with emission of radiation, to a state of lower energy. However, in the state of lowest energy, this possibility is excluded, and, according to our assumptions, the atom cannot change its state unless energy is supplied from the outside, and of such an amount as to bring the atom into one of its other stationary states. Thus, we see that the quantum theory permits us to understand at the outset certain essential properties of atoms. However, with regard to the more detailed elaboration of the theory, the fact is that almost all decisive results of a quantitative nature have been obtained by considering the extremely simple system consisting of a nucleus and a single electron, and, although convincing results of a more qualitative character have been obtained for other systems, especially in the fields of series spectra and X-ray spectra, it has at the same time been very difficult to explain many other properties of the elements. This is connected with the difficulty of accounting for the stability of an atom containing a large number of electrons; for this stability requires what might be called a coupling of the different electrons in the atom, a coupling for which the simple assumptions, made in the first sketchy attempt to develop a theory of atomic structure, by applying at the same time the nuclear model and the quantum theory, did not provide a sufficient basis. These difficulties have been strongly emphasized from different sides, and the discussion of them has given rise to a large number of papers in the last few years. Through these, the conditions that an atomic theory must satisfy, to have any chance of explaining the chemical properties of the elements, have come to stand in a clearer light. However, several of the proposals for changing these simple considerations are hardly consistent with the proven fundamental principles of the nuclear theory and the quantum theory. On the other hand, in other papers changes are proposed which do not conflict with these principles, and which undoubtedly contain hints of the direction in which we may search for improvements; however, since the proposed changes are somewhat arbitrary, one cannot avoid the impression that the difficulties have been moved to another place rather than eliminated. In what follows, I shall attempt to give a survey of the nature of the difficulties in question, and then try to show how it is possible, by certain simple considerations closely connected with the points of view developed in the treatment of the series spectra, to get around the difficulties and throw new light on certain aspects of the problem of atomic structure.

*System consisting of nucleus and single electron**

Keplerian orbit. Energy and value of the major axis in the stationary state. Spectrum. More accurate determination of the orbit in given circumstances. Influence of the relativistic modifications. Description of the orbits in Sommerfeld's theory of the fine structure. Low stability of the orbital forms except in the normal state. The hydrogen molecule, simple model, its formation, only approximate agreement. Lenz' model, formation, stability. The calculation of Kramers. Discussion of the problem of stability. The helium spectrum. Calculation of orbits. Normal state. Simple model. Lack of agreement. Lithium. Beryllium. The periodic system. 8 rings. Lack of inner stability. Kossel (Ne model). Lewis, Langmuir, Born and Landé. Spatial symmetry, several-quantic orbits. X-ray spectra. Production of K-radiation. Moseley's results. Similarity to hydrogen. Difference with respect to absorption. Kossel's considerations. Sommerfeld's theory of the fine structure of X-ray spectra. Heuristic viewpoint on several-quantic orbits.

MAIN BODY OF THE LECTURE

I shall not enter here into detail upon the assumptions underlying this theory of series spectra, but shall merely refer to a comprehensive survey of these points of view which I gave in April this year before the "Deutsche physikalische Gesellschaft" ⟨published in Z. Phys. **2** (1920) 423⟩, and in what follows I shall refer to the main results only in so far as they bear on the questions treated in the present paper. To facilitate the survey, we shall begin with the results of the theory for elements containing a small number of electrons and shall then proceed to discuss the questions presented by the properties of elements with higher atomic numbers.

* [The following paragraph and the main body of the lecture are handwritten.]

As shown in several previous papers*, and discussed in detail in the lecture mentioned, it has been possible, because of the simple periodic character of the motion in the hydrogen atom, to develop a simple, and in certain respects complete, quantum theory of the hydrogen spectrum. In fact, hydrogen consists simply of a nucleus and a single electron which, if relativity corrections are neglected and the mass of the nucleus is considered to be infinite as compared to that of the electron, will move in an ellipse with the nucleus in one of the foci. According to the theory in question, the atom possesses a series of stationary states in which it can exist without emitting radiation and for which the values of the work W necessary to remove the electron to an infinite distance from the nucleus and the major axis $2a$ of the electronic orbit are given by the following formulas:

$$W_n = \left\langle \frac{1}{n^2} \frac{2\pi^2 N^2 e^4 m}{h^2} \right\rangle, \qquad 2a_n = \left\langle n^2 \frac{h^2}{2\pi^2 N e^2 m} \right\rangle. \tag{1}$$

On the other hand, the theory makes no particular assumption about the value of the eccentricity of the electronic orbits in the stationary states. On this simple basis, it was not possible to give a detailed explanation of phenomena which depend on deviations of the motion of the electron from a simple periodic orbit, such as the fine structure and the Stark and Zeeman effects of the hydrogen lines. However, a basis for the explanation of these phenomena was attained when the application of the quantum theory was extended to the so-called conditionally-periodic systems, the motions of which may be considered as made up of a number of components, each of which can be treated in a similar way as a periodic motion. For these systems, the stationary states are, therefore, not characterized by a single number but by just as many whole numbers as there are quasi-periodic components in the motion. As is well known, on the basis of this theory, Sommerfeld and his followers were able to account for many of the details of the phenomena just mentioned. However, as shown in the lecture, it is possible to develop a simple method for these phenomena, in which we are dealing with motions of systems that are very nearly periodic, a method that is directly related to the simple theory of the hydrogen spectrum. In fact, when the motion deviates from a simple periodic motion, it is possible, from points of view similar to those underlying the latter theory, to find the additional conditions for the stationary states by direct examination of the nature of these deviations. For it turns out that these conditions must be regarded as being connected in a simple manner with the changes in shape and position suffered by the orbit in the course of time as a result of the given influence, changes which in celestial mechanics

* Phil. Mag. **26** ⟨(1913) 1, 476, 857⟩; **27** ⟨506⟩. See also N. Bohr, Abhandlungen über Atombau aus den Jahren 1913–1916, German translation by H. Stintzing, ⟨Friedr. Vieweg & Sohn⟩ Braunschweig 1921. This book will be referred to as *Abh. ü. Atombau* in what follows.

are called secular perturbations of the orbit. This method is not only capable of treating phenomena in which the stationary states can be determined from the theory of conditionally-periodic systems, but also cases in which this theory cannot be applied. From this method, which not only permits the treatment of phenomena for which the stationary states of the atom can be fixed by the theory of conditionally-periodic systems but also opens the possibility of determining these states in cases where the theory is not applicable (Note about Kramers), it follows immediately that the shape of the electronic orbit in the stationary states is, in general, ill determined or at least quite unstable, since it undergoes major changes under the influence of weak external forces, provided these are capable of producing changes in the periodic motion of the same order of magnitude as those demanded by the theory of relativity. However, the matter is different for the stationary state in which W has its maximum value, i.e., is given by eq. (1) with $n = 1$. For this state the orbit is completely defined; it is not only circular for the undisturbed atom, but, in the presence of constant external fields, it will deviate from the circular form only by small quantities proportional to the intensity of the external forces. In this state the atom will take such a position relative to the external field that there will be no secular perturbations of the electronic orbit. We are here dealing with especially striking evidence of the inherent stability of the atom, which, although mechanically inexplicable, is demanded by the quantum theory and reveals itself in the very existence of stationary states.

With regard to the closer investigation of the normal state of the hydrogen atom, we are faced with the well-known fact that hydrogen atoms join to form molecules, so that the normal state of the atom is accessible to experimental study only under special circumstances. For an attempt to explain the physical and chemical properties of hydrogen, the constitution of the hydrogen molecule is, therefore, of primary interest. As a possible model of such a molecule, the writer has proposed a simple configuration which immediately suggests itself, namely a configuration in which the two nuclei are at rest at a fixed distance apart, while the electrons move in a circular orbit perpendicular to the line connecting the nuclei and with its center at the midpoint of this line. At any instant the electrons find themselves at opposite ends of a diameter of this circle, and in the equilibrium configuration the length of this diameter is $\sqrt{3}$ times the distance between the nuclei. Moreover, it was shown how such a configuration can be formed in a simple manner by the joining of two atoms. In fact, if we imagine two hydrogen atoms at a considerable distance from each other and assume that the electrons revolve in parallel orbits in the same direction, but such that the radius vectors from nucleus to electron point in opposite directions in the two atoms, and, if we assume that the nuclei approach each other slowly, it can be shown that the electronic orbits will suffer a gradual change, in that they will approach each other faster than the nuclei, and that, when the nuclei have reached a certain distance, the electrons will move in the same orbit symmetrical to the two nuclei. By a further approach of the nuclei to each other, the equilibrium configuration mentioned will finally be reached. It turns out that the two systems will attract each other during this entire process, so we have a perfect picture of a possible spontaneous formation of a molecule by the joining together of two atoms. Moreover, from this consideration we obtain a unique method for determining the motion of the electrons in the molecular model; for the angular momentum of the electrons around the line connecting the nuclei does not change during the process and, hence, must be the same in the completed molecule as in the original state, i.e., equal to $h/2\pi$ for each electron. In this connection, it may be of interest to point out the similarity between the process considered and the adiabatic processes, introduced in the quantum theory by Ehrenfest from other points of view, and, in particular, his transformability principle for stationary states ["Adiabatenhypothese"], which has been of such great importance for the elaboration of the original quantum theory. A detailed investigation of this simple model of the hydrogen molecule has made it possible to explain approximately such properties as the ionization potential, the heat of dissociation, and the specific heat, of hydrogen; however, the calculations are not in close agreement with the measurements of these quantities. (An attempt by Debye to show that such a hydrogen molecule offers an accurate explanation of the dispersion of hydrogen can hardly be taken seriously because of the contradiction between the assumptions made in Debye's calculations and those underlying the application of the quantum theory to spectral problems. (See C. W. Oseen ⟨Phys. Zeitschr. **16** (1915) 395⟩, O. Stern; see also *Abh. ü. Atombau*, p. 138.) Objections have also been raised against this model for other reasons; thus Lenz has pointed to the difficulty of explaining the Zeeman effect of the many-line spectrum resulting from the assumption that the hydrogen molecule has a magnetic moment. Although, because of our insufficient knowledge of electrodynamics, it is difficult to account for magnetic phenomena, e.g., as they appear in the anomalous Zeeman effect, there are undoubtedly strong reasons for giving credence to a molecular model of the type proposed by Lenz, in which the two electrons move in opposite directions in a plane containing the nuclei, each electron revolving around its own nucleus, the two electrons always having one of the nuclei between them when they pass the line containing the nuclei. Because of the great difficulties involved, no detailed calculations of orbits and configurations have yet been made; however, there is so much more reason to be hopeful about this model, as it can be shown that two hydrogen atoms placed at a great distance apart, in orientations corresponding to Lenz' model, will attract each other with a force that is $1\frac{1}{2}$ times as strong as if the atoms were placed in the manner mentioned above when the formation of the geometrically simpler model was considered, a fact that indicates that Lenz' model may be the more stable. In this connection, it may be of interest to mention the models for the hydrogen molecule that have recently been proposed by Langmuir. Like Lenz' model, these possess no magnetic moment; however, the electronic motions are regarded as being of oscillatory nature, whereas in the former models they may be said to be of rotatory character. However, just this difference makes it difficult to see how molecules of the type proposed by Langmuir could be formed by the joining of two hydrogen atoms.

Turning to consider the element with atomic number 2, i.e., helium, we meet first of all an atom which contains only one electron, that may be called a singly charged positive helium ion. This is a system that is mechanically perfectly similar to the hydrogen atom, and, as is well known, it has been possible, with a theory entirely analogous to that holding for hydrogen, to account in all details for the so-called helium spark spectrum, whose series formula differs from the Balmer formula only in that the Rydberg constant is replaced by a four times larger constant. Before the present atomic theory was developed, this spectrum was ascribed to hydrogen, since there exists between it and the hydrogen spectrum a closer numerical connection than had ever been found between the properties of any two elements. However, this connection was recognized to be just the kind that was to be expected from Rutherford's nuclear theory and of which examples have later been found through Moseley's fundamental investigations of X-ray spectra.

When we turn to consider the neutral helium atom, we are faced with a far more complex problem. As discussed in the lecture mentioned above, from the existence of two complete series spectra of helium (the ortho- and parhelium spectra) of the usual Rydberg type, we are led to assume two separate classes of motions in the atom, in both of which one of the electrons moves in an orbit which is very large compared to the distance of the other electron from the nucleus, and in which this outer electron moves similarly to the electron in a hydrogen atom when it is perturbed by a central field with its center in the nucleus. Investigations of this problem, which have been carried out on the one hand by Landé and on the other hand as mentioned in the lecture by Dr. Kramers and the author, will be discussed in the following. Here we shall only mention the various attempts that have been made to determine the motion in the normal state of the atom, and thereby to explain the ordinary physical and chemical properties of helium. A simple model of the helium atom, which is closely related to the model of the hydrogen atom mentioned above, is obtained by assuming that both electrons move in a single circular orbit with the nucleus as center, each electron having an angular momentum of $h/2\pi$, and the two electrons being at opposite ends of a diameter. Because of the high energy required to remove one of the electrons from such a configuration to an infinite distance from the nucleus, this model, as has been shown by the author, seems suited to account for the chemical inertness of helium. Moreover, it gave a value for the ionization potential which is of the same order of magnitude as that found by experiments. However, according to the more accurate experiments carried out in recent years by Horton and Davies and by Franck ⟨and Knipping⟩, a value of 25.5 volts is found, and this is considerably lower than the value of about 29.3 volts calculated from the simple model in question. Thus, this model must be abandoned, a fact that is also indicated by the attempt to account for the formation of a helium atom by the successive binding of two electrons. In fact, it turns out that this simple configuration does not belong to any of the general classes of motions mentioned above. I shall later mention the conclusions about the motion in the normal state of the atom which it seems possible to draw from such considerations, and I shall discuss how it seems possible to give a simple and natural explanation of the important discovery by Franck and Knipping ⟨Should be Reiche; see Z. Phys. **1** (1920) 154⟩ of the so-called metastable state into which a helium atom can be brought by electron impact. Before leaving the question of the constitution of the helium atom, I shall just mention that is seems impossible to assume a configuration of the type proposed by Langmuir, which is similar to that proposed by him for hydrogen; for it appears that a type of motion as that proposed by Langmuir, in which the two electrons oscillate against each other, cannot arise from a process in which the two electrons are bound successively by the nucleus.

Passing to the next element, lithium, whose neutral atom contains three electrons, we have first the doubly charged positive ion, i.e., a system that is mechanically perfectly similar to a hydrogen atom and that must be expected to emit a spectrum of quite the same type as that of hydrogen, except that the Rydberg constant is replaced by a nine times larger constant. This spectrum, and that of the singly-charged lithium ion, have not been observed, in spite of strenuous and persistent attempts to do so, including some carried out in recent years in the physical laboratory in Copenhagen. Especially the latter spectrum would be of great value for testing the theoretical efforts made to interpret the spectrum of helium; for in this lithium ion we are dealing with a system that is mechanically similar to a neutral helium atom*. The difficulty of producing these spectra is undoubtedly closely related to the circumstance that two of the electrons in the normal state of the neutral lithium atom are bound much more strongly than

* This is an example of the general "Verschiebungssatz" for series spectra advanced by Sommerfeld and Kossel ⟨Verh. d. D. phys. Ges. **21** (1919)⟩, according to which a similarity must be expected between the arc spectrum of an element and the spark spectrum of the next element in the periodic system, an expectation that in many cases has been found to agree with experience.

the third electron, something that is plainly indicated by the chemical properties of lithium. This follows also from a consideration of the usual series spectrum of lithium which belongs to the well-known alkali metal spectra that have been discussed in the lecture mentioned. It was shown there that the quantum theory provides a simple explanation of the laws holding for the absorption spectra of alkali metal vapors, and for the production of radiation and ionization of such vapors by electron impact. In what follows, we shall return to the more detailed conclusions about the normal state of the lithium atom that can be drawn from its spectrum.

Proceeding to elements of higher atomic number, we should expect the presence of a large number of series spectra of different types, corresponding to the successive binding of the different electrons in the atom. However, the majority of these spectra cannot be directly observed by our present means of experimentation, since it has been impossible to produce atoms which have simultaneously lost a great number of their electrons. So far, in addition to the usual series spectra of the Rydberg type, which arise from neutral atoms, we have only been able to ascertain with certainty the existence of the so-called spark spectra, the series formula of which is of the type first investigated by Fowler. However, in addition to these spectra, there appear in elements of higher atomic number the so-called X-ray spectra, whose origin differs essentially from that of the series spectra, and whose structure is also different, in that they do not possess series of lines with an infinite number of terms whose frequencies converge to definite limits. These spectra do not arise when a number of electrons are removed from the atom, but the emission of them is connected essentially with a process in which a single one of the inner electrons moves to another orbit in the atom. This may perhaps be seen most clearly from experiments on the production of X-rays by electron bombardment. Thus, already in the early version of the quantum theory, it could be shown that the energy needed to produce the most penetrating kind of X-rays, the so-called K-radiation, is just of the order of magnitude required to remove an electron to infinity from a system consisting of a charge corresponding to the atomic number in question and a single electron in its normal state, an energy that may be expected to be of the same order of magnitude as that required to remove one of the inner-most electrons in the neutral atom. This view is also consistent with the simple laws for X-ray spectra found shortly afterwards in the fundamental investigations of Moseley. However, the formulas for the X-ray spectra given by Moseley show a greater similarity to the spectrum which, according to the theory, should be emitted by a nucleus and a single electron, than one would off-hand be inclined to expect. Thus, Moseley found that not only is the strongest line of the K-radiation given with considerable accuracy by the simple formula for the hydrogen spectrum when, in the expression for the Rydberg constant, N is put equal to the atomic number of the atom in question and $n' = 1$ and $n'' = 2$, but he also found that the strongest line of the less penetrating rays, the so-called L-radiation, was expressed, to a good approximation, by the same formula with $n' = 2$ and $n'' = 3$. However, as is shown by the widely different absorption features of the X-ray and the series spectra, it is clearly out of the question that the X-ray spectra have an origin similar to that of a series spectrum of the type corresponding to hydrogen; this is discussed in more detail in the often mentioned lecture. As mentioned there, and demanded by the theory, not all lines in a series spectrum are found to suffer selective absorption, but only a single series of lines, the so-called principal series, originating in transitions from the normal state of the atom to higher stationary states. On the other hand, in the case of the X-rays, it is found that there is selective absorption associated with both the K-radiation and the L-radiation; however, none of the K- or L-lines are absorbed themselves, but selective absorption sets in at the so-called K- and L-edges. This must be regarded as showing decisively that both the K- and the L-radiation are associated with processes in which a change occurs in the motion of an electron in the normal state of the atom, and that, as the Moseley formula at first sight might indicate, the latter kind of radiation does not correspond to a transition between two stationary states in which the motions of the electrons in both states deviate considerably from the motions in the normal state of the atom. Thus, we can hardly adopt any other explanation of the X-rays than that developed by Kossel, according to which, corresponding to the assumption that the K-radiation arises from the settling down of the atom after the removal of one of the most firmly bound electrons, the L-radiation is thought to arise from the reorganization of the atom after the removal of an electron belonging to a group lying farther from the nucleus and being less firmly bound. As shown by Kossel, an elaboration of this view leads to the prediction of certain relations between the frequencies of the individual K- and L-lines which are found to be verified with considerable accuracy by the measurements. From this view, it is also possible to account in detail for the characteristic absorption of X-rays (see Phil. Mag. ⟨**30** (1915) 394⟩, see also Millikan). Attempts to develop a more detailed theory of X-ray spectra, and especially an explanation of the differences in frequency between the individual K- and L-lines, have been given by Sommerfeld who, however, in his considerations nearly adopts the simple origin of these spectra suggested by the Moseley formula. Thus, Sommerfeld seeks to explain the differences mentioned by the deviations of the electronic motions from those expected from the usual mechanics, demanded by the theory of relativity, in a similar way as he has done for the fine structure of the hydrogen lines. In spite of the extremely interesting results obtained in this way by Sommerfeld, it is not possible, for the reasons mentioned, to carry over the conditions holding for the hydrogen spectrum directly to the X-ray spectra, and Sommerfeld has recently attempted to alter his assumptions, so that they can be brought into agreement with the absorption of X-rays, although at the same time he retains the essential basis for explaining the fine structure of X-rays. In the following, I shall return to the difficulties in the way of such an attempt and to possible means of overcoming them. Here, I shall only mention that any attempt in this direction seems to demand that the more loosely bound electrons responsible for the L-radiation must have a motion that quantum-theoretically corresponds not to that of the electron in the normal state of hydrogen but rather to that of the stationary state for which $n = 2$. To obtain a formal explanation of this surprising circumstance, Sommerfeld has tentatively advanced the interesting hypothesis that, in the normal state of an atom, only the innermost electrons are bound in the same way as the electron in the normal state of the hydrogen atom, whereas the more loosely bound electrons in the next "layer" ["Schicht"] would be bound as an electron in the second stationary state of hydrogen, and the electrons in the third layer as an electron in the third state of hydrogen, etc. However, this assumption cannot be valid; for it is definitely inconsistent with what we learn from the line spectra about the motions, in the normal state, of the electrons in the outermost layer, namely, that these motions do not differ much from one element to another in the same group (e.g., the alkali metals) of the periodic system, while, according to Sommerfeld's hypothesis, they should differ greatly, since they should correspond to different states of the hydrogen atom. However, we shall see in the following how it appears possible to obtain a rational basis for determining the motion of the electrons in the L-layer.

If we consider the other physical and chemical properties of the elements of higher atomic number, we find, as is well known, that, in contrast to the X-ray spectra, most of these properties vary in a characteristic periodic manner with the atomic number. This evidently requires an explanation of a different kind from anything that can be obtained by a simple generalization of conclusions drawn from the theory of the hydrogen spectrum. Thus, it seems impossible to explain satisfactorily this periodic variation of the properties of the elements by starting with the simple assumption that the electrons in the normal states move in circular orbits in the same or parallel planes, such as, for simplicity, it was assumed in the first tentative attempt to apply the quantum theory to the nuclear atom. Indeed, according to such an assumption, the different possible electron configurations in the atom will not have the stability that appears necessary to explain the specific properties of the elements. It is true that it seemed possible to argue that electrons preferably arrange themselves in groups of eight, as seems to follow from the chemical properties of the elements; for, from the condition that two rings can join together continuously (adiabatically) only when they

[56]

contain the same number of electrons, it follows that only electron rings with 2, 4, 8, . . . electrons can be formed adiabatically (cf. Phil. Mag. ⟨**26** (1913) 851⟩). But whatever value this argument may possess, it is neutralized by the fact that we cannot, by any similar consideration, see why rings, e.g., with eight electrons, should be formed again in the reorganization of the atom after the removal of one of the electrons, as in a process which, according to what was said above, must be assumed to lead to the emission of X-rays. A thorough discussion of the possible connection between the configuration of electrons in the atom and the chemical properties of the elements has been given by Kossel ⟨Ann. d. Phys. **49** (1916) 229, Naturwis. **7** (1919) 339, 360⟩ who has justly emphasized that the configurations tentatively suggested in the first paper of the writer do not allow us to understand the characteristic difference between the so-called electropositive and electronegative elements. Since Kossel still retains the assumption of coplanar electron orbits, his investigation does not provide any direct point of departure for a detailed quantum-theoretical treatment of the problems in question. From another side, an attempt has been made by Lewis, and elaborated by Langmuir, to explain the chemical properties of the elements by introducing the assumption that the electrons in an atom are distributed in space, and that, in particular, the appearance of groups of eight electrons corresponds to the presence of electron configurations with cubical symmetry. In the original form of this theory, it was assumed that the electrons are stationary or oscillate around definite equilibrium positions. This assumption, however, appears to be inconsistent with the basic postulates of the nuclear theory as well as with the quantum theory; hence, it is of great significance that, as pointed out by Born and Landé, there exist possible electronic motions with cubical symmetry which are consistent with the assumptions of the quantum theory and which appear to allow a determination of stationary states from quantum-theoretical principles. Born and Landé were led to the study of such possible motions not from a consideration of the chemical properties of the elements but from their investigation of crystal structure and, especially, of the compressibility of crystals. In fact, it turns out that it is not possible to account for the observed compressibility if the electrons are assumed to move in coplanar rings but only if the system of orbits has a higher degree of spatial symmetry. However, when trying to determine the stationary states corresponding to such orbits and especially to the normal state, we encounter difficulties which have already appeared prominently when simpler orbital forms were considered, namely, that the dimensions of the atoms and of the crystals become considerably smaller than calculated from measurements, provided we assume that the motions, in a generalized way, correspond to those in the normal state of the hydrogen atom. In a paper recently published by Landé, he points out that agreement in order of magnitude can be attained in many cases by introducing the assumption, similar to that used by Sommerfeld to explain the origin of the L-radiation, that the motion of the outer electrons in the normal state corresponds to a stationary state of the point atom characterized by $n = 2$. In a paper on the structure of diamond, Landé has even obtained some very promising results by assuming that in the normal state these electron orbits are not even approximately circular but strongly elliptical. After getting acquainted with this work through a lecture given by Landé before the Physical Society of Copenhagen in October this year, it has occurred to me that it might be possible from simple points of view to give a rational explanation of this at a first glance strange assumption. In fact, in the following I shall try to show that it seems possible, by pursuing these viewpoints further, to throw light on a great number of the problems mentioned in the foregoing and to create a hitherto lacking basis for an understanding of the peculiar stability of the elements, of which their specific properties so strongly bear witness.*

The point of departure of the following considerations is a simple investigation of the mechanical equilibrium of a particle (electron) moving in a circle in a central field. Denoting the necessarily attractive central force by $f(r)$, where r is the distance from the center, and denoting the mass of the particle by m and its velocity by v, we have as a condition for mechanical equilibrium in a circular orbit

$$mv^2/r = f(r).$$

For the angular momentum I about the center we further have

$$I = mvr$$

and, eliminating v, we get as a relation between I and r

$$I^2 = mr^3f(r). \tag{2}$$

Considering a central force varying as the inverse square of the distance, as that acting upon the electron in the hydrogen atom, we have $f(r) = eE/r^2$ and, hence, $I^2 = mreE$, a relation which simply determines the value of r corresponding to a given value of I. Now, putting $I = nh/2\pi$, we obtain directly values for r which are equal to those given by formula (1) for the major axis of the electron orbit in the stationary states of hydrogen. In the case considered, there is not only a value of I corresponding to any given value of r, but there is also a value of r corresponding to every value of I. However, this is not always the case. An extremely simple example is obtained by assuming that the central force, besides the main term varying inversely as the square of the distance, contains a term varying inversely

* [The preceding sentence was left uncompleted in the manuscript. However, there can be little doubt that its meaning is rendered correctly. There seems to be a gap between this sentence and what follows.]

as some higher power of the distance. To begin with, let us write, for simplicity,

$$f(r) = eE/r^2 + A/r^3,$$

where A is positive. Substituting in (2), we get

$$I^2 = m(eEr + A), \tag{3}$$

so that I has the minimum value

$$\sqrt{mA}.$$

This is related to a well-known theorem in the mechanics of central motions, which holds not only for circular orbits but for orbits of arbitrary shape, and which states that, if we compare a central motion corresponding to the force $f(r)$ with a motion for which there is an additional force inversely proportional to the third power of the distance, then there will correspond to each orbit of the former motion an orbit of the latter motion obtained by leaving the values of the radius vector unchanged but multiplying all angles formed by the radius vector with a fixed line by a constant equal to the ratio between the angular momenta of the latter and the former motions. We shall denote these by I' and I; they are connected by the relation

$$I'^2 = I^2 + mA,$$

provided the additional force, as above, is denoted by A/r^3. Thus, we see that, in the case considered, there exists a minimum value of the angular momentum which in the simple case of circular orbits corresponds to an infinitely small radius. However, we meet an even more different situation if we consider forces varying inversely as higher powers of the distance and write

$$f(r) = eE/r^2 + A'/r^3 + A''/r^4 + \ldots.$$

From (2) we obtain

$$I^2 = m(eEr + A' + A''/r + \ldots).$$

Assuming that A'', A''', ..., like A', are positive quantities, we get a minimum value for I which does not, as formerly, correspond to $r = 0$ but to a value $r_0 > 0$. Thus, for such a force field it is not possible to obtain a circular orbit with $r < r_0$. A similar limitation of the possible values of the angular momentum holds for other types of orbits, but the calculations are considerably more difficult to carry out.

We shall now turn to consider how this circumstance leads to the exclusion of certain states which beforehand would be regarded as possible, and which are just of the kind that must be considered in our search for the normal state. That this circumstance does not seem to have been sufficiently considered is because the corrections to the Coulomb force fields needed in the calculations of electron orbits in the atoms generally correspond to negative values of the coefficients A', A'', ... As long as we retain the assumption that the electrons move in the same plane or, in any case, as long as we assume that all orbits are circular and neglect the mutual perturbations among the electronic orbits which produce deviations from the circular form ...*. Let us assume that some spatially symmetrical electron configuration is moving around a nucleus, and let us investigate how its presence will affect the motion of an electron which is being caught by the atom after the formation of this configuration. A general idea of the influence of these first electrons on the motion of the last electron may be obtained by assuming that their charge is distributed over some spatial region, say, a sphere. If the orbits of the individual electrons are approximately circular, the charge on this sphere will be distributed largely over its surface; on the other hand, if the orbits have a considerable eccentricity, a large part of the charge will be found in the interior of the sphere. With regard to the question of the influence of this distribution of electricity on the motion of the last electron to be caught, we see that, as long as this electron never gets inside the surface of the sphere, the effect will be essentially the same as if the charge of the earlier-bound electrons were concentrated in the nucleus and simply tended to neutralize a part of its positive charge. On the other hand, if we imagine that the electron gets inside the sphere during part of its orbit, the influence will naturally be different. In this case, the part of the orbit lying outside the sphere will still be a Kepler ellipse, but, when the electron gets out again, it will no longer move in the same ellipse, but its orbit will be a segment of an identical ellipse whose major axis has turned an angle that will depend upon how far the electron was inside the sphere, and on the charge distribution of the sphere. If the entire orbit lies inside the sphere, the result will be a motion of a very different kind, and if, for example, we consider circular orbits, we must be prepared to meet peculiarities as those considered in the foregoing example, which may exclude certain conceivable stationary states. We get an especially drastic example by imagining that all the charge is concentrated on the surface of the sphere. In this case, when the electron is moving from the outside into the sphere, the attractive force will increase suddenly on its passage through the surface; for the charge in the sphere will have no effect inside, while exerting a considerable repulsion outside. Hence, if we imagine a circular electron orbit which, to begin with, lies outside the sphere and has a definite angular momentum, and then consider a series of circular orbits with gradually decreasing radii, we find that, as long as we are outside the sphere, the

* [This sentence is left incomplete. There seems to be a gap here.]

angular momentum will decrease continuously, but, the moment we pass through the surface of the sphere, the angular momentum and the velocity of the electron must increase considerably, if the dynamical equilibrium is to be retained. Afterwards, the angular momentum will again decrease continuously ⟨with decreasing radius⟩ and it will eventually reach and become smaller than the value corresponding to orbits immediately outside the surface of the sphere. However, these ⟨inside⟩ orbits will not have any simple mechanical relation to the orbits lying outside. Consider, e.g., a thought experiment in which we begin with a circular orbit lying outside the sphere and imagine that the electron loses angular momentum continuously, e.g., by the emission of electromagnetic radiation; the orbit would then slowly shrink at first and become a spiral, which at each instant may be regarded as practically circular. However, as soon as it touches the sphere, the orbit will assume an essentially different shape, coinciding very nearly with an ellipse of finite eccentricity which touches the sphere in an inside point. Under continued emission, and resulting loss of angular momentum and energy of the electron, the major axis of this ellipse will decrease continuously, and the orbit will more and more approach a circular form, which it will not reach, however, until its dimensions have become infinitely small, i.e., until the electron has fallen into the nucleus and has emitted an infinite amount of energy. In such a case, all stationary states will depend on the radius of the sphere. If this is smaller than the smallest distance of the electron from the nucleus in an orbit lying entirely in the outer field, i.e., less than one half of the value of the parameter obtained from the formula for $2p$ by putting $k = 1$, ⟨i.e., $h^2/4\pi^2 e^2 m$; the manuscript has $\frac{1}{4}$ instead of $\frac{1}{2}$; the formula is given in Z. Phys. **2** (1920) 451⟩ we must expect that the stationary states will be very nearly the same as if it were assumed that all the earlier bound electrons were concentrated in the nucleus. Hence, during the binding of the electron in question, a spectrum would be emitted of very nearly the same type as that of hydrogen. If the sphere were larger than the value just mentioned, orbits with $k = 1$ would be capable of penetrating into the sphere, and the energy in the stationary states with $k = 1$ would differ from those of hydrogen for the same value of n. This will appear as a decrease in the value of the energy which to begin with will be more pronounced for high values of n, since in such orbits the electron will get closer to the nucleus. When we investigate this effect for increasing values of the radius of the sphere, ⟨we find that⟩, as soon as this becomes equal to the radius of the circular orbit corresponding to $n = 1$ (i.e., equal to ⟨the manuscript has "twice as great as"⟩ the formerly mentioned critical value), the result will be that the state corresponding to the orbit will not be capable of physical existence; hence, the normal state of the atom will correspond to an orbit with $n = 2$, $k = 1$, which will have considerably less ⟨energy⟩ than the circular orbit, $n = 2$, $k = 2$, which will still lie far outside the sphere (since its radius will be four times as large as that of the sphere). During the continued increase of the radius of the sphere, there will continue to be an influence upon the stationary states corresponding to $k = 1$; their energies will deviate more and more from those of the hydrogen states, until the radius has become four times as large as the critical value ρ_0, in which case the effect will begin to appear in orbits corresponding to $k = 2$. It is interesting to note that the observed spectra of the alkali metals are very similar to the spectra which would be emitted by the imagined idealized model in the interval in which the radius of the sphere lies between $2\rho_0$ and $4\rho_0$. Thus, we find for the Li spectrum that the energy in the series of stationary states corresponding to $k = 2, 3, \ldots$ deviates very little from the states of hydrogen. In the state $n = 2$, $k = 2$, where the deviation is largest, it is no more than ca. 4%. Moreover, we find that the energy values for the series of states corresponding to $k = 1$ deviate very much from those of hydrogen. In fact, as is well known, the spectral terms corresponding to these states are given with great accuracy by

$$K/(\tau+\tfrac{1}{2})^2,$$

where τ is an integer that assumes the values 1, 2, 3, ... for the different terms; this is contrary to hydrogen where the states are simply given by

$$K/n^2.$$

Since these terms, so to speak, deviate equally much from the preceding and the following value of a hydrogen term, it has hitherto not been possible from the terms alone to know if one should correlate the values of τ with n or with $n-1$; but, since there was no reason to assume that the normal state did not correspond to $n = 1$, it was most natural, from a theoretical point of view, to assume that τ corresponds to n. However, from a mechanical consideration, it has been difficult to understand why the energy values were so much greater than for the corresponding states of hydrogen; for this difference would have to be ascribed to a much greater repulsion from the inner electrons than would be obtained if these were placed in the nucleus. Such a repulsion would be difficult to account for mechanically, since it is a general rule that mutual perturbations lead to increased attraction rather than repulsion between mechanical systems; for the perturbations cause repelling particles to tend to be as far from one another as possible, while attracting particles tend to approach one another. However, from the considerations to which we are led by the investigation of the idealized model mentioned, there is no difficulty in assuming that we are dealing with a decrease in the energy values resulting from an increase in the attraction; for this attraction can be thought to have caused the state corresponding to $n = 1$ to become physically excluded, so that, in the expression for the Li terms, we can identify τ with $n-1$.

Quite similar circumstances hold for the Na spectrum, which was used in the lecture as an example of a typical alkali spectrum, and whose spectral terms were illustrated in the figure on p. 434 ⟨Z. Phys. **2** (1920) 423⟩. However, since at that time there seemed to be no reason to exclude the state $n = 1, k = 1$, it was assumed in the schematic figure that the first spectral term of the sharp series was characterized by $n = 1$, $k = 1$, while, according to the considerations developed in the present paper, there can hardly be any doubt that this state must be characterized by $n = 2$, $k = 1$. However, this involves only a displacement of the n-values in the first line of the figure and does not entail any change at all in the general considerations that follow about the absorption and the energies of the spectral lines, or about the apparent limitation of the combination principle. In fact, these considerations concern only the question of the energy values in the stationary states, which are independent of the nomenclature, and the k-values determining the angular momentum. Only the brief consideration at the bottom of p. 460 ⟨Z. Phys. **2** (1920) 423⟩ about the ratio between the relative intensities of the different lines which can appear in a spontaneous transition from an initial state corresponding to $n = 3$, $k = 2$, can clearly not be retained in detail. In the case of the Na spectrum, however, the corrections for the other series of spectral terms are considerably greater than for Li, being, as shown by the figure, about 10% for the series of terms with $k = 2$. However, this is very little in comparison with the great changes in the energy values of the series corresponding to $k = 1$, and this deviation from what might be expected from the idealized model is hardly surprising. I shall return to discuss the great difference that may be thought to exist between the inner electron configurations in Li and Na. ...*.

In the case just mentioned, it was assumed that the additional terms in the attractive force originate from a fixed field, in the sense that the constants of the added terms are independent of the motion of the ⟨outer⟩ electron and, in particular, of its angular momentum. However, a different situation is met with when these terms depend to a considerable degree upon changes in the motion of the inner electrons resulting from forces originating in the outer electron. In fact, we are dealing with such a case in the helium atom and, in the investigations of Landé, and of Kramers and the author, the main problem has been to determine the effect of the mutual perturbations between the inner and the outer electrons. In Landé's work the investigation is limited to the case in which the orbit of the outer electron is approximately circular, but, as will be explained elsewhere, since he has applied a method of approximation that cannot be justified, his results are not even valid for this case. In our investigation, the calculation has been carried out without introducing, at the outset, any assumptions about

* [This short sentence is illegible.]

the orbit of the outer electron, and we have been led to the recognition of two classes of motions which, in our opinion, provide an understanding of the two separate helium spectra. As in Landé's work, we assume that we are dealing, in the orthohelium spectrum, with stationary states in which both electrons move in the same plane and with the same direction of revolution; we have shown that there exists in this case a general class of motions in which the shape of the orbit of the inner electron remains approximately constant under the influence of the outer electron, while this orbit turns in such a way that the direction from the nucleus to the perihelion coincides at all times with the radius vector from the nucleus to the outer electron. Now, the calculation shows that, for the solution in question, the eccentricity of the inner orbit, which to a first approximation determines the field in which the outer electron moves, depends only upon the parameter of the outer orbit and is quite independent of its eccentricity. Thus, for the eccentricity of the inner orbit, we find

$$\varepsilon = \sqrt{p/a_0}\,.$$

This, in turn, results in a field of force which to a first approximation can be expressed by

$$f(r) = eE/r^2 + \tfrac{4}{3}\varepsilon a_0 e/r^3.$$

Since this force depends on the angular momentum of the outer electron, which is proportional to p, the investigation cannot be carried out as simply as when the force field was given by $f(r) = eE/r^2 + A/r^3$. However, if we limit ourselves to circular orbits, the problem is very simple, and corresponding to relation (3) we get

$$I^2 = \langle m \rangle eEr + f(I),$$

an expression which, as may be seen, gives a minimum value for I greater than $h/2\pi$. Even though the calculation is only approximative, it is clear, nevertheless, that these considerations provide a natural explanation of the peculiar phenomenon discovered by Franck and Knipping that the minimum values of the energy in the stationary states involved in the orthohelium spectrum correspond not to $n = 1$, $k = 1$, but to $n = 2$, $k = 1$. It may be noted here that Landé, without considering the circumstance mentioned here, has calculated, from his formulas, the energy of the possible orthohelium state $n = 1$, $k = 1$, and has found a value even smaller than that obtained from the simple helium model mentioned above, a value too small even to explain the measured ionization potential of helium. In much the same way, Kramers and the author had calculated the energy in this state and had found the fundamental difficulty about the origin of the helium spectrum which disappears when the viewpoints advanced here are invoked ...*

As concluded by Franck and Reiche from their experimental results, it must be assumed that the normal state of the helium atom belongs to the states of the parhelium spectrum; thus, it may be said that in this spectrum we observe the formation of the helium atom, in the same manner as the hydrogen spectrum reveals the formation of the hydrogen atom. It seems possible to give a simple mechanical explanation of this circumstance. In Landé's investigation as well as in ours, it is assumed from essentially different points of view, however, that the two orbits connected with parhelium do not lie in the same plane; this will be explained in detail in the comprehensive paper.

When we consider stationary states in which there is less and less difference between the dimensions of the two orbits, the situation is essentially different for spatial and for coplanar orbits, since, for the former type, it is possible to obtain a gradual transition between states in which the orbits of the two electrons are very different and one in which the motions of the two electrons are of the same kind, e.g., in which the two electrons may be said to be equivalent; hence, we are enabled to form an idea of how such a configuration can be reached by successive binding of the two electrons. On the other hand, in the case of coplanar orbits, such a successive formation of the atom is impossible, since a transition between states in which the two electrons move in different positions but in equivalent orbits is excluded. For example, it will be seen that, quite apart from these considerations about the motion in the normal state, it is impossible to bring about a gradual transition from the orthohelium states to the simple model of the atom, often mentioned above, in which the two electrons move in the same circular orbit.

Applying the foregoing results, we shall now attempt to give a survey of certain main features of the atomic structure of the elements. In the foregoing, we have dealt in some detail with the atoms of hydrogen and helium. The next question concerns the electron configuration in lithium. In the first place, we must assume that the two innermost electrons have the same configuration as the electrons in the normal state of the helium atom, in which, as we have seen, we must assume that a configuration in which the motions of the two electrons are equivalent has been formed by successive binding of the electrons. Asking now about the binding of the third electron, we have direct evidence about it in the spectrum. As was discussed in the foregoing, everything indicates that this electron, in the normal state, will move in an elliptic orbit corresponding to $n = 2$, $k = 1$; thus, in most of its orbit it moves at distances from the nucleus which are large compared to those of the two inner electrons, but, when it passes

* [This sentence is left incomplete in the manuscript. It may have been completed on the next page which, however, is missing. Hence, there is a gap here.]

its perihelion, it gets as close to the nucleus as these; indeed, the foregoing consideration of the idealized case, in which the charges of the two inner electrons are evenly distributed over a sphere, gives very strong reasons for believing that, during its passage through the perihelion, the third electron will get even closer to the nucleus than the largest distance from the nucleus of the two inner electrons during their motion. In this way, it becomes possible to understand the great difference between the chemical properties of the hydrogen and the lithium atoms. This was puzzling in the preliminary version of the theory given in my first papers; for it was not clear how the inner electrons could have the strong influence upon the outer electron demanded by the chemical properties, if it was assumed that this electron moved in an approximately circular orbit in the same plane as the inner electrons, whose orbits were also assumed to be nearly circular. On the other hand, just this assumed strongly eccentric orbit of the outer electron makes possible a strong influence of the inner electrons on the outer, resulting in what might be called a direct coupling between the motions of the three electrons.

Passing to the next element, beryllium, we shall assume that the first three electrons possess a configuration, formed again by successive binding, which is similar to that of the neutral lithium atom. The question is now how the fourth electron may be assumed to be bound in the atom. In view of the strongly eccentric orbit of the third electron, it is natural to assume that the fourth electron will be bound in a similar manner; for the orbits will apparently not disturb each other very much, if we assume that they are oriented in such a way that the major axes point nearly in opposite directions, and that the electrons revolve in their orbits with such a phase difference that one of the electrons is in its aphelion when the other is in its perihelion. In this way, it seems actually possible to understand the characteristic properties of beryllium. First of all, it is seen how the coupling between the inner and the outer electrons makes it possible to explain the great difference between the properties of beryllium and helium, as evidenced by the fact that the former is an active electropositive element, while the latter is definitely a chemically inert element. There was no possibility whatever of understanding this difference in the first version of the theory, in which, although it was assumed that beryllium possesses two inner and two outer electrons, it was concluded, from a consideration of coplanar orbits, that the outer electrons in the beryllium atom were bound almost as firmly as the electrons in the helium atom. Moreover, the peculiar equivalent motions of the outer electrons offer not only the possibility of explaining the two electropositive valencies, but also the ease with which the spark spectra of such divalent elements are obtained. It is true that no data of this kind are known for beryllium itself; however, for the next element of this kind, magnesium, the spark spectrum has been determined very accurately by Fowler, and, as mentioned above, its series formula has precisely the form to be expected, if it is emitted by an atom from which one electron has been entirely removed, while another moves in an orbit whose dimensions are large compared to the distances of the inner electrons from the nucleus. For a long time, however, it was very difficult to reconcile the surprising ease with which the spark spectra of the group in question could be obtained, with the prevalent ideas of the electron configuration in the normal state. (Two-quantic orbits bound more firmly than in hydrogen, although less firmly than orbits in hydrogen with $n = 1$.)*

Passing to the next element, boron, we shall assume that the four innermost electrons are bound in the same way as the electrons in beryllium. In fact, a coupling as that described between the two inner and the two outer electrons will be the more probable the higher the charge on the nucleus; for, as this charge increases, the relative value of the repulsion between the electrons will decrease, and, hence, the dimensions of the various electron orbits will become more nearly equal. With regard to the fifth electron in the boron atom, we are naturally led to assume that it will be bound in much the same way as the third and fourth electrons; this binding corresponds, so to speak, to a falling in towards the nucleus, and it is reasonable to assume that the final orbit can be classified among the series of states which corresponds to the binding of the electron in the atom. Thus, in the normal state, we shall assume that the three outer electrons are bound in equivalent manner, such that they find themselves in the perihelion of their orbits at equally spaced instants. Whether we must assume that the orbits lie in the same plane or in other equivalent configurations is hard to decide without a detailed calculation of their mutual perturbations. From certain considerations, concerned mainly with the magnetic properties of the elements, it seems probable that the entire angular momentum of the atom in the normal state is either 0 or $h/2\pi$. (Of course, these sketchy remarks also apply to beryllium, but they must be worked out in detail.)

Passing to carbon, we are led to assume that the first five electrons are bound as in boron, while the sixth electron will be bound in such a way that the four outermost electrons form a configuration in which the motion of each electron is equivalent, and in which the electrons pass through their perihelions and get into the region of the two innermost electrons at equally spaced intervals. With regard to the detailed description of the orbits, we can hardly assume that they all lie in the same plane; for the average distance between the electrons, and, hence, their potential energy, may be expected to be less when the major axes at each instant have relative positions approximately as the lines connecting the center and the corners of a regular tetrahedron than when the major axes form the diagonals of a square. However, the detailed computation would be extremely

* [This parenthesis is in Bohr's handwriting.]

complicated because of the large perturbations (rotation of the perihelions) suffered by the orbits of the outer electrons each time these pass close by the nucleus, i.e., the perturbations upon which the coupling and the very stability of the configuration depend. When the considerations presented here are compared with those of my first papers, it is hardly necessary to discuss the great difference that these circumstances make; for all that was said earlier in the discussion of beryllium is still more valid here. On the other hand, it may be of interest to discuss more closely the difference between the electron configuration of carbon that we are led to here and the configurations with spatial cubical symmetry proposed by Born and Landé, and which especially the latter has applied successfully in his recent work on the structure of diamond. First of all, it must be emphasized that these authors have given no thought to the influence of the coupling between different electron layers in the atom and, hence, have based their arguments for possible configurations solely on symmetry properties of the field of force surrounding the atom (as in Born's considerations about the compressibility of crystals), or on the rather small differences in the potential energy of the electrons in planar and spatial configurations. However, if we consider the general properties of the elements, we see in the most striking manner that this coupling plays a decisive role for the properties of the elements. Thus, in the last case considered, carbon, it is clear that it is not in the number four of the outer electrons in this element that we must seek its specific properties, e.g., its well-known valence of four in organic chemistry, and that it forms the hardest and most compact crystal of cubical symmetry; for, if this were the case, there seems to be no reason why these properties should not be possessed in equal or even higher degree by beryllium, an element which, as is well known, has quite different properties. On the contrary, the characteristic properties of carbon must be assumed to arise from the fact that in the neutral atom there are four electrons which are coupled to the two inner electrons, so as to make their motions essentially independent of one another, in the sense that the binding energy of each of them is far less dependent on the other electrons and far smaller than would be the case in a normal state calculated on the assumption that the four electrons revolve in the same circular orbit or in orbits distributed with cubical symmetry, as suggested and investigated by Born and Landé. They assumed that the electrons not only are equivalent but, as in the simple plane motions discussed by several authors in connection with the problem in question and called "Ellipsenverein" by Sommerfeld, that they at any one time occupy corresponding positions, such that they all pass through their perihelions together and at a later moment all are at their greatest distance from the nucleus. When, as in Landé's and Born's papers, no special regard is paid to the presence of the inner electrons, it is impossible to give any reason for excluding a state analogous to the normal state of hydrogen.

If it is assumed, nevertheless, that the outer electrons move in elliptic orbits corresponding to the second stationary state of hydrogen, such as Landé did in order to obtain agreement with the observed sizes of the atoms, e.g., from crystal dimensions, then it seems unavoidable that these orbits would lead to difficulties; for, if the four electrons enter the region of the two inner electrons simultaneously, they must exert perturbing effects on these of a magnitude that is hardly compatible with the continued existence of the electron configuration considered, a configuration characterized by the sharp distinction between outer and inner electrons. (This must be worked out in more detail, with a more precise description of how exceedingly close the outer electrons get to the nucleus.)

It will be noted that the criticism contained in these remarks is of a fundamental nature and involves general viewpoints which, it seems, must form the basis for a quantum-theoretical explanation of the properties of the atoms; on the other hand, it is not meant to show any lack of appreciation of the interesting results achieved especially by Landé, results which, independently of the theoretical discussion, depended on introducing, for certain of the electrons, orbits which are not analogous to the circular ones in the normal state of hydrogen, but rather correspond to the elliptic orbits corresponding to the second stationary state of this atom.

Continued discussion of the periodic system.

X-ray spectra.

Perhaps a word about magnetism.

Passing on to the next elements in the periodic system, nitrogen, oxygen, and fluorine, which make up the last elements of the second row* of the system, similar considerations lead us to assume that the first six electrons to be bound in the atom will assume a configuration similar to that in carbon, i.e., the two inner electrons will move in spatial equivalent "one-quantic" orbits, while the four outer electrons will move in equivalent "two-quantic" orbits, coupled to the inner electrons and exhibiting perfect spatial symmetry with respect to the nucleus. If we now ask about the configuration of the electrons to be bound later, we are faced with the circumstance that five, six, or seven orbits as those of the outer electrons in lithium, beryllium, boron, and carbon, cannot be distributed in a symmetrical manner, as was the case with two, three or four electrons. However, it is difficult to decide what significance should be attributed to this fact; for, as already mentioned, we consider the existence of a coupling between the inner and the outer electrons to be of primary importance and the symmetry of the outer electrons to play a secondary role. However, with an increasing number of electrons, there are several points to consider with regard to this coupling.

* [The manuscript has "first row"; however, "second row or period" is in conformity with Bohr's later usage.]

II. ATOMIC STRUCTURE

(Nature **107** (1921) 104)*

* [See Introduction, sect. 2.]

Reprinted from NATURE, *March* 24, 1921]

Atomic Structure.

IN a letter to NATURE of November 25 last Dr. Norman Campbell discusses the problem of the possible consistency of the assumptions about the motion and arrangement of electrons in the atom underlying the interpretation of the series spectra of the elements based on the application of the quantum theory to the nuclear theory of atomic structure, and the apparently widely different assumptions which have been introduced in various recent attempts to develop a theory of atomic constitution capable of accounting for other physical and chemical properties of the elements. Dr. Campbell puts forward the interesting suggestion that the apparent inconsistency under consideration may not be real, but rather appear as a consequence of the formal character of the principles of the quantum theory, which might involve that the pictures of atomic constitution used in explanations of different phenomena may have a totally different aspect, and nevertheless refer to the same reality. In this connection he directs attention especially to the so-called "principle of correspondence," by the establishment of which it has been possible—notwithstanding the fundamental difference between the ordinary theory of electromagnetic radiation and the ideas of the quantum theory—to complete certain deductions based on the quantum theory by other deductions based on the classical theory of radiation.

In so far as it must be confessed that we do not possess a complete theory which enables us to describe in detail the mechanism of emission and absorption of radiation by atomic systems, I naturally agree that the principle of correspondence, like all other notions of the quantum theory, is of a somewhat formal character. But, on the other hand, the fact that it has been possible to establish an intimate connection between the spectrum emitted by an atomic system—deduced according to the quantum theory on the assumption of a certain type of motion of the particles of the atom—and the constitution of the radiation, which, according to the ordinary theory of electromagnetism, would result from the same type of motion, appears to me to afford an argument in favour of the reality of the assumptions of the spectral theory of a kind scarcely compatible with Dr. Campbell's suggestion. On the contrary, if we admit the soundness of the quantum theory of spectra, the principle of correspondence would seem to afford perhaps the strongest inducement to seek an interpretation of the other physical and chemical properties of the elements on the same lines as the interpretation of their series spectra; and in this letter I should like briefly to indicate how it seems possible by an extended use of this principle to overcome certain fundamental difficulties hitherto involved in the attempts to develop a general theory of atomic constitution based on the application of the quantum theory to the nucleus atom.

The common character of theories of atomic constitution has been the endeavour to find configurations and motions of the electrons which would seem to offer an interpretation of the variations of the chemical properties of the elements with the atomic number as they are so clearly exhibited in the well-known periodic law. A consideration of this law leads directly to the view that the electrons in the atom are arranged in distinctly separate groups, each containing a number of electrons equal to one of the periods in the sequence of the elements, arranged according to increasing atomic number. In the first attempts to obtain a definite picture of the configuration and motion of the electrons in these groups it was assumed that the electrons within each group at any moment were placed at equal angular intervals on a circular orbit with the nucleus at the centre, while in later theories this simple assumption has been replaced by the assumptions that the configurations of electrons within the various groups do not possess such simple axial symmetry, but exhibit a higher degree of symmetry in space, it being assumed, for instance, that the configuration of the electrons at any moment during their motions possesses polyhedral symmetry. All such theories involve, however, the fundamental difficulty that no interpretation is given why these configurations actually appear during the formation of the atom through a process of binding of the electrons by the nucleus, and why the constitution of the atom is essentially stable in the sense that the original configuration is reorganised if it be temporarily disturbed by external agencies. If we reckon with no other forces between the particles except the attraction and repulsion due to their electric charges, such an interpretation claims clearly that there must exist an intimate interaction or "coupling" between the various groups of electrons in the atom which is essentially different from that which might be expected if the electrons in different groups are assumed to move in orbits quite outside each other in such a way that each group may be said to form a "shell" of the atom, the effect of which on the constitution of the outer shells would arise mainly from the compensation of a part of the attraction from the nucleus due to the charge of the electrons.

These considerations are seen to refer to essential features of the nucleus atom, and so far to have no special relation to the character of the quantum theory, which was originally introduced in atomic problems in the hope of obtaining a rational interpretation of the stability of the atom. According to this theory an atomic system possesses a number of distinctive states, the so-called "stationary states," in which the motion can be described by ordinary mechanics, and in which the atom can exist, at any rate for a time, without emission of energy radiation. The characteristic radiation from the atom is emitted only during a transition between two such states, and this process of transition cannot be described by ordinary mechanics, any more than the character of the emitted radiation can be calculated from the motion by the ordinary theory of electro-magnetism, it being, in striking contrast to this theory, assumed that the transition is always followed by an emission of monochromatic radiation the frequency of which is determined simply from the difference of energy in the two states. The application of the quantum theory to atomic problems—which took its starting point from the interpretation of the simple spectrum of hydrogen, for which no *a priori* fixation of the stationary states of the atoms was needed—has in recent years been largely extended by the development of systematic methods for fixing the stationary states corresponding to certain general classes of mechanical motions. While in this way a detailed interpretation of spectroscopic results of a very different kind has been obtained, so far as phenomena which depend essentially on the motion of one electron in the atom were concerned, no definite elucidation has been obtained with regard to the constitution of atoms containing several electrons, due to the circumstance that the methods of fixing stationary states were not able to remove the arbitrariness in the choice of the number and configurations of the electrons in the various groups, or shells, of the atom. In fact, the only immediate consequence to which they lead is that the motion of every electron in the atom will on a first approximation correspond to one of the stationary states of a system consisting of a particle moving in a central field of force, which in their limit are represented by the various circular or elliptical stationary orbits which appear in Sommerfeld's theory of the fine structure of the hydrogen lines. A way to remove the arbitrariness in question is opened, however, by the introduction of the correspondence principle, which gives expression to the tendency in the quantum theory to see not merely a set of formal rules for fixing the stationary states of atomic systems and the frequency of the radiation emitted by the transitions between these states, but rather an attempt to obtain a rational generalisation of the electromagnetic theory of radiation which exhibits the discontinuous character necessary to account for the essential stability of atoms.

Without entering here on a detailed formulation of the correspondence principle, it may be sufficient for the present purpose to say that it establishes an intimate connection between the character of the motion in the stationary states of an atomic system and the possibility of a transition between two of these states, and therefore offers a basis for a theoretical examination of the process which may be expected to take place during the formation and reorganisation of an atom. For instance, we are led by this principle directly to the conclusion that we cannot expect in actual atoms configurations of the type in which the electrons within each group are arranged in rings or configurations of polyhedral symmetry, because the formation of such configurations would claim that all the electrons within each group should be originally bound by the atom at the same time. On the contrary, it seems necessary to seek the configurations of the electrons in the atoms among such configurations as may be formed by the successive binding of the electrons one by one, a process the last stages of which we may assume to witness in the emission of the series spectra of the elements. Now on the correspondence principle we are actually led to a picture of such a process which not only affords a detailed insight into the structure of these spectra, but also suggests a definite arrangement of the electrons in the atom of a type which seems suitable to interpret the high-frequency spectra and the chemical properties of the elements. Thus from a consideration of the possible transitions between stationary states, corresponding to the various steps of the binding of each of the electrons, we are led in the first place to assume that only the two first electrons move in what may be called one-quantum orbits, which are analogous to that stationary state of a central system which corresponds to the normal state of a system consisting of one electron rotating round a nucleus. The electrons bound after the first two will not be able by a transition between two stationary states to procure a position in the atom equivalent to that of these two electrons, but will move in what may be called multiple-quanta orbits, which correspond to other stationary states of a central system.

The assumption of the presence in the normal state of the atom of such multiple-quanta orbits has already been introduced in various recent theories, as, for instance, in Sommerfeld's work on the high-frequency spectra and in that of Landé on atomic dimensions and crystal structure; but the application of the correspondence principle seems to offer for the first time a rational theoretical basis for these conclusions and for the discussion of the arrangement of the orbits of the electrons bound after the first two. Thus by means of a closer examination of the progress of the binding process this principle offers a simple argument for concluding that these electrons are arranged in groups in a way which reflects the periods exhibited by the chemical properties of the elements within the sequence of increasing atomic numbers. In fact, if we consider the binding of a large number of electrons by a nucleus of high positive charge, this argument suggests that after the first two electrons are bound in one-quantum orbits, the next eight electrons will be bound in two-quanta orbits, the next eighteen in three-quanta orbits, and the next thirty-two in four-quanta orbits.

Although the arrangements of the orbits of the electrons within these groups will exhibit a remarkable degree of spatial symmetry, the groups cannot be said to form simple shells in the sense in which this expression is generally used as regards atomic constitution. In the first place, the argument involves that the electrons within each group do not all play equivalent parts, but are divided into sub-groups corresponding to the different types of multiple-quanta orbits of the same total number of quanta, which represents the various stationary states of an electron moving in a central field. Thus, corresponding to the fact that in such a system there exist two types of two-quanta orbits, three types of three-quanta orbits, and so on, we are led to the view that the above-mentioned group of eight electrons consists of two sub-groups of four electrons each, the group of eighteen electrons of three sub-groups of six electrons each, and the group of thirty-two electrons of four sub-groups of eight electrons each.

Another essential feature of the constitution described lies in the configuration of the orbits of the electrons in the different groups relative to each other. Thus for each group the electrons within certain sub-groups will penetrate during their revolution into regions which are closer to the nucleus than the mean distances of the electrons belonging to groups of fewer-quanta orbits. This circumstance, which is intimately connected with the essential features of the processes of successive binding, gives just that expression for the "coupling" between the different groups which is a necessary condition for the stability of atomic configurations. In fact, this coupling is the predominant feature of the whole picture, and is to be taken as a guide for the interpretation of all details as regards the formation of the different groups and their various sub-groups. Further, the stability of the whole configuration is of such a character that if any one of the electrons is removed from the atom by external agencies not only may the previous configuration be reorganised by a successive displacement of the electrons within the sequence in which they were originally bound by the atom, but also the place of the removed electron may be taken by any one of the electrons belonging to more loosely bound groups or sub-groups through a process of direct transition between two stationary states, accompanied by an emission of a monochromatic radiation. This circumstance—which offers a basis for a detailed interpretation of the characteristic structure of the high-frequency spectra of the elements—is intimately connected with the fact that the electrons in the various sub-groups, although they may be said to play equivalent parts in the harmony of the interatomic motions, are not at every moment arranged in configurations of simple axial or polyhedral symmetry as in Sommerfeld's or Landé's work, but that their motions are, on the contrary, linked to each other in such a way that it is possible to remove any one of the electrons from the group by a process whereby the orbits of the remaining electrons are altered in a continuous manner.

These general remarks apply to the constitution and stability of all the groups of electrons in the atom. On the other hand, the simple variations indicated above of the number of electrons in the groups and sub-groups of successive shells hold only for that region in the atom where the attraction from the nucleus compared with the repulsion from the electrons possesses a preponderant influence on the motion of each electron. As regards the arrangements of the electrons bound by the atom at a moment when the charges of the previously bound electrons begin to compensate the greater part of the positive charge of the nucleus, we meet with new features, and a consideration of the conditions for the binding process forces us to assume that new, added electrons are bound in orbits of a number of quanta equal to, or fewer than, that of the electrons in groups previously bound, although during the greater part of their revolution they will move outside the electrons in these groups. Such a stop in the increase, or even decrease, in the number of quanta characterising the orbits corresponding to the motion of the electrons in successive shells takes place, in general, when somewhat more than half the total number of electrons is bound. During the progress of the binding process the electrons will at first still be arranged in groups of the indicated constitution, so that groups of three-quanta orbits will again contain eighteen electrons and those of two-quanta orbits eight electrons. In the neutral atom, however, the electrons bound last and most loosely will, in general, not be able to arrange themselves in such a regular way. In fact, on the surface of the atom we meet with groups of the described constitution only in the elements which belong to the family of inactive gases, the members of which from many points of view have also been acknowledged to be a sort of landmark within the natural system of the elements. For the atoms of these elements we must expect the constitutions indicated by the following symbols:

Helium	(2_1),	Krypton	$(2_1 8_2 18_3 8_2)$,
Neon	$(2_1 8_2)$,	Xenon	$(2_1 8_2 18_3 18_3 8_2)$,
Argon	$(2_1 8_2 8_2)$,	Niton	$(2_1 8_2 18_3 32_4 18_3 8_2)$,

where the large figures denote the number of electrons in the groups starting from the innermost one, and the small figures the total number of quanta characterising the orbits of electrons within each group.

These configurations are distinguished by an inherent stability in the sense that it is especially difficult to remove any of the electrons from such atoms so as to form positive ions, and that there will be no tendency for an electron to attach itself to the atom and to form a negative ion. The first effect is due to the large number of electrons in the outermost group; hence the attraction from the nucleus is not compensated to the same extent as in configurations where the outer group consists only of a few electrons, as is the case in those families of elements which in the periodic table follow immediately after the elements of the family of the inactive gases, and, as is well known, possess a distinct electro-positive character. The second effect is due to the regular constitution of the outermost group, which prevents a new electron from entering as a further member of this group. In the elements belonging to the families which in the periodic table precede the family of the inactive gases we meet in the neutral atom with configurations of the outermost group of electrons which, on the other hand, exhibit a great tendency to complete themselves by the binding of further electrons, resulting in the formation of negative ions.

The general lines of the latter considerations are known from various recent theories of atomic constitution, such as those of A. Kossel and G. Lewis, based on a systematic discussion of chemical evidence. In these theories the electro-positive and electro-negative characters of these families in the periodic table are interpreted by the assumption that the outer electrons in the atoms of the inactive gases are arranged in especially regular and stable configurations, without, however, any attempt to give a detailed picture of the constitution and formation of these groups. In this connection it may be of interest to direct attention to the fundamental difference between the picture of atomic constitution indicated in this letter and that developed by Langmuir on the basis of the assumption of stationary or oscillating electrons in the atom, referred to in Dr. Campbell's letter.

Quite apart from the fact that in Langmuir's theory the stability of the configuration of the electrons is considered rather as a postulated property of the atom, for which no detailed *a priori* interpretation is offered, this difference discloses itself clearly by the fact that in Langmuir's theory a constitution of the atoms of the inactive gases is assumed in which the number of electrons is always largest in the outermost shell. Thus the sequence of the number of electrons within the groups of a niton atom is, instead of that indicated above, assumed to be 2, 8, 18, 18, 32, such as the appearance of the periods in the sequence of the elements might seem to claim at first sight.

The assumption of the presence of the larger groups in the interior of the atom, which is an immediate consequence of the argument underlying the present theory, appears, however, to offer not merely a more suitable basis for the interpretation of the general properties of the elements, but especially an immediate interpretation of the appearance of such families of elements within the periodic table, where the chemical properties of successive elements differ only very slightly from each other. The existence of such families appears, in fact, as a direct consequence of the formation of groups containing a larger number of electrons in the interior of the atom when proceeding through the sequence of the elements. Thus in the family of the rare earths we may be assumed to be witnessing the successive formation of an inner group of thirty-two electrons at that place in the atom where formerly the corresponding group possessed only eighteen electrons. In a similar way we may suppose the appearance of the iron, palladium, and platinum families to be witnessing stages of the formation of groups of eighteen electrons. Compared with the appearance of the family of the rare earths, however, the conditions are here somewhat more complicated, because we have to do with the formation of a group which lies closer to the surface of the atom, and where, therefore, the rapid increase in the compensation of the nuclear charge during the progress of the binding process plays a greater part. In fact, we have to do in the cases in question, not, as in the rare earths, with a transformation which in its effects keeps inside one and the same group, and where, therefore, the increase in the number in this group is simply reflected in the number of the elements within the family under consideration, but we are witnesses of a transformation which is accompanied by a confluence of several outer groups of electrons.

In a fuller account which will be published soon the questions here discussed will be treated in greater detail. In this letter it is my intention only to direct attention to the possibilities which the elaboration of the principles underlying the spectral applications of the quantum theory seems to open for the interpretation of other properties of the elements. In this connection I should also like to mention that it seems possible, from the examination of the change of the spectra of the elements in the presence of magnetic fields, to develop an argument which promises to throw light on the difficulties which have hitherto been involved in the explanation of the characteristic magnetic properties of the elements, and have been discussed in various recent letters in NATURE.

N. BOHR.

Copenhagen, February 14.

III. OUR PRESENT KNOWLEDGE OF ATOMS

UNSERE HEUTIGE KENNTNIS VOM ATOM
(Die Umschau **25** (1921) 229)

(TRANSLATION*)

* [See Introduction, sect. 3.]

The ancient idea of the atomists was that matter is not indefinitely divisible but that, when attempting to divide a body into smaller and smaller parts, we should finally reach very small, not further divisible, particles which must be considered as elementary building stones of matter and which, because of their indivisibility, were called "atoms". Until the beginning of the nineteenth century this idea was of little significance for exact natural science. However, since *Dalton*, the assumption that all matter is made up of atoms has been an indispensable working hypothesis of chemists. The chemical atoms, however, had not much in common with the atoms of the ancient atomists. Their divisibility or indivisibility was not being questioned; rather, they were small particles forming the building stones of the different elements, in the sense that each chemical element implies the existence of a certain kind of atom. However, chemistry did not allow a direct investigation of all properties of these atoms and was unable, e.g., to determine the weight and size of individual atoms; chemistry was primarily concerned with the more or less pronounced tendency of like or unlike atoms to join to form larger structures, the so-called molecules. Although also physicists, in their attempts to explain the most important properties of matter, such as the characteristic differences between the solid, liquid and gaseous states, more and more pointed to an atomistic conception as the only natural hypothesis, it is nevertheless true that in the first half of the nineteenth century the atomic theory was not of particular importance for physical research; for apparently no *physical* phenomena were known which gave direct evidence of the existence of individual atoms or of their properties. Only the laws discovered by *Faraday* on the conduction of an electric current through aqueous solutions of various materials pointed to the important fact that atoms can appear as *carriers of electricity*, and in such a way that a charged atom always carries a quite definite positive or negative charge, independently of the manner in which the experiment is performed, and independently of the nature of the element. However, in the last four decades a series of new physical phenomena has been discovered which are intimately connected with the atomistic structure of matter, and the close study of which has succeeded not only in removing any doubt about the reality of the atoms of Dalton's theory, and in determining their weights and sizes accurately, but even in forming detailed pictures of the "structure" of these atoms. In fact, these new discoveries have shown that an atom is very far from being indivisible and unchangeable; rather, *an atom has a more or less complicated structure*; it is a world in itself, a "microcosm", an "ordered aggregate in miniature", in a more literal sense than anyone had ever dreamed.

Experiments on the conduction of electricity through gases, now familiar to almost everybody from the beautiful optical phenomena in Geissler tubes, showed in the last decade of the last century that an essential constituent of all matter is an extremely small and light particle charged with negative electricity. These particles are all perfectly alike, i.e., they all possess the same mass and *negative charge*; they are called "electrons". It became clear that every atom contains one or more of these electrons, and the question arose how and where *the positive electricity* is located in the atom; for there must be such a charge in order to compensate for the negative charges of the electrons, in order that the entire atom can be electrically neutral. Information on this question has been obtained through experiments on the so-called "radioactive" elements, such as, e.g., radium, the discovery of which in recent years has played an extremely important role in physical research*. It is not feasible in a few lines to explain these investigations which we owe mainly to the great English scientist *Rutherford*, and here I shall merely sketch briefly the picture which we can now form of the structure of the atom. The positive electricity in an atom is concentrated in a "nucleus" which has a weight that is practically the same as that of the entire atom and which is surrounded by a crowd of electrons bound to it by electric attraction. Thus, an atom consists entirely of electrically charged particles, and it is clear that it can appear as a carrier of electricity as soon as the number of negative electrons is smaller or larger than needed to compensate for the positive charge of the nucleus, i.e., smaller or larger than in the neutral atom. The dimensions of the electrons, and also of the nucleus, are very small compared to the distances between electrons and nucleus. While the distance from the nucleus to the most

* The most important facts in question will be known to readers of "Die Umschau" from the article by Hahn (Radioactivity and the Study of the Elements, 1921, Nos. 1 and 2).

remote electrons is about a ten-millionth of a millimeter, the dimensions of the individual particles are perhaps no more than some billionths of a millimeter ⟨10^{-12} mm⟩. Since, because of its small size, a possible *structure of the nucleus* will not exert any appreciable influence upon the arrangement and motion of the surrounding electrons, so that these will depend only on the total charge of the nucleus and, hence, on the number of electrons in the neutral atom, the physical and chemical properties of the element will depend only on this number which is called the *atomic number*. In general, the weight of atoms increases with increasing atomic number. Thus, the atomic number of the lightest atom *hydrogen* is equal to one, i.e., a neutral hydrogen atom consists of a nucleus and a single electron circling around it; on the other hand, the heaviest known atom, the atom of the fairly rare metal *uranium*, contains no less than 92 electrons in its neutral state.

By the certain establishment of all of these facts, a true wonderland has been opened up, and the task of atomic research has thereby been divided into two sharply distinct problems. The former is to determine in detail the laws according to which the electrons move around the nucleus; the latter consists in discovering the internal structure of the nucleus itself.

With regard to the latter point, i.e., "*nuclear physics*", several extremely interesting discoveries have already been made, especially in the field of radioactivity. In fact, it is from the nucleus alone that all the characteristic radiations emitted by the radioactive elements originate. They result from a spontaneous explosion of an atomic nucleus. In such an explosion either an electron or a positive particle—which is identical with the atomic nucleus of the element with atomic number two, i.e., helium—is ejected, while the remaining part of the original nucleus then forms the nucleus of an atom with an atomic number different from that of the original atom. Of course, such a phenomenon must be ascribed to a complicated internal structure of the atomic nucleus itself, a structure about which we as yet know very little. Another remarkable result of recent research which bears on the properties of the nucleus is the discovery of the so-called "*isotopic*" *elements*. By these we mean elements which have the same nuclear charge, so that most of their physical and chemical properties are accurately the same, but for which the weight and the internal structure of their nuclei are different, so that their specific weights and radioactive properties may be different. In the last few years it has been found that isotopes occur not only among the radioactive elements but that also a considerable number of old well-known elements, such as chlorine and mercury, consist of several isotopes. This discovery has been made through ingenious experiments on a type of rays emitted in discharge tubes; these consist of positively charged atoms, and the experiments allow a direct comparison of the masses of the individual atoms.

Moreover, it may be of interest to mention that in the last few months it has been possible, by a suitable method of distillation, to *separate ordinary mercury into two modifications* for which the specific weight of one is a little larger and of the other a little smaller than that of ordinary mercury, while otherwise all the characteristic properties of this element are common to both of them.

Before leaving our considerations of the atomic nucleus itself, we shall mention the new experiments of *Rutherford* in which he has succeeded, by bombardment with rays from radium, in splitting the atomic nuclei of an element (nitrogen) and thus in changing one element into another by external influence. By this achievement, as already by the demonstration of the spontaneous disintegration of atoms of the radioactive elements, the old view that the chemical atoms are in principle unchangeable has been most clearly disproved; or, as we might perhaps rather say, in order to better emphasize the positive in these results, through these experiments we have reached a clearer perception of the chemical elements than we formerly had.

Let us now turn to the former of the above-mentioned tasks of atomic physics, namely that of analyzing the behavior of the *electrons surrounding the nucleus* and thereby obtaining an explanation of the physical and chemical properties of the elements. At first sight, this problem appears closely analogous to the astronomical problem of the motion of the planets around the sun; for as was said above, in a certain sense an atom is a kind of planetary system in which the nucleus represents the sun and the electrons the planets. At the same time, upon closer consideration there is a fundamental difference because of the fact that in a planetary system the orbits of the planets depend on the special circumstances existing when the system was originally formed, while we must assume that in an atomic system the motions that the electrons carry out under normal conditions are quite definite and *can depend only on the nuclear charge* of the atom in question. In fact, this assumption is demanded by the empirical fact that the elements have characteristic definite physical and chemical properties. Nevertheless, in the last few years it has been possible to unravel several of the laws that govern the behavior of the electrons in an atom. Above all, the theory of electrodynamics, developed in the second half of the nineteenth century, has rendered great service as a guiding principle in this effort. However, it has become clear that this theory is not directly applicable to atomic problems. In particular, to explain the above-mentioned definite character of the properties of the elements, it has become necessary to develop a view of the electromagnetic behavior of atoms which is revolutionary and entirely alien to the older views of such questions. The new laws contained in this view, which are summarized under the name "*quantum theory*", have their essential origin in the bold theory of the so-called heat radiation advanced twenty years ago by the great German theorist *Max Planck*.

I shall be content to give just a few indications*. In contrast to the pronounced continuity, supposedly possessed by all electromagnetic phenomena according to the classical theory of electricity, we must assume that processes that can occur in the "electron sheet" ("Elektronenhülle") possess a peculiar discontinuity. This is to be understood in the following manner: When an atom is forced out of its equilibrium state by external influences of any kind, e.g., bombardment by electrons, irradiation by X-rays, etc., it cannot thereby be transferred to any conceivable other state, but for every atom there exists a series of quite definite electronic motions, and the atom will always attempt to pass to one of these preferred states, so that all processes in an atom can be regarded as transitions between two such distinguished states. The state in which a neutral atom normally exists, the state we have always tacitly assumed when speaking of atoms, is the state among these distinguished states in which the arrangement and motion of the electrons are most stable. Now, with regard to the explanation of the characteristic properties of the elements, it is assumed that the majority of these properties are not connected directly with the motion of the electrons in the atom in this state, but rather that they are connected with processes occurring within the atom *during its transitions* between different characteristic states. Thus, we must assume that in such a transition process *light may be emitted or absorbed by the atom*, and the quantum theory now ties a very definite bond between the properties of this light, such as its color and intensity, on the one hand, and the motion of the electrons and the change in this motion during the transition process considered, on the other hand. In this way it becomes possible by a closer so-called spectroscopic analysis of the light emitted under suitable experimental conditions by the atoms of an element to obtain information about certain properties of the *motion of the electrons* in this atom. At the same time, the conceptions which we can form from such investigations about the orbits described by the electrons in the different atoms have been able also to offer an explanation of certain other properties of the atoms not connected with the emission of light rays but related to the chemical behavior of the elements.

For the present we are exceedingly far from being able to exploit theoretically the enormous mass of facts collected in the course of time pertaining to the chemical properties of the elements, or the no less abundant results of experimental spectroscopy. Nevertheless, we can say that it is now possible to explain qualitatively and quantitatively many individual facts and that the—partly regular and partly capricious—variations in the chemical properties as one goes from elements of lower to elements of higher atomic number, variations which are so beautifully expressed in the empirical so-called *periodic system of the elements*,

* The quantum theory is treated more fully in an article by Valentiner (Die Umschau 1921, No. 4).

[88]

are no longer such an incomprehensible secret as they were a few years ago. The solution of this enigma is now within the realm of possibility. With regard to this point, it seems very recently to have become possible to develop a point of view which permits a survey of the arrangement of the electrons in the atoms of the whole series of elements from hydrogen to uranium and which enables us in a simple manner to interpret the main features of the regularities mentioned. However, it would be impossible in this short article, which aims only at giving a general survey of our knowledge of atoms, to enter more closely into such a question.

IV. EXTRACT OF SECOND HALF OF SOLVAY REPORT

(1921*)

* [See Introduction, sect. 4.]

BRIEF EXTRACT OF CONTENTS OF SECOND HALF OF PROFESSOR BOHR'S REPORT

Applications of general considerations in first halfth to the problem of the constitution of actual atoms.

Line-spectrum of Hydrogen. Purely periodic system. Stationary states fixed by one condition

$$\mathcal{I} = nh.$$

Energy in stationary states

$$E_n = -\frac{2\pi^2 N^2 e^4 m}{n^2 h^2}, \tag{1}$$

half major axis

$$a_n = \frac{n^2 h^2}{4\pi^2 N e^2 m}; \tag{2}$$

degree of eccentricity of orbit, just as position of plane in space, arbitrary. Ne charge on nucleus, $-e$ charge electron, etc.

The spectrum gives expression to the process of binding of an electron by a positive nucleus. *Normal state* $\rightarrow n = 1$. Correspondence between spectrum and motion easily seen to be fulfilled.

General analogy with spectrum of atom containing one electron exhibited by

1) *series spectra*. Here appearance of Rydberg's constant directly interpreted by assuming *one* electron moves largest part of orbit far outside other elec-

trons. In these spectra we witness binding of last electron by a nucleus. In *spark spectra* (4 times Rydberg's constant) all electrons but two are already bound, and the last but one is being bound.

2) *Röntgen spectra*. As observed by Moseley, and Sommerfeld, there exists a general analogy in the formal representation of these spectra with spectrum of atom containing one electron. In these spectra we witness *reorganisation* of normal state of atom. The properties as to the absorption prove that there is no direct physical analogy with the process of the emission of a series-spectrum.

Application of general theory of multiple-periodic systems to *central systems* (Sommerfeld): stationary states fixed by two conditions. If central force at large distances asymptotically coincides with inverse-square law force, the motion in the stationary states may be described as *rotating* Keplerian ellipse, the *major axis* of which is given by (2), while the parameter $2p$ (diameter through nucleus, perpendicular to major axis) is fixed by

$$2p = \frac{k^2h^2}{2\pi^2Ne^2m} \tag{3}$$

where k integer ($k \leqq n$). (Impulsmoment equal to $kh/2\pi$).

Application to *fine-structure* (Sommerfeld, Paschen). Tentatively also applied by Sommerfeld to Röntgenspectra, with success (L-dublet). Detailed interpretation of this result first to be got, when we have picture of appearance of Röntgen spectra (to be discussed below).

Application to *series spectra* of elements. Only formal analogy, since inner electrons do not act exactly as central field of force. Success due to fact that effect of inner electrons will be to cause perihelion precession of orbit of outer electron, every time this electron on its way comes near the nucleus. *Understanding* of general scheme of series spectra (S, P, D, B terms); not yet explanation of triplets, doublets etc.*. *Understanding*, on account of *correspondence*, that k can

* If, as by Rydberg's formulae, the series terms can approximately be represented by

$$-\frac{1}{hc}E_{nh} = \frac{R}{n-\alpha(k)^2}$$

(R Rydberg's constant), the frequency, with which the electron every time returns to perihelion, is given by

$$\frac{1}{h}\frac{\partial E}{\partial n}$$

while the frequency of the rotation of the major axis is

$$\frac{1}{h}\frac{\partial E}{\partial k}.$$

only decrease or increase by 1 during a transition (i.e. of capriciousness in the appearance of *combination* lines).

Difficulty for S-terms in alkali spectra; solution of this obtained by assuming that the series electron in normal state describes a *two-quanta orbit* ($n = 2, k = 1$), and in this, as well as in the other S-states, possesses a large *progressive* perihelion precession, due to the fact that the outer electron penetrates in the cluster of inner electrons, and thus for a short time is exposed to a much larger attraction than in the analogous state in a hydrogen atom.

The *one-quantum orbit does not exist*, is mechanically impossible. This falling out of certain states quite simply illustrated by simple examples of central field.

These considerations strongly supported by detailed investigation of *helium spectrum*, which on the whole has been of utmost importance in order to become clear as regards binding of electrons, and as regards general view on *correspondence principle*, as formulated in first part.

Helium spectrum (i.e. spectrum emitted by the neutral atom) consists of *two* complete series spectra, between which *no* combinations are observed. Even if we apply electric fields, which cause the appearance of many new combination lines within each spectrum, such mutual combinations do not appear. Now *Bohr* and *Kramers* found two special classes of mechanical motions in He atom, where the character of motion especially simple, and for the *outer electron* of just the type as in central system, while *inner electron* describes perturbed one-quantum orbit. In one class the two electrons move in *coplanar* orbits, and calculation gives in large features agreement with *ortho-helium* spectrum (doublets). In other class the two orbits are at right angles (at any rate for k not very small), and approximate agreement with *parhelium* spectrum. Now *Franck* and *Knipping's* experiments prove that the normal state in helium corresponds to ($n = 1, k = 1$, one-quantum orbit) in *parhelium*, while the most stable orbit in orthohelium is a two-quanta orbit ($n = 2, k = 1$). The calculation gives really that for the ortho-helium states the state $n = 1$ is mechanically excluded. This state is *meta-stable*, and analogous to normal state in alkali-atoms.

Dividing these two quantities, I obtain for the angle through which the major axis has turned after a single revolution

$$\beta/2\pi = -\frac{\mathrm{d}\alpha}{\mathrm{d}k} \quad \text{(independent of } n\text{)} \quad \alpha_k = \frac{1}{2\pi}\int_k^\infty \beta\,\mathrm{d}k.$$

In actual spectra Rydberg's formula holds, as known, often with large degree of approximation; this means, that the rotation of the major axis after one revolution nearly independent of the *length* of the major axis, and this understood directly since this rotation is wholly due to the effect of the inner electron at the moment where the outer electron is nearest the nucleus and therefore independent of length of *major axis*, but only dependent on the parameter.

The normal state of helium is thus not a *ring* of two one-quantum electron. Such a ring cannot be formed by the binding of a second electron after the first has been bound. It is moreover in disagreement with the observed ionisation potential (25,2 Volt instead of 29 Volt). The binding process, which we witness in the *Parhelium* spectrum, gives rise to the actual normal state, in which both electrons move in one-quantum orbits, but not coplanar (in mechanically equivalent orbits).

Fundamental difference with Landé's view (see Bohr's letter).

After binding of two electrons one-quantum orbits by a positive nucleus, as illustrated in *helium spectrum*, we must assume that third electron only can be bound, until it in the normal state remains in two-quanta orbit, ($n = 2$, $k = 1$) as illustrated in *Lithium spectrum*.

The next three electrons will be bound in equivalent two-quanta orbits (illustrated in *double valency* of Be, *triple valency* of B, *quadruple valency* of C), so that of six electrons the first two will be bound in equivalent one-quanta orbits, the four next in equivalent two-quanta orbit the arrangement of which must be assumed to possess a high degree of central spatial symmetry, but not cubical in Born's and Landé's sense. Hereafter it must be assumed, that on account of this symmetry we may expect, that the "opportunity" of binding further electrons in orbits equivalent with those of the electrons already bound, and that there is so to say no more "room" for more two-quanta orbits of the *elliptical* type. In the normal state the next electrons will be bound in *circular* two-quanta orbits ($n = 2$, $k = 2$), and when four such electrons are bound, a configuration is found, which again exhibits high degree of central symmetry (such symmetry being exhibited not only by each sub-group of four two-quanta orbits, but also by manner by which they are "coupled" together). Illustrated by position of *neon* in periodic system.

Electrons to be bound after the first ten will for the largest part of their revolution move outside the 10 inner electrons, and in case of large nucleus charge there will clearly be no "room" for two-quanta orbits, when the 10 first such electrons moving in two-quanta orbits. There will appear, on the contrary, electrons moving in 3-quanta orbits, of which the first will be of the type ($n = 3$, $k = 1$), the electrons during their revolution penetrating in the region of the inner 10 electrons. From the periodic system we infer directly that the three-quanta orbits will have formed a compact whole, when there are 18 of them most probably in such a way that there are 6 orbits corresponding to $k = 1$, 6 to $k = 2$, 6 to $k = 3$. A probable indication of constitution and formation of this group may be obtained by observing that three sub-groups of four electronic orbits each cannot be expected to amalgamate in the same symmetrical way as two such sub-groups (corresponding to the fact, that two tetrahedrons may be arranged in quasi-tetrahedral symmetry but not three), while three sub-groups of six electrons may be more symmetrically amalgamated (corresponding to the fact that three octahedrons may be arranged in quasi-octahedral symmetry). A possible explanation is therefore suggested, that the binding of a third sub-group of three-quanta orbits corresponding to $n = 3$, $k = 3$, will disturb the symmetry of the two first sub-groups $k = 2$ and $k = 1$, if these contain only four electrons each, with the effect that new electrons will be incorporated in these groups, until they contain six each.

After the 18 three-quanta electrons, there will be no more "room" for other three-quanta orbits, and the following electrons will now only be bound as four-quanta orbits. From the periodic system, again, we infer the existence of 32 four-quanta orbits; most probably in four sub-groups of 8, corresponding to $k = 1$, $k = 2$, $k = 3$ and $k = 4$ respectively, for the formation of which corresponding arguments can be given as in the case of the 18 groups.

All these last considerations hold only for large nucleus charge. If charge small, the inner electrons will compensate so much of nucleus-attraction, that instead of an increase of the number of quanta characterising the orbits of the new electrons to be bound, this number may not increase, even decrease (due to fact, that dimensions of orbits increase by decreasing nuclear charge (see (2)). Simplest example: in *natrium*, the 10 inner electrons have compensated almost all the nucleus charge, and the valency-electron will be bound in a two-quanta orbit again, ($n = 2$, $k = 1$) just as in lithium.

General survey of arrangement of electrons got from consideration of inactive gases (see "Nature" ⟨**107** (1921) 104⟩).

He 2_1	Kr $2_1\,8_2\,18_3\,8_2$
Ne $2_1\,8_2$	Xe $2_1\,8_2\,18_3\,18_3\,8_2$
Ar $2_1\,8_2\,8_2$	Ni $2_1\,8_2\,18_3\,32_4\,18_3\,8_2$

(first two groups of 8, then two of 18, then one of 32).

General understanding of groups in periodical system. Especially the appearance of Fe Ni Co groups (establishment of new groups of 18 three-quanta electrons) and of groups of rare earth (establishment of group 32 four-quanta electrons in the inner of the atom (see "Nature")).

Coupling of the various groups of electrons between each other. Electrons of various groups penetrate during their revolution the groups lying inside.

Röntgen rays. Understanding of the general way, in which the different characteristic high-frequency lines appear. Kossel's scheme; reason of success of Sommerfeld's theory.

Magnetism. Apparent capriciousness in appearance of para-magnetism. This circumstance in suggestive connection

a) with anomalous effect of magnetic field in spectra lines, which exhibits peculiar departure of behaviour of bound electrons from classical electronic theory,

b) with the anomalous results of Einstein–de Hass experiments. (see Bohr's letter).

V. CONSTITUTION OF ATOMS

(1921*)

* [See Introduction, sect. 4.]

given by (14) will asymptotically coincide with $(n'-n'')\ \omega$, that is with the frequency of one of the harmonic elliptical vibrations, into which the variation of the electrical moment of the atom due to the revolution of the particles, may be resolved. From this correspondence it has actually been possible to deduce the theoretical expression for the factor on the right side of formula (14) before the establishment of the complete theory of the stationary states of multiple-periodic motions.

In the stationary state corresponding to $n = 1$ the electron will be bound as firmly as possible, and in this state the atom will not be capable of performing any spontaneous transition accompanied by emission of radiation. This state may therefore be considered as the normal state of the atom, which it cannot leave except by a process, caused by the action of external agencies, by which it is transferred to some other stationary state of the atom, corresponding to higher values of n, in which the electron is looser bound, or to a state in which it is "free" and capable of moving to infinite distances from the nucleus. In the case of hydrogen we get from (13) by putting $N = 1$ and $n = 1$, and introducing the empirical values for e, m and h, values for W and $2a$ corresponding to the normal state, which are of the same order of magnitude as the values for atomic energies and dimensions to be expected from measurements of ionisation potentials and from the interpretation of the properties of gases on the kinetic theory. A direct quantitative comparison with such data are, however, out of question for hydrogen, already thereby since hydrogen atoms under usual conditions combine into molecules. Quite analogous considerations, which hold for the binding of the electron in the hydrogen atom, are found to hold for the binding of a single electron by a nucleus of higher positive charge. Thus if in (14) we put $N = 2$, we get a formula, which is found closely to represent a spectrum emitted by helium, when this gas is subject to a condensed discharge; corresponding to the circumstance that an atom of helium, the atomic number of which is equal to 2, if one electron is removed from the neutral atom, forms a system of the type under consideration. For still greater values of N we obtain expressions for line spectra, which according to the theory should be emitted from elements of higher atomic numbers, when all the electrons but one are removed from the atom. Hitherto such spectra have not been observed, probably owing to the difficulty of producing such highly ionised atoms in sufficient amount in the discharge tubes.

Formula (14), however, has also for large values of N acquired a considerable interest from an examination of the Røntgen spectra, which may be assumed to appear during the reorganisation of an atom, when one of the inner electrons has been removed by the effect of some external agency. In fact, it is found that the energy necessary to excite the most penetrating of characteristic Røntgen radiation, the so-called K-ray, is with a close approximation given by formula (13), if we put $n = 1$ and for N introduce the atomic number of the element in question. Further Moseley from the basis of his famous investigation of the structure of the high frequency spectra of the elements drew the attention to the fact, that the most intensive term of the K-radiation could be approximately represented by formula (14), if we put $n'' = 2$ and $n' = 1$, and that the frequency of the less penetrating L-radiation in frequency were approximately given by the same formula, if we put $n' = 2$ and $n'' = 3$. In spite of this agreement it must be borne in mind, that there exists a most pronounced contrast between the structure of the Røntgen spectra of the elements and a series spectrum of the same type as the hydrogen spectrum, which is exhibited not only thereby, that the lines in the Røntgen spectrum, instead of forming series consisting of an infinite sequence of lines converging to a limit may be arranged in sets (K-set, L-set, M-set etc.), each containing a definite number of lines, but also by the very different laws holding for the absorption of high frequency radiation by atoms and for absorption of radiation in the optical region. In fact, while the latter absorption, at any rate for monatomic gases, extends not only over a continuous interval, starting at the end of a certain series of lines, but will also be directly related to the lines within this series, the absorption of the high frequency radiation is not directly related to any line within the characteristic Røntgen spectra, but appears always as extended over continuous intervals, which start suddenly with a definite frequency at a so-called absorption edge, of which there corresponds one edge (or several closely situated edges) to each of the mentioned sets of lines in the characteristic Røntgen spectrum.

Notwithstanding the difficulty, which due to these circumstances is apparently involved in a detailed interpretation of the Røntgen spectra, the numerical relations in question give us naturally the hope of penetrating into the problem of the constitution of the normal state of the atom by means of considerations analogous to those indulging the application of the quantum theory to the hydrogen spectrum. From the large number of electrons in the neutral atoms of most elements we might beforehand be prepared to expect a very complex structure of the characteristic high frequency spectra. The circumstance, that these spectra, however, consist of a few sets of neighbouring lines separated by large frequency intervals, suggests, that the electrons in the atom are arranged in a moderate number of distinctly separated groups in such a way, that the same (or approximately the same) amount of work must be expected for the removal of the various electrons within each group. Such a group-arrangement of the electrons in the normal state of the atom is in fact claimed not only for the interpretation of the structure of the high frequency spectra, but is also directly suggested from the peculiar way, in which most physical and chemical properties of the atom vary with the increase of atomic number, and which obtains its adequate expression in the so-called periodic law. Already before we had the knowledge of the constituent particles of the atom acquired in recent years, such an arrangement has also appeared as an essential feature in all attempts of a closer picture of the atom, as for instance in the famous theory of J. J. Thomson. As regards the question of the closer connection of the Røntgen spectra with these groups considerable light on this problem was thrown by the work of Kossel, who assumed, that the various lines belonging to one and the same set in the high frequency spectrum of an element was emitted by atomic processes, by which an electron, removed from one group within the atom, was replaced by electrons belonging to the various other groups of electrons, less firmly bound. On this view he was able by means of the general frequency condition to predict certain numerical relations, to which we shall come back in the following, between the frequencies of lines belonging to the various sets in the high frequency spectrum, and which at any rate in a slightly modefied form have been found to be fulfilled with a close approximation by the measurements. These relations form in fact the first example of an application outside the optical spectra of the general principles of combination of spectral lines, mentioned in part A. This view on the origin of the Røntgen spectra throws also light on the characteristic difference mentioned above between the absorption connected with the series spectra, and that connected with the Røntgen spectra. Thus in the first case the absorption may not only be connected with a complete removal of an electron from the atom, but may also be connected with processes, by which the atom is transferred from the normal state into some other stationary state, from which it may directly return by a process, by which the normal state is re-established, and which gives rise to the emission of a spectral line with the same frequency. On the other hand we must assume, that the absorption of the Røntgen rays is always connected with the complete removal of an electron from one of the groups within the atom, and that it can never form the inverse process of the emission of a line in the Røntgen spectra, since on the above view the latter process does not result in a complete re-establishment of the normal state.

As regards the attempts of obtaining a detailed picture of the constitution of the atom in its normal state the simplest possible assumption is presented by the hypothesis, that the electrons within each of the various groups rotate at equal angular intervals in a circular orbit as in the model of Sir J. J. Thomson. Starting from this hypothesis the author made a preliminary attempt to picture the electronic constitution of the nucleus atom by assuming, that the angular momentum round the nucleus of each of the rotating electrons was equal to $h/2\pi$ corresponding to the value for the angular momentum of the electron in the normal state of the hydrogen atom, if it is assumed, that its orbit is circular. Although in this way it was apparently possible to account as regards the order of magnitude for the firmness, with which the electrons in atoms are bound by the nucleus, it was not found possible, however, by means of this simple picture to account for the properties of the elements in any finer details. Attempts by other authors to modify this picture by introducing the assumption, suggested by the empirical laws, governing the Røntgen spectra, that in some of the rings the angular momentum of every electron would correspond to higher multiple of $h/2\pi$, corresponding to intermediate stages in the binding of the electron in the hydrogen atom, can also not be said to have led to satisfactory results. Even as regards the interpretation of the Røntgen rays it remained a paradox, how to reconcile the almost complete absence in these spectra of such variations with the atomic number as according to the periodic law appear in most other properties of the elements. The fundamental difficulty, involved in the assumption of a ring arrangement, consisted, however, therein that the picture offered no sufficient explanation of the essential stability of the atoms, since it presented no basis for an apriori fixation of a distribution of the electrons on the various rings. In order to avoid the difficulties involved in a ring-constitution it has from various sides been proposed to assume a more shell-like constitution of the atom, in which the arrangement of the electrons within the various groups possessed a higher degree of spatial symmetry, exhibiting not as a "ring" the symmetry of a regular plane polygon, but a symmetry of the same type as a regular polyhedron. On one hand Born and Landé have from an examination of the properties of crystals deduced arguments in favour of this view in order to account for the character of the field of force between atoms necessary for the elasticity properties of the crystal-lattice. On the other hand a similar view has from chemical considerations been advocated by Lewis and Langmuir, who draw the attention to the possibility, from an arrangement of cubical symmetry of the electrons in the groups to arrive at a simple interpretation of the prominent part the number 8, as well known, plays in the periodic law. Although certainly these arguments involve the germ of an essential progress, they cannot be said to bring us at once over the difficulties in question. Thus the fundamental difficulty remains of an understanding of the inherent stability of the configuration of the electrons in the atom, which shows itself thereby, that as soon as one electron is removed from its position, a spontaneous process of reorganisation takes place, as the result of which the normal distribution of the electrons within the groups is re-established. An explanation of this stability claims the existence of an intimate coupling between the electrons in the various groups, which is at first sight no more exhibited by a shell-arrangement than by a ring-arrangement. In the following we shall see, however, that it seems possible on the basis of the general principles of the quantum theory to obtain a picture of the process of formation of an atom by successive binding of electrons, which leads us to a view of the constitution of the atom, in which such a coupling between the various groups just appears as the essential feature.

For an investigation as that just mentioned the simple theory of the hydrogen spectrum offers no sufficient guidance, but fortunately we possess in the series spectra of the other elements a means of studying the binding process more closely. As well known, the structure of these spectra exhibits a higher degree of complexity than the hydrogen spectrum, although it has been possible to represent the frequencies of the lines of these spectra by means of empirical formulae, which exhibit a remarkable analogy with a formula holding for the latter spectrum. Thus looking for a moment apart from the complex structure of the series lines (doublets, triplets) the frequencies of the lines of the series spectra of many elements may be represented by a formula of the type

$$\nu = f_{k''}(n'') - f_{k'}(n') \tag{15}$$

where $f_{k'}$ and $f_{k''}$ are two among a set of functions $f_1, f_2, f_3, \ldots$ characteristic for the element, and where n' and n'' are positive integers. The various "series" are represented by (15), if the first term on the right side is kept constant, and if in the second term we let the argument n of a given function $f_{k'}$ take a successive sequence of integer values. Formula (15) is seen to include as a special case the series spectrum of hydrogen, where there enters only one function $f(n)$, which has the simple form

$$f = R/n^2 \tag{16}$$

where, looking apart from the small influence due to the motion of the nucleus relative to the centre of gravity, R as shown in the preceding can be represented by the theoretical expression $2\pi^2 e^4 m/h^3$.

The functions $f_{k'}$ appearing in the impirical formulae for the series spectra of other elements are not so simple as in the case of hydrogen, but it has been found, that for series spectra excited under usual conditions, the so-called arc spectra, these functions can be brought in the form

$$f_k(n) = R/(n+\phi_k(n))^2 \tag{17}$$

where in many cases $\phi_k(n)$ is very nearly constant, and in all simple series spectra converges to a finite limit for increasing n, while R to a high degree of approximation is the same constant as that appearing in the formula for the hydrogen spectrum. This last circumstance obtains its immediate interpretation by the assumption, that in the stationary states involved in this emission of these spectra the atom is neutral, and that in the states corresponding to large n's one electron is moving in an orbit of such dimensions, that it at any rate during the greater part of its revolution moves at a distance from the nucleus, large compared with the distances of the other electrons from the nucleus. In fact, while in that case the motion of the inner electrons may be expected to differ only little from their motion in the normal state of the atom, in the absence of the outer electron, the orbit of the latter electron will during the greater part of the motion differ only very little from a possible simple Keplerian orbit of the electron in the neutral hydrogen atom. This general view on the appearance of the series spectra of usual type, which has obtained a most direct support from experiments on excitation of spectral lines and ionisation of atoms by impact of electrons as those, referred to in part A, offers an immediate understanding of the fact, that many elements, when subject to powerful discharge, are found to emit a line spectrum, which equally may be arranged in a scheme represented by formula (17), but which differ from the series spectra, excited under ordinary conditions, thereby that the constant R appearing in the formula (16) is replaced by a constant 4 times larger. In fact, this is just what should be expected, if these spectra were emitted by atoms, from which one electron is completely removed, while a second electron during the greater part of the revolution moves at a distance from the nucleus, large compared with the distances of remaining $N-2$ electrons. This electron will consequently be subject to forces, which differ only little from the forces on the electron in a helium ion, containing only one electron. This point of view, according to which the spark spectrum of helium possesses exactly the same relation to the spark spectra of the other elements as the hydrogen spectrum to the usual series spectra, may be traced still further, and leads to the expectation that there in general will be a pronounced similarity between the details of the structure of the arc spectrum of an element of given atomic number and the spark spectrum of the element following next in the periodic table, as in such cases the inner system contains the same total number of electrons and therefore in general may be expected to exhibit an analogous structure. This point has been emphasised by Kossel and Sommerfeld, who have called attention to the experimental evidence in favour of this "displacement law of spectra". From the above general considerations we must, besides arc and spark spectra, expect the possibility of series spectra, for which the constant R in the formula is replaced by a constant k^2R, where k is an integer larger than 2, and which are emitted by atoms, from which $k-1$ electrons are removed, while a kth electron is moving in an orbit of dimensions large compared with the distance of the remaining $N-k$ electrons. Hitherto no definite experimental evidence has been obtained for the appearance of such spectra, which may be denoted as series spectra of the order k, a circumstance which is probably due to the experimental difficulties involved in the excitation of such spectra with sufficient intensity.

Already from a purely theoretical point of view the existence of these spectra is a problem of great interest, as they must be considered as bearing witness of the various stages of the formation of atoms by the successive binding of electrons, in such a way that the ordinary arc-spectra or series spectra of the first order correspond to the last stage of this process, which final result is the establishment of the normal state of the neutral atom, while spark spectra or series spectra of the second order correspond to the last stage but one in this process and so on.

While the above considerations rest essentially on the pronounced analogy existing between the spectra accompanying the binding of a single electron by a nucleus and the series spectra, which are emitted during the binding of an electron by an atom, in which the nucleus has already bound one or more electrons, a closer examination of the binding process in the latter case leads us directly to the problem of the characteristic difference between the two types of spectra in question. The origin of this difference may be directly sought in the more complex motion of the outer electron in the atoms of other elements, compared to the motion of the electron in the hydrogen atom. Thus the simple structure of the hydrogen spectrum is, as we have seen, just related to the simple periodic character of the latter motion, and it was not before the theory of stationary states of multiple periodic systems was developed, that it was possible to throw light on the characteristic features of the structure of the series spectra. The starting point for the closer interpretation of these spectra was in fact offered by Sommerfeld's theory of the fixation of the stationary states of a system, consisting of an electrified particle moving in a central field of force, which presents an especially simple example of a multiple periodic motion.

The so-called central orbits, described by the particle in such a central system, may, as well known, be considered as a plane purely periodic motion, on which is superposed a uniform rotation round the centre in its plane. Asking for the harmonic constituents of such a motion it will first be observed, that the periodic motion, considered for itself, may be resolved in a sequence of elliptical harmonic vibrations in the plane of the orbit with frequencies equal to an entire multiple of the frequency ω of the periodic motion. Next it will be seen, that the influence of the superposed uniform rotation is simply obtained, if we imagine every elliptical vibration as composed of two circular rotations of the same frequency but with opposite sense of revolution, and observe that the frequencies of these rotations will be increased or decreased by the frequency of the superposed rotation, according as the revolution has the same or the opposite sense as this, while their amplitudes remain unaltered. Denoting the frequency of the periodic motion by ω and that of the superposed uniform rotation by σ—which two frequencies may be considered as fundamental frequencies of the multiple periodic motion of the particle—we get therefore, that this motion may be resolved in two sets of harmonic circular rotations with frequencies $\tau\omega+\sigma$ and $\tau\omega-\sigma$, where τ is a positive integer. According to the theory of multiple periodic motions, exposed in part A, the stationary states will now be fixed by two conditions, which in Sommerfeld's theory—which rests upon the possibility of separation of variables in polar coordinates—are written as

$$\mathscr{I}_1 = \int q_1 \mathrm{d}q_1 = n_1 h \qquad \mathscr{I}_2 = \int p_2 \mathrm{d}q_2 = n_2 h, \tag{18}$$

where n_1 and n_2 are positive integers, and where q_1 and q_2 are the length of the radius vector from the centre to the particle and the angular distance of this radius-vector from a fixed direction in the plan of the orbit respectively, while p_1 and p_2 are the conjugated harmonic momenta. The first integral is to be extended over the period of the radial motion and the second over an entire revolution of the radius vector. Since p_2 is constant during the motions, we see that $\mathscr{I}_2$ is simply equal to the angular momentum of the particle round the centre multiplied by 2π. Instead of the formula (17) of the conditions for the stationary states we shall in the following present these conditions in a modified form, more suited for the exposition of the problems under consideration. Thus we shall write

$$\mathscr{I} = nh, \qquad \mathscr{P} = kh \tag{19}$$

where $\mathscr{I} = \mathscr{I}_1 + \mathscr{I}_2$, while $\mathscr{P}$ is identical with $\mathscr{I}_2$, and where n and k are again positive integers, which clearly are subject to the condition $n \geqq k$.

With this choice of conditions the fixation of stationary states in the present case appears as a direct generalisation of this fixation of these states in the case of purely periodic systems. In fact if we imagine an adiabatic transformation of our central system, by which the central character is preserved, but the result of which is, that through the change of the law of attraction the motion from multiple periodic is transformed to simply periodic, the quantity $\mathscr{I}$ will just coincide with the quantity defined by the relation (2), the fixation of which by the condition (5) constitutes the only condition to which the stationary states of a periodic system are subject. Further it will be seen, how the quantities $\mathscr{I}$ and $\mathscr{P}$ are intimately connected with the choice of fundamental frequencies used in the above description of the motion and its resolution in harmonic constituents. Thus we have for the difference δE of the energy in two neighbouring motions of the system, for which the values of $\mathscr{I}$ and $\mathscr{P}$ differ by $\delta\mathscr{I}$ and $\delta\mathscr{P}$ respectively

$$\delta E = \omega\delta\mathscr{I} + \sigma\delta\mathscr{P} \tag{20}$$

On one hand this equation directly illustrates, why the second of the conditions (19) becomes ineffective, as soon as the motion is purely periodic, because in this case, where σ becomes zero, $\mathscr{P}$ does not enter explicitly in the expression of the energy of the system, which will depend on $\mathscr{I}$ only. On the other hand it will be seen, that as soon as the central orbit is no longer purely periodic, $\mathscr{P}$ will enter in the expression for the energy in such a way, that its fixation by means of the second of the conditions (19) will have the effect, that the spectrum calculated by the general frequency-condition will continue to reflect the motion of the system in a way consistent with the correspondence principle. As regards the possibility of spontaneous transitions between the stationary states, it follows next that the possibility of a transition from a state characterised by $n = n'$, $k = k'$ to a state $n = n''$, $k = k''$ is to be sought in the occurrence in the motion of constituent harmonic vibration with frequency $(n'-n'')\omega+(k'-k'')\sigma$. From the above analysis of the motion we may therefore conclude, that only such transitions must be regarded as possible, for which k changes by one unit.

An interesting support for this conclusion may be derived from an examination of the moment of momentum round the axis of the system possessed by the radiation emitted during a transition. In fact the assumption lies at hand, that this radiation will possess the same constitution as the radiation, which according to the classical electronic theory would be emitted by an electric particle performing a motion of the same type as the harmonic constituents of the actual motion corresponding with the transition. In case of a circular harmonic rotation this gives for the ratio between the angular momentum and energy of the emitted radiation a value equal to 2ν, where ν is the frequency of the radiation. As will be seen, this is just equal to the ratio between the change of angular momentum $h/2\pi$ and the loss of energy $h\nu$ suffered by the system during a transition, in which k changes by one unit.*

In case the law of attraction towards the centrum differs only slightly from the law of the inverse square of the distance, the orbit may be said to be formed by a Keplerian ellipse, the major axis of which undergoes a slow uniform precession. As regards the dimensions of this ellipse with stationary states the length of the major axis will approximately be expressed by the formula (13), as it follows immediately from the first of the conditions (19). On the other hand it is easily seen, that the second of these conditions claims, that the shape of the preceding ellipse is fixed in such a way, that the ratio between the minor and major axis is equal to k/n. In the applications, however, it is more often convenient instead of the minor axis to consider the so-called "parameter", i.e. the cord through

* Compare D. Kgl. Danske Vidensk. Selsk. Skr. 8. Raekke IV, **1**. p. 34, 1918. Independently of the correspondence principle, the importance of a consideration of conservation of angular momentum as a means of obtaining information about the possibility of transitions between stationary states and the constitution of the radiation emitted has been pointed out by Rubinovicz (Phys. ZS. **19**, p. 441, 465, 1918). As regards a closer discussion the relation of such considerations with the correspondence principle compare N. Bohr, (ZS. f. Phys. **2**. p. 423, 1920 and especially **6**, p. 1, 1921).

the focus perpendicular to the major axis. As a simple calculation shows, the length of this cord—which as regards the order of magnitude gives a measure for the shortest distance, within which the electron during its revolution approaches the centre, just as the major axis gives a measure for the largest distance of the electron from the centre—is given by an expression similar to that holding for the major axis with the only difference, that $\mathscr{I}$ is replaced by $\mathscr{P}$, or—in the stationary states—n is replaced by k. Denoting the length of the parameter by $2p$ we have thus

$$2p = \frac{\mathscr{P}^2}{2\pi^2 Ne^2 m} = k^2 \frac{h^2}{2\pi^2 Ne^2 m} \qquad \langle(20a)\rangle$$

if it is assumed, that the central force differs only little from the attraction to a nucleus of charge Ne. A very instructive application of the stationary states of central systems is offered by the detailed interpretation of the so-called fine-structure of the hydrogen lines, which formed the starting point of Sommerfeld's work on the spectral theory. As well known the interpretation of this phenomenon rests upon the circumstance, that the orbit of the electron round its nucleus, if the small variations of its mass with the velocity claimed by the theory of relativity is taken into account, will not be exactly periodic, but may, just as in the above case, where it was assumed, that the law of attraction differs slightly from the inverse square, be described as a Keplerian orbit, the major axis of which undergoes a slow precession. As the closer calculation shows, the frequency of this precession σ possesses a ratio to the frequency of revolution in the Keplerian orbit, which in first approximation is given by

$$\sigma/\omega = \frac{Ne^2}{2mc^2}\frac{1}{p} = \frac{1}{2}\left(\frac{2\pi Ne^2}{\mathscr{P}c}\right)^2 *$$

Due to this deviation from a purely periodic motion the energy of the atom will not, as assumed in the simple theory of the hydrogen spectrum, depend on $\mathscr{I}$ only, but will also vary slightly with varying $\mathscr{P}$. Thus the calculation gives in first approximation

$$E = -\frac{2\pi^2 N^2 e^4 m}{\mathscr{I}^2}\left\{1 + \left(\frac{2\pi Ne^2}{c}\right)^2\left(-\frac{3}{4\mathscr{I}^2} + \frac{1}{\mathscr{I}\mathscr{P}}\right)\right\}.$$

This formula offers a simple illustration of the application of the relation (19). In fact, it will be seen that in first approximation the partial differential coefficient $\partial E/\partial\mathscr{P}$ possesses a ratio to $\partial E/\partial\mathscr{I}$, which coincides with the above expression for σ/ω. For the value of the energy in the stationary states we get

* Cf. Q.o.L. p. 65, eq. (69).

by application of (19)

$$E_{n,k} = -\frac{2\pi^2 N^2 e^4 m}{n^2 h^2}\left\{1+\left(\frac{2\pi^2 e^2}{hc}\right)^2\left(-\frac{3}{4n^2}+\frac{1}{nk}\right)+\ldots\right\}$$

which as shown by Sommerfeld by application of (1) leads to values for the frequencies of the radiation emitted by the atom, which are in remarkable agreement with the measurements of the fine-structure of the lines of the hydrogen spectrum and of the analogous helium spark spectrum, investigated by Paschen.

Comparing the theory of the fine structure of the hydrogen lines with the simple theory of the hydrogen spectrum, exposed in the beginning of this chapter, we meet with a difference as regards the fixation of the stationary states, the essential feature of which is not so much, that the values of the energies are slightly different, but which must rather be thought therein, that in the first mentioned case the shape, of the orbit in these states was completely determined, while in the second case only the major axis was determined, the eccentricity being left entirely undetermined. It may not be thought, however, that this want of definition of the shape of the orbit in the latter case is merely to be regarded as a consequence of the simplified assumptions used in the deductions. On the contrary if the radiating atoms are exposed to the action of external agencies, producing perturbations in the orbits of the same order of magnitude as the slight deviations from a Keplerian motion, to which the small modifications in the laws of mechanics claimed by the theory of relativity give rise, the orbits in the stationary states will be completely different from those of the undisturbed atom, although the spectral lines will only be influenced in a measure not exceeding the order of magnitude of the fine structure of the undisturbed lines itself. This circumstance yields not only the explanation of the instability of the fine structure phenomenon in the presence of such small disturbing fields as are usually present in discharge tubes, but appears in clear light in the detailed theoretical interpretation, obtained in recent years on the basis of the quantum theory, of the effect of electric fields on the lines of the hydrogen spectrum.

On the basis of this theory Sommerfeld has now obtained an explanation of the characteristic difference between the structure of the other series spectra and that of the hydrogen spectrum, by assuming that the inner electrons in the other atoms besides their partial compensation of the effect of the nucleus charge would act upon the electron as a field of force of central symmetry, which at large distances converges more rapidly to zero than an electrostatic attraction from a point charge. Assuming for example with Sommerfeld that, in case of an atom emitting a series spectrum of the order k, the additional central force on the outer electron may be represented by a term of the type

$$F = Ne^2/r^2 + A/r^3 + B/r^4 + \ldots \tag{21}$$

it is possible by application of the condition (18) to find the values $E(k, n)$ for the energy in the stationary states as functions of n and k, and by dividing these terms by h to obtain a set of spectral terms, which exhibit a pronounced similarity with the sets of empirical spectral terms. In the special case where we confine ourselves to the first term in the above expression of the force, which varies as the inverse cube, the result may be expressed in the simple formula

$$E_{n,k} = -\frac{2\pi^2 k^2 e^4 m}{h^2} \cdot \frac{1}{\left(n - \left(\sqrt{k^2 - \frac{Am}{h^2}} + k\right)\right)^2} \tag{22}$$

which expression is seen just to correspond to Rydberg's formula (17) in the case, where ϕ_k is independent of n. In case we take higher terms in (21), the expression for $E_{n,k}$ becomes more complex, the term corresponding to ϕ_k depending also on n in a way, which—as shown by Sommerfeld—is similar to the modifications of the simple Rydberg formula proposed by Ritz. The analogy between the multitude of terms of the observed series spectra and those deduced in the above way appears not only in the expressions for the different sequences of terms, but first of all in the general structure of this multitude. Thus in the formulae (15) and (17) the letter k appeared only as an index, introduced in order to discriminate between the various simple sequences of terms, but as a consequence of the present application of the quantum theory it may be claimed, that within each such sequence k can be identified with a definite positive integer, in such a way that n in this sequence is always equal to or larger than k, and that—due to the geometrical significance as determining the smallest distance, at which the outer electron approaches the nucleus—the deviations from the corresponding hydrogen term decreases for increasing k. Now it has actually been found possible to identify the various observed sequences in a manner consistent with this condition. In fact, these sequences may as well for arc spectra as for such spark spectra, which hitherto have been differently distangled, be numerated 1, 2, 3, ... k in such a manner, that the deviations from the corresponding hydrogen terms are rapidly decreasing, and that apart from an uncertainty entering in the first sequence due to the especially large deviations from the hydrogen terms, the value of n characterising the first term is just equal to the number k characterising the sequence, while the values of ϕ_k diminish rapidly with increasing k. In spectroscopical literature these sequences are often termed as the sequence of sharp terms or S-sequence ($k = 1$), the sequence of "principal" terms or P-sequence ($k = 2$), the sequence of "diffuse" terms or D-sequence ($k = 3$), the sequence of fundamental (or Bergmann) terms or F-sequence ($k = 4$). Further terms of other sequences, which differ extremely little from the simple hydrogen (or the helium spark) terms, and about the arrangement of which in the scheme of series terms there had, just of this reason, been some uncertainty, must according to the present theory be ascribed to values of k larger than 4.

A very strong additional argument for the central symmetrical character of the orbit of the outer electron in the stationary states, involved in the emission of the series spectra, is obtained from a consideration based on the correspondence principle. Thus we have seen in the preceding that due to the character of the motion we can only expect the possibility of spontaneous transitions between such pairs of stationary states, in which k differs by one unit. This expectation is found to be fulfilled in a striking way by the experimental evidence. In fact, all the series appearing under ordinary conditions in flames or discharge tubes correspond to combinations between two spectral terms within two consecutive sequences, when arranged in the above mentioned way. The interesting observation by Stark and others of the appearance of other series of combination lines under such conditions of excitation, where the atoms are subject to strong electric fields, is also just what was to be expected on the correspondence principle, since in the presence of such fields the motion will undergo perturbations with the effect of the appearance of new harmonic components in the motion of the outer electron, the frequencies of which are given by $\tau\omega+\mu\sigma$, where μ is an integer, which differs from ± 1.

While the above considerations may be said to offer a convincing interpretation of the characteristic structure of the series spectra of the other elements, at least as long as we look apart from the complexity of the single series terms, which form a characteristic feature for the spectra of most elements, it is not possible to account with any closer approximation for the simultaneous dependence of the spectral terms on n and k for the spectrum of a given element by means of a simple expression for the force exerted by the inner system on the outer electron of the type given by (21). It is also clear, that such an assumption of a central force could only offer a very rough measure for the effect of the inner electrons on the motion of the outer. In fact, an accurate interpretation of the series spectrum of an element will naturally claim not only a detailed picture of the constitution of the atom in the absence of the outer electron, but also a detailed investigation of the mutual mechanical interaction of the motion of the inner electron with that of the outer. In the following we shall, when discussing the properties of special elements, have occasion to come back to this problem. For the moment we shall confine ourselves to consider some points of general character, which are of essential importance for the latter discussion.

The first of these points deals with the difficulty referred to above which appears for many spectra in the closer interpretation of the terms of the S-sequence, and which is clearly brought to light for instance through the uncertainty involved in the characterisation of these terms by definite values of n. This uncertainty is due to the very great deviations from the hydrogen terms given by (16), exhibited by the observed values for the terms within this sequence, as a consequence of which a term of the sequence often does not lie essentially closer to the one than to the other of the two successive hydrogen terms, between which it is lying, in contrast to the terms of the P, D and F sequences, for which there usually can be no doubt as to which value of n they must be ascribed. In spectroscopic literature this circumstance has given rise to a terminology, in which the terms of the S-sequence are characterised by numbers 1.5, 2.5 ..., while the terms of the other sequences always are characterised by whole numbers, a notation which, however, may be considered as wholly inadequate from the point of view of the present state of the spectral theory. At first sight it might appear, that the principles of the quantum theory directly afford a definite interpretation of the series terms in question. In fact from experiments on selective absorption in vapours of the alkali metals it is known, that the first observed term of the S-sequence in the spectra of these elements, which in the usual notation is characterised by 1.5 S, corresponds to the normal state of the atom, and it seems therefore necessary to assume, that this term should correspond to $n = 1$, since another assumption would apparently involve an inexplicable absence of a state corresponding to $n = 1$*. The problem appears in a different light, however, when we enter on a closer consideration of the interaction between the various electrons in the atom. While in the sequences of states corresponding to higher values of k we may be sure, that the orbit of the outer electron lies quite outside the region, where the electrons move, this will not be so for the states corresponding to $k = 1$. On the contrary the large deviations from the corresponding hydrogen terms bears direct evidence of the fact, that the outer electrons in these states penetrate into distances from the nucleus, which are of the same order as the dimensions of the orbit of the inner electron. This, however, will be seen to effect a large difference between the character of a possible state corresponding to $n = 1$, and of states corresponding to the larger value $n = 2$. While in the latter case the outer electron under the greater part of its revolution will move at distances large compared with the dimensions of the inner system, it will only during very short intervals pass within such short distances of the nucleus, the electron would in the former case during the whole time move within distances from the nucleus, which are

* Such a view comes for instance to expression in the formal scheme, illustrating the origin of the sodium spectrum, used by the writer in a lecture in the Physical Society in Berlin (ZS. f. Phys. **2** p. 423, 1920). On the other hand Roschdestwensky (See "Naturwissenschaften") has from an examination of the empirical material thought it possible to find evidence in favour of the assumption, that in the alkali metals the normal state is characterised by $n = 2$, without, however, attempting an explanation of the absence of the state $n = 1$.

of the same order of magnitude as the dimensions of the orbits previously bound. In the consideration of the possibility of a final state of the binding process corresponding to $n = 1$ in the equation (18) we meet therefore with the question of the constitution and the stability of the inner system. This problem, however, can according to its nature only be approached by an examination of the formation of actual atoms by successive binding. As we for instance shall see in the following, we must for the atoms of the alkali metals assume configurations for the normal state of the inner electrons, which exclude the possibility of binding an additional electron in a state corresponding to $n = 1$, and which naturally force us to assume, that in the normal state of the atom the electron last bound will move in an orbit $n = 2$, $k = 1$. At the same time we shall see, how the comparative simple structure of the alkali-spectra is connected with a remarkable spherical symmetry of the configuration of the inner electrons in the elements of this family, and how the more complex structure of other spectra is connected with the less symmetrical character of the configuration of the inner system. This circumstance comes not only to light in the very intricate spectra of certain families of elements, the disentanglement of which has first been attacked with success in recent years, but also thereby that certain elements possess several series spectra of the same order, the terms of which do not form mutual combinations, and which may be ascribed to various performances of the binding process, as regards the interaction between the outer electron and the inner system. With a typical case of this we shall meet already in helium, which is subsequent to hydrogen in the periodic table. Even in the spectra of the alkali metals we meet, however, with a characteristic feature, for which the simple analogy with central systems offers no interpretation, namely the well known doubling of the series terms. As pointed out by Sommerfeld and Landé an explanation of this complexity may be sought in a small deviation of the motion of the outer electron from a simple plane central orbit, so that the multitude of stationary states does not bear a complete analogy with that of a particle moving in a central field of force. Such an effect might in actual atoms be expected to appear, if the system of inner electrons in their normal state in the absence of the outer possessed a resultant angular momentum round an axis. In fact, it might be expected to involve, that the effect of the inner system on the outer electron will not exhibit complete central symmetry, but rather be comparable with that of a field of force possessing axial symmetry round the mentioned axis. In the following chapters we shall have occasion to examine, to what extent such a consideration may offer an interpretation of the complexity of the term of the alkali-spectra and of the spectra of the alkaline earths. Here we shall only for sake of the later discussion briefly expose the principles, in which the fixation of the stationary states in such a case rests. This fixation may be obtained in immediate connection with the theory of central systems by considering the influence on the motion caused by the deviation of the outer electron from a central orbit. In fact, in case as that under consideration the orbit of the outer electron may be described as a central orbit, the plane of which is subject to a slow and uniform rotation round the invariable impulse axis of the whole atom, and as a consequence of the additional fundamental frequency, which due to this rotation enters in the motion, the fixation of the stationary states will involve an additional condition. Thus these states will be fixed by three conditions, which may be written in the form

$$\mathcal{I} = nh, \quad \mathcal{P} = kh, \quad \mathcal{Q} = lh \tag{23}$$

where the two first conditions are analogous to the condition, which characterises the stationary states of a simple central system, and may be derived from these by means of an imagined adiabatic transformation, by which the deviation from central symmetry is made to vanish, while the third condition fixes the total angular momentum of the whole atom in the stationary states, the quantity $\mathcal{Q}$ being equal to 2π times this momentum. Just as in the case of a simple central system the second condition may be said to determine of the rotating deformed Keplerian orbit, the major axis of which was determined by the first condition, while its shape in case of a purely periodic orbit was left undetermined; the third of the conditions (23) may be said to fix the orientation of the plane of the outer orbit relative to the inner system, which in case of complete central symmetry remained completely undetermined. In fact, the orientation of the outer orbit follows directly from the two last conditions of the angular momentum if the inner system in the absence of the outer electron is known. Now this momentum must again be assumed, multiplied by 2π, to be equal to an entire multiple of h, so that we may write $2\pi A = jh$, where j is a positive integer. Denoting the angle between the rotating plane of the outer orbit and the invariable axis of the system by ϑ, we get thus for the stationary states

$$\cos\vartheta = \frac{j^2+l^2-k^2}{2jl}$$

since naturally the numbers k, j and l are subject to the conditions $n \geqq k \geqq |l-j|$. We see, that to each state corresponding to a given value for n correspond not only n-states of different states of the outer electron, but that there to each of these correspond $2j+1$ or $2k+1$ states with different orientations according as $k > j$ or $k < j$ respectively.

As regards the question of the correspondence between the spectrum and the motion we have in the first place, that the difference δE in the energy between two neighbouring states, which correspond to values for $\mathcal{I}$, $\mathcal{P}$, $\mathcal{Q}$ which differ by $\delta\mathscr{I}$, $\delta\mathscr{P}$ and $\delta\mathscr{Q}$ respectively

$$\delta E = \omega\delta\mathscr{I} + \sigma\delta\mathscr{P} + \rho\delta\mathscr{Q} \tag{24}$$

where in analogy to the notation in case of central systems ω determines the frequency of the revolving periodic orbit of the outer electron, and σ the frequency of the precession of the major axis of this orbit, relative to the line of intersection between its plane and the invariable plane of the atom through the nucleus perpendicular to the invariable axis, while finally ρ denotes the frequency of rotation of this line. Next, as regards the revolution of the motion of the outer electron in its harmonic constituents it will easily be seen, that this motion may be considered as a superposition of a set of linear harmonic vibrations of frequencies $\tau\omega \pm \sigma$ parallel to the invariable axis and two sets of circular harmonic rotations in opposite directions round this axis of frequencies $\tau\omega \pm \sigma + \rho$ and $\tau\omega \pm \sigma - \rho$ respectively. From this we conclude, that also in the case of the more complex motion under consideration only such transitions are possible, by which k changes by one unit. While in the case of a simple central system each such transition corresponds to a radiation of definite frequency, which may be expected to possess circular polarisation round an axis perpendicular to the plane of the orbit, we meet, however, in the present case with two types of transitions, in which k changes by one unit. In the first of these l will remain unchanged, and the emitted radiation will be linearly polarised parallel with the invariant axis of the atom. In the second l has changed by one unit, and the radiation will show circular polarisation round this axis. Just as in the case of an atom of complete central symmetry this polarisation is naturally not subject to direct observation. In view of a consideration of conservation of angular momentum analogous to that referred to in the former case, the correspondence between spectrum and motion obtains, however, enlarged importance in the present more complex case, as from this we may directly conclude, that the angular momentum of the atom during transitions must either remain invariant or change by $h/2\pi$, and thereby obtain a support of the conditions (23), deduced on the theory of multiple-periodic systems, but an independent argument for concluding that in actual atoms, where the direct application of this theory may involve complications, the total angular momentum must always be equal to an entire multiple of $h/2\pi$.

As mentioned we shall postpone the question of applications of the above considerations to the interpretation of the complex structure of the lines of the series spectra. Before we conclude this chapter, however, we shall still have to mention the important development of the theory of the hydrogen spectrum, by which Sommerfeld was able to explain in detail the fine structure of the hydrogen lines observed by instruments of high dispersive power, a theory which in fact forms the starting point for Sommerfeld's work on the spectral problems. The

[118]

interpretation of this phenomenon rests upon the circumstance, that due to the small variation of the mass of the electron with its velocity claimed by the theory of relativity the orbit of the electron round the nucleus will not—as assumed in the simple theory given in the beginning of this chapter—be exactly periodic, but may, just as the orbit of a central system, for which the law of attraction differs slightly from that of the inverse square, be described as a slightly deformed Keplerian orbit, the major axis of which undergoes a slow precession. Due to this deviation from a purely periodic motion the stationary states of the atom will not be fixed by one condition only, but the motion in these states will be defined by two conditions of the form (19). The application of these conditions gives now for the energy in the stationary states, instead of the simple expressions (10) and (13), a more complex expression, which in first approximation may be written

$$E = -\frac{2\pi^2 N^2 e^4 m}{\mathcal{I}^2}\left\{1+\left(\frac{2\pi N e^2}{c}\right)^2\left(-\frac{3}{4\mathcal{I}^2}+\frac{1}{\mathcal{I}\mathcal{P}}\right)\right\}$$

$$= -\frac{2\pi^2 N^2 e^4 m}{n^2 h^2}\left\{1+\left(\frac{2\pi N e^2}{hc}\right)^2\left(-\frac{3}{4n^2}+\frac{1}{nk}\right)\right\}$$

Since the numerical value of the factor $(2\pi^2 e^2/hc)$ is very small (about $7{,}24 \times 10^{-3}$), this expression leads for every value of n to a set of values, which all lie very near to the value given by (13). By application of the frequency relation (1) we get therefore for each line represented by (14) a number of components lying close together, which were found to be in remarkable agreement with the measurements of the fine structure of the lines of the hydrogen spectrum and of the analogous helium spark spectrum, investigated by Paschen.

Comparing this theory of the fine structure of the hydrogen lines with the simple theory of the hydrogen spectrum, exposed in the beginning of this chapter, it will be noticed that the difference as regards the fixation of the stationary states appears not only in the slightly different values of the energy, but that an essential feature of this difference lies therein, that in the fine structure theory the shape of the orbit in these states is completely determined, while in the simple theory only the major axis was determined, the eccentricity being left entirely undetermined. It may be thought at first sight, that this want of definition of the shape of the orbit in the latter case is merely to be regarded as a formal consequence of the simplified treatment. A closer consideration, however, brings out the justification of considering the shape of the orbit as actually undetermined as far as the motion may be treated as periodic. In fact, if the radiating atoms are exposed to the action of external agencies, producing perturbations in the orbits of the same order of magnitude as the slight deviations from a Keplerian motion, to which the small modifications in the laws of mechanics claimed by the theory of relativity give rise, the orbits in the stationary states will be completely different from those of the undisturbed atom, although the spectral lines will only be influenced in a measure not exceeding the order of magnitude of the fine structure of the undisturbed lines itself. This circumstance yields not only the explanation of the marked instability of the fine structure phenomenon in the presence of such small disturbing fields as are usually present in discharge tubes, but appears in clear light in the detailed theoretical interpretation, obtained on the basis of the quantum theory, of the effect of electric fields on the lines of the hydrogen spectrum. In this connection it is of importance to notice the fundamental difference between the part played by the relativity modifications in the theory of the hydrogen spectrum and in the series spectra of other elements. As we have seen, we have in the latter case with the motion of an outer electron in the atom, the orbit of which due to the presence of the inner system will be subject to considerable deviations from a simple periodic motion, the motion in first approximation being comparable with a central orbit, in consequence of which each component of the fine structure of a hydrogen line appears as analogous to a certain member in one of the series composing the spectrum of the other elements. The neglect of the relativity modifications involved in the above treatment of the latter spectra will not result in any alteration of the general character of the orbit and its resolution in harmonic constituents, but only in a slight alteration in the amplitudes and frequencies of these constituents, accompanied by minute alterations in the frequencies of the spectral lines, the character of this spectrum remaining the same.

While thus the variability of the mass of the electrons with their velocity is of no essential importance for the interpretation of the series spectra of the elements, apart from hydrogen, the problem stands differently as regards the interpretation of the Røntgen spectra. Thus Sommerfeld was able to show, that the intimate numerical relation mentioned above between the frequencies of the prominent lines in the Røntgen spectrum of an element, and frequencies of lines of the simple spectrum belonging to an atom, containing only one electron, could be traced still further, as the observed complexity of certain of the Røntgen lines in question and of the conjugated absorption edges, exhibited as regards frequency differences of the components, a close numerical relation with the relativity fine structure to be expected for the latter spectrum. At the same time as this result undoubtedly affords a most striking evidence of the similarity of the motion of the inner electrons in the atom and the motion in the stationary states corresponding to the various stages of the process of binding of a single electron by a positive nucleus, the extended analogy under consideration brings on the other hand the paradoxes, we meet in an interpretation of the Røntgen spectra, consistent with a conception of the atom suitable for the explanation of the periodic law, to appear in still sharper light. In fact on the view, exposed in the preceding, according to which the absorption and emission of Røntgen rays are connected with processes of removal of one of the electrons from the atom and of subsequent reorganisation of the normal state, we must from the analogy in question not only conclude, that in the normal state of the atom electrons are present, which move in orbits corresponding to the various stages of the binding process of a single electron by the nucleus, considered in the simple theory of the hydrogen spectrum, but we must even claim the simultaneous presence of electronic orbits of the various shapes occurring in the theory of the fine structure of the hydrogen lines. As we shall see in the following chapters, a consideration of the formation of atoms by successive binding of the electrons leads us to a view as regards the constitution and stability of the normal state of the atom, which, at the same time as it seems to offer a suitable basis for the interpretation of the periodic law of the chemical properties of the elements, actually fulfills this claim.

§ 2. Helium.

Helium, which is the first element after hydrogen, and the neutral atom of which contains two electrons, is in chemical respects especially marked by its inactive character. Helium, being in fact the first of the family of the inactive gases, is distinguished by having the lowest critical temperature and highest ionisation potential of all elements. These properties of helium bear witness of an intimate interaction in the normal state of the neutral atom between the two electrons. For this it has been tried to account by the simple assumption, that the two electrons in this state moved in the same circular orbit round the nucleus, each with the same angular momentum $h/2\pi$. Such a model of the helium atom offered apparently a possible explanation of the chemical inactivity and of the order of magnitude of the ionisation potential, which is about double as high as that corresponding to the removal of an electron from the neutral hydrogen atom which is equal to 13,6 Volt. An accurate agreement with the observations could not be obtained amongst others since the ionisation potential corresponding to the model was found 2,125 times that corresponding to the hydrogen atom, which would give about 29 Volt, while the recent accurate measurements have given only 25 Volt. At first sight this disagreement appears as a fundamental difficulty for the simple applications of the quantum theory, because seemingly no argument could be offered, which allows to exclude the mentioned simple configuration from the possible stationary states, and it would therefore seem difficult to assume a normal state with lower ionisation potential than 29 Volt. As we shall see, it appears, however, possible from a study of the helium spectrum in the light of the correspondence principle to throw light on this point. As mentioned in the former §, helium possesses, apart from a series spectrum of the second order, which is completely analogous to the hydrogen* spectrum, two complete series spectra of the first order, which do not form mutual combinations. The one of these spectra consists of single lines and is usually referred to as the Parhelium spectrum, while the other consists of narrow doublets and will be denoted as the Orthohelium spectrum. While the spectrum of second order is due to the binding of one electron by the nucleus, the two last spectra are due to the binding of the second electron by the atom and bear witness about the fact, that this process can take place in two essentially different ways. A theoretical investigation of this problem clearly claims a close examination of the mutual interaction of the motions of the two electrons round the nucleus, which might represent stationary states, corresponding to the various stages of the binding process.

First of all we must assume, that the electron first bound in all these states moves in an orbit, which differs only little from the stationary circular one-quantum orbit, it will describe in the absence of the second electron. On the other hand the orbit of the outer will, at any rate at the initial stages of the binding process, move in an orbit differing only little from a possible stationary orbit described round a single positive nucleus of unit charge. The smaller the dimensions of this orbit become during the binding, however, the more the mutual perturbations of the electrons will influence the motion in the stationary states, producing at the same time deviations of the spectral terms from the corresponding terms in the hydrogen spectrum.

A very interesting attempt to base the interpretation of the helium spectra of the first order on the study of these mutual perturbations of the orbits of the inner and outer electron has been published by Landé. The starting point of Landé's work forms an examination of the mutual perturbation, in case the orbits of the two electrons, the orbit of the outer electron being large compared to that of the inner electron, are situated in the same plane and are circular in first approximation. As result of this calculation Landé found, that if the two electrons move in the same direction, the mean effect of the perturbations as regards the inner orbit might be described as that of a displacement of the centre of this orbit in a direction opposite to the radius vector from the nucleus to the outer electron. Due to this fact the mean effect of the inner system on the outer electron was found to deviate from the attraction of a single nucleus of unit charge by a small quantity, which varies with a lower power of the distance than the effect calculated on the assumption, that the mean effect of the inner electron on the

* [The manuscript has "helium", but this is clearly an error.]

outer one was equivalent with that of an electronic charge uniformly distributed along the undisturbed circular orbit of the inner electron.

In order to apply this calculation to the interpretation of the spectra under consideration Landé makes use of the theory of multiple-periodic systems comparing the effect of the inner electrons in its approximately circular orbit by the effect of an inner system, which produces a field of force of simple axial symmetry, and which possesses an angular momentum round the axis of this field. As mentioned the stationary states will in such a case be fixed by three conditions, of which the first two determine the dimensions and the shape of the orbit of the outer electron, and the third the angle between the plane of this orbit and that of the axis of the inner system in such a way, that the total angular momentum of the atom becomes equal to an entire multiple of $h/2\pi$. In fact, Landé's calculations form a first example of a treatment as that discussed on page [126].

Since in the case under consideration the angular momentum of the inner system is equal to $h/2\pi$, we get for every set of values for n and k only three possible values for l*, namely $k+1$, k and $k-1$. Of these the first and the last correspond to states, in which the two electrons move in the same plane in the same or opposite direction respectively, while the second case corresponds to states, where the planes of the two orbits make an angle with each other, which for large value of k converges rapidly to 90°. Discarding now the possibility of $l = k-1$, assuming that these states for some unknown reason are physically impossible, Landé referred the interpretation of the orthohelium spectrum to the stationary states, in which the two electrons revolve in "coplanar" orbits in the same direction, and the interpretation of the parhelium spectrum to "intersecting" orbits, and obtain in this way for both spectra an agreement as regards the order of magnitude of the deviations of the observed spectral terms from the corresponding hydrogen terms.

This interpretation of the helium spectra, however, is open to serious objections as regards the underlying general arguments as well as the elaboration of the details of the calculations. In the first place Landé's assumptions seem not to be consistent with the absence of lines corresponding to combinations of terms belonging to the ortho- and the parhelium spectra. In fact, the assumed analogy between the motion of the outer electron and the motion, which would appear in a multiple-periodic system as that described, would claim the appearance of such combination lines, which would correspond to the linear vibrations parallel to the axis of the atom, which form part of the constituent harmonic vibrations of the motion. These vibrations would certainly in general possess smaller amplitudes than the circular rotations round this axis, but this circum-

* [For the definition of l, see eq. (23), p. [117].]

stance would by no means explain the marked absence of any such combination lines, which holds not only, if the helium spectra are excited under usual conditions, but also for the spectra observed when the atoms are subject to strong electric fields, in which case as mentioned a large number of new combination lines between pairs of terms of the orthohelium as well as the parhelium spectrum appear. This last circumstance would seem to show, that the direct comparison between the motion of the atoms and that of a multiple-periodic system, which underlies Landé's interpretation of the ortho- and parhelium spectra, involves an idealisation which is hardly legitimate. One must rather be prepared to expect, that the mutual perturbations of the deformable orbits of the two electrons in general would give rise to an exchange of energy and momentum of a nature far more intricate than would correspond to the assumption of an inner system of simple axial symmetry. In view of such a consideration the lecturer has already some years ago, in collaboration with Dr. Kramers, been occupied with a closer investigation of the mechanical problem of the mutual perturbations of the orbits of the two electrons in the helium atom. Our results have not yet been published, but as they seem to throw light on the problem under consideration, they may be briefly exposed here. The calculation deals with the general mechanical problem of the mutual influence of the orbits of the two electrons, with the sole restriction that the dimensions, and consequently also the period of revolution of the outer orbit, are large compared with the dimensions, and the period respectively, of the inner orbit. In consequence of this assumption the motion of the inner orbit may in first approximation be described as a Keplerian orbit round a nucleus of two units charge, which undergoes slow variations as regards shape and position due to the influence of the outer electron, while the motion of the latter in first approximation will be a Keplerian ellipse round a nucleus of one unit charge, which equally undergoes slow variations in shape and position, due to the fact that the mean effect of the inner electron will be slightly different from that of an electronic charge placed at the nucleus, in a manner depending on the shape and orientation of the inner orbit. The investigation of these variations now shows, that in general the variations of the orbit are so complex, that the motions of the electrons will not possess a character, which allows a direct application of the theory of the stationary states of a multiple periodic system. Among the multitude of possible motions there are two cases, however, in which the variations of the orbits exhibit certain simple periodicity properties. In the first of these cases the electrons move through all times in one and the same plane, and in the second the planes of the orbits form with a high degree of approximation constantly a right angle with each other, the major axis of the inner orbit remaining perpendicular to the plane of the outer orbit. While in both cases the shape and orientation of the inner orbit, apart from the conditions just mentioned, may at a given moment be chosen arbitrarily, and in general may undergo considerable changes during the revolution of the outer electron, the orbit of the latter electron will, as the calculation shows, as a consequence of these conditions possess a character similar to the variations of a Keplerian orbit perturbed by a simple central field, i.e. in first approximation the shape of this orbit will remain unvaried, while its major axis will undergo a slow and uniform precession. These kinds of orbits present themselves therefore as suitable for a selection of stationary states, responsible for the emission of series spectra of the type observed. As regards the fixation of these states the direct application of the theory of multiple-periodic systems involves still a certain amount of arbitrariness, which may be overcome, however, from a consideration of the establishment of these states during the binding process of the outer electron, as it seems possible among the motions of the atom in the two classes mentioned to select families of motion, distinguished by fulfilling certain conditions, which this establishment seemingly would claim.

If we first consider the case of coplanar orbits, where in general the shape of the inner orbit and its orientation relative to the position of the outer electron will undergo periodical changes due to the perturbing effect, as well as the motion of this electron, it can be shown, that there exists a type of distinguished solutions, where the shape of the inner orbit remains invariant, and where only the direction of its major axis is subject to changes, in such a way, that at any moment it points away, or towards, the momentary position of the outer electron; the perturbing effect in this case, consisting in a precession of the major axis of the inner orbit, which for any position of the outer electron in its elliptical orbit is just equal to the angular velocity of this electron relative to the nucleus. In the first approximation the value of the eccentricity ε of the inner orbit corresponding to this type of solution may be expressed by

$$\varepsilon = \tfrac{3}{2}\sqrt{a_i/2p_o}$$

where $2a_i$ is the major axis of the inner orbit and $2p_o$ the parameter of the outer orbit*. While such a type of mechanical solution exists for the motion of the

* This eccentricity of the inner orbit gives rise to a precession of the major axis of the outer orbit which can be simply calculated in first approximation if it is taken into account that at large distances the effect of an electron revolving in a Keplerian orbit coincides with that of an electronic charge at a point which may be denoted as electrical centre and which is at a distance from the attracting centre equal to $\frac{3}{2}\varepsilon a$, where ε is the eccentricity and $2a$ the major axis. In fact, taking this into account, one gets by a simple application of the theory of perturbation that the ratio between the frequency σ of the precession of the outer orbit and the frequency of revolution of this orbit $\omega : \sigma/\omega = \pm\frac{3}{2}\varepsilon a_i/p_o = \pm(9/4\sqrt{2})(a_i/p_o)^{\frac{3}{2}}$, a quantity of lower order of magnitude than that which would be produced by an electronic charge uniformly distributed along a circle of radius a_i which in fact would give rise to a precession of the major axis of the outer orbit, the ratio of which to ω is easily seen to be of the same order as $(a_i/p_o)^2$.

atom in the absence of external disturbances, it may be proved that also in the case this motion undergoes variations consisting in a slow change of energy and angular momentum of the system due to the effect of external agencies acting on the outer electron, a similar solution exists, in which the inner orbit will continuously adjust itself in such a way, that its major axis always points along the radius vector from the nucleus to the outer electron, and that its eccentricity is always related to the momentary value of the parameter of the outer orbit through the above expression. During this adjustment as well as during the undisturbed motion of the atom the mean value of the sum of the kinetic energy of the inner electron and its potential energy with respect to the nucleus will to a high degree of approximation remain unaltered, while the mean value of the angular momentum of the inner electron, which is constant during the undisturbed motion, during the adjustment will undergo gradual changes corresponding to the slow variation of the eccentricity of the inner orbit.

From these properties of the distinguished solutions of the coplanar problem it will be seen, that the motion of the outer electron in several respects exhibits a pronounced analogy with the behaviour of a system consisting of a single particle moving in a fixed central field. Since the inner electron as well in the undisturbed states of the atom as during such processes which may correspond to transitions between stationary states accompanied by emission or absorption of radiation so to say plays a passive part, its adjustment having only the effect of preserving the central character of the mean force to which the outer electron is subject. We are therefore presented directly with a fixation of the stationary states, which would seem to form a rational application of the theory of multiple periodic systems. Thus by fixing in the first place the motion of the inner electron by the condition that the mean value of the sum of its kinetic energy and its potential energy relative to the nucleus is equal to the energy in the normal state of the atom in the absence of the outer electron, we obtain that the inner system during the binding process as well as in the stationary states, as during the transitions between these, may be considered as remaining in its normal state, under the varying influence of the outer electron, which may be compared with the effect of an external field, undergoing adiabatic changes. Next the motion of the outer electron in the stationary states may be fixed in direct analogy with the theory of central system, if only due attention is paid to the effect of the exchange of energy and angular momentum between the two electrons, in the stationary states as well as during the transition between these. For this fixation we may start from a consideration of the following expression for the difference in the total energy for two neighbouring states of the system, which belong to the family under consideration, and for which the mean value of the "inner" energy is the same.

$$\delta E = \omega\delta\mathscr{I}^x + \sigma\delta\mathscr{P}^x$$

where ω represents the radial frequency of the outer electron and σ the frequency of the precession of the major axis, while $\mathscr{P}^x$ denotes the total angular momentum of the atom, and $\mathscr{I}^x$ the sum of this angular momentum and an integral of the type $\int p dq$ taken over one period of the radial motion of the outer electron, all in intimate analogy with the expression (20) for the energy difference between two neighbouring motions of a simple central system. In analogy with the fixation of the stationary states in the latter case we may therefore assume, that in the present case the stationary states are fixed by the two conditions

$$\mathscr{I}^x = nh, \quad \mathscr{P}^x = (k \pm 1)h, \quad n^x = n \pm 1,$$

where the upper and lower sign in the last condition correspond to the case, where the two electrons revolve in the same or opposite direction round the nucleus. By this fixation we secure a correspondence between the spectrum and the motion of exactly the same kind as for multiple periodic systems; as well as the conservation of angular momentum during the transitions, expected from the constitution of the emitted radiation claimed from the character of the resolution of the motion of the outer electron in constituent harmonic vibrations. Moreover, we see, that for increasing values of n and k the motion of the outer electron converges more and more to the possible stationary states of the electron in the hydrogen atom, which is perturbed by some agency producing a uniform progression of the major axis, just as the modifications in the laws of mechanics claimed by the theory of relativity*.

When next we consider the second case, where the orbits of the inner and the outer electrons are approximately perpendicular to each other, the major axis of the inner pointing along the normal to the plane of the outer, we meet with a different problem, when asking for the selection of stationary states. Also in this case a distinguished type of mechanical solutions exists, for which the shape of the inner orbit remains invariant, and where the orbit preserves a constant orientation with respect to the radius vector of the outer electron, the normal of the plane of the inner orbit pointing always to the momentary position of the latter electron. For the present type of solution the eccentricity will, using the same notation as in the former case of coplanar orbits, be given by

$$\sqrt{1-\varepsilon^2} = \frac{3}{2\sqrt{2}}\sqrt{\frac{a_i}{p_o}}$$

* The inner orbit on the other hand will be of more and more circular shape corresponding to the normal state of the undisturbed atom with one electron. It may be remarked that we in the foregoing have looked apart from the relativity modifications [one word illegible] which for a large period of revolution of the outer orbit will exert perturbations of the orbit of the same order of magnitude as those produced by the presence of the outer electron. It is of importance to observe that also these perturbations will disappear at the limit when the dimension of the outer orbit becomes infinitely large due to the fact that the inner orbit in this limit will be circular.

In contrast to the co-planar case, where the eccentricity converges to zero for increasing p_o, we see that in this case the eccentricity will increase for increasing p_o and converge to 1 in case of very large dimensions of the outer orbit. Quite apart from the fact that in this limit the solution becomes illusory due to the preponderant influence of the relativity modifications on the inner orbit compared with the effect of the outer electron, we may from this character of the solution conclude, that the orbits in question are not suited for representing stationary states corresponding to stages of the process of binding of the outer electron. On the contrary we must assume, that these stationary states must be selected between such types of solutions for which the inner orbit for increasing dimensions of the outer orbit will tend to be circular. As the study of the motion shows, this condition excludes a solution of so simple a character as regards the shape of the inner orbit and its orientation with respect to the outer electron as in the solutions described above, but still a distinguished solution of a rather simple character exists, which would seem suitable for the representation of stationary states. In this solution the variation in the shape of the inner orbit may be described as a slow harmonic oscillation of its centre along the normal to the plane of the outer orbit and a slow rotation of its plane round this normal. The calculation gives for the amplitude of the oscillation of the centre $a_i(3/2\sqrt{2})\sqrt{a_i/p_o}$, and for the ratio between the mean frequency of the rotation of the inner orbit and the frequency of revolution of the outer

$$\sigma/\omega = \tfrac{9}{16}a_i/p_o \tag{25}$$

Comparing with the case of coplanar orbits the effect of the inner system on the precession of the major axis of the outer orbit will, however, be much smaller on account of the fact, that displacement of the electrical centre of the inner orbit due to the mutual perturbations will not be directed along the radius vector of the outer electron, but perpendicular to it. This last circumstance holds accurately, however, only in the limit, where the dimensions of the outer orbit are very large compared with those of the inner orbit. For orbits where p_o becomes smaller, the angle between the planes of the orbits will gradually be altered, due to the varied conditions for the preservation of the motions of distinguished type. For the closer insight in this question the present investigation of the mutual perturbations does not, however, offer sufficient guidance. Nevertheless it seems possible to give certain definite statements as regards the fixation of the stationary states, using the analogy with a central system in a way corresponding to that mentioned for coplanar orbits. Thus from the correspondence between the spectrum and the motion, and the condition to be fulfilled in the limit, where the outer orbit is very large compared to the inner we may conclude, that two conditions must hold, of the same form as (19), where in the last of these conditions $\mathscr{P}$ will now denote the total angular momentum of the atom.

When now we ask for the energy corresponding to the possible stationary states corresponding to the various distinguished solutions described, we find, that the contribution to the energy of the atom, which appears as a consequence of the binding of the second electron, in case of coplanar orbits deviates from the values of the energy in the hydrogen atom by terms of a lower order of magnitude than the effect calculated by neglecting the perturbations in the inner orbit, and estimating the effect of the inner electron on the motion of the outer one as that of an electronic distribution over the circumference of a circle, while in the case of intersecting orbits the deviations are only of the same order of magnitude as that, which would be obtained on such an assumption. Neglecting small quantities of the same order of the latter deviations we may deduce the expression for the energy in the stationary states with coplanar orbits by direct use of equation (20) and the expression (00) for σ*. In fact in first approximation we may for a_i and p_o use the formulae (13) and (20a) in the former §, and we may therefore write:

$$\sigma = \omega \frac{9}{16} \frac{h^3}{\mathscr{P}^3} \tag{26}$$

From (20) we get therefore

$$\delta E = \omega \left(\delta \mathscr{I} \pm \frac{9}{16} \frac{h^3}{\mathscr{P}^3} \delta \mathscr{P} \right) = \omega \delta \mathscr{I}' \tag{27}$$

where

$$\mathscr{I}' = \mathscr{I} \mp \frac{9}{32} \frac{h^3}{\mathscr{P}^2}$$

From (27) it follows, that E, as well as ω, is a function of $\mathscr{I}'$ only, and it may be simply shown, that with the approximation under consideration the relation between E, ω and $\mathscr{I}'$ is the same as the relation between E, ω and $\mathscr{I}$ holding for a simple Keplerian motion, and given by (10). For the energy we get therefore

$$E = -2\pi^2 e^4 m \Big/ \left(\mathscr{I} \mp \frac{9}{32} \frac{h^3}{\mathscr{P}^2} \right)^2 \tag{28}$$

which for the energy in the stationary states gives

$$E_{n,k} = -(2\pi^2 e^4 m) \big/ h^2 \left(n \mp \frac{9}{32} \frac{1}{k^2} \right)^2 \tag{29}$$

a formula, which according to the approximative procedure followed, involves

* [Cf. eqns. (73) and (74) in the fourth lecture held in Göttingen, pp. [383] and [384].]

the neglect of small terms, which relative to the main terms are of the order of magnitude $1/nk^2$ and higher orders.

Comparing formula (29) with the observations it is now found possible to obtain an approximate agreement with the ortho-helium spectrum, if the upper sign in the denominator is used. For the parhelium spectrum on the other hand the numerical values of the deviations of the spectral terms from the hydrogen terms are so much smaller, that in an attempt to account for them we are presented with the assumption, that the corresponding stationary states belong to the class of orbits of intersecting planes. As regards the origin of the ortho- and parhelium spectrum we are then led to conclusions, which as regards the description of the motion in the stationary states exhibit features corresponding to Landé's interpretation*. On the entirely different way, on which the discussion of the selection of stationary states is based, it seems possible, however, to escape from the fundamental difficulty, which the absence of mutual combinations between the ortho- and parhelium spectrum presents for the latter interpretation, which is based on the formal assumption of the direct applicability of the theory of stationary states of multiple-periodic system. In fact, although in our considerations the motion in the stationary states is certainly assumed to be of multiple-periodic type, it is not assumed, that these motions belong to one and the same continuously connected class of such motions. On the contrary they appear as members in two separated families of such motions, the gap between which cannot be overbridged by motions of multiple periodic type, and for transitions between which no basis for the application of the theory of such motions is therefore offered. At the same time it would seem, that our point of view might suggest an explanation of the non-appearance during the binding process of stationary states, corresponding to coplanar orbits of opposite sense of revolution. Thus it must be imagined, that the establishment during the binding of the second electron by the helium atom an adjustment will take place, which cannot be described on the basis of the ordinary laws of mechanics, and by which the orbits of the two electrons—the planes of which before the process must be assumed to have an arbitrary relative orientation—are brought to fulfill the conditions

* As regards the mathematical expression for the spectral terms, the formula derived by Landé for the orthohelium case is essentially different from (29), since the derivation from the hydrogen terms appearing as an entirely different function of n and k. This difference is due to the way in which the perturbations are calculated in the simple case of two approximately circular coplanar orbits which form the starting point of Landé's calculations. Thus, in Landé's procedure the inner orbit is considered as a perturbed circular orbit, although the value of the eccentricity corresponding to his result is much larger than could be produced by the perturbing forces during a single revolution of the inner orbit. As a consequence of this, an approximative procedure must, in order to obtain the correct result as first approximation, consider the inner orbit as a perturbed Keplerian ellipse with finite eccentricity, as done in the calculations which form the basis of our investigation.

necessary for the appearance of well defined stationary states. Now the cause of this adjustment may be sought in the irregular variation of the inner orbit, which may give rise to the appearance of a radiation process, by which a more regular motion of the inner electron is brought about, and the assumption lies therefore at hand, that the adjustment can only result in a decrease of the energy of the inner electron relative to the outer. Now the case of coplanar orbits with opposite sense of revolution corresponds to an absolute maximum of this relative energy for given outer orbit, and it will therefore be seen that we are presented with an argument for assuming that the probability for the establishment of stationary states of this type is vanishing small. Before concluding these remarks about the possible interpretation of the helium spectrum it may once more be emphasised, that the considerations—being based on an investigation of the mutual interaction between the two electrons only for large values of the ratios of dimensions and periods of the outer and inner orbits—do not in this form allow an accurate evaluation of the energy in the stationary states and thereby a definitive numerical test of the theoretical arguments. For a closer discussion of the limit of accuracy, which may be obtained by the method of approximation used, and of the procedure by which this accuracy may be increased by taking not only the variation of the orbits as such into consideration, but by tracing the finer harmony between the periodicity properties of the motion of the inner and outer electron themselves, the reader may be referred to a more detailed account of our investigation, which it is hope to publish soon.

After this discussion of the origin of the helium spectra of first order we shall now return to the problem mentioned in the beginning of this chapter of the normal state of the neutral helium atom, which may be expected to appear in the final state resulting from the binding process, of which these spectra bear witness. In the first place it will be clear, that corresponding to the appearance of the two spectra, which do not show mutual combinations, we must be prepared to find two states, both of which must be assumed to present the properties characteristic of a normal state of being able of permanent existence in the absence of external agencies. Now it is of the utmost interest, that it has actually been possible by experiments of Franck and Knipping, on reaction of atoms against collision with free electrons, to transfer the helium atom from its usual normal state into another state, which just possesses this property. This state has been called "metastable" by Franck and Reiche, who had in fact anticipated the existence of such a state from the result of Paschen's remarkable experiments of resonance in luminous helium gas, before it was detected by Franck and Knipping by experiments on collision between helium atoms and free electrons, by which it was shown that a modification of ordinary helium gas appeared, which in the absence of impurities was not liable of spontaneous retransformation under the emission of radiation. The state under consideration corresponded to the largest observed spectral term within the ortho helium spectrum, which in the notation adopted in this paper is characterised by $n = 2$, $k = 1$, and it was thus to be concluded, that the ordinary normal state of the neutral helium atom corresponded to the member of the sequence of the parhelium spectrum, characterised by $n = 1$, $k = 1$, the existence of which had hitherto not been liable to detection by optical experiments, but which quite recently has been verified by experiments of Lyman and Fricke* by means of a vacuum spectrograph through the detection of an ultra-violet line 585 Å, which corresponds to the transition between the parhelium terms $n = 2$, $k = 1$ and $n = 1$, $k = 1$, on the assumption that the energy for removing the electron from this state just corresponds to the ionisation potential 25,2 Volt.

In this connection it will be noted, that the transition mentioned, as k remains unaltered, is of a kind which does not possess any finite probability for an undisturbed atom. On the contrary since the term $n = 2$, $k = 1$ is the largest but one of all parhelium terms, it might at first sight be anticipated, that the atom also in this state was capable of a permanent existence. There is, however, that great difference between a normal state, which as mentioned may be considered as "stable", or a "meta-stable" state on one hand, and a state of the mentioned type on the other hand, that the latter would acquire a certain, although small, probability of transition in the presence of an external electric field, however small. In the terminology of the quantum theory such a state may therefore conveniently be called "a-stable" in contrast to the ordinary stationary states, which are inherently "un-stable" in that sense that they under all circumstances possess a definite probability for transitions to other stationary states within a given time-interval. In contrast to the problem of the properties of the other states mentioned, for which an immediate formal interpretation is offered by the correspondence principle, the existence of the meta-stable state of the helium atom, which claims the non-existence of a state corresponding to a term $n = 1$, $k = 1$ in the ortho-helium sequence, is at first sight surprising, as it would seem that neither Landé's theory nor our investigations offers a plausible argument. A closer examination gives, however, a different result, and in helium we seem in fact to meet with an instructive example of the falling out of a stationary state, due to the mechanical impossibility of a motion consistent with the formal conditions of the quantum theory, a phenomenon to which we have already had occasion briefly to refer in the former § in connection with the interpretation of the series spectra of the alkali metals.

* See p. [16].

For this circumstance we may obtain indication already from the approximate treatment of the corresponding type of orbit mentioned above, although it must be borne in mind, that these considerations find only direct application when the dimensions of the orbit of the outer electron are large compared with those of the inner, and we must be prepared to meet with difficulties. Thus we get, if—in conformity with the approximation underlying the calculation of the energy in the stationary states—which was found to offer an approximate interpretation of the orthohelium terms, we compare the effect of the inner electron on the motion of the outer with that of an electronic charge placed at the nucleus, on which is superposed the effect of an electronic dipole of moment $\sim(3/2)a_i\varepsilon$, which always points away from the momentary position of the outer electron, the following expression for the major axis and parameter of the preceding orbit of this electron in the stationary states:

$$2a = 2a_o\left(n-\frac{9}{32}\frac{1}{k^2}\right)^2, \qquad 2p = 2a_o\left(k-\frac{9}{32}\frac{1}{k^2}\right)^2 \tag{30}$$

where, as in the former §, a_o denotes the radius of the circular orbit of the electron in the normal state of the hydrogen atom. Quite apart from the fact, that the approximation, which is only effective in case the dimensions of the outer orbit are large compared with that of the inner, is of a very preliminary character for such orbits where $k = 1$, and where the parameter of the outer orbit, if the mutual perturbations could be neglected, would be only double as large as the major axis of the inner orbit, which itself would only be equal to a_o, we see these approximations in case $k = 1$ clearly before the mind. Even if the solution also in the latter cases claims, that the "outer" electron during its revolution comes within distances from the nucleus of the same order as the dimensions of the inner orbit, it will in the greater part of the orbit move at much larger distances from the nucleus, and due to this circumstance there is no hindrance for the existence of stationary states, for which at the same time as the motion preserves the essential features exhibited by the family of orbits, which we for large n and k have adopted for the motion corresponding to the orthohelium terms, a high degree of harmony exists between the periodicity properties of the motion of the inner and outer electron. In fact, the assumption presents itself, that such states may be obtained, if the motion is so adjusted that at the moment of pericentime-passage* of the outer electron the inner electron is situated on the opposite side of the nucleus and possess its largest distance from it. If on the other hand we consider the case $n = 1$, $k = 1$, where the outer orbit in first approximation should be a circle, we cannot in a similar way obtain a well defined regular motion,

* [Should undoubtedly read "perihelion passage".]

as the "outer" electron through all times remains within distances from the nucleus, which are of the same order as the mean distance of the inner electron. Indeed, the rough approximation represented by (30) gives a value for the major axis of the outer orbit, which is nearly coinciding with that of the major axis of the inner orbit.

From a closer consideration of the difficulties of establishing a motion of the orthohelium type corresponding to $n = 1$ and $k = 1$, which are intimately connected with the large exchange of angular momentum between the outer and inner electron, which takes place within a single revolution of the latter, in case the dimensions of the outer orbit are of the same order of magnitude as those of the inner, we are in fact left to the view, that the only coplanar motion of regular periodicity properties, in which the two electrons move in orbits of the same order of magnitude, is represented by the simple case mentioned in the beginning of this chapter, where the two electrons move with constant angular momentum in one and the same circular orbit round the nucleus. Such a motion, however, where the two electrons play equivalent part, cannot be expected to represent a stationary state corresponding to the term in the orthohelium sequence. In fact, in all these states we may expect, that a sharp separation is contained between the orbit of the outer electron and that of the inner, although during the progress of the binding process of the outer electron the difference in the dimensions of the two orbits becomes less and less. Thus we may say, that in all these states the orbit of the inner electron so to say severs the outer electron from the nucleus, this effect being brought about thereby, that the inner orbit for decreasing parameter of the outer orbit becomes more and more eccentric and clearly does not* in the tendency of the binding process to result in a state of motion, where the two electrons play equivalent parts. Viewing the problem of the possible establishment of such a "ring" through a process of spontaneous transition from some stationary state, we are from the point of view of the correspondence principle, as long as we confine ourselves to coplanar orbits, on the contrary led to the conclusion, that this would claim, that also in the initial state the electrons played equivalent parts, and—tracing the binding process further back—to the conclusion, that the electrons should be bound by the nucleus simultaneously and not in succession, such as we must assume for the formation of actual atoms. Before entering more closely on this problem we shall first consider the case of the normal state corresponding to the intersecting orbits, which may be assumed to correspond to the emission of the parhelium spectrum.

In the first place we see, that in case of intersecting orbits there is no question of a severance of outer electron and nucleus by inner orbit of a kind similar to

*. [A word appears to have fallen out.]

what we meet with in case of coplanar orbits, and we may naturally expect, that in the case $n = 1$, $k = 1$ we have to do with a motion, where the two electrons move in perfectly equivalent orbits.

The mechanical problem of such equivalent non-coplanar orbits* has now quite recently been investigated by Kramers, who has found that there exists a simple solution of distinguished type, in which the two electrons move in orbits, which may be considered as deformed circular orbits, the planes of which form equal angles with the variable impulse-axis of the system, round which they turn slowly with the same uniform frequency. The motion of the electrons in these orbits is further so adjusted, that the line connecting their momentary positions is perpendicular to the axis, and divided by this in two equal parts. Further the motion parallel to the axis is perfectly symmetrical with respect to the invariable plane through the nucleus. By an imagined continuous transformation of the nucleus charge the motion can be shown to preserve these properties and can therefore be continuously mechanically connected with a motion corresponding to a high value for this charge, where the mutual interaction of the motion of the electrons is small compared with the attraction of the nucleus, and the motion of each electron therefore will tend to a simple undeformed circular orbit. According to Ehrenfest's principle we may therefore expect, that a stationary state to the parhelium term $n = 1$, $k=1$ may be derived by the consideration of such a transformation from a state corresponding to a high nucleus charge, where the motion of each electron will be a circular orbit with angular momentum $h/2\pi$, and where the normals of these planes of this circle—corresponding to a value of the total angular momentum of the system of equally $h/2\pi$—each will form an angle of 60° with the invariable axis. By a procedure of successive approximations Kramers has in this way derived a value for the energy of the system, which for a nucleus charge of Ne is given by the expression

$$E = -W_0[2N^2 - 1.52N + 0.42 + \ldots] \tag{31}$$

where W_0 as in the former § represents the energy necessary for the removal of the electron from the hydrogen atom in the normal state. For $N = 2$, corresponding to helium, (31) gives

$$E = -5.85\,W_0.^{**}$$

giving for the energy necessary for the removal of one electron from the atom $5.85\,W_0 - 4\,W_0 = 1.85\,W_0$ corresponding to an ionisation potential of $1{,}85 \times 13{,}53$ Volt = 25,2 Volt***. Since this does not differ more from the observed value 25,3

* Such an idea of the constitution of the normal state of the helium atom is also advocated in a paper by Kemble which has just appeared.

** [This does not agree with (31). Some pencilled corrections are illegible.]

*** [$25.2 = 1.85 \times 13.6$.]

more than was to be expected from the degree of approximation, with which the calculation has hitherto been carried through, this result may be said to offer a most interesting support of the general ideas, as regards the constitution of the normal state of the helium atom, and the interpretation of the ortho- and par-helium spectrum, exposed in the preceding*. For closer information as regards the description of the motion and the details of the calculation the reader may be referred to a paper of Dr. Kramers, which will appear in a near future.

The above considerations may be taken as a typical illustration of the laws, which govern the normal state of an atom, laws according to which this state may not be looked for as the state of smallest energy corresponding to motions which apparently fulfill the formal conditions claimed by stationary states, but may be rather traced from a consideration of its establishment by a possible process of binding of the electron by the nucleus. Before concluding this chapter it may be of interest therefore to examine a point related with the normal state of the helium atom somewhat more closely. In the above we have met with two families of mechanical solutions, in which the electrons play equivalent parts. In the first the electrons move in coplanar orbits and are always situated on a line through the nucleus at equal distances from this, each electron moving in an elliptical orbit with the nucleus at the common focus. In the other the orbits of the electrons may be considered as two deformed circular orbits of intersecting planes, which rotate slowly round an invariable axis of the system, which makes equal angles with these planes. While the actual normal state as mentioned may be considered to belong to the last family, the simple ring model appears as a member of each family. The question now arises: What are the conditions holding for transitions between possible stationary states within each of these families, and especially, what argument may be found for the non-possibility claimed by the observations of a transition between the motion, adopted for the actual normal state, and the motion previously proposed for this state, in which the two electrons move within the same circular orbit, each with angular momentum $h/2\pi$. For simplicity let us first consider the family of equivalent coplanar orbits. Here one finds directly, that the relation between the energy difference between two neighbouring states and the frequencies of revolution of the electrons is given by

$$\delta E = 2\omega\delta\mathscr{I} = \omega\delta 2\mathscr{I} \tag{32}$$

where $\mathscr{I}$ is the quantity defined by the expression (⟨2⟩) taken over a revolution of one of the electrons. As regards the vibrations corresponding to possible

* [The final value of Kramers' calculations was $E = -5.525\ W_0$, while the empirical value is $E = -5.907\ W_0$.]

transitions we meet here with the circumstance, that the electrical moment of the system remains zero during the motion, and that therefore the radiation, which would be emitted according to the classical laws of electrodynamics, would in first approximation be vanishing small, compared with that which would be due to a single electron, which circumstance leads us to imagine, that spontaneous transitions between the stationary states of such a system will, if possible at all in the absence of external agencies, possess a very small probability compared with the case of stationary states of less symmetrical character. At any rate it will be seen, that it would not be possible to establish a correspondence between the motion and transitions, for which $2\mathscr{I}$ would alter by an odd multiple of h, as might at first sight be anticipated from the last form, in which δE is written in (32), because the fundamental frequency which appears in the external electromagnetic effect of the atom, due to its motion, will not be ω but 2ω. This circumstance becomes still more lucid, if we consider such transitions, the appearance of which may be brought about by the effect of external fields because also here 2ω will be the frequency in the variation of the electrical moment, brought about by such fields. This consideration might therefore seem to offer an argument for the assumption, that the stationary states of the system belonging to the type in question are characterised by values of $\mathscr{I}$ for each electron, equal to an entire multiple of h. If next we consider the case of equivalent orbits of intersecting planes we get by a simple consideration the following expression for the energy difference in two neighbouring states

$$\delta E = 2\omega\delta\mathscr{I} + \sigma\delta\mathscr{P}$$

where ω is the frequency of oscillation of the electrons parallel to the invariable axis of the system, and σ the small frequency with which the planes of their orbits rotate round this axis. Further $\mathscr{P}$ is the total angular momentum, and $\mathscr{I}$ a quantity which is not altered during the adiabatic transformation corresponding to a change of the nucleus charge, and which in the limit of high values for this charge coincide with the quantity $\mathscr{I}$ defined by (⟨2⟩) for a simple Keplerian motion. As regards the harmonic constituents of the motion it will be seen, that the motion of each of the electrons can be resolved in a set of linear vibrations parallel to the axis of frequency $\tau\omega$, and of circular rotations round this axis with frequencies $\tau\omega \pm \sigma$. Of these, however, only the first frequencies will appear in the variation of the electric moment of the system. In the first place we may be prepared that for the undisturbed atom only transitions will be possible, by which $\mathscr{P}$ remains unchanged, and by such a multiple of transitions corresponding to the factor 2, $\mathscr{I}$ will change by τh corresponding to a process appearing in the expression for the energy, by which so to speak each electron will emit a quantum $h\nu$, as it might also be expected from the equivalent part, which these elec-

[138]

trons play in the motion concerned, in contrast to motions involved in the emission of series spectra. In case the motion of the atom is disturbed by external electric fields, the motion will be changed, and we may expect possibilities of other types of transitions, but just corresponding to what holds for the family of co-planar equivalent orbits, it may be assumed, that these can result only in a change of $\mathscr{P}$ by an even multiple of $h/2\pi$, since due to the symmetry only the double value of the frequency σ will enter in the change of the electric moment of the atom. Of all reasons we are therefore on the principle of correspondence not presented by any argument, which would make a transition from a state $\mathscr{I} = 1$, $\mathscr{P} = 1$, to a "ring"-state corresponding to $\mathscr{I} = 1$, $\mathscr{P} = 2$ necessary, although this transition would be accompanied with a strengthening of the binding of the electrons.

§ 3. On the elements within the second period of the periodic system.

Similar considerations as those in the former chapters, regarding the formation of the hydrogen and helium atom by binding of the electrons by the nuclei, may be expected to hold for the binding of a first and a second electron by the nucleus of any atom. Thus we must, as already mentioned, assume that the binding of a single electron by a nucleus of charge Ne will be accompanied by the emission of a simple spectrum, which on the first approximation is expressed by formula (14), resulting in a final state in which the electron describes a circular orbit round the nucleus with angular momentum $h/2\pi$, the diameter of which in first approximation is given by formula (13). Next we may assume, that the binding of the second electron will give rise to series spectra of the same type as the helium spectra of the first order, and that the binding process will result in a normal state, in which the electrons may be expected to revolve in equivalent orbits, which may be described as two deformed circular orbits, each with angular momentum $h/2\pi$, the planes of which form an angle $\frac{2}{3}\pi$ with each other to the effect, that the total angular momentum is again just equal to $h/2\pi$. While the energy involved in the binding process of the first electron to the first approximation is given by (10), the energy involved in the subsequent binding of the second electron will with approximation be given by (⟨29?⟩).

When next we ask for the binding of a third electron by the atom, we meet with a new problem, about which we may obtain information from an examination of the lithium series spectrum of the first order. Compared with the helium spectrum, the lithium spectrum exhibits a greater simplicity, the multitude of empirical series terms forming a simple example of a series scheme of the Rydberg type, so that the spectrum, looking apart from the circumstance that the lines are narrow doublets, exhibits complete analogy with the spectrum to be expected from a simple central system. This simplicity of the spectrum of lithium, compared with that of helium, may obtain a simple interpretation in the circumstance, that the motion of the two first bound electrons in the lithium atom in their normal state is of a multiple-periodic character and therefore exhibits an essentially higher stability against the influence of external forces than the periodic orbit of the first bound electron in the helium atom. Entering on a closer examination of the lithium spectrum we remark in the first place, that the empirical terms of the S-sequence, which with a close approximation can be written $R/(1.6)^2$, $R/(2.6)^2$, $R/(3.6)^2$ . . . , deviate very much from the terms of the hydrogen spectrum, while the terms of the other sequences deviate very little from the corresponding hydrogen terms, the values of the function $\phi_k(n)$ in (17) being already here the first member in the P-sequence ($n = 2$, $k = 2$) with approximation given by $\phi = -0.04$. For this property of the series scheme, characteristic for the lithium spectrum, we obtain an immediate explanation, when we compare the dimensions of the outer orbit in the various sequences with the dimensions of the inner system. Thus while the dimensions of the orbits of the inner electrons are nearly three times smaller than those of the orbit of the electron in the normal state of the hydrogen atom, we get, if for the moment we consider the effect of the inner system as that of a unit point charge, that the parameter of the orbit of the outer electron in a stationary state corresponding to a given value of k is equal to $k^2 \times 2a_0$, where according to the notation used above $2a_0$ denotes the diameter of the orbit in the normal state of the hydrogen atom. Already in the states corresponding to the P-sequence ($k = 2$) we get therefore, that the circular orbit corresponding to $n = 2$ is about 12 times larger than the inner orbits, while in the states corresponding to larger values of n the smallest distance, to which the outer electron during the revolution in its elliptical orbit approaches the nucleus is almost 6 times larger than the radius of the orbit of the inner electrons. We can thus easily understand, that all lithium terms except those of the S-sequence exhibit only very small deviations from the hydrogen terms. As regards the latter sequence, the first question which arises is about the values of n to be conjugated with the various terms. Also here conclusions may be drawn directly from a rough examination of the dimensions of the orbits. In fact, if the first term should correspond to $n = 1$, we would, in order to account for the large positive value of ϕ in the S-sequence, have to assume that the inner system—besides the attraction due to its resultant charge—should exert a considerable additional repulsion on the outer of the same order of magnitude as that of a dipole with axis pointing towards this electron and possessing an electric moment corresponding to that of two opposite charges placed at a distance from each other equal to the radius of the orbits of the inner electrons in their normal state. The existence of such a repulsion, however, is completely out of question, even if we take possible perturbations of the inner system, due to the presence of the outer electron, into consideration. In fact, if we calculate for instance the radius of the circular orbit, which on the view under consideration would correspond to the first member of the S-sequence, we find, taking the mentioned repulsion into account, a value for its radius, which is about double as large as a_0, or six times larger than the radius of the inner orbits, at what distance the effect, exerted by the outer electron on the orbits of the inner electrons, would be very small compared with the mutual perturbing influence of the orbits of these electrons, responsible for the stability of the inner system. On the contrary we may from a simple consideration expect, that, at any rate if its distance from the nucleus is large compared with the dimensions of the inner system, the outer electron will induce in this system an electric moment, which possesses a direction so as to increase the attraction of the inner system on the electron, and which will have a magnitude proportional to the intensity of the electric field due to its charge. The presence of such an excess of attraction would also seem necessary in order to account for the negative values of ϕ in the terms of the P- and D-sequence. As soon as we take an effect of this kind into account, we are now led to a very different view as regards the interpretation of the S-sequence from that discussed above. In the first place, as regards the dimensions of an eventual state corresponding to $n = 1$ we should, instead of an orbit of larger dimensions than those of the electron in the normal state of the hydrogen atom, expect an orbit of considerably smaller dimensions. In fact, estimating the moment of the induced dipole from the observed value of ϕ in the state $n = 2$, $k = 2$, we would find, that the excess attraction for the state $n = 1$, $k = 1$ will be large enough to drag the outer electron into distances from the nucleus of the same order as those of the electrons previously bound*. The existence of such a state, however, would mean, that the third electron would be very firmly bound by the nucleus, what would be irreconcilable with the general physical and chemical properties of lithium and especially with the evidence from experiments on selective absorption in lithium vapour, which shows that the first observed term in the S-sequence $R/(1.6)^2$ corresponds to the normal state of the lithium atom. On the other hand theoretical arguments for the non-appearance of a state as that in consideration may also be derived from a con-

* A dipole of the same magnitude as that to be expected from a consideration of the dielectric constant of helium in which the configuration in the normal state is analogous with the inner configuration in lithium. (Possibly a reference to Kramers' work; however, perhaps already a discussion of the problem of the dielectric constant under helium.)

sideration similar to that, on which the discussion of the normal state of the helium atom was based. In fact, in the choice of a motion, in which all three electrons revolve in orbits round the nucleus of the same order of magnitude, and which possess such periodicity properties as may be considered necessary for the definition of a stationary state, we are naturally limited to such motions, where all three electrons play equivalent parts. In analogy with the case of the helium atom the simplest imaginable configuration of this kind, consistent with a formal application of the principles of the quantum theory is one, in which the three electrons revolve at equal angular intervals in one and the same circular orbit, each possessing an angular momentum $h/2\pi$. Further also non-coplanar configurations present themselves, in which the orbit of each electron may be considered as a circular orbit deformed under the effect of their mutual repulsions, and which would correspond to a somewhat smaller firmness of the binding. In contrast to the case of helium we would in lithium, however, not only have to reject a coplanar configuration of the first type, but also configurations of the latter type, from a consideration of formation of the atom by a spontaneous process of transition accompanied by radiation. In fact, no process—which bears any resemblance with a possible transition between two stationary states of a multiple periodic system—can be imagined, by which the configurations of the two first bound electrons would allow a third electron coming from outside to enter as an equivalent partner in the motion round the nucleus.

By the exclusion of a state corresponding to $n = 1$, $k = 1$ the interpretation of the observed terms of the S-sequence comes at once in a very clear light. In fact, if in first approximation we compare the effect of the inner system with that of a unit point charge, we get, that the outer electron in these states will describe an elliptical orbit, for which the mean value for the radius vector will be large compared to the dimensions of the inner system, at the same time as the minimum value of the radius vector for all states will only differ very little from $\frac{1}{2}a_0$. While during the greater part of the revolution of the outer electron the inner system will act approximately as a point charge, we must for the parts of the orbit, where the electron passes closest to the nucleus, expect a very different effect of the inner system. Not only will the outer electron come within distances from the inner electron of the same order as the mutual distances of the particles of the inner system, but the outer electron will further in this part of the orbit possess a velocity, which is not smaller than that of the inner electrons. Due to a great similarity of the part closest to the nucleus of the Keplerian orbit of the outer electron in all states corresponding to $k = 1$, deduced under the assumption that the inner system acts as a point charge, we may, however, expect, that even if the motion in this part of the orbit as a consequence of the interaction with the inner system will deviate considerably from the corresponding part of a Keplerian orbit, this deviation will be very similar for all the states in question. Under these circumstances the orbit of the outer electron may adequately be described as composed of a number of equal loops, each of which is nearly a Keplerian orbit, and which are joined together in a way allowing the major axis of successive loops to be turned relative to each other by a considerable angle, which will be the same for all the states corresponding to the terms within the sequence under consideration. Now it can be simply shown, that this property of the orbit will just give rise to a sequence of energy values in the stationary states, which exhibit the character of the spectral terms observed. In fact, if we for one of the states in question calculate the quantity $\mathcal{I}_1$ given by formula (18) for a period of the radial motion relative to the nucleus, we get an expression, which consists of two terms, of which the one is the same as that corresponding to a Keplerian motion with the same value for the energy and for $\mathcal{I}_2$, while with a high degree of approximation the other is identical for all states within the sequence. Denoting this last term by αh the energy may consequently be expressed by*

$$E = -\frac{4\pi^2 e^4 m}{2(\mathcal{I}_1+\mathcal{I}_2-\alpha h)^2} = -\frac{2\pi^2 e^4 m}{(\mathcal{I}-\alpha h)^2} = -\frac{R}{(n-\alpha)^2}$$

In the last consideration as well in the estimate of the effect of the inner system in the states of the P- and D-sequences we have compared the motion of the outer electron with a simple central orbit without paying attention to the fact, that the inner system due to its resultant angular momentum is of the type considered in Chapter II in connection with the discussion of the possible origin of the observed complexity of the terms of many series spectra. From the considerations referred to we might in fact for each of the states corresponding to given values for k and n in (19) for a simple central motion expect three states with different orientations of the outer electron relative to the inner system. In two of these the plane of the outer orbit is perpendicular to the axis of the inner system, the angular momentum of the outer orbit having the same and opposite direction as the resultant angular momentum of the inner system respectively, while in the third case the plane of the outer orbit makes an angle with the axis of the inner system, in such a way that the total angular momentum of the atom becomes equal to $k(h/2\pi)$. As to the magnitude of the effect on the spectral terms of a difference in relative orientation of the plane of the outer electron and the axis of the inner system it should beforehand be expected to be largest for states corresponding to $k = 1$, but just for these states the observed terms are single, and we will therefore have to look for reasons, which may allow the preference of one orientation before the two others. Now a closer examination of this point

* [The constant R in this formula is equal to the constant R in (16) and (17) multiplied by h.]

indicates, that just the more symmetrical orientation of the plane of the outer orbit relative to the configuration of the inner system may bring with it a greater irregularity in the effect on the outer orbit during its closest passage to the nucleus, since due to the synchronism of the motion of the two inner electrons in a direction parallel to the axis of the system they may happen during the passage of the outer electron to be both on one and the same side of the plane, in which this electron moves, producing thereby irregular disturbances in the orientation of this plane and in the precession of the major axis. Due to this fact the less symmetrical orientation in the third case may bring with it a considerably greater regularity of the action of the inner system on the orbit of the outer electron resulting in a greater similarity with a simple multiple-periodic motion of the electrons. This circumstance may in fact offer an argument for the assumption, that in the states corresponding to the S-sequence, and especially in the normal state, the relative orientation of inner system and outer orbit will be such, that the atom possesses a resultant angular momentum equal to $h/2\pi$. As regards the other sequences of states the effect of a different orientation may be expected to be much smaller and may possibly offer an interpretation of the observed complexity of the term of the first member of the P-sequence, analogous with the interpretation for the characteristic doubling of the terms of the P- and D-sequences in the other members of the family of the alkali metal, proposed by Sommerfeld and Landé, and to which we shall come back later.

We have entered on this discussion of the lithium spectrum in order to investigate the binding process of a third electron, and it may be emphasised, that the arguments developed do not only have direct bearing to the problem of the constitution of the lithium atom, but may also be used to throw light on the more complex problem of the constitution of atoms of higher atomic number. As regards the problem of the lithium atom the main purpose of the considerations consisted naturally in finding a basis for an interpretation of the remarkable looseness, with which the last electron is bound by this atom, compared with the firmness by which the electrons are held in the hydrogen and helium atom, and which is responsible for the characteristic electro-chemical properties of lithium. As we have seen, such a basis may actually be found in the nature of the coupling between the electron last bound and the two first electrons, which brings with it the exclusion of a state, in which this electron moves in an orbit analogous with that of the electrons in the normal state of the helium and hydrogen atom, with the effect that in the normal state of lithium the orbit of the last bound electron will correspond to an intermediate stage of the binding of an electron in one of these atoms. This last result can now be expected to apply equally to the binding of a third electron by a nucleus of higher charge. In fact, for higher nucleus charge corresponding mechanical considerations hold with the only modification, that due to the smaller dimensions of the orbit of the third electron in corresponding stages of the process of its binding, compared with those of the inner system, we may expect an interaction between this system and the third electron, which exhibits the features discussed in case of lithium to a still higher degree. Thus for large values of the nucleus charge Ne already a first approximation, where the effect of the inner system is compared with that of a point charge $(N-2)e$, gives for the dimensions of the orbit in an eventual state $n = 1$, $k = 1$ values closely coinciding with those of the inner system, while the electron in the state corresponding to $k = 1$ and values of n larger than 1, will penetrate into distances from the nucleus, which are almost two times smaller than the radii of the orbits of the two first electrons in the absence of the third.

As regards the binding of a 4th, 5th and 6th electron by an atom we have unfortunately no direct spectral evidence, as no spectra of beryllium, boron or carbon are hitherto observed or disentangled. From the considerations about the binding of the 3rd electron, discussed above, it seems possible, however, to arrive to a view as regards the binding of these electrons, which lends itself to an interpretation of the characteristic chemical and physical properties of the named elements. Considering the binding of the 4th electron we may in the first place assume, that the binding takes place in steps in such a way, that the orbit in the earlier stages, where its dimensions are large compared with those of the orbits of the electrons previously bound, will differ only little from that in the stationary states corresponding to the binding of an electron by a nucleus of a charge $(N-3)e$. In the later stages on the other hand we may expect an intricate interaction of the 4th electron not only with the 3rd, but also with the two electrons first bound. In fact considering the dimensions of the orbit corresponding to the normal state ($n = 1$) of the binding of an electron by a nucleus of charge $(N-3)e$, we find that this would be several times smaller than the dimensions of the orbit of the 3rd electron, and that consequently the motion of an electron rotating in a possible orbit corresponding to ($n = 1$) would be only little affected by the presence of this electron, but would depend primarily on the immediate interaction with the inner configuration of the two electrons first bound. From considerations analogous with those holding for the binding of the 3rd electron we may therefore directly reject the existence of a state, in which the 4th electron was bound in a way corresponding to the final state of the process of binding of the electron by a simple nucleus, and may in the normal state of the atom look for an orbit of the 4th electron, which, as that of the 3rd, corresponds to one of the intermediate stages of this process. As regards such an orbit we see next, that the eccentricity of the orbit of the 3rd electron—as a consequence of which it remains in the immediate neighbourhood of the orbits of the two first electrons only within a short fraction of its period of revolution—offers a possibility of a type of configuration, in which the 4th electron in first approximation may be said to be coupled to the inner system independently of the 3rd, in that sense that the main interaction between the inner system and the 4th electron takes place at a moment, where the 3rd electron is situated at the remoter part of its orbit, where its effect on this interaction in first approximation may be neglected. Especially we are led to assume a configuration in the normal state, where the 3rd and 4th electron move in equivalent orbits, each of which may be described as a deformed central orbit of the same type as that performed by the 3rd electron in the absence of the 4th*.

* [This manuscript is incomplete. The remaining elements of the second period are discussed on ca. 25 pages which are handwritten in pencil and now largely illegible. The conceptions of "coupling" and "penetrating orbits" occur there.]

§ 4. The correspondence principle.

Without possessing means of directly attacking the problem of the mechanism underlying transitions between stationary states, we shall now see, how the closer examination of the relations of the quantum theory, as developed for multiple periodic systems, discloses a certain suggestive connection between the motion of the atom and the occurrence of the transitions, given rise to the spectrum. Consider a multiple-periodic system, for which the stationary states are fixed by the three conditions (6). The connection alluded to consists then in the circumstance, that it is possible to conjugate the appearance of a transition between two states, where the numbers n_1, n_2, n_3 have the values n'_1, n'_2, n'_3 and n''_1, n''_2, n''_3 respectively, with the appearance in the motion of the system of a harmonic vibration, the frequency of which in terms of the fundamental frequencies ω_1, ω_2, ω_3 may be expressed by $(n'_1-n''_1)\omega_1+(n'_2-n''_2)\omega_2+(n'_3-n''_3)\omega_3$. This peculiar connection reveals itself most clearly, if we consider such transitions, for which the motion in the stationary states and consequently the values of the fundamental frequencies differ only comparatively little from each other, a condition which in general will be fulfilled, if we consider a transition between stationary states, for which the value of each of the quantum numbers n_1, n_2, n_3 is large compared with the difference between the values of the number in the two states. In that case we have according to (7) approximately:

$$E(n'_1\,n'_2\,n'_3)-E'(n''_1\,n''_2\,n''_3) \sim \omega_1(\mathcal{I}'_1-\mathcal{I}''_1)+\omega_2(\mathcal{I}'_2-\mathcal{I}''_2)+\omega_3(\mathcal{I}'_3-\mathcal{I}''_3)$$

which by use of the general frequency relation (1) and the state relations (6) gives

[148]

us the following asymptotical relation:

$$\nu \sim (n'_1 - n''_1)\omega_1 + (n'_2 - n''_2)\omega_2 + (n'_3 - n''_3)\omega_3$$

Notwithstanding this intimate quantitative connection in the region of large quantum numbers between the frequencies of the spectral lines emitted during the various transitions between the stationary states and the frequencies of the constituent harmonic vibrations, into which the motion of the system may be resolved, it must be borne in mind, however, that so far there is no question of a gradual approach for increasing quantum numbers between the character of the radiation process on the quantum theory and the classical ideas of radiation. In fact, the consideration is entirely based on the postulate, that radiation is always emitted as single trains of harmonic waves, and that accordingly the various trains of waves, which in frequency coincide with the frequencies of the constituent harmonic components are not emitted simultaneously by the atomic system, but by a number of independent processes, by which the system undergoes transitions between various sets of stationary states. Just through this circumstance we are in the above result led to perceive a general connection between the appearance of the various types of transitions and the various constituent vibrations of the motion of the atom, which hides itself at first sight in the region of states, where the quantum numbers are small, and where a direct relation between the values of the frequencies of the spectral lines and the frequencies occurring in the motion of the atom is excluded already thereby, that the values of the latter frequencies may be quite different in the two stationary states, between which the transitions take place.

Now in view of the internal consistency and wide field of successful application of the classical theory of electrodynamics the idea lies at hand, that the direct connection between spectrum and motion, disclosed for a multiple periodic system in the region of large quantum numbers, will hold not only as regards the frequencies of the spectral lines, but that in this region the spectrum in a more intimate way will bear witness of the properties of the motion. Thus we are led to conclude, that within the region of large quantum numbers the probability for the various possible spontaneous transitions between the stationary states of a multiple-periodic system and the constitution of the radiation emitted are connected with the character of the corresponding harmonic vibrations, in such a manner that the spectrum reflects the motion in the atom in exactly the same way as in the classical theory, according to which the intensity and polarisation of the different components of the radiation, emitted by a system the variation of the electric moment of which can be resolved in constituent harmonic vibrations, is simply related to the amplitude of these vibrations in the different directions in space. By these considerations we are next led to conclude, that this extended correspondence may be traced also in the region of small quantum numbers, although it is here naturally excluded just as in the case of the frequencies to obtain an exact quantitative direct connection between the probabilities of the various transitions and the motion, because the amplitudes of the corresponding harmonic vibration as well as their frequencies will in general differ essentially in the two stationary states, involved in the process. In fact, we are led to consider the possibility of the occurrence of a transition between any two given stationary states of a multiple-periodic system, as conditioned by the appearance in the motion of the corresponding harmonic vibration, and for instance to expect the occurrence of lines in the spectrum corresponding to all imaginable transitions only in the case, where the motion of the atomic system contains harmonic constituence corresponding to all sets of values for the n's in the relation (6). Further as regards the state of polarisation of the radiation corresponding to the various transitions we shall in general expect an elliptical polarisation in accordance with the fact, that in the general case the constituent harmonic vibrations of a multiple-periodic motion possess an elliptical character and possess an eccentricity dependent on the values of the $\mathscr{I}$'s. On the other hand in many atomic systems appearing in nature, which possess a motion of multiple-periodic character, the constituent harmonic vibrations will due to some symmetry of the atom be of linear or circular type for all motions of the system, and we may therefore in such cases expect, that the radiation emitted during the various types of transitions will be either linear or circular. As we shall see in the next chapter, the examination of atomic problems, hitherto treated by means of the theory of multiple-periodic systems, has given an unrestricted and convincing support for the above view, which may be termed the correspondence principle. It may here be especially emphasised, that the establishment of this principle has offered an immediate interpretation of the apparent capriciousness, involved in the application of the principle of combination of spectral lines, which consists in the circumstance, that only a small part of the spectral lines, which might be anticipated from an unrestricted application of this principle, are actually observed in the experiments. It may also be mentioned, that in case of atoms containing only one electron the principle has allowed to account for the intensities and polarisations of the components, into which spectral lines are resolved, when the atoms are subject to external electric and magnetic forces.

As regards the problem of the constitution of atoms, containing several electrons, the principle of correspondence will further be shown to offer a guidance not only as regards the interpretation of the spectra of the elements, but especially as regards the question about the various possible types of transitions between stationary states, which may take place within the atom, and about the occurrence of which these spectra bear witness. Thus the emission of the so-called series spectra must be assumed as accompanying the binding of an additional electron by a system consisting of a nucleus and a number of electrons previously bound, while the so-called characteristic Røntgen-spectra are emitted during a process, which may be described as the reorganisation of the normal state of an atomic system, when one of the constituent electrons are removed by some external agency.

Although, as mentioned in § 1, the multitude of conceivable motions of the particles in an atom, containing several electrons, is far more intricate than that of the possible motions of a multiple-periodic system, it will nevertheless be found, that it is possible from considerations analogous with the application of the correspondence principle to such systems to obtain a guidance for the examination of the interatomic process under consideration. By these considerations we are led to restrict the choice of stationary states, which may occur as a result of actual processes of formation and reorganisation of atoms. As we shall see, this leads again to a picture of atomic constitution, which, at the same time as it offers a basis for a formal interpretation of the spectra of the elements, presents aspects suitable for the interpretation of the specific properties of these elements, in accordance with the results, embodied in the periodic table.

CHAPTER V.1.

Principles

Never orbits belonging to two different groups or subgroups of an atom with the same two quantum numbers k and n. Should perhaps already be mentioned to some extent in Chapters I and II. Probably only a closer discussion of the principles a little later in Chapter V, namely after the third group of the periodic system*.

V_2 2a

The dimensions of configurations of 10 ⟨electrons⟩. To be calculated from the value of $\mathscr{I}$ for the outer "loops". If we imagine that these move in a field that has $(N-7)e$, ⟨and⟩ if we put $\mathscr{I} = 1.7$, we get ca. $1.9a_{\mathrm{H}}$.

Elliptical orbits in neon corresponding to ⟨$\mathscr{I} =$⟩1.7 have a ratio between minor and major axes of $\sqrt{1.7^2 \times 1^2/1.7^2} = 1/1.7$**. An elliptical orbit corresponding to a threefold charge and $\mathscr{I} = 1.7$ has a value for W equal to $(9/1.7^2)W_{\mathrm{H}} = 3.1\ W_{\mathrm{H}}$. This value, which corresponds to an ionization potential of ca. 40 volts, does not signify any serious difficulty for explaining the ionization potential of neon of ca. 20 volts, provided we consider the change in the energy of the configuration of the remaining 9 electrons when one of the outer electrons is removed. For example, if we assume that one of the outer electrons during a $1/x$'th part of its revolution is inside the sphere, we obtain that the shielding for each of the four circular orbits to a first approximation is diminished by e/x. If we figure that the field in which the electrons move corresponds to ca. 6 e, this represents a change in the energy of the four electrons of $(4 \times 6^2/3x) \times \frac{1}{4}W = 12W/x$. If x is 10, this already means ca. 16 volts. If we attempt for the moment to estimate the influence of the energy resulting from the change in the binding of the two innermost electrons, we get, assuming that one of the outer electrons during a $1/y$'th part of its revolution is inside the two innermost ⟨electrons⟩, that the effect is ca. $(2 \times 10^2/5y)W_{\mathrm{H}} = (40/y)W_{\mathrm{H}}$. If y is 100, this becomes ca. 5 volts***.

* [This is in Danish.]

** [This sentence is written in Bohr's own hand.]

*** [This page is in Danish.]

§ 5. On the General Interpretation of the Periodic Table.*

Before we enter on the problems of this chapter, the content of which will be the outline of the theory of the constitution of the atom, which lends itself naturally to the interpretation of the laws of the variation of the physical and chemical properties of the elements with the atomic number embodied in the periodic table. We shall briefly state the principles on which the preceding considerations lie, the tendency and foundation of which will already appear as indicated by the considerations of the preceding chapters. When, in the rest of the chapter, proceeding with the discussion of elements of higher atomic number, we shall only consider the details as far as they may be considered as illustrative of these principles.

As mentioned in the former chapter, we may expect that, in the normal state of an atom with ten electrons, we meet with a configuration which is complete, in the sense that not only the inner group of orbits $n = 1$, but also the outer group of orbits $n = 2$, cannot take up further electrons by a process corresponding to the stages of binding of an eleventh electron. In fact, we may assume that each subgroup in the outer group containing orbits characterized by $k = 1$ and $k = 2$, respectively, was completed, in the sense that the inclusion of a further electron would not show any resemblance with a process of transition between two stationary states of a multiple periodic system. As regards the configuration of the eleventh electron, we may therefore look for orbits of a new type.

* [As mentioned in the Introduction, sect. 4, p. [18], § 5 and the preceding short paragraphs have been transcribed from a handwritten manuscript.]

Now, as regards the binding of this electron, we are so fortunate to possess detailed evidence in the sodium spectrum which, in a similar way as the lithium spectrum, exhibits a marked simplicity. In contrast to the lithium spectrum, however, not only the spectral terms of the S-sequence, but also those of the P-sequence, differ considerably from the hydrogen terms. Thus, the latter terms do not only exhibit a much greater doubling than the Li terms, but they show also deviations from the hydrogen terms which are considerably larger in the opposite sense as those of the Li terms. On the other hand, the terms of the D- and F-series coincide, just as the Li terms, very closely with the terms of the hydrogen spectrum. We shall now see how all this is to be simply interpreted on the picture of the first 10 electrons bound by the atom.

Regarding, in the first place, the terms of the S-sequence, we may expect that these correspond to stationary states in which the orbit of the outer electron during its revolution penetrates into the region of the configuration of the electrons previously bound. In fact, the dimensions of this configuration may be estimated from the fact that the outer loops of the ellipses possess a value of $\mathscr{I}$ which must be somewhat smaller than 2 (ca. 1.8) and move in a field which, taking the repulsing effect of the electrons moving in the four circular orbits and the three other elliptical orbits approximately equal to $4e$ $(11-2-4-1=4)$, giving for the radius ca. 1.6 a_H and is thus considerably larger than the parameter of an orbit of an electron revolving in an orbit corresponding to $k=1$ round a unit charge. We must therefore in the S-sequence expect an orbit of the last electron which during its motion penetrates right into the region of the orbits of the inner electrons; indeed, a simple consideration shows that it will, just as the electrons previously bound which revolve in elliptical orbits, penetrate into distances from the nucleus which are smaller than the orbits of the two electrons first bound. Now, as mentioned in Chapter II, § 4, we must, from experiments on selective absorption in sodium vapor, conclude that the normal state of the atom corresponds to the spectral term usually denoted by 1.5 S. Now, from arguments perfectly analogous to those used in the preceding chapter as regards the configuration of the lithium and the nitrogen atoms, we must reject orbits corresponding to $k=1$ and for which n should be equal to 1 or 2, and we must accordingly assume that the normal state corresponds to $k=1$, $n=3$. As regards the closer interpretation of the numerical values for the observed spectral terms, which for the sharp series are found to be nearly given by $T_H/1.6^2$, $T_H/2.6^2$, ... corresponding to values for $\mathscr{I}$ for the outer Keplerian loops of the orbit of the electron which for a given value of n is equal to $n-1.4$, which is of the order of magnitude to be expected from a rough estimate of the contribution to the integral $\mathscr{I}_1$ corresponding to the part of the orbit situated within the region of the inner electrons.

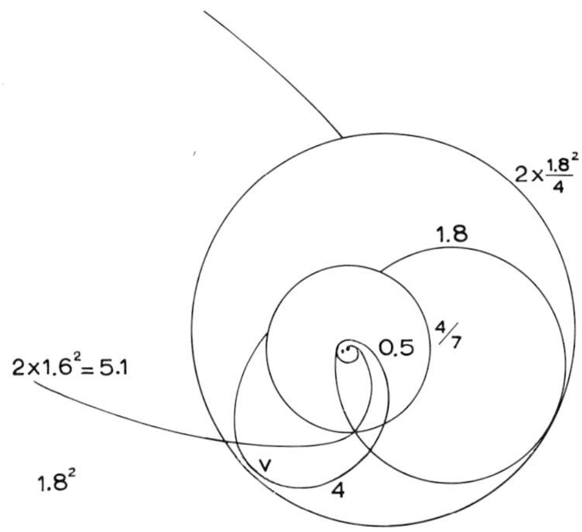

As regards the interpretation of the terms of the P-sequence, we meet with a problem of a somewhat different kind. In fact, the parameter of an orbit of an electron revolving round a unit charge corresponding to $k = 2$, which is $4a_{\text{H}}$, is more than twice as large as the region of the 10 orbits previously bound. It might therefore be expected that the orbits of the electrons in the P-sequence would remain wholly outside the configuration of the inner electrons. This, however, would lead to a serious difficulty as regards the interpretation of the numerical values for the P-terms which, looking apart from the doubling, are with a close approximation given by $T_{\text{H}}/2.1^2$, $T_{\text{H}}/3.1^2$ ... This would mean that the electron, if it remained outside, should, besides the attraction from the resulting unit charge of the inner system, be subject to a considerable repulsion varying as the inverse cube of the distance from the nucleus. Just as in case of the discussion of the Li-spectrum, we must assume however that such a repulsion is out of question and that we must rather expect that the inner configuration would exert an additional attraction on the outer electron, on account of the perturbations in the orbit of the inner orbit induced by the presence of the outer electron. In fact, such an attraction would also seem necessary to account for the terms of the diffuse series which, although they coincide closely with the terms of the hydrogen spectrum, are decisively greater than these, corresponding to a greater firmness of the binding. Thus, the term 3D is about 1% greater than $T_{\text{H}}/3^2$. On the other hand, a field of force of the repulsive kind, necessary on the above assumption to account for the P-terms, would claim for the D-terms a deviation from the hydrogen terms in the opposite sense, and the magnitude of which for the term 3D should be somewhat more than 6%. Although an estimate of the excess of attraction, based on the observed spectral terms of the D-sequence, does not in the present case, as we found with a similar estimate in the discussion of the Li-spectrum, offer a sufficient reason for concluding that the possible orbit corresponding to $n = 2$, $k = 2$ would be drawn within the region of the orbits previously bound and therefore is to be rejected on account of the completed character of the configuration of orbits with the same number of quanta. Nevertheless, on account of the far less symmetrical character of the inner configuration as compared with that of the two inner electrons of the Li-atom, which in the first place arises from the phase difference with which the elliptical orbits in the sodium atom are performed, and on account of the fact that the velocity with which the outer part of these orbits is performed, and on account of the fact that, in striking contrast with the case of the Li-atom, the velocity of the electrons in the outer part of the orbits will be of the same order of magnitude as that of an electron in a possible outer circular orbit corresponding to $n = 2$, it would seem that such an estimate would hardly be conclusive and that an outer orbit of the kind under consideration still was to be rejected. (This point is of greater importance as regards the interpretation of the sodium spectrum than as regards the constitution of the atom, as the question in the latter respect is not concerned with the normal state of the atom but only with one of the earlier stages of its formation by binding of the eleventh electron.) Indeed, as we shall see, we use in the ...* in general in the normal state not to be left with any uncertainty as regards the character of the various kinds of orbits of the electrons.

If now we assume that—due to some excess of attraction of the inner system arising from perturbations of the inner orbits as wants of uniformity of the distribution of the charges of the inner system—the orbit 2, 2 is actually excluded and that the orbits corresponding to $k = 2$ and entire values of n will penetrate into the region of the configuration of the inner electrons, we are at once presented with a suggestive explanation of the spectral terms of the P-sequence. In fact, an estimate of the dimensions of an orbit of the latter kind, for which the inner loop will penetrate within the circular orbits $n = 2$, $k = 2$ but not within the region of the innermost orbits $n = 1$, $k = 1$, ...** leads to the expectation that the contributions to the integral $\mathscr{I}_1$ of this loop will be almost equal to one, so that the orbits in question, as regards the value of $\mathscr{I}$ for the outer loop, will only differ little from that of a Keplerian orbit round a unit charge characterized by $n-1$, 2, although it will possess somewhat smaller value for W, as claimed by the spectral terms observed.

As regards the interpretation of the D- and F-sequences, we must, on the other hand, as it follows from the above considerations, assume that they correspond

* [Two words are illegible.]

** [A half dozen, partly illegible, words are crossed out.]

[156]

to orbits for which $k = 3$ and 4, which never approach the nucleus within distances comparable with the dimensions of the inner system.

Hitherto, we have looked apart from the interpretation of the observed well-known doubling of the terms of the P- and D-sequences of the sodium spectrum. This is immediately connected with the fact that we have in the above looked apart from the question of the closer orientation of the orbits of the outer electron with respect to the axis of the inner system. In fact, as mentioned in Chapter 2, we must, as pointed out by Sommerfeld and Landé, expect that the doubling is connected with the departure of the configuration of the inner electrons from a purely spherical symmetry, resulting in the stationary states of certain definite orientations of the outer orbit relative to the axis of the inner system, distinguished by the condition that the angular momentum of the whole atom is equal to an entire multiple of $h/2\pi$. On first sight, we should, on such a view, expect a complexity not only of the P- and D-terms but also of the S-terms, but, on a consideration of the definite configurations of the inner electrons to which we are led, we see that there exists that essential difference between the orbits corresponding to the various sequences that only the orbits of the S-terms penetrate right into the region of the configurations of the innermost orbits, which is the primary* seat of the asymmetry of the atom. Just as mentioned in the case of Li, we may therefore in the case of the S-orbits find arguments for the exclusion of such orientations of the orbits where its plane stands perpendicular to the axis of the system, so that we may explain that we are left only with an orientation in which the total angular momentum of the atom is equal to $h/2\pi$. As regards the orbits corresponding to the P- and D-sequences, we may assume that the effect of the orientation is to be sought in the orientation relative to the orbits of the electrons in the second group of 8 electrons moving in orbits corresponding to $n = 2$ and that any such effect therefore will depend only indirectly on the asymmetry of the first group, which is ultimately responsible for the asymmetry of the second group through the close interaction between this group and the elliptical orbits of the second group.

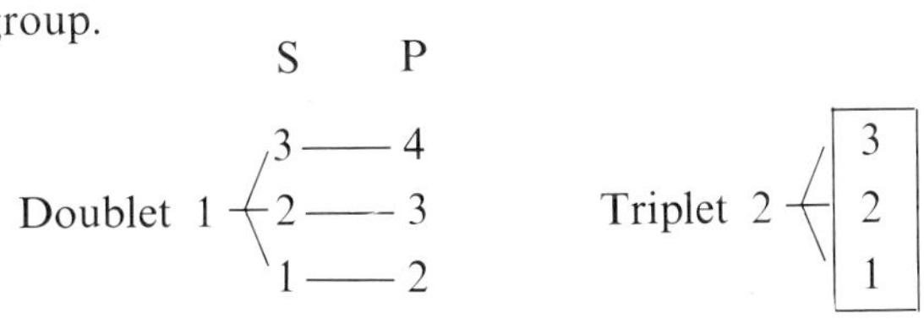

We may therefore expect that here we have no reason to exclude any orientation of the orbit of the outer electron relative to the inner configuration, if it satisfies the condition that the total angular momentum of the atom is equal to an entire multiple of $h/2\pi$, and if it is reconcilable with the motion of the outer group of 8

* [The manuscript has "preliminary".]

electrons. We may therefore understand that the terms of the P- and D-sequences are double*.

We should also get the same result if the energy in the two orbits, which, apart from the direction of revolution, were oriented alike with regard to the axis of the inner system, were the same. No difficulty in understanding the doublet in the principal series. Only question about the interpretation of the fine structure in the diffuse series. Variation of the doublets with the serial number**.

Polar symmetry of innermost group. Axial symmetry of second group. General way in which doublet distance varies for members within the series is directly accounted for by the fact that the inner loop for orbits of increasing n amounts to a smaller part of the orbit. (May be simplest expressed by noticing that doublet members are separately expressed by Rydberg formula.)

From the above discussion of the *sodium spectrum*, it seems possible to throw light on the binding of the eleventh electron of the atom, at the same time as we obtain from this process further information of the configuration of the ten electrons previously bound. Still more information about the binding process of the 11 electrons of a very direct kind we obtain from the spark spectrum of magnesium. For this, the main features are closely analogous to those of the sodium spectrum, not only through the fact that the terms of the P- and D-sequences are doublets, but, in the first place, through the fact that the terms, when divided by a factor 4 (corresponding to the appearance of 4 times the Rydberg constant in spark spectra), show a relation to the terms of the hydrogen spectrum of the same kind as the terms of the sodium spectrum, differing only thereby that the deviations are still more exaggerated. Thus, the terms of the sharp series are closely given by $4T_H/1.8^2$, $4T_H/2.8^2$, ..., the terms of the P-series by $4T_H/2.4^2$, ..., the terms of the D-series by $4T_H/3.03^2$ ... This may simply be accounted for by considering the dimensions of an orbit rotating outside the configuration of the first 10 electrons when the nuclear charge is $12e$. In fact, the larger deviations of the P-sequence from the hydrogen terms is to be referred to the smaller departure of the apparent quantum number from the real quantum number $n = 3$ of the outer loop due to the comparatively larger dimensions of the inner loop. Similar holds for the states corresponding to the P-sequence; here we may further note that a circular orbit $n = 2$, lying outside the inner configuration, would quite clearly be impossible, since the radius of such an orbit, calculated already in first approximation, would hardly fall outside the region of the second group, leaving very little doubt at any rate for the interpretation of the terms in question of the magnesium spectrum. Finally, the larger value of the

* [Several lines are crossed out here.]

** [The preceding sketchy paragraph is written in Danish.]

corrections of the D-terms is just what should be expected from the smaller value, if compared with that of the inner system, of the dimensions of the orbit of the outer electron, which in this case, just as in sodium, must be expected to lie wholly outside the region of the inner system.

When we proceed to an examination of the problem of the binding of a 12th electron, we may expect that this will result in a configuration in which an 11th and 12th electron move in equivalent orbits corresponding to $n = 3$, $k = 1$, and, just as was the case with the 3rd and 4th electron of the Be-atom, we may expect that in the neutral magnesium atom the planes of the orbits will make the same angle with the axis of the inner system, and be so oriented relative to this that the total angular momentum of the atom is equal to $2h/2\pi$. This is also confirmed in a very direct way from a consideration of the magnesium spark spectrum.

In the first place, the terms of the S-, P- and D-sequences are considerably more firmly bound, which may be accounted for by the circumstance that, in the configuration of 12 electrons, the 12th moves outside the others in a central orbit composed of very eccentric loops, so that the effect of the inner system, for orbits which do not correspond to large values of k, will differ considerably from that of unit charge, the resultant attraction being much larger. Next, the terms of the P- and D-sequences are not doublets as in sodium but triplets. This may be explained by noticing that the region of the Mg atom, within which the inner loops of the orbits corresponding to these sequences lie, does not, in contrast to the sodium atom, possess a simple non-polar axial symmetry, but, due to the angle between the plane of the orbit with the invariable axis of the atom, the system will exhibit a pronounced polar character, so that we may expect that orbits, the planes of which coincide but which differ in sense of revolution, will give rise to an essential difference in the spectral terms.

Scheme s p d

4
3 3
2 < 2 2
1

$k = 1$ $k = 2$ $k = 3$
$i = 1$ $i = 1$ $i = 1$

Relation to Landé and Sommerfeld ...* in absolute values which is immediately connected with the atom model.

* [Word illegible.]

BINDING OF 13 ELECTRONS

As regards the binding of the 13th electron, we must assume that this results in a configuration in which the atom contains an outer group of three equivalent electrons rotating in orbits corresponding to $n = 3$, $k = 1$. The orientation of these orbits may take place in two different ways, resulting in the total angular momentum of the atom of, respectively, $h/2\pi$ or $2h/2\pi$. This may offer an interpretation of the fact that the arc spectrum of aluminium presents a more complicated structure than those of sodium and magnesium.

(Perhaps the doublets might be explained by the assumption that certain orientations of the outer orbit in the P- and D-states are excluded, because configurations for which the plane is perpendicular to the axis of the system might assume a too special position with respect to the orbits of the 11th and 12th electrons. The difficulty of this argument is that, to a certain degree, we meet similar conditions already in sodium. However, the tetrahedral symmetry might make an essential difference.)

(If this explanation is permissible, the P-terms might correspond to 3, 2, and 1 times $h/2\pi$, and one might therefore think that the doublet system corresponded to an S-term, for which the angular momentum corresponded to $h/2\pi$, and we should then expect another Al spectrum consisting of triplets corresponding to an S-term with $2h/2\pi$.)*

BINDING OF THE 14TH ELECTRON

As regards the binding of the 14th electron, we must further expect that the four outer electrons, moving in orbits corresponding to $n = 3$**, $k = 1$, will possess an orientation relative to each other and to the axis of the inner system corresponding to that of the four outer electrons in the carbon atom. However, we must expect some greater looseness in the binding (compare the spectra of Na and Li) which could explain the difference in the electrochemical character of Si and C (compare also Li, Be, B with Na, Mg, Al)***.

Proceeding now to the problem of the binding of the 15th electron, we may expect that this will result in the appearance of an orbit of a new type, since, as was the case with the binding of the 7th electron, the electrons previously bound that, due to the pronounced central symmetry of the configuration of the elec-

* [The preceding two parentheses are in Danish.]

** [The manuscript has $n = 1$, but that is clearly an error.]

*** [The last sentence, with the two parentheses, is in Danish.]

trons previously bound, the binding process cannot result in an inclusion of an additional electron as an equivalent partner to the groups of electrons previously bound, by which inclusion the harmony of the motions within these groups would be essentially disturbed. In contrast to the case of the binding of the seventh electron, we cannot now expect that the new electron will be bound in a circular orbit. In fact, a circular orbit $n = 2$ is excluded for reasons analogous to those holding for a circular orbit $n = 1$ and further will an elliptical orbit $n = 3, k = 2$ be bound very much more firmly than electrons moving in a circular orbit $n = 3$. In fact, the four elliptical orbits $n = 3$, $k = 1$* will, although somewhat larger, be of the same order of magnitude as the configuration of the four orbits $n = 2$, $k = 1$** in the nitrogen atom. While in the latter case a circular orbit $n = 2$ would fall within the region of the elliptical orbits, this will not be the case of a circular orbit in the case of P. In fact, a simple estimate shows that such an orbit must fall outside the region of the elliptical orbits and will therefore correspond to a stationary state in which the electron moves in an elliptical orbit $n = 3$, $k - 2$, which in first approximation would correspond to the binding of an electron moving in an orbit corresponding to $n = 2$ in the field corresponding to that existing outside the configuration of the first 10 electrons. We may therefore, in the normal state of the atom, expect that the 15th electron moves in an orbit $n = 3$, $k = 2$ which, just as the circular orbit $n = 2$, $k = 2$ in the nitrogen atom is situated wholly within the region of the group of elliptical orbits $n = 2$, $k = 1$, and which, due to the circumstance that it penetrates within the region of the 8 electrons of orbits corresponding to $n = 2$, will in its outer part possess an eccentricity very much smaller than that of a simple Keplerian orbit $n = 3$, $k = 2$.

As regards the binding of the 16th, 17th and 18th electrons, we meet now with features closely analogous to those of the binding of the 8th, 9th, and 10th electrons with the difference that we, instead of circular orbits in the normal state, must assume that these electrons must be bound in central orbits $n = 3$, $k = 2$. Thus, we must in the case of 18 electrons expect a configuration in which the outer 8 electrons, in a similar way as the outer 8 electrons in the neon atom, possess a configuration which exhibits a high degree of symmetry and harmony and which consists of two groups, in each of which the orbits of the electrons are oriented in a manner which exhibits tetrahedral symmetry, and which are oriented in a symmetrical way relative to the axis of the atom which coincides with the axis of the configuration of the first two electrons.

* [The manuscript has $n = 1$, $k = 1$.]
** [The manuscript has $n = 1$, $k = 1$.]

(Electronegative character of phosphorus, sulphur and chlorine. Compare with nitrogen, oxygen and fluorine. Weaker electronegativity corresponding to weaker binding of the entire group. No knowledge about the magnetism of the atoms, but of the ions. Compare Ne with A. Larger atomic volume of the latter)*.

Binding of the 19th electron. New type of orbit. Information from the K spectrum shows that the normal state must correspond to $n = 4, k = 1$, and the first term in the P-sequence to $n = 4$, $k = 2$ and the first D-term probably to $n = 4$, $k = 3$. However, the last is uncertain from a direct consideration, since the circumstances are still more unfavorable than in sodium. (This is wrong, however. The increase in the D-terms of K, Rb and Cs is the opposite of what occurs with the S- and P-terms. This shows that the orbit with $n = 3$, $k = 3$ still lies outside (?) the inner configuration. Must investigate if the D-terms form a good Rydberg series, or if the first term falls outside.)

We must therefore assume that in the normal state the 19th electron is bound in an orbit with $n = 4$, $k = 1$. As well the smaller values of the S-terms, and the greater electropositive character of the potassium atom as compared with sodium, are easily understood when it is considered that the dimensions of the inner system as compared to the dimensions of the outer orbit, calculated as a Kepler ellipse round a nucleus with unit charge and for a given apparent quantum number, is larger in potassium than in sodium.

Something quite similar holds for *calcium* for which all conditions as regards spectrum and chemical properties are analogous, only the electropositivity is greater, and the spectral terms are smaller, as well in the arc as in the spark spectrum. Thus, in the normal state of the neutral calcium atom, we must expect that the 19th and 20th electrons are bound in equivalent orbits with quantum numbers $n = 4$, $k = 1$. However, if we proceed farther in the periodic system, we soon meet changed circumstances, in that the elements become more and more different from those corresponding to an eight units lower atomic number, until finally, in the iron group, we get a number of strongly electropositive elements rather than an inert gas. This is just what we should be prepared for, however, on account of the essentially different configurations of the outer groups in the neon and argon atoms, in spite of their great similarity. While the structure of neon is such that we must expect it to be repeated in all atoms with higher atomic number than neon, this is not the case with argon, since this atom is not completed in the same manner. In fact, with increasing nuclear charge, we arrive at a point where a circular orbit with $n = 3$, $k = 3$ could be located inside the group of 8 electrons bound in 3-quanta orbits, and where the electron in question will be bound more firmly than an electron moving in a central orbit with $n = 4$, $k = 1$,

* [This parenthesis, and the following nine handwritten pages, are in Danish.]

which only partly comes within the group mentioned. When we have reached this point, the 19th electron, rather than being bound in an orbit with $n = 4$, $k = 1$, will be bound in an $n = 3$, $k = 3$ orbit; for, such an orbit, which differs in its classification from the orbits of the previously bound electrons, cannot be rejected by arguments such as we have applied to explain the impossibility in certain cases of additional electrons being bound in orbits characterized by the same quantum numbers as previously bound electrons. If this occurs, and if the atomic number is sufficiently large, the group of 3-quanta orbits will continue to grow as the succeeding electrons are bound. Now, without a thorough calculation of the possible orbits, it is difficult to predict in detail the individual steps of this process of transformation, which the group will undergo until it has reached such a degree of symmetry and harmony that the additional binding in 3-quanta orbits will no more depend on the firmness of the binding, but must be rejected by considerations based on the correspondence principle. However, it seems possible by simple arguments to reach rather definite conclusions about the general course of the transformation process and its outcome.

In the first place, it is clear that the binding of the first electrons in circular orbits with $n = 3$ will not increase, but will rather decrease, the symmetry of the group of 3-quanta orbits already present, in that it will introduce polarity. However, it is perhaps useful to point out at the outset that this is not at variance with the tendency of the previous considerations. While, from the correspondence principle, we had to reject the additional binding of electrons with the same quantum numbers n and k as those previously bound whenever the symmetry of the group of individual electrons in question was essentially disturbed, no such argument applies to the binding by the atom of electrons in orbits with new quantum numbers, even if these orbits might act ever so disturbing on previous groups; for, we must assume that only one electron is effectively involved with the radiation emitted during a certain transition, and that we therefore from correspondence considerations must confine ourselves essentially to the proper motion, during the process of binding, of the electron in question. Now, if nothing further happens during the continued binding of circular orbits, the symmetry of the subgroup in question will continue to increase until, after the binding of the fourth electron in a circular orbit, we might well assume a configuration characterized by tetrahedral symmetry. However, it is not at all certain that this will be the case and that such a tetrahedral configuration can be developed in this region of the atom where there already is a configuration of 8 electrons which does not, as the configuration of four electrons in carbon and silicon, directly leave a place open for the additional binding.

We must therefore be prepared that such a tetrahedral configuration is not developed at all but that, when the circular orbits appear, the previous configuration of 3-quanta orbits is disturbed to such a high degree that it no more behaves as completed as regards the question of the inclusion of new electrons with quantum numbers 3, 1 and 3, 2*, but rather will be completed by further binding of electrons in such orbits. Even if the individual steps in the transformation of the group, as mentioned above, can hardly be determined at the present time, it is nevertheless possible to give strong *a priori* arguments for understanding that the final configuration of 3-quanta orbits will not contain 12 electrons, distributed in three tetrahedral groups of 4 electrons each, but will rather contain 18 electrons; as we shall see, this is necessary in order to explain the period of 18 elements in the periodic system which follows after the two previous periods with 8 electrons each, which we have been concerned with in the preceding. In fact, it can be shown that, while it is hardly possible to obtain a configuration of 12 electrons sufficiently harmonious for completion of the group, a simple configuration of 18 electrons, distinguished by a high degree of harmony and spatial symmetry, offers itself.

In fact, we obtain such a configuration if we assume that the group includes six 3,1 orbits, six 3,2 orbits and six 3,3 orbits, and further that each subgroup, as regards the symmetry of the orbits, is built up in such a way that the normals of the orbital planes lie pairwise in the same plane as the axis of the inner system, and in this plane form equal angles with a normal to the axis, and further that the normals to the orbits are distributed in three planes which form equal angles with one another as shown in the accompanying simple stereographic projection, and if finally we assume that the three subgroups are placed relative to one another as shown in the following second projection.

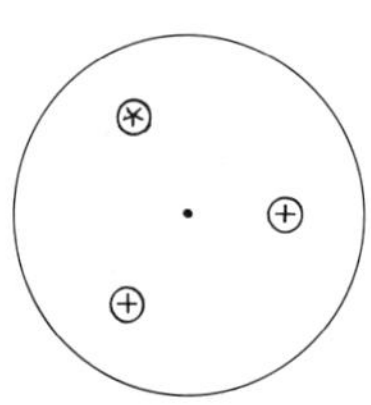

1.

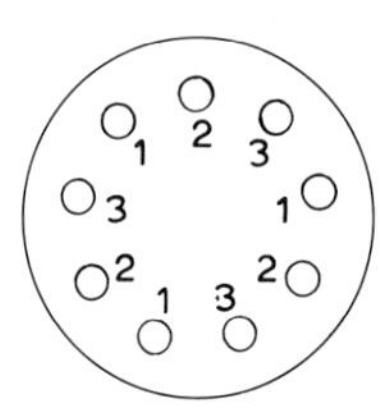

2.

This configuration is not only distinguished by a considerable symmetry, but it is of such a nature that we may assume that it can be formed by a gradual completion of the group of eight 3-quanta orbits occurring in the argon atom, which we may expect to occur as a stage in the formation of any atom with a higher atomic charge in which only 18 electrons are bound. In fact, as has been mentioned more closely when the constitution of the carbon atom was discussed, the in-

* [That is, $n = 3$, $k = 1$ and $n = 3$, $k = 2$. This notation is used in the following.]

dividual subgroups differ from tetrahedral configurations, in that the normals to the orbits lie grouped in pairs in equivalent planes containing the axis, while in a tetrahedral configuration the normals to the orbits lie distributed in planes containing the axis in such a way that they in turn make an acute or an obtuse angle with the latter.

As we saw when discussing the configuration of 8 outer electrons in the neon and argon atoms, we must now expect that, during the binding of the next 4 electrons, these must either move in circular orbits 2,2 or in elliptical orbits 3,2, and, to avoid coincidence of planes, we must assume that these, as well as the previously bound elliptical orbits corresponding to $k = 1$, will no more move in pure tetrahedral configurations but in configurations which may be regarded as deformed tetrahedrons, in which the deformation consists in a turning of the planes containing the normals of orbits making acute angles with the axis relative to those making obtuse angles with the axis. However, as long as the configurations contain 4 electrons each, this turning cannot be so great that it results in a coincidence of the normal planes in pairs, since, as is immediately seen, that would bring about a coincidence of orbital planes. However, as soon as more electrons are being bound in the subgroups, so that each contains 6 electrons, the matter is different, since then such a coincidence of the normal planes in question is possible, while a distribution in which they bisect the angles between one another, contrary to what was the case with the tetrahedral configuration of the first 4 electrons, ...*. Moreover, we see that, if the subgroups contain 6 electrons and have a suitable configuration, we can just arrange three such without their disturbing one another, while it would not be possible to have two or four without bringing about a coincidence of orbital planes. We shall return to the significance of the last circumstance when we proceed to discuss the still larger groups of 4-quanta orbits, which must be expected to exist in the interior of atoms with high atomic numbers.

After these considerations, we shall proceed to indicate briefly how the formation of such a group of 18 electrons in 3-quanta orbits appears to offer a basis for a detailed interpretation of the characteristic properties of the elements in the fourth group of the periodic system.

A. FACTA. SURVEY.

Potassium and calcium already discussed. Scandium, vanadium and titanium doubtful. Chromium and manganese definitely different from sulphur and

* [This sentence is left incomplete.]

chlorine in their electrochemical properties and in the magnetic properties of their ions. The quite new electrochemical and magnetic properties of the iron group. Gradual approach of copper and zinc to the alkali metals and the group of alkaline earths. Thereafter, close kinship with the last of the elements in the preceding groups of 8.

B. INTERPRETATION.

1) Beginning possibility of the appearance of circular 3-quanta orbits can, from simple calculations, be expected just in atoms with atomic numbers 21, 22 or 23.

2) The beginning appearance of 3-quanta circular orbits ...* as regards electrochemical properties, on account of the small difference in the binding between these and orbits 4,1. (Perhaps not correct.)

3) The number of electrons bound in the 3-quanta groups must, to begin with, be expected to increase by 1 with increasing atomic number, since the inclusion of an electron in the inner group must be expected to diminish essentially the firmness of the binding of succeeding electrons.

[There follows about three quarters of a page with largely illegible notes and crude sketches by Bohr himself. Practically all of the remaining manuscript is in English.]

As regards the formation of such a group of 18 electrons, it is not only difficult to account for its establishment in detail but, without laborious calculations, it is also difficult to calculate *a priori* at what atomic number we may expect that its establishment is completed, so that any further electron will be bound in a 4-quantum orbit. From a simple estimate, we may, however, expect that, in the neighborhood of the atomic number 22, the 19th electron will be bound with nearly the same firmness whether it moves in an orbit 4,1 or 3,3. In fact, at a nucleus charge of 22, the dimensions of the region of the configuration of the 18 electrons first bound will possess a diameter of approximately the same size as the diameter of the electron in the hydrogen atom, denoted by $2a_H$. Further, an electron bound in an orbit 4,1 will be bound with an energy which is nearly 3 times W_H, which will again be expected tó be approximately the same as an electron moving in a circular orbit 3,3 in the outskirts of the configuration of elliptical orbits 3,1 and 3,2. Next, we may expect, when we first have arrived to a nucleus charge for which one electron moving in an orbit 3,3 may just be in-

* [One or two words are illegible.]

cluded in the configuration of 3-quanta orbits, that, for a nucleus charge of one unit higher, not only will the configuration in question contain one electron moving in an orbit 3,3 but that at least a second electron will be included in this configuration. In fact, we must, from a simple estimate of the forces within the atom, conclude that the inclusion of, let us say, the N'th electron in this configuration for a nucleus charge N will be easier than the inclusion of the $N-1$'th electron in an atom of a nucleus charge $N-1$, and to this comes the fact that, the more the group of three quanta orbits is nearing its completion, the greater will the tendency be of including new electrons, so as to form a distinguished symmetrical configuration. In fact, this effect may just be expected to be analogous with the tendency of completion of the first group of 8 electrons in the atom which gives rise to the appearance in the periodic table of elements of electronegative valency; the only difference is that in the latter case the question is about the formation of a system of resulting negative charge, while in the former case we have to do with a completion of a configuration of a resulting positive charge, brought about by the larger dimensions of a circular three-quaŋtum orbit compared with a 2-quantum orbit moving in the same field. Due to the increasing relative tendency under consideration of including electrons in a group of three-quanta orbits, we may expect that this group will be completed at an atomic number which shows a difference from the number where the transformation of the group of 18 electrons starts which is smaller than the included number of electrons ⟨or⟩ equal to the total number of ⟨additional⟩ electrons included ⟨in the 3-quanta group⟩. In other words, we may expect that the neutral atom of the last element for which the group of 8 still exists intact, compared with the neutral atom of the first element for which the transformation of this group to a group of 18 is completed, contains a larger number of electrons moving in four-quanta orbits round the ...* groups.

Now, all this corresponds most closely and suggestively with the way in which the properties of the elements within the fourth period of the periodic table change with the atomic number. As we have seen already, the first elements (K, Ca) within this period possess properties which are closely analogous with those of the preceding periods of 8 elements and may be expected in the neutral atom to possess a configuration in which two inner electrons move in one-quanta orbits, 8 electrons move in 2-quanta orbits, and 8 electrons in 3-quanta orbits, forming a configuration round which a few electrons, the valency electrons, move in 4-quanta orbits. Now, with increasing atomic number, the properties of the elements within the period under consideration depart more and more from the corresponding members of the preceding periods, due to the relative increase in the

* [A word is illegible.]

dimensions of the configuration of 3-quanta orbits compared with that of the configuration of one- and two-quanta orbits which determines the dimension of the "inner" system of atomic elements of the preceding periods. Proceeding still further within the group, this departure, however, becomes so pronounced that it must be ascribed to an inherent transformation of the configuration of 3-quanta orbits, corresponding to the inclusion of electrons in the group moving in circular orbits 3,3. Instead of elements of decided electronegative character, as those of the latter half of the preceding group, we meet with a sequence of elements which rather may be characterized as electropositive and whose (?) properties, interpreted in the terminology of valency, exhibit very complex features, and among which the elements of the so-called iron family may be taken as typical specimens. This complexity of the valency properties is just what might be expected from the interpretation given of the appearance of these elements. In fact, this interpretation rests on the assumption that the successive stages of the completion of the group of 3-quanta orbits take place, so to speak, by the gradual displacement of the balance between 3- and 4-quanta orbits, as regards the firmness by which additional elements are bound by the atom. Through this balance, however, a definite distinction between what may be considered as valence electrons is not possible, in the simple sense as in the families of electropositive elements within the preceding periods of the periodic table where, for instance in the monovalent elements Li and Na, the energy needed to remove the electron last bound is several times smaller than that necessary to remove a second electron. Next, due to the same balance, it is difficult, not only from the properties of the elements, to ascertain the exact atomic numbers where the completion of the group starts and where it stops, but it is even difficult to obtain a rigorous definition of these points. If, for instance, we consider copper which possesses an atomic number 29, and which follows immediately after the elements of the iron family, we are naturally led to assume that in the neutral atom the group under consideration is completed, so that the constitution of the atom may be represented as 29 $(2_1 8_2 18_3 1_4)$*. In fact, this would offer an immediate interpretation of the circumstance that copper, in contrast to the preceding elements, may in combination with electronegative elements appear as monovalent and may in solution form monovalent ions. On the other hand, the circumstance that copper usually appears in combination as a divalent electropositive element bears witness of the fact that, even if this group ⟨of 18⟩ may be completed in the neutral atom and in singly charged positive ions, it does not hold its electrons with a force comparable with that of the inner groups in the Na and K atoms.

* [The parenthesis signifies: 2 one-quantum, 8 two-quanta, 18 three-quanta and one four-quanta electron.]

Proceeding to elements of higher atomic number, we meet already in Zn ($N = 30$) an element of pronounced divalent properties, bearing witness of the decided greater firmness by which the first 28 electrons are bound compared with the two last. Proceeding still further, the properties of the elements resemble more and more the elements in the third period in the periodic table, when we compare elements of atomic numbers $28 + a$ with elements of numbers $10 + a$. In fact, we meet with a number of electropositive elements followed by a number of elements of decided electronegative character, stopping by bromine, which resembles chlorine in a remarkable degree and which is followed by krypton ($N = 36$) which, as argon ($N = 18$), belongs to the family of inactive gases. This increasing resemblance between the elements of the third and fourth groups, when nearing the end of the latter group, is due not only to the appearance of a complete additional group of 18 electrons within the atom, but may be accounted for more closely by estimating the dimensions of the region of this group for increasing ⟨atomic number⟩ and the firmness with which the electrons within the outer group of 4-quanta orbits are bound. In fact, while for atomic number 30 (Zn) the dimensions of the group of 18 electrons may be expected to ...*.

(Comparison between the dimensions of the group of 18 in zinc and the group of 8 in Mg. Comparison between the binding of the outer electrons in Zn and Mg, partly theoretically and partly from the spectra. Comparison between the dimensions of argon and krypton.)

Constitution of krypton 36 ($2_1 8_2 18_3 8_4$). Description of the outer group of 8 electrons compared with the outer groups in Ne and A.

Before we proceed further with the discussion of the following group of the periodic table, it may be remarked how a most suggestive evidence in favor of the interpretation given of the general chemical character of the properties of the elements within the fourth group may be derived from the magnetic properties of these elements. As mentioned, we meet already in the second period of the periodic table with elements which under circumstances may exhibit paramagnetic properties, and for which the interpretation of the experimental evidence was very difficult, due to the fact that the magnetic properties of the elements in question were different under different circumstances, and especially very different in such combinations where the element appears as a neutral atom or ion, as it may also naturally be expected due to the difference in the character of the configuration of the electrons in such cases. Due to our very small knowledge of so-called homopolar combinations, evidence which does not concern the properties of ions is very difficult at the present state to interpret.

* [This sentence is left incomplete. The following parenthesis and the two first (incomplete) sentences of the next short paragraph are in Danish.]

As regards the elements of the former periods, they do not show paramagnetic properties in the ionic state, a circumstance which is just what should be expected (however, see the discussion under He)* from the marked non-polar symmetry of the groups which enter just in the constitution of ions, as well electropositive as electronegative. Now, the number of elements of the fourth group are distinguished, as regards their magnetic properties, thereby that they show decided paramagnetism in the ionic state. Now, this is just what should be expected if the atoms of such elements contain an inner group of non-completed asymmetrical character, as we have been led to assume, and, in fact, the experimental evidence as regards magnetism gives a most illustrative evidence just regarding the stages of completion of this group. As already called attention to by Kossel, the experimental evidence which at first sight appears very complex—each element possessing different atomic properties in its combinations of different valencies—exhibits a remarkable simplicity if, for different elements, the magnetism of ions which contain the same total number of electrons is compared. For instance, if the magnetism of Mn ($N = 25$) in divalent combinations is compared with that of iron ($N = 26$) in trivalent combinations and with that of Cr ($N = 24$) in the monovalent state ...**. If a comparison is made in this way, it is found that the ions with the same total number of electrons in different elements, with a single exception which may be due to an experimental error, possess the same magnetic moment, and that this moment varies in a remarkably simple way with this number, rising to a maximum and then continuously decreasing to zero. As regards the first result, Kossel points out that this may be explained by the assumption of an analogous constitution of atomic systems possessing the same total number of electrons***.

Now, on the present consideration, it appears that this interpretation can actually be traced to a certain point. In the first place, it deals in Kossel's arrangement† of the magnetic properties of ions with a total number of electrons between 21 and 28, derived from elements of atomic numbers between 24 and 30. Now, in the first place, these elements lie in the region within which we have seen that we must expect the transformation of the group of 3-quanta orbits to have begun, so that we, in the ions under consideration, must expect to have to do with configurations corresponding to stages of this transformation which are analogous in ions containing the same total number of electrons. Further, the figures show that the magnetism disappears just by ions containing 28 electrons,

* [This parenthesis is in Danish.]

** [This incomplete sentence is inserted in Bohr's own hand. It is only partly legible.]

*** [Some 13 lines are crossed out.]

† [This word is in Danish ("Sammenstilling").]

that is, just at the point where we shall expect the group under consideration to be completed and possess a symmetrical non-polar character. Although, unfortunately, the experimental data do not allow to trace the number of electrons in an ion where the first appearance of magnetism takes its beginning. Still, indications may be found from figures at hand which materially support a tracing of the transformation process in detail. In fact, the maximum of magnetism corresponds to a number of 23 electrons (corresponding to the divalent Fe ion), which number lies just in the midst between the number 28 and the number 18 which should correspond to an ion of constitution N $(2_1 8_2 8_3)$ which, just as little as the atom of argon or the ions of K and Ca, should exhibit any paramagnetism. The maximum will therefore be seen to correspond to a state of the process where asymmetry might be expected to possess an especially marked character.

FIFTH PERIOD OF THE PERIODIC TABLE.

As regards the interpretation of the chemical properties of the elements within this period, as well as the configurations of the atoms of these elements, the characteristic features are essentially the same as those of the fourth group, or rather the configurations in question possess a relation to those involved in the discussion of the fourth group which is quite analogous to the relation between the considerations involved in the discussion of the elements within the second and the third groups. Thus, for the first element within the group under consideration, rubidium ($N = 37$), we shall expect a configuration given by 37 $(2_1 8_2 18_3 8_4 1_5)$, and for the last element, xenon, 54 $(2_1 8_2 18_3 18_4 8_5)$, so that the only essential difference between the features of this and the preceding period is that, in the formation of the group of 18, we have not to deal with orbits of 3-quanta but such of 4-quanta ($n = 4$). Just as we in the third group have to do with the formation of a configuration of eight 3-quanta orbits, and not with eight 2-quanta orbits, as in the second group. This brings, however, no essential difference with it, as far as the elements under consideration are concerned. In fact, the second 18 group in Xe is formed by the inclusion of elliptical orbits 4,3 instead of circular orbits 3,3 in the former period, just as the second 8 group in argon is formed by inclusion of elliptical orbits ⟨3,1 and⟩ 3,2 instead of circular orbits 2,2. Just as in the latter case, orbits 3,3 possess so large dimensions that they do not come into consideration in the normal state of the atom and, therefore, as far as a third period of elements is concerned, do not play any part in considerations of symmetry and number of electrons in the group in question. We get, as regards the elements within the fifth period, that orbits 4,4 will not have to be taken into account. In the meanwhile, just as the occurrence of orbits 3,3 could only be said to be postponed as regards the elements within the third period, and the occurrence of such orbits in fact plays the predominant feature as regards the interpretation of the properties within the fourth group, so we would have to be prepared that, by increasing atomic number, we will reach such a point where the group of 4-quanta orbits, which in the fourth and fifth period arrive at successive stages of its completion, will undergo a final transformation by inclusion of 4,4 orbits. Now, there can hardly be any doubt that it is just such a transformation we witness in the appearance of the group of rare earths, which falls within the sixth period of the periodic table.

Just as we have seen with the formation of the group of 18 in the fourth period, we shall expect that the transformation of the second group of 18 will start at the point where the binding of an electron moving in an orbit 4,4 will be greater than that of electrons moving in orbits corresponding to $n = 5$. Now, just as was the case with the discussion of the analogous problem as regards the formation of the group of 18 electrons in the fourth period, it is difficult without laborious calculations to predict *a priori* exactly where this will take place. Nevertheless, it does seem possible from an estimate to account approximately for the point within the periodic table where the start of the transformation is claimed in order to account for the variation in the chemical properties of the elements with increasing atomic number. In fact, for an atomic number of 58, the first five-quanta orbits bound after the configuration 58 ($2_1 8_2 18_3 18_4$) may, by close approximation, be described as "apparent" 3-quanta orbits moving in a field of an "effective" nucleus charge of about $12e$, while an electron moving in a circular orbit in the outskirts of the mentioned configuration would be held by a nuclear charge of about $16[12^2/9 = 16 = 16^2/16]$. (Closer numerical investigation.)*

Next, as regards the constitution of a complete group of orbits $n = 4$, we may apply similar considerations as used in the discussion of the formation and constitution of the complete group of 3-quanta orbits. Thus, just as the inclusion of orbits 3,3 would affect the disturbance of the harmony of the previous group of eight 3-quanta orbits, leading to the completion not only of the whole group but also of the subgroups of four electrons within the latter configuration, we may expect that the inclusion of orbits 4,4 will lead to a transformation of the subgroups in the 18 group. Moreover (?), for the fact that the final result, as claimed by the interpretation of the structure of the sixth group in the periodic table, is the formation of a group of 32 electrons, arguments of similar nature may be given, as those given for the formation of a group of 18 electrons in the discussion of the elements within the fourth period. Thus, as we have seen already

* [This parenthesis is in Danish.]

in the mentioned discussion, it is not possible to obtain a group consisting of four subgroups of six electrons, which presents similar features as regards harmony as that of a group consisting of 3 subgroups of 6 electrons. Just the argument there used would lead, moreover, to expect that the subgroups of a group of 4-quanta orbits would contain a number which, divided by 2, would be even such that a configuration of 32 electrons presents itself. (Each subgroup possesses probably the same structure, as regards the direction of the normals, as the group of 8 in neon, twisted, however, through an angle in a similar way as the tetrahedral subgroups in neon.)

The next point on which it seems possible to throw a certain light is the marked difference between the variation of the chemical properties of the elements within the fourth and the sixth group the atoms, which may be considered as containing stages of the final completion of the configurations of 3-quanta and 4-quanta orbits. This difference shows itself in that the corresponding elements within the latter period differ markedly less than the elements within the former period. This circumstance, however, may have its natural explanation therein that, due to the larger number of electrons within the 4-quanta group, and the greater resultant change of the atom at the moment where the balance between the tendency of inclusion of electrons in the group and that of binding electrons in higher quanta orbits takes place, the variation in the resultant charge, when proceeding from one element to the next, will, compared with the effect of the ...* of the internal constitution of the groups ...* by this inclusion, play a relative larger part in the formation of the 4-quantum group than in the 3-quantum group. We may, for the first group, therefore expect, in the normal state of atoms of successive elements, in general to find configurations of 4-quanta orbits containing a number of electrons higher by one unit, with the effect that such elements will possess the same number of outer electrons bound with nearly the same firmness and, according to this, very nearly the same chemical properties. A decrease in the valency of the elements under consideration, corresponding to the tendency of finishing the completion in its final stages, which was so marked with the elements within the fourth group, however, does also take place in this group, as a closer examination shows**. (Look into this more closely.)

The only significant difference between the individual rare earths, however, must be expected in their magnetic properties, as is actually the case.

* [Corrections made by Bohr are largely illegible.]

** [Part of this sentence, the parenthesis, and the remaining two short paragraphs are in Danish.]

Explanation of how the rare earths in no way diminish the simplicity of the periodic system, but merely occur as an epoch in a group of quite the same constitution as the previous ones. Constitution of cesium and niton: 55 $(2_1 8_2 18_3 18_4 8_5 1_6)$, 86 $(2_1 8_2 18_3 32_4 18_5 8_6)$.

VI. ATOMIC STRUCTURE

(Nature **108** (1921) 208)*

* [See Introduction, sect. 5.]

Reprinted from NATURE, *October* 13, 1921]

Atomic Structure.

IN connection with the problem of the constitution of the atom discussed in my letter to NATURE of March 24 last (vol. 107, p. 104), I should like to add a few complementary remarks about the manner in which the orbits of the electrons in the atom are characterised.

According to this view of atomic constitution, the electrons in the atom are arranged in groups in such a way that the orbit of every electron within one and the same group is characterised by the same total number of quanta. Since, however, for orbits characterised by more than one quantum there exist several types of orbits possessing the same total number of quanta, the electrons within each group do not in general play equivalent parts, but are divided into a number of sub-groups corresponding to different types of possible orbits. Now it is a salient feature of this picture that the atom cannot be said to be composed of a number of well-defined spherical shells of electrons moving in sharply separated regions of the atom. In fact, although the electrons of a given group mainly move within one and the same shell-shaped region of the atom, the electrons, at any rate of certain sub-groups, will in their revolution penetrate into the region of the orbits of the electrons of inner groups. This gives rise to a coupling between the various groups, which is an essential feature of the interpretation of the stability of the atom. As a consequence of this, the orbit of an electron may be considered from different points of view, according as attention is mainly paid (1) to the larger part of the orbit which lies outside the region of inner groups, and which nearly coincides with an almost closed Keplerian ellipse, or (2) to the mechanical properties of the whole orbit, regarded as a type of central orbit composed of loops which only in their outer part possess an approximately Keplerian character.

Now in the classification described in my former letter the orbits were regarded from the first, and more superficial, point of view. The numbers of quanta characterising the orbits of the electrons in the different groups correspond to Keplerian ellipses, which coincide approximately with the outer parts of the orbits of the electrons in question. It has since been possible, by a detailed examination of the parts of the orbital loops situated within the region of inner groups, to classify the orbits from the second, and more fundamental, point of view, leading to a simple and unambiguous result. In fact, we are led to a classification in which, when we proceed outwards from the nucleus, the number of quanta characterising a certain group of orbits is always larger by one unit than that of the preceding group. For the groups in the inner region of the atom, where the attraction of the nucleus preponderates, this new rigorous classification coincides with the old one of my former letter. But it departs from the old for groups in which the orbits of the electrons mainly fall in the outer region of the atom, where the attraction of the nucleus is largely compensated for by the repulsion of the electrons in the inner groups. For these groups the quantum numbers of the orbits given in my former letter were equal to, or even smaller than, those of inner groups.

Notwithstanding the essential progress made by this modification in the classification of the orbits, the main features of this model of the atom remain the same. For instance, my former statements of the numbers of electrons in the various groups and subgroups in the atom hold unaltered for all groups. In fact, in fixing these numbers by the correspondence principle we find them to depend on the harmony of the motion of the electrons within each single group. They depend, therefore, primarily on the relative dimensions of the approximately Keplerian loops, and only secondarily on the way in which these loops are joined together to form complete central orbits. Thus the previous model of the atoms of the inert gases holds unaltered also as regards the outer groups, provided that the numbers stated as defining the number of quanta of the orbits in the various groups are considered instead as defining the number of sub-groups within the corresponding groups. Moreover, the numbers in question offer an approximate estimate of the spatial extension of the regions of the orbits of the electrons in the different groups in the atom. For instance, the orbits in the outermost "shell" in the Niton atom must be characterised as six-quanta orbits instead of as two-quanta orbits; but the dimensions of the orbital loops will by no means be of the same order of magnitude as those of the orbit of an electron revolving in a Keplerian orbit characterised by six quanta in the region outside that of the orbits of the electrons in the five inner groups; they will rather be of the same order as those of a similar Keplerian orbit characterised by only two quanta.

From these remarks it will be seen that my former applications of the theory to the interpretation of the physical and chemical properties of the elements remain in substance unaltered. At the same time the elaboration of the theoretical considerations sketched in this letter throws a good deal more light on the interpretation of many details. For instance, it is possible to account for the appearance in the atom, with increasing atomic number, of groups of new constitution in such a way that we obtain a natural interpretation not only, as before, of the existence of such families of elements in the periodic table as those of the iron metals and the rare earths, but also of the almost complete absence of any effect on the Röntgen-ray spectra of the appearance of such groups. This absence is explained by the fact that in these families we do not witness any sudden change with increasing atomic number in the total number of quanta of the orbits of the electrons of certain groups. On the contrary, we may be assumed to witness in the appearance of each of these families the completion of a group by the inclusion of further electrons moving in orbits characterised by the same number of quanta. This addition to the group is brought about by a change in the interaction between the various possible types of orbits with this number of quanta caused by the alteration in the dimensions of the orbital loops and in the "apparent" number of quanta which may be said to characterise these loops.

I have confined myself here to these points of general character. For details of the theory and its applications I must refer the reader to a paper in preparation for publication by the Royal Danish Academy of Science.

N. BOHR.

Copenhagen, September 16.

PRINTED IN GREAT BRITAIN BY THE CORNWALL PRESS, LTD., PARIS GARDEN, STAMFORD STREET, LONDON, S.E. 1.

VII. THE STRUCTURE OF THE ATOM AND THE PHYSICAL AND CHEMICAL PROPERTIES OF THE ELEMENTS

(1921–1922)

DANISH TEXT AND TRANSLATION*

* [See Introduction, sect. 7.]

Atomernes Bygning

og

Stoffernes fysiske og kemiske Egenskaber

Af

N. Bohr

København
Jul. Gjellerups Forlag
1922

Hr. Professor Dr. phil. Niels Nielsen
med venlig Hilsen
fra Forfatteren

Atomernes Bygning

og

Stoffernes fysiske og kemiske Egenskaber

Af

N. Bohr

Foredrag holdt i Fysisk Forening den 18. Oktober 1921

København
Jul. Gjellerups Forlag
1922

NAAR jeg er gaaet ind paa Forlagets venlige Opfordring til at lade dette i Fysisk Forening holdte Foredrag, der er offentliggjort i Fysisk Tidsskrift, udkomme som et særskilt Hefte, skyldes det den Betragtning, at Foredraget paa Grund af dets Indhold og Form maaske vilde have Interesse for andre end Fysisk Tidsskrifts sædvanlige Læsekreds. Foredragets Hensigt vil fremgaa af de Bemærkninger, hvormed det indledes; jeg vil imidlertid her gerne tilføje et Par Ord for at lette Læserne Oversigten over Indholdet. Idet Foredragsformen er bibeholdt ved Trykningen, er der ikke benyttet en ved indholdsangivende Overskrifter betegnet Inddeling; men en vis Inddeling er alligevel søgt antydet ved Tegn, der angiver større naturlige Pauser i Foredraget. Den første af disse Pauser findes saaledes paa Side 5 efter den egentlige Indlednings Afslutning. Det paafølgende Afsnit af Foredraget, der strækker sig til Pausen nederst paa Side 18, indeholder en Oversigt over nogle af de vigtigste af de Resultater, man har naaet ved Arbejdet paa Spørgsmaalet om Atombygningen i Løbet af de senere Aar, samt en Angivelse af de Problemer, der vil blive søgt belyst i det følgende. I Foredragets næste Afsnit, der strækker sig til Pausen paa Side 31, gøres der dernæst nærmere Rede for de teoretiske Synspunkter, ud fra hvilke denne Belysning søges opnaaet. Dersom Læseren skulde finde Vanskelighed ved enkelte Betragtninger i Slutningen af dette Afsnit, vil jeg gerne allerede her gøre opmærksom paa, at flere af de omhandlede Enkeltheder fuldt saa meget er medtagne for Fuldstændighedens Skyld som for Forstaaelsen af det følgende. Saaledes vil der for Forstaaelsen af det paafølgende Hovedafsnit af Foredraget, hvori der gøres et Forsøg paa i Omrids at give en Fortolkning af Grundstoffernes kemiske Egenskaber, kun kræves en Tilegnelse af Hovedtrækkene af de teoretiske Synspunkter. I lignende om end mindre Grad gælder dette ogsaa for Foredragets næste og sidste Afsnit, der begynder ved Pausen paa Side 61. I dette Afsnit søges de i det foregaaende udviklede Resultater anvendt paa den teoretiske Forklaring af Røntgenstraalespektrene, hvis eksperimentelle Undersøgelse har givet saa vigtige Bidrag til Studiet af Spørgsmaalet om Atombygningen.

København, December 1921.

N. Bohr.

De Forestillinger om Atombygningen, som jeg i Aften skal forsøge at udvikle for Dem, har jeg i Hovedtræk allerede antydet i et Foredrag her i Fysisk Forening for omtrent et Aar tilbage. I Mellemtiden har disse Forestillinger imidlertid antaget fastere Former, og i to Breve til »Nature«, offentliggjort for nogle Maaneder siden, har jeg skitseret Teoriens nærmere Udformning. De Resultater, jeg skal forelægge for Dem, har vel ikke nogen afsluttet Karakter, men jeg haaber at kunne give Dem et Indtryk af, hvorledes det synes muligt ud fra de omhandlede Forestillinger om Atombygningen at belyse Sammenhængen mellem Stoffernes forskellige Egenskaber paa en Maade, der tillader at komme ud over visse Vanskeligheder, som hidtil tilsyneladende har stillet sig i Vejen for en simpel, sammenhængende Fortolkning af disse Egenskaber. Jeg maa paa Forhaand bede om Overbærenhed, dersom jeg i Foredragets første Halvdel kommer til at omtale adskillige Ting, der turde være mange af Dem velkendte; men for at føre Dem ind i den Tankebygning, det drejer sig om, bliver jeg nødt til først kort at omtale nogle af de vigtigste Resultater, som man er naaet til ved Arbejdet med Atomspørgsmaalene i Løbet af de senere Aar.

Det Billede af Atomet, som jeg skal lægge til Grund for alt det følgende, er det saakaldte K e r n e a t o m, efter hvilket man forestiller sig, at et Atom bestaar af en Kerne omgivet af et Antal Elektroner, hvis Afstande fra hinanden og fra Kernen er meget store i Forhold til Partiklernes egne Dimensioner. Kernen besidder næsten

1

hele Atomets Masse og en positiv elektrisk Ladning af en saadan Størrelse, at Antallet af Elektroner i et neutralt Atom er lig det paagældende Grundstofs Nummer i det periodiske System, det saakaldte Atomnummer. Dette Billede af Atomet, til hvilket man som bekendt er naaet væsentlig gennem R u t h e r f o r d's grundlæggende Undersøgelser over radioaktive Stoffer, fremviser overordentlig simple Træk, men netop denne Simpelhed rummer i første Øjeblik tilsyneladende Vanskeligheder for en Forklaring af Stoffernes Egenskaber. Hvis vi behandler dette Spørgsmaal ud fra vore sædvanlige Begreber om Mekanik og Elektrodynamik, er det nemlig ikke muligt i dette Billede at finde Udgangspunkter for en Forklaring af de udprægede Egenskaber, som de forskellige Stoffer besidder, ja ikke engang for Stoffernes Bestandighed. Paa den ene Side vil der øjensynligt ikke eksistere stabile Ligevægtstilstande af Atomets Partikler, paa den anden Side maa vi vente, at enhver Bevægelse, der kunde komme i Betragtning, vilde give Anledning til en Udsendelse af elektromagnetisk Straaling, der ikke vilde høre op, før hele Systemets Energi var udstraalet og alle Elektronerne faldne ind i Kernen. En Udvej til at komme ud over disse Vanskeligheder har man imidlertid fundet ved Benyttelse af Forestillinger hentede fra den saakaldte K v a n t e t e o r i, hvortil Grunden blev lagt ved P l a n c k s berømte Arbejder over Varmestraalingsloven, og som betegner et afgørende Brud med tidligere Forestillinger, idet der for første Gang ved Formuleringen af almindelige Naturlove indførtes Antagelser om Optræden af Diskontinuiteter.

I den Form for Kvanteteorien, i hvilken den har fundet Anvendelse paa Spørgsmaalet om Atomernes Bygning, hviler den paa to Postulater, der netop tager Sigte paa de ovenfor berørte Vanskeligheder. Efter det første Postulat findes der visse særegne Tilstande af Atomet, i hvilke dette kan eksistere uden Udsendelse af Straaling til Trods for, at Partiklerne tænkes at befinde sig i accelereret Bevægelse i Forhold til hinanden. Disse saakaldte s t a t i o n æ r e T i l s t a n d e antages endvidere at besidde en særegen Stabilitet af en saadan Art, at det ikke er muligt at tilføre eller fratage et Atom Energi undtagen ved en Proces, der bestaar i en Overførelse af Atomet til en anden af disse Tilstande. Efter det andet Postulat vil enhver Straaling, der udsendes fra Atomet, og som altsaa skal tænkes at være knyttet til en Overgangsproces mellem to stationære Tilstande, altid bestaa af en Rækkefølge af rent harmoniske Bølger. Svingningstallet for disse Bølger afhænger ikke direkte af Atomets Bevægelse, men er bestemt af den saakaldte Frekvensbetingelse, hvorefter Svingningstallet multipliceret med den af Planck indførte universelle Konstant er lig med den totale ved Processen udstraalede Energi. For en Overgang mellem to stationære Tilstande, for hvilken Atomets Energi før og efter Udstraalingsprocessen er henholdsvis E' og E'' har man saaledes

$$h\nu = E' - E'', \tag{1}$$

hvor h er Plancks Konstant og ν Svingningstallet for den udsendte Straaling. Tiden vil ikke her tillade mig at give en systematisk Fremstilling af Kvanteteorien, hvis senere Udvikling er gaaet Haand i Haand med dens Anvendelse paa Atombygningen; jeg skal derfor straks gaa over til at betragte de Anvendelser af Teorien, der har direkte Betydning for de i det følgende omhandlede Spørgsmaal.

* *
*

Vi skal da begynde med at se paa det simplest tænkelige Atom, der blot bestaar af en Kerne og en Elektron. Dersom Kernens Ladning svarer til en enkelt Elektrons Ladning, og Systemet altsaa er neutralt, har vi at gøre med et Brintatom, og den nøjere Udformning af Anvendelsen af Kvanteteorien paa Atombygningen har netop taget sit Udgangspunkt i Redegørelsen for det bekendte simple Spektrum, som udsendes fra Brint. Dette Spektrum bestaar af en Række Spektrallinier, for hvilke Svingningstallet er givet ved den overordentlig simple saakaldte Balmerske Formel

$$\nu = K\left(\frac{1}{(n'')^2} - \frac{1}{(n')^2}\right), \tag{2}$$

hvor n' og n'' er to hele Tal. Efter Kvanteteorien tænker vi os nu, at Atomet besidder en Række stationære Tilstande karakteriserede ved Rækken af hele Tal, og vi ser, hvorledes de ved Formel (2) givne Svingningstal fremgaar af Frekvensbetingelsen, dersom vi antager, at en Brintlinie svarer til en Straaling, der udsendes ved en Overgang imellem to af disse Tilstande karakteriserede ved Tallene n' og n''.

og dersom vi antager, at Energien i den $n$te Tilstand paa nær en vilkaarlig Konstant er givet ved

$$E_n = -\frac{hK}{n^2}. \tag{3}$$

Benyttelsen af det negative Tegn hænger sammen med, at Atomets Energi simplest maales ved det Arbejde, der skal til for at fjerne Elektronen til en uendelig Afstand fra Kernen, og vi skal antage, at den numeriske Værdi af Udtrykket paa højre Side af Formel (3) netop er lig med dette Arbejde.

Hvad den nærmere Beskrivelse af de stationære Tilstande angaar, finder vi, dersom vi ser bort fra Kernens Medbevægelse, at Elektronen med meget stor Tilnærmelse vil beskrive en Ellipse med Kernen i det ene Brændpunkt. Storaksen af denne Ellipse hænger paa simpel Maade sammen med Atomets Energi, og svarende til de ved Ligning (3) givne Energiværdier af de stationære Tilstande finder vi en Række Værdier for Elektronbanens Storakse $2a$ givne ved

$$2a_n = \frac{n^2 e^2}{hK}, \tag{4}$$

hvor e er den numeriske Værdi af Elektronens og Kernens Ladning.

Sammenfattende kan vi sige, at Brintspektret belærer os om den Maade, hvorpaa Brintatomet dannes, idet de stationære Tilstande tør betragtes som forskellige Stadier af en Proces, ved hvilken Elektronen under Udsendelse af Straaling bindes i Baner af mindre og mindre Dimensioner, svarende til Tilstande med aftagende Værdier for n. Det vil ses, hvorledes dette Billede har visse Træk tilfælles med Bindingsprocessen mellem Elektron og Kerne, saaledes som det maatte ventes efter den sædvanlige Elektrodynamik; men at vort Billede afviger fra den sidstnævnte paa en Maade, som gør det muligt at gøre Regnskab for Brintens iagttagne Egenskaber. Specielt ser vi, hvorledes Bindingsprocessens Slutresultat fører til en ganske bestemt Tilstand af Atomet, nemlig den stationære Tilstand, for hvilken $n = 1$. Denne Tilstand, der svarer til Atomets mindste Energiindhold, skal i det følgende betegnes som Atomets N o r m a l t i l s t a n d. Det kan allerede her bemærkes, at de Værdier for Atomets Energi og Elektronbanens Storakse, som man finder, dersom man i Formlerne (3) og (4) sætter

$n = 1$, har samme Størrelsesorden som de Værdier for Fastheden af Elektronbindinger og for Atomdimensioner, som man har udledt af Undersøgelser over Luftarters elektriske og mekaniske Egenskaber. Om en nøje Prøvelse af Formlerne (3) og (4) ved en saadan Sammenligning er der dog ikke Tale, fordi Brint under de Forhold, hvorunder saadanne Undersøgelser kan udføres, ikke optræder som enkelte Atomer, men som Molekyler, i hvilke Atomerne er parvis sammensluttede i indgribende Vekselvirkning.

Kvanteteoriens formelle Grundlag bestaar nu imidlertid ikke alene i Frekvensbetingelsen, men ogsaa i Betingelser der tillader at bestemme de stationære Tilstande for Atomsystemer, og som ligesom Frekvensbetingelsen kan opfattes som naturlige Generalisationer af de Antagelser om Vekselvirkningen mellem simple elektrodynamiske Systemer og et omgivende elektromagnetisk Straalingsfelt, der ligger til Grund for Plancks Varmestraalingsteori. Jeg skal ikke her komme nærmere ind paa disse Betingelsers Art, men blot sige, at de stationære Tilstande gennem dem karakteriseres ved et Antal hele Tal, de saakaldte Kvantetal. For en simpel periodisk Bevægelse, som den vi har antaget i Brintatomet, indgaar der i Fastlæggelsen af de stationære Tilstande kun eet Kvantetal. Dette fastlægger Atomets Energi og dermed Elektronbanens Storakse, men ikke Graden af dennes Ekscentricitet og derfor ikke Banens Form. Den nærmere Beregning giver, at Energien i de forskellige Tilstande vil være bestemt ved Udtrykket

$$E_n = -\frac{2\pi^2 N^2 e^4 m}{n^2 h^2}, \tag{5}$$

hvor $-e$ og m er henholdsvis Elektronens elektriske Ladning og dens Masse, og hvor vi for de følgende Anvendelsers Skyld har betegnet Kernens Ladning med Ne. Det hele Tal n er Kvantetallet, der karakteriserer de forskellige Tilstande.

For Brintatomet er $N = 1$, og Sammenligning med Ligning (3) fører til det følgende teoretiske Udtryk for Konstanten K i Formel (2):

$$K = \frac{2\pi^2 e^4 m}{h^3}, \tag{6}$$

hvilket Udtryk indenfor den Nøjagtighed, hvormed de indgaaende Størrelser kan bestemmes, har vist sig at stemme overens med den empiriske Værdi for Konstanten i Brintspektret. Dersom vi i den ovenstaaende Formel sætter $N = 2$, svarende til et Atom bestaaende af en Elektron, der bevæger sig om en Kerne med dobbelt Ladning, faar vi en Værdi for K, der er fire Gange saa stor som den for Brintspektret. For det Spektrum, der skulde udsendes under Dannelsen af et saadant Atom, faar vi derfor følgende Udtryk:

$$\nu = 4K\left(\frac{1}{(n'')^2} - \frac{1}{(n')^2}\right). \tag{7}$$

Denne Formel fremstiller visse Linier, som i længere Tid har været kendt, og som man tidligere paa Grund af den store Lighed mellem Formlerne (2) og (7) har tilskrevet Brint, fordi man aldrig havde iagttaget, at to forskellige Stoffer kunde vise Egenskaber, der lignede hinanden saa meget. Vi maa imidlertid efter Teorien vente, at Formel (7) fremstiller et Spektrum, som under visse Omstændigheder kan udsendes fra Helium, idet et Heliumatom, der har mistet en Elektron, netop bestaar af en Kerne med dobbelt Ladning og en Elektron. Denne Opfattelse har ogsaa vist sig at stemme overens med senere Erfaringer, og bl. a. har det været muligt at frembringe det omhandlede Spektrum i ren Helium. Jeg har opholdt mig ved dette Punkt for at vise, hvorledes denne nøje Forbindelse mellem to Grundstoffers Egenskaber, som i første Øjeblik kunde synes overraskende, er at betragte som et umiddelbart Udtryk for den for Kerneatomet karakteristiske, simple Bygning. Kort Tid efter Klarlæggelsen af det omhandlede Spørgsmaal fik man imidlertid nyt Vidnesbyrd af overordentlig Interesse om en saadan Lighed mellem Grundstoffernes Egenskaber, nemlig gennem Moseleys grundlæggende Arbejder over Stoffernes Røntgenspektre. Som velkendt fandt Moseley, at disse Spektre varierer paa overordentlig simpel Maade, naar man indenfor det periodiske System gaar fra Stof til Stof. Røntgenspektrenes Linier kan, som bekendt, deles i Grupper, og for den stærkeste Linie i den saakaldte K-Gruppe, der omfatter de mest gennemtrængende Røntgenstraaler, som et givet Stof kan udsende, naar det udsættes for Bombardement af Katodestraaler, fandt Moseley, at Svingningstallet for alle undersøgte Stoffer kunde udtrykkes ved Hjælp af en Formel, der med en ringe Forenkling kan skrives

$$\nu = N^2 K \left(\frac{1}{1^2} - \frac{1}{2^2}\right), \tag{8}$$

hvor K er den samme Konstant som i Brintspektret, og hvor N er Nummeret af det paagældende Element i det periodiske System. Denne Opdagelses store Betydning var først og fremmest, at den førte til almindelig Anerkendelse af den Antagelse, at dette saakaldte Atomnummer netop var lig Antallet af Elektroner i det paagældende Stofs Atomer, en Antagelse der allerede tidligere var lagt til Grund for Arbejder over Atombygningen, og som først var udtalt af van den Broek. Medens man straks var klar over denne Side af de Moseleyske Resultaters Betydning, har det paa den anden Side været vanskeligere tilfulde at forstaa den ganske overordentlige Lighed med Brintspektret, som Røntgenspektrene ifølge disse Resultater udviser. Denne Lighed fremtræder ikke alene for Linierne i K-Gruppen, men ogsaa for Grupperne af mindre gennemtrængende Røntgenstraaler. Saaledes fandt Moseley, at Svingningstallene for den stærkeste Linie i den saakaldte L-Gruppe for alle undersøgte Stoffer kunde angives ved en Formel, der med en Forenkling svarende til den, der er indført i Formel (8), kan skrives

$$\nu = N^2 K \left(\frac{1}{2^2} - \frac{1}{3^2}\right). \tag{9}$$

I begge Tilfælde faar vi saaledes et Udtryk for Svingningstallet, der svarer til en Linie i et Spektrum, der vilde udsendes under en Elektrons Binding af en Kerne med Ladning Ne.

Denne Lighed imellem Bygningen af Røntgenspektrene og Brintspektret blev yderligere uddybet paa overordentlig interessant Maade af Sommerfeld gennem hans betydningsfulde Udvidelse af Teorien for Brintspektret. Den ovenfor omtalte Fastsættelse af Energien i de stationære Tilstande af Brintsystemet, hvor hver Tilstand er mærket ved et enkelt Kvantetal, hviler paa den Antagelse, at Elektronens Bane i Atomet er rent periodisk. Dette gælder imidlertid kun med Tilnærmelse. Det viser sig nemlig, at Banen, naar man tager den Ændring, som Elektronens Masse undergaar med Hastigheden, i Betragtning, ikke bliver en simpel Ellipse, men at Elektronbanen derimod kan beskrives som en Centralbevægelse fremkommen ved en Overlejring af en jævn langsom Drejning paa en rent periodisk Bevægelse i en Bane, der kun afviger meget lidt fra en Ellipse. For en saadan Centralbevægelse vil de stationære Tilstande være karakteriseret ved to Kvantetal. Af disse kan dog i det her betragtede Tilfælde det ene vælges saaledes, at det med stor Tilnærmelse bestemmer Atomets Energi, paa samme Maade som det tidligere benyttede Kvantetal antoges at bestemme Energien under den Antagelse, at Elektronbanen i Brintatomet var en simpel Ellipse. Dette Kvantetal, som vi stadig skal kalde n, skal vi derfor betegne som »Hovedkvantetallet«. Foruden denne Bestemmelse, der med stor Tilnærmelse fastlægger Storaksen i den roterende næsten elliptiske Bane, vil for en Centralbane de stationære Tilstande være underlagt endnu en Betingelse, der udsiger, at Momentet af Elektronens Bevægelsesmængde omkring Centret, det saakaldte Impulsmoment, er lig med et helt Multiplum af Plancks Konstant divideret med 2π. Det hele Tal, der optræder som Faktor til denne Størrelse, kan betragtes som det andet Kvantetal og skal betegnes ved k. Denne sidstnævnte Betingelse bestemmer i det her betragtede Tilfælde Ekscentriciteten af den roterende næsten elliptiske Bane nærmere. Det kan i denne Forbindelse bemærkes, at den mulige Betydning af Impulsmomentet i Kvanteteorien var fremhævet af Nicholson allerede før den første Anvendelse af denne Teori paa Brintspektret, samt at en lignende Fastsættelse af de stationære Tilstande for et Brintatom som den af Sommerfeld benyttede omtrent samtidig var foreslaaet af Wilson, uden at denne dog opnaaede at give en fysisk Fortolkning af sine Resultater.

Den simpleste Beskrivelse af Formen af den roterende, næsten elliptiske Elektronbane i Brintatomet faas ved at betragte den paa Storaksen vinkelrette Korde gennem Brændpunktet, den saakaldte Parameter. Længden af denne Parameter vil nemlig med stor Tilnærmelse være givet ved et Udtryk, der har ganske samme Form som Udtrykket for Storaksen, blot med den Forskel, at n er ombyttet med k. Med de samme Betegnelser som i det foregaaende har vi saaledes

$$2a = n^2 \frac{h^2}{2\pi^2 N e^2 m}, \quad 2p = k^2 \frac{h^2}{2\pi^2 N e^2 m}. \qquad (10)$$

For hver af de stationære Tilstande, som vi tidligere havde karakteriseret ved en given Værdi af n, faar vi derfor et Sæt af n Tilstande svarende til Værdier for k fra 1 til n. Ved at tage Hensyn til Elektronmassens Afhængighed af Hastigheden fandt Sommerfeld nu for Energien i de stationære Tilstande, i Stedet for den simple Formel (5), et mere kompliceret Udtryk, som afhænger saavel af n som af k, og som med Tilnærmelse kan skrives

$$E_{n,k} = -\frac{2\pi^2 N^2 e^4 m}{n^2 h^2}\left[1 + \frac{4\pi^2 N^2 e^4}{h^2 c^2}\left(-\frac{3}{4n^2} + \frac{1}{nk}\right)\right]. \tag{11}$$

Idet det andet Led indenfor Parentesen er meget lille, faar vi, svarende til enhver af de ved den enkle Formel (5) givne Energiværdier for de stationære Tilstande af Brintatomet, n Værdier, der kun afviger meget lidt fra hinanden. Ved Anvendelse af den almindelige Frekvensbetingelse (1) faar vi saaledes for enhver Brintlinie, givet ved den simple Formel (2), et Antal Svingningstal, der er meget nær lige store. Sommerfeld kunde nu vise, at Beregningen virkelig stemte overens med Maalingerne af Komponenterne i den saakaldte Finstruktur af Brintlinierne, der iagttages, naar Brintspektret undersøges med Spektrografer med meget stor Opløsningsevne. Denne Overensstemmelse gjaldt ikke alene for Brintliniernes Finstruktur, hvor Maalinger paa Grund af Komponenternes overordentlig nære Beliggenhed er meget vanskelige, men navnlig var det muligt i Enkeltheder at gøre Regnskab for Finstrukturen af de ved Formel (7) givne Heliumlinier, som har været meget nøje undersøgt af Paschen. I Forbindelse med denne Teori gjorde Sommerfeld nu opmærksom paa, at den samme Formel (11) ogsaa tillod en Anvendelse paa Røntgenspektrene, idet han kunde vise, at der saavel i K- som L-Grupperne optraadte Liniepar, hvis Svingningstals Differenser ved Hjælp af Frekvensbetingelsen med Tilnærmelse kunde bestemmes ud fra dette Udtryk for Energien i de stationære Tilstande, svarende til en Elektrons Binding af en Kerne med Ladning Ne.

Trods den store formelle Lighed mellem Røntgenspektrene og Brintspektret, som disse Resultater aflægger Vidnesbyrd om, maa der dog antages at være en dybtgaaende Forskel imellem de Processer, der giver Anledning til Fremkomsten af de nævnte Typer af Spektre. Medens Udsendelsen af Brintspektret ligesom Udsendelsen af de sædvanlige optiske Spektre af andre Stoffer maa antages at være knyttet til en Proces, hvorved en Elektron indfanges og bindes af Atomet, viser Iagttagelser over Frembringelsen og Absorptionen af Røntgenspektre paa tydelig Maade, at disse Spektre er knyttet til en Proces, hvorved et færdigt dannet Atom reorganiseres efter en Forstyrrelse i dets Indre, frembragt ved Indgreb af ydre Kræfter. Vi maa saaledes tænke os, at Røntgenspektrenes Fremkomst ikke alene bestemmes af den direkte Vekselvirkning mellem en enkelt Elektron og Atomkernen, men at den er paa det nøjeste forbunden med den Maade, hvorpaa Elektronerne i det færdigt dannede Atom er bundet.

Ved Betragtning af dette sidste Problem besidder vi en Ledetraad af største Værdi i den ejendommelige Maade, hvorpaa mange af Stoffernes Egenskaber varierer med Atomnummeret, og som finder sit Udtryk i Grundstoffernes saakaldte periodiske System. Til Brug for det følgende er en simpel Oversigt over dette System givet i Tabellen paa næste Side. Tallene ud for ethvert Grundstof betegner Atomnummeret, og Stofferne indenfor de forskellige Rammer udgør som angivet Systemets »Perioder«, indenfor hvilke de kemiske Egenskaber varierer paa den velkendte karakteristiske Maade.

For at opnaa en Forstaaelse af denne ejendommelige Lovmæssighed, ligger det nær at antage en udpræget Gruppedeling af Elektronerne i Atomet, paa en saadan Maade, at Gruppeordningen af Grundstofferne i det periodiske System tilskrives disse Elektrongruppers gradvise Udformning med stigende Atomnummer. En Gruppedeling af Elektronerne i Atomet har da ogsaa været et Hovedtræk i alle nøjere udformede Forestillinger om Atombygningen, lige fra J. J. Thomsons berømte Forsøg paa en Fortolkning af det periodiske System paa Grundlag af en Undersøgelse af forskellige Elektronkonfigurationers Stabilitetsforhold. Skønt det Billede af Atomet, Thomson benyttede, ikke er foreneligt med vore siden vundne Erfaringer, har hans Arbejde paa Grund af de mange originale Tanker, der er nedlagt deri, øvet stor Indflydelse paa den senere Udvikling af Atomteorien.

Med Benyttelsen af de Oplysninger om Elektroners Binding af

I PERIODE
1 Brint
2 Helium

II PERIODE
3 Lithium
4 Beryllium
5 Bor
6 Kulstof
7 Kvælstof
8 Ilt
9 Fluor
10 Neon

III PERIODE
11 Natrium
12 Magnium
13 Aluminium
14 Silicium
15 Fosfor
16 Svovl
17 Klor
18 Argon

IV PERIODE
19 Kalium
20 Kalcium
21 Scandium
22 Titan
23 Vanadin
24 Krom
25 Mangan
26 Jern
27 Kobolt
28 Nikkel
29 Kobber
30 Zink
31 Gallium
32 Germanium
33 Arsen
34 Selen
35 Brom
36 Krypton

V PERIODE
37 Rubidium
38 Strontium
39 Yttrium
40 Zirkon
41 Niobium
42 Molybdæn
43 —
44 Ruthenium
45 Rhodium
46 Palladium
47 Sølv
48 Kadmium
49 Indium
50 Tin
51 Antimon
52 Tellur
53 Jod
54 Xenon

VI PERIODE
55 Cæsium
56 Barium
57 Lanthan
58 Cerium
59 Praseodym
60 Neodym
61 —
62 Samarium
63 Europium
64 Gadolinium
65 Terbium
66 Dysprosium
67 Holmium
68 Erbium
69 Thulium
70 Ytterbium
71 Lutetium
72 —
73 Tantal
74 Wolfram
75 —
76 Osmium
77 Iridium
78 Platin
79 Guld
80 Kvægsølv
81 Thallium
82 Bly
83 Vismut
84 Polonium
85 —
86 Niton

VII PERIODE
87 —
88 Radium
89 Actinium
90 Thorium
91 Protactinium
92 Uran

Atomkernen, som blev vundet gennem Teorien for Brintspektret, forsøgte Foredragsholderen allerede i samme Afhandling, hvori denne Teori fremsattes, at udkaste et Billede af Atombygningen i store Træk, idet det blev antaget, at enhver Elektron i Atomets Normaltilstand vilde bevæge sig paa en Maade, der var analog med Bevægelsen af en Elektron i det sidste Stadium af dennes Binding af en Kerne. Ligesom i den Thomsonske Teori antoges det endvidere, at Elektronerne bevægede sig i cirkulære Baner, og at Elektronerne indenfor de forskellige Grupper under denne Bevægelse indtog Stillinger i Forhold til hverandre som Vinkelspidserne i plane regulære Polygoner. En saadan Ordning er ofte betegnet som en Fordeling af Elektronerne i »Ringe«. Ved disse Antagelser viste det sig muligt at opnaa en Redegørelse for Størrelsesordenen af Atomernes Dimensioner, samt af den Fasthed, hvormed Elektronerne i Atomet er bundne, og for hvilken sidste man opnaar et Maal ved Forsøg over Frembringelse af de forskellige Typer af Spektre. En detailleret Forklaring af Stoffernes Egenskaber var man imidlertid paa denne Maade ikke i Stand til at opnaa, selv efter at man fra Moseleys Resultater og Sommerfelds Arbejder havde faaet den Oplysning, at det skitserede, simple Billede af Atomet, saaledes som blandt andre Vegard har fremhævet, maatte modificeres derhen, at der foruden Elektronbaner svarende til Normaltilstanden af Brintatomet, i det Indre af det færdigt dannede Atom forekom Elektronbaner med højere Kvantetal svarende til tidligere Stadier af Brintatomets Dannelse. Vanskeligheden ved paa denne Maade at opnaa et tilfredsstillende Billede af Atomet er paa det nøjeste knyttet sammen med Vanskeligheden ved at gøre Rede for den udprægede »Stabilitet«, som en Forklaring af Stoffernes Egenskaber kræver af Atombygningen. Som vi saa ved Omtalen af Spørgsmaalet om Brintatomets Dannelse, tager Indførelsen af Kvanteteoriens Postulater netop Sigte paa dette Punkt, men de Resultater, som paa den omtalte Maade blev opnaaet for et Atom, der kun indeholder een Elektron, tillader ikke umiddelbart at kaste Lys over et saadant Spørgsmaal som Gruppedelingen af Elektronerne i et Atom med flere Elektroner. Forestiller vi os som i det omhandlede simple Billede, at Elektronerne i Atomets Grupper i hvert enkelt Øjeblik er orienterede relativt til hverandre som Vinkelspidserne i regulære Polygoner, giver de nævnte Postulater — hvad enten vi tænker os de enkelte Elektroners Baner som cirkulære eller elliptiske — os saaledes ikke tilstrækkelige Holdepunkter til at belyse den eventuelle Forskel i Stabilitet hos Elektronkonfigurationer med forskellige Antal af Elektroner i Grupperne.

De af Stoffernes Egenskaber krævede, ejendommelige Stabilitetsforhold hos Atombygningen kommer klart frem ved en Betragtning saavel af Røntgenspektrene som af de kemiske Egenskaber, der, som allerede nævnt, maa antages at hænge nøje sammen med Gruppedelingen af Elektronerne i Atomet. Dette er paa interessant Maade blevet belyst gennem to betydningsfulde Arbejder af Kossel. I det første Arbejde viser han, paa hvilken Maade man nærmere kan forklare Røntgenstraalernes Fremkomst som knyttet til Gruppebygningen af Atomet. Han antager, at en Linie i Røntgenspektret kommer frem ved en Proces, hvorved, efter at en Elektron ved Indgreb udefra er fjernet fra Atomet, en Elektron i en af de andre Grupper indtager dennes Plads; en Pladsbytning, der antages at kunne finde Sted paa lige saa mange Maader, som der findes Grupper med løsere bundne Elektroner. Denne Opfattelse af Røntgenstraalernes Fremkomst førte til en simpel Forklaring af Røntgenstraalernes ejendommelige Absorptionsforhold, og har — ved at føre til Forudsigelsen af visse simple Relationer mellem Svingningstallene af Røntgenlinier for et og samme Stof — vist sig egnet til at danne Grundlag for en Udredelse af de i første Øjeblik meget indviklede Forhold, som det senere, indgaaende eksperimentelle Studium af Røntgenspektrenes Bygning, der navnlig skyldes Siegbahn, har bragt for Dagen. Til Trods for det store Arbejde, som en Række Forskere i Tilknytning til Sommerfelds grundlæggende Undersøgelser har udført paa dette Omraade, har det dog hidtil været vanskeligt fuldtud at drage Nytte af de Kosselske Forestillinger om Røntgenspektrenes Fremkomst. Som vi imidlertid i Slutningen af Foredraget skal se, synes det muligt udfra de ny Synspunkter, som vi skal lægge til Grund for Betragtningen af Atomernes Stabilitetsforhold, paa utvungen Maade at bringe de forskellige Resultater i naturlig Forbindelse med hverandre.

I det andet af Kossels ovennævnte Arbejder undersøger han Betingelsen for Forklaringen af det periodiske System udfra Atomteorien. Uden at gaa nærmere ind paa Spørgsmaalet om den dybere Aarsag til Gruppedelingen af Elektronerne eller de forskellige Elektronfigurationers forskellige Stabilitet fremhæver han i Tilslutning til Tanker, der allerede spillede en Rolle i Thomsons før omtalte Teori, hvorledes det periodiske System aflægger Vidnesbyrd om en periodisk Fremkomst af Elektronkonfigurationer af særlig stabil Natur. Disse Konfigurationer optræder i de neutrale Atomer hos Grundstofferne af de inaktive Luftarters Familie, der i den foranstaaende Tabel danner Afslutningen paa Systemets Perioder, og den omhandlede Stabilitet antages ikke alene at forklare disse Stoffers Mangel paa Evne til at indgaa kemiske Forbindelser, men ogsaa at forklare de karakteristiske aktive kemiske Egenskaber hos Grundstofferne i de Familier indenfor Systemet, der gaar forud for eller følger umiddelbart efter den nævnte Familie. Betragter vi f. Eks. en inaktiv Luftart som Argon, der har Atomnummer 18, skal vi antage, at de 18 Elektroner i Atomet danner en særlig regelmæssig Konfiguration med udpræget Stabilitet. Ser vi derimod paa det foregaaende Element, Klor, tænkes dets udtalte elektronegative Karakter forklaret derved, at det neutrale Atom, der jo kun indeholder 17 Elektroner, besidder en Tendens til at indfange en yderligere Elektron, hvorved der dannes en negativ Klorion, der indeholder en Elektronkonfiguration af 18 Elektroner med lignende Egenskaber som den, der optræder i det neutrale Argonatom. Paa den anden Side tænker vi os den udprægede elektropositive Karakter hos Kalium forklaret derved, at en af de 19 Elektroner i det neutrale Atom saa at sige er overflødig i de første 18 Elektroners Sammenspil, og at denne derfor let mistes under Dannelsen af en positiv Kaliumion af en lignende Konstitution som Argonatomet. Paa tilsvarende Maade gøres Rede for den elektronegative og elektropositive Karakter af saadanne Stoffer som Svovl og Kalcium med Atomnumrene 16 og 20, der dog i Modsætning til Klor og Kalium er divalente, idet der for at danne den stabile Konfiguration af 18 Elektroner for Svovlatomet kræves Optagelse af 2 Elektroner og for Kalciumatomet Fraspaltning af 2 Elektroner. Ved at udvikle disse Tanker nærmere har Kossel ikke alene formaaet at opnaa en overordentlig interessant Belysning af et stort Antal kemiske Erfaringer, men er ogsaa for Elementerne af de første Perioder i det periodiske System blevet ført til Antagelser om en Gruppedeling af Elektronerne i Atomet, der i formel Henseende frembyder Lighed med dem, som vi i det følgende skal omtale. Kossels Arbejde er siden paa interessant Maade blevet videreført af Ladenburg med særligt Henblik paa Gruppedelingen af Elektronerne i Atomerne hos Grundstofferne i de senere Perioder af det periodiske System. Som vi skal se, frembyder ogsaa Ladenburgs Resultater Lighedspunkter med dem, vi senere skal omtale.

Paa Spørgsmaalet om den dybere Aarsag til den udprægede Stabilitet af visse Elektronkonfigurationer fremfor andre har det hidtil ikke været muligt at opnaa et tilfredsstillende Svar ud fra en konsekvent Anvendelse af Kvanteteorien paa Kerneatomet. Dog har man i nogen Tid været klar over, at en Løsning maatte søges ved at undersøge Mulighederne for en rumlig Fordeling af Elektronernes Baner i Atomet i Stedet for at indskrænke sig til at betragte Konfigurationer, hvor alle Elektronerne indenfor en og samme Gruppe i Atomet bevæger sig i samme Plan, saaledes som det ved de første Forsøg paa at opnaa et Overblik over Atombygningen for Simpelheds Skyld blev antaget. Denne Nødvendighed af at antage rumlige Elektronkonfigurationer har været paapeget fra forskellige Sider. Born og Landé har i Tilknytning til Undersøgelser over Krystallernes Bygning og Egenskaber paapeget, hvorledes Antagelsen af saadanne Konfigurationer synes krævet for at gøre Rede for forskellige af Krystallernes Forhold, og Landé har forfulgt dette Spørgsmaal nærmere og har, som det vil blive omtalt i det følgende, foreslaaet forskellige saadanne »rumlige Atommodeller«, i hvilke Elektronerne i Atomets Grupper hvert Øjeblik indtager Stillinger relativt til hverandre, der udviser en lignende Symmetri som regulære Polyedre. Disse Modeller betyder i visse Henseender et væsentligt Fremskridt, selv om der dog derved ikke er naaet afgørende Resultater med Hensyn til Forstaaelsen af Atombygningens Stabilitetsforhold.

Endvidere har Lewis og Langmuir udfra Betragtninger over Stoffernes kemiske Egenskaber paapeget Betydningen af rumlige Elektronkonfigurationer. Saaledes har Lewis, der paa mange Punkter uafhængig har naaet de samme Resultater som Kossel, fremhævet, at Tallet 8, der karakteriserer Længden af de første Grupper i det periodiske System, kunde tænkes at antyde en Konstitution af Atomets Grupper, hvor Elektronerne var anbragt i Forhold til hinanden i Hjørnerne af en Kubus, og har fremhævet hvorledes en Konfiguration af denne Beskaffenhed fører til anskuelige Forestillinger om Bygningen af Molekyler af kemiske Forbindelser. Nu er dertil at sige, at en saadan »statisk« Elektronkonfiguration ikke vil være mulig, naar man antager, at Kræfterne i Atomet hidrører udelukkende fra Partiklernes elektriske Ladninger. Langmuir, der har forsøgt at udarbejde Lewis' Forestillinger videre og at gøre Rede ikke alene for Fremkomsten af de første Oktaver i det periodiske System, men ogsaa for de længere Perioder indenfor dette, tænker sig da ogsaa, at Atomernes Bygning og Stabilitet beherskes af Kræfter, hvis Natur vi ikke nøje kender. Han forestiller sig, at Atomet besidder en »Cellestruktur«, saaledes at der paa Forhaand er anvist enhver Elektron Plads i en Celle, og at disse Celler er ordnede i Lag paa en saadan Maade, at de forskellige Lag, regnet fra Atomkernen og udefter, netop indeholder lige saa mange Pladser som Perioderne i det periodiske System, optalt i Retning af stigende Atomnummer. Langmuirs Arbejder har tiltrukket sig stor Interesse indenfor Kemikernes Kreds, idet de har bidraget til at anskueliggøre de Begreber, hvormed den empiriske kemiske Videnskab opererer. Idet efter en saadan Teori Forklaringen af de forskellige Stoffers Egenskaber henføres til visse for dette Formaal opstillede Postulater angaaende de paagældende Atomers Bygning, adskiller den sig dog paa Forhaand principielt fra Bestræbelserne for at forklare Stoffernes specifike Egenskaber ved Hjælp af almindelige Love, gældende for Vekselvirkningen mellem Partiklerne i ethvert Atom. Min Hovedopgave i dette Foredrag skal imidlertid være at forsøge at vise, hvorledes Arbejdet paa det Program, der udtrykkes i disse Bestræbelser, ingenlunde synes haabløst, men hvorledes det tvertimod synes muligt ud fra en konsekvent Benyttelse af Kvanteteoriens almindelige Postulater at opnaa en Indsigt i Atomernes Bygning og Stabilitetsforhold, naar man — støttet paa disse Postulater — angriber dette Problem ved at stille sig Spørgsmaalet: Hvordan kan et Atom tænkes dannet ved Indfangning og Binding af Elektronerne en for en i det Kernen omgivende Kraftfelt?

* * *

Før vi forsøger at besvare dette Spørgsmaal, maa vi dog først se lidt nærmere paa, hvad Kvanteteorien lærer os om Bindingsprocessens almindelige Karakter. For Brintatomets Vedkommende har vi allerede set, hvorledes Brintspektret tør antages at give bestemte Oplysninger om Forløbet af den Proces, hvorved Elektronen bindes af Kernen. Spørger vi om Dannelsen af andre Stoffers Atomer, har vi ogsaa i disses Spektre Kilder til Oplysning om Dannelsesprocessen, men de direkte Oplysninger, som vi faar her, er i Almindelighed ikke saa udtømmende som ved Brintatomet. For et Stof med Atomnummer N kan vi nemlig tænke os Dannelsesprocessen som forløbende i N Afsnit, svarende til den successive Indfangning i Kernens Felt af N Elektroner. Til hver af disse Indfangningsprocesser maa der tænkes at svare et Spektrum; men kun for de to første Elementer, Brint og Helium, besidder vi udtømmende Kendskab til disse Spektre. For de andre Stoffers Vedkommende, hvor flere Spektre vilde være knyttet til Dannelsesprocessen, kender vi hidtil i intet Tilfælde mere end to Typer af saadanne Spektre, der ofte benævnes som henholdsvis »Buespektre« og »Gnistspektre«, med Henblik paa de Forhold, hvoraf deres eksperimentelle Iagttagelse afhænger. Selv om disse Spektre viser en betydelig mere kompliceret Struktur end Brintspektret, givet ved Formel (2) og det ved Formel (7) udtrykte Heliumspektrum, har det i mange Tilfælde været muligt at finde simple Love for Svingningstallene, der besidder en nøje Analogi med de ved disse Formler udtrykte Love.

Ser vi for Simpelheds Skyld bort fra den Struktur, som Linierne i de fleste Spektre udviser (Fremkomst af Dubletter, Tripletter o. s. v.), kan saaledes Svingningstallene for Linierne af mange B u e s p e k t r e med stor Tilnærmelse fremstilles paa den saakaldte R y d b e r g s k e Formel

$$\nu = \frac{K}{(n'' + \alpha_{k''})^2} - \frac{K}{(n' + \alpha_{k'})^2}, \qquad (12)$$

hvor n' og n'' er hele Tal og K den samme Konstant som i Brintspektret, medens $\alpha_{k'}$ og $\alpha_{k''}$ er to blandt et Sæt af Konstanter, karakteristiske for det betragtede Stof. Et Spektrum med en Bygning af denne Art kaldes ligesom Brintspektret et S e r i e s p e k t r u m, idet

2

Linierne kan ordnes i Rækker, de saakaldte »Spektralserier«, indenfor hvilke Svingningstallene konvergerer til bestemte Grænseværdier. Disse Rækker fremstilles f. Eks. ved Formel (12), dersom man ved Benyttelsen af to bestemte Konstanter for $\alpha_{k''}$ og $\alpha_{k'}$ lader n'' forblive uforandret, medens n' efter hinanden antager en Række stadig voksende Værdier.

Som nævnt gælder Formel (12) kun med Tilnærmelse, men altid finder man, at Svingningstallene for Spektrenes Linier ligesom i Formlerne (2) og (12) kan skrives som en Differens mellem to Funktioner af hele Tal. Saaledes gælder den sidste Formel nøjagtigt, dersom Størrelserne α_k ikke betragtes som Konstanter, men som Repræsentanter for et Sæt af Talrækker $\alpha_k(n)$, karakteristiske for det betragtede Stof, hvis Værdier for voksende n indenfor hver enkelt Række hurtigt nærmer sig en konstant Grænseværdi. Den Omstændighed, at Svingningstallene for Spektrene altid optræder som Differens mellem to Led, de saakaldte »Spektraltermer«, af hvis Kombinationer det fuldstændige Spektrum fremgaar, er navnlig blevet fremhævet af Ritz, der ved Opstilling af det saakaldte Kombinationsprincip i høj Grad har befordret Studiet af Spektrene. For dette Princip tilbyder nu Kvanteteorien en umiddelbar Fortolkning, idet vi efter Frekvensbetingelsen ganske som i Brintspektret føres til at opfatte Linierne som fremkommende ved Overgange mellem stationære Tilstande af Atomet, blot at vi for de andre Stoffers Spektre ikke har at gøre med en enkelt Række stationære Tilstande, men med et Sæt af saadanne Rækker. Fra Formel (12) faar vi saaledes for et Buespektrum, idet vi foreløbig ser bort fra de enkelte Liniers Struktur, Oplysning om en Samling stationære Tilstande, for hvilke Energien af Atomet i den $n$te Tilstand i den $k$te Række kan angives ved

$$E_k(n) = -\frac{Kh}{(n+\alpha_k)^2} \tag{13}$$

i nøje Lighed med den simple Formel (3) for Energien i Brintatomets stationære Tilstande.

Hvad Gnistspektrene, hvis Struktur navnlig er klarlagt igennem Undersøgelser af Fowler, angaar, har det været muligt for mange Stoffers Vedkommende med Tilnærmelse at udtrykke Svingningstallene ved en Formel af ganske samme Type som (12), kun med den Forskel, at Konstanten K, ligesom i det ved Formel (7) givne Heliumspektrum, er ombyttet med en Konstant, der er 4 Gange saa stor. For Gnistspektrene vil derfor Energiværdierne i de tilsvarende stationære Tilstande af Atomet være givet ved et Udtryk af samme Type som (13), kun med den Forskel, at K er ombyttet med 4 K.

Denne ejendommelige Lighed mellem Strukturen af de nævnte Typer af Spektre og de simple Spektre givet ved (2) og (7) fortolkes simpelt ved at antage, at Buespektrene er knyttet til det sidste Afsnit af Dannelsen af det neutrale Atom, bestaaende af Indfangningen og Bindingen af den *Nte* Elektron, medens Gnistspektrene er knyttet til det næstsidste Afsnit af Atomets Dannelse, nemlig Bindingen af den $(N - 1)$te Elektron. I disse Tilfælde vil nemlig den paagældende Elektron — i det mindste paa de tidligere Stadier af Bindingsprocessen, hvor Elektronen under største Del af sit Omløb bevæger sig i Afstande fra Kernen, der er store i Forhold til Dimensionerne af de tidligere bundne Elektroners Baner — bevæge sig i et Kraftfelt stammende fra Kernen og de $N - 1$ eller $N - 2$ tidligere indfangne Elektroner, der vil være meget nær det samme som det, der omgiver Kernen i et Brint- eller Heliumatom. Analogt med hvad der gjaldt for Formel (3) vedrørende de stationære Tilstande af Brintatomet, skal vi saaledes tænke os, at den numeriske Værdi af Udtrykket paa højre Side af (13) vil være lig det Arbejde, der kræves for fra Atomet at fjerne den sidst indfangne Elektron, til hvis Binding Udsendelsen af det omhandlede Stofs Buespektrum er knyttet.

Medens disse Forhold naturligt fortolkedes ud fra Foredragsholderens simple Teori for Brintspektret, gav først S o m m e r f e l d s Teori for Brintliniernes Finstruktur en klar Forstaaelse af den karakteristiske Forskel mellem Brintens Spektrum og Heliums Gnistspektrum paa den ene Side og de andre Stoffers Bue- og Gnistspektre paa den anden Side. Naar vi betragter Bindingen ikke af den første, men af de paafølgende Elektroner i et Atom, vil nemlig — i det mindste for de senere Stadier af Bindingsprocessens Vedkommende, hvor den sidst fangne Elektron under sit Omløb kommer i indgribende Vekselvirkning med de tidligere indfangne Elektroner, — den paagældende Elektrons Bane ikke mere med stor Tilnærmelse være en lukket Ellipse, men derimod i første Tilnærmelse en Centralbane af

2*

lignende Type som den vi allerede møder i Brintatomet, naar vi tager Elektronmassens Foranderlighed med Hastigheden i Betragtning; kun vil den overlejrede Rotation her være forholdsvis langt hurtigere og den periodiske Banes Afvigelse fra en Ellipse langt større. For en Bane af denne Type vil de stationære Tilstande ligesom i Teorien for Finstrukturen være bestemt ved to Kvantetal, som vi stadig skal betegne ved n og k, og som paa simpelt angivelig Maade hænger sammen med Banens kinematiske Egenskaber. Jeg skal for det følgendes Skyld blot nævne, at medens Kvantetallet k i alle Tilfælde paa den tidligere angivne simple Maade er forbundet med Værdien af Elektronens konstante Impulsmoment omkring Centret, kræver Fastsættelsen af Hovedkvantetallet n en Undersøgelse af Banens hele Forløb og vil for en vilkaarlig Centralbane ikke hænge simpelt sammen med den roterende periodiske Banes Dimensioner, dersom denne afviger væsentlig fra en Keplerellipse.

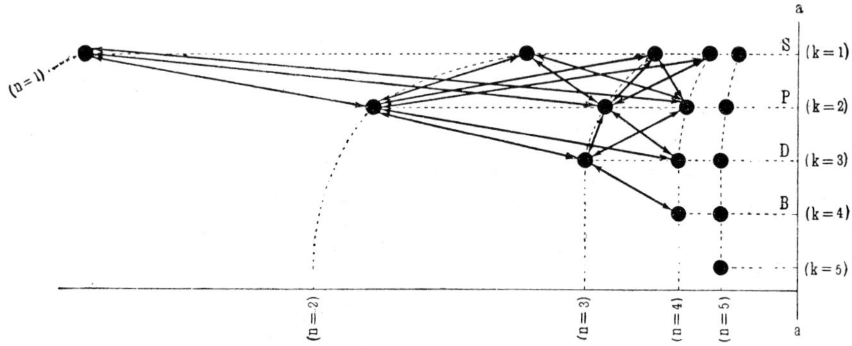

Skematisk Oversigt over Natriumspektrets Fremkomst.

De omtalte Resultater er skematisk fremstillet paa Figuren, der er en Gengivelse af en Tavle, som jeg har benyttet ved en tidligere Lejlighed, og som giver en Oversigt over Forholdene ved Natriumspektrets Fremkomst. De sorte Punkter repræsenterer de stationære Tilstande svarende til forskellige Rækker af Spektraltermer, kendetegnede ved de til højre staaende Bogstaver *S, P, D* og *B*, svarende til de i den spektroskopiske Litteratur sædvanlig benyttede Betegnelser, der tager Sigte paa Egenskaber hos de til Termernes Kombinationer knyttede Spektralserier (skarpe Serie, principale Serie, diffuse Serie o. s. v.). De enkelte Punkters Afstand fra den lodrette Linie *aa* yderst til højre paa Figuren er proportionale med den numeriske Værdi for Atomets Energi, udtrykt ved Formel (13). De skraa, sorte Pile antyder endelig de Overgange mellem stationære Tilstande, der giver Anledning til Fremkomsten af Linier i det sædvanligt iagttagne Natriumspektrum. De til de forskellige Tilstande føjede Værdier for n og k er ment som angivende de Kvantetal, der efter Sommerfelds Teori ud fra en foreløbig Betragtning kunde tænkes at kendetegne den ydre Elektrons Bane. For Oversigtens Skyld er de Tilstande, der er tænkt at svare til samme Værdi for n, forbundne ved de punkterede Kurver, der er dragne saaledes, at deres lodrette Asymptoter svarer til saadanne Termer i Brintspektret, der hører til samme Værdi for Hovedkvantetallet. Kurvernes Forløb er tænkt at illustrere, hvorledes Afvigelsen fra Brinttermerne maa ventes at aftage med voksende Værdi af k, svarende til Tilstande, hvor den mindste Afstand, hvortil Elektronen under sit Omløb nærmer sig Kernen, bliver stadig større.

Ved Betragtning af Figuren maa det straks bemærkes, at selv om Teorien lader Hovedtrækkene i Spektrenes Struktur paa tydelig Maade komme for Dagen, har det ikke hidtil været muligt ved Undersøgelser af Elektronbaner, der kan optræde i et simpelt Kraftfelt med Centralsymmetri, at gøre Rede for noget Stofs Spektrum i Enkeltheder. Jeg tænker ikke her saa meget paa Spørgsmaalet om de enkelte Spektralliniers sammensatte Bygning, der f. Eks. for Natriumspektrets Vedkommende kommer til Syne ved Optræden af to nær sammenfaldende Spektraltermer for hvert Led i P- og D-Rækkerne. Dette Forhold, som vi skal komme tilbage til i det følgende, skyldes utvivlsomt mindre Afvigelser fra Centralsymmetri af det fra det indre System stammende Kraftfelt, der vil bevirke, at Banen ikke nøjagtig bliver en Centralbane, men har en noget mere indviklet Karakter, og derfor i de stationære Tilstande maa karakteriseres ved mere end to Kvantetal, paa analog Maade som Optræden af Afvigelser i Elektronbanen i Brintatomet fra en simpel periodisk Bane som nævnt bevirker, at de stationære Tilstande af dette Atom maa karakteriseres ved mere end eet Kvantetal. Det, der i første Række er af Betydning for den følgende Diskussion af Atomernes Dannelse, er imidlertid den entydige Fastlæggelse af de to for Beskrivelsen af en simpel Centralbane fordrede

Kvantetal n og k, der svarer til de forskellige stationære Tilstande, om hvis Eksistens Spektrene lærer os. Allerede ved Fastlæggelse af disse Tal er man nemlig stødt paa Vanskeligheder af dybtliggende Art, der — som vi skal se — paa det nøjeste er knyttet til Spørgsmaalet om Atombygningens ejendommelige Stabilitetsforhold. Jeg skal her kun bemærke, at de Værdier for Kvantetallet n, der er angivne paa Figuren, utvivlsomt ikke kan opretholdes, hverken for S-Rækkens eller for P-Rækkens Vedkommende. Hvad derimod de benyttede Værdier for Kvantetallet k angaar, kan man med Sikkerhed sige, at den Tydning af Egenskaberne ved Banerne, som de angiver, er rigtig. Et Støttepunkt for Undersøgelsen af dette Spørgsmaal har man opnaaet ved Betragtninger af en væsentlig anden Karakter end dem, jeg hidtil har haft Lejlighed til at omtale, og som har gjort det muligt at opnaa en Forbindelse mellem Bevægelsen i Atomet og Fremkomsten af Spektrets Linier.

Hvad Kvanteteoriens P r i n c i p p e r angaar, har det hidtil været den gennemgribende Afvigelse fra vore sædvanlige Forestillinger om mekaniske og elektrodynamiske Fænomener, som det har haft Betydning at fremhæve. Som Foredragsholderen gennem Arbejder i de sidste Aar har forsøgt at vise, synes det imidlertid muligt for Anvendelsen af Kvanteteorien at lægge et Synspunkt til Grund, der holder os det Haab for Øje, ikke desto mindre at lade denne Teori fremtræde som en rationel Udvidelse af vore sædvanlige Forestillinger. Som det fremgaar af Kvanteteoriens Postulater og specielt af Frekvensbetingelsen, er en direkte Forbindelse mellem Spektrene og Bevægelsen som den, den klassiske elektrodynamiske Teori vilde fordre, paa Forhaand udelukket, men samtidig leder disse Postulaters Form os paa Sporet efter en anden Forbindelse af ejendommelig Karakter. Dersom vi betragter et elektrodynamisk System og ud fra vore sædvanlige Forestillinger spørger om Arten af den Straaling, som Systemets Bevægelse vilde foraarsage, tænker vi os Bevægelsen opløst i saakaldte rene harmoniske Svingninger og antager, at Straalingen vil bestaa i den samtidige Udsendelse af et Antal elektromagnetiske Bølgesystemer, der besidder samme Svingningstal som disse harmoniske Komponenter og Intensiteter, der afhænger af disse Komponenters Amplituder. En nærmere Undersøgelse af Kvanteteoriens formelle Grundlag viser nu, at det er muligt at føre Spørgsmaalet om Fremkomsten af de Straalingsprocesser, der ledsager de forskellige Overgange mellem et Atoms stationære Tilstande tilbage til en Undersøgelse af de forskellige harmoniske Svingninger, der optræder i Atomets Bevægelse, saaledes at Muligheden for, at en bestemt Overgang skal kunne finde Sted, kan anses for at være betinget af Optræden i Bevægelsen af en bestemt angivelig »korresponderende« Svingningskomponent. Dette Synspunkt tillader samtidig at kaste Lys over det i det foregaaende flere Gange berørte Spørgsmaal om Sammenhængen mellem Antallet af de Kvantetal, der maa benyttes til Beskrivelsen af et Atoms stationære Tilstande, og Typen af Elektronernes Baner, hvis Klassificering simplest føres tilbage netop til en Opløsning af Bevægelsen i harmoniske Svingningskomponenter. Tiden tillader mig imidlertid ikke at komme nærmere ind paa dette sidste Spørgsmaal, og jeg skal her væsentlig indskrænke mig til at omtale de Slutninger, som det antydede K o r r e s p o n d e n s p r i n c i p tillader at drage angaaende Fremkomsten af Overgange mellem forskellige Par af stationære Tilstande, og som er af afgørende Betydning for det følgende.

Det simpleste Eksempel paa en saadan Slutning faar vi, dersom vi betragter et Atomsystem, der indeholder en Partikel, som beskriver en ren periodisk Bane, og hvor som nævnt de stationære Tilstande vil være karakteriserede ved et enkelt Kvantetal n. I dette Tilfælde kan Bevægelsen efter F o u r i e r s Teorem opløses i en enkelt Række harmoniske Svingninger, hvis Svingningstal kan skrives $\tau\omega$, hvor τ er et helt Tal og ω er Omløbstallet i Banen. Det viser sig nu, at en Overgang imellem to stationære Tilstande, for hvilken Værdierne for Kvantetallet er lig henholdsvis n' og n'', vil korrespondere med en harmonisk Svingningskomponent, for hvilken $\tau = n' - n''$. Dette tillader at kaste Lys over den ejendommelige Forskel, der findes imellem Mulighederne for Overgange mellem de stationære Tilstande af et Brintatom paa den ene Side og af et simpelt System bestaaende af en elektrisk Partikel, der kan udføre rent harmoniske Svingninger omkring en Ligevægtsstilling, paa den anden Side. For det sidste System, der ofte kaldes en P l a n c k s Oscillator, bestemmes Energien i de stationære Tilstande ved den velkendte Formel $E = nh\omega$, og ved Anvendelse af Frekvensbetingelsen faar vi derfor for den Straaling, der vilde udsendes ved en Overgang mellem to stationære Tilstande,

$\nu = (n' - n'') \omega$. Nu er det imidlertid ikke alene en væsentlig Antagelse i **Plancks** Varmestraalingsteori, men ogsaa en Antagelse der synes krævet for at gøre Rede for Molekylers Absorption i det ultrarøde Straalingsomraade, at en saadan harmonisk Oscillator kun vil emittere og absorbere en Straaling, for hvilken Svingningstallet ν er lig Oscillatorens Svingningstal ω. Vi tvinges derfor til at antage, at for Oscillatoren kan Overgange kun finde Sted mellem stationære Tilstande, karakteriserede ved umiddelbart paa hinanden følgende Kvantetal, medens det, som vi saa, for Forklaringen af Brintspektret givet ved Formel (2) netop var en væsentlig Betingelse, at alle mulige Overgange mellem de stationære Tilstande, for hvilke Atomets Energi var givet ved Formel (5), kunde finde Sted. Ud fra det omhandlede Korrespondenssynspunkt ser vi imidlertid, hvorledes denne tilsyneladende Vanskelighed umiddelbart fortolkes derigennem, at der i den elliptiske Bevægelse af Elektronen i Brintatomet, til Forskel fra Oscillatorens Bevægelse, findes Svingningskomponenter svarende til Værdier af τ forskellig fra 1; eller ved Benyttelse af en velkendt Betegnelse fra Lydlæren, at der i Brintatomets Bevægelse findes Overtoner.

Et andet simpelt Eksempel paa Anvendelsen af Korrespondensprincippet faar vi nu netop i Strukturen af de ovenfor betragtede Seriespektre. Betragter vi igen Figuren med den skematiske Fremstilling af Natriumspektrets Fremkomst, ser vi, at de sorte Pile, der svarer til de Spektrallinier, der fremkommer under sædvanlige Udladningsbetingelser, kun forbinder Punktpar i hinanden følgende Rækker. Denne ejendommelige Begrænsning i Optræden af Kombinationer mellem Spektraltermerne viser sig nu at kunne fortolkes naturligt ud fra en Undersøgelse over de harmoniske Svingningskomponenter, der optræder i en Centralbevægelse af den beskrevne Art. En saadan Bevægelse kan nemlig, som man let viser, opløses i to Rækker af harmoniske Svingningskomponenter, hvis Svingningstal kan udtrykkes henholdsvis ved $\tau\omega + \sigma$ og $\tau\omega - \sigma$, hvor τ er et helt Tal, og hvor ω er Omløbstallet i den roterende periodiske Bane, medens σ er Omløbstallet for den overlejrede Rotation. Disse Svingningskomponenter korresponderer netop med Overgange, hvor Hovedkvantetallet n aftager med τ Enheder, medens Kvantetallet k henholdsvis aftager og vokser med 1 Enhed, altsaa netop med saadanne Overgange, der svarer til de sorte Pile paa Figuren. En yderligere Støtte og interessant Uddybning af denne Betragtning kan man endvidere vinde gennem en Fortolkning af Iagttagelser, gjort af Stark og hans Medarbejdere, over Virkningen af elektriske Felter paa Seriespektre.

For Brintspektrets Vedkommende bestaar, som bekendt, den af Stark opdagede Virkning af et elektrisk Felt i en Opløsning af de enkelte Linier i et Antal polariserede Komponenter. Jeg beklager, at Tiden her ikke tillader mig nærmere at omtale den indgaaende Fortolkning af denne Starkeffekt paa Brintlinierne, som det, kort efter Fremkomsten af Sommerfelds Teori for Brintliniernes Finstruktur, lykkedes Epstein og Schwarzschild at give, og som ikke alene danner en af de smukkeste Anvendelser af Kvanteteorien paa Atomspørgsmaal, men som ogsaa har bidraget væsentlig til denne Teoris nøjere Udformning. Jeg skal blot nævne, at den Fastlæggelse af stationære Tilstande, som det her drejer sig om, er væsentlig forskellig fra den, der benyttes i Teorien for Finstrukturen, idet de Perturbationer, som Elektronens næsten periodiske Bane undergaar under de ydre elektriske Kræfters Indflydelse, har en ganske anden Karakter end den langsomme Drejning af Banen, der som Følge af Elektronmassens Afhængighed af Hastigheden optræder i det uforstyrrede Atom, men som ganske undertrykkes af de elektriske Kræfter, for hvilke Atomerne udsættes i Starks Forsøg. Dette bevirker, at det Kvantetal, der — foruden Hovedkvantetallet, som i første Tilnærmelse bestemmer Atomets Energi og den perturberede Banes Storakse — optræder ved Fastsættelsen af de stationære Tilstande, har en ganske anden Betydning end det Kvantetal k, der optræder i Teorien for Finstrukturen. Dette Punkt har det været muligt at belyse ved Hjælp af Korrespondensprincippet, der, som Foredragsholderen har vist, tillader en simpel Fortolkning af Epsteins og Schwarzschilds Resultater ud fra en direkte Undersøgelse af Baneperturbationerne. I Forbindelse med Starkeffekten paa Brintlinierne maa det dog navnlig nævnes, at det har været muligt at benytte dette Princip til, som udført af Kramers, i Enkeltheder at gøre Rede for den ejendommelige Intensitetsfordeling, hvormed Opspaltningskomponenterne optræder, ud fra en Undersøgelse af Amplituderne af de med disse Komponenter korresponderende harmoniske Svingninger i det perturberede Atoms Bevægelse. For andre Stoffers Spektre møder vi væsentlig andre Forhold, fordi her som oftest de ydre Kræfter ikke vil være i Stand til fuldstændig at forstyrre Centralbanekarakteren af den paagældende Elektrons Bevægelse, men kun til at deformere denne Bevægelse i ringe Grad. Derigennem opnaar man en umiddelbar Fortolkning af den svage Virkning, som, ifølge Forsøgene, elektriske Kræfter har paa de fleste af Spektrenes Linier. Hvad der imidlertid i Sammenhæng med de ovenstaaende Betragtninger er af særlig Interesse, er den Iagttagelse, at der i et Spektrum som Natriumspektret under de omhandlede Forhold optræder helt nye Spektralserier svarende til andre Overgange mellem stationære Tilstande end de, der er angivet paa Figuren. Dette er netop, hvad vi ud fra Korrespondensprincippet maatte vente, fordi der i Elektronernes Bevægelse som Følge af de Perturbationer, som de ydre elektriske Kræfter frembringer, foruden de Svingningskomponenter, der optræder i en uforstyrret Centralbane, i Banen vil optræde nye Typer af Svingningskomponenter med Amplituder proportionale med Intensiteten af de ydre Kræfter.

Jeg beklager ligeledes, at Tiden ikke tillader mig at komme nærmere ind paa Spørgsmaalet om Virkningen af magnetiske Felter paa Spektrenes Linier, der jo blev opdaget af Zeeman for netop 25 Aar siden. Jeg skal blot lige nævne, at det har været muligt ved Hjælp af Korrespondensprincippet at uddybe den af Sommerfeld og Debye paa Kvanteteoriens Grund opbyggede Teori for den »normale« Zeemaneffekt paa en saadan Maade, at Teorien kan siges at udvise saa stor en Lighed med den berømte, af Lorentz umiddelbart efter Zeemans Opdagelse givne Teori, som det overhovedet kan ventes, naar man holder sig den store Forskel mellem den klassiske elektromagnetiske Teori og Kvanteteorien for Øje. Jeg skal imidlertid kort omtale et Spørgsmaal, der har Betydning for det følgende, og som har en nøje Tilknytning til Spørgsmaalet om Virkningen af magnetiske Kræfter paa Spektrallinier, nemlig den sammensatte Bygning, som mange Seriespektres Linier udviser, og som vi allerede ovenfor har hentydet til. Denne saakaldte Komplexstruktur af Spektrenes Linier maa tænkes at hidrøre fra en ringe Afvigelse fra Centralsymmetri hos de Kræfter, som den sidst indfangne Elektron er underkastet under dens Binding i Atomet. En saadan Afvigelse, der maa tænkes at have sin Grund i Konfigurationen af de tidligere bundne Elektroners Bane, vil nemlig bevirke, at den paagældende Elektrons Bevægelse ikke vil foregaa i en fast Plan, men at Baneplanen stadig vil ændre sin Stilling paa en saadan Maade, at Bevægelsen kan beskrives som en Centralbevægelse, paa hvilken der er overlejret en jævn, langsom Drejning om en fast Akse, af lignende Art som den Drejning, som ifølge Larmors Teorem efter den klassiske elektromagnetiske Teori vilde blive overlejret et Atoms Bevægelse under Indvirkning af et homogent ydre Magnetfelt. Foruden de to Kvantetal n og k, som med stor Tilnærmelse vil fastlægge Elektronens Bevægelse i den roterende Centralbane i Atomets stationære Tilstande, svarende til Stadier af den sidst indfangne Elektrons Binding, vil der i Fastlæggelsen af disse Tilstande derfor tillige indgaa et tredie Kvantetal j, der vil bestemme Baneplanens Orientering relativt til de tidligere bundne Elektroners Konfiguration paa en saadan Maade, at Atomets resulterende Impulsmoment multipliceret med 2π vil være lig et helt Multiplum af Plancks Konstant. I Forbindelse med Spørgsmaalet om Forklaringen af Spektralliniernes Komplexstruktur har Sommerfeld paapeget Muligheden for en formel Fortolkning af Iagttagelserne ved Indførelsen af et tredie ikke nærmere defineret »indre« Kvantetal ved de stationære Tilstandes Fastsættelse. Ved at forfølge denne Tanke nærmere har Landé vist, hvorledes Indførelsen af et tredie Kvantetal med den angivne Betydning aabner Mulighed for at udvikle en formel, paa Frekvensbetingelsen bygget Teori for den saakaldte »anomale« Zeemaneffekt, som Komponenterne af de omhandlede komplexe Serielinier udviser, og hvis »Type«, som bekendt, viser nøje Forbindelse med Karakteren af Liniernes Struktur. Spørger vi nu om den Indflydelse, som en Afvigelse af de indre Elektroners Konfiguration fra simpel radial Symmetri vil have paa Fremkomsten af de forskellige Typer af Overgange mellem stationære Tilstande, finder vi ved Anvendelse af Korrespondensprincippet, at den omhandlede Virkning paa den ydre Elektrons Bane ikke kan ventes at give Anledning til Fremkomst af nye Typer af Overgange, forsaavidt det drejer sig om Ændringer i Kvantetallene n og k. I Modsætning til de Perturbationer, der f. Eks. frembringes af et ydre elektrisk Felt, vil den omhandlede, overlejrede Drejning af Banen nemlig ikke frembringe nye Svingningskomponenter, men blot bevirke en Opspaltning af hver af Komponenterne i Centralbevægelsen i et Antal Svingninger med nær sammenfaldende Svingningstal, der vil korrespondere med et Antal Overgange mellem stationære Tilstande af Atomet, for hvilke n og k forandrer sig paa samme Maade, medens Kvantetallet j ændrer sig paa forskellig Maade. Vi forstaar derfor, at den Afvigelse fra Centralsymmetri i Atombygningen, der giver Anledning til Komplexstrukturen af Seriespektrenes Linier, i Lighed med Virkningen af ydre magnetiske Felter, men i Modsætning til Virkningen af ydre elektriske Felter, ikke giver Anledning til Fremkomsten af andre Typer af Overgange mellem stationære Tilstande end de, der kan ventes for en ren Centralbevægelse. [Tilføjelse efter Foredraget: I et netop udkommet Hefte af det hollandske Tidsskrift »Physica«, udgivet til Minde om Zeemans Opdagelse, har Paschen og Back meddelt den interessante Iagttagelse, at der i et Magnetfelt fremkommer nye Komponenter i Serieliniernes Struktur, svarende til Overgange, hvor det tredie Kvantetal ændrer sig paa anden Maade end for det uforstyrrede Atom. For dette faar man efter Korrespondensprincippet en simpel Fortolkning, naar man antager, at Karakteren af de Ændringer, som den ydre Elektrons Baneplan i det uforstyrrede Atom vil undergaa, vil ændres af et Magnetfelts Tilstedeværelse paa anden Maade end gennem en simpel overlejret Drejning af den Art, som den klassiske Elektrodynamik efter Larmors Teorem vilde kræve. At et saadant Forhold tør betragtes som den egentlige Aarsag til Zeemaneffektens »Anomalitet« synes ogsaa tydeligt at fremgaa af den af Paschen og Back i et tidligere, betydningsfuldt Arbejde gjorte Opdagelse, at Zeemaneffekten paa Spektrallinier af sammensat Bygning, med voksende Intensitet af de magnetiske Kræfter, undergaar en gradvis Forandring paa en saadan Maade, at Opspaltningsbilledet mere og mere nærmer sig til en normal Zeemaneffekt paa en Spektrallinie af usammensat Bygning. En saadan Virkning maa nemlig ventes at indtræde, naar det ydre magnetiske Felts Indflydelse paa Elektronbanens Karakter er af samme Størrelsesorden som den fra de indre Elektroners Konfiguration stammende Afvigelse fra en plan Centralbevægelse; thi det vil simpelt forstaas, at denne sidste Afvigelse under saadanne Omstændigheder ikke vil kunne gøre sig gældende som bestemmende for Elektronbanens Konfiguration relativt til det indre System].

I Forbindelse med disse Spørgsmaal tør det være af Interesse at gøre opmærksom paa, at en Undersøgelse af Begrænsningen for Overgangsmuligheder mellem stationære Tilstande, baseret paa en simpel Betragtning af Ændringen i Atomets Impulsmoment under Straalingsprocessen, i Modsætning til hvad man tidligere har ment, ikke er egnet til at kaste Lys over den ved Figuren illustrerede, ejendommelige Struktur af Seriespektrene. I Tilknytning til de ovennævnte Arbejder af Sommerfeld og Landé maa vi nemlig antage, at »Komplexiteten« af Spektraltermerne, svarende til givne Værdier for n og k, der kommer til Syne i Finstrukturen af Spektrenes Linier, maa tilskrives Tilstande, svarende til forskellige Værdier for dette Impulsmoment, i hvilke Elektronbanens Plan er orienteret paa forskellig Maade relativt til Konfigurationen af de tidligere bundne Elektroner i Atomet. Betragtninger af Ændringen af Impulsmomentet under Straalingsprocessen — saaledes som de er fremdragne af Foredragsholderen til Støtte for visse Slutninger fra Korrespondensprincippet og, uafhængig af dette Princip, samtidig er udviklede af Rubinowicz — kan derfor ikke belyse den Begrænsning af Kombinationsmulighederne, der giver sig Udtryk i den omhandlede simple Struktur af Seriespektrene, og den Ændring, som denne Struktur undergaar ved ydre Kræfters Indflydelse. Saadanne Betragtninger, der er af betydelig principiel Interesse, kan i Sammenhæng med Seriespektrene kun bidrage til en Forstaaelse af den Begrænsning i Kombinationsmuligheder, der giver sig til Kende i de ejendommelige Love, der gælder for det Komponenttal, hvormed de enkelte Spektrallinier optræder. Hvad dette sidste Spørgsmaal angaar, tilbyder Betragtninger over Impulsmomentets Ændring under Straalingsprocessen en direkte Støtte for de ovenfor antydede Konsekvenser af Korrespondensprincippet.

* *
*

I det følgende skal vi se, hvorledes den for Seriespektrene paaviste Korrespondens mellem Bevægelsen af den sidst indfangne Elektron og Fremkomsten af Overgange mellem de til forskellige Stadier af Bindingsprocessen svarende stationære Tilstande giver os et Udgangspunkt for at vælge imellem de i første Øjeblik overordentlig righoldige Muligheder, der frembyder sig ved en Betragtning af Atomernes Dannelse ved successiv Indfangning og

Binding af Elektronerne. Vi ledes nemlig til blandt de tænkelige Processer, der efter Kvanteteoriens Postulater kunde ventes at finde Sted i Atomet, at udskyde saadanne, hvis Optræden ikke kan anses for forenelig med en Korrespondens af den krævede Beskaffenhed.

Ved Betragtning af Atomernes Dannelse behøver vi ikke at beskæftige os længe med Spørgsmaalet om den første Elektrons Binding. Som det fremgaar af det foregaaende, maa vi antage, at Slutresultatet af Bindingsprocessen vil være en Tilstand, hvor Atomets Energi er givet ved Formlen (5), naar vi sætter $n = 1$, eller nøjere ved Formlen (11), naar vi sætter $n = 1$ og $k = 1$, samt at Elektronens Bane vil være en Cirkel, hvis Dimensioner er givne ved Formel (10), dersom vi sætter $n = 1$ og $k = 1$. En saadan Bane skal vi betegne som en 1-kvantet Bane, idet vi i Almindelighed skal betegne en Bane, for hvilken Hovedkvantetallet har en given Værdi n som en n-kvantet Bane. For Anvendelser, hvor det gælder om at skelne imellem Baner svarende til forskellige Værdier af Kvantetallet k, skal vi betegne en Centralbane karakteriseret ved givne Værdier for Kvantetallene n og k som en n_k Bane.

Betragter vi Bindingen af den anden Elektron, møder vi straks et langt mere indviklet Spørgsmaal. Oplysninger om Bindingsprocessen kan vi imidlertid faa ud fra Heliums Buespektrum. Dette bestaar i Modsætning til de fleste andre simple Spektre ikke af eet, men af to fuldstændige Systemer af Linier, hvis Svingningstal er udtrykt ved Formler af Typen (12). Denne Omstændighed har bevirket, at man paa et tidligere Tidspunkt har antaget, at Helium var en Blanding af to Luftarter, »Orthohelium« og »Parhelium«, men nu ved vi, at de to Spektre blot betyder, at den anden Elektrons Binding kan finde Sted paa to forskellige Maader. En teoretisk Klarlæggelse af Hovedtrækkene af Heliumspektrets Bygning er givet i et betydningsfuldt Arbejde af Landé. Han tænker sig, at Orthoheliumspektrets Udsendelse skyldes Overgange mellem stationære Tilstande, hvor begge Elektronerne bevæger sig i samme Plan og i samme Omløbsretning; den sidst indfangne i en Bane udenom den først bundne. Parheliumspektret derimod tænker han sig som knyttet til stationære Tilstande, hvor Elektronbanerne danner Vinkler med hinanden. En nærmere Undersøgelse af den Maade, hvorpaa de to Baner i de stationære Tilstande af Heliumatomet maa tænkes at indvirke paa hinanden, er udført af Foredragsholderen i Samarbejde med Dr. Kramers. Resultaterne af vor Undersøgelse, der var paabegyndt længere Tid før Offentliggørelsen af Landés Arbejde, har endnu ikke været offentliggjort. Uden at komme ind paa Detailler kan jeg sige, at selv om vore Resultater paa flere Punkter skiller sig væsentligt fra Landés, er vi enige med ham i de ovennævnte almindelige Slutninger angaaende Orthohelium- og Parheliumspektrets Oprindelse.

I nøjeste Forbindelse med Betragtninger over Heliumspektrets Fremkomst er Spørgsmaalet om Slutresultatet af Bindingen af den anden Elektron i Atomet blevet belyst paa betydningsfuld Maade ved Forsøg af Franck og hans Medarbejdere, der som bekendt i de sidste Aar ved Undersøgelser over Virkningen af Bombardement af Atomer med Elektroner med forskellige Hastigheder har formaaet at kaste overordentlig interessant Lys over vigtige Træk af Atombygningens Stabilitet og Spektrenes Fremkomst. For kort Tid siden har disse Forsøg saaledes ledt til den Opdagelse, at Heliumatomer, der udsættes for et Bombardement af Elektroner, kan bringes over i en Tilstand, som Franck benævner »metastabil«, og fra hvilken Atomet ikke kan vende tilbage til Normaltilstanden ved en simpel Overgangsproces ledsaget af Udsendelse af Straaling, men kun ved en Proces der kan betegnes som en kemisk Reaktion, idet den er betinget af Tilstedeværelsen af Atomer af andre Stoffer. Dette Resultat staar i nøjeste Forbindelse dermed, at Bindingen af den anden Elektron i Heliumatomet, saaledes som Spektret lærer os, kan foregaa paa to forskellige Maader. Det fremgaar saaledes af Francks Forsøg, at Atomets normale Tilstand svarer til Slutresultatet af den til Udsendelse af Parheliumspektret knyttede Bindingsproces, hvorved den sidst indfangne Elektron ligesom den først indfangne vil bindes i en 1_1 Bane. Den nævnte metastabile Tilstand derimod svarer til Slutresultatet af den til Fremkomsten af Orthoheliumspektret knyttede Proces, ved hvilken den sidst indfangne Elektron i Modsætning til den første vil bevæge sig i en 2_1 Bane, svarende til en Fasthed af Bindingen, der er ca. 7 Gange mindre end for Elektronen i Atomets Normaltilstand.

Hvis vi nu forsøger nærmere at undersøge dette i første Øjeblik overraskende Resultat, viser det sig, at en Forklaring, eller rettere en Forstaaelse, deraf kan opnaas paa Basis af det i det ovenfor omtalte

Korrespondenssynspunkt. Man indser nemlig let, at der er en stor Forskel paa de to Bindingsprocesser, der netop kommer tydeligt frem ved Undersøgelse af disse Processers sidste Stadium. Betragter vi saaledes Orthoheliumbanerne, hvor de to Elektroner bevæger sig i samme Plan, viser det sig, at den ydre Elektrons Bane har den samme Karakter som en Bane af en Elektron der bevæger sig i et simpelt centralsymmetrisk Kraftfelt, og at derfor Fremkomsten af Overgange mellem de stationære Tilstande, paa den i det foregaaende omtalte Maade, direkte kan henføres til Egenskaber ved denne Elektrons Bevægelse. Dette gælder dog netop kun, naar vi indskrænker os til Baner, for hvilke Hovedkvantetallet er større end 1. Vilde vi tænke os Muligheden af en Orthoheliumbane, for hvilken baade n og k var lig 1, maatte vi nemlig, som den nøjere Undersøgelse af Banerne viser, antage, at de to Elektroners Baner var »ækvivalente«, og da de skal ligge i samme Plan, vilde vi blive ført til en Tilstand, hvor de to Elektroner bevægede sig om Kernen i en og samme cirkulære Bane, saaledes at de hvert Øjeblik befandt sig i Endepunkterne af en Diameter. Denne meget enkle Ringkonfiguration kunde efter en simpel Anvendelse af Kvanteteorien ventes at svare til den fastest mulige Binding af Elektronerne i Atomet, og var derfor i mit første Arbejde over Atombygningen foreslaaet som en Model for et neutralt Heliumatom. Spørger vi imidlertid efter Muligheden for en Overgang fra en af Orthoheliumtilstandene til en Konfiguration af den nævnte Art, møder vi Forhold, der er ganske forskellige fra dem, der gælder for indbyrdes Overgange mellem to af Orthoheliumbanerne. For en Overgangsproces, hvorved den sidst indfangne Elektron fra at bevæge sig »udenom« den anden blev optaget som jævnbyrdig Deltager i Bevægelsen omkring Kernen, kan vi nemlig ikke forestille os en Række simple Mellemformer for de to Elektroners Bevægelse, under hvilke den sidst indfangne Elektrons Bane udviste en saadan Lighed med en Centralbevægelse, at der for denne Overgang kunde blive Tale om en Korrespondens af den efterlyste Art. Det bliver derfor forstaaeligt, at den sidst indfangne Elektron ved en Proces, hvorunder de to Elektroner bevæger sig i samme Plan, ikke kan bindes fastere end i en 2_1 Bane. Ser vi derimod paa den Bindingsproces, der ledsager Parheliumspektrets Udsendelse, og hvor Elektronerne i de stationære Tilstande bevæger sig i Baner, hvis Planer danner Vinkler med hinanden, møder vi væsentlig andre Forhold. Her kræves ikke en tilsvarende, gennemgribende Forandring i Vekselspillet mellem den sidst indfangne Elektron og den tidligere bundne for at opnaa, at de to Elektroners Plads i Atomet bliver ligestillet, og vi kan derfor forestille os, at det sidste Afsnit af Bindingsprocessen kan foregaa paa en lignende Maade som de Afsnit, der er betegnet ved Overgange mellem Baneformer, karakteriserede ved større Værdier for Kvantetallene n og k.

I Heliumatomets Normaltilstand maa de to Elektroner tænkes at bevæge sig i ækvivalente 1_1 Baner, der i første Tilnærmelse kan beskrives som to Cirkelbaner, hvis Planer — i Overensstemmelse med den i det foregaaende omtalte Betingelse, der efter Kvanteteorien er paalagt Impulsmomentet for et Atom — danner en Vinkel paa 60 Grader med hinanden, men iøvrigt paa Grund af Vekselspillet mellem de to Elektroner ikke vil besidde en fast Orientering i Rummet. Udfra Synspunkter, der afviger væsentlig fra dem, jeg her har fremsat, er en tilsvarende Model for Heliumatomet fornylig foreslaaet af Kemble, der, hvad det finere Vekselspil mellem Elektronerne angaar, samtidig har gjort opmærksom paa en mulig Bevægelsestype af udpræget Symmetri og af en saadan Beskaffenhed, at Elektronerne under deres hele Omløb indtager symmetriske Stillinger i Forhold til en fast Akse. Kemble har dog ikke undersøgt denne Bevægelse nærmere. Allerede før Fremkomsten af hans Arbejde havde imidlertid Dr. Kramers paabegyndt en nøjere Gennemregning af netop denne Bevægelsestype, for at forsøge hvorvidt det var muligt ud fra en saadan Beregning at gøre nøjere Rede for den Fasthed, hvormed Elektronerne er bundne i Heliumatomet, og som kan maales ved en Undersøgelse af det saakaldte Ioniseringspotential. Tidligere Maalinger af dette Potential havde givet Værdier for Fastheden af Bindingen, der omtrentlig svarede til den, der vilde fremgaa af den ovenfor omtalte Ringkonfiguration, der for Fjernelsen af en Elektron kræver et Arbejde lig $^{17}/_{8}$ af det Arbejde, der kræves for Fjernelsen af Elektronen i Brintatomets Normaltilstand. Idet den teoretiske Værdi for dette sidste Arbejde — som vi i det følgende for Kortheds Skyld skal betegne ved W — svarer til et Ioniseringspotential af 13,53 Volt, skulde saaledes Ioniseringspotentialet for Helium efter denne Model ventes at være 28,8 Volt. Nyere, nøjagtigere Maalinger har imidlertid givet en Værdi for Heliums Ioniseringspotential, der er betydelig lavere og ligger i Nærheden af 25 Volt. Ganske uafhængig af de ovenstaaende Betragtninger var Uholdbarheden af den omhandlede Ringkonfiguration som fremstillende Normaltilstanden af et Heliumatom allerede derved fastslaaet. I Modsætning til, hvad der gælder for denne Konfiguration, kræver en nøjere Undersøgelse af den beskrevne rumlige Atomkonfiguration et vidtløftigt Beregningsarbejde, og K r a m e r s Regninger har endnu ikke givet afsluttede Resultater. Med den Tilnærmelse hvormed de hidtil er gennemførte, peger imidlertid Beregningerne paa Muligheden af at opnaa Overensstemmelse med Forsøgsresultaterne, og det endelige Resultat maa afventes med stor Interesse, fordi det giver os Mulighed for i det simplest tænkelige Tilfælde at opnaa en Prøvelse af de Synspunkter, ud fra hvilke man forsøger at fastlægge stationære Tilstande for Atomer indeholdende mere end een Elektron.

Som det fremgaar af Oversigten over Grundstoffernes System, danner Brint og Helium tilsammen d e n f ø r s t e P e r i o d e i Systemet, idet Helium jo er den første inaktive Luftart. Den store Forskel i de kemiske Egenskaber af Brint og Helium hænger nøje sammen med den store Forskel i Elektronbindingens Beskaffenhed, som Undersøgelsen af Spektrene og Ioniseringspotentialerne direkte lærer os om, og som vi har set, det ogsaa synes muligt at fortolke ved direkte Anvendelse af Kvanteteorien. Medens Helium har det højeste Ioniseringspotential af alle Stoffer, er Elektronen i Brintatomet ikke bundet fastere, end at vi kan forstaa Brintens Tilbøjelighed til at optræde som positiv Ion i vandige Opløsninger og kemiske Forbindelser. En nærmere Betragtning af dette Spørgsmaal kræver imidlertid en Sammenligning mellem Elektronkonfigurationernes Art og Fasthed hos andre Stoffers Atomer og kan derfor ikke diskuteres, før vi har betragtet dette Spørgsmaal nærmere.

Naar vi nu gaar over til at betragte Atombygningen hos Stoffer, der i det neutrale Atom indeholder mere end to Elektroner, skal vi først og fremmest antage, at alt, hvad vi ovenfor har sagt om Heliumatomets Dannelse, i Hovedtræk vil gælde ogsaa for Indfangningen og Bindingen af de to første Elektroner i saadanne Atomer. Disse Elektroner maa saaledes i Normaltilstanden antages at bevæge sig i ækvivalente Baner, karakteriserede ved Kvantebetegnelsen 1_1.

Spørger vi nu om Bindingen af den 3die Elektron, faar vi direkte Oplysning fra en Undersøgelse af Lithiumspektret. Dette Spektrum lærer os om en Række stationære Tilstande, hvor den Fasthed, hvormed den sidst indfangne Elektron er bundet, er meget nær den samme som i de stationære Tilstande svarende til Stadier af Brintatomets Dannelse. Foruden disse Rækker af Tilstande, der svarer til Baner, hvor k er større eller lig 2, og hvor den sidst indfangne Elektron bevæger sig helt udenfor det Omraade, hvori de to første Elektroner bevæger sig, lærer Spektret os om en Række Tilstande, der svarer til $k=1$, og hvor Energien afviger væsentlig fra de tilsvarende stationære Tilstande af Brintatomet. I disse Tilstande, hvortil Atomets Normaltilstand — saaledes som det fremgaar af Forsøg over Absorption i Lithiumdamp — hører, maa den sidst indfangne Elektron, selv om den under Størstedelen af Omløbet befinder sig i større Afstande fra Kernen, under et vist Tidspunkt af Omløbet antages at trænge ind til Afstande fra denne, der er af samme Størrelsesorden som Dimensionerne af de tidligere bundne Elektroners Baner. Paa Grund af denne Omstændighed vil Elektronen være bunden med en Fasthed, der er betydelig større end den, hvormed Elektronen er bunden i stationære Tilstande af Brintatomet, svarende til samme Værdi for n. I Normaltilstanden, hvor, som det fremgaar af Spektret, det Arbejde, der kræves for Elektronens Fjernelse fra Atomet, kun er 0,396 W, maa vi derfor antage, at denne bevæger sig i en 2_1 Bane, selv om den er bunden omtrent halvanden Gang saa fast som i den tilsvarende Bane i Brintatomet, hvor det for Fjernelsen krævede Arbejde er ¼ W. Hvis vi her spørger om Aarsagen til, at en Overgang fra en af de stationære Tilstande, hvorom Spektret belærer os, til en Tilstand, hvor den sidst indfangne Elektron vilde bevæge sig i en 1_1 Bane, er udelukket, møder vi Forhold, der frembyder Lighed med Fortolkningen af Heliums metastabile Tilstand. En saadan Overgang, der, som en nærmere Undersøgelse viser, maatte tænkes at føre til en Tilstand af Atomet, hvor alle Elektronerne bevægede sig i ækvivalente Baner, vilde nemlig være af en ganske anden Type end de til Udsendelsen af Lithiumspektret knyttede Overgange mellem stationære Tilstande og vilde i Modsætning til disse ikke udvise Korrespondens med harmoniske Komponenter i Atomets Bevægelse. Vi faar derfor et Billede af Lithiumatomets Dannelse og Bygning, der tilbyder en naturlig Forklaring for de fra Heliums og Brints stærkt afvigende, kemiske Egenskaber, som Lithium udviser, idet vi opnaar en Basis for at forstaa, hvorfor den sidst indfangne Elektron i Lithiumatomet er næsten 5 Gange løsere bunden end Elektronerne i Heliumatomet, og lidt mere end 2 Gange løsere end Elektronen i Brintatomet.

Hvad her er sagt gælder ikke alene om Dannelsen af Lithiumatomet, men maa antages at gælde om Bindingen af den 3die Elektron i ethvert Atom, der altsaa i Modsætning til de to første Elektroner, som bevæger sig i 1_1 Baner, maa tænkes at bevæge sig i en 2_1 Bane. Tilsvarende vil nu gælde om den 4de, 5te og 6te Elektron. Jeg skal ikke her komme nærmere ind paa de Forestillinger, man kan gøre sig om den successive Indfangning og Binding af disse Elektroner, men blot nævne, at Aarsagen til, at Bindingen af de første af disse ikke vil forhindre Indfangningen af de paafølgende i ækvivalente 2_1 Baner, maa antages at hænge sammen med, at Banerne ikke er cirkulære, men stærkt ekscentriske, og at saaledes f. Eks. den 3die Elektron ikke kan holde de senere Elektroner borte fra det indre System paa tilsvarende Maade, som Tilstedeværelsen af de to først bundne Elektroner i Lithiumatomet hindrer den 3die i at bindes i en 1_1 Bane. Vi maa saaledes vente, at den 4de, 5te og 6te Elektron paa tilsvarende Maade som den 3die hver især under visse Øjeblikke af deres Omløb trænger ind i det Omraade, hvor de to først bundne Elektroner bevæger sig. Vi skal imidlertid ikke tænke os, at disse Besøg hos det indre System for de fire Elektroners Vedkommende finder Sted paa samme Tid, men at de fire Elektroner besøger Kernen enkeltvis med lige store Tidsmellemrum. Ved tidligere Arbejder over Atombygningen har man tænkt sig, at Elektronerne i Atomets forskellige Grupper bevægede sig i adskilte Omraader indenfor dette, samt at Elektronerne i hver enkelt Gruppe hvert Øjeblik indtog Konfigurationer, der besad en Symmetri af lignende Art som en regelmæssig Polygon eller Polyeder, hvad blandt andet medførte, at Elektronerne i Gruppen antoges samtidig at befinde sig i de Punkter af deres Baner, hvor de er Kernen nærmest. En saadan Bygning af Atomet kan betegnes som en, hvor Bevægelsen af Elektronerne indenfor Grupperne er koblet sammen paa en Maade, der er uafhængig af Vekselvirkningen mellem de forskellige Grupper. Det karakteristiske Træk hos en Bygning for Atomet som den her foreslaaede er derimod den intime K o b l i n g mellem Elektronbevægelserne i de forskellige Grupper, samt den større U a f h æ n g i g h e d, som kommer for Dagen i Bindingsarten indenfor en og samme Gruppe af Elektroner med ækvivalente Baner karakteriserede ved samme Kvantetal. Ved at fremhæve dette sidste Træk tænker jeg ikke alene paa den langt ringere Indflydelse af Tilstedeværelsen af tidligere bundne Elektroner indenfor samme Gruppe paa Fastheden af Bindingen af de paafølgende Elektroner, som den beskrevne Bevægelsesmaade bringer med sig, men fuldt saa meget paa den Maade hvorpaa Bevægelserne af Elektronerne indenfor Gruppen afspejler Uafhængigheden af de Processer, gennem hvilke Gruppen kan dannes ved de enkelte Elektroners Indfangning, samt kan reorganiseres ved Pladsbytning af de enkelte Elektroner i Atomet efter en Forstyrrelse frembragt ved ydre Kræfter. Det sidste Spørgsmaal skal vi komme nærmere ind paa ved Omtalen af Fortolkningen af Røntgenspektrenes Fremkomst og Beskaffenhed; foreløbig skal vi gaa videre med at betragte, hvilken Bygning af Atomet vi føres til ved at undersøge de til Elektronernes successive Indfangning knyttede Processer.

De foregaaende Betragtninger giver et Grundlag for at forstaa, at de paa Lithium følgende Stoffer, B e r y l l i u m og B o r, i Forbindelser med andre Stoffer kan optræde som elektropositive med henholdsvis to og tre Valenser, idet, ligesom den 3die Elektron i Lithiumatomet, de senere indfangne Elektroner i disse Stoffers Atomer vil være langt løsere bundne end de to første Elektroner. Samtidig forstaar vi, at den elektropositive Karakter dog for disse Stoffer er mindre udpræget end hos Lithium, idet Elektronerne i 2_1 Banerne paa Grund af det stærkere Felt, hvori de bevæger sig, vil være langt fastere bundne i Beryllium- og Bor-Atomet end i Lithium-Atomet. Hos det næste Grundstof, K u l s t o f, møder vi nye Forhold, idet dette Stof i sine typiske kemiske Forbindelser ikke kan antages at optræde som Ion, men snarere som neutralt Atom. Dette maa antages ikke alene at hænge sammen med Elektronbindingens store Fasthed, men tillige at være væsentlig betinget af Symmetriegenskaber hos Elektronkonfigurationen.

Ved Bindingen af den 4de, 5te og 6te Elektron i 2_1 Baner maa den rumlige Symmetri af den relative Konfiguration af de paagældende Elektronbaner antages stadig at forøges, indtil ved Bindingen af den 6te Elektron Banerne af de fire sidst bundne Elektroner maa ventes at indtage en overordentlig symmetrisk Konfiguration, hvori Normalerne til Baneplanerne indtager en Stilling relativt til hverandre, meget nær som Linierne fra Centret til Hjørnespidserne i et regulært Tetraeder. En saadan Konfiguration af Gruppen af 2-kvantede Baner i Kulstofatomet synes nu virkelig egnet til at danne Grundlag for Forestillinger om Bygningen af de organiske Forbindelser. Jeg skal imidlertid ved denne Lejlighed ikke komme nærmere ind paa dette Spørgsmaal, hvis nøjere Undersøgelse vil kræve et indgaaende Studium af Vekselvirkningen mellem Elektronbevægelsen i de i Molekylet indgaaende Atomer. Det skal i denne Forbindelse blot nævnes, at de Typer af Elektronkonfigurationer, som man ud fra de her fremsatte Betragtninger naturligt føres til, bliver væsentlig forskellige fra saadanne Molekylmodeller, som blev foreslaaet i Foredragsholderens første Arbejder, og hvor de kemiske »Valensbaand« fremstilledes ved »Elektronringe« af lignende Type som de, der antoges at udgøre Elektrongrupperne indenfor de enkelte Atomer. Naar det, uden at komme ind paa saadanne Spørgsmaal, dog er muligt, som jeg skal forsøge her, at give en Oversigt over Fortolkningen af Grundstoffernes kemiske Egenskaber ud fra en Undersøgelse af disse Stoffers Atomer, ligger dette i, at Betragtningen af Molekyldannelser af den Art, som optræder i Sammenslutningen af Atomer af et og samme Grundstof og i mange organiske Forbindelser, ikke spiller den Rolle for en saadan Oversigt som Betragtningen af Molekyldannelser, hvori de enkelte Atomer tør anses at indgaa som elektrisk ladede Ioner. Den sidste Art af Forbindelser, som ofte benævnes »heteropolare«, og hvortil de simple uorganiske Forbindelser for den største Del hører, besidder nemlig en langt mere typisk Karakter end de førstnævnte Forbindelser, der benævnes »homoeopolare«, og hvis Egenskaber i en ganske anden Grad er individuelle for de indgaaende Stoffer. Hovedopgaven i det følgende vil derfor blive at betragte den Tilpasning, som Elektronkonfigurationerne i de forskellige Atomer frembyder for Iondannelsen.

Før vi forlader Spørgsmaalet om Kulstofatomets Bygning, maa jeg imidlertid nævne, at en Model af et Kulstofatom, hvor Konfigurationen af Banerne af de fire løsest bundne Elektroner besidder en udtalt Tetraeder Symmetri, allerede har været foreslaaet af Landé, der for at gøre Rede for det eksperimentelle Materiale vedrørende Atomernes Størrelse ligeledes har benyttet den Antagelse, at de paagældende Elektroner bevægede sig i Baner, karakteriserede ved Kvantebetegnelsen 2_1. Der er imidlertid den Væsensforskel mellem Landés Betragtninger og de her anførte, at medens Kulstofatomets karakteristiske Forhold hos Landé søges baseret alene paa en Undersøgelse af de simpleste Bevægelsesformer med rumlig Symmetri, som fire Elektroner kan udføre, er Betragtningerne her ført tilbage til det hele Atoms Stabilitetsforhold, idet, som vi har set, Antagelserne om de paagældende Elektroners Baner er direkte baseret paa en Undersøgelse af Vekselvirkningen mellem disse Elektroner og de to først bundne. Denne Forskel kommer til Syne gennem de Antagelser om Koblingen mellem Bevægelserne af Elektronerne i de to Grupper med ækvivalente Baner og om Vekselspillet mellem Bevægelserne indenfor hver af disse, som Betragtninger af Atomets Dannelse har ledt os til, og som giver vor Model for Kulstofatomet dynamiske Egenskaber, der er væsentlig forskellige fra Egenskaberne af Landés Model.

For at gøre Rede for de nærmest paafølgende Grundstoffers Egenskaber kommer det, som nævnt i det foregaaende ved Omtalen af Kossels Arbejder, i første Række an paa at forstaa, hvorfor den Konfiguration af 10 Elektroner, der optræder i det neutrale Neonatom besidder Stabilitetsforhold af særegen Art. Ved Betragtninger over dette Spørgsmaal har man hidtil lagt den Antagelse til Grund, at denne Konfigurations Egenskaber var betinget af Vekselspillet mellem 8 Elektroner, der i ækvivalente Baner bevægede sig udenom Kernen og en indre Gruppe af to Elektroner svarende til Konfigurationen af Elektronerne i Heliumatomet. Vi skal imidlertid se, at Løsningen maa antages at skulle søges i en væsentlig forskellig Retning. Spørger vi om Indfangningen og Bindingen af den 7ende Elektron i Atomet, kan vi nemlig ikke vente, at denne vil blive bundet i en 2_1 Bane ækvivalent med de fire foregaaende Elektroners Baner. En Optræden af 5 saadanne Baner i Atomet vilde nemlig paa saa afgjort Maade forstyrre Symmetrien i disse Elektroners Vekselspil, at vi ikke kan forestille os, at en Proces resulterende i Optagelsen af en femte Elektron i den paagældende Gruppe vilde besidde den af Korrespondensprincippet krævede Analogi med en Proces, hvormed Udsendelsen af Straaling efter vore sædvanlige Forestillinger vil være forbunden. Vi maa tvertimod tænke os, at de fire Elektroner i deres overordentlig symmetriske Banekonfiguration vil holde senere tilkomne Elektroner ude fra deres Gruppe, med det Resultat at disse Elektroner vil blive bundne i Baner karakteriserede ved højere Kvantetal.

De Baner, der kommer i Betragtning ved Bindingen af den 7ende Elektron i Kvælstofatomet og den 7ende, 8nde, 9ende og 10ende Elektron hos Atomerne af de paafølgende Elementer, vil være cirkulære Baner af Typen 2_2, der — selv om deres Diameter vil være betydelig større end Banediametrene for de to første Elektroner —, dog vil forløbe indenfor det Omraade, i hvilket de fire næste Elektroner bevæger sig, saaledes at de yderste Dele af de ekscentriske 2_1 Baner vil rage et Stykke udenfor de cirkulære 2_2 Baner. Jeg skal ikke her komme ind paa Spørgsmaalet om de paagældende Elektroners successive Indfangning og Binding, for hvilke en nærmere Undersøgelse af Vekselspillet mellem Bevægelserne af Elektronerne i de 2 Typer af 2-kvantede Baner er nødvendig. Jeg skal blot nævne, at i Neonatomet, hvor vi skal antage, at der findes fire Elektroner i 2_2 Baner, disse Baners Planer ikke alene maa tænkes at indtage en Stilling i Forhold til hverandre, der udmærker sig ved en høj Grad af rumlig Symmetri, men ogsaa at besidde en harmonisk Konfiguration relativt til de fire elliptiske 2_1Baner. Et Vekselspil af denne Art kan imidlertid uden Sammenfald af Baneplaner kun opnaas ved, at Banekonfigurationerne i begge Undergrupper viser en systematisk Afvigelse fra Tetraeder-Symmetri, der vil bevirke, at Elektrongruppen med 2-kvantede Baner i Neonatomet kun vil besidde en enkelt Symmetriakse, der maa antages at falde sammen med Symmetriaksen for Konfigurationen af den inderste Gruppe med de to Elektroner i 1-kvantede Baner.

Med Henblik paa Forklaringen af den udprægede elektronegative Karakter af de forud for Neon gaaende Elementer, Ilt og Fluor, med Atomnummer 8 og 9 maa det her fremhæves, at den Tendens, som disse Stoffers neutrale Atomer besidder for ved yderlig Indfangning af Elektroner at danne negative Ioner med en lignende Konstitution som det neutrale Neonatom, maa tilskrives ikke alene den Forøgelse af Symmetrien og den dermed følgende Forøgelse af Elektronbygningens Fasthed, som Kompletteringen af den paagældende Elektrongruppe vil medføre, men lige saa meget den Omstændighed, at de sidst indfangne Elektroners Baner vil finde deres Plads indenfor det Omraade, hvori de tidligere indfangne Elektroner i 2_1 Baner bevæger sig. Dette Forhold betinger nemlig en Forstaaelse af den store Forskel, som Stofferne i den sidste Halvdel af den anden Periode af det periodiske System udviser fra Stofferne i den første Halvdel, i hvis Atomer i Modsætning til de førstnævnte Stoffers der kun findes en Type af 2-kvantede Baner.

Vi skal nu gaa over til at undersøge Spørgsmaalet om Bygningen af Atomerne for Stofferne i den tredie Periode i Grundstoffernes System og stilles da straks overfor Spørgsmaalet om, hvorledes den 11te Elektron i Atomet bindes. Her møder vi Forhold, der i visse Henseender er analoge med dem, vi omtalte ved Bindingen af den 7ende Elektron. I Lighed med hvad der gjaldt for Konfigurationen i Kulstofatomet vilde Symmetrien af Konfigurationen i Neonatomet blive væsentlig, for ikke at sige fuldstændig, forstyrret ved Optagningen af endnu en Elektron i Atomet i en Bane af samme Type som den, hvori de sidst indfangne Elektroner blev bundne. Ligesom ved den 3die og 7ende Elektron maa vi derfor vente ved den 11te Elektron at møde en ny Type af Baner i Atomet, og de Baner, der denne Gang melder sig, er 3_1 Baner. En Elektron, der bevæger sig i en saadan Bane, vil vel for Størstedelen af Bevægelsen befinde sig udenfor Konfigurationen af de 10 første Elektroners Baner; men i visse Øjeblikke af dens Omløb vil den ikke alene trænge ind i Omraadet for 2_1 og 2_2 Banerne, men vil ligesom de første af disse Baner endda trænge helt ind til Afstande fra Kernen, der er mindre end Radierne af de 1-kvantede Baner af de to først bundne Elektroner. Dette Forhold, der har den største Betydning for Forstaaelsen af Atomets Stabilitet, fører for Bindingen af den betragtede 11te Elektron til et ejendommeligt Resultat. I Natriumatomet vil saaledes for de ydre Banedeles Vedkommende denne Elektron bevæge sig i et Felt, der kun afviger meget lidt fra det, der omgiver Kernen i Brintatomet, men alligevel vil Dimensionerne af de omhandlede Dele af Banen være væsentlig forskellige fra Dimensionerne af de tilsvarende Dele af en 3_1 Bane i

Brintatomet. Dette hidrører fra, at selv om Elektronen kun i korte Øjeblikke af sit Omløb kommer indenfor den indre Konfiguration af 10 Elektroner, vil dog denne Del af Banen være væsentlig bestemmende for Fastlæggelsen af Hovedkvantetallet, hvad der hænger simpelt sammen med, at Elektronens Bevægelse for denne Banedels Vedkommende kun afviger lidt fra den Bevægelse, som hver af de tidligere bundne Elektroner i 2_1 Baner vil udføre under et helt Omløb i disse Baner. Med denne Omstændighed hænger den Usikkerhed sammen, der har hersket angaaende Fastlæggelsen af Kvantetallene for de stationære Tilstande svarende til et Spektrum som Natriumspektret, og som vi har hentydet til i det foregaaende. Dette Spørgsmaal har være underkastet Drøftelse fra forskellig Side. Ud fra Sammenligninger mellem Spektraltermerne for de forskellige Alkalimetaller har Roschdestwensky ment at kunne drage den Slutning, at Normaltilstanden ikke, som man paa Forhaand kunde være tilbøjelig til at vente, svarer til en 1_1 Bane — en Opfattelse, der f. Eks. kommer til Udtryk i de Talværdier for n, hvormed i Figuren paa Side 22 de stationære Tilstande er kendetegnet — men at denne Tilstand svarer til en 2_1 Bane. I et Forsøg paa at gøre Rede for den store Forskel imellem S-Termerne og Termerne i P- og D-Rækkerne hos Alkalispektrene, der viser sig i Værdierne for Konstanterne α_k i Formlen (12), er Schrödinger kommet til et tilsvarende Resultat ud fra den Antagelse, at den »ydre« Elektron i de Tilstande der svarer til S-Termerne, i Modsætning til de, der svarer til P- og D-Termerne, under sit Omløb forløber delvis indenfor Omraadet for de indre Elektroners Baner. Disse Undersøgelser indeholder vel betydningsfulde Fingerpeg, men som vi skal se, maa Forholdene i Virkeligheden antages at ligge væsentlig forskelligt for de forskellige Alkalispektre. I Modsætning til Lithiumspektret, hvor de første Termer af S- og P-Rækkerne svarer henholdsvis til 2_1 Baner og 2_2 Baner, maa vi saaledes for Natriumspektrets Vedkommende ikke alene antage, at det første Spektralterm i S-Rækken svarer til en 3_1 Bane, men en nøjere Betragtning viser tilmed, at det første Term i P-Rækken ikke svarer til en 2_2 Bane, som antydet paa Figuren, men til en 3_2 Bane. Den Opfattelse, der ligger til Grund for Figurens Betegnelser, vilde blandt andet føre med sig, at P-Termerne skulde være mindre end Brinttermerne, svarende til samme Hovedkvantetal. En

Forklaring heraf vilde imidlertid kræve, at Middelvirkningen af de indre Elektroners Tilstedeværelse kunde beskrives som en Frastødning, større end den der vilde fremkomme, hvis deres samlede elektriske Ladning var forenet i Atomets Centrum. Ud fra en Forestilling om Atombygningen som den her omhandlede kan et Forhold som dette imidlertid ikke ventes at forekomme. Den Omstændighed, at den sidst indfangne Elektron, i hvert Fald for lave Værdier af *k*, forløber delvis indenfor de tidligere indfangne Elektroners Bane, vil tværtimod bevirke, at disse Elektroners Tilstedeværelse vil give Anledning til en virtuel Frastødning, der er betydelig mindre end den, der vilde skyldes deres forenede Ladning. I Stedet for de fra højre til venstre forløbende Kurver, der paa Figuren er trukket mellem Punkter, der tænkes at repræsentere stationære Tilstande, svarende til samme Værdi for Hovedkvantetallet, faar vi derfor Kurver, der forløber fra venstre til højre, og et saadant Forhold finder vi ikke alene for Natriumspektret, men ogsaa for andre Stoffers Spektre. I Forbindelse med Spørgsmaalet om Fortolkningen af Natriumspektret ud fra det omhandlede Billede af Atomet kan det nævnes, at dette ikke alene synes at tilbyde en Forklaring af Størrelserne af Termerne, men at ogsaa Komplexiteten af Termerne i *P*- og *D*-Rækkerne forklares naturligt ud fra den Afvigelse fra ren Centralsymmetri, som Konfigurationen af de 10 først bundne Elektroner udviser. Denne Mangel paa Symmetri har sit Udspring i de to allerinderste Elektroners Konfiguration og »forplanter« sig ud til de ydre Dele af Atombygningen netop paa Grund af den Omstændighed, at 2_1 Banerne delvis trænger ind paa disse Elektroners Omraade.

De ovenstaaende Betragtninger angaaende Natriumspektrets Fremkomst giver samtidig en umiddelbar Fortolkning af Natriums stærke elektropositive Egenskaber, idet som Spektret lærer os, og som Atommodellen gør forstaaeligt, den sidst bundne Elektron i Natriumatomet er endnu løsere bunden end den sidst indfangne Elektron i Lithiumatomet. Det kan i denne Forbindelse anføres, at ogsaa en Betragtning af Atomvolumenets Tiltagen med stigende Atomnummer indenfor Alkalimetallernes Familie fortolkes simpelt ud fra den stedse løsere Binding af Valenselektronen. I Tilknytning til sine Arbejder over Røntgenspektrene har Sommerfeld paa et tidligere Tidspunkt ment i denne Tiltagen at Atomvolumenerne at kunne finde

Støttepunkter for den Antagelse, at det Hovedkvantetal, der svarer til Valenselektronernes Bane, ligesom det her er antaget, vil vokse med en Enhed, hver Gang vi gaar fra et Metal i den omhandlede Familie til det næste. Sommerfeld har dog senere i Sammenhæng med Undersøgelser over Seriespektrene bestemt taget Afstand fra denne Antagelse, der i første Øjeblik ogsaa kunde synes at medføre en Tiltagen af Atomvolumenerne langt større end den iagttagne. En simpel Forstaaelse af disse Forhold opnaar man imidlertid ved, som ovenfor beskrevet, at tage Hensyn til, at den paagældende Elektrons Bane vil delvis forløbe indenfor de indre Banekonfigurationers Omraade, og at derfor de »effektive« Kvantetal, der svarer til Banens ydre, næsten elliptiske Sløjfer, vil være langt mindre end det Hovedkvantetal, hvormed den hele Centralbane beskrives. Det maa i denne Forbindelse ogsaa bemærkes, at Vegard, ligeledes i Tilknytning til Undersøgelser over Røntgenspektrene, har foreslaaet Antagelsen af stedse stigende Kvantetal for Elektronbanerne i Atomets Grupper, regnet fra Kernen og udefter, og i denne Sammenhæng indført Antagelser angaaende Forbindelsen mellem Elektrontallene i Atomets Grupper og Periodelængderne i Grundstoffernes System, der udviser visse formelle Lighedspunkter med de i det følgende fremsatte Resultater. Heller ikke Vegards Betragtninger tilbyder imidlertid Udgangspunkter for en nærmere Undersøgelse af Gruppernes Udformning og Stabilitet og derfor ikke noget Grundlag for en indgaaende Fortolkning af Stoffernes Egenskaber.

Betragter vi de paa Natrium følgende Stoffer i den 3die Periode af Grundstoffernes System møder vi ved Bindingen af den 12te, 13de og 14de Elektron Forhold, der er nøje analoge med dem, vi mødte ved Spørgsmaalet om Bindingen af den 4de, 5te og 6te Elektron i Atomet. Vi skal saaledes antage, at vi i det neutrale Siliciumatom har at gøre med en Konfiguration, hvor de fire sidst indfangne Elektroner bevæger sig i 3_1 Baner, der ligesom 2_1 Banerne hos Kulstof maa tænkes at danne en Konfiguration med saadanne Symmetriegenskaber, at en Optagen i Atomerne af de følgende Elementer af en yderligere Elektron i en 3_1 Bane er udelukket. Den 15de Elektron hos Stofferne af højere Atomnummer vil derfor være bunden i en ny Banetype; men her vil vi ikke, som ved Indfangningen af den 7ende Elektron, møde cirkulære Baner, men en roterende ekscentrisk Bane af Typen 3_2. Dette hænger paa det nøjeste sammen med de Forhold, som jeg før omtalte, og som bevirker, at ikke-cirkulære Baner vil svare til en fastere Binding end cirkulære Baner med samme Værdi for Hovedkvantetallet, idet for de første Baners Vedkommende Elektronen i visse Øjeblikke vil trænge langt dybere ind i Atomets Indre. En Bane 3_2 vil vel ikke komme helt ind i den inderste Konfiguration af 1_1 Baner, men den vil trænge ind til Afstande fra Kernen, der er betydelig mindre end Radierne i de cirkulære 2_2 Baner. For den 16de, 17de og 18de Elektrons Vedkommende stiller Forholdene sig paa tilsvarende Maade som for den 15de, og vi maa for Argon vente en Konfiguration, hvori de ti inderste Elektroner bevæger sig i Baner af samme Typer som Elektronerne i Neonatomet, medens de 8 sidste Elektroner vil danne en Konfiguration af fire 3_1 Baner og fire 3_2 Baner, der besidder Symmetriegenskaber, der maa tænkes nøje at svare til Konfigurationen af 2-kvantede Baner i Neonatomet. Samtidig med at denne Forestilling giver en Forstaaelse af den Lighed i kvalitativ Henseende, som de kemiske Egenskaber af Stofferne i den sidste Del af henholdsvis den anden og tredie Periode i Grundstoffernes System udviser, aabner den en Mulighed for en naturlig Forklaring af den betydelige Forskel, der i kvantitativ Henseende findes hos Egenskaberne af disse Stoffer.

Gaar vi nu over til den 4de Periode af Grundstoffernes System, møder vi til at begynde med Stoffer, som i kemisk Henseende ligner Stofferne i Begyndelsen af de to foregaaende Perioder. Dette er ogsaa, hvad vi maatte vente. Vi maa saaledes forestille os, at den 19de Elektron vil blive bundet i en Bane af ny Type, og den nærmere Betragtning viser, at dette maa være en 4_1 Bane. De Forhold, som vi omtalte ved Bindingen af Elektronen i Natriumatomet, viser sig her endnu mere udprægede paa Grund af det større Kvantetal, hvormed de indre Elektroners Baner er kendetegnede. Saaledes vil, som Følge af den Omstændighed at den paagældende Bane, hvad dens indre Sløjfer angaar, falder meget nær sammen med en 3_1 Bane, Dimensionerne af de ydre Dele af Banen hos den 19de Elektron i Kaliumatomet ikke alene afvige stærkt fra Dimensionerne af en 4_1 Bane i Brintatomet, men maa ventes at falde nær sammen med en Brintbane af Typen 2_1, hvis Dimensioner er ca. fire Gange mindre end den førstnævnte Brintbane. Dette Resultat tillader umiddelbart i Hovedtræk at gøre Rede for Kaliums spektrale og kemiske Egenskaber. Tilsvarende Resultater gælder for Calcium, i hvis neutrale Atom vi skal tænke os to Valenselektroner i ækvivalente 4_1 Baner. Gaar vi til Grundstoffer med højere Atomnummer, afviger imidlertid, som bekendt, Egenskaberne hos Stoffernes i den 4de Periode af Grundstoffernes System mere og mere fra tilsvarende Stoffer i de første Perioder, indtil vi i Jernmetallernes Familie møder Stoffer, der viser Egenskaber, som er væsentlig forskellige fra Stofferne i disse Perioder. Naar vi gaar til endnu højere Atomnumre, træffer vi paany ændrede Forhold, idet vi møder en Række Stoffer, der i deres kemiske Egenskaber nærmer sig mere og mere til Stofferne i den sidste Del af de forudgaaende Perioder, indtil vi endelig ved Atomnummeret 36 igen møder en inaktiv Luftart, nemlig Krypton.

Forhold af denne Art er nu netop, hvad vi maatte være forberedt paa at træffe. Medens Betragtningerne over Dannelsen og Stabiliteten af Atomkonfigurationerne for Stofferne af de tre første Perioder af Grundstoffernes System medfører, at enhver af de første 18 Elektroner i Atomet ogsaa i de følgende Elementer stadig vil blive bundet i en Bane af samme Type som den, hvormed den paagældende Elektron første Gang optræder, gælder dette, som man simpelt ser, ikke mere for den 19de Elektron. Med stigende Kerneladning og dermed følgende Formindskelse af Forskellen mellem de Kraftfelter, der hersker udenfor og indenfor Omraadet for Banerne af de 18 først bundne Elektroner, vil nemlig Dimensionerne af de Dele af en 4_1 Bane, der falder udenfor den nævnte Konfiguration mere og mere nærme sig til Dimensionerne af en 4-kvantet Bane beregnet ud fra den Antagelse, at man kan se bort fra Vekselvirkningen mellem Elektronerne i Atomet. Med stigende Atomnummer kommer der derfor snart et Øjeblik, hvor en 3_3 Bane vil svare til en fastere Binding af den 19de Elektron end en 4_1 Bane, og dette indtræffer allerede i Begyndelsen af den fjerde Periode. Medens vi ud af Kaliumspektret direkte kan læse, at Banen 4_1 vil svare til en mere end dobbelt saa fast Binding som Banen 3_3, der giver Anledning til det første Spektralterm i *D*-Rækken, stiller Forholdene sig allerede meget forskellige i Calcium. Her skal vi ikke se paa Buespektret, der udsendes under Indfangningen af den 20nde Elektron, men paa Gnistspektret, der svarer til Indfangningen og Bindingen af den 19de Elektron. Medens Magniums Gnistspektrum, der — bortset fra, at der i Formlen (12) optræder en Konstant fire Gange saa stor som den Rydbergske Konstant — viser stor Lighed med Natriumspektret, hvad Værdierne for Spektraltermerne i de forskellige Rækker angaar, møder vi i Calciums Gnistspektrum det ejendommelige Forhold, at det første Term af D-Rækken er større end det første Term af P-Rækken og kun er meget lidt mindre end det første Term af S-Rækken, der maa tænkes at svare til Bindingen af den 19de Elektron i Calciumatomets Normaltilstand. Allerede for det næste Element, Scandium, maa vi derfor være forberedt paa, at en 3_3 Bane vil svare til en fastere Binding af den 19de Elektron end en 4_1 Bane. Den 19de Elektron i Scandiumatomet, der saaledes maa ventes at bindes i en 3_3 Bane, vil dog, som det følger af det foregaaende, være bundet langt løsere end de første 18 Elektroner i Atomet, hvilket ogsaa svarer til, at Scandium i kemiske Forbindelser optræder elektropositivt med tre Valenser. Gaar vi til de følgende Stoffer, vil der i Atomet optræde et større Antal af Elektroner i 3_3 Baner i Atomets Normaltilstand, idet Antallet af saadanne Elektronbaner vil afhænge af Fastheden af deres Binding, sammenlignet med Fastheden af Elektronbindingen i en 4_1 Bane, i hvilken Type af Baner vi maa tænke os, at i det mindste den sidst indfangne Elektron i Atomet vil bevæge sig. Vi møder derfor Forhold vedrørende Atomernes Stabilitet og Dannelse, der er væsensforskellige fra dem, vi har betragtet ved Omtalen af de forudgaaende Perioder af Grundstoffernes System, og som bevirker, at vi her, i Modsætning til tidligere, har at gøre med den successive Udvikling med stigende Atomnummer af en af de indre Elektrongrupper i Atomet, i vort Tilfælde af Gruppen af Elektroner i 3-kvantede Baner. Først naar Udbygningen af denne Gruppe vil være fuldendt, kan vi vente igen at faa en tilsvarende Ændring i Stoffernes Egenskaber med stigende Atomnummer, som den vi finder i de foregaaende Perioder i Grundstoffernes System. En Betragtning af Egenskaberne af Stofferne i den sidste Del af den fjerde Periode viser nu direkte, at den vedkommende Gruppe i sin færdige Form maa besidde 18 Elektroner, idet vi f. Eks. maa tænke os, at Krypton foruden Grupperne af 1, 2 og 3-kvantede Baner besidder en udpræget symmetrisk Konfiguration af 8 Elektroner i 4-kvantede Baner bestaaende af fire 4_1 Baner og fire 4_2 Baner.

Spørgsmaalet er nu: Hvordan maa vi nærmere tænke os Gruppen af Elektroner i 3-kvantede Baner udformet? I Lighed med Konstitutionen af Elektrongruppen af 2-kvantede Baner kunde man maaske ved første Betragtning være tilbøjelig til at vente, at den fuldstændige Gruppe med 3-kvantede Baner vilde bestaa af tre Undergrupper med hver fire Elektroner med Baner henholdsvis af Typerne 3_1, 3_2 og 3_3, altsaa ialt kun 12 Elektroner i Stedet for 18, saaledes som det kræves for at gøre Rede for Grundstoffernes iagttagne Egenskaber. En nærmere Betragtning viser imidlertid, at en saadan Forventning heller ikke vilde være berettiget. Stabiliteten af den i Neon optrædende Konfiguration af 8 Elektroner med 2-kvantede Baner maa nemlig ikke alene tilskrives den symmetriske Konfiguration af Elektronbanerne i de to Undergrupper med henholdsvis 2_1 og 2_2 Baner, men ligesaa fuldt Muligheden for at bringe Banerne indenfor disse Undergrupper i harmonisk Forhold til hverandre. Anderledes stiller Forholdet sig imidlertid for Elektrongruppen med 3-kvantede Baner; her kan tre Undergrupper med hver fire Baner ikke ventes paa tilsvarende simpel Maade at kunne bringes i indbyrdes Sammenspil. Vi maa tværtimod tænke os, at Optagelsen af Elektroner i 3_3 Baner vil formindske Harmonien i Konfigurationen af Banerne indenfor de to første Undergrupper med 3-kvantede Baner, i det mindste naar vi ved stigende Atomnummer er naaet til et Punkt, hvor den 19de Elektron ikke mere som hos Scandium er bundet væsentlig løsere end de tidligere bundne Elektroner i 3-kvantede Baner, men er trukket saa dybt ind i Atomet, at den forløber væsentlig indenfor det samme Omraade af dette, hvor disse Elektroner bevæger sig. Vi maa nu tænke os, at denne Formindskelse af Harmonien saa at sige vil »aabne« den tidligere »lukkede« Konfiguration af Elektroner i 3_1 og 3_2 Baner og derved skabe Muligheden for Binding af yderligere Elektroner i Baner af disse Typer. Hvad Slutresultatet angaar, peger Tallet 18 paa, at der efter Gruppens endelige Udformning vil være tre Undergrupper med seks Elektroner hver. Selv om det ikke paa det nuværende Tidspunkt har været muligt indgaaende at følge de forskellige Trin i Gruppens Dannelse, støttes denne Slutning dog paa interessant Maade af den Omstændighed, at det er muligt at placere tre Konfigurationer hver med seks Elektronbaner paa simpel Maade relativt til hverandre. Denne Konfiguration af Undergrupperne udviser imidlertid ikke en saaden tetraedral Symmetri som Gruppen af 2-kvantede Baner i Kulstof, men derimod en Symmetri der, hvad Banenormalernes relative Konfiguration angaar, kan betegnes som trigonal.

Trods den store Forskel i de paagældende Grundstoffers Egenskaber kan Fuldstændiggørelsen af Gruppen med 18 Elektroner i 3-kvantede Baner for en nærmere Betragtning siges at ytre sig paa en Maade analog med Fuldstændiggørelsen af Gruppen af 2-kvantede Baner, der, som vi har set, ikke alene betinger Neons Egenskaber som inaktiv Luftart, men ogsaa maa antages at betinge de foregaaende Stoffers elektronegative og de paafølgende Stoffers elektropositive Egenskaber. At vi ikke træffer en inaktiv Luftart med en ydre Gruppe af 18 Elektroner i 3-kvantede Baner forklares simpelt ved de langt større Dimensioner, som en 3_3 Bane har i Sammenligning med en 2_2 Bane, der bevæger sig under Indflydelse af samme Kraftfelt. Dette bevirker nemlig, at den førstnævnte Gruppe ikke kan optræde som yderste Gruppe i et neutralt Atom, men kun i positivt ladede Ioner. I det karakteristisk Fald i Valens hos Kobber i Sammenligning med de foregaaende Grundstoffer, som Optræden af Cupro-Ioner betegner, tør vi imidlertid se en lignende Tendens til Afslutning af en symmetrisk Elektronkonfiguration, som vi har lært at kende i den udprægede elektronegative Karakter af et Stof som Fluor. At vi i Cupro-Ionen virkelig har at gøre med en fuldstændig Gruppe af 3-kvantede Baner, faar vi direkte Oplysning om gennem Kobberspektret, der i Modsætning til de yderst komplicerede Spektre hos de foregaaende Stoffer, der maa tilskrives den usymmetriske Karakter af det indre System, besidder en simpel Bygning af en Type meget nær svarende til Natriumspektrets. Dette maa utvivlsomt tilskrives en lignende simpel symmetrisk Struktur hos Cupro-Ionen som hos Natrium-Ionen, selv om den forskellige Konstitution af den ydre Elektrongruppe i disse Ioner gør sig gældende i den betydelige Forskel i Spektraltermernes Værdier for Kobberspektret sammenlignet med Natriumspektret saavel som i Dubletafstanden i P-Termerne for de to Spektre. Paa den anden Side lærer Optræden af Cupri-Forbindelserne os, at Gruppen af 3-kvantede Baner ikke i Kobberatomet besidder en tilsvarende Fasthed med Hensyn til Elektronbindingen som Gruppen af 2-kvantede Baner i Natriumatomet. Først hos det næste Grundstof, Z i n k, der optræder udelukkende som divalent, er Gruppens Elektroner saa fast bundne, at de ikke kan afgives ved sædvanlige kemiske Processer.

De her fremsatte Forestillinger om Dannelsen og Bygningen af Atomerne hos Stofferne af den fjerde Periode i Grundstoffernes System tillader ikke alene i store Træk at gøre Rede for de kemiske og spektrale Egenskaber af disse Stoffer, men den faar ogsaa Støtte ved Betragtning af Egenskaber af anden Art end dem, vi hidtil har betragtet. Som det er vel kendt, udmærker Grundstofferne i den 4de Periode sig paa forskellig Maade fra Stofferne i de foregaaende Perioder, nemlig dels ved deres magnetiske Egenskaber, dels ved de karakteristiske Farver, som Forbindelserne af disse Stoffer udviser. Vel optræder der Paramagnetisme og Farver hos Stoffer i de foregaaende Perioder, men ikke i saadanne simple Forbindelser, hvor de paagældende Atomer optræder som Ioner. Mange Stoffer i den 4de Periode derimod viser netop i dissocierede, vandige Opløsninger udprægede paramagnetiske Egenskaber og karakteristiske Farver. Betydningen af disse Forhold har bl. a. været betonet af L a d e n b u r g i Forbindelse med hans føromtalte Betragtninger over Fortolkningen af Egenskaberne hos Stofferne i de senere Perioder i Grundstoffernes System. I Modsætning til L a n g m u i r, — der for at gøre Rede for Forskellen mellem den 4de Periode og de forudgaaende Perioder i Systemet simpelthen antager, at Atomet udenom de Lag af Celler med hver 8 Elektroner, der benyttes til at anskueliggøre Egenskaberne hos Elementerne i de sidstnævnte Perioder, besidder et Lag af Celler med Plads til 18 Elektroner, der først er fuldt besat hos Krypton, som danner Afslutningen af den 4de Periode — tænker L a d e n b u r g sig, at der i Løbet af denne Periode af en eller anden ikke nærmere angivet Aarsag mellem den indre Elektronkonfiguration i Atomet, der allerede optræder i Argon, og den yderste Gruppe af Valenselektroner, udvikler sig et »Mellemlag«, hvis Dannelse begynder ved Scandium og er afsluttet netop ved Afslutningen af Jernmetallernes Familie. Til Støtte for denne Antagelse anfører Ladenburg ikke alene Betragtninger over de kemiske Egenskaber hos Stofferne indenfor den 4de Periode, men han henviser ogsaa til de ovenfor omtalte Egenskaber hos disse Stoffer, idet han fremhæver, at Optræden af Paramagnetisme og Farve netop findes hos de Stoffer, hos hvis Atomer dette Mellemlag skulde være under Udvikling. Som det vil ses, udviser L a d e n b u r g s Opfattelse visse formelle Lighedspunkter med den ovenfor givne Fortolkning af Fremkomsten af den 4de Periode i Grundstoffernes System, og det er nu af Interesse at se, at vore Betragtninger, baseret paa en direkte Undersøgelse af Betingelserne for Atomernes Dannelse, netop tilbyder Mulighed for Forstaaelse af de af L a d e n b u r g fremdragne Forhold.

Hvad de m a g n e t i s k e E g e n s k a b e r angaar, har vi i vore sædvanlige, elektrodynamiske Forestillinger vel ikke et tilstrækkeligt Grundlag til at gøre Rede for Atommagnetismen, hvad der dog ikke tør undre os, naar vi tænker paa, at disse Forestillinger ikke har vist sig i Stand til at gøre Rede for Straalingsfænomenerne, der netop er betinget af den intime Vekselvirkning mellem de fra Elektronernes Bevægelse hidrørende elektriske og magnetiske Kraftvirkninger. Hvordan disse Vanskeligheder end skal løses, ligger imidlertid den Antagelse nær, at en Optræden af Magnetisme som den, vi møder hos Stofferne i den 4de Periode i Grundstoffernes System, er betinget af en Mangel paa Symmetri i Atombygningen, hvorved de fra Elektronernes Bevægelse hidrørende magnetiske Kræfter hindres i at danne et System af lukkede, indenfor Atomerne selv forløbende Kraftlinier. Medens vi for Stofferne indenfor de forudgaaende Perioder i Systemet har antaget, at Ionerne, hvad enten de er positivt eller negativt ladede, indeholder Elektronkonfigurationer af udpræget symmetrisk Karakter, maa vi netop være forberedt paa at møde en udtalt Mangel paa Symmetri i Elektronkonfigurationen hos Ionerne af de Stoffer indenfor den 4de Periode, der indeholder en Elektrongruppe med 3-kvantede Baner, der befinder sig paa et Overgangstrin mellem en symmetrisk Konfiguration med henholdsvis 8 og 18 Elektroner. Som paapeget af K o s s e l, udviser de eksperimentelle Resultater en overordentlig Simpelhed, idet det magnetiske Moment hos Ionerne af de paagældende Grundstoffer er det samme hos Ioner, der indeholder det samme Antal Elektroner; f. Eks. udviser Ferri-Ioner samme Atommagnetisme som Mangano-Ioner, medens Mangani-Ioner udviser samme Atommagnetisme som Kromo-Ioner. I smukkeste Overensstemmelse med, hvad der ovenfor blev antaget om Bygningen af Atomerne hos Kobber og Zink, viser det sig nu, at Magnetismen netop forsvinder hos Ioner med 28 Elektroner, der som nævnt maa antages at indeholde en fuldstændig Gruppe af 3-kvantede Baner. I det hele giver Betragtninger af de magnetiske Egenskaber hos Stofferne indenfor den 4de Periode os en levende Forestilling om, hvordan vi hos de paagældende Stoffer iagttager et Saar i Atomernes ellers meget symmetriske indre Bygning, der, naar vi vandrer fra Stof til Stof, ses saavel at udvikles som at heles, og det maa forventes, at den nærmere Undersøgelse af de magnetiske Forhold vil kunne give en Ledetraad for den nøjere Forfølgning af den Maade, hvorpaa Udviklingen af Elektrongruppen af 3-kvantede Baner foregaar Skridt for Skridt.

Hvis vi vender os til Spørgsmaalet om Ionernes Farve, faar vi her en tilsvarende direkte Støtte for den udviklede Opfattelse af Atombygningen. Efter Kvanteteoriens Postulater maa vi tænke os, at Absorption saavel som Emission af Straaling foregaar ved Overgang mellem to stationære Tilstande. Optræden af Farver, d. v. s. Absorption af Lys svarende til det synlige Spektralomraade, lærer os derfor om Overgangsprocesser, der kan foregaa i Atomet, og som er ledsaget af Energiændringer af samme Størrelsesorden som de, vi iagttager i Stoffernes sædvanlige optiske Spektre. I Modsætning til Ionerne hos Elementerne af de foregaaende Perioder, hvor alle Elektronerne maa antages at være meget fast bundne, er Muligheden af saadanne Processer netop, hvad vi kan vente for Ionerne af de omhandlede Stoffer i den 4de Periode; thi, som vi har set, vil Udviklingen og Kompletteringen af Gruppen af Elektroner med 3-kvantede Baner foregaa saa at sige under Kappestrid med en Elektronbinding i højere-kvantede Baner, idet Antallet af optrædende 3-kvantede Baner netop er betinget af, at Elektronerne i disse Baner er bundne fastere end Elektroner i 4_1 Baner, og Udviklingen af den paagældende Gruppe vil derfor foregaa lige til det Punkt, hvor der saa at sige er Ligevægt mellem Bindingen i de to Slags Baner. Dette Forhold maa antages ikke alene at have nøjeste Forbindelse med Ionernes Farver, men ogsaa med de paagældende Stoffers Tilbøjelighed til at danne Ioner med flere forskellige Ladninger i Modsætning til Stofferne af de første Perioder, hvor Ladningen af Ioner i vandige Opløsninger altid er den samme for et og samme Stof.

For de paafølgende Perioder af Grundstoffernes System udvikler alle Betragtningerne sig nu af sig selv i fuldkommen Analogi med det, der hidtil er sagt. Ser vi saaledes paa de første Stoffer indenfor den 5te Periode, maa vi tænke os, at den 37te og 38te Elektron bindes i 5_1 Baner, hvad ogsaa en Undersøgelse af R u b i d i u m s Buespektrum og S t r o n t i u m s Gnistspektrum bekræfter. Som det sidste Spektrum viser, melder Baner 4_3 sig dog hurtigt, og i den paagældende Periode, der ligesom den 4de indeholder 18 Grundstoffer, maa vi tænke os, at vi er Vidne til Tilbagelæggelsen af et Stadium i Udviklingen af Elektrongruppen med 4-kvantede Baner. Denne Gruppe, hvis Dannelses første Stadium kan siges at være tilbagelagt i Krypton ved Optræden af en symmetrisk Konfiguration af 8 Elektroner bestaaende af to Undergrupper med hver fire Elektroner i 4_1 og 4_2 Baner, maa i S ø l v tænkes at have opnaaet en anden foreløbig Afslutning ved Optræden af en symmetrisk Konfiguration af 18 Elektroner, bestaaende af tre Undergrupper med hver seks Elektroner i Baner af Typerne 4_1, 4_2 og 4_3. Alt hvad vi har sagt om Omdannelsen af Elektrongruppen med 3-kvantede Baner gælder nemlig uforandret for det omhandlede Stadium af Omdannelsen af Gruppen med 4-kvantede Baner, idet vi paa intet Punkt har gjort Brug af de absolute Værdier af Kvantetallene eller af nærmere Antagelser om Banernes Form, men blot om Antallet af mulige Banetyper, der kunde komme i Betragtning. Paa samme Tid er det af Interesse at nævne, at de paagældende Elementers Egenskaber, sammenlignet med dem af den foregaaende Periode, dog viser en Forskel, der netop svarer til, hvad man skulde vente efter den antagne Forskel i Banetyperne. I den femte Periode kommer saaledes Forskellen fra de for Stofferne i den anden og tredie Periode karakteristiske Valensforhold senere for Dagen end hos Stofferne i den fjerde Periode; medens et Stof som T i t a n i den fjerde Periode allerede viser en udpræget Tilbøjelighed for at optræde med forskellige Valenser, er et Stof som Z i r k o n endnu rent firevalent ligesom Kulstof i den anden Periode og Silicium i den tredie. En simpel Undersøgelse af Elektronbanernes kinematiske Forhold viser nu netop, at en Elektron i en ekscentrisk 4_3 Bane hos et Stof i den femte Periode vil være betydelig løsere bundet end en Elektron i en cirkulær 3_3 Bane hos det tilsvarende Stof i den fjerde Periode, medens Elektroner, som i disse Stoffer er bundne i ekscentriske Baner af henholdsvis 5_1 og 4_1 Typen vil svare til en omtrentlig lige fast Binding.

Ved Afslutningen af den 5te Periode maa vi tænke os, at vi i Xenon, der har Atomnummeret 54, møder en Konfiguration, hvor der foruden de omtalte Konfigurationer af to 1-kvantede, otte 2-kvantede, atten 3-kvantede og atten 4-kvantede Baner optræder en symmetrisk Konfiguration af 8 Elektroner i Baner, bestaaende af to Undergrupper med hver 4 Elektroner i henholdsvis 5_1 og 5_2 Baner. Betragter vi Atomerne hos Stoffer med endnu højere Atomnumre, maa vi først og fremmest, som Spektrene af Cæsium og Barium lærer os, antage, at i disse Metallers Atomer den 55nde og 56nde Elektron bindes i 6_1 Baner; men vi maa nu være forberedt paa snart at træffe helt nye Forhold. Vi maa nemlig vente, at der med stigende Kerneladning ikke alene vil komme et Øjeblik, hvor en Elektron i en 5_3 Bane vil være bunden fastere end i en 6_1 Bane; men vi maa ogsaa vente, at det Øjeblik vil komme, hvor under Atomets Dannelse den 47nde Elektron ikke mere vil indfanges i en 5_1 Bane, men hvor en 4_4 Bane vil repræsentere en fastere Binding af denne Elektron, ganske paa samme Maade som vi ved Stofferne af den tredie Gruppe mødte det Øjeblik, hvor for første Gang den 19de Elektron bandtes i en 3_3 Bane i Stedet for i en 4_1 Bane. Er vi naaet til dette Punkt, maa vi vente med stigende Atomnummer at møde en Række paa hinanden følgende Stoffer, der ligesom Stofferne i Jernmetallernes Familie har meget nær de samme Egenskaber, kun at dette Forhold vil fremtræde i endnu højere Grad, fordi vi her har at gøre med den successive Omdannelse af en Elektronkonfiguration, der befinder sig dybere inde i Atomet. De vil allerede have forstaaet, at hvad jeg sigter til er en simpel Fortolkning af Optræden af de saakaldte sjældne Jordarters Familie i Begyndelsen af den 6te Periode indenfor Grundstoffernes System. Ligesom ved den tidligere betragtede Omdannelse og Komplettering af Gruppen med 3-kvantede Baner i Systemets 4de Periode og den delvise Komplettering af Gruppen med 4-kvantede Baner i den 5te Periode, kan vi her umiddelbart ud fra Længden af den 6te Periode paa Forhaand være sikre paa Antallet af Elektroner, nemlig 32, der indgaar i den betragtede Gruppe af Elektroner med 4-kvantede Baner efter denne Gruppes endelige Udvikling. I Analogi med hvad der gjaldt for Gruppen af 3-kvantede Baner, ligger den Slutning nær, at den færdige Gruppe vil indeholde 8 Elektroner i hver af de fire Undergrupper. Selv om det endnu ikke har været muligt at følge Gruppens Udvikling Skridt for Skridt, kan vi imidlertid ogsaa her allerede ud fra simple Betragtninger finde Støttepunkter for en teoretisk Forstaaelse af Optræden af en symmetrisk Konfiguration med netop dette Elektrontal. Jeg skal blot nævne, at det ikke er muligt uden Sammenfald af Baneplaner at opnaa et lignende simpelt Vekselspil mellem fire Undergrupper med hver seks Elektroner i en Banekonfiguration af trigonal Symmetri, som det der tilbyder sig for tre saadanne Undergrupper, samt at de Vanskeligheder, vi møder herved, gør det sandsynligt, at et harmonisk Sammenspil netop kan opnaas for fire Grupper med hver 8 Elektroner, hvis Banekonfigurationer udviser aksial Symmetri.

Ligesom det var Tilfældet ved Jernmetallernes Familie i den 3die Gruppe, støttes den fremsatte Fortolkning af Fremkomsten af de sjældne Jordarters Familie paa interessant Maade af en Undersøgelse af disse Stoffers magnetiske Egenskaber. Til Trods for den store Lighed i kemisk Henseende viser saaledes Medlemmerne af den sidstnævnte Familie meget forskellige magnetiske Forhold, og medens nogle af dem kun viser ringe Magnetisme, viser andre et større magnetisk Moment pr. Atom end noget andet undersøgt Stof. Ogsaa for de ejendommelige Farver, hvormed Forbindelserne af de omhandlede Stoffer optræder, opnaar vi Mulighed for en simpel Fortolkning analog med de ved Omtalen af den 4de Periode fremsatte Betragtninger.

Foruden den endelige Udformning af Gruppen med 4-kvantede Baner iagttager vi hos Stofferne i den 6te Periode af Grundstoffernes System det andet Stadium af Udformningen af Gruppen med 5-kvantede Baner i Fremkomsten af P l a t i n m e t a l l e r n e s Familie, samt det første foreløbige Stadium i Udformningen af en Elektrongruppe med 6-kvantede Baner, nemlig hos den radioaktive, kemisk inaktive Luftart N i t o n, der danner Afslutningen paa denne Periode. I Atomet af dette Stof maa vi, foruden Elektrongrupper med henholdsvis to 1-kvantede, otte 2-kvantede, atten 3-kvantede, to og tredive 4-kvantede og atten 5-kvantede Baner, tænke os en ydre symmetrisk Konfiguration af otte Elektroner i 6-kvantede Baner, bestaaende af to Undergrupper med hver fire Elektroner i henholdsvis 6_1 og 6_2 Baner.

Rent bortset fra den fundamentale Forskel, der ligger deri, at vi i Modsætning til Langmuirs statiske Modeller betragter Konfigurationer af Elektroner, der bevæger sig med store Hastigheder mellem hverandre paa en saadan Maade, at Elektronerne i de »ydre« Grupper under deres Omløb trænger ind paa Omraadet for Banerne af Elektronerne i de »indre« Grupper, vil det ses, at den Gruppedeling i Atomet, som vi ved direkte Betragtning af Atomernes Dannelse er blevet ført til f. Eks. for Niton, er væsentlig forskellig fra den Gruppedeling, der i Langmuirs Teori benyttes i Forsøget paa at fortolke Stoffernes Egenskaber. Paa lignende Maade som vi nævnte ved Spørgsmaalet om Forskellen mellem de kemiske Egenskaber af Stofferne i den 4de Periode af Grundstoffernes System og Egenskaberne hos Stofferne i de foregaaende Perioder antager saaledes Langmuir for at forklare Egenskaberne hos Stofferne i den 6te Periode, at Atomet udenom de indre Cellelag med henholdsvis 2, 8, 8, 18 og 18 Elektroner, der benyttes til at gøre Rede for Egenskaberne hos Stofferne i de tidligere Perioder, besidder et Lag af Celler med Plads til 32 Elektroner, der netop er udfyldt i Niton. [Tilføjelse efter Foredraget: Det er i denne Forbindelse af Interesse at nævne et nyligt udkommet Arbejde af Bury, som jeg lige har faaet Lejlighed til at gøre mig bekendt med, og som indeholder en interessant Gennemgang af Grundstoffernes kemiske Egenskaber, belyst ud fra lignende Forestillinger om Atombygningen som Lewis' og Langmuirs. Bury fremhæver med Styrke, at en Redegørelse for Egenskaberne hos Stofferne i de senere Perioder af Grundstoffernes System, langt naturligere end den af Langmuir foreslaaede, opnaas ved at antage, ikke alene at Fremkomsten af den 4de og 5te Periode i Systemet — i Lighed med hvad allerede Ladenburg havde foreslaaet — hidrører fra Komplettering af indre Elektrongrupper, men at ogsaa Fremkomsten af de sjældne Jordarters Familie i den 6te Periode hidrører fra en Komplettering af en indre Gruppe i Atomet; en Opfattelse som iøvrigt allerede Vegard har gjort sig til Talsmand for i Sammenhæng med sine Undersøgelser over Røntgenspektrene. Bury omtaler min første Note til »Nature«, hvori det her udviklede Billede af Atomet fremsattes i Hovedtræk, uden at han dog tager Stilling til Spørgsmaalet om Aarsagen til Fremkomsten og Stabiliteten af de forskellige Grupper i Atomet.]

Betragter vi endelig den 7ende og sidste Periode i Grundstoffernes System, maa vi vente Optræden i Atomernes Normaltilstand af 7-kvantede Baner. I det neutrale Radium-Atom vil der saaledes, foruden en Elektronkonfiguration svarende til den i Niton, optræde to Elektroner i 7_1 Baner, der under deres Omløb ikke alene vil trænge ind i Omraadet for Elektronbanerne med lavere Værdi for Hovedkvantetallet, men endda helt ind til Afstande fra Kernen mindre end Baneradierne af de inderste 1-kvantede Baner. Hvad Egenskaberne hos Stofferne i den 7ende Periode angaar, minder de meget om Egenskaberne hos Stofferne i den 5te Periode. I Modsætning til hvad der gjaldt for den 6te Periode optræder der i Begyndelsen af den 7ende Periode saaledes ikke Stoffer, hvis Egenskaber ligner hinanden paa en Maade, der svarer til Ligheden i kemisk Henseende mellem Medlemmerne af de sjældne Jordarters Familie. I nøje Analogi med hvad vi har sagt om Forholdet imellem Egenskaberne hos Stofferne i den 4de og 5te Periode, fortolkes dette Forhold simpelt derved, at en ekscentrisk 5_4 Bane vil svare til en betydelig løsere Binding af den 79nde Elektron i Atomet hos et Stof i den 7ende Periode end Bindingen af den 47nde Elektron i en cirkulær 4_4 Bane hos det tilsvarende Stof i den 6te Periode, medens der vil være en langt mindre Forskel paa Fastheden af Bindingen af de nævnte Elektroner i Baner af henholdsvis Typerne 6_1 og 5_1.

Som bekendt er den 7ende Periode hidtil ikke afsluttet, idet intet Stof er fundet, der har et Atomnummer større end 92, hvad der antagelig hænger sammen med, at de sidste Stoffer i Systemet er radioaktive, og at Atomkerner med en samlet Ladning af mere end 92 ikke vil være tilstrækkelig stabile til at eksistere under Forhold, hvor Elementernes Egenskaber kan iagttages. Det kunde være fristende at søge at gøre sig Forestillinger om Bygningen af Atomer, dannede ved Indfangning og Binding af Elektroner omkring Atomkerner med saadanne højere elektriske Ladninger, og derigennem om de Egenskaber, der kunde ventes hos de dertil svarende hypotetiske Stoffer. Jeg skal imidlertid ikke komme nærmere ind herpaa, idet Arten af de Forestillinger, hvortil vi vilde blive ført, vil være Dem klar af de Betragtninger, jeg her har fremsat om Fortolkningen af de virkelig iagttagne Stoffers Egenskaber. En Oversigt over Resultaterne af disse Betragtninger faar De maaske bedst ved følgende Tabel, der giver en skematisk Fremstilling af Forestillingerne om Atombygningen hos de inaktive Luftarter, der danner Afslutningen

Grundstof	Atomnum.	Antal af Elektroner i n_k Baner.																							
		1_1	2_1	2_2	3_1	3_2	3_3	4_1	4_2	4_3	4_4	5_1	5_2	5_3	5_4	5_5	6_1	6_2	6_3	6_4	6_5	6_6	7_1	7_2	7_3
Helium .	2	2																							
Neon . .	10	2	4	4																					
Argon . .	18	2	4	4	4	4	—																		
Krypton	36	2	4	4	6	6	6	4	4	—	—														
Xenon. .	54	2	4	4	6	6	6	6	6	6	—	4	4	—	—	—									
Niton . .	86	2	4	4	6	6	6	8	8	8	8	6	6	6	—	—	4	4	—	—	—	—			
? . .	118	2	4	4	6	6	6	8	8	8	8	8	8	8	8	—	6	6	6	—	—	—	4	4	—

paa de seks første Perioder i Grundstoffernes System. For at understrege Tendensen er yderligere angivet den formodentlige Elektronfiguration af det næste Atom, der vilde besidde lignende Egenskaber som Atomerne af de omhandlede Stoffer.

Før jeg forlader Spørgsmaalet om Fortolkningen af Grundstoffernes kemiske Egenskaber ud fra vort Billede af Atombygningen, vil jeg gerne endnu engang minde om, at de fremsatte Betragtninger hviler paa en Undersøgelse — baseret paa Kvanteteoriens Postulater — af de Processer, der maa tænkes knyttet til Dannelsen af et Atom ved Indfangningen af Elektronerne en efter en. Det Hovedsynspunkt, der er lagt til Grund for denne Undersøgelse, finder sit Udtryk i det saakaldte Korrespondensprincip, der bag Kvanteteoriens Anvendelser lader ane en indre Sammenhæng af en lignende Art som den, der findes i de klassiske Forestillinger om elektrodynamiske Processer, der til Trods for deres udstrakte Anvendelighed paa mange fysiske Fænomener dog har vist sig ude af Stand til at gøre Rede for den af Stoffernes Egenskaber krævede Stabilitet hos Atombygningen. Selv om det har været muligt paa mange Punkter at uddybe Betragtningerne i betydelig højere Grad, end Tiden her har tilladt mig at give Dem noget Indtryk af, har man dog endnu ikke ved en direkte Anvendelse af Korrespondensprincippet kunnet forfølge alle

Stadier i Atomernes Dannelse paa en saadan Maade, at man f. Eks. kan sige, at de Resultater, der fremtræder i ovenstaaende Tabel, i alle Enkeltheder kan betragtes som det entydige Resultat af en saadan Anvendelse. Paa den anden Side synes det, at allerede den Tendens, der ligger i Betragtningerne, stiller de empiriske Data i et Lys, der næppe tillader nogen væsentlig anden Fortolkning af Stoffernes Egenskaber udfra Kvanteteoriens Postulater. Dette gælder ikke alene om Seriespektrene og deres nøje Sammenhæng med Stoffernes kemiske Egenskaber, til Betragtning af hvilke vi direkte føres ved Undersøgelse af de Processer, der er knyttet til Atomernes Dannelse, men i fuldt saa høj Grad om Røntgenspektrene, hvis Betragtning fører os ind paa Undersøgelsen af indre-atomiske Processer af en helt anden Art. Som vi har nævnt, maa vi nemlig tænke os, at Udsendelsen af de sidste Spektre er knyttet til Processer, hvorved det færdigt dannede Atom reorganiseres efter en senere Forstyrrelse, frembragt i Atomets Indre ved ydre Kræfters Indgreb.

* *
*

For Røntgenspektrenes Vedkommende har det ligesom for Seriespektrene været muligt at opfatte Svingningstallene for de enkelte Linier som Differenser (Kombinationer) mellem en Række Spektraltermer karakteristiske for det betragtede Stof. Tydet ud fra Kvanteteoriens almindelige Frekvensbetingelse (1) skal vi derfor tænke os, at enhver Røntgenlinie fremkommer ved en Overgang mellem to blandt et vist Antal af stationære Tilstande af Atomet. Værdierne for Atomets Energi svarende til disse Tilstande benævner man ofte som de til Røntgenspektrene svarende »Energiniveauer«. Som allerede nævnt i Begyndelsen af Foredraget kommer den store Forskel imellem Røntgenspektrenes og Seriespektrenes Oprindelse imidlertid tydeligt for Dagen ved de forskellige Love, der gælder for Stoffernes Absorption af Straaling i Røntgenomraadet og i det optiske Omraade. Medens Absorptionen i det sidstnævnte Omraade under sædvanlige Forhold er knyttet til saadanne Linier i Seriespektrene, der svarer til Kombinationer mellem to Spektraltermer, hvoraf det ene er det størst forekommende — og derfor Absorptionen maa tilskrives en Proces, hvorved den sidst indfangne Elektron i Atomet fra sin Bane i dettes Normaltilstand bringes over i en Bane, svarende til et tidligere

Stadium af Bindingsprocessen — viser Absorptionen i Røntgenomraadet sig at være knyttet ikke til selve Linierne i Røntgenspektrene, men til udstrakte Spektralomraader, der tager deres Begyndelse ved de saakaldte »Absorptionskanter«, hvis Svingningstal, i det mindste med stor Tilnærmelse, har vist sig netop at svare til de Termer, af hvis Kombination de enkelte Linier i Spektret fremgaar. Vi maa derfor tænke os, at Atomet ved Absorption fra Normaltilstanden kan bringes over til enhver blandt de stationære Tilstande, ved Overgang mellem hvilke Røntgenspektrets Linier udsendes, og for hvilke Atomets Energi svarer til de ovennævnte Niveauer. Vi skal nu se, hvorledes det Billede af Atombygningen, der er udviklet i det foregaaende, tillader en simpel Fortolkning af disse Forhold, naar vi spørger om, hvilke Ændringer i Atomets Tilstand der kan ventes frembragt ved Absorption af Straaling, og hvilke efterfølgende Udstraalingsprocesser en saadan Ændring kan give Anledning til.

At vi ved Bestraaling overhovedet kan ændre Bevægelsen af en Elektron i Atomets Indre maa først og fremmest tænkes at staa i nøjeste Forbindelse med den Karakter af Vekselspillet mellem Elektronerne indenfor de enkelte Grupper, som vi blev ført til ved Betragtning af Muligheden for Atomernes Dannelse ved Elektronernes successive Binding, og som — i Modsætning til Bevægelsesformer ved hvilke Elektronernes Stilling hvert Øjeblik udviser polygonal eller polyhedral Symmetri, — medfører, at de harmoniske Svingningskomponenter i de enkelte Elektroners Bevægelse i Almindelighed er repræsenteret i Atomets resulterende elektriske Moment. Dette bevirker nemlig, at det vil være muligt ved en Proces, der besidder den krævede Analogi med det Billede af en Absorptionsproces, som vilde svare til vore sædvanlige elektrodynamiske Forestillinger, at udløse en enkelt Elektron af Vekselspillet med de andre Elektroner i samme Gruppe. De Synspunkter, som vi har lagt til Grund for Fortolkningen af Elektrongruppernes Udformning og Afslutning under Atomets Dannelse, fører paa den anden Side med sig, at en Ændring af Atomets Tilstand, frembragt ved Bestraaling, lige saa lidt som en til Udsendelse af Straaling knyttet Ændring i Elektronernes Bevægelse, kan resultere i Optagelsen af en ny Elektron i nogen Gruppe i Atomet af forud afsluttet Karakter. Dette betyder, at en Elektron, der tilhører en af Atomets indre Grupper, ved en Absorptionsproces — dersom den ikke fuldstændigt frigøres fra Atomet — kun kan overføres til en Gruppe af uafsluttet Karakter eller til en Bane, hvor Elektronen under Størstedelen af sit Omløb bevæger sig i en Afstand fra Kernen, der er stor i Forhold til de andre Elektroners Afstande. Paa Grund af de særegne Stabilitetsforhold, der, som vi har set, behersker Optræden af ufuldstændige Grupper i det Indre af Atomet, vil imidlertid den Energi, der er nødvendig for Overførelsen af en Elektron til en saadan Gruppe, i Almindelighed kun afvige meget lidt fra den, der kræves til at fjerne vedkommende Elektron fuldstændigt fra Atomet. I første Tilnærmelse kan vi derfor sige, at de til Absorptionskanterne svarende Energiniveauer lærer os om det Arbejde, der skal til for at fjerne en Elektron i en af de indre Grupper fuldstændigt fra Atomet. Hvad dernæst Fremkomsten af Røntgenspektrenes Emissionslinier ved Overgange mellem de til disse Niveauer svarende Tilstande i Atomet angaar, giver de omhandlede Synspunkter os et tilsvarende Grundlag for Forstaaelsen af Iagttagelserne. Den ovenfor beskrevne Form af Vekselspillet mellem Elektronerne i Atomets Grupper fører nemlig med sig, at enhver Elektron i Atomet saa at sige er forberedt paa, uafhængig af de andre Elektroner i samme Gruppe, at gribe enhver Lejlighed, der tilbyder sig for en fastere Binding ved Optagelse i en Elektrongruppe med Baner svarende til mindre Værdier for Hovedkvantetallet. Ud fra de Synspunkter, vi har lagt til Grund for Atomets Opbygning, er det imidlertid indlysende, at en saadan Lejlighed er til Stede, saasnart en Elektron er fjernet fra en af disse Grupper.

Samtidig med, at det omhandlede Billede af Atomet saaledes utvungent fører til en Opfattelse af Røntgenspektrenes Fremkomst og Absorptionsforhold, der stemmer nøje overens med de Synspunkter, ud fra hvilke Kossel paa formel Maade har søgt at fortolke Iagttagelserne, tillader det at give en simpel Forstaaelse af de kvantitative Relationer gældende for Røntgenliniernes Svingningstal, som Moseleys og Sommerfelds Undersøgelser har lært os at kende. Som vi har set, bragte disse Undersøgelser en ejendommelig og overraskende Lighed for Dagen mellem Røntgenspektret af et givet Stof og det Spektrum, der maatte ventes at fremkomme ved Atomkernens Binding af en enkelt Elektron. Denne Lighed forstaas umiddelbart, naar vi tænker paa, at der i Atomets Normaltilstand vil findes Elektroner i Baner, der, med en vis Begrænsning, svarer til alle Stadier af en saadan Bindingsproces, samt at der, naar en Elektron er fjernet fra sin oprindelige Plads i Atomet, indenfor det forstyrrede Atom vil kunne udløses Processer, der svarer til alle af Korrespondensprincippet tilladte Overgange mellem disse Stadier. Disse Omstændigheder bringer os med et Slag ud over de Vanskeligheder, som man tilsyneladende møder, dersom man forsøger at gøre Rede for Røntgenspektrenes Fremkomst ved Hjælp af et Billede af Atomet, egnet til en Fortolkning af Grundstoffernes periodiske System. Denne Vanskelighed har været følt i den Grad, at den f. Eks. har ført Sommerfeld til i sine senere Arbejder at antage, at Elektronfigurationerne i de forskellige Atomer af et og samme Stof allerede under sædvanlige Omstændigheder kan være forskellige. Idet han i Modsætning til vore Forestillinger tænker sig, at alle Elektronerne i Atomets Hovedgrupper bevæger sig i ækvivalente Baner, tvinges han nemlig til at antage, at disse Grupper i de forskellige Atomer er forskelligt udformede, svarende til forskellige mulige Typer af Baneformer. En saadan Antagelse forekommer imidlertid uforenelig med en Fortolkning af den bestemte Karakter af Stoffernes karakteristiske fysiske og kemiske Egenskaber og staar, som det vil ses, i den mest iøjnefaldende Modstrid med de Synspunkter om Atomernes Stabilitetsforhold, der er lagt til Grund for det her fremsatte Billede af Atombygningen.

Det er i denne Forbindelse af Interesse at fremhæve, at Gruppedelingen af Elektronerne i Atomet, — som i Overensstemmelse med vort Billede af Atombygningen ligger til Grund saavel for Fortolkningen af det periodiske System som af Klassificeringen af Linierne i Røntgenspektrene, — iøvrigt ytrer sig paa helt forskellig Maade i de kemiske Egenskaber og de røntgenspektroskopiske Fænomener. Medens den karakteristiske Ændring af de kemiske Egenskaber med Atomnummeret skyldes den gradvise Udformning og Komplettering af Grupperne af de løsest bundne Elektroner, har den for Røntgenspektrene karakteristiske Mangel paa saa at sige ethvert Spor af en periodisk Ændring med Atomnummeret sin Grund ikke alene deri, at Elektronkonfigurationen for de afsluttede Gruppers Vedkommende, hvad de optrædende Banetyper angaar, gentages uforandret for stigende Atomnummer, men lige saa meget i den gradvise Maade, hvorpaa de uafsluttede Grupper ifølge vore Betragtninger udformes, og som har til Følge, at en Banetype, fra det Øjeblik hvor den for første

Gang optræder i Normaltilstanden af det neutrale Atom, stadig vil optræde i denne Tilstand og vil svare til en Binding af de paagældende Elektroner, hvis Fasthed vokser jævnt med Atomnummeret. Den Udformning af Grupperne i Atomet med stigende Atomnummer, der behersker Stoffernes kemiske Egenskaber, kommer i Røntgenspektrene derfor kun direkte for Dagen ved Optræden af nye Linier i Spektrene, efterhaanden som der i Atomet optræder Elektronbaner af nye Typer. Til en Forbindelse af denne Art mellem det periodiske System og Røntgenspektrene har allerede Swinne henvist i Tilknytning til Kossels Teori. [Tilføjelse efter Foredraget: Som Hr. Coster venligt har meddelt mig, har han i den sidste Tid i Sammenhæng med Fortsættelsen af sine nedenfor omtalte, betydningsfulde Undersøgelser over Røntgenspektrenes formelle Udredning opnaaet Resultater, der paa interessant Maade synes at støtte de Antagelser om den gradvise Udformning af Elektrongrupperne i Atomet, som faar deres Udtryk i den i Tabellen paa Side 60 antydede Opbygning af de inaktive Luftarters Atomer]. Iøvrigt kan vi kun for saadanne røntgenspektroskopiske Fænomener, hvor Forholdene ved Atomets »Overflade« kommer direkte i Betragtning, vente nogen nærmere Forbindelse mellem disse Fænomener og Stoffernes kemiske Egenskaber. I Overensstemmelse med hvad der er bragt for Dagen ved Undersøgelser over Absorptionen af Røntgenstraaler i Stoffer af lavere Atomnummer, som i de sidste Aar er udførte i det fysiske Laboratorium i Lund, forstaar vi saaledes umiddelbart, at Absorptionskanternes Beliggenhed og eventuelle Struktur i nogen Grad afhænger af den fysiske og kemiske Tilstand, hvori det undersøgte Stof befinder sig, medens en saadan Afhængighed ikke gør sig gældende for de karakteristiske Emissionslinier i Røntgenspektrene.

Ved et Forsøg paa en nærmere Fortolkning af Iagttagelserne ud fra vore Forestillinger om Atombygningen møder vi Spørgsmaalet om, hvilken Indflydelse Tilstedeværelsen af de andre Elektroner i Atomet vil have paa Fastheden af Bindingen af en Elektron i en bestemt Banetype. Denne Indflydelse vil, som man umiddelbart ser, være mindst for de indre Dele af Atomet, hvor for enhver Elektron Kernens Tiltrækning er stor i Forhold til Frastødningen fra de andre Elektroners Side. Det maa i denne Forbindelse ogsaa erindres, at medens den relative Indflydelse af de andre Elektroners Tilstedeværelse paa Bindingens Fasthed med voksende Kerneladning vil aftage, vil Virkningen af Elektronmassens Foranderlighed med Hastigheden paa Bindingens Fasthed, saaledes som det fremgaar af Sommerfelds Formel (11), vokse stærkt, naar Kerneladningen tiltager. Medens vi derfor for de Niveauer, der svarer til Fjernelsen af en af de inderste Elektroner i Atomet, faar en nogenlunde god Overensstemmelse allerede ved Brug af den simple Formel (11), er det imidlertid for blot en tilnærmet Redegørelse for de Niveauer, der svarer til en Fjernelse af en Elektron fra en af de ydre Grupper i Atomet, nødvendigt at tage de andre Elektroners Indflydelse i Betragtning. Til Gengæld byder imidlertid netop denne Omstændighed os en Mulighed for ved Hjælp af Røntgenspektrene at faa Oplysninger om Elektronfigurationerne i Atomernes Indre. Over dette Spørgsmaal er der anstillet talrige indgaaende Undersøgelser saavel af Sommerfeld og hans Elever, som af Debye, Vegard og andre. I denne Forbindelse kan det ogsaa bemærkes, at de Broglie og Dauvillier i et nylig offentliggjort Arbejde har ment i det eksperimentelle Materiale at kunne finde Støttepunkter for Antagelser om Elektrontallene i Atomets Grupper, til hvilke Dauvillier var blevet ført ved lignende Betragtninger over det periodiske System, som de der er blevet fremsat af Langmuir og Ladenburg. Ved de i Sammenhæng med disse Undersøgelser udførte Beregninger er det antaget, at Elektronerne i de forskellige Grupper af Atomet bevæger sig i adskilte koncentriske Omraader af dette, saaledes at Virkningen af Tilstedeværelsen af Elektronerne i indre Grupper paa Bevægelsen af Elektronerne i ydre Grupper i første Tilnærmelse maa ventes at bestaa i en simpel »Afskærmning« af den fra Kernen udgaaende Tiltrækning. Paa Grundlag af et Billede af Atombygningen som det her fremsatte stiller Forholdene sig imidlertid væsentlig anderledes, idet det for Beregningen af Elektronbindingens Fasthed er nødvendigt at tage Hensyn til, ikke alene at Elektronerne i de løsere bundne Grupper under Størstedelen af deres Omløb bevæger sig udenom de fastere bundne Elektroners Baner, men ogsaa at Elektronerne i Almindelighed under en vis Brøkdel af deres Omløb vil trænge ind i Omraadet for Elektronbanerne i de fastere bundne Grupper. Paa Grund af denne sidstnævnte Omstændighed, hvis Indflydelse vi har set Eksempler paa ved Betragtningen af Seriespektrene, kan vi ikke vente at gøre Regnskab for Fastheden af Bindingen af de enkelte Elektroner ved en simpel »Afskærmningskorrektion«, bestaaende i et Fradrag af en konstant Størrelse i Værdien for N i saadanne Formler som (5) og (11). Hertil kommer endvidere, at man ved Beregning af det til Energiniveauerne svarende Arbejde maa tage Hensyn ikke alene til Vekselvirkningen mellem Elektronerne i Atomets Normaltilstand, men ligeledes til de Ændringer i de øvrige Elektroners Konfiguration og Vekselvirkning, der uden Udsendelse af Straaling indstiller sig automatisk under den paagældende Elektrons Fjernelse fra sin Plads i Atomet. Selv om saadanne Beregninger endnu ikke er gennemført med nogen større Tilnærmelse, har en foreløbig Undersøgelse dog allerede vist, at det er muligt tilnærmelsesvis at gøre Rede for de eksperimentelle Resultater ud fra de Antagelser om Elektronkonfigurationernes Beskaffenhed, som Betragtningerne over Atomernes Dannelse har ført os til.

Uafhængig af Spørgsmaalet om Udformningen af bestemte Forestillinger om Atombygningen er det i den allersidste Tid lykkedes ved en formel Anvendelse af Kossels og Sommerfelds Teorier paa detailleret Maade at gøre Rede for det store Iagttagelsesmateriale angaaende Røntgenspektrene og de til disse svarende Niveauer, der først og fremmest er fremskaffet ved de nøjagtige Maalinger af disse Spektre, der i de sidste Aar er udført af Siegbahn og hans Medarbejdere. Ved denne Udredelse af Erfaringsmaterialet, i hvilken foruden Sommerfeld og hans Elever, navnlig Smekal og Coster har taget Del, er man saaledes ført til Erkendelsen af Muligheden for en fuldstændig Klassification af de til Røntgenspektrene svarende Energiniveauer, baseret paa en formel Henføring til Banetyper, karakteriseret ved to Kvantetal n og k, samt til Opstillingen af bestemte Regler for Kombinationsmulighederne mellem de saaledes klassificerede Niveauer. Herigennem er der naaet flere Resultater af stor Interesse for den nærmere Belysning af Røntgenspektrenes Fremkomst. For det første har det ikke alene været muligt at finde Niveauer, der indenfor visse Grænser svarer til alle mulige Talpar for n og k, men det har vist sig, at der i Almindelighed til hvert saadant Talpar maa tilskrives mere end et Niveau. Dette Resultat, der i første Øjeblik kunde forekomme overraskende ud fra vort Billede af Atombygningen, tillader imidlertid ved nærmere Betragtning en simpel Fortolkning. Vi maa nemlig, som ovenfor berørt, huske paa, at Niveauerne ikke alene er betinget af Atomets Konstitution i Normaltilstanden, men lige saa meget af de Elektronkonfigurationer, der optræder efter Fjernelsen af en af Elektronerne i Atomets Indre, og som i Modsætning til Normaltilstanden netop ikke besidder en entydig afsluttet Karakter. Betragter vi saaledes en Proces, hvorved en af Elektronerne i en Gruppe (Undergruppe) fjernes, maa vi derfor være forberedt paa, at Banerne af de resterende Elektroner i denne Gruppe efter Processen vil kunne være orienteret paa mere end en Maade i Forhold til hverandre i Konfigurationer, der opfylder de af Kvanteteorien til stationære Tilstande stillede Krav. En saadan Opfattelse af Niveauernes Komplexitet tillader, som en nærmere Betragtning viser, ogsaa netop at gøre Rede for den Maade, hvorpaa Afstandene mellem sammenhørende Niveauer varierer med Atomnummeret. Uden at forsøge at danne sig nogen nærmere udformet Forestilling om Atomets Bygning har allerede S m e k a l diskuteret Muligheden af at gøre Rede for Mangfoldigheden af de iagttagne Niveauer. Foruden at henvise til den Mulighed, at de enkelte Elektroner i Atomets Hovedgrupper, ligesom det her er antaget, ikke bevæger sig i ækvivalente Baner, foreslaar S m e k a l Indførelsen af tre Kvantetal til Beskrivelsen af Elektronkonfigurationerne i Atomets forskellige Grupper, men angiver ikke nærmere, hvorvidt disse Kvantetal skal opfattes som karakteriserende en Komplexitet i Bygningen af Grupperne i Normaltilstanden selv eller af saadanne Grupper af ufuldstændig Karakter, der optræder i Atomer, hvorfra en Elektron er fjernet ved ydre Indgreb.

Den her fremsatte Opfattelse af R ø n t g e n n i v e a u e r n e s K o m p l e x i t e t vil ses at udvise nøje Analogi med den i det foregaaende omtalte Fortolkning af Komplexiteten af Seriespektrenes Termer. Mellem den med Niveauernes Komplexitet sammenhængende finere Struktur af Røntgenspektrene og den omhandlede sammensatte Bygning af Linierne i Seriespektrene er der dog den Forskel, at der i Røntgenspektrene — i Modsætning til hvad der under sædvanlige Omstændigheder gælder for de sidstnævnte Spektre — ikke alene forekommer Kombinationer mellem Spektraltermer (Niveauer), for hvilke Værdien af k varierer med en Enhed, men ogsaa imellem Termer, der svarer til samme Værdi af k. Dette maa antages at hid-

[254]

røre fra, at vi i Røntgenspektrene i Modsætning til Seriespektrene har at gøre med Overgange mellem stationære Tilstande, hvor baade før og efter Overgangen den Elektron, til hvis Bevægelsesændring Straalingsudsendelsen er knyttet, deltager i intimt Vekselspil med andre Elektroner i Baner med samme Hovedkvantetal. Selv om dette Vekselspil maa antages at være af saadan Beskaffenhed, at de Svingningskomponenter, der vilde fremtræde i en enkelt af Elektronernes Bevægelser uden de andres Nærværelse, i Almindelighed ogsaa vil optræde i Atomets resulterende Moment, maa vi nemlig vente, at Vekselvirkningen mellem Elektronerne vil give Anledning til Optræden i dette Moment af Svingningskomponenter af nye Typer. Samtidig med at dette Forhold giver en Forstaaelse af, at det, som Niveauernes Komplexitet viser os, er muligt ved en Straalingsproces paa mere end en Maade at udløse en Elektron af Vekselspillet med de andre i samme Undergruppe i ækvivalente Baner bundne Elektroner, tilbyder det os et Udgangspunkt for at gøre Rede for de iagttagne Kombinationer mellem Niveauerne.

Jeg skal her indskrænke mig til disse Antydninger af den Fortolkning, som de fremsatte Betragtninger over Atombygningen synes at tilbyde for de røntgenspektroskopiske Fænomener, men idet jeg afslutter dette Foredrag, vil jeg gerne endnu en Gang understrege den nøje Forbindelse mellem denne Fortolkning og de forudgaaende Betragtninger vedrørende Fortolkningen af Stoffernes karakteristiske kemiske Egenskaber, saaledes som de kommer til Syne i Grundstoffernes periodiske System. Den Omstændighed, at det er en ganske tilsvarende Anvendelse af Kvanteteorien paa Kerneatomets Stabilitetsforhold, som vi ved Fortolkningen af de to Slags Fænomener har gjort Brug af, er nemlig et Punkt af største Vægt for Bedømmelsen af Betragtningernes Realitet; thi Berettigelsen af at benytte Betragtninger over Atomernes D a n n e l s e, ved den første Indfangning af Elektronerne en efter en, som en Hovedledetraad ved Udforskningen af Atombygningen kunde paa Forhaand forekomme tvivlsom, dersom saadanne Betragtninger ikke lod sig bringe i utvungen Overensstemmelse med Betragtninger over Atomets R e o r g a n i s a t i o n efter en senere Forstyrrelse i det Indre af det færdigt dannede Atom frembragt ved ydre Indgreb. Selv om der maaske saaledes kan paapeges

5*

en vis indre Sammenhæng i de Forestillinger om Atombygningen, som jeg har søgt at give Dem et Indblik i, behøver jeg dog neppe at fremhæve Betragtningernes uafsluttede Karakter ikke alene hvad Udarbejdelsen af Enkeltheder, men ogsaa hvad Underbygningen af de almindelige Synspunkter angaar. Der synes imidlertid ikke at tilbyde sig nogen anden Vej til Fremtrængen paa Atomspørgsmaalets Omraade end den, som man hidtil har fulgt, nemlig at lade Arbejdet paa Udviklingen af vore Forestillinger i de to nævnte Retninger følges ad Haand i Haand.

The Theory of Spectra
and
Atomic Constitution

THREE ESSAYS

BY

NIELS BOHR

Professor of Theoretical Physics in the University of Copenhagen

SECOND EDITION

CAMBRIDGE
AT THE UNIVERSITY PRESS
1924

PREFACE

THE three essays which here appear in English all deal with the application of the quantum theory to problems of atomic structure and refer to the different stages in the development of this theory.

The first essay "On the spectrum of hydrogen" is a translation of a Danish address given before the Physical Society of Copenhagen on the 20th of December 1913, and printed in *Fysisk Tidsskrift,* XII. p. 97, 1914. Although this address was delivered at a time when the formal development of the quantum theory was only at its beginning, the reader will find the general trend of thought very similar to that expressed in the later addresses, which form the other two essays. As emphasized at several points the theory does not attempt an "explanation" in the usual sense of this word, but only the establishment of a connection between facts which in the present state of science are unexplained, that is to say the usual physical conceptions do not offer sufficient basis for a detailed description.

The second essay "On the series spectra of the elements" is a translation of a German address given before the Physical Society of Berlin on the 27th of April 1920, and printed in *Zeitschrift für Physik,* VI. p. 423, 1920. This address falls into two main parts. The considerations in the first part are closely related to the contents of the first essay; especially no use is made of the new formal conceptions established through the later development of the quantum theory. The second part contains a survey of the results reached by this development. An attempt is made to elucidate the problems by means of a general principle which postulates a formal correspondence between the fundamentally different conceptions of the classical electrodynamics and those of the quantum theory. The first germ of this correspondence principle may be found in the first essay in the deduction of the expression for the constant of the hydrogen spectrum in terms of Planck's constant and of the quantities which in Rutherford's atomic model are necessary for the description of the hydrogen atom.

The third essay "The structure of the atom and the physical and chemical properties of the elements" is based on a Danish address, given before a joint meeting of the Physical and Chemical Societies of Copenhagen on the 18th of October 1921, and printed in *Fysisk Tidsskrift*, XIX. p. 153, 1921. While the first two essays form verbal translations of the respective addresses, this essay differs from the Danish original in certain minor points. Besides the addition of a few new figures with explanatory text, certain passages dealing with problems discussed in the second essay are left out, and some remarks about recent contributions to the subject are inserted. Where such insertions have been introduced will clearly appear from the text. This essay is divided into four parts. The first two parts contain a survey of previous results concerning atomic problems and a short account of the theoretical ideas of the quantum theory. In the following parts it is shown how these ideas lead to a view of atomic constitution which seems to offer an explanation of the observed physical and chemical properties of the elements, and especially to bring the characteristic features of the periodic table into close connection with the interpretation of the optical and high frequency spectra of the elements.

For the convenience of the reader all three essays are subdivided into smaller paragraphs, each with a headline. Conforming to the character of the essays there is, however, no question of anything like a full account or even a proportionate treatment of the subject stated in these headlines, the principal object being to emphasize certain general views in a freer form than is usual in scientific treatises or text books. For the same reason no detailed references to the literature are given, although an attempt is made to mention the main contributions to the development of the subject. As regards further information the reader in the case of the second essay is referred to a larger treatise "On the quantum theory of line spectra," two parts of which have appeared in the Transactions of the Copenhagen Academy (*D. Kgl. Danske Vidensk. Selsk. Skrifter*, 8. Række, IV. 1, I and II, 1918), where full references to the literature may be found. The proposed continuation of this treatise, mentioned at several places in the second essay, has for various reasons been delayed, but in the near future the work will be completed by the publication of a third part. It is my intention to deal more fully with the problems discussed in the third essay by a larger systematic account of the application of the quantum theory to atomic problems, which is under preparation.

As mentioned both in the beginning and at the end of the third essay, the considerations which it contains are clearly still incomplete in character. This holds not only as regards the elaboration of details, but also as regards the development of the theoretical ideas. It may be useful once more to emphasize, that—although the word "explanation" has been used more liberally than for instance in the first essay—we are not concerned with a description of the phenomena, based on a well-defined physical picture. It may rather be said that hitherto every progress in the problem of atomic structure has tended to emphasize the well-known "mysteries" of the quantum theory more and more. I hope the exposition in these essays is sufficiently clear, nevertheless, to give the reader an impression of the peculiar charm which the study of atomic physics possesses just on this account.

I wish to express my best thanks to Dr A. D. Udden, University of Pennsylvania, who has undertaken the translation of the original addresses into English, and to Mr C. D. Ellis, Trinity College, Cambridge, who has looked through the manuscript and suggested many valuable improvements in the exposition of the subject.

N. BOHR.

COPENHAGEN,
May 1922.

PREFACE TO SECOND EDITION

Since the appearance of the first edition of these essays work has been published by which the main ideas of the theory of atomic constitution would seem to have obtained further support. At the same time this work has revealed difficulties at various points, to the effect of emphasizing still more strongly the incomplete character of the theory. Although this might make desirable certain changes in the form of the essays, especially of the third, I have nevertheless decided to retain the original text and confine myself to giving a brief account of the later development in an appendix at the end of the book.

N. B.

COPENHAGEN,
May 1924.

ESSAY III

THE STRUCTURE OF THE ATOM AND THE PHYSICAL AND CHEMICAL PROPERTIES OF THE ELEMENTS

APPENDIX

ESSAY III*

THE STRUCTURE OF THE ATOM AND THE PHYSICAL AND CHEMICAL PROPERTIES OF THE ELEMENTS

I. PRELIMINARY

In an address which I delivered to you about a year ago I described the main features of a theory of atomic structure which I shall attempt to develop this evening. In the meantime this theory has assumed more definite form, and in two recent letters to *Nature* I have given a somewhat further sketch of the development†. The results which I am about to present to you are of no final character; but I hope to be able to show you how this view renders a correlation of the various properties of the elements in such a way, that we avoid the difficulties which previously appeared to stand in the way of a simple and consistent explanation. Before proceeding, however, I must ask your forbearance if initially I deal with matters already known to you, but in order to introduce you to the subject it will first be necessary to give a brief description of the most important results which have been obtained in recent years in connection with the work on atomic structure.

The nuclear atom. The conception of atomic structure which will form the basis of all the following remarks is the so-called nuclear atom according to which an atom is assumed to consist of a nucleus surrounded by a number of electrons whose distances from one another and from the nucleus are very large compared to the dimensions of the particles themselves. The nucleus possesses almost the entire mass of the atom and has a positive charge of such a magnitude that the number of electrons in a neutral atom is equal to the number of the element in the periodic system, the so-called *atomic number*. This idea of the atom, which is due principally to Rutherford's fundamental researches on radioactive substances, exhibits extremely simple features, but just this simplicity appears at first sight to present difficulties in explaining the properties of the elements. When we treat this question on

* Address delivered before a joint meeting of the Physical and Chemical Societies in Copenhagen, October 18, 1921.

† *Nature*, March 24, and October 13, 1921.

the basis of the ordinary mechanical and electrodynamical theories it is impossible to find a starting point for an explanation of the marked properties exhibited by the various elements, indeed not even of their permanency. On the one hand the particles of the atom apparently could not be at rest in a state of stable equilibrium, and on the other hand we should have to expect that every motion which might be present would give rise to the emission of electromagnetic radiation which would not cease until all the energy of the system had been emitted and all the electrons had fallen into the nucleus. A method of escaping from these difficulties has now been found in the application of ideas belonging to the quantum theory, the basis of which was laid by Planck in his celebrated work on the law of temperature radiation. This represented a radical departure from previous conceptions since it was the first instance in which the assumption of a discontinuity was employed in the formulation of the general laws of nature.

The postulates of the quantum theory. The quantum theory in the form in which it has been applied to the problems of atomic structure rests upon two postulates which have a direct bearing on the difficulties mentioned above. According to the first postulate there are certain states in which the atom can exist without emitting radiation, although the particles are supposed to have an accelerated motion relative to one another. These *stationary states* are, in addition, supposed to possess a peculiar kind of stability, so that it is impossible either to add energy to or remove energy from the atom except by a process involving a transition of the atom into another of these states. According to the second postulate each emission of radiation from the atom resulting from such a transition always consists of a train of purely harmonic waves. The frequency of these waves does not depend directly upon the motion of the atom, but is determined by a *frequency relation*, according to which the frequency multiplied by the universal constant introduced by Planck is equal to the total energy emitted during the process. For a transition between two stationary states for which the values of the energy of the atom before and after the emission of radiation are E' and E'' respectively, we have therefore

$$h\nu = E' - E'', \quad \ldots\ldots\ldots\ldots\ldots\ldots\ldots(1)$$

where h is Planck's constant and ν is the frequency of the emitted radiation. Time does not permit me to give a systematic survey of the quantum theory, the recent development of which has gone hand in hand with its applications to atomic structure. I shall therefore immediately proceed to the consideration of those applications of the theory which are of direct importance in connection with our subject.

Hydrogen atom. We shall commence by considering the simplest atom conceivable, namely, an atom consisting of a nucleus and one electron. If the charge on the nucleus corresponds to that of a single electron and the system consequently is neutral we have a hydrogen atom. Those developments of the quantum theory which have made possible its application to atomic structure started with the interpretation of the well-known simple spectrum emitted by hydrogen. This spectrum consists of a series of lines, the frequencies of which are given by the extremely simple Balmer formula

$$\nu = K\left(\frac{1}{(n'')^2} - \frac{1}{(n')^2}\right), \quad \text{.....................(2)}$$

where n'' and n' are integers. According to the quantum theory we shall now assume that the atom possesses a series of stationary states characterized by a series of integers, and it can be seen how the frequencies given by formula (2) may be derived from the frequency relation if it is assumed that a hydrogen line is connected with a radiation emitted during a transition between two of these states corresponding to the numbers n' and n'', and if the energy in the nth state apart from an arbitrary additive constant is supposed to be given by the formula

$$E_n = -\frac{Kh}{n^2}. \quad \text{.......................(3)}$$

The negative sign is used because the energy of the atom is measured most simply by the work required to remove the electron to infinite distance from the nucleus, and we shall assume that the numerical value of the expression on the right-hand side of formula (3) is just equal to this work.

As regards the closer description of the stationary states we find that the electron will very nearly describe an ellipse with the nucleus in the focus. The major axis of this ellipse is connected with the energy of the atom in a simple way, and corresponding to the energy values of the stationary states given by formula (3) there are a series of values for the major axis $2a$ of the orbit of the electron given by the formula

$$2a_n = \frac{n^2 e^2}{hK}, \quad\ldots\ldots\ldots\ldots\ldots\ldots\ldots\ldots (4)$$

where e is the numerical value of the charge of the electron and the nucleus.

On the whole we may say that the spectrum of hydrogen shows us the *formation of the hydrogen atom,* since the stationary states may be regarded as different stages of a process by which the electron under the emission of radiation is bound in orbits of smaller and smaller dimensions corresponding to states with decreasing values of n. It will be seen that this view has certain characteristic features in common with the binding process of an electron to the nucleus if this were to take place according to the ordinary electrodynamics, but that our view differs from it in just such a way that it is possible to account for the observed properties of hydrogen. In particular it is seen that the final result of the binding process leads to a quite definite stationary state of the atom, namely that state for which $n = 1$. This state which corresponds to the minimum energy of the atom will be called the *normal state* of the atom. It may be stated here that the values of the energy of the atom and the major axis of the orbit of the electron which are found if we put $n = 1$ in formulae (3) and (4) are of the same order of magnitude as the values of the firmness of binding of electrons and of the dimensions of the atoms which have been obtained from experiments on the electrical and mechanical properties of gases. A more accurate check of formulae (3) and (4) can however not be obtained from such a comparison, because in such experiments hydrogen is not present in the form of simple atoms but as molecules.

The formal basis of the quantum theory consists not only of the frequency relation, but also of conditions which permit the determination of the stationary states of atomic systems. The latter conditions, like that assumed for the frequency, may be regarded as natural generalizations of that assumption regarding the interaction between simple electrodynamic systems and a surrounding field of

[266]

electromagnetic radiation which forms the basis of Planck's theory of temperature radiation. I shall not here go further into the nature of these conditions but only mention that by their means the stationary states are characterized by a number of integers, the so-called *quantum numbers*. For a purely periodic motion like that assumed in the case of the hydrogen atom only a single quantum number is necessary for the determination of the stationary states. This number determines the energy of the atom and also the major axis of the orbit of the electron, but not its excentricity. The energy in the various stationary states, if the small influence of the motion of the nucleus is neglected, is given by the following formula:

$$E_n = -\frac{2\pi^2 N^2 e^4 m}{n^2 h^2}, \quad \ldots\ldots\ldots\ldots\ldots\ldots\ldots(5)$$

where e and m are respectively the charge and the mass of the electron, and where for the sake of subsequent applications the charge on the nucleus has been designated by Ne.

For the atom of hydrogen $N=1$, and a comparison with equation (3) leads to the following theoretical expression for K in formula (2), namely

$$K = \frac{2\pi^2 e^4 m}{h^3}. \ldots\ldots\ldots\ldots\ldots\ldots\ldots\ldots\ldots(6)$$

This agrees with the empirical value of the constant for the spectrum of hydrogen within the limit of accuracy with which the various quantities can be determined.

Hydrogen spectrum and X-ray spectra. If in the above formula we put $N=2$ which corresponds to an atom consisting of an electron revolving around a nucleus with a double charge, we get values for the energies in the stationary states, which are four times larger than the energies in the corresponding states of the hydrogen atom, and we obtain the following formula for the spectrum which would be emitted by such an atom:

$$\nu = 4K\left(\frac{1}{(n'')^2} - \frac{1}{(n')^2}\right). \quad \ldots\ldots\ldots\ldots\ldots\ldots(7)$$

This formula represents certain lines which have been known for some time and which had been attributed to hydrogen on account of the great similarity between formulae (2) and (7) since it had never been anticipated that two different substances could exhibit properties so closely resembling each other. According to the theory we may, however, expect that the emission of the spectrum given by (7) corresponds to the *first stage of the formation of the helium atom*, i.e. to the binding of a first electron by the doubly charged nucleus of this atom. This interpretation has been found to agree with more recent information. For instance it has been possible to obtain this spectrum from pure helium. I have dwelt on this point in order to show how this intimate connection between the properties of two elements, which at first sight might appear quite surprising, is to be regarded as an immediate expression of the characteristic simple structure of the nuclear atom. A short time after the elucidation of this question, new evidence of extraordinary interest was obtained of such a similarity between the properties of the elements. I refer to Moseley's fundamental researches on the X-ray spectra of the elements. Moseley found that these spectra varied in an extremely simple manner from one element to the next in the periodic system. It is well known that the lines of the X-ray spectra may be divided into groups corresponding to the different characteristic absorption regions for X-rays discovered by Barkla. As regards the K group which contains the most penetrating X-rays, Moseley found that the strongest line for all elements investigated could be represented by a formula which with a small simplification can be written

$$\nu = N^2 K\left(\frac{1}{1^2} - \frac{1}{2^2}\right). \quad \text{.....................(8)}$$

K is the same constant as in the hydrogen spectrum, and N the atomic number. The great significance of this discovery lies in the fact that it would seem firmly to establish the view that this atomic number is equal to the number of electrons in the atom. This assumption had already been used as a basis for work on atomic structure and was first stated by van den Broek. While the significance of this aspect of Moseley's discovery was at once clear to all, it has on the other hand been more difficult to understand the very great similarity between the spectrum of hydrogen and the X-ray spectra. This similarity is shown, not only by the lines of the K group, but also by groups of less penetrating X-rays.

[268]

Thus Moseley found for all the elements he investigated that the frequencies of the strongest line in the L group may be represented by a formula which with a simplification similar to that employed in formula (8) can be written

$$\nu = N^2 K\left(\frac{1}{2^2} - \frac{1}{3^2}\right). \quad \text{......................(9)}$$

Here again we obtain an expression for the frequency which corresponds to a line in the spectrum which would be emitted by the *binding of an electron to a nucleus, whose charge is Ne.*

The fine structure of the hydrogen lines. This similarity between the structure of the X-ray spectra and the hydrogen spectrum was still further extended in a very interesting manner by Sommerfeld's important theory of the fine structure of the hydrogen lines. The calculation given above of the energy in the stationary states of the hydrogen system, where each state is characterized by a single quantum number, rests upon the assumption that the orbit of the electron in the atom is simply periodic. This is, however, only approximately true. It is found that if the change in the mass of the electron due to its velocity is taken into consideration the orbit of the electron no longer remains a simple ellipse, but its motion may be described as a *central motion* obtained by superposing a slow and uniform rotation upon a simple periodic motion in a very nearly elliptical orbit. For a central motion of this kind the stationary states are characterized by *two quantum numbers.* In the case under consideration one of these may be so chosen that to a very close approximation it will determine the energy of the atom in the same manner as the quantum number previously used determined the energy in the case of a simple elliptical orbit. This quantum number which will always be denoted by n will therefore be called the "principal quantum number." Besides this condition, which to a very close approximation determines the major axis in the rotating and almost elliptical orbit, a second condition will be imposed upon the stationary states of a central orbit, namely that the angular momentum of the electron about the centre shall be equal to a whole multiple of Planck's constant divided by 2π. The whole number, which occurs as a factor in this expression, may be regarded as the second quantum number and will be denoted by k. The latter condition fixes the excentricity of the rotating orbit which in the case of a simple periodic orbit was undetermined. It should be mentioned that the possible importance of the angular momentum in the quantum theory was pointed out by Nicholson before the application of this theory to the spectrum of hydrogen, and that a determination of the stationary states for the hydrogen atom similar to that employed by Sommerfeld was proposed almost simultaneously by Wilson, although the latter did not succeed in giving a physical application to his results.

The simplest description of the form of the rotating nearly elliptical electronic orbit in the hydrogen atom is obtained by considering the chord which passes through the focus and is perpendicular to the major axis, the so-called "parameter." The length $2p$ of this parameter is given to a very close approximation by an expression of exactly the same form as the expression for the major axis, except that k takes the place of n. Using the same notation as before we have therefore

$$2a = n^2 \frac{h^2}{2\pi^2 N e^2 m}, \quad 2p = k^2 \frac{h^2}{2\pi^2 N e^2 m}. \quad \ldots\ldots\ldots(10)$$

For each of the stationary states which had previously been denoted by a given value of n, we obtain therefore a set of stationary states corresponding to values of k from 1 to n. Instead of the simple formula (5) Sommerfeld found a more complicated expression for the energy in the stationary states which depends on k as well as n. Taking the variation of the mass of the electron with velocity into account and neglecting terms of higher order of magnitude he obtained

$$E_{n,k} = -\frac{2\pi^2 N^2 e^4 m}{n^2 h^2}\left[1 + \frac{4\pi^2 N^2 e^4}{h^2 c^2}\left(-\frac{3}{4n^2} + \frac{1}{nk}\right)\right], \quad \ldots(11)$$

where c is the velocity of light.

Corresponding to each of the energy values for the stationary states of the hydrogen atom given by the simple formula (5) we obtain n values differing only very little from one another, since the second term within the bracket is very small. With the aid of the general frequency relation (1) we therefore obtain a number of components with nearly coincident frequencies instead of each hydrogen line given by the simple formula (2). Sommerfeld has now shown that this calculation actually agrees with measurements of the fine structure. This agreement applies not only to the fine structure of the hydrogen lines which is very difficult to measure on account of the extreme proximity of the components, but it is also possible to account in detail for the fine structure of the helium lines given by formula (7) which has been very carefully investigated by Paschen. Sommerfeld in connection with this theory also pointed out that formula (11) could be applied to the X-ray spectra. Thus he showed that in the K and L groups pairs of lines appeared the differences of whose frequencies could be determined by the expression (11) for the energy in the stationary states which correspond to the binding of a single electron by a nucleus of charge Ne.

Periodic table. In spite of the great formal similarity between the X-ray spectra and the hydrogen spectrum indicated by these results a far-reaching difference must be assumed to exist between the processes which give rise to the appearance of these two types of spectra. While the emission of the hydrogen spectrum, like the emission of the ordinary optical spectra of other elements, may be assumed to be connected with the binding of an electron by an atom, observations on the appearance and absorption of X-ray spectra clearly indicate that these spectra are connected with a process which may be described as a *reorganization of the electronic arrangement* after a disturbance within the atom due to the effect of external agencies. We should therefore expect that the appearance of the X-ray spectra would depend not only upon the direct interaction between a single electron and the nucleus, but also on the manner in which the electrons are arranged in the completely formed atom.

The peculiar manner in which the properties of the elements vary with the atomic number, as expressed in the periodic system, provides a guide of great value in considering this latter problem. A simple survey of this system is given in fig. 1. The number preceding each element indicates the atomic number, and the elements within the various vertical columns form the different "periods" of the system. The lines, which connect pairs of elements in successive columns, indicate homologous properties of such elements. Compared with usual representations of the periodic system, this method, proposed more than twenty years ago by Julius Thomsen, of indicating the periodic variations in the properties of the elements is more suited for comparison with theories of atomic constitution. The meaning of the frames round certain sequences of elements within the later periods of the table will be explained later. They refer to certain characteristic features of the theory of atomic constitution.

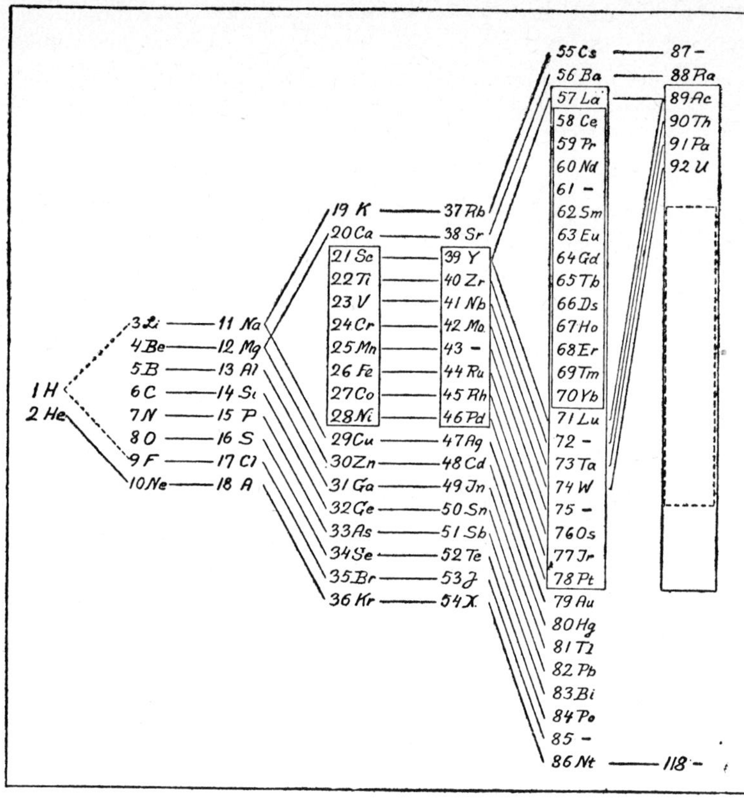

Fig. 1.

In an explanation of the periodic system it is natural to assume a division of the electrons in the atom into distinct groups in such a manner that the grouping of the elements in the system is attributed to the gradual formation of the groups of electrons in the atoms as the atomic number increases. Such a grouping of the electrons in the atom has formed a prominent part of all more detailed views of atomic structure ever since J. J. Thomson's famous attempt to explain the periodic system on the basis of an investigation of the stability of various electronic configurations. Although Thomson's assumption regarding the distribution of the positive electricity in the atom is not consistent with more recent experimental evidence, nevertheless his work has exerted great influence upon the later development of the atomic theory on account of the many original ideas which it contained.

With the aid of the information concerning the binding of electrons by the nucleus obtained from the theory of the hydrogen spectrum I attempted in the same paper in which this theory was set forth to sketch in broad outlines a picture of the structure of the nucleus atom. In this it was assumed that each electron in its normal state moved in a manner analogous to the motion in the last stages of the binding of a single electron by a nucleus. As in Thomson's theory, it was assumed that the electrons moved in circular orbits and that the electrons in each separate group during this motion occupied positions with reference to one another corresponding to the vertices of plane regular polygons. Such an arrangement is frequently described as a distribution of the electrons in "rings." By means of these assumptions it was possible to account for the orders of magnitude of the dimensions of the atoms as well as the firmness with which the electrons were bound by the atom, a measure of which may be obtained by means of experiments on the excitation of the various types of spectra. It was not possible, however, in this way to arrive at a detailed explanation of the characteristic properties of the elements even after it had become apparent from the results of Moseley and the work of Sommerfeld and others that this simple picture ought to be extended to include orbits in the fully formed atom characterized by higher quantum numbers corresponding to previous stages in the formation of the hydrogen atom. This point has been especially emphasized by Vegard.*

The difficulty of arriving at a satisfactory picture of the atom is intimately connected with the difficulty of accounting for the pronounced "stability" which the properties of the elements demand. As I emphasized when considering the formation of the hydrogen

* [Vegard, Lars (1880–1963). Norwegian physicist. Became professor at the University of Oslo in 1918.]

atom, the postulates of the quantum theory aim directly at this point, but the results obtained in this way for an atom containing a single electron do not permit of a direct elucidation of problems like that of the distribution in groups of the electrons in an atom containing several electrons. If we imagine that the electrons in the groups of the atom are orientated relatively to one another at any moment, like the vertices of regular polygons, and rotating in either circles or ellipses, the postulates do not give sufficient information to determine the difference in the stability of electronic configurations with different numbers of electrons in the groups.

The peculiar character of stability of the atomic structure, demanded by the properties of the elements, is brought out in an interesting way by Kossel in two important papers. In the first paper he shows that a more detailed explanation of the origin of the high frequency spectra can be obtained on the basis of the group structure of the atom. He assumes that a line in the X-ray spectrum is due to a process which may be described as follows: an electron is removed from the atom by some external action after which an electron in one of the other groups takes its place; this exchange of place may occur in as many ways as there are groups of more loosely bound electrons. This view of the origin of the characteristic X-rays afforded a simple explanation of the peculiar absorption phenomena observed. It has also led to the prediction of certain simple relations between the frequencies of the X-ray lines from one and the same element and has proved to be a suitable basis for the classification of the complete spectrum. However it has not been possible to develop a theory which reconciles in a satisfactory way Sommerfeld's work on the fine structure of the X-ray lines with Kossel's general scheme. As we shall see later the adoption of a new point of view when considering the stability of the atom renders it possible to bring the different results in a natural way in connection with one another.

In his second paper Kossel investigates the possibilities for an explanation of the periodic system on the basis of the atomic theory. Without entering further into the problem of the causes of the division of the electrons into groups, or the reasons for the different stability of the various electronic configurations, he points out in connection with ideas which had already played a part in Thomson's theory, how the periodic system affords evidence of a periodic appearance of especially stable configurations of electrons. These configurations appear in the neutral atoms of elements occupying the final position in each period in fig. 1, and the stability in question is assumed in order to explain not only the inactive chemical properties of these elements but also the characteristic active properties of the immediately preceding or succeeding elements. If we consider for instance an inactive gas like argon, the atomic number of which is 18, we must assume that the 18 electrons in the atom are arranged in an exceedingly regular configuration possessing a very marked stability. The pronounced electronegative character of the preceding element, chlorine, may then be explained by supposing the neutral atom which contains only 17 electrons to possess a tendency to capture an additional electron. This gives rise to a negative chlorine ion with a configuration of 18 electrons similar to that occurring in the neutral argon atom. On the other hand the marked electropositive character of potassium may be explained by supposing one of the 19 electrons in the neutral atom to be as it were superfluous, and that this electron therefore is easily lost; the rest of the atom forming a positive ion of potassium having a constitution similar to that of the argon atom. In a corresponding manner it is possible to account for the electronegative and electropositive character of elements like sulphur and calcium, whose atomic numbers are 16 and 20. In contrast to chlorine and potassium these elements are divalent, and the stable configuration of 18 electrons is formed by the addition of two electrons to the sulphur atom and by the loss of two electrons from the calcium atom. Developing these ideas Kossel has succeeded not only in giving interesting explanations of a large number of chemical facts, but has also been led to certain general conclusions about the grouping of the electrons in elements belonging to the first periods of the periodic system, which in certain respects are in conformity with the results to be discussed in the following paragraphs. Kossel's work was later continued in an interesting manner by Ladenburg with special reference to the grouping of the electrons in atoms of elements belonging to the later periods of the periodic table. It will be seen that Ladenburg's conclusions also exhibit points of similarity with the results which we shall discuss later.

Recent atomic models. Up to the present time it has not been possible to obtain a satisfactory account based upon a consistent application of the quantum theory to the nuclear atom of the ultimate cause of the pronounced stability of certain arrangements of electrons. Nevertheless it has been apparent for some time that the solution should be sought for by investigating the possibilities of a *spatial distribution of the electronic orbits* in the atom instead of limiting the investigation to configurations in which all electrons belonging to a particular group move in the same plane as was assumed for simplicity in my first papers on the structure of the atom. The necessity of assuming a spatial distribution of the configurations of electrons has been drawn attention to by various writers. Born and Landé, in connection with their investigations of the structure and properties of crystals, have pointed out that the assumption of spatial configurations appears necessary for an explanation of these properties. Landé has pursued this question still further, and as will be mentioned later has proposed a number of different "spatial atomic models" in which the electrons in each separate group of the atom at each moment form configurations possessing regular polyhedral symmetry. These models constitute in certain respects a distinct advance, although they have not led to decisive results on questions of the stability of atomic structure.

The importance of spatial electronic configurations has, in addition, been pointed out by Lewis and Langmuir in connection with their atomic models. Thus Lewis, who in several respects independently came to the same conclusions as Kossel, suggested that the number 8 characterizing the first groups of the periodic system might indicate a constitution of the outer atomic groups where the electrons within each group formed a configuration like the corners of a cube. He emphasized how a configuration of this kind leads to instructive models of the molecular structure of chemical combinations. It is to be remarked, however, that such a "static" model of electronic configuration will not be possible if we assume the forces within the atom to be due exclusively to the electric charges of the particles. Langmuir, who has attempted to develop Lewis' conceptions still further and to account not only for the occurrence of the first octaves, but also for the longer periods of the periodic system, supposes therefore the structure of the atoms to be governed by forces whose nature is unknown to us. He conceives the atom to possess a "cellular structure," so that each electron is in advance assigned a place in a cell and these cells are arranged in shells in such a manner, that the various shells from the nucleus of the atom outward contain exactly the same number of places as the periods in the periodic system proceeding in the direction of increasing atomic number. Langmuir's work has attracted much attention among chemists, since it has to some extent thrown light on the conceptions with which empirical chemical science is concerned. On his theory the explanation of the properties of the various elements is based on a number of postulates about the structure of the atoms formulated for that purpose. Such a descriptive theory is sharply differentiated from one where an attempt is made to explain the specific properties of the elements with the aid of general laws applying to the interaction between the particles in each atom. The principal task of this lecture will consist in an attempt to show that an advance along these lines appears by no means hopeless, but on the contrary that with the aid of a consistent application of the postulates of the quantum theory it actually appears possible to obtain an insight into the structure and stability of the atom.

II. SERIES SPECTRA AND THE CAPTURE OF ELECTRONS BY ATOMS

We attack the problem of atomic constitution by asking the question: "How may an atom be formed by the successive capture and binding of the electrons one by one in the field of force surrounding the nucleus?"

Before attempting to answer this question it will first be necessary to consider in more detail what the quantum theory teaches us about the general character of the binding process. We have already seen how the hydrogen spectrum gives us definite information about the course of this process of binding the electron by the nucleus. In considering the formation of the atoms of other elements we have also in their spectra sources for the elucidation of the formation processes, but the direct information obtained in this way is not so complete as in the case of the hydrogen atom. For an element of atomic number N the process of formation may be regarded as occurring in N stages, corresponding with the successive binding of N electrons in the field of the nucleus. A spectrum must be assumed to correspond to each of these binding processes; but only for the first two elements, hydrogen and helium, do we possess a detailed knowledge of these spectra. For other elements of higher atomic number, where several spectra will be connected with the formation of the atom, we are at present acquainted with only two types, called the "arc" and "spark" spectra respectively, according to the experimental conditions of excitation. Although these spectra show a much more complicated structure than the hydrogen spectrum, given by formula (2) and the helium spectrum given by formula (7), nevertheless in many cases it has been possible to find simple laws for the frequencies exhibiting a close analogy with the laws expressed by these formulae.

Arc and spark spectra. If for the sake of simplicity we disregard the complex structure shown by the lines of most spectra (occurrence of doublets, triplets etc.), the frequency of the lines of many arc spectra can be represented to a close approximation by the Rydberg formula

$$\nu = \frac{K}{(n'' + \alpha_{k''})^2} - \frac{K}{(n' + \alpha_{k'})^2}, \qquad \text{.............(12)}$$

where n' and n'' are integral numbers, K the same constant as in the hydrogen spectrum, while $\alpha_{k'}$ and $\alpha_{k''}$ are two constants belonging to a set characteristic of the element. A spectrum with a structure of this kind is, like the hydrogen spectrum, called a series spectrum, since the lines can be arranged into series in which the frequencies converge to definite limiting values. These series are for example represented by formula (12) if, using two definite constants for $\alpha_{k''}$ and $\alpha_{k'}$, n'' remains unaltered, while n' assumes a series of successive, gradually increasing integral values.

Formula (12) applies only approximately, but it is always found that the frequencies of the spectral lines can be written, as in formulae (2) and (12), as a difference of two functions of integral numbers. Thus the latter formula applies accurately, if the quantities α_k are not considered as constants, but as representatives of a set of series of numbers $\alpha_k\ (n)$ characteristic of the element, whose values for increasing n within each series quickly approach a constant limiting value. The fact that the frequencies of the spectra always appear as the difference of two terms, the so-called "spectral terms," from the combinations of which the complete spectrum is formed, has been pointed out by Ritz, who with the establishment of the combination principle has greatly advanced the study of the spectra. The quantum theory offers an immediate interpretation of this principle, since, according to the frequency relation we are led to consider the lines as due to transitions between stationary states of the atom, just as in the hydrogen spectrum, only in the spectra of the other elements we have to do not with a single series of stationary states, but with a set of such series. From formula (12) we thus obtain for an arc spectrum—if we temporarily disregard the structure of the individual lines—information about an ensemble of stationary states, for which the energy of the atom in the nth state of the kth series is given by

$$E_k\,(n) = -\frac{Kh}{(n+\alpha_k)^2} \qquad (13)$$

very similar to the simple formula (3) for the energy in the stationary states of the hydrogen atom.

As regards the spark spectra, the structure of which has been cleared up mainly by Fowler's investigations, it has been possible in the case of many elements to express the frequencies approximately by means of a formula of exactly the same type as (12), only with the difference that K, just as in the helium spectrum given by formula (7), is replaced by a constant, which is four times as large. For the spark spectra, therefore, the energy values in the corresponding stationary states of the atom will be given by an expression of the same type as (13), only with the difference that K is replaced by $4K$.

This remarkable similarity between the structure of these types of spectra and the simple spectra given by (2) and (7) is explained simply by assuming the arc spectra to be connected with the *last stage in the formation of the neutral atom* consisting in the capture and binding of the Nth electron. On the other hand the spark spectra are connected with the *last stage but one in the formation of the atom*, namely the binding of the $(N-1)$th electron. In these cases the field of force in which the electron moves will be much the same as that surrounding the nucleus of a hydrogen or helium atom respectively, at least in the earlier stages of the binding process, where during the greater part of its revolution it moves at a distance from the nucleus which is large in proportion to the dimensions of the orbits of the electrons previously bound. From analogy with formula (3) giving the stationary states of the hydrogen atom, we shall therefore assume that the numerical value of the expression on the right-hand side of (13) will be equal to the work required to remove the last captured electron from the atom, the binding of which gives rise to the arc spectrum of the element.

Series diagram. While the origin of the arc and spark spectra was to this extent immediately interpreted on the basis of the original simple theory of the hydrogen spectrum, it was Sommerfeld's theory of the fine structure of the hydrogen lines which first gave us a clear insight into the characteristic difference between the hydrogen spectrum and the spark spectrum of helium on the one hand, and the arc and spark spectra of other elements on the other. When we consider the binding not of the first but of the subsequent electrons in the atom, the orbit of the electron under consideration —at any rate in the latter stages of the binding process where the electron last bound comes into intimate interaction with those previously bound—will no longer be to a near approximation a closed ellipse, but on the contrary will to a first approximation be a central orbit of the same type as in the hydrogen atom, when we take into account the change with velocity in the mass of the electron. This motion, as we have seen, may be resolved into a plane periodic motion upon which a uniform rotation is superposed in the plane of the orbit; only the superposed rotation will in this case be comparatively much more rapid and the deviation of the periodic orbit from an ellipse much greater than in the case of the hydrogen atom. For an orbit of this type the stationary states, just as in the theory of the fine structure, will be determined by two quantum numbers which we shall denote by n and k, connected in a very simple manner with the kinematic properties of the orbit. For brevity I shall only mention that while the quantum number k is connected with the value of the constant angular momentum of the electron about the centre in the simple manner previously indicated, the determination of the principal quantum number n requires an investigation of the whole course of the orbit and for an arbitrary central orbit will not be related in a simple way to the dimensions of the rotating periodic orbit if this deviates essentially from a Keplerian ellipse.

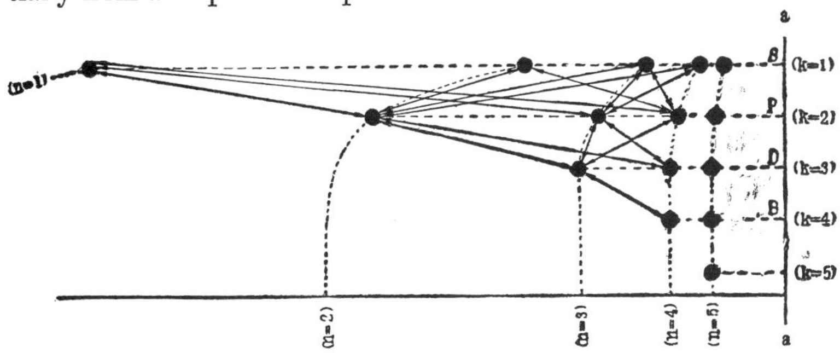

Fig. 2.

These results are represented in fig. 2 which is a reproduction of an illustration I have used on a previous occasion (see Essay II, p. 30), and which gives a survey of the origin of the sodium spectrum. The black dots represent the stationary states corresponding to the various series of spectral terms, shown on the right by the letters S, P, D and B. These letters correspond to the usual notations employed in spectroscopic literature and indicate the nature of the series (sharp series, principal series, diffuse series, etc.) obtained by combinations of the corresponding spectral terms. The distances of the separate points from the vertical line at the right of the figure are proportional to the numerical value of the energy of the atom given by equation (13). The oblique, black arrows indicate finally the transitions between the stationary states giving rise to the appearance of the lines in the commonly observed sodium spectrum. The values of n and k attached to the various states indicate the quantum numbers, which, according to Sommerfeld's theory, from a preliminary consideration might be regarded as characterizing the orbit of the outer electron. For the sake of convenience the states which were regarded as corresponding to the same value of n are connected by means of dotted lines, and these are so drawn that their vertical asymptotes correspond to the terms in the hydrogen spectrum which belong to the same value of the principal quantum number. The course of the curves illustrates how the deviation from the hydrogen terms may be expected to decrease with increasing values of k, corresponding to states, where the minimum distance between the electron in its revolution and the nucleus constantly increases.

It should be noted that even though the theory represents the principal features of the structure of the series spectra it has not yet been possible to give a detailed account of the spectrum of any element by a closer investigation of the electronic orbits which may occur in a simple field of force possessing central symmetry. As I have mentioned already the lines of most spectra show a complex structure. In the sodium spectrum for instance the lines of the principal series are doublets indicating that to each P-term not one stationary state, but two such states correspond with slightly different values of the energy. This difference is so little that it would not be recognizable in a diagram on the same scale as fig. 2. The appearance of these doublets is undoubtedly due to the small deviations from central symmetry of the field of force originating from the inner system in consequence of which the general type of motion of the external electron will possess a more complicated character than that of a simple central motion. As a result the stationary states must be characterized by more than two quantum numbers, in the same way that the occurrence of deviations of the orbit of the electron in the hydrogen atom from a simple periodic orbit requires that the stationary states of this atom shall be characterized by more than one quantum number. Now the rules of the quantum theory lead to the introduction of a third quantum number through the condition that the resultant angular momentum of the atom, multiplied by 2π, is equal to an entire multiple of Planck's constant. This determines the orientation of the orbit of the outer electron relative to the axis of the inner system.

In this way Sommerfeld, Landé and others have shown that it is possible not only to account in a formal way for the complex structure of the lines of the series spectra, but also to obtain a promising interpretation of the complicated effect of external magnetic fields on this structure. We shall not enter here on these problems but shall confine ourselves to the problem of the fixation of the two quantum numbers n and k, which to a first approximation describe the orbit of the outer electron in the stationary states, and whose determination is a matter of prime importance in the following discussion of the formation of the atom. In the determination of these numbers we at once encounter difficulties of a profound nature, which—as we shall see—are intimately connected with the question of the remarkable stability of atomic structure. I shall here only remark that the values of the quantum number n, given in the figure, undoubtedly can not be retained, neither for the S nor the P series. On the other hand, so far as the values employed for the quantum number k are concerned, it may be stated with certainty, that the interpretation of the properties of the orbits, which they indicate, is correct. A starting point for the investigation of this question has been obtained from considerations of an entirely different kind from those previously mentioned, which have made it possible to establish a close connection between the motion in the atom and the appearance of spectral lines.

Correspondence principle. So far as the principles of the quantum theory are concerned, the point which has been emphasized hitherto is the radical departure of these principles from our usual conceptions of mechanical and electrodynamical phenomena. As I have attempted to show in recent years, it appears possible, however, to adopt a point of view which suggests that the quantum theory may, nevertheless, be regarded as a rational generalization of our ordinary conceptions. As may be seen from the postulates of the quantum theory, and particularly the frequency relation, a direct connection between the spectra and the motion of the kind required by the classical dynamics is excluded, but at the same time the form of these postulates leads us to another relation of a remarkable nature. Let us consider an electrodynamic system and inquire into the nature of the radiation which would result from the motion of the system on the basis of the ordinary conceptions. We imagine the motion to be decomposed into purely harmonic oscillations, and the radiation is assumed to consist of the simultaneous emission of series of electromagnetic waves possessing the same frequency as these harmonic components and intensities which depend upon the amplitudes of the components. An investigation of the formal basis of the quantum theory shows us now, that it is possible to trace the question of the origin of the radiation processes which accompany the various transitions back to an investigation of the various harmonic components, which appear in the motion of the atom. The possibility, that a particular transition shall occur, may be regarded as being due to the presence of a definitely assignable "corresponding" component in the motion. This principle of correspondence at the same time throws light upon a question mentioned several times previously, namely the relation between the number of quantum numbers, which must be used to describe the stationary states of an atom, and the types to which the orbits of the electrons belong. The classification of these types can be based very simply on a decomposition of the motion into its harmonic components. Time does not permit me to consider this question any further, and I shall confine myself to a statement of some simple conclusions, which the correspondence principle permits us to draw concerning the occurrence of transitions between various pairs of stationary states. These conclusions are of decisive importance in the subsequent argument.

The simplest example of such a conclusion is obtained by considering an atomic system, which contains a particle describing a *purely periodic orbit*, and where the stationary states are characterized by a single quantum number n. In this case the motion can according to Fourier's theorem be decomposed into a simple series of harmonic oscillations whose frequency may be written $\tau\omega$, where τ is a whole number, and ω is the frequency of revolution in the orbit. It can now be shown that a transition between two stationary states, for which the values of the quantum number are respectively equal to n' and n'', will correspond to a harmonic component, for which $\tau = n' - n''$. This throws at once light upon the remarkable difference which exists between the possibilities of transitions between the stationary states of a hydrogen atom on the one hand and of a simple system consisting of an electric particle capable of executing simple harmonic oscillations about a position of equilibrium on the other. For the latter system, which is frequently called a Planck oscillator, the energy in the stationary states is determined by the familiar formula $E = nh\omega$, and with the aid of the frequency relation we obtain therefore for the radiation which will be emitted during a transition between two stationary states $\nu = (n' - n'')\,\omega$. Now, an important assumption, which is not only essential in Planck's theory of temperature radiation, but which also appears necessary to account for the molecular absorption in the infra-red region of radiation, states that a harmonic oscillator will only emit and absorb radiation, for which the frequency ν is equal to the frequency of oscillation ω of the oscillator. We are therefore compelled to assume that in the case of the oscillator transitions can occur only between stationary states which are characterized by quantum numbers differing by only one unit, while in the hydrogen spectrum represented by formula (2) all possible transitions could take place between the stationary states given by formula (5). From the point of view of the principle of correspondence it is seen, however, that this apparent difficulty is explained by the occurrence in the motion of the hydrogen atom, as opposed to the motion of the oscillator, of harmonic components corresponding to values of τ, which are different from 1; or using a terminology well known from acoustics, there appear overtones in the motion of the hydrogen atom.

Another simple example of the application of the correspondence principle is afforded by a *central motion*, to the investigation of which the explanation of the series spectra in the first approximation may be reduced. Referring once more to the figure of the sodium spectrum, we see that the black arrows, which correspond to the spectral lines appearing under the usual conditions of excitation, only connect pairs of points in consecutive rows. Now it is found that this remarkable limitation of the occurrence of combinations between spectral terms may quite naturally be explained by an investigation of the harmonic components into which a central motion can be resolved. It can readily be shown that such a motion can be decomposed into two series of harmonic components, whose frequencies can be expressed by $\tau\omega + \sigma$ and $\tau\omega - \sigma$ respectively, where τ is a whole number, ω the frequency of revolution in the rotating periodic orbit and σ the frequency of the superposed rotation. These components correspond with transitions where the principal number n decreases by τ units, while the quantum number k decreases or increases, respectively, by one unit, corresponding exactly with the transitions indicated by the black arrows in the figure. This may be considered as a very important result, because we may say, that the quantum theory, which for the first time has offered a simple interpretation of the fundamental principle of combination of spectral lines has at the same time removed the mystery which has hitherto adhered to the application of this principle on account of the apparent capriciousness of the appearance of predicted combination lines. Especially attention may be drawn to the simple interpretation which the quantum theory offers of the appearance observed by Stark and his collaborators of certain new series of lines, which do not appear under ordinary circumstances, but which are excited when the emitting atoms are subject to intense external electric fields. In fact, on the correspondence principle this is immediately explained from an examination of the perturbations in the motion of the outer electron which give rise to the appearance in this motion—besides the harmonic components already present in a simple central orbit—of a number of constituent harmonic vibrations of new type and of amplitudes proportional to the intensity of the external forces.

It may be of interest to note that an investigation of the limitation of the possibility of transitions between stationary states, based upon a simple consideration of conservation of angular momentum during the process of radiation, does not, contrary to what has previously been supposed (compare Essay II, p. 62), suffice to throw light on the remarkably simple structure of series spectra illustrated by the figure. As mentioned above we must assume that the "complexity" of the spectral terms, corresponding to given values of n and k, which we witness in the fine structure of the spectral lines, may be ascribed to states, corresponding to different values of this angular momentum, in which the plane of the electronic orbit is orientated in a different manner, relative to the configuration of the previously bound electrons in the atom. Considerations of conservation of angular momentum can, in connection with the series spectra, therefore only contribute to an understanding of the limitation of the possibilities of combination observed in the peculiar laws applying to the number of components in the complex structure of the lines. So far as the last question is concerned, such considerations offer a direct support for the consequences of the correspondence principle.

III. FORMATION OF ATOMS AND THE PERIODIC TABLE

A correspondence has been shown to exist between the motion of the electron last captured and the occurrence of transitions between the stationary states corresponding to the various stages of the binding process. This fact gives a point of departure for a choice between the numerous possibilities which present themselves when considering the formation of the atoms by the successive capture and binding of the electrons. Among the processes which are conceivable and which according to the quantum theory might occur in the atom we shall reject those whose occurrence can not be regarded as consistent with a correspondence of the required nature.

First Period. Hydrogen—Helium. It will not be necessary to concern ourselves long with the question of the constitution of the hydrogen atom. From what has been said previously we may assume that the final result of the process of *binding of the first electron* in any atom will be a stationary state, where the energy of the atom is given by (5), if we put $n=1$, or more precisely by formula (11), if we put $n=1$ and $k=1$. The orbit of the electron will be a circle whose radius will be given by formulae (10), if n and k are each put equal to 1. Such an orbit will be called a 1-quantum orbit, and in general an orbit for which the principal quantum number has a given value n will be called an n-quantum orbit. Where it is necessary to differentiate between orbits corresponding to various values of the quantum number k, a central orbit, characterized by given values of the quantum numbers n and k, will be referred to as an n_k orbit.

In the question of the constitution of the helium atom we meet the much more complicated problem of the *binding of the second electron.* Information about this binding process may, however, be obtained from the arc spectrum of helium. This spectrum, as opposed to most other simple spectra, consists of two complete systems of lines with frequencies given by formulae of the type

(12). On this account helium was at first assumed to be a mixture of two different gases, "orthohelium" and "parhelium," but now we know that the two spectra simply mean that the binding of the second electron can occur in two different ways. A theoretical explanation of the main features of the helium spectrum has recently been attempted in an interesting paper by Landé. He supposes the emission of the orthohelium spectrum to be due to transitions between stationary states where both electrons move in the same plane and revolve in the same sense. The parhelium spectrum, on the other hand, is ascribed by him to stationary states where the planes of the orbits form an angle with each other. Dr Kramers and I have made a closer investigation of the interaction between the two orbits in the different stationary states. The results of our investigation which was begun several years before the appearance of Landé's work have not yet been published. Without going into details I may say, that even though our results in several respects differ materially from those of Landé (compare Essay II, p. 56), we agree with his general conclusions concerning the origin of the orthohelium and parhelium spectra.

The final result of the binding of the second electron is intimately related to the origin of the two helium spectra. Important information on this point has been obtained recently by Franck and his co-workers. As is well known he has thrown light upon many features of the structure of the atom and of the origin of spectra by observing the effect of bombarding atoms by electrons of various velocities. A short time ago these experiments showed that the impact of electrons could bring helium into a "metastable" state from which the atom cannot return to its normal state by means of a simple transition accompanied by the emission of radiation, but only by means of a process analogous to a chemical reaction involving interaction with atoms of other elements. This result is closely connected with the fact that the binding of the second electron can occur in two different ways, as is shown by the occurrence of two distinct spectra. Thus it is evident from Franck's experiments that the normal state of the atom is the last stage in the binding process involving the emission of the parhelium spectrum by which the electron last captured as well as the one first captured will be bound in a 1_1 orbit. The metastable state, on the contrary, is the final stage of the process giving the orthohelium spectrum. In this case the second electron, as opposed to the first, will move in a 2_1 orbit. This corresponds to a firmness of binding which is about six times less than for the electron in the normal state of the atom.

If we now consider somewhat more closely this apparently surprising result, it is found that a clear grasp of it may be obtained from the point of view of correspondence. It can be shown that the coherent class of motions to which the orthohelium orbits belong does not contain a 1_1 orbit. If on the whole we would claim the existence of a state where the two electrons move in 1_1 orbits in the same plane, and if in addition it is claimed that the motion should possess the periodic properties necessary for the definition of stationary states, then there seems that no possibility is afforded other than the assumption that the two electrons move around the nucleus in one and the same orbit, in such a manner that at each moment they are situated at the ends of a diameter. This extremely simple ring-configuration might be expected to correspond to the firmest possible binding of the electrons in the atom, and it was on this account proposed as a model for the helium atom in my first paper on atomic structure. If, however, we inquire about the possibility of a transition from one of the orthohelium states to a configuration of this type we meet conditions which are very different from those which apply to transitions between two of the orthohelium orbits. In fact, the occurrence of each of these transitions is due to the existence of well-defined corresponding constituent harmonic vibration in the central orbits which the outer electron describes in the class of motions to which the stationary states belong. The transition we have to discuss, on the other hand, is one by which the last captured electron is transferred from a state in which it is moving "outside" the other to a state in which it moves round the nucleus on equal terms with the other electron. Now it is impossible to find a series of simple intermediate forms for the motion of those two electrons in which the orbit of the last captured electron exhibits a sufficient similarity to a central motion that for this transition there could be a correspondence of the necessary kind. It is therefore evident, that where the two electrons move in the same plane, the electron captured last can not be bound firmer than in a 2_1 orbit. If, on the other hand, we consider the binding process which accompanies the emission of the parhelium spectrum and where the electrons in the stationary states move in orbits whose planes form angles with one another we meet essentially different conditions. A corresponding intimate change in the interaction between the electron last captured and the one previously bound is not required here for the two electrons in the atom to become equivalent. We may therefore imagine the last stage of the binding process to take place in a manner similar to those stages corresponding to transitions between orbits characterized by greater values of n and k.

In the *normal state of the helium atom* the two electrons must be assumed to move in equivalent 1_1 orbits. As a first approximation these may be described as two circular orbits, whose planes make an angle of 120° with one another, in agreement with the conditions which the angular momentum of an atom according to the quantum theory must satisfy. On account of the interaction between the two electrons these planes at the same time turn slowly around the fixed impulse axis of the atom. Starting from a distinctly different point of view Kemble has recently suggested a similar model for the helium atom. He has at the same time directed attention to a possible type of motion of very marked symmetry in which the electrons during their entire revolution assume symmetrical positions with reference to a fixed axis. Kemble has not, however, investigated this motion further. Previous to the appearance of this paper Kramers had commenced a closer investigation of precisely this type of motion in order to find out to what extent it was possible from such a calculation to account for the firmness with which the electrons are bound in the helium atom, that is to account for the ionization potential. Early measurements of this potential had given values corresponding approximately to that which would result from the ring-configuration already mentioned. This requires 17/8 as much work to remove a single electron as is necessary to remove an electron from the hydrogen atom in its normal state. As the theoretical value for the latter amount of work—which for the sake of simplicity will be represented by W—corresponds to an ionization potential of 13·53 volts, the ionization potential of helium would be expected to be 28·8 volts. Recent and more accurate determinations, however, have given a value for the ionization potential of helium which is considerably lower and lies in the neighbourhood of 25 volts. This showed therefore the untenability of the ring-configuration quite independently of any other considerations. A careful investigation of the spatial atomic configuration requires elaborate calculation, and Kramers has not yet obtained final results. With the approximation to which they have been so far completed the calculations point to the possibility of an agreement with the experimental results. The final result may be awaited with great interest, since it offers in the simplest case imaginable a test of the principles by which we are attempting to determine stationary states of atoms containing more than one electron.

Hydrogen and helium, as seen in the survey of the periodic system given in fig. 1, together form the first period in the system of elements, since helium is the first of the inactive gases. The great difference in the chemical properties of hydrogen and helium is closely related to the great difference in the nature of the binding of the electron. This is directly indicated by the spectra and ionization potentials. While helium possesses the highest known ionization potential of all the elements, the binding of the electron in the hydrogen atom is sufficiently loose to account for the tendency of hydrogen to form positive ions in aqueous solutions and chemical combinations. Further consideration of this particular question requires, however, a comparison between the nature and firmness of the electronic configurations of other atoms, and it can therefore not be discussed at the moment.

Second Period. Lithium—Neon. When considering the atomic structure of elements which contain more than two electrons in the neutral atom, we shall assume first of all that what has previously been said about the formation of the helium atom will in the main features also apply to the capture and binding of the first two electrons. These electrons may, therefore, in the normal state of the atom be regarded as moving in equivalent orbits characterized by the quantum symbol 1_1. We obtain direct information about the *binding of the third electron* from the spectrum of lithium. This spectrum shows the existence of a number of series of stationary states, where the firmness with which the last captured electron is bound is very nearly the same as in the stationary states of the hydrogen atom. These states correspond to orbits where k is greater than or equal to 2, and where the last captured electron moves entirely outside the region where the first two electrons move. But in addition this spectrum gives us information about a series of states corresponding to $k=1$ in which the energy differs essentially from the corresponding stationary states of the hydrogen atom. In these states the last captured electron, even if it remains at a considerable distance from the nucleus during the greater part of its revolution, will at certain moments during the revolution approach to a distance from the nucleus which is of the same order of magnitude as the dimensions of the orbits of the previously bound electrons. On this account the electrons will be bound with a firmness which is considerably greater than that with which the electrons are bound in the stationary states of the hydrogen atom corresponding to the same value of n.

Now as regards the lithium spectrum as well as the other alkali spectra we are so fortunate (see p. 32) as to possess definite evidence about the normal state of the atom from experiments on selective absorption. In fact these experiments tell us that the first member of the sequence of S-terms corresponds to this state. This term corresponds to a strength of binding which is only a little more than a third of that of the hydrogen atom. We must therefore conclude that the outer electron in the normal state of the lithium atom moves in a 2_1 orbit, just as the outer electron in the metastable state of the helium atom. The reason why the binding of the outer electron can not proceed to an orbit characterized by a smaller value for the total quantum number may also be considered as analogous in the two cases. In fact, a transition by which the third electron in the lithium atom was ultimately bound in a 1_1 orbit would lead to a state in the atom in which the electron would play an equivalent part with the two electrons previously bound. Such a process would be of a type entirely different from the transitions between the stationary states connected with the emission of the lithium spectrum, and would, contrary to these, not exhibit a correspondence with a harmonic component in the motion of the atom.

We obtain, therefore, a picture of the formation and structure of the lithium atom which offers a natural explanation of the great difference of the chemical properties of lithium from those of helium and hydrogen. This difference is at once explained by the fact that the firmness by which the last captured electron is bound in its 2_1 orbit in the lithium atom is only about a third of that with which the electron in the hydrogen atom is held, and almost five times smaller than the firmness of the binding of the electrons in the helium atom.

What has been said here applies not alone to the formation of the lithium atom, but may also be assumed to apply to the binding of the third electron in every atom, so that in contrast to the first two electrons which move in 1_1 orbits this may be assumed to move in a 2_1 orbit. As regards the *binding of the fourth, fifth and sixth electrons* in the atom, we do not possess a similar guide as no simple series spectra are known of beryllium, boron and carbon. Although conclusions of the same degree of certainty can not be reached it seems possible, however, to arrive at results consistent with general physical and chemical evidence by proceeding by means of considerations of the same kind as those applied to the binding of the first three electrons. In fact, we shall assume that the fourth, fifth and sixth electrons will be bound in 2_1 orbits. The reason why the binding of a first electron in an orbit of this type will not prevent the capture of the others in two quanta orbits may be ascribed to the fact that 2_1 orbits are not circular but very excentric. For example, the 3rd electron cannot keep the remaining electrons away from the inner system in the same way in which the first two electrons bound in the lithium atom prevent the third from being bound in a 1-quantum orbit. Thus we shall expect that the 4th, 5th and 6th electrons in a similar way to the 3rd will at certain moments of their revolution enter into the region where the first two bound electrons move. We must not imagine, however, that these visits into the inner system take place at the same time, but that the four electrons visit the nucleus separately at equal intervals of time. In earlier work on atomic structure it was supposed that the electrons in the various groups in the atom moved in separate regions within the atom and that at each moment the electrons within each separate group were arranged in configurations possessing symmetry like that of a regular polygon or polyhedron. Among other things this involved that the electrons in each group were supposed to be at the point of the orbit nearest the nucleus at the same time. A structure of this kind may be described as one where the motions of the electrons within the groups are coupled together in a manner which is largely independent of the interaction between the various groups. On the contrary, the characteristic feature of a structure like that I have suggested is the *intimate coupling between the motions of the electrons in the various groups* characterized by different quantum numbers, as well as the *greater independence in the mode of binding within one and the same group of electrons* the orbits of which are characterized by the same quantum number. In emphasizing this last feature I have two points in mind. Firstly the smaller effect of the presence of previously bound electrons on the firmness of binding of succeeding electrons in the same group. Secondly the way in which the motions of the electrons within the group reflect the independence both of the processes by which the group can be formed and by which it can be reorganized by change of position of the different electrons in the atom after a disturbance by external forces. The last point will be considered more closely when we deal with the origin and nature of the X-ray spectra; for the present we shall continue the consideration of the structure of the atom to which we are led by the investigation of the processes connected with the successive capture of the electrons.

The preceding considerations enable us to understand the fact that the two elements beryllium and boron immediately succeeding lithium can appear electropositively with 2 and 3 valencies respectively in combination with other substances. For like the third electron in the lithium atom, the last captured electrons in these elements will be much more lightly bound than the first two electrons. At the same time we understand why the electropositive character of these elements is less marked than in the case of lithium, since the electrons in the 2-quanta orbits will be much more firmly bound on account of the stronger field in which they are moving. New conditions arise, however, in the case of the next element, carbon, as this element in its typical chemical combinations can not be supposed to occur as an ion, but rather as a neutral atom. This must be assumed to be due not only to the great firmness in the binding of the electrons but also to be an essential consequence of the symmetrical configuration of the electrons.

With the binding of the 4th, 5th and 6th electrons in 2_1 orbits, the spatial symmetry of the regular configuration of the orbits must be regarded as steadily increasing, until with the binding of the 6th electron the orbits of the four last bound electrons may be expected to form an exceptionally symmetrical configuration in which the normals to the planes of the orbits occupy positions relative to one another nearly the same as the lines from the centre to the vertices of a regular tetrahedron. Such a configuration of groups of 2-quanta orbits in the carbon atom seems capable of furnishing a suitable foundation for explaining the structure of organic compounds. I shall not discuss this question any further, for it would require a thorough study of the interaction between the motions of the electrons in the atoms forming the molecule. I might mention, however, that the types of molecular models to which we are led are very different from the molecular models which were suggested in my first papers. In these the chemical "valence bonds" were represented by "electron rings" of the same type as those which were assumed to compose the groups of electrons within the individual atoms. It is nevertheless possible to give a general explanation of the chemical properties of the elements without touching on those matters at all. This is largely due to the fact that the structures of combinations of atoms of the same element and of many organic compounds do not have the same significance for our purpose as those molecular structures in which the individual atoms occur as electrically charged ions. The latter kind of compounds, to which the greater number of simple inorganic compounds belong, is frequently called "heteropolar" and possesses a far more typical character than the first compounds which are called "homoeopolar," and whose properties to quite a different degree exhibit the individual peculiarities of the elements. My main purpose will therefore be to consider the fitness which the configurations of the electrons in the various atoms offer for the formation of ions.

Before leaving the carbon atom I should mention, that a model of this atom in which the orbits of the four most lightly bound electrons possess a pronounced tetrahedric symmetry had already been suggested by Landé. In order to agree with the measurements of the size of the atoms he also assumed that these electrons moved in 2_1 orbits. There is, however, this difference between Landé's view and that given here, that while Landé deduced the characteristic properties of the carbon atom solely from an investigation of the simplest form of motion which four electrons can execute employing spatial symmetry, our view originates from a consideration of the stability of the whole atom. For our assumptions about the orbits of the electrons are based directly on an investigation of the interaction between these electrons and the first two bound electrons. The result is that our model of the carbon atom has dynamic properties which are essentially different from the properties of Landé's model.

In order to account for the properties of *the elements in the second half of the second period* it will first of all be necessary to show why the configuration of ten electrons occurring in the neutral atom of neon possesses such a remarkable degree of stability. Previously it has been assumed that the properties of this configuration were due to the interaction between eight electrons which moved in equivalent orbits outside the nucleus and an inner group of two electrons like that in the helium atom. It will be seen, however, that the solution must be sought in an entirely different direction. It can not be expected that *the 7th electron* will be bound in a 2_1 orbit equivalent to the orbits of the four preceding electrons. The occurrence of five such orbits would so definitely destroy the symmetry in the interaction of these electrons that it is inconceivable that a process resulting in the accession of a fifth electron to this group would be in agreement with the correspondence principle. On the contrary it will be necessary to assume that the four electrons in their exceptionally symmetrical orbital configuration will keep out later captured electrons with the result that these electrons will be bound in orbits of other types.

The orbits which come into consideration for the 7th electron in the nitrogen atom and the 7th, 8th, 9th and 10th electrons in the atoms of the immediately following elements will be circular orbits of the type 2_2. The diameters of these orbits are considerably larger than those of the 1_1 orbits of the first two electrons; on the other hand the outermost part of the excentric 2_1 orbits will extend some distance beyond these circular 2_2 orbits. I shall not here discuss the capture and binding of these electrons. This requires a further investigation of the interaction between the motions of the electrons in the two types of 2-quanta orbits. I shall simply mention, that in the atom of neon in which we will assume that there are four electrons in 2_2 orbits the planes of these orbits must be regarded not only as occupying a position relative to one another characterized by a high degree of spatial symmetry, but also as possessing a configuration harmonizing with the four elliptical 2_1 orbits. An interaction of this kind in which the orbital planes do not coincide can be attained only if the configurations in both subgroups exhibit a systematic deviation from tetrahedral symmetry. This will have the result that the electron groups with 2-quanta orbits in the neon atom will have only a single axis of symmetry which must be supposed to coincide with the axis of symmetry of the innermost group of two electrons.

Before leaving the description of the elements within the second period it may be pointed out that the above considerations offer a basis for interpreting that tendency of the neutral atoms of oxygen and fluorine for capturing further electrons which is responsible for the marked electronegative character of these elements. In fact, this tendency may be ascribed to the fact that the orbits of the last captured electrons will find their place within the region, in which the previously captured electrons move in 2_1 orbits. This suggests an explanation of the great difference between the properties of the elements in the latter half of the second period of the periodic system and those of the elements in the first half, in whose atoms there is only a single type of 2-quanta orbits.

Third Period. Sodium—Argon. We shall now consider the structure of atoms of elements in the third period of the periodic system. This brings us immediately to the question of *the binding of the* 11*th electron* in the atom. Here we meet conditions which in some respects are analogous to those connected with the binding of the 7th electron. The same type of argument that applied to the carbon atom shows that the symmetry of the configuration in the neon atom would be essentially, if not entirely, destroyed by the addition of another electron in an orbit of the same type as that in which the last captured electrons were bound. Just as in the case of the 3rd and 7th electrons we may therefore expect to meet a new type of orbit for the 11th electron in the atom, and the orbits which present themselves this time are the 3_1 orbits. An electron in such an orbit will for the greater part of the time remain outside the orbits of the first ten electrons. But at certain moments during the revolution it will penetrate not only into the region of the 2-quanta orbits, but like the 2_1 orbits it will penetrate to distances from the nucleus which are smaller than the radii of the 1-quantum orbits of the two electrons first bound. This fact, which has a most important bearing on the stability of the atom, leads to a peculiar result as regards the binding of the 11th electron. In the sodium atom this electron will move in a field which so far as the outer part of the orbit is concerned deviates only very little from that surrounding the nucleus in the hydrogen atom, but the dimensions of this part of the orbit will, nevertheless, be essentially different from the dimensions of the corresponding part of a 3_1 orbit in the hydrogen atom. This arises from the fact, that even though the electron only enters the inner configuration of the first ten electrons for short intervals during its revolution, this part of the orbit will nevertheless exert an essential influence upon the determination of the principal quantum number. This is directly related to the fact that the motion of the electron in the first part of the orbit deviates only a little from the motion which each of the previously bound electrons in 2_1 orbits executes during a complete revolution. The uncertainty which has prevailed in the determination of the quantum numbers for the stationary states corresponding to a spectrum like that of sodium is connected with this. This question has been discussed by several physicists. From a comparison of the spectral terms of the various alkali metals, Roschdestwensky has drawn the conclusion that the normal state does not, as we might be inclined to expect a priori, correspond to a 1_1 orbit as shown in fig. 2 on p. 79, but that this state corresponds to a 2_1 orbit. Schrödinger has arrived at a similar result in an attempt to account for the great difference between the S terms and the terms in the P and D series of the alkali spectra. He assumes that the "outer" electron in the states corresponding to the S terms—in contrast to those corresponding to the P and D terms—penetrates partly into the region of the orbits of the inner electrons during the course of its revolution. These investigations contain without doubt important hints, but in reality the conditions must be very different for the different alkali spectra. Instead of a 2_1 orbit as in lithium we must thus assume for the spectrum of sodium not only that the first spectral term in the S series corresponds to a 3_1 orbit, but also, as a more detailed consideration shows, that the first term in the P series corresponds not to a 2_2 orbit as indicated in fig. 2, but to a 3_2 orbit. If the numbers in this figure were correct, it would require among other things that the P terms should be smaller than the hydrogen terms

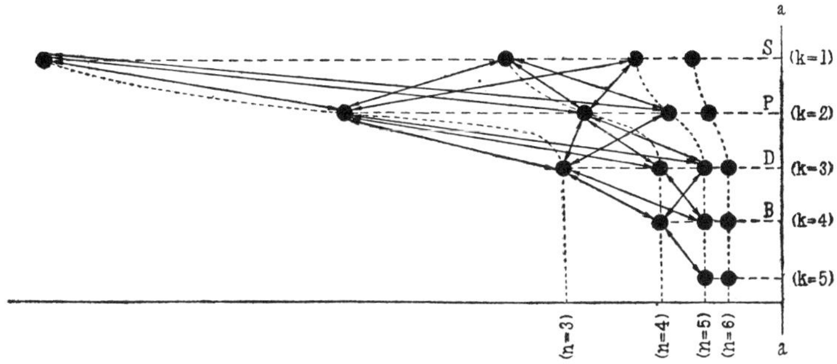

Fig. 3.

corresponding to the same principal quantum number. This would mean that the average effect of the inner electrons could be described as a repulsion greater than would occur if their total electrical charge were united in the nucleus. This, however, can not be expected from our view of atomic structure. The fact that the last captured electron, at any rate for low values of k, revolves partly inside the orbits of the previously bound electrons will on the contrary involve that the presence of these electrons will give rise to a virtual repulsion which is considerably smaller than that which would be due to their combined charges. Instead of the curves drawn between points in fig. 2 which represent stationary states corresponding to the same value of the principal quantum number running from right to left, we obtain curves which run from left to right, as is indicated in fig. 3. The stationary states are labelled with quantum numbers corresponding to the structure I have described. According to the view underlying fig. 2 the sodium spectrum might be described simply as a distorted hydrogen spectrum, whereas according to fig. 3 there is not only distortion but also complete disappearance of certain terms of low quantum numbers. It may be stated, that this view not only appears to offer an explanation of the magnitude of the terms, but that the complexity of the terms in the P and D series finds a natural explanation in the deviation of the configuration of the ten electrons first bound from a purely central symmetry. This lack of symmetry has its origin in the configuration of the two innermost electrons and "transmits" itself to the outer parts of the atomic structure, since the 2_1 orbits penetrate partly into the region of these electrons.

This view of the sodium spectrum provides at the same time an immediate explanation of the pronounced electropositive properties of sodium, since the last bound electron in the sodium atom is still more loosely bound than the last captured electron in the lithium atom. In this connection it might be mentioned that the increase in atomic volume with increasing atomic number in the family of the alkali metals finds a simple explanation in the successively looser binding of the valency electrons. In his work on the X-ray spectra Sommerfeld at an earlier period regarded this increase in the atomic volumes as supporting the assumption that the principal quantum number of the orbit of the valency electrons increases by unity as we pass from one metal to the next in the family. His later investigations on the series spectra have led him, however, definitely to abandon this assumption. At first sight it might also appear to entail a far greater increase in the atomic volume than that actually observed. A simple explanation of this fact is however afforded by realizing that the orbit of the electron will run partly inside the region of the inner orbit and that therefore the "effective" quantum number which corresponds to the outer almost elliptical loop will be much smaller than the principal quantum number, by which the whole central orbit is described. It may be mentioned that Vegard in his investigations on the X-ray spectra has also proposed the assumption of successively increasing quantum numbers for the electronic orbits in the various groups of the atom, reckoned from the nucleus outward. He has introduced assumptions about the relations between the numbers of electrons in the various groups of the atom and the lengths of the periods in the periodic system which exhibit certain formal similarities with the results presented here. But Vegard's considerations do not offer points of departure for a further consideration of the evolution and stability of the groups, and consequently no basis for a detailed interpretation of the properties of the elements.

When we consider the elements following sodium in the third period of the periodic system we meet in *the binding of the* 12*th*, 13*th and* 14*th electrons* conditions which are analogous to those we met in the binding of the 4th, 5th and 6th electrons. In the elements of the third periods, however, we possess a far more detailed knowledge of the series spectra. Too little is known about the beryllium spectrum to draw conclusions about the binding of the fourth electron, but we may infer directly from the well-known arc spectrum of magnesium that the 12th electron in the atom of this element is bound in a 3_1 orbit. As regards the binding of the 13th electron we meet in aluminium an absorption spectrum different in structure to that of the alkali metals. In fact here not the lines of the principal series but the lines of the sharp and diffuse series are absorption lines. Consequently it is the first member of the P terms and not of the S terms which corresponds to the normal state of the aluminium atom, and we must assume that the 13th electron is bound in a 3_2 orbit. This, however, would hardly seem to be a general property of the binding of the 13th electron in atoms, but rather to arise from the special conditions for the binding of the last electron in an atom, where already there are two other electrons bound as loosely as the valency electron of aluminium. At the present state of the theory it seems best to assume that in the silicon atom the four last captured electrons will move in 3_1 orbits forming a configuration possessing symmetrical properties similar to the outer configuration of the four electrons in 2_1 orbits in carbon. Like what we assumed for the latter configuration we shall expect that the configuration of the 3_1 orbits occurring for the first time in silicon possesses such a completion, that the addition of a further electron in a 3_1 orbit to the atom of the following elements is impossible, and that *the* 15*th electron* in the elements of higher atomic number will be bound in a new type of orbit. In this case, however, the orbits with which we meet will not be circular, as in the capture of the 7th electron, but will be rotating excentric orbits of the type 3_2. This is very closely related to the fact, mentioned above, that the non-circular orbits will correspond to a firmer binding than the circular orbits having the same value for the principal quantum number, since the electrons will at certain moments penetrate much farther into the interior of the atom. Even though a 3_2 orbit will not penetrate into the innermost configuration of 1_1 orbits, it will penetrate to distances from the nucleus which are considerably less than the radii of the circular 2_2 orbits. In the case of the 16th, 17th and 18th electrons the conditions are similar to those for the 15th. So for argon we may expect a configuration in which the ten innermost electrons move in orbits of the same type as in the neon atom while the last eight electrons will form a configuration of four 3_1 orbits and four 3_2 orbits, whose symmetrical properties must be regarded as closely corresponding to the configuration of 2-quanta orbits in the neon atom. At the same time, as this picture suggests a qualitative explanation of the similarity of the chemical properties of the elements in the latter part of the second and third periods, it also opens up the possibility of a natural explanation of the conspicuous difference from a quantitative aspect.

Fourth Period. Potassium—Krypton. In the fourth period we meet at first elements which resemble chemically those at the beginning of the two previous periods. This is also what we should expect. We must thus assume that *the 19th electron* is bound in a new type of orbit, and a closer consideration shows that this will be a 4_1 orbit. The points which were emphasized in connection with the binding of the last electron in the sodium atom will be even more marked here on account of the larger quantum number by which the orbits of the inner electrons are characterized. In fact, in the potassium atom the 4_1 orbit of the 19th electron will, as far as inner loops are concerned, coincide closely with the shape of a 3_1 orbit. On this account, therefore, the dimensions of the outer part of the orbit will not only deviate greatly from the dimensions of a 4_1 orbit in the hydrogen atom, but will coincide closely with a hydrogen orbit of the type 2_1, the dimensions of which are about four times smaller than the 4_1 hydrogen orbit. This result allows an immediate explanation of the main features of the chemical properties and the spectrum of potassium. Corresponding results apply to calcium, in the neutral atom of which there will be two valency electrons in equivalent 4_1 orbits.

After calcium the properties of the elements in the fourth period of the periodic system deviate, however, more and more from the corresponding elements in the previous periods, until in the family of the iron metals we meet elements whose properties are essentially different. Proceeding to still higher atomic numbers we again meet different conditions. Thus we find in the latter part of the fourth period a series of elements whose chemical properties approach more and more to the properties of the elements at the end of the preceding periods, until finally with atomic number 36 we again meet one of the inactive gases, namely krypton. This is exactly what we should expect. The formation and stability of the atoms of the elements in the first three periods require that each of the first 18 electrons in the atom shall be bound in each succeeding element in an orbit of the same principal quantum number as that possessed by the particular electron, when it first appeared. It is readily seen that this is no longer the case for the 19th electron. With increasing nuclear charge and the consequent decrease in the difference between the fields of force inside and outside the region of the orbits of the first 18 bound electrons, the dimensions of those parts of a 4_1 orbit which fall outside will approach more and more to the dimensions of a 4-quantum orbit calculated on the assumption that the interaction between the electrons in the atom may be neglected. *With increasing atomic number a point will therefore be reached where a 3_3 orbit will correspond to a firmer binding of the 19th electron than a 4_1 orbit,* and this occurs as early as at the beginning of the fourth period. This cannot only be anticipated from a simple calculation but is confirmed in a striking way from an examination of the series spectra. While the spectrum of potassium indicates that the 4_1 orbit corresponds to a binding which is more than twice as firm as in a 3_3 orbit corresponding to the first spectral term in the *D* series, the conditions are entirely different as soon as calcium is reached. We shall not consider the arc spectrum which is emitted during the capture of the 20th electron but the spark spectrum which corresponds to the capture and binding of the 19th electron. While the spark spectrum of magnesium exhibits great similarity with the sodium spectrum as regards the values of the spectral terms in the various series—apart from the fact that the constant appearing in formula (12) is four times as large as the Rydberg constant—we meet in the spark spectrum of calcium the remarkable condition

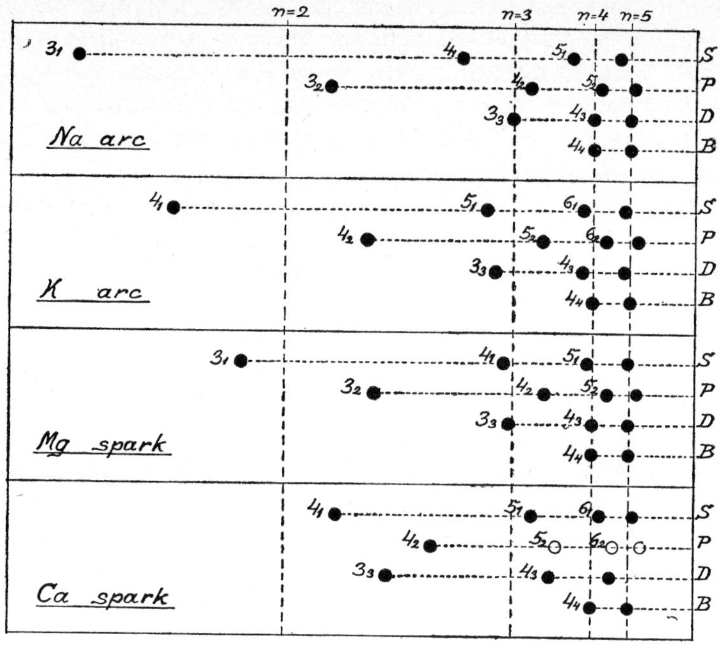

Fig. 4.

that the first term of the D series is larger than the first term of the P series and is only a little smaller than the first term of the S series, which may be regarded as corresponding to the binding of the 19th electron in the normal state of the calcium atom. These facts are shown in figure 4 which gives a survey of the stationary states corresponding to the arc spectra of sodium and potassium. As in figures 2 and 3 of the sodium spectrum, we have disregarded the complexity of the spectral terms, and the numbers characterizing the stationary states are simply the quantum numbers n and k. For the sake of comparison the scale in which the energy of the different states is indicated is chosen four times as small for the spark spectra as for the arc spectra. Consequently the vertical lines indicated with various values of n correspond for the arc spectra to the spectral terms of hydrogen, for the spark spectra to the terms of the helium spectrum given by formula (7). Comparing the change in the relative firmness in the binding of the 19th electron in a 4_1 and 3_3 orbit for potassium and calcium we see that we must be prepared already for the next element, scandium, to find that the 3_3 orbit will correspond to a stronger binding of this electron than a 4_1 orbit. On the other hand it follows from previous remarks that the binding will be much lighter than for the first 18 electrons which agrees that in chemical combinations scandium appears electropositively with three valencies.

If we proceed to the following elements, a still larger number of 3_3 orbits will occur in the normal state of these atoms, since the number of such electron orbits will depend upon the firmness of their binding compared to the firmness with which an electron is bound in a 4_1 orbit, in which type of orbit at least the last captured electron in the atom may be assumed to move. We therefore meet conditions which are essentially different from those which we have considered in connection with the previous periods, so that here we have to do with *the successive development of one of the inner groups of electrons in the atom,* in this case with groups of electrons in 3-quanta orbits. Only when the development of this group has been completed may we expect to find once more a corresponding change in the properties of the elements with increasing atomic number such as we find in the preceding periods. The properties of the elements in the latter part of the fourth period show immediately that the group, when completed, will possess 18 electrons. Thus in krypton, for example, we may expect besides the groups of 1, 2 and 3-quanta orbits a markedly symmetrical configuration of 8 electrons in 4-quanta orbits consisting of four 4_1 orbits and four 4_2 orbits.

The question now arises: In which way will the gradual formation of the group of electrons having 3-quanta orbits take place? From analogy with the constitution of the groups of electrons with 2-quanta orbits we might at first sight be inclined to suppose that the complete group of 3-quanta orbits would consist of three subgroups of four electrons each in orbits of the types 3_1, 3_2 and 3_3 respectively, so that the total number of electrons would be 12 instead of 18. Further consideration shows, however, that such an expectation would not be justified. The stability of the configuration of eight electrons with 2-quanta orbits occurring in neon must be ascribed not only to the symmetrical configuration of the electronic orbits in the two subgroups of 2_1 and 2_2 orbits respectively, but fully as much to the possibility of bringing the orbits inside these subgroups into harmonic relation with one another. The situation is different, however, for the groups of electrons with 3-quanta orbits. Three subgroups of four orbits each can not in this case be expected to come into interaction with one another in a correspondingly simple manner. On the contrary we must assume that the presence of electrons in 3_3 orbits will diminish the harmony of the orbits within the first two 3-quanta subgroups, at any rate when a point is reached where the 19th electron is no longer, as was the case with scandium, bound considerably more lightly than the previously bound electrons in 3-quanta orbits, but has been drawn so far into the atom that it revolves within essentially the same region of the atom where these electrons move. We shall now assume that this decrease in the harmony will so to say "open" the previously "closed" configuration of electrons in orbits of these types. As regards the final result, the number 18 indicates that after the group is finally formed there will be three subgroups containing six electrons each. Even if it has not at present been possible to follow in detail the various steps in the formation of the group this result is nevertheless confirmed in an interesting manner by the fact that it is possible to arrange three configurations having six electrons each in a simple manner relative to one another. The configuration of the subgroups does not exhibit a tetrahedral symmetry like the groups of 2-quanta orbits in carbon, but a symmetry which, so far as the relative orientation of the normals to the planes of the orbits is concerned, may be described as trigonal.

In spite of the great difference in the properties of the elements of this period, compared with those of the preceding period, the completion of the group of 18 electrons in 3-quanta orbits in the fourth period may to a certain extent be said to have the same characteristic results as the completion of the group of 2-quanta orbits in the second period. As we have seen, this determined not only the properties of neon as an inactive gas, but in addition the electronegative properties of the preceding elements and the electropositive properties of the elements which follow. The fact that there is no inactive gas possessing an outer group of 18 electrons is very easily accounted for by the much larger dimensions which a 3_3 orbit has in comparison with a 2_2 orbit revolving in the same field of force. On this account a complete 3-quanta group can not occur as the outermost group in a neutral atom, but only in positively charged ions. The characteristic decrease in valency which we meet in copper, shown by the appearance of the singly charged cuprous ions, indicates the same tendency towards the completion of a symmetrical configuration of electrons that we found in the marked electronegative character of an element like fluorine. Direct evidence that a complete group of 3-quanta orbits is present in the cuprous ion is given by the spectrum of copper which, in contrast to the extremely complicated spectra of the preceding elements resulting from the unsymmetrical character of the inner system, possesses a simple structure very much like that of the sodium spectrum. This may no doubt be ascribed to a simple symmetrical structure present in the cuprous ion similar to that in the sodium ion, although the great difference in the constitution of the outer group of electrons in these ions is shown both by the considerable difference in the values of the spectral terms and in the separation of the doublets in the P terms of the two spectra. The occurrence of the cupric compounds shows, however, that the firmness of binding in the group of 3-quanta orbits in the copper atom is not as great as the firmness with which the electrons are bound in the group of 2-quanta orbits in the sodium atom. Zinc, which is always divalent, is the first element in which the groups of the electrons are so firmly bound that they can not be removed by ordinary chemical processes.

The picture I have given of the formation and structure of the atoms of the elements in the fourth period gives an explanation of the chemical and spectral properties. In addition it is supported by evidence of a different nature to that which we have hitherto used. It is a familiar fact, that the elements in the fourth period differ markedly from the elements in the preceding periods partly in their *magnetic properties* and partly in the *characteristic colours* of their compounds. Paramagnetism and colours do occur in elements belonging to the foregoing periods, but not in simple compounds where the atoms considered enter as ions. Many elements of the fourth period, on the contrary, exhibit paramagnetic properties and characteristic colours even in dissociated aqueous solutions. The importance of this has been emphasized by Ladenburg in his attempt to explain the properties of the elements in the long periods of the periodic system (see p. 73). Langmuir in order to account for the difference between the fourth period and the preceding periods simply assumed that the atom, in addition to the layers of cells containing 8 electrons each, possesses an outer layer of cells with room for 18 electrons which is completely filled for the first time in the case of krypton. Ladenburg, on the other hand, assumes that for some reason or other an intermediate layer is developed between the inner electronic configuration in the atom appearing already in argon, and the external group of valency electrons. This layer commences with scandium and is completed exactly at the end of the family of iron metals. In support of this assumption Ladenburg not only mentions the chemical properties of the elements in the fourth period, but also refers to the paramagnetism and colours which occur exactly in the elements, where this intermediate layer should be in development. It is seen that Ladenburg's ideas exhibit certain formal similarities with the interpretation I have given above of the appearance of the fourth period, and it is interesting to note that our view, based on a direct investigation of the conditions for the formation of the atoms, enables us to understand the relation emphasized by Ladenburg.

Our ordinary electrodynamic conceptions are probably insufficient to form a basis for an explanation of atomic magnetism. This is hardly to be wondered at when we remember that they have not proved adequate to account for the phenomena of radiation which are connected with the intimate interaction between the electric and magnetic forces arising from the motion of the electrons. In whatever way these difficulties may be solved it seems simplest to assume that the occurrence of magnetism, such as we meet in the elements of the fourth period, results from a lack of symmetry in the internal structure of the atom, thus preventing the magnetic forces arising from the motion of the electrons from forming a system of closed lines of force running wholly within the atom. While it has been assumed that the ions of the elements in the previous periods, whether positively or negatively charged, contain configurations of marked symmetrical character, we must, however, be prepared to encounter a definite lack of symmetry in the electronic configurations in ions of those elements within the fourth period which contain a group of electrons in 3-quanta orbits in the transition stage between symmetrical configurations of 8 and 18 electrons respectively. As pointed out by Kossel, the experimental results exhibit an extreme simplicity, the magnetic moment of the ions depending only on the number of electrons in the ion. Ferric ions, for example, exhibit the same atomic magnetism as manganous ions, while manganic ions exhibit the same atomic magnetism as chromous ions. It is in beautiful agreement with what we have assumed about the structure of the atoms of copper and zinc, that the magnetism disappears with those ions containing 28 electrons which, as I stated, must be assumed to contain a complete group of 3-quanta orbits. On the whole a consideration of the magnetic properties of the elements within the fourth period gives us a vivid impression of how a wound in the otherwise symmetrical inner structure is first developed and then healed as we pass from element to element. It is to be hoped that a further investigation of the magnetic properties will give us a clue to the way in which the group of electrons in 3-quanta orbits is developed step by step.

Also the colours of the ions directly support our view of atomic structure. According to the postulates of the quantum theory absorption as well as emission of radiation is regarded as taking place during transitions between stationary states. The occurrence of colours, that is to say the absorption of light in the visible region of the spectrum, is evidence of transitions involving energy changes of the same order of magnitude as those giving the usual optical spectra of the elements. In contrast to the ions of the elements of the preceding periods where all the electrons are assumed to be very firmly bound, the occurrence of such processes in the fourth period is exactly what we should expect. For the development and completion of the electronic groups with 3-quanta orbits will proceed, so to say, in competition with the binding of electrons in orbits of higher quanta, since the binding of electrons in 3-quanta orbits occurs when the electrons in these orbits are bound more firmly than electrons in 4_1 orbits. The development of the group will therefore proceed to the point where we may say there is equilibrium between the two kinds of orbits. This condition may be assumed to be intimately connected not only with the colour of the ions, but also with the tendency of the elements to form ions with different valencies. This is in contrast to the elements of the first periods where the charge of the ions in aqueous solutions is always the same for one and the same element.

Fifth Period. Rubidium—Xenon. The structure of the atoms in the remaining periods may be followed up in complete analogy with what has already been said. Thus we shall assume that the 37th and 38th electrons in the elements of the fifth period are bound in 5_1 orbits. This is supported by the measurements of the arc spectrum of rubidium and the spark spectrum of strontium. The latter spectrum indicates at the same time that 4_3 orbits will soon appear, and therefore in this period, which like the 4th contains 18 elements, we must assume that we are witnessing a *further stage in the development of the electronic group of 4-quanta orbits.* The first stage in the formation of this group may be said to have been attained in krypton with the appearance of a symmetrical configuration of eight electrons consisting of two subgroups each of four electrons in 4_1 and 4_2 orbits. A second preliminary completion must be regarded as having been reached with the appearance of a symmetrical configuration of 18 electrons in the case of silver, consisting of three subgroups with six electrons each in orbits of the types 4_1, 4_2 and 4_3. Everything that has been said about the successive formation of the group of electrons with 3-quanta orbits applies unchanged to this stage in the transformation of the group with 4-quanta orbits. For in no case have we made use of the absolute values of the quantum numbers nor of assumptions concerning the form of the orbits but only of the number of possible types of orbits which might come into consideration. At the same time it may be of interest to mention that the properties of these elements compared with those of the foregoing period nevertheless show a difference corresponding exactly to what would be expected from the difference in the types of orbits. For instance, the divergencies from the characteristic valency conditions of the elements in the second and third periods appear later in the fifth period than for elements in the fourth period. While an element like titanium in the fourth period already shows a marked tendency to occur with various valencies, on the other hand an element like zirconium is still quadri-valent like carbon in the second period and silicon in the third. A simple investigation of the kinematic properties of the orbits of the electrons shows in fact that an electron in an excentric 4_3 orbit of an element in the fifth period will be considerably more loosely bound than an electron in a circular 3_3 orbit of the corresponding element in the fourth period, while electrons which are bound in excentric orbits of the types 5_1 and 4_1 respectively will correspond to a binding of about the same firmness.

At the end of the fifth period we may assume that xenon, the atomic number of which is 54, has a structure which in addition to the two 1-quantum, eight 2-quanta, eighteen 3-quanta and eighteen 4-quanta orbits already mentioned contains a symmetrical configuration of eight electrons in 5-quanta orbits consisting of two subgroups with four electrons each in 5_1 and 5_2 orbits respectively.

Sixth Period. Caesium—Niton. If we now consider the atoms of elements of still higher atomic number, we must first of all assume that the 55th and 56th electrons in the atoms of caesium and barium are bound in 6_1 orbits. This is confirmed by the spectra of these elements. It is clear, however, that we must be prepared shortly to meet entirely new conditions. With increasing nuclear charge we shall have to expect not only that an electron in a 5_3 orbit will be bound more firmly than in a 6_1 orbit, but we must also expect that a moment will arrive when during the formation of the atom a 4_4 orbit will represent a firmer binding of the electron than an orbit of 5 or 6-quanta, in much the same way as in the elements of the fourth period a new stage in the development of the 3-quanta group was started when a point was reached where for the first time the 19th electron was bound in a 3_3 orbit instead of in a 4_1 orbit. We shall thus expect in the sixth period to meet with a new stage in the development of the group with 4-quanta orbits. Once this point has been reached we must be prepared to find with increasing atomic number a number of elements following one another, which as in the family of the iron metals have very nearly the same properties. The similarity will, however, be still more pronounced, since in this case we are concerned with the successive transformation of a configuration of electrons which lies deeper in the interior of the atom. You will have already guessed that what I have in view is a simple explanation of the occurrence of the *family of rare earths* at the beginning of the sixth period. As in the case of the transformation and completion of the group of 3-quanta orbits in the fourth period and the partial completion of groups of 4-quanta orbits in the fifth period, we may immediately deduce from the length of the sixth period the number of electrons, namely 32, which are finally contained in the 4-quanta group of orbits. Analogous to what applied to the group of 3-quanta orbits it is probable that, when the group is completed, it will contain eight electrons in each of the four subgroups. Even though it has not yet been possible to follow the development of the group step by step, we can even here give some theoretical evidence in favour of the occurrence of a symmetrical configuration of exactly this number of electrons. I shall simply mention that it is not possible without coincidence of the planes of the orbits to arrive at an interaction between four subgroups of six electrons each in a configuration of simple trigonal symmetry, which is equally simple as that shown by three subgroups. The difficulties which we meet make it probable that a harmonic interaction can be attained precisely by four groups each containing eight electrons the orbital configurations of which exhibit axial symmetry.

Just as in the case of the family of the iron metals in the fourth period, the proposed explanation of the occurrence of the family of rare earths in the sixth period is supported in an interesting manner by an investigation of the magnetic properties of these elements. In spite of the great chemical similarity the members of this family exhibit very different magnetic properties, so that while some of them exhibit but very little magnetism others exhibit a greater magnetic moment per atom than any other element which has been investigated. It is also possible to give a simple interpretation of the peculiar colours exhibited by the compounds of these elements in much the same way as in the case of the family of iron metals in the fourth period. The idea that the appearance of the group of the rare earths is connected with the development of inner groups in the atom is not in itself new and has for instance been considered by Vegard in connection with his work on X-ray spectra. The new feature of the present considerations lies, however, in the emphasis laid on the peculiar way in which the relative strength of the binding for two orbits of the same principal quantum number but of different shapes varies with the nuclear charge and with the number of electrons previously bound. Due to this fact the presence of a group like that of the rare earths in the sixth period may be considered as a direct consequence of the theory and might actually have been predicted on a quantum theory, adapted to the explanation of the properties of the elements within the preceding periods in the way I have shown.

Besides *the final development of the group of 4-quanta orbits* we observe in the sixth period in the family of the platinum metals *the second stage in the development of the group of 5-quanta orbits.* Also in the radioactive, chemically inactive gas niton, which completes this period, we observe the first preliminary step in the development of a group of electrons with 6-quanta orbits. In the atom of this element, in addition to the groups of electrons of two 1-quantum, eight 2-quanta, eighteen 3-quanta, thirty-two 4-quanta and eighteen 5-quanta orbits respectively, there is also an outer symmetrical configuration of eight electrons in 6-quanta orbits, which we shall assume to consist of two subgroups with four electrons each in 6_1 and 6_2 orbits respectively.

Seventh Period. In the seventh and last period of the periodic system we may expect the appearance of 7-quanta orbits in the normal state of the atom. Thus in the neutral atom of radium in addition to the electronic structure of niton there will be two electrons in 7_1 orbits which will penetrate during their revolution not only into the region of the orbits of electrons possessing lower values for the principal quantum number, but even to distances from the nucleus which are less than the radii of the orbits of the innermost 1-quantum orbits. The properties of the elements in the seventh period are very similar to the properties of the elements in the fifth period. Thus, in contrast to the conditions in the sixth period, there are no elements whose properties resemble one another like those of the rare earths. In exact analogy with what has already been said about the relations between the properties of the elements in the fourth and fifth periods this may be very simply explained by the fact that an excentric 5_4 orbit will correspond to a considerably looser binding of an electron in the atom of an element of the seventh period than the binding of an electron in a circular 4_4 orbit in the corresponding element of the sixth period, while there will be a much smaller difference in the firmness of the binding of these electrons in orbits of the types 7_1 and 6_1 respectively.

It is well known that the seventh period is not complete, for no atom has been found having an atomic number greater than 92. This is probably connected with the fact that the last elements in the system are radioactive and that nuclei of atoms with a total charge greater than 92 will not be sufficiently stable to exist under conditions where the elements can be observed. It is tempting to sketch a picture of the atoms formed by the capture and binding of electrons around nuclei having higher charges, and thus to obtain some idea of the properties which the corresponding hypothetical elements might be expected to exhibit. I shall not develop this matter further, however, since the general results we should get will be evident to you from the views I have developed to explain the properties of the elements actually observed. A survey of these results is given in the following table, which gives a symbolical representation of the atomic structure of the inactive gases which complete the first six periods in the periodic system. In order to emphasize the progressive change the table includes the probable arrangement of electrons in the next atom which would possess properties like the inactive gases.

The view of atomic constitution underlying this table, which involves configurations of electrons moving with large velocities between each other, so that the electrons in the "outer" groups penetrate into the region of the orbits of the electrons of the "inner' groups, is of course completely different from such statical models of the atom as are proposed by Langmuir. But quite apart from this it will be seen that the arrangement of the electronic groups in the atom, to which we have been lead by tracing the way in which each single electron has been bound, is essentially different from the arrangement of the groups in Langmuir's theory. In order to explain the properties of the elements of the sixth period Langmuir assumes for instance that, in addition to the inner layers of cells containing 2, 8, 8, 18 and 18 electrons respectively, which are employed to account for the properties of the elements in the earlier periods, the atom also possesses a layer of cells with room for 32 electrons which is just completed in the case of niton.

Element	Atomic number	Number of Electrons in n_k-orbits																							
		1_1	2_1	2_2	3_1	3_2	3_3	4_1	4_2	4_3	4_4	5_1	5_2	5_3	5_4	5_5	6_1	6_2	6_3	6_4	6_5	6_6	7_1	7_2	7_3
Helium	2	2																							
Neon	10	2	4	4																					
Argon	18	2	4	4	4	4	–																		
Krypton	36	2	4	4	6	6	6	4	4	–	–														
Xenon	54	2	4	4	6	6	6	6	6	6	–	4	4	–	–	–									
Niton	86	2	4	4	6	6	6	8	8	8	8	6	6	6	–	–	4	4	–	–	–	–			
?	118	2	4	4	6	6	6	8	8	8	8	8	8	8	8	–	6	6	6	–	–	–	4	4	–

In this connection it may be of interest to mention a recent paper by Bury, to which my attention was first drawn after the deliverance of this address, and which contains an interesting survey of the chemical properties of the elements based on similar conceptions of atomic structure as those applied by Lewis and Langmuir. From purely chemical considerations Bury arrives at conclusions which as regards the arrangement and completion of the groups in the main coincide with those of the present theory, the outlines of which were given in my letters to *Nature* mentioned in the introduction.

Survey of the periodic table. The results given in this address are also illustrated by means of the representation of the periodic system given in fig. 1. In this figure the frames are meant to indicate such elements in which one of the "inner" groups is in a stage of development. Thus there will be found in the fourth and fifth periods a single frame indicating the final completion of the electronic group with 3-quanta orbits, and the last stage but one in the development of the group with 4-quanta orbits respectively. In the sixth period it has been necessary to introduce two frames, of which the inner one indicates the last stage of the evolution of the group with 4-quanta orbits, giving rise to the rare earths. This occurs at a place in the periodic system where the third stage in the development of an electronic group with 5-quanta orbits, indicated by the outer frame, has already begun. In this connection it will be seen that the inner frame encloses a smaller number of elements than is usually attributed to the family of the rare earths. At the end of this group an uncertainty exists, due to the fact that no element of atomic number 72 is known with certainty. However, as indicated in fig. 1, we must conclude from the theory that the group with 4-quanta orbits is finally completed in lutetium (71). This element therefore ought to be the last in the sequence of consecutive elements with similar properties in the first half of the sixth period, and at the place 72 an element must be expected which in its chemical and physical properties is homologous with zirconium and thorium. This, which is already indicated on Julius Thomsen's old table, has also been pointed out by Bury. [Quite recently Dauvillier has in an investigation of the X-ray spectrum excited in preparations containing rare earths, observed certain faint lines which he ascribes to an element of atomic number 72. This element is identified by him as the element celtium, belonging to the family of rare earths, the existence of which had previously been suspected by Urbain. Quite apart from the difficulties which this result, if correct, might entail for atomic theories, it would, since the rare earths according to chemical view possess three valencies, imply a rise in positive valency of two units when passing from the element 72 to the next element 73, tantalum. This would mean an exception from the otherwise general rule, that the valency never increases by more than one unit when passing from one element to the next in the periodic table.] In the case of the incomplete seventh period the full drawn frame indicates the third stage in the development of the electronic group with 6-quanta orbits, which must begin in actinium. The dotted frame indicates the last stage but one in the development of the group with 5-quanta orbits, which hitherto has not been observed, but which ought to begin shortly after uranium, if it has not already begun in this element.

With reference to the homology of the elements the exceptional position of the elements enclosed by frames in fig. 1 is further emphasized by taking care that, in spite of the large similarity many elements exhibit, no connecting lines are drawn between two elements which occupy different positions in the system with respect to framing. In fact, the large chemical similarity between, for instance, aluminium and scandium, both of which are trivalent and pronounced electropositive elements, is directly or indirectly emphasized in the current representations of the periodic table. While this procedure is justified by the analogous structure of the trivalent ions of these elements, our more detailed ideas of atomic structure suggest, however, marked differences in the physical properties of aluminium and scandium, originating in the essentially different character of the way in which the last three electrons in the neutral atom are bound. This fact gives probably a direct explanation of the marked difference existing between the spectra of aluminium and scandium. Even if the spectrum of scandium is not yet sufficiently cleared up, this difference seems to be of a much more fundamental character than for instance the difference between the arc spectra of sodium and copper, which apart from the large difference in the absolute values of the spectral terms possess a completely analogous structure, as previously mentioned in this essay. On the whole we must expect that the spectra of elements in the later periods lying inside a frame will show new features compared with the spectra of the elements in the first three periods. This expectation seems supported by recent work on the spectrum of manganese by Catalán, which appeared just before the printing of this essay.

Before I leave the interpretation of the chemical properties by means of this atomic model I should like to remind you once again of the fundamental principles which we have used. The whole theory has evolved from an investigation of the way in which electrons can be captured by an atom. The formation of an atom was held to consist in the successive binding of electrons, this binding resulting in radiation according to the quantum theory.

8—2

According to the fundamental postulates of the theory this binding takes place in stages by transitions between stationary states accompanied by emission of radiation. For the problem of the stability of the atom the essential problem is at what stage such a process comes to an end. As regards this point the postulates give no direct information, but here the correspondence principle is brought in. Even though it has been possible to penetrate considerably further at many points than the time has permitted me to indicate to you, still it has not yet been possible to follow in detail all stages in the formation of the atoms. We cannot say, for instance, that the above table of the atomic constitution of the inert gases may in every detail be considered as the unambiguous result of applying the correspondence principle. On the other hand it appears that our considerations already place the empirical data in a light which scarcely permits of an essentially different interpretation of the properties of the elements based upon the postulates of the quantum theory. This applies not only to the series spectra and the close relationship of these to the chemical properties of the elements, but also to the X-ray spectra, the consideration of which leads us into an investigation of interatomic processes of an entirely different character. As we have already mentioned, it is necessary to assume that the emission of the latter spectra is connected with processes which may be described as a reorganization of the completely formed atom after a disturbance produced in the interior of the atom by the action of external forces.

IV. REORGANIZATION OF ATOMS AND X-RAY SPECTRA

As in the case of the series spectra it has also been possible to represent the frequency of each line in the X-ray spectrum of an element as the difference of two of a set of spectral terms. We shall therefore assume that each X-ray line is due to a transition between two stationary states of the atom. The values of the atomic energy corresponding to these states are frequently referred to as the "energy levels" of the X-ray spectra. The great difference between the origin of the X-ray and the series spectra is clearly seen, however, in the difference of the laws applying to the absorption of radiation in the X-ray and the optical regions of the spectra. The absorption by non-excited atoms in the latter case is connected with those lines in the series spectrum which correspond to combinations of the various spectral terms with the largest of these terms. As has been shown, especially by the investigations of Wagner and de Broglie, the absorption in the X-ray region, on the other hand, is connected not with the X-ray lines but with certain spectral regions commencing at the so-called "absorption edges." The frequencies of these edges agree very closely with the spectral terms used to account for the X-ray lines. We shall now see how the conception of atomic structure developed in the preceding pages offers a simple interpretation of these facts. Let us consider the following question : What changes in the state of the atom can be produced by the absorption of radiation, and which processes of emission can be initiated by such changes ?

Absorption and emission of X-rays and correspondence principle. The possibility of producing a change at all in the motion of an electron in the interior of an atom by means of radiation must in the first place be regarded as intimately connected with the character of the interaction between the electrons within the separate groups. In contrast to the forms of motion where at every moment the position of the electrons exhibits polygonal or polyhedral symmetry, the conception of this interaction evolved from a consideration of the possible formation of atoms by successive binding of electrons has such a character that the harmonic components in the motion of an electron are in general represented in the resulting electric moment of the atom. As a result of this it will be possible to release a single electron from the interaction with the other electrons in the same group by a process which possesses the necessary analogy with an absorption process on the ordinary electrodynamic view claimed by the correspondence principle. The points of view on which we based the interpretation of the development and completion of the groups during the formation of an atom imply, on the other hand, that just as no additional electron can be taken up into a previously completed group in the atom by a change involving emission of radiation, similarly it will not be possible for a new electron to be added to such a group, when the state of the atom is changed by absorption of radiation. This means that an electron which belongs to one of the inner groups of the atom, as a consequence of an absorption process—besides the case where it leaves the atom completely—can only go over either to an incompleted group, or to an orbit where the electron during the greater part of its revolution moves at a distance from the nucleus large compared to the distance of the other electrons. On account of the peculiar conditions of stability which control the occurrence of incomplete groups in the interior of the atom, the energy which is necessary to bring about a transition to such a group will in general differ very little from that required to remove the particular electron completely from the atom. We must therefore assume that the energy levels corresponding to the absorption edges indicate to a first approximation the amount of work that is required to remove an electron in one of the inner groups completely from the atom. The correspondence principle also provides a basis for understanding the experimental evidence about the appearance of the emission lines of the X-ray spectra due to transitions between the stationary states corresponding to these energy levels. Thus the nature of the interaction between the electrons in the groups of the atom implies that each electron in the atom is so to say prepared, independently of the other electrons in the same group, to seize any opportunity which is offered to become more firmly bound by being taken up into a group of electrons with orbits corresponding to smaller values of the principal quantum number. It is evident, however, that on the basis of our views of atomic structure, such an opportunity is always at hand as soon as an electron has been removed from one of these groups.

At the same time that our view of the atom leads to a natural conception of the phenomena of emission and absorption of X-rays, agreeing closely with that by which Kossel has attempted to give a formal explanation of the experimental observations, it also suggests a simple explanation of those quantitative relations holding for the frequencies of the lines which have been discovered by Moseley and Sommerfeld. These researches brought to light a remarkable and far-reaching similarity between the Röntgen spectrum of a given element and the spectrum which would be expected to appear upon the binding of a single electron by the nucleus. This similarity we immediately understand if we recall that in the normal state of the atom there are electrons moving in orbits which, with certain limitations, correspond to all stages of such a binding process and that, when an electron is removed from its original place in the atom, processes may be started within the atom which will correspond to all transitions between these stages permitted by the correspondence principle. This brings us at once out of those difficulties which apparently arise, when one attempts to account for the origin of the X-ray spectra by means of an atomic structure, suited to explain the periodic system. This difficulty has been felt to such an extent that it has led Sommerfeld for example in his recent work to assume that the configurations of the electrons in the various atoms of one and the same element may be different even under usual conditions. Since, in contrast to our ideas, he supposed all electrons in the principal groups of the atom to move in equivalent orbits, he is compelled to assume that these groups are different in the different atoms, corresponding to different possible types of orbital shapes. Such an assumption, however, seems inconsistent with an interpretation of the definite character of the physical and chemical properties of the elements, and stands in marked contradiction with the points of view about the stability of the atoms which form the basis of the view of atomic structure here proposed.

X-ray spectra and atomic structure. In this connection it is of interest to emphasize that the group distribution of the electrons in the atom, on which we have based both the explanation of the periodic system and the classification of the lines in the X-ray spectra, shows itself in an entirely different manner in these two phenomena. While the characteristic change of the chemical properties with atomic number is due to the gradual development and completion of the groups of the loosest bound electrons, the characteristic absence of almost every trace of a periodic change in the X-ray spectra is due to two causes. Firstly the electronic configuration of the completed groups is repeated unchanged for increasing atomic number, and secondly the gradual way in which the incompleted groups are developed implies that a type of orbit, from the moment when it for the first time appears in the normal state of the neutral atom, always will occur in this state and will correspond to a steadily increasing firmness of binding. The development of the groups in the atom with increasing atomic number which governs the chemical properties of the elements shows itself in the X-ray spectra mainly in the appearance of new lines. Swinne has already referred to a connection of this kind between the periodic system and the X-ray spectra in connection with Kossel's theory. We can only expect a closer connection between the X-ray phenomena and the chemical properties of the elements, when the conditions on the surface of the atom are concerned. In agreement with what has been brought to light by investigations on absorption of X-rays in elements of lower atomic number, such as have been performed in recent years in the physical laboratory at Lund, we understand immediately that the position and eventual structure of the absorption edges will to a certain degree depend upon the physical and chemical conditions under which the element investigated exists, while such a dependence does not appear in the characteristic emission lines.

If we attempt to obtain a more detailed explanation of the experimental observations, we meet the question of the influence of the presence of the other electrons in the atom upon the firmness of the binding of an electron in a given type of orbit. This influence will, as we at once see, be least for the inner parts of the atom, where for each electron the attraction of the nucleus is large in proportion to the repulsion of the other electrons. It should also be recalled, that while the relative influence of the presence of the other electrons upon the firmness of the binding will decrease with increasing charge of the nucleus, the effect of the variation in the mass of the electron with the velocity upon the firmness of the binding will increase strongly. This may be seen from Sommerfeld's formula (11). While we obtain a fairly good agreement for the levels corresponding to the removal of one of the innermost electrons in the atom by using the simple formula (11), it is, however, already necessary to take the influence of the other electrons into consideration in making an approximate calculation of the levels corresponding to a removal of an electron from one of the outer groups in the atom. Just this circumstance offers us, however, a possibility of obtaining information about the configurations of the electrons in the interior of the atoms from the X-ray spectra. Numerous investigations have been directed at this question both by

Sommerfeld and his pupils and by Debye, Vegard and others. It may also be remarked that de Broglie and Dauvillier in a recent paper have thought it possible to find support in the experimental material for certain assumptions about the numbers of electrons in the groups of the atom to which Dauvillier had been led by considerations about the periodic system similar to those proposed by Langmuir and Ladenburg. In calculations made in connection with these investigations it is assumed that the electrons in the various groups move in separate concentric regions of the atom, so that the effect of the presence of the electrons in inner groups upon the motion of the electrons in outer groups as a first approximation may be expected to consist in a simple screening of the nucleus. On our view, however, the conditions are essentially different, since for the calculation of the firmness of the binding of the electrons it is necessary to take into consideration that the electrons in the more lightly bound groups in general during a certain fraction of their revolution will penetrate into the region of the orbits of electrons in the more firmly bound groups. On account of this fact, many examples of which we saw in the series spectra, we can not expect to give an account of the firmness of the binding of the separate electrons, simply by means of a "screening correction" consisting in the subtraction of a constant quantity from the value for N in such formulae as (5) and (11). Furthermore in the calculation of the work corresponding to the energy levels we must take account not only of the interaction between the electrons in the normal state of the atom, but also of the changes in the configuration and interaction of the remaining electrons, which establish themselves automatically without emission of radiation during the removal of the electron. Even though such calculations have not yet been made very accurately, a preliminary investigation has already shown that it is possible approximately to account for the experimental results.

Classification of X-ray spectra. Independently of a definite view of atomic structure it has been possible by means of a formal application of Kossel's and Sommerfeld's theories to disentangle the large amount of experimental material on X-ray spectra. This material is drawn mainly from the accurate measurements of

Siegbahn and his collaborators. From this disentanglement of the experimental observations, in which besides Sommerfeld and his students especially Smekal and Coster have taken part, we have obtained a nearly complete classification of the energy levels corresponding to the X-ray spectra. These levels are formally referred to types of orbits characterized by two quantum numbers n and k, and certain definite rules for the possibilities of combination between the various levels have also been found. In this way a number of results of great interest for the further elucidation of the origin of the X-ray spectra have been attained. First it has not only been possible to find levels, which within certain limits correspond to all possible pairs of numbers for n and k, but it has been found that in general to each such pair more than one level must be assigned. This result, which at first may appear very surprising, upon further consideration can be given a simple interpretation. We must remember that the levels depend not only upon the constitution of the atom in the normal state, but also upon the configurations which appear after the removal of one of the inner electrons and which in contrast to the normal state do not possess a uniquely completed character. If we thus consider a process in which one of the electrons in a group (subgroup) is removed we must be prepared to find that after the process the orbits of the remaining electrons in this group may be orientated in more than one way in relation to one another, and still fulfil the conditions required of the stationary states by the quantum theory. Such a view of the "complexity" of the levels, as further consideration shows, just accounts for the manner in which the energy difference of the two levels varies with the atomic number. Without attempting to develop a more detailed picture of atomic structure, Smekal has already discussed the possibility of accounting for the multiplicity of levels. Besides referring to the possibility that the separate electrons in the principal groups do not move in equivalent orbits, Smekal suggests the introduction of three quantum numbers for the description of the various groups, but does not further indicate to what extent these quantum numbers shall be regarded as characterizing a complexity in the structure of the groups in the normal state itself or on the contrary characterizing the incompleted groups which appear when an electron is removed.

It will be seen that the complexity of the X-ray levels exhibits a close analogy with the explanation of the complexity of the terms of the series spectra. There exists, however, this difference between the complex structure of the X-ray spectra and the complex structure of the lines in the series spectra, that in the X-ray spectra there occur not only combinations between spectral terms, for which k varies by unity, but also between terms corresponding to the same value of k. This may be assumed to be due to the fact, that in the X-ray spectra in contrast to the series spectra we have to do with transitions between stationary states where, both before and after the transition, the electron concerned takes part in an intimate interaction with other electrons in orbits with the same principal quantum number. Even though this interaction may be assumed to be of such a nature that the harmonic components which would appear in the motion of an electron in the absence of the others will in general also appear in the resulting moment of the atom, we must expect that the interaction between the electrons will give rise to the appearance in this moment of new types of harmonic components.

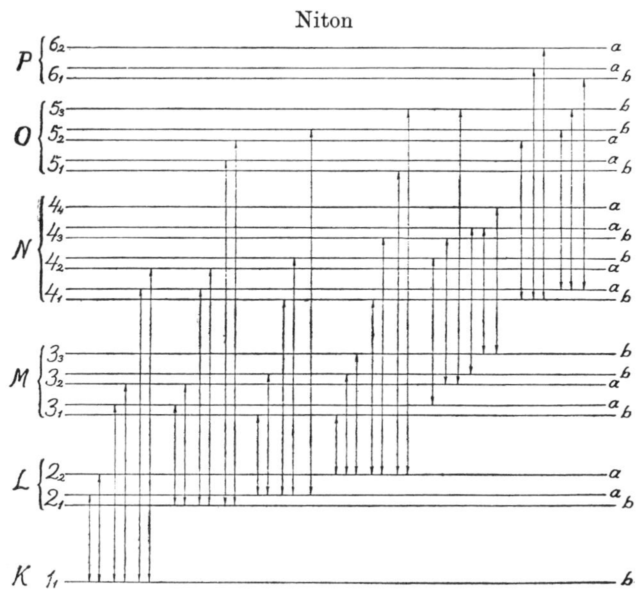

Fig. 5.

It may be of interest to insert here a few words about a new paper of Coster which appeared after this address was given, and in which he has succeeded in obtaining an extended and detailed connection between the X-ray spectra and the ideas of atomic structure given in this essay. The classification mentioned above was based on measurements of the spectra of the heaviest elements, and the results in their complete form, which were principally due to independent work of Coster and Wentzel, may be represented by the diagram in fig. 5, which refers to elements in the neighbourhood of niton. The vertical arrows

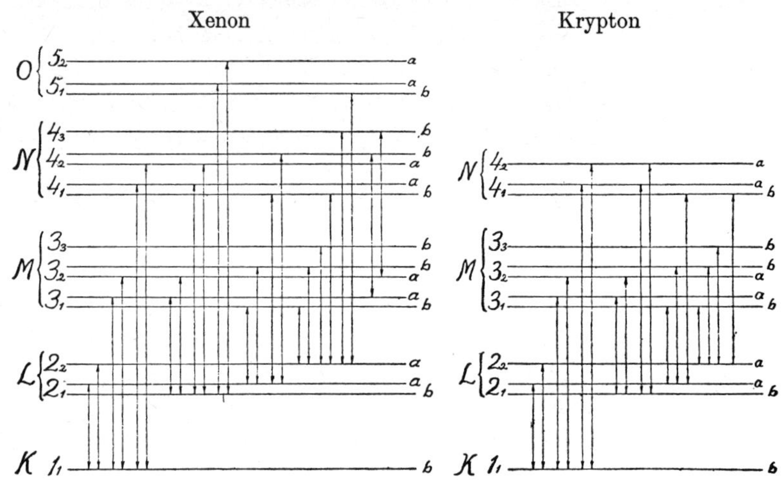

Fig. 6.

represent the observed lines arising from combinations between the different energy levels which are represented by horizontal lines. In each group the levels are arranged in the same succession as their energy values, but their distances do not give a quantitative picture of the actual energy differences, since this would require a much larger figure. The numbers n_k attached to the different levels indicate the type of the corresponding orbit. The letters a and b refer to the rules of combination which I mentioned. According to these rules the possibility of combination is limited (1) by the exclusion of combinations, for which k changes by more than one unit, (2) by the condition that only combinations between an a- and a b-level can take place. The latter rule was given in this form by Coster; Wentzel formulated it in a somewhat different way by the formal introduction of a third quantum number. In his new paper Coster has established a similar classification for the lighter elements. For the elements in the neighbourhood of xenon and krypton he has obtained results illustrated by the diagrams given in fig. 6. Just as in fig. 5 the levels correspond exactly to those types of orbits which, as seen from the table on page 113, according to the theory will be present in the atoms of these elements. In xenon several of the levels present in niton have disappeared, and in krypton still more levels have fallen away. Coster has also investigated in which elements these particular levels appear for the last time, when passing from higher to lower atomic number. His results concerning this point confirm in detail the predictions of the theory. Further he proves that the change in the firmness of binding of the electrons in the outer groups in the elements of the family of the rare earths shows a dependence on the atomic number which strongly supports the assumption that in these elements a completion of an inner group of 4-quanta orbits takes place. For details the reader is referred to Coster's paper in the *Philosophical Magazine*. Another important contribution to our systematic knowledge of the X-ray spectra is contained in a recent paper by Wentzel. He shows that various lines, which find no place in the classification hitherto considered, can be ascribed in a natural manner to processes of reorganization, initiated by the removal of more than one electron from the atom; these lines are therefore in a certain sense analogous to the enhanced lines in the optical spectra.

Conclusion

Before bringing this address to a close I wish once more to emphasize the complete analogy in the application of the quantum theory to the stability of the atom, used in explaining two so different phenomena as the periodic system and X-ray spectra. This point is of the greatest importance in judging the reality of the theory, since the justification for employing considerations, relating to the formation of atoms by successive capture of electrons, as a guiding principle for the investigation of atomic structure might appear doubtful if such considerations could not be brought into natural agreement with views on the reorganization of the atom after a disturbance in the normal electronic arrangement. Even though a certain inner consistency in this view of atomic structure will be recognized, it is, however, hardly necessary for me to emphasize the incomplete character of the theory, not only as regards the elaboration of details, but also so far as the foundation of the general points of view is concerned. There seems, however, to be no other way of advance in atomic problems than that which hitherto has been followed, namely to let the work in these two directions go hand in hand.

APPENDIX

Classification of electronic orbits. Since the third essay of this book was written considerable work on the development of the theory has been done, and it has especially been possible in certain respects to obtain a close test of the theoretical conclusions by means of experimental data. These conclusions concern primarily the classification of the electron orbits in the atom by means of the quantum symbol n_k which is suited to the description of the stationary states of a central motion. A survey of this classification is given in the scheme on the following page.

As seen, the scheme includes for the most part only those elements which are found in the beginning of the periods of the natural system. This is due to the fact that it is mainly as regards the first occurrence of orbits of new types that the theory allows a comparison with experiment. The problem of the exact way in which the groups of electrons of orbits of types already present are successively built up is one which still affords unsolved difficulties. In fact the present state of the quantum theory hardly provides an unambiguous basis for conclusions as to the distribution of the electrons among the different subgroups of a completed or partially completed electron group and for testing such conclusions by comparison with experiment. In the scheme this state of affairs is indicated by placing the numbers of these electrons in square brackets every first time where such a group appears in the scheme. Further, the bracketing in certain cases of the number of electrons in a subgroup of the group with highest quantum number means that the experimental material at hand is as yet insufficient for fixing these numbers with certainty.

Series spectra. As an illustration of the way in which use is made of the experimental material obtained from the investigation of series spectra a number of diagrams are given below which are exactly analogous to the diagram on p. 102 of the text in their

N \ n_k	1_1	$2_1 2_2$	$3_1 3_2 3_3$	$4_1 4_2 4_3 4_4$	$5_1 5_2 5_3 5_4 5_5$	$6_1 6_2 6_3 6_4 6_5 6_6$	$7_1 7_2$
1 H	1						
2 He	2						
3 Li	2	1					
4 Be	2	2					
5 B	2	2 1					
— —	–	– –					
10 Ne	2	[4 4]					
11 Na	2	4 4	1				
12 Mg	2	4 4	2				
13 Al	2	4 4	2 1				
— —	–	– –	– –				
18 A	2	4 4	[4 4]				
19 K	2	4 4	4 4	1			
20 Ca	2	4 4	4 4	2			
21 Sc	2	4 4	4 4 1	(2)			
22 Ti	2	4 4	4 4 2	(2)			
— —	–	– –	– – –	–			
29 Cu	2	4 4	[6 6 6]	1			
30 Zn	2	4 4	6 6 6	2			
31 Ga	2	4 4	6 6 6	2 1			
— —	–	– –	– – –	– –			
36 Kr	2	4 4	6 6 6	[4 4]			
37 Rb	2	4 4	6 6 6	4 4	1		
38 Sr	2	4 4	6 6 6	4 4	2		
39 Y	2	4 4	6 6 6	4 4 1	(2)		
40 Zr	2	4 4	6 6 6	4 4 2	(2)		
— —	–	– –	– – –	– – –	–		
47 Ag	2	4 4	6 6 6	[6 6 6]	1		
48 Cd	2	4 4	6 6 6	6 6 6	2		
49 In	2	4 4	6 6 6	6 6 6	2 1		
— —	–	– –	– – –	– – –	– –		
54 X	2	4 4	6 6 6	6 6 6	[4 4]		
55 Cs	2	4 4	6 6 6	6 6 6	4 4	1	
56 Ba	2	4 4	6 6 6	6 6 6	4 4	2	
57 La	2	4 4	6 6 6	6 6 6	4 4 1	(2)	
58 Ce	2	4 4	6 6 6	6 6 6 1	4 4 1	(2)	
59 Pr	2	4 4	6 6 6	6 6 6 2	4 4 1	(2)	
— —	–	– –	– – –	– – – –	– – –	–	
71 Cp	2	4 4	6 6 6	[8 8 8 8]	4 4 1	(2)	
72 Hf	2	4 4	6 6 6	8 8 8 8	4 4 2	(2)	
— —	–	– –	– – –	– – – –	– – –	–	
79 Au	2	4 4	6 6 6	8 8 8 8	[6 6 6]	1	
80 Hg	2	4 4	6 6 6	8 8 8 8	6 6 6	2	
81 Tl	2	4 4	6 6 6	8 8 8 8	6 6 6	2 1	
— —	–	– –	– – –	– – – –	– – –	– –	
86 Em	2	4 4	6 6 6	8 8 8 8	6 6 6	[4 4]	
87 —	2	4 4	6 6 6	8 8 8 8	6 6 6	4 4	1
88 Ra	2	4 4	6 6 6	8 8 8 8	6 6 6	4 4	2
89 Ac	2	4 4	6 6 6	8 8 8 8	6 6 6	4 4 1	(2)
90 Th	2	4 4	6 6 6	8 8 8 8	6 6 6	4 4 2	(2)
— —	–	– –	– – –	– – – –	– – –	– – –	–
118 ?	2	4 4	6 6 6	8 8 8 8	[8 8 8 8]	[6 6 6]	[4 4]

method of representation. Figure 7 gives the terms of the arc spectra of the alkalis. In this diagram, as well as in the following, the number in brackets refers to the atomic number of the element, while the index in Roman type indicates the degree of ionization of the atomic residue. Considering the diagram one must remember

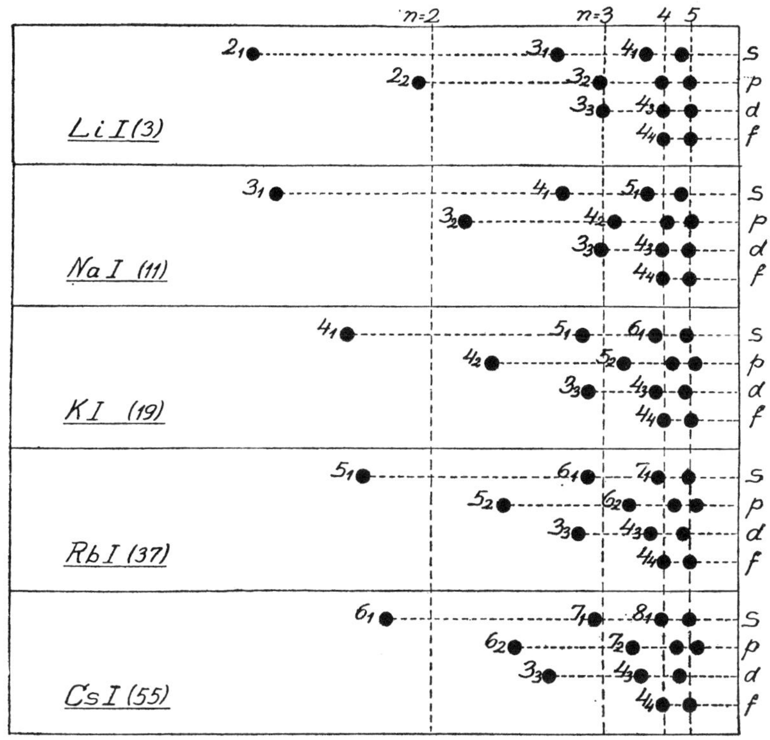

Fig. 7.

the difference in character between the orbits corresponding to the *s*- and *p*-terms and those corresponding to the *d*- and *f*-terms. While the former, at least from Na on, penetrate into the inner region of the atomic residue, the latter lie completely in the outer region of the atom. As a consequence of this the first terms of the *s*- and *p*-series in the sequence of these elements do not increase nearly as rapidly as would hydrogen terms with the same principal quantum numbers. The difference in the behaviour of the two types of orbits is clearly brought to light by the continuous decrease of these terms from element to element in contrast to the *d*- and *f*-terms, which remain constant or slowly increase.

Figure 8 refers to the spark spectra of the alkaline earths. As will be remembered spark spectra of this type are distinguished from the arc spectra by the replacement of the Rydberg constant

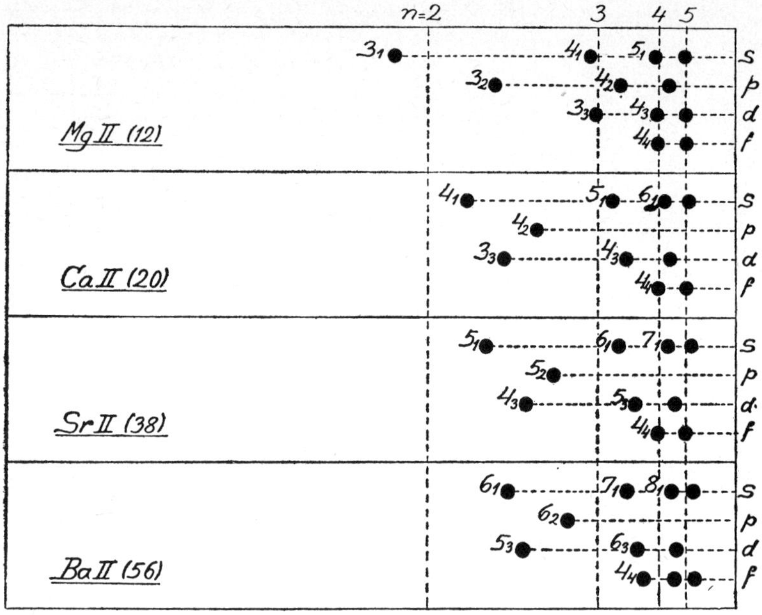

Fig. 8.

K by a constant $4K$, due to the double charge of the atomic residue. Just as in the diagram on p. 102 the scale for representing the terms of these spectra is therefore taken as a quarter of that used in figure 7. As far as the *s*-, *p*- and *f*-terms are concerned the state of affairs is similar to that in the arc spectra of the alkalis The *d*-terms, however, show a different behaviour which, as discussed in detail in the essay, is intimately connected with the commencement of a new stage in the building up of one of the inner electron groups in the elements following the alkaline earths in the later periods of the natural system.

Figure 9 is of particular interest, as it contains the results of the recent work of Paschen and Fowler, in which new types of series spectra were discovered which, instead of the Rydberg constant K, contain $9K$ and $16K$ respectively. According to the theory these spectra are ascribed to atoms in which an electron revolves around an atomic residue with three or four unit charges. The diagram

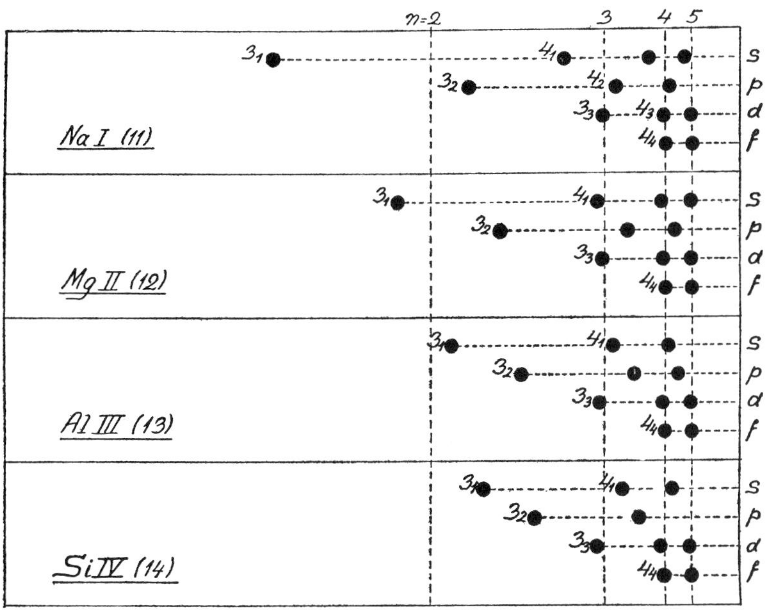

Fig. 9.

gives the terms of the spectra emitted by the binding of the eleventh electron in the atoms of Na, Mg, Al, and Si. In conformity with the representation in the other diagrams the scale is in the latter two cases 1/9 and 1/16 respectively of that used in arc spectra. As was to be expected, one sees how, with increasing charge in the atomic residue, the strength of binding of the 3_1 and 3_2 orbits approaches more and more nearly to that of an electron in a 3-quanta orbit corresponding to a nuclear charge equal to that of the atomic residue. The behaviour of the *d*- and *f*-terms also corresponds exactly to the theoretical expectations.

Figure 10 refers to the arc spectra of certain elements where according to the scheme there occurs for the first time a new type

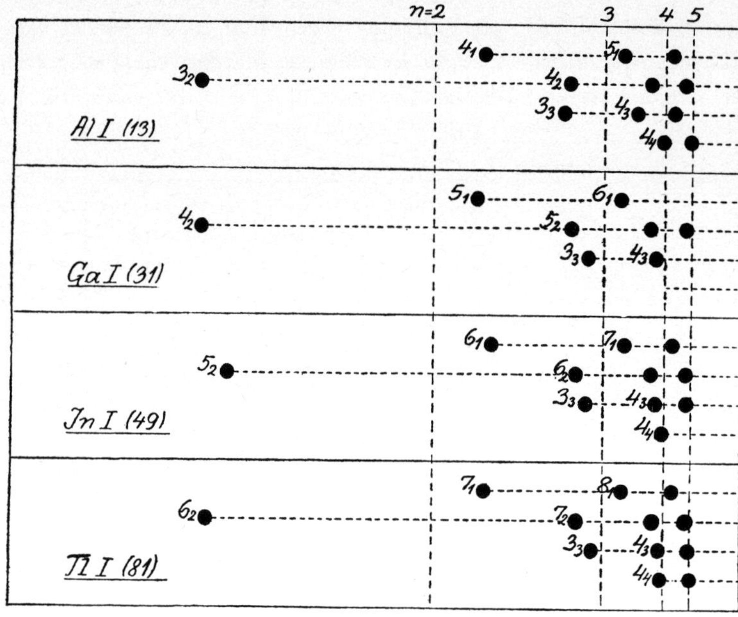

Fig. 10.

of orbit in the normal state, in which the quantum number k is equal to 2. In striking contrast to the spectra discussed above, the first p-terms are here considerably larger than the first s-terms, and a principal quantum number is assigned to them which is smaller by one unit than that of the largest s-term.

Finally in figure 11 are given the results of the investigations just published by Fowler on a spark spectrum of carbon corre-

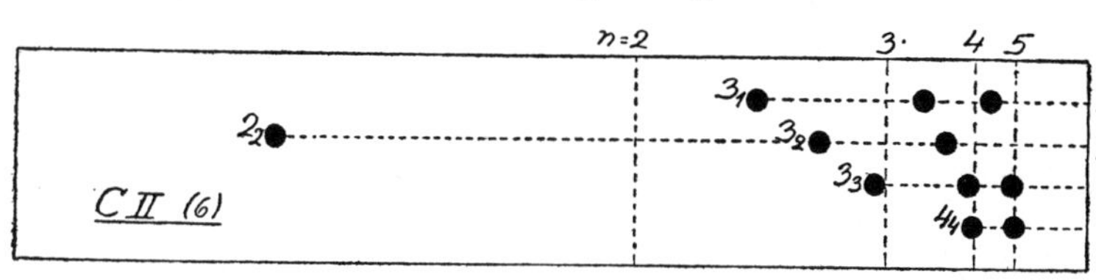

Fig. 11.

sponding to the binding of the 5th electron in the C atom. The interpretation of this spectrum involves a point in which a correction is necessary in the representation given in the third essay. In fact we must assume that the result of the binding process is an electronic orbit of the type 2_2, and not a 2_1 orbit, as assumed in the essay, for the 5th electron in an atom. As the progress of the theoretical work had already indicated, such a behaviour was also to be expected from the general regularities shown by the spectra of homologous elements in the system.

As in the diagrams of the text the manifold of spectral terms reproduced in the diagrams given here is simplified for the sake of clarity in the representation by the neglect of the complexity of the series terms, a mean value of the term components being substituted. As mentioned in Essay III, the origin of this complexity is to be sought for in a deviation of the electron orbits from central orbits. The further pursuit of this question brings us to the problem of the finer coupling of the orbits of the different electrons in the atom and has disclosed difficulties of a fundamental nature which can hardly be satisfactorily solved in the present state of the quantum theory, as has been particularly shown by the analysis of the Zeeman effect. We shall not go more deeply into these questions here, which are closely connected with the above mentioned difficulties confronting a closer discussion of the building up of electron groups in the atom. For these questions, as well as for the closer comparison of the regularities of series spectra with the picture of atomic structure, the reader is referred to a recent paper by the author ("Linienspektren und Atombau," *Annalen der Physik*, LXXI. p. 228, 1923).

X-ray spectra. Another region in which the conclusions of the theory may be tested by means of experimental evidence is that of X-ray spectra. As mentioned at the end of the third essay the work of Coster, in particular, has furnished new empirical material of great importance in this respect. In figure 12 is given a schematical survey of the regularities shown by the values of the X-ray terms. In the figure the ordinates are the square roots of the X-ray terms and the abcissae the atomic numbers. The vertical lines at the bottom of the figure mark the places where, according to the theory, the different orbital types appear for the first time in the normal state of the atom. The horizontal lines attached to

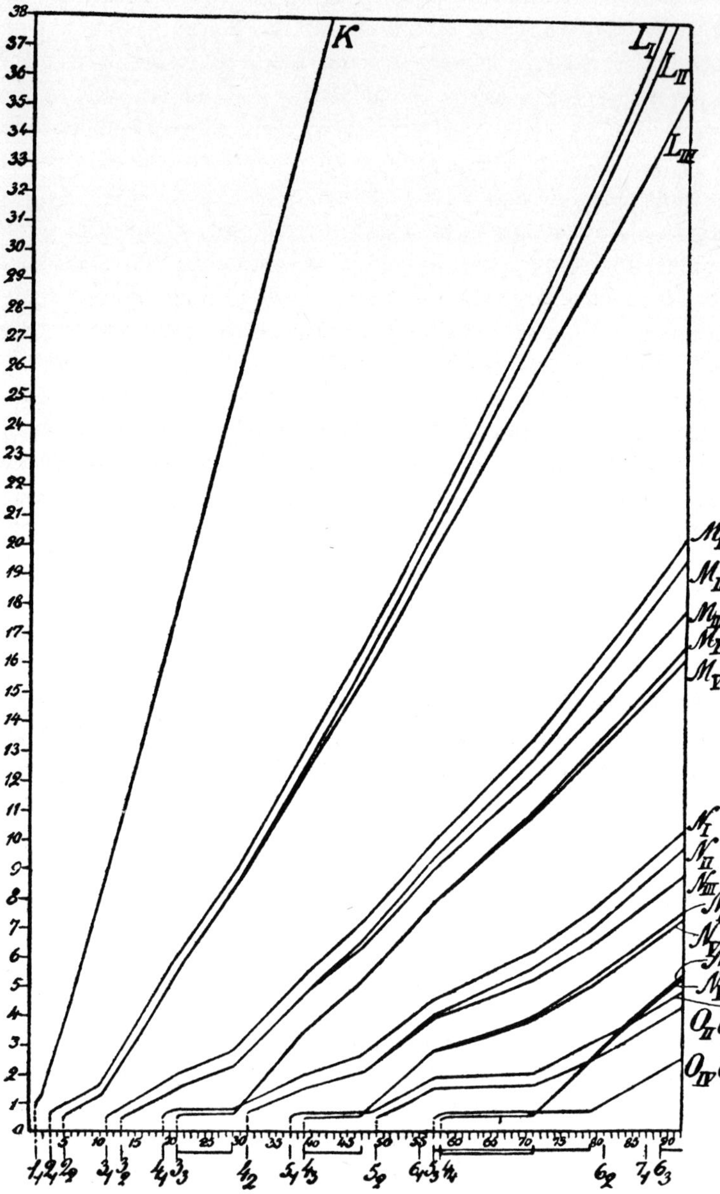
K
L_I
L_{II}
L_{III}
M_I
M_{II}
M_{III}
M_{IV}
M_V
N_I
N_{II}
N_{III}
N_{IV}
N_V
N_{VI}
N_{VII}
O_I
O_{II}
O_{III}
O_{IV}
O_V

Fig. 12.

some of these vertical lines mark those places in the system of the elements where a building up of inner electron groups takes place in the atom. In the first place it appears, as indicated by the curves, that it has been possible to correlate each term with the occurrence of electron orbits of a given type n_k in the normal state of the atom. Although the data at hand are not yet sufficiently complete to fix the course of the curves with the accuracy desired, especially in the region where the terms are small and can only be indirectly determined, the survey given by the figure may be considered as a general confirmation of the theoretical conclusions as regards the successive building up of atoms. Thus as indicated on the diagram in a schematical way it appears possible everywhere to trace the development of an inner electron group by its effect on the rate of variation of the X-ray terms with the atomic number. In particular, the marked irregularities in the course of the N- and O-curves may be regarded as direct evidence of that stage in the development of the electron group of 4-quanta orbits which gives rise to the occurrence of the rare earths in the system of the elements. On the whole the analysis of the new experimental data shows that Moseley's simple relation between X-ray spectra and atomic number constitutes a first approximation, and that we can now trace an intimate connection between these spectra and the general relationships of the elements within the natural system. At the same time it must be kept in mind that a satisfactory explanation of the complexity of the manifold of terms, which consists in the fact that in general each type of orbit n_k corresponds to more than one curve, is confronted at the present time with great difficulties. Clearly the origin of this complexity is to be sought for in the coupling of the different subgroups belonging to an electron group with given principal quantum number. As mentioned above, however, the quantum theory in its present form does not permit us to draw definite conclusions regarding this coupling. For this question reference may be made to a paper by Coster and the author ("Röntgenspektren und periodisches System," *Zs. für Phys.* XII. p. 342, 1923), which contains a general discussion of the empirical material of X-ray spectra from the standpoint of the atomic theory.

Chemical relationship. It may also be of interest in this appendix to enter briefly on the subject of the explanation of chemical relationship. As emphasized in the third essay, the theory permits only conclusions as to the appearance of the so-called heteropolar compounds. It is assumed that in the molecules of these compounds each electron may be regarded as belonging to a particular atomic nucleus, the molecules being held together by the electrostatic forces between the ions, each of which consists of a nucleus with the electrons belonging to it. For the stability of these compounds the factor of primary importance is the size of such ions and the energy which is necessary for the formation of the ions from the neutral atoms, both of which quantities are obviously related directly to the dimensions of the orbits of electrons in the atom and the strength of their binding. In this connection the theory has revealed a close relation between the spectral and chemical properties of the elements, and has even made it possible in certain cases to draw conclusions from the picture of atomic structure as to the chemical properties of a substance.

On the whole the theory has, in this respect, proved to be in accord with experiment. Only in one special case was there, for a time, an apparent difficulty of a fundamental nature. This had to do with the chemical properties of the long unknown element of atomic number 72. As already mentioned in the insertion on p. 114 Dauvillier concluded from an investigation of the X-ray spectrum of a preparation of rare earths published about two years ago that this element was identical with a new member of the family of rare earths, the existence of which had previously been suspected by Urbain and for which the name celtium had been proposed. Obviously the correctness of this conclusion would be in contradiction to the theoretical survey of the system of the elements given in connection with the theory on p. 70 of Essay III. According to this scheme the element 71 should be the last member of the family of rare earths and the element 72 should be the homologue of zirconium (40), just as the element 73 is the homologue of niobium (41). While according to the theory the 68th electron in all these elements should belong to the completed group of 4-quanta orbits, the 69th electron should move in a 5_3 orbit with a strength of binding increasing with increasing atomic number. If actually the element 72 showed chemical properties analogous to those of the trivalent rare earths, we should be compelled, however, to assume that the 69th electron was more strongly bound in this element than in the next element, tantalum, which shows a pronounced quintuple valency. Under these circumstances it was a confirmation of the theory when Coster and Hevesy succeeded in proving by X-ray spectroscopic investigation that the element of atomic number 72 forms an essential constituent of most zirconium minerals. This element for which the name hafnium was proposed proved to be markedly different in its chemical properties from the rare earths, but in every way similar to zirconium. In fact it is only due to the considerable difficulty of separating the new element from zirconium that it is possible to understand that it has so long escaped detection by the chemist, although it is present in minerals in appreciable quantities. The optical spectrum of the new element proved to be completely different from the spectrum previously ascribed by Urbain to the hypothetical celtium, the lines of which have since been shown to belong to the element 71. The closer investigation of the X-ray spectra of the elements of atomic numbers in the neighbourhood of 72 has led to further results of particular interest for the theory. Thus it seems that the magnitude of the terms denoted by N_{VI}, N_{VII} remains approximately constant up to and including the element with atomic number 71, while starting from this element a rapid increase of these terms takes place. In other words the curve representing these terms shows actually a sharp nick at the atomic number 71, which indicates that at this element the stepwise development of the 4-quanta electron group has come to a conclusion.

As to the constitution of the so-called homoeopolar compounds, in the molecules of which a number of electrons are assumed to be shared by several nuclei, the theory permits, at the present time, of only few definite statements. The beautiful regularities brought to light not least by organic chemistry give unquestionable evidence of certain pronounced symmetry properties in the structure of the molecules of a large class of such compounds. It is hardly justifiable, however, to draw conclusions from this as to definite symmetry properties of the configurations of the electron orbits in the isolated atom. It must be particularly emphasized that in the neutral carbon atom it is hardly possible to speak of a tetrahedral symmetry of the orbital configurations of the 2-quanta electron group. As mentioned above, the new investigations of Fowler on the spark spectrum of carbon have proved that the simply charged carbon ion in its normal state besides two electrons in 2_1 orbits possesses one electron moving in a 2_2 orbit. The conclusion that the neutral carbon atom, in contrast to the representation given in Essay III, contains 2_2 orbits as well as 2_1 orbits can therefore hardly be avoided.

The difficulties which confront the development of definite pictures of the structure of the molecules of such compounds depend primarily on the fact that we must here be prepared to meet such large deviations from central motion, that an adequate classification of the electron orbits by means of the simple quantum symbol n_k can no longer be maintained. For just this reason it seems at present hardly possible to proceed, with the aid of the ideas of the quantum theory, far beyond the more qualitative considerations with which the chemist has explained the regularities of organic chemistry with so great success.

CAMBRIDGE: PRINTED BY W. LEWIS AT THE UNIVERSITY PRESS

VIII. SEVEN LECTURES ON THE THEORY OF ATOMIC STRUCTURE

GÖTTINGEN 1922
(TRANSLATION*)

* [See Introduction, sect. 7.]

FIRST LECTURE. JUNE 12, 1922.

The present state of physics is characterized by the fact that we not only are convinced about the reality of atoms but also believe that we possess a detailed knowledge of their building stones. I shall not discuss here the development of our conceptions of these building stones, nor Rutherford's discovery of the atomic nuclei which initiated a new epoch in physics.

The picture of the atom that we shall discuss here is the nuclear atom to which Rutherford's discovery has led. We suppose that an atom consists of a positively charged nucleus around which a number of electrons revolve. It is a remarkable fact that the number of electrons in an atom, i.e., the number of elementary charges on the nucleus, ["Kernladungszahl"], is equal to the ordinal number of the atom in the periodic system of the elements. When I speak of a detailed knowledge of the building stones, I mean a relatively detailed knowledge, just as the degree of knowledge in physics is always relative. In fact, we know nothing about the internal structure of electrons, and we do not know what forces hold them together. About the nuclei, we know that they exhibit a complex structure. This follows from the existence of radioactive nuclei. Moreover, Rutherford has succeeded in splitting ⟨certain⟩ nuclei, thus, for the first time, achieving an artificial transmutation of a chemical element. It is of essential interest to us that the spatial extension of the building stones is very small compared to the dimensions of the atoms; for this reason, we believe that, for a large number of properties of atoms, the building stones may be regarded as point charges and mass points.

A further feature characterizing present-day atomic physics is that we are convinced that, on the basis of this picture, it is not possible to make any progress with classical electrodynamics. We cannot comprehend the peculiar stability of atoms, which we are forced to assume, in order to understand the properties of the elements. The classical theory fails especially when we attempt to use it to explain the radiation phenomena. We must expect that any motion that may have to be considered would give rise to an emission of electromagnetic radiation, which would not cease until the total energy of the system was radiated away and all electrons had fallen into the nucleus. All physicists know that these difficulties can be overcome with the aid of the quantum theory. I shall not develop the quantum theory in detail, but shall only state the main results upon which all the following depends.

Planck's investigation of radiation was based on the use of the harmonic oscillator as a simple model of an atom. The motion of the quasi-elastically bound charge of an oscillator is given by

$$\rho = C \cos 2\pi(\omega_0 t + \delta), \tag{1}$$

where I denote the frequency by ω_0, since it is assumed to be independent of the energy of the system but is determined only by its nature. Planck investigated the radiative equilibrium between a large number of such atoms and a radiation field. In the statistical treatment of this problem, it appeared that not all energy values of the oscillator are to be considered, but only those that are whole multiples of an energy quantum given by $h\omega_0$:

$$E_n = nh\omega_0 \quad (n = 0, 1, 2, \ldots). \tag{2}$$

This result reveals features which are essentially different from the conceptions of the hitherto existing physics, in that here, for the first time, whole numbers play a preferential role in the formulation of a physical law. This result leads us to attempt, also for other systems, to distinguish certain motions which possess stability of a peculiar kind. In his treatment on the radiation phenomena, Planck based his work on the assumption, hitherto always adhered to, that the frequency ν of the emitted or absorbed radiation is identical with the frequency ω_0 of the moving charge:

$$\nu \equiv \omega_0.$$

However, Einstein has pointed out, for the first time, that Planck's result not only teaches us something about the motion in the atom but also expresses something directly about the radiation process. He remarks that Planck's result can also be described by saying that energy is emitted or absorbed only in quanta of amount $h\nu$, so that the frequency of the emitted or absorbed radiation is determined by the equation

$$\Delta E = h\nu. \tag{3}$$

This idea led Einstein to his theory of the photoelectric effect. The speed of the emitted electrons depends only on the frequency ν and not on the intensity of the incident light which is absorbed in quanta of amount $h\nu$. The kinetic energy of the emitted electrons is given by

$$\tfrac{1}{2}mv^2 = h\nu - P,$$

where P is the work necessary to remove an electron from the metal. This result of Einstein is particularly important, because here Planck's idea is applied, for the first time, to a problem of a non-statistical nature. The difficulty resulting from this assumption is that the emitted radiation has an atomistic structure; the radiation is not emitted as spherical waves but only in a single direction. Yet, only on this assumption could the photoelectric effect be understood. On the other hand, such an assumption of the nature of light cannot be reconciled with the explanation of the interference phenomena. We obtain the frequency from the picture of the classical theory by analyzing the emitted spectrum. The theory of light quanta gives only a definition of the quantity ν which enters in eq. (3), but offers no explanation of what we shall understand by wavelength.

That such a theory was adopted illustrates the impossibility of preserving an inner cohesion in the theory of radiation. Here, we simply state that no consistent picture of the radiation processes exists, and we shall perhaps return to this point later.

In what follows we shall stick to facts. The difficulties are overcome with the aid of two fundamental assumptions. The first one reads: Of the possible mechanical motions in a nuclear atom, certain definite ones are distinguished in which the atom, without emitting radiation, possesses a peculiar stability. We shall call these distinguished states "stationary states." A change of energy occurs only in such a manner that the atom goes from one stationary state to another stationary state. According to the second assumption, in this transition a train of harmonic waves is emitted having a frequency given by the relation

$$E' - E'' = h\nu. \tag{4}$$

I shall later attempt to express these postulates more precisely and discuss their scope and limitation. Today I shall mention only the simplest applications. First, we shall use these principles to interpret the hydrogen spectrum which possesses an especially simple structure. The series spectrum of hydrogen is represented by the formula

$$\nu = K\left(\frac{1}{(n'')^2} - \frac{1}{(n')^2}\right) \tag{5}$$

which, in this general form, was first stated by Ritz. The factor K is a constant. If we set $n'' = 2$ and let n' run through the values 3, 4, 5, ..., we obtain the Balmer series. The series found by Paschen in the infrared is represented by $n'' = 3$, $n' = 4, 5, 6, \ldots$, the ultraviolet Lyman series, discovered a few years ago, is represented by $n'' = 1$, $n' = 2, 3, 4, \ldots$, and recently Hughes has found a series in the far infrared for which $n'' = 4$ and $n' = 5, 6, 7, \ldots$. Our problem is to connect this spectrum with the picture of the hydrogen atom which, according to Rutherford, consists of a singly charged positive nucleus and an electron. To determine the motion, we apply ordinary mechanics, from which we can calculate the motion to a high degree of approximation. We obtain thus a purely periodic motion, the well-known Kepler motion, whose components we can express by Fourier series of the form

$$\rho = \sum_{\tau=1}^{\infty} C_\tau \cos 2\pi(\tau\omega t + \delta_\tau). \tag{6}$$

According to the well-known Keplerian laws, the major axis ⟨2⟩a and the period of revolution ω are connected with the energy W by the formulas

$$2a = -Ne^2/W, \tag{7}$$

$$\omega = \sqrt{2W^3(m+M)/\pi^2N^2e^4mM}. \tag{8}$$

Here N is the atomic number ["Kernladungszahl"], which we introduce for the sake of generality; m and M are the masses of the electron and the nucleus, respectively. If we now assume that

$$W_n = -hK/n^2 \tag{9}$$

and substitute this in the frequency relation (4), we obtain the Ritz formula (5). Moreover, we have

$$2a_n = n^2\frac{Ne^2}{hK} = n^2 \times 1.1 \times 10^{\langle - \rangle 8} \text{ cm},$$

$$\omega_n = \frac{1}{n^3}\sqrt{\frac{2K^3h^3(m+M)}{\pi^2N^2e^4mM}} = \frac{1}{n^3} \times 6.2 \times 10^{15} \text{ sec}^{-1}.$$

What we can relate to experience is merely transitions from one stationary state to another. First, we shall consider the orders of magnitude of the atoms in the stationary states. These depend very much upon the value of n. However, a distinguished state is that of lowest possible energy, i.e., the most stable state corresponding to $n = 1$. For this ⟨state⟩ we obtain the value of 0.55×10^{-8} cm for half the major axis. This is the otherwise assumed order of magnitude for the atomic radius. Also, ω is of the order of magnitude of the characteristic frequencies occurring in the dispersion theory. However, there can be no question of a direct comparison.

We shall now consider the question of the passage from the quantum theory to the classical theory. As we let n become larger and larger, the ratio between the frequencies of revolution of consecutive states approaches unity:

$$\omega_n/\omega_{n-1} \to 1 \quad \text{as} \quad n \to \infty.$$

The frequency emitted in a transition

$$\nu_{n' \to n''} = K\left(\frac{1}{(n'')^2} - \frac{1}{(n')^2}\right) \tag{10}$$

approaches the value

$$(n' - n'')\frac{2K}{n^3}.$$

On the other hand, the frequency emitted according to the classical theory is

$$\omega = (n' - n'')\omega_0,$$

where $n' - n'' = 1$. If we now, in order to obtain agreement, attempt to put

$$2K/n^3 = \omega_0,$$

we obtain the formula

$$K = \frac{2\pi^2 N^2 e^4 m}{h^3(1 + m/M)}. \tag{11}$$

For hydrogen $K_H = 3.26 \times 10^{15}\ \text{sec}^{-1}$, and we get

$$\frac{2\pi^2 e^4 m}{h^3} = 3.29 \times 10^{15}\ \text{sec}^{-1}.$$

Thus, in this limit of high quantum numbers, we obtain a connection between the emitted radiation and the motion. This result shows that there exists a very close connection between our conception of the structure of the atom and the observed spectrum, but this does not at all mean a merging of classical and quantum theory. We shall now consider more closely this connection which is of a remarkable nature. A transition in which a high quantum number changes by one unit gives a frequency of the emitted radiation equal to the fundamental frequency of the mechanical motion. According to the classical theory, all the different overtones are emitted simultaneously; however, here we obtain the different overtones by quite different processes. We have obtained no connection between the conceptions, but only a numerical agreement. If this is not to be regarded as mere accident, there must also, besides the approach of the frequencies, occur an approach of the amplitudes of the radiation, which we must then claim to be just as essential as that of the frequencies. In the classical theory, the intensity of a frequency $\tau\omega$ depends on the amplitude C_τ; in the quantum theory it depends on the frequency with which the transitions occur. What this depends on is a perfectly open question, which we have no basis for answering. When developing the theory of heat radiation with the aid of the postulates used here, Einstein emphasized these difficulties and overcame them by introducing probabilities for the transitions. This raises the serious question of whether we must rest content with statements of probabilities for individual processes. As matters stand at present, we are so far from being able to give a real description of these processes that we may well assume that Einstein's mode of treatment may actually be the most appropriate. In that case, however, the transition probabilities for the different processes must be related to the amplitudes of the corresponding overtones. So far, we have traced this connection only for large quantum numbers. However, since the connection is of a purely qualitative nature, it is natural to assume a connection between the transition probabilities and the presence of overtones ⟨in the motion⟩ also for small quantum numbers. From this point of view, we may regard H_β as the octave of H_α, since H_β corresponds to a quantum jump of 2 and H_α to a quantum jump of 1. It is true that H_β does not have twice the frequency of H_α, but it corresponds to the octave. This relationship we call the "correspondence principle." To each transition there corresponds a harmonic component of the mechanical motion. We shall attempt to apply this to the Planck oscillator. Application of the frequency relation gives

$$\nu = (n' - n'')\,\omega_0\,. \tag{12}$$

In order that $\nu \equiv \omega_0$, we must have $n' - n'' = 1$. Hence, only transitions to an immediately adjacent state are possible. This is essentially different from the case of the hydrogen spectrum, in which all possible jumps occur. To begin with, this gave rise to serious doubt. The reason for the difference is that the oscillator, contrary to the hydrogen atom, possesses no overtones. A diatomic molecule, e.g., HCl, is an oscillator found in nature. Here, the atoms can oscillate relative to each other around their stable equilibrium positions. This motion can be regarded as harmonic only for very small amplitudes. Since the masses of the atoms are large compared to that of an electron, at least the first few quantum

[348]

states will correspond to very small oscillations. To a first approximation, only a single absorption line actually appears (for simplicity, we neglect the rotation here). Kemble has found that this is not strictly the case but that also the octave occurs; to be sure, very much more weakly. The reason is that the vibrations are not purely harmonic. This entails that there will also exist a small probability for higher quantum jumps; this would, of course, also follow from the classical theory; however, the superiority of the quantum theory shows up immediately. The overtone does not lie exactly at the calculated place but is shifted a little, so that also here the matter is much as in the hydrogen spectrum, although there the frequency corresponding to the quantum jump 2 deviates greatly from the octave and here only very little. Next time, we shall discuss the conditions which govern the stationary states. We shall then also formulate these conditions for more complex systems. In the time remaining today, we shall mention briefly some simple applications of the previous considerations.

We ask ourselves the question if the formula

$$\nu = K\left(\frac{1}{(n'')^2} - \frac{1}{(n')^2}\right)$$

really represents all the lines of the hydrogen series. This has not always been assumed. Lines were found in stellar spectra which were regarded as hydrogen lines but were represented by the formula

$$\nu = K'\left(\frac{1}{2^2} - \frac{1}{(n+\frac{1}{2})^2}\right)$$

and not by the Balmer formula. They were assumed to be hydrogen lines because no other case had ever occurred in which different elements had exhibited properties as similar as those expressed by these spectral formulas. The latter formula can also be written

$$\nu = K'\left(\frac{1}{(\frac{1}{2}n'')^2} - \frac{1}{(\frac{1}{2}n')^2}\right) \quad \text{or} \quad \nu = 4K'\left(\frac{1}{(n'')^2} - \frac{1}{(n')^2}\right), \tag{13}$$

where, in particular, $n'' = 4$. Now, the observations show that, apart from a small deviation, K' is equal to K. This indicates strongly that these lines should not be ascribed to hydrogen but to He^+. This conclusion is supported by a closer consideration of the small difference between K and K'. We find

$$K'/K_H = 1.000408.$$

This small difference from unity is large compared to the experimental error. It agrees with the correction obtained from formula (11) because of the difference between the nuclear masses of H and He:

$$(1+m/M_{\mathrm{H}})/(1+m/M_{\mathrm{He}}) = 1.000409.$$

We have gone into this to show that this surprising relationship between the properties of different elements is a consequence of the picture of the nuclear atom.

A year after these considerations were made, Moseley's fundamental work on the X-ray spectra appeared. He found that the first line in the K-series could be represented by

$$\nu = K(N-\alpha)^2\left(\frac{1}{1^2}-\frac{1}{2^2}\right) \tag{14}$$

and the first line in the L-series by

$$\nu = K(N-\beta)^2\left(\frac{1}{2^2}-\frac{1}{3^2}\right). \tag{15}$$

The significance of this discovery by Moseley lies in the fact that these formulas are very similar to the formula for the hydrogen spectrum. Because of the fundamental difference between the X-ray spectra and the hydrogen spectrum, this is very peculiar. While the hydrogen spectrum gives information about the formation of the hydrogen atom which contains only a single electron, the X-ray spectra have their origin in the disturbance and reorganization of a complete atom containing many electrons. It is also difficult to relate this behavior of X-ray lines to the other properties of the elements which unquestionably exhibit periodicities. The main purpose of these lectures will be to present a picture of atomic structure which permits us to explain, at the same time, this peculiarity and the periodicities that find their expression in the periodic system.

SECOND LECTURE. JUNE 13, 1922.

The quantum theory, which may be regarded as a rational generalization of Planck's original theory, represents a sharp break with the conceptions of classical electrodynamics. This places us in a very peculiar situation. So far, only concepts developed in the classical theories, such as those of the electron and electric and magnetic forces, are available to us for describing the natural phenomena; however, we assume at the same time that the picture of the classical theories is invalid. Now, the question arises if there is any possibility at all of uniting the classical concepts with the quantum theory without contradiction. We are not in a position to settle this question; however, physicists hope that the ideas of both theories possess a certain reality. When attempting to formulate the principles of the quantum theory, we encounter these difficulties, which are formidable, indeed. In such a situation, the most cautious procedure is always to stick to the applications of thc principles. As we have seen, the applications of the quantum theory rest on two fundamental postulates:

(1) Among all mechanically possible motions, a series of particular states are distinguished by their stability. A change of the energy can be brought about only by a transition from one stationary state to another.

(2) When such a transition is accompanied by radiation, this will consist of a train of harmonic waves, the frequency of which is given by the condition

$$E' - E'' = h\nu. \tag{16}$$

When attempting to consider these postulates more closely and to form a picture of the stationary states, we encounter not only the difficulty that no motion of especial stability is distinguished in the classical theory but also that every motion is connected with the appearance of radiation, so that the classical theory cannot be applied without change. The question is now how far-reaching this change needs to be. In all processes to which the quantum theory has been applied, the radiation which, according to the classical theory, is emitted during a period (a revolution of the electron) is but small. Hence, it is a very natural assumption that, in calculating the stationary states, we may neglect the influence of the radiation upon the motion, i.e., that we may treat the system as conservative. We emphasize, however, that such a procedure represents only an approximation; later on, we shall consider more closely the significance of this approximation. That we consider a system as conservative means that we can describe its motion by means of the ordinary equations of motion which, in their canonical form, are as follows

$$\frac{\mathrm{d}p_l}{\mathrm{d}t} = -\frac{\partial E}{\partial q_l}, \quad \frac{\mathrm{d}q_l}{\mathrm{d}t} = \frac{\partial E}{\partial p_l} \quad (l = 1, 2, \ldots s). \tag{17}$$

Here s is the number of degrees of freedom of the system, the q_l are its generalized spatial coordinates, the p_l are the canonically conjugate momenta (generalized momenta), and the energy E is to be regarded as a function of the q_l and p_l. Since the system is conservative, its energy remains constant during the motion. We shall treat here only the motion relative to a coordinate system in which the center of mass of the atom is at rest. To a first approximation, the energy is then given by the relative positions of the building blocks and the kinetic energy. However, we cannot decide definitely whether the system can be treated as conservative, or to what extent the radiation must be taken into account. The general solution of the equations of motion is generally so complicated that it is impossible to single out the stationary states. It appears that certain periodicity properties of the motion are needed to ensure the peculiar stability possessed by the stationary states and to make possible the application of the correspondence principle to these states. The necessary periodicity properties of the motion consist in the motion of the atomic particles being of simple or multiply periodic character, i.e., that it can be represented as a superposition of periodic motions with the fundamental frequencies $\omega_1, \ldots \omega_u$. The number u of these fundamental frequencies, which must be smaller than or equal to the number s of the degrees of freedom of the system, we shall call the degree of periodicity of the motion. A coordinate ζ of the system is then represented by a u-fold Fourier series of the form

$$\zeta = \sum_{-\infty}^{+\infty} C_{\tau_1 \ldots \tau_u} \cos 2\pi[(\tau_1 \omega_1 + \ldots + \tau_u \omega_u)t + \gamma_{\tau_1 \ldots \tau_u}] \tag{18}$$

in which the $\tau_1 \ldots \tau_u$ take on all positive and negative whole numbers, including zero. In order that the representation in terms of the fundamental frequencies may be unique, no relations of the form

$$m_1 \omega_1 + \ldots + m_u \omega_u = 0 \quad (m_l \ldots m_u \text{ integers}) \tag{19}$$

may exist. Then we are always in a position to fix the stationary states by the conditions

$$\mathscr{I}_{\omega_1} = n_1 h, \ldots, \mathscr{I}_{\omega_u} = n_u h. \tag{20}$$

The $\mathscr{I}_\omega$, the definitions of which we shall consider later, are closely connected with the periodicity properties of the motion, as indicated by the indices ω.

We now raise the question of what happens to the periodicity properties of the motion when the atom is subjected to external forces. We shall first consider the simple case that the atom is situated in a constant external field of force. We shall assume, moreover, that the motion is of multiply periodic nature, also in this field of force. A condition for the stability is then that the $\mathscr{I}$'s are of such a nature that they do not vary appreciably with time when the field is increased slowly. On the other hand, when the atom is exposed to a rapidly varying field, e.g., if it is in a radiation field, we must assume that the motion can no longer be described by ordinary mechanics. This is also the case in collisions between atoms. The process itself cannot be described with the aid of ordinary mechanics, since the stationary states are determined by the periodicity properties of the motion. We assume, however, that the atom is in a stationary state before and after the collision, an assumption the consequences of which have been verified by the interpretation of the experiments of Franck and Hertz and their followers. We have just remarked that ordinary mechanics is incapable of describing what happens in a collision. The reason for this fact is easily perceived. In mechanics, the state of a system determines directly only states that are infinitesimally adjacent in time; here, however, the final state is determined beforehand and can be defined only by a consideration of the entire motion.

It is very difficult to visualize this result. In a simple case, we can more or less imagine that an atom, when it suddenly enters a field, is unable to adjust itself to the altered conditions. The matter is different when the field varies slowly. Then the atom gets prepared to adjust its motion to the new conditions. This is called an "adiabatic transformation" of the motion. According to Ehrenfest's principle, it can be described with the aid of ordinary mechanics; of course, the degree of approximation achieved with mechanics is the same as in other cases.

After these reflections, we shall turn to consider the detailed nature of the conditions holding for the stationary states, and we shall first deal with systems of one degree of freedom. In this case, the conditions become particularly simple, since we have to do with the variation of only a single variable. We exclude from the consideration motions in which the particles move infinitely far from each other, since in such cases the motion has no periodicity properties. Thus, we consider only periodic motions. Then $\mathscr{I}$ must be defined by the equation

$$\mathscr{I} = \oint p\,dq.$$

Here the integral is to be extended over a period of the motion. This integral, which we shall term the action integral, has many simple properties. It is independent of the coordinate system chosen, and it is adiabatically invariant, i.e., its value does not change when a change in the motion by slowly varying fields is described by means of ordinary mechanics. If we consider two different neighboring states of motion, the change in the energy is

$$\delta E = \omega \delta \mathscr{I}, \tag{21}$$

where ω is the fundamental frequency of the motion. For a Planck harmonic oscillator, the motion is represented by

$$\zeta = C \cos 2\pi(\omega t + \delta), \tag{22}$$

where $\omega = \omega_0$ is constant. The energy is

$$E_n = nh\omega_0. \tag{23}$$

Hence, in order that eq. (21) may be satisfied, it follows that

$$\mathscr{I}_n = E/\omega_0 = nh. \tag{24}$$

According to Ehrenfest's principle, this equation must hold for all systems that can be obtained from the Planck oscillator by an adiabatic transformation. The equation $\mathscr{I} = nh$ appears as a rational generalization of Planck's fundamental assumption, not only from the point of view of the adiabatic hypothesis but also from the point of view of the correspondence principle. Indeed,

$$\nu = \frac{1}{h}(E_{n'} - E_{n''}) = \frac{1}{h}\int \delta E = \frac{1}{h}\int \omega \delta \mathscr{I},$$

and, for neighboring stationary states for which ω can be considered nearly constant, we get with the aid of $\mathscr{I} = nh$ that

$$\nu = \frac{1}{h}\int \omega \delta \mathscr{I} \sim (n' - n'')\omega, \tag{25}$$

as demanded by the correspondence principle.

We now proceed to systems with several degrees of freedom. Here, the circumstances are in general very complicated; however, they become simple when the motion can be composed of mutually independent components. As an example, we may consider an anisotropic oscillator with two degrees of freedom, the motion of which is made up of two perpendicular components which can be treated independently of each other. Then, the above equations hold for each degree of freedom separately. The same is true of the correspondence. But, an added feature is that there are two periods in the motion. If the motions are simple harmonic, only the quantum number of one of these motions can change in a radiative process and, indeed, only by one unit; the quantum numbers of the two motions can never change simultaneously; for, the combination $\tau_1\omega_1+\tau_2\omega_2$ does not occur in the representation of the motion, but only the frequencies ω_1 and ω_2 separately. Just as the linear harmonic oscillator is approximately realized by a diatomic molecule, as discussed yesterday, it is now a triatomic molecule that corresponds approximately to the anisotropic harmonic oscillator with two degrees of freedom. In fact, Hettner found for the spectrum of water vapor that essentially only such processes take place in which a single quantum number changes by one unit. However, since the motion is not purely harmonic, integral combinations of the fundamental frequencies occur also in the motion with small amplitudes, so that there is a small probability for transitions in which both quantum numbers simultaneously change by one unit. Also here, there is a characteristic difference between the classical theory and the quantum theory. When the binding is anharmonic, "overtones", primarily sum and difference tones, occur also in the classical theory; and, indeed, also when the molecule vibrates about its equilibrium position, as is the case in absorption processes. Actually, however, only sum tones and not difference tones are observed in absorption, i.e., only the simultaneous change of both quantum numbers by $+1$. The reason for this is that the molecule before absorbing is in the state distinguished by the lowest quantum numbers. In emission, differences in the quantum numbers can also occur. I owe this remark to a personal communication from Einstein. As in the case of diatomic molecules, the higher overtones do not lie precisely at the positions demanded by the classical theory.

We now meet cases in which a system of several degrees of freedom has especially simple properties, namely, when separation of variables is possible. In this case it is not necessary that the motion is capable of being represented in the above-mentioned manner as a superposition of independent harmonic motions, but it is possible, nevertheless, to choose special coordinates such that we may speak of a kind of independent parts of the motion. In these cases, each p_l depends only on the conjugate coordinate q_l, so that we can form the quantities

$$\mathcal{J}_l = \oint p_l \mathrm{d}q_l .$$

An especially simple case of this kind appears in the central motion. If the force field is not a Coulomb one, one does not obtain a periodic motion but a motion consisting of a uniform rotation superposed on a simple periodic motion. Such a central motion is separable in polar coordinates. The stationary states are fixed by the quantum conditions

$$\mathscr{I}_r = \oint p_r \mathrm{d}r = n_r h, \qquad \mathscr{I}_\varphi = \int_0^{2\pi} p_\varphi \mathrm{d}\varphi = 2\pi P = n_\varphi h. \tag{26}$$

A minor mathematical difficulty is that the coordinate φ, unlike r, has no natural turning point but increases continually during the motion; it is natural, then, to take 2π as the period of φ and to extend the integral $\mathscr{I}_\varphi$ over this interval. Besides, the evaluation of the second integral is particularly simple because the momentum p_φ is independent of φ and equal to P. These equations are Sommerfeld's well-known conditions, with which he obtained the theory of fine structure. We shall later discuss this theory in some detail but shall now pursue a little further, and in a general way, the method of separation of variables. What about the application of Ehrenfest's principle in the case considered? Since p_φ remains constant, we are dealing formally with a system of one degree of freedom and, hence, $\mathscr{I}_r$ is adiabatically invariant. We shall discuss later the question of how this motion can be changed adiabatically into one that can be represented as the superposition of two harmonic oscillations.

One generally encounters difficulties with the separation of variables when the degree of periodicity u is smaller than the number s of the degrees of freedom. The system is then separable in several different coordinate systems and, since they depend on the coordinate system chosen as basis, the equations $\mathscr{I}_l = \oint p_l \mathrm{d}q_l$ can, therefore, not tell anything specific about the system. The smaller the degree of periodicity, the greater is the arbitrariness in the choice of the system of separation variables. The reason is that, in the case that $u < s$, the periodicity properties do not manifest themselves when the separation method is employed. However, when we turn to the correspondence principle, we encounter the periodicity properties of the system which contain no indication of the particular choice of separation coordinates. The motions mentioned here (with $u < s$) form a family within the general class of multiply periodic systems. There exists a beautiful mathematical theory for them, which I shall not develop in detail; I shall only call your attention to a few essential points. In place of the q_l, p_l, it is possible to introduce a number of "analytical" variables, of which a set $w_1, \ldots w_u$ are usually called "angle coordinates" because of special cases in celestial mechanics where they are of the nature of angles. It seems to me that it would be more reasonable to call them "uniformizing variables". Their principal property is

[356]

that the q_l are multiply periodic functions of the $w_1, \ldots w_u$:

$$q_l = \sum_{-\infty}^{+\infty} C_{\tau_1 \ldots \tau_u} \cos 2\pi(\tau_1 w_1 + \ldots + \tau_u w_u + \gamma_{\tau_1 \ldots \tau_u}). \tag{27}$$

The energy depends only on the variables $\mathscr{I}_{\omega_1} \ldots \mathscr{I}_{\omega_u}$ which are canonically conjugate to the w_l.

Besides the w_l and $\mathscr{I}_{\omega_l}$, there are two sets, each of $(s-u)$ canonically conjugate quantities $\beta_1, \ldots \beta_{s-u}$ and $\alpha_1, \ldots \alpha_{s-u}$, so that the entire set of new coordinates is

$$w_1, \ldots, w_u, \quad \beta_1, \ldots, \beta_{s-u}$$

and the new momenta

$$\mathscr{I}_{\omega_1}, \ldots, \mathscr{I}_{\omega_u}, \quad \alpha_1, \ldots, \alpha_{s-u}.$$

Now, it turns out that the $\mathscr{I}_{\omega_l}$, β_l and α_l are all constant during the motion; the β_l, α_l are the "orbital parameters" of the motion. Now, since E depends only on the $\mathscr{I}_{\omega_l}$, it follows that the

$$\omega_l = \delta E/\delta \mathscr{I}_{\omega_l} \quad (l = 1, 2, \ldots u) \tag{28}$$

are also constant and, hence, that

$$w_l = \omega_l t + \delta_l \quad (l = 1, 2, \ldots u). \tag{29}$$

Thus, the uniformizing variables are linear functions of the time. Substituting (29) into (27), we obtain the coordinates q_l and, hence, also the displacement variables ζ of the particles as functions of the time in the form

$$\zeta = \sum_{-\infty}^{+\infty} C'_{\tau_1 \ldots \tau_u} \cos 2\pi\{(\tau_1 \omega_1 + \ldots + \tau_u \omega_u)t + \delta_1 + \ldots + \delta_u\}. \tag{30}$$

The energy difference between neighboring motions is given by

$$\delta E = \omega_1 \delta \mathscr{I}_{\omega_1} + \ldots + \omega_u \delta \mathscr{I}_{\omega_u}. \tag{31}$$

We still need to determine the zero values of the $\mathscr{I}_{\omega_l}$. They can be fixed by demanding that the $\mathscr{I}_{\omega}$ be adiabatically invariant. As shown by Burgers, the conditions for the constants can be obtained by forming

$$\int_{t_0}^{t_1} \sum \mathscr{I}_{\omega_l} \mathrm{d}w_l = (t_1 - t_0) \sum \mathscr{I}_{\omega_l} \omega_l.$$

This becomes

$$(t - t_0) \sum \mathscr{I}_{\omega_l} \omega_l = \int_{t_0}^{t_1} \sum p_l \mathrm{d}q_l + \text{Per.}\,(w_1, \ldots, w_u), \tag{32}$$

where Per. is a periodic function of the variables indicated. This equation, which must be satisfied for every value of t_1, can be used to determine the zero values of $\mathcal{I}_{\omega_1}, \ldots \mathcal{I}_{\omega_\mu}$. This manner of treating a multiply periodic system originated with Schwarzschild and has been elaborated by Epstein, Burgers, and Kramers. So much for the adiabatic properties of the conditions

$$\mathcal{I}_{\omega_l} = n_{\omega_l} h. \tag{33}$$

What is now the connection of these conditions with the correspondence principle? If we express the emitted frequency by means of the frequency relation, and introduce the energy changes as functions of the angle variables, we get

$$\nu = \frac{1}{h} \int_{(')}^{('')} \sum \omega_l \delta \mathcal{I}_{\omega_l}. \tag{34}$$

Passing to the limit of large quantum numbers, we may regard the ω_l as constants in the integrals and may write

$$\nu \sim \frac{1}{h} \sum \omega_l (\mathcal{I}'_{\omega_l} - \mathcal{I}''_{\omega_l}) = \sum (n' - n'') \omega_l. \tag{35}$$

Thus, the situation is quite analogous to what it was in the case of a single degree of freedom. As in that case, we demand that only such changes in the quantum numbers are possible which correspond to a harmonic component of the motion in such a way that

$$\tau_l = n'_l - n''_l. \tag{36}$$

We shall not again discuss this correspondence in detail but shall only recall that we assume that probabilities of transition can be calculated from this correspondence first of all in the limit of high quantum numbers; for lower quantum numbers, the correspondence principle gives only a clue. The question now arises if it is possible to find a general expression for transition probabilities. This problem is not as hopeless as it might appear at a first glance. Introducing a parameter λ such that $\mathcal{I}_{\omega_l}$ may be written in the form

$$\mathcal{I}_{\omega_l}(\lambda) = h(n''_l + \lambda(n'_l - n''_l)),$$

we have

$$\nu = \frac{1}{h} \int_{(')}^{('')} \sum \omega_l \delta \mathcal{I}_{\omega_l} = \frac{1}{h} \int_0^1 \sum \omega_l \frac{\mathrm{d}\mathcal{I}_{\omega_l}}{\mathrm{d}\lambda} \mathrm{d}\lambda = \int_0^1 \sum (n'_l - n''_l) \omega_l \mathrm{d}\lambda.$$

We see, as has been pointed out by Kramers, that ν is obtained by averaging the ω_l over all intermediate states. After all, the ω_l themselves have no direct physical significance. Kramers has endeavored to find a similar general expression for the average values of amplitudes and to use it in the calculation of transition probabilities. Of course, we cannot claim that this goal can be reached; however, it does not seem impossible.

The general theory of multiply periodic motions sketched here allows us to master these motions completely. As an example, we shall mention the adiabatic transformation, given by Burgers, of a central motion into one composed of two independent harmonic motions, i.e., the motion of an anisotropic harmonic oscillator. If such an oscillator is brought into a magnetic field, its motion cannot be treated by means of the method of separation of variables. However, it can be proved to be an especially simple case of a multiply periodic motion. The general motion may be obtained by the superposition of two elliptical vibrations taking place in opposite directions. If we now let the magnetic field increase so much that the force exerted by it is large compared to the degree of anisotropy, then the two elliptical vibrations will more and more approach circular vibrations without degeneracy occurring. However, as is well known, two circular oscillations in opposite directions and with different periods are equivalent to an elliptic motion with precession of the perihelion, i.e., to a central motion. In this way, it is possible to carry the motion of an anisotropic [the manuscript has "anharmonic"] oscillator adiabatically into a central motion, and vice versa.

As in this special example, the resolution into harmonic components is in general essential. Thus, a central motion can be regarded as the superposition of a periodic motion and a slow uniform rotation. Each of the harmonic components in the periodic motion can be represented by circular motions going in opposite directions. These are superposed on the uniform rotation. Thus, a component is represented by the formula

$$\zeta = \sum C_{\tau, \pm 1} \cos 2\pi\{(\tau\omega \pm \sigma)t + \delta_{\tau, \pm 1}\}. \tag{37}$$

Here, ω is the frequency [the manuscript has "Periode"] of the periodic motion, σ that of the superposed rotation. When we treat this motion with the aid of the theory developed above, we obtain as a relation between the energy and the ω, σ the formula

$$\delta E = \omega\delta\mathcal{I}_\omega + \sigma\delta\mathcal{I}_\sigma. \tag{38}$$

The stationary states are fixed by the conditions

$$\mathcal{I}_\omega = n_\omega h, \quad \mathcal{I}_\sigma = n_\sigma h. \tag{39}$$

The $\mathcal{I}_\omega$ and $\mathcal{I}_\sigma$ are expressed in the following way in terms of the momenta $\mathcal{I}_r$ and $\mathcal{I}_\varphi$ obtained when polar coordinates are used

$$\mathcal{I}_\omega = \mathcal{I}_r + \mathcal{I}_\varphi, \quad \mathcal{I}_\sigma = \mathcal{I}_\varphi. \tag{40}$$

If $\mathscr{I}_\omega$ and $\mathscr{I}_\sigma$ are chosen to fix the stationary states, they are related in a simple way to the periodicity properties of the system, contrary to $\mathscr{I}_r$ and $\mathscr{I}_\varphi$.

We shall now briefly discuss Sommerfeld's fine structure theory and then offer some general considerations about the degree of approximation to which the motion in the stationary states is described.

Sommerfeld's theory of the fine structure rests on the fact that, when the relativistic correction is taken into consideration, the motion is not strictly periodic but is a central motion that can be represented by the superposition of a very slow rotation upon a purely periodic Kepler motion.

For the purely periodic Kepler motion, the quantity

$$\mathscr{I} = \oint p\,\mathrm{d}q$$

is independent of the coordinate system chosen and is closely related to the period. That means that in formula ⟨32⟩ the periodic function is absent. The energy E is then expressed in terms of $\mathscr{I}$ in the form

$$E = -2\pi^2N^2e^4m/\mathscr{I}^2 \tag{41}$$

and

$$\nu = \frac{2\pi^2N^2e^4m}{h}\left(\frac{1}{(\mathscr{I}''_\omega)^2} - \frac{1}{(\mathscr{I}'_\omega)^2}\right). \tag{42}$$

However, the expression for E depends strictly, i.e., when relativity is taken into account, not only upon $\mathscr{I}_\omega$ but also slightly upon $\mathscr{I}_\sigma$; in fact

$$E = -\frac{2\pi^2N^2e^4m}{\mathscr{I}^2_\omega}\left\{1+\left(\frac{2\pi Ne^2}{c}\right)^2\left(-\frac{3}{4\mathscr{I}^2_\omega}+\frac{1}{\mathscr{I}_\omega\mathscr{I}_\sigma}\right)+\dots\right\}. \tag{43}$$

Hence, the frequency becomes

$$\nu = \frac{1}{h}\left\langle\left[2\pi^2N^2e^4m\left\{\left(\frac{1}{(\mathscr{I}''_\omega)^2}-\frac{1}{(\mathscr{I}'_\omega)^2}\right)+\left(\frac{2\pi Ne^2}{c}\right)^2\left(\frac{1}{(\mathscr{I}''_\omega)^2}\left(\frac{-3}{4(\mathscr{I}''_\omega)^2}+\frac{1}{\mathscr{I}''_\omega\mathscr{I}''_\sigma}\right)\right.\right.\right.\right.$$
$$\left.\left.\left.\left.-\frac{1}{(\mathscr{I}'_\omega)^2}\left(\frac{-3}{4(\mathscr{I}'_\omega)^2}+\frac{1}{\mathscr{I}'_\omega\mathscr{I}'_\sigma}\right)\right)\right\}\right]\right\rangle.$$

While, when relativity is disregarded, only the major axis of the ellipse is determined, and this by $\mathscr{I}_\omega$ alone, now also its shape is determined by the two quantities $\mathscr{I}_\omega$ and $\mathscr{I}_\sigma$. We call n_ω the "principal quantum number". There are a number of motions for which n_ω has the same value, while n_σ can assume any value $\leqq n_\omega$. For example, if $n_\omega = 3$, n_σ can have any one of the values 1, 2, or 3. Certain transitions are excluded, according to the correspondence principle, for, since

σ occurs only with the factors $+1$ or -1 in the expansion of the motion, only such transitions are possible for which $n'_\sigma - n''_\sigma = \pm 1$. If external forces are present having an influence comparable to the changes demanded by the theory of relativity in the motion calculated by classical mechanics, the fine structure is lost. Then only a single period is obtained. The essential point in the treatment of fine structure is the consideration of the multiple periodicity. If the fine structure disappears, it is appropriate that the stationary states be fixed by means of a single condition. This is a very essential point for the formal coherence of the theory; we shall consider it later when discussing the effect of electric and magnetic forces on the motion.

We shall now turn to the second contemplated point, namely, the degree of approximation to which the stationary states are fixed. In fact, the entire description of the stationary states is only approximative, since we have not taken into account any radiation in the stationary states. Thus, in view of the fact that the values of E' and E'' are not determined exactly, because of our neglect of this radiation, the question arises as to the accuracy of the equation $E' - E'' = h\nu$. If we ask ourselves what we are to understand by ν, we must assume that in reality there exists no strictly monochromatic radiation. Rather, ν must depend on the number of emitted vibrations, i.e., on the duration of the emission process; and about this we know nothing definite. We can only obtain an upper limit for it from experiments on the time of decay of the luminescence of canal ray particles. For reasons of correspondence, we believe that this time is of the same order of magnitude as the time of emission calculated by the classical theory. Actually, it has been found that the duration of luminescence agrees in order of magnitude with that calculated classically. This means that the energy and the frequency are determined to just the same degree of approximation. What this agreement really means is hard to say. This may be a point that might be suited to throw light on various problems of the quantum theory. In any case, this result contributes much to round off the applications of the quantum theory mentioned here.

I should like to make two more brief remarks. The first one refers to a paper by Ehrenfest and Breit [Z. Phys. **9** (1922) 207]. It deals with a special case which, at a first glance, appears very paradoxical. Ehrenfest considers a system in which a particle revolves in a circular orbit without being subjected to any forces. However, by a suitable kinematic device, it is provided that the particle, after a definite large number of revolutions (which, for simplicity, we may take to be integral), is suddenly reflected elastically. Thus, there are two periods involved here: the "micro period" of the revolution and the "macro period", i.e., the time between two consecutive moments of reversal. The macro period must then, of course, be quantized, However, the stationary states thus fixed do not alone suffice to characterize the properties of the system, since the particle has behaved as free during the long intervening times. If the motion is resolved into a Fourier series, the overtone corresponding to the micro period will occur very strongly, and when the macro period is very long even more strongly than the fundamental frequency. This means that high quantum jumps of order τ predominate greatly, where τ is the (integral) ratio between the macro period and the micro period; thus, the system behaves optically as if the micro period were quantized and the particle rotated freely, However, a paradox is obtained when the thermal properties are considered. These depend only on the magnitude of the energy element, i.e., on the macro period. If this is increased, thus making its ν smaller and smaller, the multiples of the energy quantum approach one another more and more, and the system behaves as if the energy can take on all values continuously (as in classical statistics). Thus, with increasing macro period, the micro period appears more and more clearly in the optical properties, but can apparently not be detected at all in the thermal properties. The paradox of this result is cleared up when we consider that, with very large macro period, the vibrations cannot be undamped; one should, in principle, not neglect the emission of radiation. In fact, the system will have lost all its energy during a macro period well before the next impact. I do *not* believe that the rules of quantum theory can be applied to such a case. They hold only when the classically calculated emission is inappreciably small during many periods. At least, the difficulty pointed out by Ehrenfest is met with only if the quantum conditions are regarded as exact.

This point is very essential, since numerous attempts have been made to quantize all motions whatever. The atom represents an especially sharp case of the quantum theory; however, when we are not dealing with a closed system, we possess no formal quantum conditions.

My second remark concerns the problem of wireless telegraphy. Here we are dealing with special circumstances; for we believe we are able to describe in all details the radiation processes that occur here. What about the application of the quantum theory to these processes? If we were able to quantize the systems in question, we should have an immense number of quantum numbers. However, I do not think that that is the essential thing. Rather, I believe that the quantum theory fails completely here, and probably because of the immense magnitude of the radiative forces involved. The emitted energy is so large that it must arise from an immense number of individual processes. Since even a tiny fraction of a wavelength contains a very large amount of energy, the magnitude of the radiative forces, which are large compared to the forces that otherwise fix the stationary states, makes the very concept of stationary states meaningless.

THIRD LECTURE. JUNE 14, 1922.

In yesterday's lecture I attempted to explain some main points of the theory of determining the stationary states for simply and multiply periodic motions. Today I shall present some applications of this theory.

My main endeavor yesterday was to mention some physical viewpoints that are of special importance for the theory, namely, the adiabatic principle and the correspondence principle. We saw that, with the aid of ordinary mechanics, we can describe, to a certain degree of approximation, not only the motions in the stationary states but also an adiabatic transformation of the system. The adiabatic principle is particularly well suited to describe the influence of external fields upon an atomic system. If we have a system in a stationary state and let an external field increase slowly, we ask what the motion of the system will become. Analytical mechanics gives the following answer to this question: Even for small external forces, the motion cannot be described in terms of the uniformized variables [angle coordinates] of the unperturbed motion. The energy will no longer remain constant during the motion. On the other hand, it can be proved that when the external force is considered to be constant during a time interval that is long compared to the periodicities of the motion, and when the average value of the energy is taken over this time interval, during which the resultant motion develops, this average value will be independent of the field strength to a first approximation, i.e., to within quadratic terms in the field strength. However, the field strength is actually not constant. Now, analytic mechanics proves that the change in the energy of the system caused by this fact can be put equal to the average of the potential of the external forces taken over the unperturbed motion. The total value of the energy is obtained by adding this latter amount to the above-mentioned energy which is constant to a first approximation. Thus, we need not investigate the perturbation in detail but only the average value of the energy.

However, the application of this method leads to difficulties in certain cases, namely, when the degree of periodicity is greater when external forces are present than when they are not. Hence, in the special case in which the degree of periodicity of the unperturbed system is equal to the number of degrees of freedom, this difficulty cannot arise. If the degree of periodicity is smaller than the number of degrees of freedom, the orbital parameters α, β will in general not remain constant under the perturbing forces. If the α, β are independent of the perturbing forces to a first approximation, the degree of periodicity of the motion remains unchanged, and the above-mentioned method can be used. On the other hand, if the α, β change with the perturbing force, new periodicities will appear in the motion. These are the periods of the so-called *secular perturbations*. In such cases, the adiabatic hypothesis fails in the first instance, because the stationary states of the perturbed system are fixed by more conditions than those of the unperturbed system. Thus, from among all the motions which can be reached adiabatically, the stationary motions must be selected by an investigation of the secular perturbations. As can be shown, the α, β satisfy canonical equations. These must be treated by means of new uniformization variables [angle variables] to give the new quantum conditions. The number of these is equal to the difference between the degrees of periodicity of the perturbed and the unperturbed motions.

I do not expect that these indications have been understood by all; nevertheless, I wanted to give them, in order to convey at least some idea about these matters. The essential thing is that we do not need to investigate the perturbed system entirely anew, but must only add new conditions to those which fix the stationary states of the unperturbed system.

Now, in order to clarify what has been discussed, we shall treat some examples. First, consider a purely periodic motion, i.e., a motion with a single degree of periodicity; its components can be represented in the form

$$\zeta = \sum C_\tau \cos 2\pi(\tau\omega t + \gamma_\tau).$$

The stationary states are fixed by the condition

$$\mathcal{I} = \oint \sum p\,\mathrm{d}q = nh.$$

The energy difference between neighboring states is

$$\delta E = \omega \delta \mathscr{I}.$$

With the help of the frequency relation, it follows from this that for high quantum numbers

$$\nu \sim (n' - n'')\omega.$$

In case of a Kepler motion, the energy is expressed in terms of $\mathscr{I}$ as follows

$$E = -2\pi^2 N^2 e^4 m / \mathscr{I}^2.$$

How will such an atom behave in an electric or a magnetic field? As we saw yesterday, because of relativity, the motion of a hydrogen atom is not purely periodic. Hence, to take relativity into account, an additional condition is needed to fix the stationary states. The modification caused by relativity in the motion calculated by classical mechanics can be regarded as a perturbation. When the perturbation calculation is carried out, the formulas given yesterday are obtained.

Under the conditions existing in experiments, the perturbations caused by the electric or the magnetic field are large compared to the relativistic correction. Hence, it appears justified to neglect the latter to begin with.

We consider first the simplest case, a hydrogen atom in a homogeneous magnetic field. According to Larmor's well-known theorem, the sole effect of the magnetic field is the superposition upon the motion taking place without the field of a uniform rotation around the direction of the field having a frequency ζ_H given by

$$\zeta_H = eH/4\pi mc, \tag{44}$$

where H is the absolute value of the magnetic field strength. It may be remarked here that Larmor's theorem is not peculiar to the hydrogen atom but holds for any nuclear atom.

It is now easy to see what the resulting change in the motion will be. In fact, every elliptic harmonic motion can be resolved into a harmonic linear motion parallel to the direction of the field and two circular motions going in opposite directions at right angles to the field. It is seen, therefore, that the component of motion in the direction of the field is represented by

$$\xi_{||} = \sum C_{\tau,0} \cos 2\pi(\tau\omega t + \gamma_{\tau,0}) \tag{45}$$

and the components of motion perpendicular to the direction of the field by

$$\eta_\perp = \sum C_{\tau,\pm 1} \cos 2\pi\{(\tau\omega \pm \zeta_H)t + \gamma_{\tau,\pm 1}\}. \tag{46}$$

Thus, there appear two fundamental frequencies in the motion, the frequency ω

of the original motion and the frequency ζ_H of the superposed rotation (precession). Hence, the stationary states are fixed by two conditions:

$$\mathscr{I}_\omega = n_\omega h, \qquad \mathscr{I}_\zeta = n_\zeta h.$$

The determination of the energy as a function of $\mathscr{I}_\omega$ and $\mathscr{I}_\zeta$ turns out to be very simple, since the frequency ζ is independent of the energy, as in the case of the Planck oscillator for which we had

$$\xi = C \cos 2\pi(\omega_0 t + \gamma), \qquad E = \omega_0 \mathscr{I} = nh\omega_0.$$

Here, we have

$$\delta E = \omega \delta \mathscr{I}_\omega + \zeta_H \delta \mathscr{I}_\zeta.$$

The energy is, therefore,

$$E = -(2\pi^2 N^2 e^4 m / \mathscr{I}_\omega^2) + \zeta_H \mathscr{I}_\zeta. \tag{47}$$

This problem was first treated by Sommerfeld and Debye with the aid of the separation method. As we see, the problem needs not at all to be treated *ab initio*; a second quantum condition is merely added to that of the unperturbed problem.

The expression for the energy obtained here is not sufficient to explain the Zeeman effect completely; in order to do that we must bring in the correspondence principle. For large quantum numbers we have

$$\nu \sim (n'_\omega - n''_\omega)\omega + (n'_\zeta - n''_\zeta)\zeta_H.$$

The appearance of a definite transition depends on the appearance of the corresponding harmonic component in the motion. The difference $n'_\omega - n''_\omega$ can assume all integral values, since all overtones of ω occur in the expansions of ξ and η. On the other hand, $n'_\zeta - n''_\zeta$ can only be ± 1 or 0, since other multiples of ζ_H do not occur in the expansions. However, these facts alone do not explain the Zeeman effect completely; some statements about the polarization of the emitted radiation are still lacking. We realize that these are also determined by the correspondence; for we assume that the polarization is similar to what would result in the classical theory from the vibration corresponding to the quantum jump; according to the classical theory, the components divide themselves into some that are linearly polarized and some that are circularly polarized. The quantum jump $n'_\zeta - n''_\zeta = 0$ corresponds to a vibration that is linearly polarized in the direction of the field, while each of $n'_\zeta - n''_\zeta = \pm 1$ corresponds to a circularly polarized vibration perpendicular to the field. The first case gives the middle component lying in the original position, the other cases give the side components of the normal Zeeman effect. This way of dealing with the Zeeman effect is similar to Lorentz' treatment which was based on classical electrodynamics. The differences are as small as could possibly be expected, in view of the dissimilarity between the classical theory and the quantum theory. However, the difference between the two theories reveals itself immediately in the fact that no sharp lines are to be expected on the classical theory. When we consider the total polarization, we find a further difference. According to the classical theory, all orientations of the atom relative to the direction of the field occur. Hence, the total polarization, obtained by the superposition of all individual processes, will be zero. On the other hand, according to the quantum theory, only definite discrete orientations occur. If P is the angular momentum of the electron, we have

$$\mathcal{J}_\zeta = n_\zeta h = 2\pi P. \tag{48}$$

From this it follows that the angular momentum can only assume discrete values. On the other hand, since the angular momentum is quantized in the direction of the normal to the orbit, it follows from the two quantum conditions that the angle between the direction of the field and the plane of the orbit can assume only discrete values.

This result has been confirmed by experiments of Stern and Gerlach who investigated the motion of paramagnetic atoms in inhomogeneous magnetic fields and showed that only definite orientations relative to the field occur. Hence, we must be prepared to find that the total polarization may not vanish; observations of Voigt actually seem to show that in certain cases the total polarization of a spectral line changes in a magnetic field. Formerly, attempts were always made to explain this phenomenon away; for no explanation could be given of it on the classical theory. On the other hand, we may now regard it as a support for the quantum theory.

I shall now discuss what can be said about the frequencies appearing in the Zeeman effect from another point of view, i.e., by applying the law of the conservation of angular momentum. When we consider radiation with a definite frequency, produced, according to the classical theory, by an electron which executes a harmonic motion with this same frequency, we find that we must ascribe angular momentum to the emitted radiation. The angular momentum of the radiation from an electron moving in an elliptic harmonic orbit is zero when the orbit degenerates into a straight line, and is a maximum when the orbit becomes circular. When the angular momentum of an atom is changed in a quantum jump, we must demand that the change appear again in the angular momentum of the radiation. Calculating the relation between angular momentum and energy, we find that

$$\frac{\Delta P}{\Delta E} \leq \frac{1}{2\pi\nu}.$$

Moreover, since $\Delta E = h\nu$, we obtain,

$$\Delta P \leqq h/2\pi. \tag{49}$$

It follows that n_ζ can only change by 0 or ± 1. It is possible, also, to derive the correct value for the polarization in this manner. This consideration is due to Rubinowicz.

We now go on to ask about the influence of an electric field. Epstein and Schwarzschild have given a theoretical explanation of the phenomena appearing here. They solved the problem by separation of variables. Indeed, it can be shown that the equations of motion of an atom in a constant electric field can be separated when parabolic coordinates are introduced. In many respects, this theory is very beautiful, but there is, nevertheless, something unsatisfactory about it. The possibility of carrying out the entire theory depends on the choice of these coordinates, and we may well ask: how can the electron know that we solve its differential equations with the help of parabolic coordinates? However, it is also possible to treat the problem very easily by applying the quantum theory to the unperturbed atom and investigate what new conditions must be added. To calculate the energy, it is necessary to determine the average value of the potential energy with respect to the external force field. This average value can be put equal to the potential energy of the electron placed in the so-called "electric center of gravity" of its orbit. Since this energy must be constant, the electric center can only move at right angles to the direction of the field. It can be proved that this point executes a harmonic motion in a plane perpendicular to the direction of the field. The frequency of the motion of this center is

$$\rho_F = F \frac{3\mathscr{I}}{8\pi^2 Nem}, \tag{50}$$

where F is the absolute value of the electric field strength. The quantity $\mathscr{I}$ determines the stationary state of the unperturbed system; thus, ρ_F is the same for all states of the perturbed system which are obtained from a single state of the unperturbed system. We must now add a new condition

$$\mathscr{I}_\rho = n_\rho h,$$

so that, as in the case of the Zeeman effect, we have only two conditions

$$\mathscr{I}_\omega = n_\omega h, \qquad \mathscr{I}_\rho = n_\rho h.$$

The energy is then

$$E = -\frac{2\pi^2 N^2 e^4 m}{\mathscr{I}_\omega^2} + F \frac{3\mathscr{I}_\omega \mathscr{I}_\rho}{8\pi^2 Nem}. \tag{51}$$

[368]

This is the same expression as that obtained by Epstein and Schwarzschild; however, an essential difference is that we have only two quantum conditions, whereas Epstein uses three conditions.

We shall now consider the conclusions which can be drawn from the correspondence principle. The component of the motion parallel to the field is

$$\xi_{||} = \sum C_{\tau_1, \tau_2} \cos 2\pi\{(\tau_1 \omega + \tau_2 \rho)t + \gamma_{\tau_1, \tau_2}\}, \tag{52}$$

where $\tau_1 + \tau_2$ is even; the components perpendicular to the field are

$$\eta_{\perp} = \sum C_{\tau_1, \tau_2} \cos 2\pi\{(\tau_1 \omega + \tau_2 \rho)t + \gamma_{\tau_1, \tau_2}\}, \tag{53}$$

where $\tau_1 + \tau_2$ is odd. Also,

$$\delta E = \omega \delta \mathscr{I}_{\omega} + \rho \delta \mathscr{I}_{\rho};$$

hence, for large quantum numbers

$$\nu \sim (n'_{\omega} - n''_{\omega})\, \omega + (n'_{\rho} - n''_{\rho})\rho.$$

Thus, such transitions will appear for which the corresponding harmonic components are present. As regards the polarization, we must expect that transitions for which $(n'_{\omega} - n''_{\omega}) + (n'_{\rho} - n''_{\rho})$ is even will give lines polarized parallel to the field, while such for which $(n'_{\omega} - n''_{\omega}) + (n'_{\rho} - n''_{\rho})$ is odd will give lines polarized perpendicular to the field. This result has been confirmed experimentally. According to the correspondence principle, the transition probability, on which the intensity of a given line depends, is determined by the amplitude of the corresponding overtone in the motion. Kramers has calculated the coefficients C in the expansions and has also found that the calculated intensities agree with experience. The Stark effect offers an enormous experimental material, for which the quantum theory can account in all details. It would be difficult to mention a field in which the classical theory has achieved something similar.

We shall now also consider the influence of external fields upon a relativistic hydrogen atom. So far, we have neglected the relativity correction, because we knew that this would not make any essential change in the coarser features of the phenomena. If we compare the result of the theory of the relativistic ionized helium atom in a field-free space with the experimental results (Paschen), we find small deviations in the intensity of the components arising from the fact that in the experiment small external fields are always present. It is seen, therefore, that it is necessary to work out the theory of an atom in an external field by also taking relativity into account.

If the external field is a magnetic field, the circumstances are very simple, because Larmor's theorem remains valid also for the relativistic Kepler motion.

Thus, there is simply superposed a precession of frequency

$$\zeta = eH/4\pi mc$$

The components of the motions parallel (ξ) and perpendicular (η) to the field are represented by

$$\begin{aligned} \xi_{||} &= \sum C_{\tau, \pm 1, 0} \cos 2\pi\{(\tau\omega \pm \sigma)t + \gamma_\tau\}, \\ \eta_{\perp} &= \sum C_{\tau \pm 1, \pm 1} \cos 2\pi\{(\tau\omega \pm \sigma \pm \zeta)t + \gamma_\tau\}. \end{aligned} \tag{54}$$

The stationary states are fixed by the conditions

$$\mathscr{I}_\omega = n_\omega h, \quad \mathscr{I}_\sigma = n_\sigma h, \quad \mathscr{I}_\zeta = n_\zeta h.$$

The energy can be written

$$E = E(\mathscr{I}_\omega, \mathscr{I}_\sigma) + \zeta \mathscr{I}_\zeta;$$

furthermore,

$$\delta E = \omega \delta \mathscr{I}_\omega + \sigma \delta \mathscr{I}_\sigma + \zeta \delta \mathscr{I}_\zeta.$$

Hence, we obtain for large quantum numbers

$$\nu \sim (n'_\omega - n''_\omega)\omega + (n'_\sigma - n''_\sigma)\sigma + (n'_\zeta - n''_\zeta)\zeta.$$

The correspondence principle yields the result that there is no limitation on the value of $(n'_\omega - n''_\omega)$. On the other hand, $(n'_\sigma - n''_\sigma)$ is always ± 1, so that, with regard to this quantum number, nothing is changed by the magnetic field. The difference $(n'_\zeta - n''_\zeta)$ can be 0 or ± 1. A change in n_ζ of 0 gives linear polarization parallel to the direction of the field, a change of ± 1 gives circular polarization perpendicular to the direction of the field. Thus, the result is that each component of the fine structure splits up into a normal Lorentz triplet. This result was first obtained by Sommerfeld by means of the method of separation of variables. There was at first some doubt about whether this agreed with experience, because we are dealing with the influence of a magnetic field upon lines having a complex structure, in which case we are used to finding an anomalous Zeeman effect. The experimental investigation of this effect is very difficult; for the fine structure is easily disturbed by external fields which are difficult to avoid; nevertheless, the normal splitting of the individual fine structure component seems to be verified by experiments of Hansen.

While the problem of the relativistic Zeeman effect is simple, because the period of the Larmor precession is simply superposed on the original motion, the problem of the relativistic Stark effect is much more complicated. The change in the orbit caused by the electric field strength is not simply added to ["wirkt ... nicht zusammen mit"] the changes in the motion caused by the relativistic correction. When the influence of the electric field is large compared to the relativistic correction, this correction is suppressed, and we get the Stark effect as given on the basis of classical mechanics. When the influence of the electric field is small compared to the relativistic correction, the electric field will change the motion very little. However, quantum jumps will occur which would not appear in the absence of the electric field; for, because of the electric field, components $\tau_2\sigma$ occur, whereas without the field only $\pm\sigma$ occur. If the influence of the electric field is of the same order of magnitude as the relativistic correction, the problem can no longer be treated by the method of separation; nevertheless, an investigation of the secular perturbations leads to the goal, which amounts to solving a problem of a smaller number of degrees of freedom. These calculations have been carried out by Kramers. They lead to the following result. The components of the motion parallel (ξ) and perpendicular (η) to the field can be expressed in the form

$$\begin{aligned} \xi_{\|} &= \sum C_{\tau_1\tau_2} \cos 2\pi\{(\tau_1\omega+\tau_2\varepsilon)t+\gamma_{\tau_1\tau_2}\}, \\ \eta_{\perp} &= \sum C_{\tau_1\tau_2\pm 1} \cos 2\pi\{(\tau_1\omega+\tau_2\varepsilon\pm\zeta)t+\gamma_{\tau_1\tau_2\pm 1}\}. \end{aligned} \tag{55}$$

Also,

$$\delta E = \omega\delta\mathscr{I}_{\omega}+\varepsilon\delta\mathscr{I}_{\varepsilon}+\zeta\delta\mathscr{I}_{\zeta}.$$

The quantum conditions are

$$\mathscr{I}_{\omega} = n_{\omega}h, \qquad \mathscr{I}_{\varepsilon} = n_{\varepsilon}h, \qquad \mathscr{I}_{\zeta} = n_{\zeta}h.$$

Furthermore, the energy is

$$E = -(2\pi^2N^2e^4m/\mathscr{I}_{\omega}^2)+\Psi(\mathscr{I}_{\omega}, \mathscr{I}_{\varepsilon}, \mathscr{I}_{\zeta}). \tag{56}$$

For large numbers, we have approximately

$$\nu \sim (n'_{\omega}-n''_{\omega})\omega+(n'_{\varepsilon}-n''_{\varepsilon})\varepsilon+(n'_{\zeta}-n''_{\zeta})\zeta.$$

Application of the correspondence principle yields the result that $n'_{\omega}-n''_{\omega}$ and $n'_{\varepsilon}-n''_{\varepsilon}$ are not subject to any limitations. On the other hand, n_{ζ} can only change by 0 or ± 1; a change of 0 corresponds to linear polarization parallel to the field, a change of ± 1 corresponds to circular polarization perpendicular to the field.

No experiments have yet been performed on the transition of the fine structure to the usual Stark effect by the gradual increase of an electric field. The quantum theory yields very many details of the phenomenon to be expected. Even if we really should not be unprepared to find that the quantum theory is false, it would surprise us very much if such a detailed picture obtained from the quantum theory should not be valid; for our belief in the formal reality of the quantum conditions is so strong that we should wonder very much if experiments were to give a different answer than what is demanded by the theory.

FOURTH LECTURE. JUNE 19, 1922.

All the applications of the theory of multiply periodic systems considered here have dealt with the simplest case, an atom with a single electron. In other cases, the solutions of the equations of motion are not of such a simple character. The main problem to which we shall now proceed is the treatment of atoms with more than one electron. Our consideration of the atom with only one electron may be regarded as the first step in attacking this problem. Of course, it has not just been a question of the hydrogen atom here but in general of the binding of the first electron by an atomic nucleus of arbitrary atomic number. As we have seen, in this case the motion can, to a first approximation, be regarded as purely periodic, so that its components can be represented by

$$\zeta = \sum C_\tau \cos 2\pi(\tau\omega t + \delta_\tau).$$

The stationary states are fixed by the condition

$$\mathscr{I} = \oint p\,\mathrm{d}q = nh.$$

The difference between neighboring states is

$$\delta E = \omega\delta\mathscr{I},$$

and in the limit of large quantum numbers

$$\nu \sim (n'-n'')\,\omega.$$

[372]

For the energy and the major axis of the elliptic orbit we found

$$E = -(2\pi^2N^2e^4m/\mathscr{I}^2_{\omega}), \qquad 2a = \mathscr{I}^2/2\pi^2Ne^2m. \tag{57}$$

With these formulas the observed terms in the spectrum of hydrogen are obtained.

Because of the relativity correction, the motion of the electron in the hydrogen atom is actually not purely periodic but is a central motion which can be represented as the superposition of a slow rotation on a Kepler motion. A component of the motion is then represented in the form

$$\zeta = \sum C_{\tau, \pm 1} \cos 2\pi\{(\tau\omega \pm \sigma)t + \gamma_{\tau, \pm 1}\},$$

where σ is the frequency of the precession. We have

$$\delta E = \omega\delta\mathscr{I}_{\omega} + \sigma\delta\mathscr{I}_{\sigma}.$$

When polar coordinates are used, the $\mathscr{I}_{\omega}$ and $\mathscr{I}_{\sigma}$ are expressed by phase integrals as follows

$$\mathscr{I}_{\omega} = \oint p_r \mathrm{d}r + \oint p_{\varphi}\mathrm{d}\varphi,$$

$$\mathscr{I}_{\sigma} = \oint p_{\varphi}\mathrm{d}\varphi \langle = 2\pi P\rangle,$$

where P is the total angular momentum. The quantum conditions are

$$\mathscr{I}_{\omega} = n_{\omega}h = nh, \qquad \mathscr{I}_{\sigma} = n_{\sigma}h = kh, \tag{58}$$

where n is written for n_{ω} and k for n_{σ}. For large quantum numbers, we have approximately

$$\nu \sim (n'-n'')\omega + (k'-k'')\sigma.$$

The energy in the stationary states depends essentially on the frequencies ω and σ. However, since $\sigma \ll \omega$, the energy and the frequency ν will depend mainly on $\mathscr{I}_{\omega}$, i.e., on the quantum number n which we, therefore, shall call the "principal quantum number". Actually,

$$E = -\frac{2\pi^2N^2e^4m}{\mathscr{I}^2_{\omega}}\left\{1 + \left(\frac{\pi^2e^2N}{c}\right)^2\left(-\frac{3}{\mathscr{I}^2_{\omega}} + \frac{4}{\mathscr{I}_{\omega}\mathscr{I}_{\sigma}}\right) + \dots\right\}.$$

Thus, for one and the same principal quantum number n, there exists a series of stationary states corresponding to different values of k and having nearly the same energy. The value of k determines the shape of the slowly precessing ellipse. Only the period of this precession fixes the shape of the ellipse. When this precession is disturbed, the stability of the ellipse is lost, and only the single condition for the purely periodic motion remains. This is essential for what follows, already when we proceed to atoms with two electrons, i.e., first of all to the atom of helium. The arc spectrum of helium throws light on this problem, while the spark spectrum gives information about the binding of the first electron. The helium arc spectrum is very simple; it was one of the first spectra to be unravelled by Paschen and Runge. However, its structure exhibits a certain complexity which is of the greatest significance. However, before we enter upon details, we shall mention some of the main features of the spectra of the other elements.

I assume that most of you are familiar with these matters, but I shall, nevertheless, emphasize the main points. As we have seen, the spectrum of hydrogen can be represented by the formula

$$\nu = K\left(\frac{1}{(n'')^2} - \frac{1}{(n')^2}\right).$$

Similarly, the spectra of the other elements can be represented by formulas of the following kind

$$\nu = f_{k''}(n'') - f_{k'}(n'), \tag{59}$$

where, in the place of K/n^2, there thus appear a number of functions $f_k(n)$ of two integers k and n. According to the quantum theory, we interpret this formula by assuming that we are dealing with quantum transitions between states having energies given by

$$E_{n,k} = -hf_k(n).$$

The terms $f_k(n)$ are similar to those of the hydrogen spectrum but are of more complex structure. It is possible to write them in the Rydberg form

$$f_k(n) = +\frac{K}{[n+\varphi_k(n)]^2}, \tag{60}$$

where K is the same Rydberg constant which appears in the formula for the hydrogen spectrum. The $\varphi_k(n)$ are functions which only depend little upon n and, with increasing n, asymptotically approach constant values

$$\varphi_k(n) \underset{n\to\infty}{\longrightarrow} \alpha_k.$$

It is easy to see that an energy expression of the form $-hf_k(n)$ is obtained, if it is assumed that one electron moves at so great a distance from the other electrons that the influence of the inner electrons can be approximately represented by a point charge, and that the small deviation from the field of a point charge can be represented by a field of central symmetry. This conception alone, on the basis of which Sommerfeld first derived the Rydberg form of the term, throws

[374]

very much light on the structure of such spectra. Sommerfeld's theory provides just two quantum numbers which correspond to the two quantum numbers for the relativistic motion of an electron. As in that theory, one quantum number corresponds to the period of the Kepler motion and the second quantum number to the period of the superposed precession. However, we must remark at once that, although this conception led Sommerfeld to the Rydberg form for the terms, he could not obtain any quantitative agreement with the experimental values of the terms, as could hardly be expected with such a simplifying assumption. The reason is that most of the orbits actually do not lie far outside the other electrons during their entire course, so that the deviation from a Kepler motion cannot be regarded as caused simply by a field of central symmetry. We shall go into this matter later on; now, I should like to discuss some properties of atoms which appear in the absorption and emission of radiation, and which can be understood simply on the basis of the view advocated here. For example, sodium vapor absorbs selectively certain spectral lines, but only those whose initial term is the largest of all existing terms and, therefore, corresponds to the state of lowest energy. It is significant that this can be explained by the quantum theory. It tells us that the initial state belonging to these lines is the normal state of the atom. This is confirmed by experiments on electron impact, after the method of Franck and Hertz, which show that no energy can be transferred to an atom, unless it is large enough to carry the atom from its normal state to the adjacent quantum state. This method also permits the determination of the ionization energy. This is the work required to remove an electron completely from the atom. It is found that the required energies agree with the values given by the spectral terms. Now, in the structure of the spectra, we find the peculiarity that not all conceivable combinations among the terms occur, but only such which correspond to adjacent values of k. This is just what should be expected according to the correspondence principle. For, even if the picture of the motion of the electron which Sommerfeld took as a basis is not quite correct, the fact remains that the motion of the electron possesses two frequencies ω and σ, of which σ appears only with coefficients ± 1 in an expansion of the components of the motion; σ may, therefore, be regarded as the frequency of the precession of a periodic motion which differs from a Kepler motion. From this we must conclude that the corresponding quantum number k can only change by ± 1. This is a very significant result which allows us to understand the peculiar limitation of the conceivable combinations. When the atom is not disturbed by external forces, it is not simply that not all possible combinations are observed. Apparently capriciously, some of the combinations excluded by the limitations on k do actually appear. However, this is due to external perturbations which make the motion more complex, so that the frequency σ appears in the Fourier expansions of the motion also with coefficients other than ± 1.

If we now compare the hydrogen spectrum, taking the fine structure into account, with the spectra of Li and Na, we find a far-reaching similarity of structure. In the alkali spectra we also find combination lines corresponding to a change in k larger than ± 1. The ease with which these lines appear must depend on the stability of the motion, i.e., on the frequency of the rotation of the perihelion. If $\sigma \ll \omega$, the influence of any existing external force upon the orbits will be strong, since the force, so to speak, has time to influence the motion. If σ is larger, the motion is only influenced very little by an external field. Now, we can estimate the magnitude of σ from the dependence of the energy on the quantum numbers; for we have

$$\delta E = \omega \delta \mathscr{I}_{\omega} + \sigma \delta \mathscr{I}_{\sigma}.$$

Thus, the deviation of a spectral term from the corresponding hydrogen term provides a measure of the magnitude of the rotation of the perihelion. Hence, according to the above-mentioned, we must expect the combination lines to appear readily when the deviations of the terms from the hydrogen terms are small. In fact, we perceive this when considering the lithium spectrum. In numerous experiments, Stark and his co-workers have investigated the appearance of combination lines also in the spectra of other elements and have found a behavior of these combination spectra for which the correspondence principle is able to account in its entirety. Moreover, in comparing the alkali spectra with the hydrogen spectrum, we have neglected the fact that the alkali spectra are more complex, in that their lines are doublets. This complexity must be caused by the circumstance that the motion has a higher degree of periodicity than so far assumed. We have already pointed out above that the field of the atomic core ["Atomrest"] can be regarded only to a first approximation as centrally symmetric.

Rather, we must assume that the atomic core possesses axial symmetry. This disturbs the central motion. If this perturbation is small, it will have the effect that the motion of the last electron to be bound will not take place in a fixed plane; rather, the plane of its orbit will alter its position in such a way that the motion can be described as a central motion with a superposed uniform slow rotation around the fixed axis of the total angular momentum of the atom. Thus, the motion is influenced in a similar manner as in the Zeeman effect. Each harmonic component of the central motion is, therefore, resolved into three components: a linear component (ξ) parallel to the fixed direction with the same frequency as the undisturbed motion and two opposite circular components (η) perpendicular to the fixed direction and having frequencies which are increased or diminished, respectively, by the frequency of precession. We have, therefore,

$$\xi_{||} = \sum C_{\tau,\pm 1} \cos 2\pi\{(\tau\omega \pm \sigma)t + \gamma_{\tau,\pm 1}\},$$

$$\eta_{\perp} = \sum C_{\tau,\pm 1,\pm 1} \cos 2\pi\{(\tau\omega \pm \sigma \pm \zeta)t + \gamma_{\tau,\mp 1,\pm 1}\}.$$

Thus, there appear three fundamental frequencies in the motion, ω, σ, and ζ, and

$$\delta E = \omega\delta\mathscr{I}_{\omega} + \sigma\delta\mathscr{I}_{\sigma} + \zeta\delta\mathscr{I}_{\zeta}.$$

If we imagine the perturbation removed adiabatically, e.g., by a magnetic field of appropriate strength and in the direction of the total angular momentum, the precession will disappear and

$$\mathscr{I}_{\omega} \rightarrow \mathscr{I}, \qquad \mathscr{I}_{\sigma} \rightarrow 2\pi P.$$

$\mathscr{I}_{\zeta}$ determines the orientation of the orbital plane relative to the previously bound electrons such that $\mathscr{I}_{\zeta} = 2\pi Q$, where Q, as shown by a simple calculation based on the analytical theory, is the total angular momentum of the atom. Thus, we see that the angular momentum of the atomic core is not at all fixed in space but only the total angular momentum of the atom. Rather, according to the principle *actio* = *reactio*, the electron turns under the influence of the core, and the core under the influence of the electron, in just such a manner that the total angular momentum of electron and core remains constant in magnitude and direction.

The quantum conditions are

$$\mathscr{I}_{\omega} = n_{\omega}h = nh, \qquad \mathscr{I}_{\sigma} = n_{\sigma}h = kh, \qquad \mathscr{I}_{\zeta} = n_{\zeta}h = jh. \qquad (61)$$

According to the correspondence principle, k can only change by ± 1, i.e., $k'-k'' = \pm 1$, while j can change by 0 or ± 1, i.e.,

$$j'-j'' = \begin{cases} 0 & \| \\ \pm 1 & \perp \end{cases}.$$

Here $j'-j'' = 0$ corresponds to linear polarization in a direction fixed in space and $j'-j'' = \pm 1$ correspond to two opposite circular polarizations perpendicular to the fixed direction. These results are supported by considerations based on the law of the conservation of angular momentum; for from this it follows that

$$|\Delta Q| \leqq h/2\pi,$$

and this requirement about the change in the angular momentum agrees with our claim that $j'-j'' = 0, \pm 1$. Contrary to older views, which I also have advocated, the angular momentum tells us nothing about the change in k, i.e., about the limitation of the combination principle; as we have seen, it is here a question of the stability of the orbits, on which only the correspondence principle can throw light.

Let us now proceed to more complex spectra and, say, consider the spectrum of mercury. Later we shall go into details. Now, we mention only that in this spectrum there appear several series of terms corresponding to the same value of k. All the combinations that occur satisfy the condition $k'-k'' = \pm 1$; but there are still other limitations which are explained by the conditions $j'-j'' = 0$ or ± 1. Before we go more closely into this, let us consider the influence of external fields on the possibilities of combinations. In order to get information about these, Hansen and Werner, in Copenhagen, have investigated the influence of electric fields upon the spectrum of mercury. It appears that the various lines are influenced to very different degrees and, as we should expect, only the unstable terms corresponding to very small values of σ are affected. In comparison with the hydrogen spectrum, the main difference is that no splittings occur but only new possibilities of combination. In fact, under the influence of the external forces, new harmonic components appear in the motion which, according to the correspondence principle, correspond to new transitions. Matters are similar for other spectra; mercury was chosen because it permits particularly clean-cut work.

Now, what about the change in j? In the mercury spectrum there are a number of narrow doublets which are not influenced by electric fields. This, too, is to be expected from the theory. This fine structure is essentially different from the fine structure of the hydrogen lines, since it depends on a precession around an axis fixed in space and not on a slow rotation of the perihelion. Because of the fast rotation of the perihelion in this atom, the plane of the central motion cannot, to a first approximation, be rotated by a homogeneous electric field; for the electric center revolves in this plane with a frequency that is so large compared to the frequency ζ of the precession around the fixed axis that the average value of the electric moment vanishes. From this we see that, if we want to find out something about the change in the quantum number j, we must investigate the very close relation between the fixing of the stationary states and the periodicity properties. A direct influence of the precession is possible by a magnetic field which produces a rotation of the orbital plane. In all cases in which we have singlet terms, we must assume that we are dealing with a pure central motion, and accordingly, we find a normal Zeeman effect. Wherever we find complex terms, an anomalous Zeeman effect also appears, i.e., we have orbits whose planes execute a precession even in the absence of a magnetic field. Then a conflict takes place, so to speak; the two rotations are not simply superposed on each other. Then the conditions $j'-j'' = 0$ or ± 1 break down, and transitions can occur which do not appear in the absence of a magnetic field. In fact, Paschen and Back have observed the appearance in a magnetic field of new lines of this fine structure. In recent years, many significant contributions to the theory of the anomalous Zeeman effect have been made, mainly by Sommerfeld and Landé, who use the designation "inner quantum number" for j, which is not quite appropriate, according to our conceptions. We shall delay a closer discussion of the anomalous Zeeman effect until we have gained some more detailed conceptions of the interaction between the outer electron and the core.

We now turn to the discussion of the *helium spectrum*. We are here dealing with a very simple spectrum, about which we have very accurate information. The lines of this spectrum are either single or very close doublets. We have two systems of spectra, each of which conforms quite closely to the pattern of a simple spectrum. No combinations occur between these two systems, so that we, so to speak, have two quite separate spectra. Hence, it was believed earlier that helium consists of two elements, orthohelium and parhelium; we know now that these are only two different states of helium. Thus, we see that the two electrons in helium can be bound in different ways. How can we account for this? The arc spectrum of helium yields information about the binding of the second electron by the helium ion; thus, we must consider an atom with an inner and an outer electron. The simplest possible approach to the calculation would be to regard the charge of the inner electron as distributed evenly over a circular orbit. This was attempted by Sommerfeld but led to a result that did not agree with experience; for, in this way, the deviations of the terms from the hydrogen terms come out much too small. Actually, the mutual perturbations of the two electrons, which were neglected here, give rise to much greater deviations. Landé has attacked the problem in a different way. He investigated the case in which both orbits are circular to a first approximation. Perturbations are then obtained which are of the required order of magnitude. The parhelium terms deviate only little from the corresponding hydrogen terms, those of orthohelium deviate more. According to Landé, this can be understood. Landé extrapolates from the simple picture of circular orbits to elliptical orbits. This is, indeed, a rational manner of attacking the problem, but one cannot understand in this way why two quite different spectra appear which do not combine with each other. Hence, we must search for the reason for the greater difference, since it appears that a formal extrapolation is not adequate. Rather, we must investigate the mutual perturbations in detail. This problem has been attacked by Kramers and the speaker; the results obtained so far have not yet been published; here I should like to give, in simple outline, a survey of the mode of attack and the results obtained.

The periodicity properties are essential for the magnitude of the perturbation. The motion of the first electron alone is almost purely periodic. In an atom with two electrons, the influence of the forces between the electrons is much larger than the relativity effect; in the investigation of the perturbations, the motion of the first electron can, therefore, be regarded as purely periodic, and, consequently, the secular perturbations will be very large. The eccentricity of the inner orbit is fixed by these, and the frequency of revolution in this orbit is of a higher order of magnitude than the precession. Landé's method cannot be used, because the inner system is degenerate to begin with. This explains that, in our calculation, the energy depends on quite different powers of the quantum numbers than in Landé's.

We must now investigate in what manner the orbits perturb each other. To do that, we assume, as a very first approximation, that one of the orbits is very large compared to the other. Then the revolution of the inner electron will be much faster than that of the outer. The inner electron moves in the field of a doubly charged nucleus, the outer electron, to a very first approximation, in the field of a singly charged nucleus. Considering the inner electron, we can, in the first approximation, regard the outer electron as being at rest, and, for the outer electron, we can at the same time, to a first approximation, regard the charge of the inner electron as distributed over its orbit. Since we have seen that the orbit of the inner electron is an elliptic orbit, and that, to a first approximation, we can regard the outer electron to be at rest, we must first investigate how such a Kepler ellipse behaves in a constant external field of force. To do this, it is convenient to introduce the following variables. We denote by β the angle between the direction of the field and the major axis and introduce an angle α defined by the equation

$$\varepsilon = \sin \alpha, \tag{62}$$

where ε is the numerical eccentricity of the ellipse. Then

$$P = \frac{1}{2\pi} \mathscr{I} \cos \alpha, \tag{63}$$

where P is the angular momentum of the inner electron and $\mathscr{I}$ is the phase integral corresponding to the period of its motion. We are dealing with a motion in a plane and ask how the major axis and the periodicity change under the influence of the external force. It is simplest to resolve the force into components, since the influence of these is simply superposed. First, consider the influence of the component lying in the direction of the major axis. This produces a secular precession without changing the eccentricity of the orbit or the angular momentum; for the major axis is already fixed by the first quantum condition and, when the major axis is fixed, a force in the direction of the major axis cannot change the eccentricity.

The angular velocity of the precession becomes

$$\frac{\mathrm{d}\beta}{\mathrm{d}t} = F \cdot \frac{3}{2} \frac{\mathscr{I}}{2\pi Nem} \cot \alpha \cos \beta, \tag{64}$$

where $F \cos \beta$ is the component of a field strength of absolute magnitude F in the direction of the major axis. Now, we consider the influence of the component perpendicular to the major axis. This does not produce any precession but changes the eccentricity and, hence, also the angular momentum of the orbit. This is easy to see, since the electric center is at a distance of $\frac{3}{2}\,a\varepsilon$ from the nucleus. Consequently, the ⟨rate of⟩ change in the angular momentum will be

$$\frac{\mathrm{d}P}{\mathrm{d}t} = F \sin \beta \cdot e \cdot \tfrac{3}{2}a\varepsilon. \tag{65}$$

Then we get

$$\frac{\mathrm{d}\alpha}{\mathrm{d}t} = F \cdot \frac{3}{2} \frac{\mathscr{I}}{2\pi Nem} \sin \beta. \tag{66}$$

It remains now to investigate the influence of the force component perpendicular to the plane of the orbit. This will only change the orbital plane, i.e., the direction of the angular momentum. When γ is the angle between the normal to the orbit and the direction of the force component which, to begin with, is at right angles to the orbital plane, we have

$$\langle P \rangle \frac{\mathrm{d}\gamma}{\mathrm{d}t} = Fe\tfrac{3}{2}a\varepsilon$$

or

$$\langle P \rangle \frac{\mathrm{d}\gamma}{\mathrm{d}t} = F \frac{3}{2} \frac{\mathscr{I}}{2\pi Nem} \tan \alpha. \tag{67}$$

In the actual computation, the position of the orbit must be expressed in terms of definite angles, the so-called orbital parameters. The accurate calculation leads to the result stated. Actually, the outer electron is not at rest; hence, the external field F is not constant, and the changes in the orbit just calculated are themselves functions of time. In general, this gives rise to very complicated motions which, I believe, are no more of multiply periodic nature and, hence, cannot be considered as possible quantum orbits. When we assume, as we have done so far, that the orbit of the outer electron is large compared to that of the inner electron, there are only two special cases in which the motions are of multiply periodic character, namely, when the two orbits either lie in the same plane or are at right angles to each other.

Let us first consider the case of coplanar orbits which, as we shall see later, corresponds to the orthohelium spectrum. Here, the major axis of the inner orbit will always point towards the outer electron, so that the frequency of rotation of the perihelion of the inner orbit agrees with the frequency of revolution of the outer electron. There are then two possibilities for the direction of revolution of the electrons; they may traverse their orbits in the same or in opposite directions. It appears that, of these two cases, only that of the same sense of revolution needs to be considered. If we denote the angular momentum of the outer electron by P', and its polar coordinates by r and φ, we have

$$P' = mr^2 \frac{\mathrm{d}\varphi}{\mathrm{d}t} \tag{68}$$

and

$$F = \frac{e}{r^2} = \frac{em}{P'} \frac{\mathrm{d}\varphi}{\mathrm{d}t}, \tag{69}$$

since we can put the distance between the two electrons approximately equal to the distance of the outer electron from the nucleus. Now, since the angular velocity of the outer electron must be equal to the rotation of the perihelion of the orbit of the inner electron, we must have, at every instant,

$$\frac{\mathrm{d}\beta}{\mathrm{d}t} = \frac{\mathrm{d}\varphi}{\mathrm{d}t}.$$

With the use of eqs. (64) and (69), this gives for the eccentricity

$$\tan \alpha = \frac{3\pi}{N} \frac{\mathscr{I}}{P'}, \tag{70}^*$$

since $\cos \beta$ is equal to one in this case. Thus, with increasing P', the inner orbit approaches the circular form. The eccentricity of the inner orbit is, therefore, fixed by the condition that the angular velocity of the outer electron be equal to the rate of rotation of the perihelion of the inner electron. Now we also understand why we cannot extrapolate from a circular to an elliptical orbit of the inner electron; for the rotation of the perihelion of the inner orbit turns out to be of essential importance, and it is undetermined for a circle.

In the case of mutually perpendicular orbital planes, the normal of the inner electron's orbit must point toward the outer electron. Hence, at every instant

$$\frac{\mathrm{d}\gamma}{\mathrm{d}t} = \frac{\mathrm{d}\varphi}{\mathrm{d}t},$$

and, as above, this fixes the eccentricity by the relation

$$\cot \alpha = \frac{3\pi}{N} \frac{\mathscr{I}}{P'}. \tag{71}$$

* [The numerical factor here and in eq. (71) should evidently be $3/4\pi$ rather than 3π. This does not affect the conclusions drawn in the text.]

[382]

Contrary to the first case, when P' increases, the eccentricity of the inner orbit approaches more and more the value 1, so that the orbit finally degenerates into a straight line. For this reason, the case of mutually perpendicular orbits must be excluded. We shall later discuss another solution which we arrive at when we drop the assumption that the orbit of one of the electrons is large compared to that of the other.

We now return to the coplanar case. If the distance of the outer electron is kept constant, the rotation of the perihelion of the inner electron will become faster the smaller its eccentricity is. Different values of the eccentricity correspond to different values of the angular momentum P' of the outer electron; the corresponding orbits can be transformed adiabatically into one another. Since $\mathscr{I}$ is adiabatically invariant, P will change whenever P' is changed, so that an automatic exchange of angular momentum takes place between the inner and the outer electron. With respect to its periodicity properties, the motion of the inner electron behaves just as in a central motion.

Before proceeding to the determination of the stationary states, we have yet to consider the perturbations produced by the inner electron on the orbit of the outer electron. The force acting on the outer electron is given by

$$\frac{(N-1)e^2}{r^2}+\frac{3e^2a\varepsilon}{r^3},$$

where the second term represents the action of a dipole whose positive charge is in the nucleus and whose negative charge is in the electric center of the inner orbit. If P' is large, we can put

$$\varepsilon = \sin\alpha = \tan\alpha$$

and then,

$$\frac{(N-1)e^2}{r^2}+\frac{3e^2a\varepsilon}{r^3} = \frac{(N-1)e^2}{r^2}+\frac{9}{16\pi^3N^2m}\frac{\mathscr{I}^3}{P'r^3}. \tag{72}$$

Thus, the orbit of the outer electron is a Kepler ellipse with rotating perihelion. The frequency σ' of the rotation of the perihelion is

$$\sigma' = \omega'\frac{9}{32\pi^3N^2}\frac{\mathscr{I}^3}{P'^3}, \tag{73}$$

where ω' is the frequency of revolution of the outer electron in its Kepler ellipse.

The difference in energy between neighboring states is given by the equation

$$\delta E = \omega'\delta\mathscr{I}'_\omega+\sigma'\delta\mathscr{I}'_\sigma.$$

From the dynamical relation between the orbits, it follows that $\mathscr{I}'_\sigma$ is the total angular momentum of the atom. In the approximation used here, P' can be replaced by $\mathscr{I}'_\sigma/2\pi$. Hence, we get

$$\delta E = \omega'\delta\left(\mathscr{I}'_\omega - \frac{9}{8N^2}\frac{\mathscr{I}^3}{\mathscr{I}'^2_\sigma}\right). \tag{74}$$

Thus, we obtain a formal resemblance to the corresponding formula for a purely periodic motion. For an orbit that is not too small, the frequency depends on the total energy, as in the case of a Kepler motion. Hence,

$$E = -2\pi^2(N-1)^2e^4m\Bigg/\left(\mathscr{I}'_\omega - \frac{9}{8N^2}\frac{\mathscr{I}^3}{\mathscr{I}'^2_\sigma}\right)^2. \tag{75}$$

The quantum conditions are

$$\mathscr{I} = h, \qquad \mathscr{I}'_\omega = nh, \qquad \mathscr{I}'_\sigma = kh.$$

The energy in the stationary states as a function of the quantum numbers is

$$E_{n,k} = (N-1)^2Kh\Bigg/\left(n - \frac{9}{8N^2}\frac{1}{k^2}\right)^2. \tag{76}$$

If the terms given by this expression are compared with the observed terms, the agreement is worse than might be hoped for. For $N = 2$, we find from eq. (76) the following values for the deviation

$$\alpha_k = \frac{9}{8N^2}\frac{1}{k^2} \tag{77}$$

of the square root of the numerator from a whole number:

$$\begin{aligned} &\text{for } k = 1 \quad \text{(s-term)} \quad \alpha_1 = 0.280 \\ &\text{for } k = 2 \quad \text{(p-term)} \quad \alpha_2 = 0.070 \\ &\text{for } k = 3 \quad \text{(d-term)} \quad \alpha_3 = 0.031, \end{aligned}$$

while observations give the following values

$$\begin{aligned} \alpha_1 &= 0.31 \\ \alpha_2 &= 0.065 \\ \alpha_3 &= 0.003. \end{aligned}$$

The rather good agreement for the s-term is surprising, since the assumption that the orbit of the outer electron is large compared to that of the inner electron is no longer valid. For the p-terms, the agreement is more or less satisfactory. The very large discrepancy in case of the d-term may perhaps be explained by

[384]

the circumstance that, for large orbits of the outer electron, the relativity correction for the inner electron becomes comparable to the perturbation exerted by the outer electron. It is possible that this is perceptible already for $k = 3$. In the orthohelium spectrum, narrow doublets are observed. It is possible to imagine reasons which make this complexity of the terms comprehensible. Far less can be said about the parhelium spectrum than about the orthohelium spectrum. The motion is very much more complicated here and cannot be treated by the method of secular perturbations. We shall return to this later.

Now, we ask about the normal state of helium. Since the selective absorption of helium must lie in the far ultraviolet, and has not yet been observed, this state has not yet been ascertained by optical experiments. However, the question about the normal state of helium has been solved by Franck and his co-workers by the method of electron impact; in fact, the ionization potential is just equal to the energy of the last electron to be bound in the normal state. It appears certain that the normal state of the atom corresponds to the end result of the binding process leading to the emission of the parhelium spectrum, in which the last electron caught, just as the first one, is bound in a 1_1-orbit. Moreover, the experiments lead to the result that helium atoms can be brought by electron impact into a state which Franck calls metastable and from which the atom cannot of itself return to the normal state with emission of radiation. The metastable state belongs to the largest term of orthohelium. The discovery of the metastable state is one of the greatest developments of recent years. It is comparable to the discoveries which established the essential features of atomic structure. The question now arises as to how this peculiar result must be interpreted. The calculations indicated above may contribute to the understanding of this phenomenon. In orthohelium the orbit of the outer electron has the same character as the orbit of an electron in a centrally symmetric force field. Thus, the spectrum of orthohelium has the properties of a spectrum arising from the motion of an electron in a central field. However, this is only true if we limit ourselves to orbits for which the principal quantum number is at least equal to two. On the other hand, when $n = k = 1$, the dimensions of the orbit become smaller than those of the hydrogen atom, and it is no longer possible to distinguish between an inner and an outer electron. There is no longer any question of secular perturbations, and one must attack the problem *ab initio*. If we ask for a state in which both electrons move in the same plane and in 1_1-orbits, it appears that only a single simple motion of this kind offers itself, namely, that in which both electrons move in the same circle around the nucleus, being at every instant at the opposite end points of a diameter. The lecturer proposed such a model in his first paper on atomic structure. However, we know now, not only from the determination of the ionization energy, which for this model comes out ca. 4 volts too high, but also from the application of the correspondence principle, that such a motion is impossible. In fact, this motion is not simply related to the family of orthohelium orbits; while the latter are connected adiabatically, so that the outer electron can be transferred, by a slow change of the energy, from one orthohelium orbit to another, it is not possible to bring the atom adiabatically from an orthohelium state to a state in which the two electrons carry out equivalent motions in one and the same circle around the nucleus. One must, so to speak, demand a readiness on the part of the outer electron to come to an understanding with the inner electron. However, such a process is not to be expected from the periodicity properties of the orthohelium orbits. Thus, we cannot conceive of a series of simple intermediate forms for orbits of the two electrons; hence, according to the correspondence principle, no transition is possible. Is this entire perception really compatible with the ideas of the quantum theory? I believe that the principles of the quantum theory are often expressed in too sharp a form. I should like to employ a simple physical picture. Corresponding to the quantum orbits and the electrons, let us imagine a number of bowls into which we are to throw balls. If we were to depend on classical mechanics, it would not be easy to get a ball into a bowl. According to the quantum theory, it seems that the ball must necessarily land in a bowl, and that is very strange. However, when we consider that the quantum states, i.e., the places where the bowls are located, as well as the processes which cause the transitions, are determined by the same periodicity properties, then we need no more wonder so much. It is, therefore, quite natural that it is impossible to get from an orthohelium orbit to the doubly occupied circular orbit; for such a transition would be extraneous to the principles of quantum theory, and it seems that all the processes that occur are closely connected with the properties of multiply periodic motions. We must assume that, in the normal state, the two electrons move in nearly circular orbits, whose planes make an angle of 120° with each other, and, in fact, move in such a way that they are at any instant in symmetrical positions with respect to an axis which is perpendicular to the straight intersection of the two orbital planes and bisect the angle between their normals. The two orbital planes execute a precession about this axis. Since the method of secular perturbations is not sufficient here, this motion is very difficult to treat. Kemble has proposed this model for the helium atom, and Kramers, even before the appearance of Kemble's paper, had started accurate calculations of just this motion. He obtains the following expression for the energy of the normal state

$$W = W_0\left(2N^2 - 1.373N + 0.0271 + \frac{a}{N} + \dots\right), \tag{78}$$

where W_0 is the energy of the hydrogen atom in the one-quantum state. So far, this result is not very satisfactory; the energy comes out too small. This is perhaps not so strange, since such an expansion of the energy in powers of the nuclear charge may converge well for large values of N, but not for $N = 2$.

So far, we have progressed very slowly; nevertheless, we have treated the entire first period of the periodic system. Even if that takes care of only two elements, we have already achieved a great deal; for, as it will appear, our progress will be easier the farther we proceed into the periodic system. Now, we have not only a clear path for each step, but we shall also obtain better and better results.

FIFTH LECTURE. JUNE 20, 1922.

Yesterday we considered, and were tantalized by, the problem of the formation and structure of the helium atom. But, in doing so, we took, after all, a big step forward and finished off the *first period* of the periodic system of the elements. Fig. 1* shows the representation of the periodic system given by Julius Thomsen, which is particularly well adapted to our purpose. The lines drawn between individual elements connect the homologous elements which, in the usual representation, stand in vertical columns. I shall not enter here upon the many investigations which have dealt with the problem of the periodic system but shall remind you only of the attempt to interpret it made by J. J. Thomson. He assumed that the periods of the system are related to periodic fluctuations of the stability of the electron configurations in the atoms. His work contains many fruitful ideas, but his starting point has not proved valid. We now view the structure of atoms quite differently. Above all, we now know the precise number of electrons in an atom, especially from investigations of X-rays, and we know that this number is equal to the number of the atom in the periodic system. Much work has been done, also more recently, on the problem of the periodic system, most successfully by Kossel, who has carried the theoretical conceptions forward, while adapting them to the new ideas. He has shown that it is possible, by simple assumptions about the formation of electron shells, to explain many chemical properties of the elements, and, at least in broad outline, to establish a connection with the X-ray spectra. We shall now, from the standpoint of the quantum theory, throw light on these questions, which in the investigations mentioned were still treated

* [See p. [412].]

in a rather formal manner. In doing so, there can be no question of deriving the experimental material from completely developed principles; rather, we must develop the principles of quantum theory from the experimental results. The solution of this problem does not appear hopeless; I should like to show you that the theory can be carried through without exerting any constraint upon the general principles.

After helium follows, as the first element of the *second period*, *lithium*, about the atomic structure of which we know a great deal from experiments. The binding of the two first electrons occurs much as in helium; the lithium spectrum, which has a particularly simple structure (see fig. 3), gives us information on the binding of the third electron. From absorption experiments, it is known that the largest s-term corresponds to the normal state of the lithium atom. The p-, d-, and b-terms are very nearly equal to the terms of hydrogen, the positions of which are indicated by broken lines in the figure; only the s-terms exhibit large deviations. This can be explained by a rough picture of the atom. The dimension of the orbits of the two electrons first bound is about one third of the orbital diameter of a hydrogen atom in its normal state. The larger the outer orbits are, the better can the influence of the nucleus and the two inner electrons, i.e., the core, be replaced by a simple point charge. From this it can be concluded that the p-, d-, and b-terms correspond to large orbits. What about the strongly deviating s-terms? And what quantum numbers must be ascribed to them? We ask, in particular, if the largest s-term corresponds to an 1_1-orbit. It is seen that the largest p-term, with quantum numbers 2_2, already deviates considerably more from the corresponding hydrogen term than does the largest d-term, with quantum numbers 3_3. We can relate this deviation to the polarizability of the atomic core. The atomic core is polarized in the electric field of the outer electron and, to a first approximation, this polarization can be considered proportional to the field strength. If A is the electric moment caused by unit field strength, and r the distance from the core, the outer electron produces a dipole in the core of moment $A\ e/r^2$. The attractive force exerted by this upon the electron will, therefore, be

$$A\frac{e}{r^2}\cdot\frac{2e}{r^3} = A\frac{2e^2}{r^5}.$$

If this picture is more or less correct, the value of A can be calculated from the deviation of the 2_2-term from the corresponding hydrogen term. In this way, the value $A = 0.4 \times 10^{-24}$ cm^3 is obtained. It is not possible to check this value by another, more direct, method; however, from the dielectric constant of helium, it is possible to determine the polarizability of the helium atom, which has a structure similar to the lithium core. In this way, we find for helium the value $A = 0.2 \times 10^{-24}$ cm^3, i.e., a value of the same order of magnitude as that found for the polarizability of the lithium core. Since the two electrons in the lithium core are more strongly bound by the trebly charged nucleus, we should really expect that the polarizability of the helium atom would be larger than that of the lithium core. The large value of A for the lithium core can perhaps be understood from the assumption that the finite extension of the lithium core gives rise to a deviation from a Coulomb field which cannot be neglected in comparison with the deviation resulting from the dipole.

If we now attempt to calculate an 1_1-orbit for the third electron, we find that such an orbit is not possible on the outside; for the attraction is so strong that the electron would be pulled entirely inside the core. Thus, the orbit lying outside the core and closest to it must be a 2_1-orbit. In the normal state the third electron moves in this orbit which corresponds to the largest s-term. This orbit comes very close to the core, and this explains the large deviation from the corresponding hydrogen term, which means a tighter binding. The question may be raised why three 1_1-orbits at the same distance need not be considered for the normal state, i.e., why not assume three such orbits similar to the two orbits in parhelium? This would give a binding that would be much too strong; in fact, about ten times as strong as observed. We must assume that, much as in metastable helium, a transition to such a state is excluded by the correspondence principle.

The lithium spectrum is not quite so simple as we have represented it up to now. Complex terms occur; the 2_2-state is double. How are we to explain that? We saw yesterday that when the field of an atomic core has axial symmetry, as we must assume to be the case for lithium, the motion of the outer electron can be regarded as a central motion whose plane executes a slow precession about the axis. In this case there are three quantum conditions, corresponding to the three periods. Then only certain distinguished orientations of the orbital plane are possible. Now, what about the situation in lithium. First, we must explain why the s-term is single. The angular momentum of the core is 1, the angular momentum of the outer electron in an s-orbit is also 1, so the total angular momentum can be 0, 1, or 2. We should like to believe that the case of zero total angular momentum must be excluded. The value 1 seems also to be excluded. The angular momentum 1 of the core is the resultant of the angular momenta of the orbits of the two inner electrons which make an angle of 120° with each other. If we should place another orbit with angular momentum 1 in such a way that the total angular momentum would also be 1, the normal of the orbit of the outer electron, since the orbit precesses, would at certain instants in time coincide with the normal of one of the inner orbits. Thus, only the single value 2 is possible for the total angular momentum. For a p-term, the angular momentum of the outer orbit is 2, so that the total angular momentum can be 3, 2, or 1. The value 1 occurs when the angular momentum of the outer orbit is opposite that of the core. We should like to believe that this case is also excluded, although we can give less compelling reasons for it than for helium. Thus, there remain the two possibilities 3 and 2, which may be responsible for the double nature of the p-terms. However, I must emphasize that this argument is quite uncertain.

I should like to say a few more words about the anomalous Zeeman effect which is very closely related to the complexity of the terms. The anomalous Zeeman effect points to very complicated circumstances which have been discussed from various sides. From the experimental material, Landé has succeeded in deriving the terms of the anomalous Zeeman effect, i.e., the energy values appearing in a magnetic field. In these terms, not only integral but also fractional multiples of the normal splitting occur. How to explain this is a very difficult question. We must conclude, from the occurrence of the anomalous Zeeman effect, that the classical theory is inadequate. On the basis of peculiar assumptions about the types of occurring orbits, Heisenberg, in a very interesting paper, has attempted to arrive at Landé's results. He assumes that the orbital planes are arranged parallel or perpendicular to the axis of the inner system. He assumes, further, that not only whole numbers must be inserted into the quantum conditions, as customary, but that k can also have the values $\frac{1}{2}$, $\frac{3}{2}$, ... The total angular momentum 1 of the s-state is distributed evenly between the outer electron and the core; in the p-, d-, ... orbits, the angular momentum of the core is always $\frac{1}{2}$. It is difficult to justify Heisenberg's assumptions. A change in angular momentum, i.e., in the orientation of the orbit, means a very great change in energy, and this is not considered in the theory; it is perhaps a thousand to ten thousand times as large as the difference between terms. Nevertheless, Heisenberg's paper is very promising and has contributed much to the formal interpretation of the experimental material. In the following, we shall not enter very much into the question of the orientation of the orbits; I regard this as a perfectly open question. Our main concern here is to show that it is possible to assign the quantum numbers n and k to the terms without ambiguity.

Proceeding further in the periodic system, we come to *beryllium*. For this and the following elements we encounter the difficulty that no simple spectra are known. We assume that the fourth electron is bound in an orbit similar to that of the third electron in lithium. Because of the higher nuclear charge, the orbits of the two inner electrons will lie much closer to the nucleus than in lithium. Another 1_1-orbit would be pulled much more into the interior than in lithium. We must, therefore, assume that the fourth electron is bound in a 2_1-orbit. We may imagine that the first 2_1-orbit shields the inner system very poorly, because the third electron revolves very slowly; this makes it comprehensible that the fourth electron is also bound in a 2_1-orbit. Consequently, the interplay between the third and the fourth electrons is not quite symmetrical. The electrons do not move in an "Ellipsenverband" ⟨see A. Sommerfeld, Atombau, 1919, p. 366⟩, but we must assume that one electron is very far removed from the nucleus at the same time that the other is very close to it. From a consideration of the magnitude of the forces, it follows that the outer electrons at certain times come closer to the nucleus than do the inner electrons.

Proceeding further in the periodic system, we come to *boron* and *carbon*. We must assume that we encounter difficulties when we add the third outer electron to boron. We shall return to this question later. Because of the symmetry properties of the atom, I believe that we must assume that in carbon the four outer electrons are bound in 2_1-orbits which are oriented relative to one another in such a way that their normals are directed towards the corners of a regular tetrahedron with its midpoint in the nucleus. This model is somewhat similar to one given previously by Landé. It has already been emphasized by Born and Landé that, for reasons originating in the theory of crystals, we must assume that the orbits are arranged with a high degree of symmetry. Landé has also shown that we must assume 2_1-orbits, in order to represent correctly the size of the carbon atom. Otherwise, Landé's considerations are essentially different from those presented here. According to Landé, the interplay of the four electrons in spatial orbits occurs under the assumption that the inner system can be replaced by a point charge in the nucleus; here the appearance of this interplay is assumed to depend essentially on the presence of the inner system. Attempts have previously been made to make the symmetry of the atom as pretty as possible, and not only spatial but also temporal symmetry has been assumed, such that all four electrons find themselves in the neighborhood of the nucleus at the same time. On the other hand, we assume that the four electrons visit the nucleus in turn, and with equal time intervals; we do this for reasons obtained from considerations of the stability properties of the entire atom. I should like to illustrate the difference between these two conceptions with a simple example. Consider the simple old atomic model, in which a number of equivalent electrons move in one and the same circle. We shall assume a nucleus with a four-fold charge and shall compare the energies and dimensions of such a model in its state of lowest energy when we place 1, 2, ... electrons in the circle. If only one electron is present, the energy is sixteen times as large as in the hydrogen atom and the radius of the circle is one quarter the radius of the hydrogen atom. If p electrons are placed in a circle, such that they at each instant are at the corners of a regular polygon, the resultant of the forces exerted by $p-1$ electrons on the p'th will always pass through the nucleus, so that the repulsive force of these electrons may be regarded as a simple reduction of the nuclear charge. If the effective nuclear charge is denoted by N_{eff}, the energy will be $p(N_{\text{eff}})^2$ times the energy of a hydrogen atom. The radius a of the circle will become $a = a_1/N_{\text{eff}}$, where a_1 is the radius of the hydrogen atom.

[392]

The effective nuclear charge, the energy and the radius for different values of p become:

p	N_{eff}	$p(N_{\text{eff}})^2$	a/a_1
1	4	16	0.25
2	3.75	28	0.267
3	3.433	35.4	0.292
4	3.043	37.0	0.329
5	2.623	34.4	0.382

Thus, the last electron is seen to be bound less firmly the more the ring is occupied, and, already for the fifth electron, energy must be expended to place the electron on the circle. In contrast to the binding energy, the radius only varies very little with the number of electrons, contrary to what we must demand. The binding energy is equally distributed over all electrons. However, in our present model, one electron is always at the verge of breaking away. In this way, we obtain a much closer connection between binding energy and atomic size. However, to carry through the accurate calculation is a very difficult task. In the next lectures, we shall present some reasons which support our views.

We know no simple spectrum of *nitrogen*, and matters are also complicated with regard to *oxygen*. We now approach a critical point, namely the end of the second period. In *neon*, a particularly stable state must have been reached. According to Kossel, the electronegative properties of *fluorine* must be explained by the fact that one electron is lacking to form the stable electron configuration of neon, while the electropositive properties of *sodium*, which comes after neon, indicate that here one electron is, so to speak, superfluous. How can this manner of looking at things be harmonized with our quantum-theoretical considerations? It is usually assumed that the orbits of the eight electrons in neon are equivalent. However, I believe that it is absolutely necessary to assume that there are 2_2-orbits as well as 2_1-orbits. This follows not only from the X-ray spectra, as I shall show later, but it is also possible to give reasons for this assumption from a consideration of the binding of the fifth outer electron. With the binding of the fourth of these electrons, we have reached the limit of the gradual increase in symmetry. Hence, it is very improbable that additional electrons be bound in the same manner. It is also possible to understand that an electron is bound very firmly in a 2_2-orbit; for, at times, electrons in 2_1-orbits move farther from the nucleus than electrons in 2_2-orbits. I believe that the electronegative properties of the elements in the second half of the second period depend on this circumstance. Thus, according to these conceptions, we have in neon four electrons in

2_1-orbits and four electrons in 2_2-orbits. We are hardly justified in imagining that the orbits of these eight outer electrons are arranged in two configurations of tetrahedral symmetry, since then at certain instants two orbital planes would coincide. As already mentioned in the case of lithium, we must assume that the occurrence of such motions is impossible. I believe, rather, that we have two deformed tetrahedral configurations, so that the electron groups with two-quantum orbits in the neon atom possess only a simple symmetry axis, which must be assumed to coincide with the symmetry axis of the innermost group of two electrons in one-quantum orbits.

Without encountering new difficulties, we now proceed to the *third period.* For *sodium*, first of all, we possess a very detailed knowledge of the spectral terms. From absorption experiments, we know that the largest s-term corresponds to the normal state. As in the case of lithium, the deviations from the hydrogen terms are very large for the s-terms and become smaller and smaller as we proceed to the p-, d-, and b-terms. The question now arises what quantum numbers are we to assign to the terms. From a formal comparison of the terms of the various alkali spectra with the terms of the hydrogen spectrum, Roschdestwensky believed that he could conclude that the quantum numbers 2_1 must belong to the normal state. Schrödinger has attempted to explain the large deviations of the s-terms from the hydrogen terms by the assumption that all s-orbits reach into the region of the innermost electron group. Also he arrived at the conclusion that the normal state is characterized by the quantum numbers 2_1. However, as with the binding of the third and seventh electrons, we must assume that, because of the peculiar stability of the electron configuration already present, the eleventh electron will now be bound in a new type of orbit, namely, in a 3_1-orbit. Considerations of the firmness of the binding of this last electron lead us to the same conclusion. It is true that, in the outer parts of its orbit, the outermost electron moves in a field that differs little from that of a simple point charge. Nevertheless, its energy is determined essentially by the parts of its orbit that lie in the region of the inner electrons. There the orbit will only differ little from a 2_1-orbit. However, in the outer parts the orbit must be larger than in the inner parts; hence, it must at least be a 3_1-orbit. By considering the polarizability of the atomic core, as in the case of lithium, we can estimate its value for a definite principal quantum number and compare it with the polarizability of neon as calculated from its dielectric constant. The best agreement between these polarizabilities is obtained when the principal quantum number 3 is assigned to the largest p-term. The view previously advocated, that the largest p-term must be assigned the principal quantum number 2, would lead to the largest p-term being smaller than the corresponding hydrogen term. One would then have to assume that the average influence of the inner system could be described as a repulsion, which would not be consistent with our view of the polarizability of the core. As in the case of lithium, the doublet nature of the p- and d-terms must be assumed to be caused by different orientations of the orbit relative to the atomic core.

The terms of the alkali spectra can be represented very well in the Rydberg form

$$K/(m+\alpha_k)^2;$$

for, from fig. 4, in which the square roots of the term denominators are listed, it is seen that the deviations of these values from whole numbers are nearly constant. In this expression, we have denoted the whole number by m, since it will appear later that it is not the principal quantum number n. We shall now show how it can be understood that α_k is nearly independent of n. In a sodium atom, for example, a 3_1-orbit has a course much as shown in fig. ( )*. Calling the part of the orbit which lies outside the region of the inner electrons a, and calling the interior loop, which is similar to a 2_1-orbit, s, we complete the outer part of the orbit with a fictitious segment g, so as to obtain an ellipse, approximately. We can then determine the energy of the real orbit by calculating the energy of this substitute ellipse. In fact, the electron enters or emerges with the same speed whether it moves in the real orbit or in this substitute ellipse, and the potential energy at the places of entry or emergence is the same in both cases. If $\mathscr{I}_r+\mathscr{I}_\varphi = \mathscr{I}$ is the sum of the phase integrals extended over the substitute ellipse, the energy of the orbit will be

$$E = -Kh^2/\mathscr{I}^2.$$

We can evaluate the integral

$$\mathscr{I} = \int^{(a+g)} p_r \mathrm{d}r + \int^{(a+g)} p_\varphi \mathrm{d}\varphi$$

in the following way. We write

$$\mathscr{I} = \int^{(a+s)} p_r \mathrm{d}r + \int^{(a+g)} p_\varphi \mathrm{d}\varphi - \int^{(s)} p_r \mathrm{d}r + \int^{(g)} p_r \mathrm{d}r.$$

Because of the constancy of the angular momentum

$$\int^{(a+g)} p_\varphi \mathrm{d}\varphi = \int^{(a+s)} p_\varphi \mathrm{d}\varphi = \mathscr{I}_\sigma.$$

Also,

$$\int^{(a+s)} p_r \mathrm{d}r = \mathscr{I}_r$$

* [This figure is no longer extant.]

and

$$\mathscr{I}_r + \mathscr{I}_\sigma = nh.$$

Consequently,

$$\mathscr{I} = nh - {}^{(s)}\mathscr{I} + {}^{(g)}\mathscr{I},$$

where ${}^{(s)}\mathscr{I}$ and ${}^{(g)}\mathscr{I}$ are the radial phase integrals over the paths indicated. Now, both ${}^{(s)}\mathscr{I}$ and ${}^{(g)}\mathscr{I}$ are nearly independent of n; this is first of all true of ${}^{(g)}\mathscr{I}$, because for constant k the form of the ellipse in the neighborhood of the focus is nearly independent of n; then also of ${}^{(s)}\mathscr{I}$, since the loop depends very little upon the outer part of the orbit. Thus, $-{}^{(s)}\mathscr{I} + {}^{(g)}\mathscr{I}$ is approximately constant, and it is seen that $\mathscr{I}/h$, i.e., the square root of a term denominator, always increases by one when n does. Without the assumption that the outer orbits penetrate into the atomic core, it would not be possible to understand this behavior of the term denominator; for then the orbits for different values of n would take quite different courses, while, on this assumption, very important parts of the orbits nearly coincide. Already the p-orbits of sodium penetrate into the inner system. As shown by fig. 4, the difference $p_1 - p_2$ remains nearly constant. This can be explained simply by assuming that the inner loops of a p_1-orbit and a p_2-orbit are slightly different. We shall not enter here into the fundamental difficulty of understanding the doublet structure.

SIXTH LECTURE. JUNE 21, 1922.

Yesterday we finished our discussion of the elements of the second period and considered the first element of the third period. Before proceeding further, I should like to say a few more words about the closing of the second period. We have sought the reason for this closing in the peculiar symmetry of the arrangement of the first electrons which prevent a newly arriving electron from gaining a place in the interplay of the other electrons, making necessary the formation of new types of orbits. At the present state of the theory, such considerations appear very uncertain, so that it may be justified to ask how much weight to attach to these conclusions. A large amount of experimental material has contributed to the shaping of our view; it is a matter of taste, I believe, whether to put the main emphasis on the general considerations or on the bare experimental facts. In order to make more clear the power of the general considerations, I shall repeat briefly our views about the binding of the eleventh electron, the last electron to be bound in sodium. We know from absorption experiments that the largest s-term belongs to the normal state. We have seen how we can form a rough picture of the occurrence of all states. On the basis of these conceptions, the normal state must be a 3_1-orbit. In its outer part, such an orbit is very nearly a Kepler ellipse; in the interior of the atomic core, the electron moves in a loop that is very nearly of the same shape for different principal quantum numbers. By assuming such a shape for the orbits of the s- and p-terms, we could explain, for each series of terms, the constancy of α_k in the Rydberg formula $K/(m+\alpha_k)^2$. If we consider a representation of the sodium spectrum, we observe not only a simple distortion of the hydrogen spectrum but also that the states with principal quantum numbers $n = 1$ and 2 have fallen out, contrary to earlier views. According to our present view, all spectra, with the single exception of the parhelium spectrum, correspond to a firmer binding than in the case of the analogous hydrogen terms. While it was formerly questionable whether the deviations from the hydrogen terms were positive or negative, all deviations are now positive; negative deviations occur only in the parhelium spectrum, but this occupies an exceptional position.

The Mg spark spectrum gives information about the binding of the eleventh electron in *magnesium*. It is very similar to the arc spectrum of sodium, also with regard to the doublet structure; the main difference is that $4K$ must be substituted for K in the Rydberg term formula. Hence, to make a comparison more convenient, the scale of the terms of the spark spectrum has been reduced to $\frac{1}{4}$ in fig. 6, while the corresponding hydrogen terms are drawn to the original scale. The similarity between the spark spectrum of Mg and the arc spectrum of Na is caused by the fact that the inner system is the same in both cases; only the nuclear charge is one unit larger, which gives rise to the appearance of $4K$ instead of K. This gives a simple explanation of the spectroscopic displacement theorem of Sommerfeld and Kossel. That the binding is increased less relative to the binding in the corresponding states of hydrogen than in the sodium arc spectrum arises from the fact that the relative difference between the force fields outside and inside the system of the ten first electrons becomes smaller the larger the nuclear charge is.

The arc spectrum of magnesium gives information about the binding of the twelfth electron. It is considerably more complex than the spark spectrum. Besides singlet terms, there also appear triplet terms; this complexity must arise because the system of the eleven electrons already present is more complex than the atomic core in sodium. I shall only touch briefly upon this question which is still an open one. I believe that the singlet and multiplet terms arise because the direction of revolution of the eleventh electron can be different, so that the angular momentum of the system of the first eleven electrons can be either zero or $2 \cdot h/2\pi$. To be sure, a zero angular momentum may not be possible for the entire atom, but, I believe, it may be possible for the atomic core. We shall not enter today on the connection of this question with the anomalous Zeeman effect. The normal state is the first S-state, which corresponds to a very firm binding. We assume that the orbit belonging to it is a 3_1-orbit. The appearance of two groups of spectral terms (s, p, d and S, P, D) is reminiscent of the helium spectrum and points to the difficulties which always arise whenever a new electron is bound in the same type of orbit as the preceding one. The s-terms can be represented very well by the Rydberg formula; in the p-terms, the α_k depend appreciably upon n.

Proceeding to the binding of the thirteenth electron in *aluminum*, we encounter again the difficulty just mentioned. The term corresponding to the firmest binding is here no s-term but, as established by McLennan from the absorption spectrum, a p-term, to which we must assign a 3_2-orbit. The admittance of another electron into a 3_1-orbit seems not to be possible here. Proceeding to *silicon*, we must ask ourselves if there here, like in carbon, occurs an interplay of four electrons in 3_1-orbits. It is not yet possible to answer this question, but it is not impossible that such a configuration is formed. As the nuclear charge increases, the orbital dimensions become smaller and, hence, the mutual perturbations larger, and it is possible that a once closed group is again broken, thus offering the possibility that new electrons are admitted to a formerly closed group. It is not yet possible to say anything definite about the next elements, *phosphorus*, *sulphur* and *chlorine*. In any case, at the end of the period, in *argon*, we must expect a configuration similar to that in neon, i.e., we must have four electrons each in 3_1- and 3_2-orbits. Thus, a difference from the neon configuration consists in the fact that there are no circular orbits here. That we have assumed no 3_3-orbits is not arbitrary; in fact, such orbits would lie far outside and, hence, be bound far too loosely. Besides, the assumption of 3_3-orbits would give an atomic diameter that would be far too large compared to that determined from the kinetic theory of gases.

The fourth period allows a still sharper test of our assumptions than did the preceding. This period is longer than the preceding; in the middle of it is located the family of iron metals, in which adjacent elements differ less in chemical properties than is normally the case. The spectrum of potassium is very similar to that of sodium. From absorption experiments we know that the largest s-term again corresponds to the normal state of potassium. We assume that the 19th electron is bound in a 4_1-orbit. It would not be a 3_1-orbit, since such an orbit is not possible outside the atomic core. The 4_1-orbit looks much like a 3_1-orbit in sodium. To calculate its inner loop would be extremely complicated; however, we can employ the following considerations. In the case of sodium, we saw that the inner loops of 3_1- and 2_1-orbits ⟨the manuscript has "4_1-orbits", but this is clearly an error⟩ are very similar. We now resume these considerations. The inner loop is very similar to an ellipse. Hence, it makes little difference for the calculation of the energy whether we let the electron move in an ellipse or in the loop, as long as only the angular momentum is the same in both cases. In sodium, the loops of 3_1-, 4_1- and 5_1-orbits, therefore, coincide approximately with the 2_1-orbit; in potassium, the loops of the 4_1-, 5_1- and 6_1-orbits coincide nearly with the 3_1-orbit. Thus, for example, to compute the outer orbit of potassium, we can replace the inner loop of a 4_1-orbit with a 3_1-orbit; in general, we can replace the inner loop of the n_1-orbit belonging to a normal state with the m_1-loop, where $m = n-1$. We saw yesterday that the square root of the term denominator is given by

$$\mathscr{I} = \oint \sum p\,dq,$$

where the integral is taken over the substitute orbit which is obtained by joining the ellipse-like outer orbit (a) with a fictitious orbital segment (g) in the interior. We found

$$\mathscr{I} = nh + {}^{(g)}\mathscr{I} - {}^{(s)}\mathscr{I},$$

where ${}^{(g)}\mathscr{I}$ and ${}^{(s)}\mathscr{I}$ are the radial phase integrals over the paths indicated. If we replace the loop of the n_1-orbit with the m_1-ellipse, we get approximately

$${}^{(s)}\mathscr{I} = (m-1)h$$

and we obtain

$$\mathscr{I} = nh - (m-1)h + {}^{(g)}\mathscr{I},$$

or, since $m = n-1$,

$$\mathscr{I} = 2h + {}^{(g)}\mathscr{I}.$$

Thus, the effective quantum number in the normal state is, to a first approximation, equal to two. The larger n becomes, the smaller is the error we commit by thus replacing the loop by an ellipse. Hence, the calculation of the term becomes more and more accurate as we proceed from potassium over rubidium to cesium, for which the terms can be estimated in this manner to within a very few percent.

As shown by the spark spectrum, the 19th electron is bound in a 4_1-orbit also in *calcium*. The arc spectrum of Ca, which gives information about the binding of the 20th electron, is, in its main features, similar to that of Mg. We assume that the largest S-term, which belongs to the normal state, corresponds to a 4_1-orbit. However, we notice a great change in the D-terms which shift in the direction of firmer binding. This arises from the fact that, with increasing nuclear charge, the orbits are attracted more and more strongly. This peculiarity appears plainly in the capture of the 19th electron. In the spark spectrum of Ca, the 3_3-orbit corresponds to a considerably firmer binding than in potassium. The 3_3-orbit is shifted more into the interior. This is a point of quite particular significance. Because of the decrease, with increasing nuclear charge, of the relative difference between the force fields inside and outside the region of the 18 first electrons, the dimensions of the parts of a 4_1-orbit lying outside this region will more and more approach the dimensions of a 4_1-orbit calculated with neglect of the interaction between the electrons. Hence, with increasing atomic number, the energy of a 3_3-orbit will more and more approach that of a 4_1-orbit, until it finally corresponds to a firmer binding than the 4_1-orbits. From this moment, the 3_3-orbit is the normal state for the 19th electron. This reversal has occurred already in scandium. Unfortunately, no series spectrum of Sc^{++} is known, so a direct verification is still pending.

A related peculiarity is the appearance of the new lines observed by Götze which do not fit into the general series scheme; in fact, we are dealing with the appearance of combinations between the normal p- and d-terms and new p′- and d′-terms. We have here the only case of an exception to the selection principle. We can understand this peculiarity from the advancing of the 3_3-terms. According to the selection principle, only such transitions occur in which k changes by ± 1. Thus, in the absence of external disturbances, the transition from a 3_3-orbit to a 4_1-orbit is impossible; but this means that the 3_3-orbit may be metastable. I should like to say a few words here about the concepts "stable" and "unstable". The normal state can be called stable, because it will persist as long as the entire energy required for a transition is not added. Apart from the metastable states, the other quantum states are also in a certain sense stable; however, they possess a probability for spontaneous transitions to states of lower energy and, thus, they are really quite unstable. The metastable states, discovered in helium, are not as stable as the normal state but are, nevertheless, stable in themselves, as long as the perturbations are sufficiently small; with stronger external influence, a transition to the normal state is possible. Thus, a metastable state is really quite unstable. It appears that the Götze lines can be explained by just such a metastable state. As a result of the disturbance occurring when the 20th electron is captured, the 19th electron, which first is in the metastable 3_3-state, can go over into a 4_1-orbit. The energy difference resulting from the circumstance that the 19th electron can be either in a 4_1- or a 3_3-orbit is within a few percent of the order of magnitude required to explain the Götze lines.

The peculiarities appearing in the capture of the 19th electron enable us to understand the greater length of the fourth period. In the elements following after scandium, the still empty 3-quantic orbits are gradually occupied. It is hard to decide if only the last or the two last electrons are bound in 4_1-orbits. In *titanium* we must assume that at least two electrons are found in 3_3-orbits. We now understand why the fourth period is quite different from the preceding periods. In the representation given of the periodic system (fig. 1)*, the elements in which the 3-quantic group is being completed are placed in a frame. No connecting line is drawn between aluminum and scandium, because the spectra are essentially different. Now, to what final state does the development of the 3-quantic group lead? At the end of the fourth period we have *krypton*, the outer structure of which must be similar to those of neon and argon. Thus, here four electrons must be bound in each of the 4_1- and 4_2-orbits. From the number of electrons present,

* [See p. [412].]

we calculate that the fully developed 3-quantic group must consist of 18 electrons. To explain this is a difficult problem. We shall content ourselves with an indication of the peculiar properties of such a group, needed for the explanation. We must assume that the development of the 3_3-orbits results in a disturbance of the interaction among the four 3_1- and the four 3_2-orbits present. The symmetry of these orbits, by which we had explained the chemical inertness of neon, is thereby disturbed, and new electrons can be admitted into 3_1- and 3_2-orbits, until finally six electrons are present in each of the 3_1-, 3_2- and 3_3-orbits. In a sense, we are dealing with a very malicious inside attack on the symmetry of the 3-quantic orbits already present. It is possible to form an idea, albeit rather uncertain, of why there appear just three groups of six electrons each. By means of a stereographic projection, it is easy to see that only in such an arrangement can the coincidence of orbits be avoided. A consideration of the X-ray spectra will offer a possibility of testing our assumption.

As already mentioned, we cannot yet follow in detail the development of the 3-quantic group. The spectrum of *manganese*, which has been investigated by Catalan, is essentially different from those of the alkalies and alkaline earths. The reason for this is that the angular momentum of the inner system is here considerably larger. A comparison of the arc and spark spectra shows that the spark spectrum of manganese is very similar to the arc spectrum. This is just what we should expect from our assumption that an inner group is being completed in the fourth period. We do not encounter simple conditions again until we reach *copper*, the spectrum of which exhibits a great similarity to that of sodium. We must assume, therefore, that in the cupro ion we are dealing with a complete group of 3-quantic orbits. That the preceding element *nickel* is not similar to the inert gases, with an outer group of 18 electrons in 3-quantic orbits, is explained simply by the circumstance that a 3_3-orbit is too large for such a group to occur as the outermost group in a neutral atom. The spectrum of *zinc* corresponds perfectly to that of calcium; only the Götze lines are missing, as was to be expected.

The peculiarity of the fourth period as compared with the preceding ones has already been discussed from other points of view. The conception of simple shells developed by Kossel is not adequate. Some years ago, Ladenburg expressed the view that, in the elements between Sc and Ni, we are dealing with the formation of an intermediate shell. This corresponds altogether to our considerations. However, we do not assume a definite intermediate shell as something lying between the outermost and the inner orbits; rather, we have found a real reason for the peculiarity of the fourth period. When we employ the quantum theory to explain the periodic system at all, we must be prepared beforehand that something new may happen here; for, from the theory of the hydrogen atom, it is impossible to exclude the occurrence of 3_3-orbits.

As pointed out by Ladenburg in this connection, the elements of the fourth period between scandium and nickel distinguish themselves, both by the colors exhibited by their compounds and by their magnetic properties. It is true that paramagnetism and color occur elsewhere, but not in compounds in which the atoms in question occur as ions. Oxygen, for example, is strongly paramagnetic in certain compounds, but not in any in which it is bound as ion. This difference can be simply explained by the consideration given here; for example, the sodium ion, like all the other ions that we have to do with in the first three periods, is a closed structure. The same is true of the electronegative ions, such as the chloride ion. That a compound is colored means that it absorbs in the visible part of the spectrum, i.e., that processes occur in which energy differences are involved which, according to the frequency condition, correspond to the visible spectrum. In closed structures, only processes with large energy differences are possible. On the other hand, we are dealing here with non-closed systems in which the electron configuration is still under development. Thus, processes involving small energy differences, i.e., colors, can appear. The increase in the strength of binding occurs only quite gradually here, so that ions with different valencies can be formed.

In paramagnetism, we encounter a perfectly open question. For example, helium and neon are diamagnetic; they do not exhibit any paramagnetism, in spite of the fact that the symmetrical group of their electrons possesses an angular momentum. It is not at all possible to understand this on the classical theory. This needs not surprise us when we realize that the classical theory is equally unable to explain the radiation phenomena, which depend just on the intimate interaction of the force effects arising from the motion of the electrons. In the fourth period, we are dealing with unsymmetrical inner systems. We may, therefore, regard it as natural and very nice that paramagnetism occurs here. As Kossel has emphasized, the experimental results exhibit a great simplicity in that all ions of different elements with the same number of electrons possess the same magnetic moment. Thus, for example, the ferri ion (Fe^{+++}) with 23 electrons has the same atomic magnetism as the mangano ion (Mn^{++}), also with 23 electrons; the mangani ion (Mn^{+++}) with 22 electrons has the same atomic magnetism as the chromo ion (Cr^{++}) which also has 22 electrons. In the most beautiful agreement with our assumptions about the structure of the elements of the fourth period, it turns out that magnetism disappears just in the cupro ion (Cu^{+}), in which we assume a complete group of 3-quantic orbits, while the cupri ion (Cu^{++}) is still magnetic. The magnetic properties of the elements of the fourth group reveal an injury occurring to the otherwise very symmetrical inner structure of the atoms, the beginning and recovery of which we witness as we proceed through the series of the elements.

In the fifth period matters are very much as in the fourth period, which is readily understood from our theory. The 37th electron in *rubidium* is bound in a 5_1-orbit. The spark spectrum of *strontium* shows that also here the 37th electron is bound in a 5_1-orbit; however, we see that the 4_3-orbits show up here, just as the 3_3-orbits do in the spectrum of calcium. The arc spectrum of strontium shows that also the 38th electron is bound in a 5_1-orbit. The new lines observed by Götze appear in Sr, as they do in Ca. We must assume that in the fifth period we witness the further development of a group of 4-quantic orbits which reaches its temporary completion in *silver*. The same reasons, as those given previously for the 3-quantic group, indicate that the 4-quantic group is built up of six electrons in each of the 4_1-, 4_2- and 4_3-orbits. The only difference is that the 3_3-orbits are circular, while the 4_3-orbits are elliptical. Certain differences between the fourth and the fifth group can be traced to this fact; they consist essentially in that the deviations from the valency properties, characteristic of the elements of the second and third periods, appear later than in the fourth period. At the end of the fifth period, we encounter *xenon* in which we must assume an outer group of four electrons in each of 5_1- and 5_2-orbits.

In the *sixth period* we meet still other phenomena. Here occur the rare earths, which all possess very similar chemical properties and whose separation has, therefore, caused the chemists great difficulties. At the beginning of the period, everything is very much as in the preceding period. The spectra of *cesium* and *barium* show that in these metals the 55th and 56th electrons are bound in 6_1-orbits, and the spark spectrum of barium shows that the 5_3-orbits are here shifted in the direction of firmer binding, as were the 3_3- and 4_3-orbits in Ca and Sr. However, we find something quite novel. The 4_4-orbits appear. We can ascertain by a calculation that, with increasing nuclear charge, there will not only come a moment when an electron is bound more firmly in a 5_3-orbit than in a 6_1-orbit but that, from a certain nuclear charge onwards, a 4_4-orbit will correspond to a firmer binding than a 5_1-orbit. From this point on we must expect, with increasing atomic number, to meet a series of successive elements which, even to a greater extent than the family of iron metals, all have very nearly the same properties; for here we are dealing with the step-wise formation of an electron configuration that is situated more deeply in the interior of the atom. We have here a simple explanation of the appearance of the rare earths in the sixth period. From the length of the period, we conclude that, in the fully developed group of 4-quantic orbits, there are 8 electrons bound in each of the 4_1-, 4_2-, 4_3- and 4_4-orbits. However, the reasons indicating this arrangement are still weaker than in the case of the 3-quantic group and the preliminary closure of the 4-quantic group in silver. In the *platinum metals*, we observe, besides the final development of the 4-quantic group, the second stage of formation of the 5-quantic group, which we must assume to be completed in *gold*, with 6 electrons in each of the 5_1-, 5_2- and 5_3-orbits. At the end of the period we encounter the radioactive inert gas *niton*, in which we must assume an outer group of 4 electrons in each of 6_1- and 6_2-orbits.

Vegard and Lewis have developed ideas similar to those of Ladenburg for explaining the appearance of the rare earths. The quantum theory automatically points to such a family of elements.

From a counting of the elements, on the one hand, and the available orbits to be filled, on the other hand, we must conclude, contrary to the customary assumption, that the family of rare earths, which begins with *praseodymium*, is completed with *lutetium*. If our ideas are correct, the not yet discovered element with atomic number 72 must have chemical properties similar to those of zirconium and not to those of the rare earths. It is not without interest that already J. Thomsen, in his representation of the periodic system, connected zirconium with element 72.

The *seventh and last* period is not complete. The reason is the instability of the nuclei, expressing itself in the radioactive properties of these elements. The spectrum of radium, compiled by Werner from the papers of Fues, shows that the 87th and 88th electrons are bound in 7_1-orbits. A difference from the sixth period is that no family similar to the rare earths occurs in the known part of the seventh period. This difference is due to the circumstance that an eccentric 5_4-orbit corresponds to a considerably looser binding of the 79th electron in the atom of an element of the seventh period than the binding of the 47th electron in a circular 4_4-orbit in the corresponding element of the sixth period, while there is a much smaller difference in the strength of the binding of these electrons in orbits of types 6_1 and 7_1, respectively. Wherever such a shift from a circular to an elliptical orbit occurs, we encounter peculiar circumstances. This manifests itself also, for example, in ion radii. When ion radii are plotted against atomic numbers, as was done by Grimm, the curve shows a bend where, according to our ideas, a change-over occurs from a circular to an elliptical orbit. It also shows up very plainly in the spectral terms, when a circular orbit takes the place of an elliptical orbit.

If there existed an atomic nucleus with that number, the seventh period would end with a noble gas atom with 118 electrons; there would then be 8 electrons each in 5_1-, 5_2-, 5_3- and 5_4-orbits, 6 in each of 6_1-, 6_2- and 6_3-orbits, and four each in 7_1- and 7_2-orbits*.

If we consider the arrangement of the noble gas configurations (see table VIII), we see that, as we proceed to higher and higher elements, there appear more and more incomplete groups. We might proceed further than done in this table and

* [The manuscript has "3 in each of 6_1-, 6_2- and 6_3-orbits" and "two in each of 7_1- and 7_2-orbits". The corrected numbers agree with those of table VIII on p. [419].]

construct hundreds or thousands of elements; however, that is not the task of physics, which deals only with things that can be put to experimental test.

I hope that I have succeeded in showing that matters become simpler and simpler the farther we proceed in the periodic system. Formerly, I believed that the difficulties would become increasingly greater the more electrons there are in the atoms. However, if we do not demand too much, it actually seems that we encounter simpler problems, and fewer new ones. I hardly need to emphasize how incomplete and uncertain everything still is.

SEVENTH LECTURE. JUNE 22, 1922.

In our investigations, so far, we have attempted to penetrate into the problem of the structure of atoms by considering the successive capture of electrons. In fact, this is a rational manner of approach, but is not sufficient in itself; for the stability of atoms is subjected to quite different tests in nature. As examples, we need only call to mind the influence of X-rays or β-rays upon atoms. The stability of atoms must also be maintained against such terrible disturbances. Hence, we shall turn to a consideration of X-ray spectra. That our assumptions, as we shall see, suffice to explain the stability conditions also for X-ray spectra, is for me perhaps the strongest support for the correctness of our views.

The X-ray spectra are of great importance for the theory. We have only postponed the discussion of them so long to obtain a more comprehensive view. I mentioned already in the first lecture that Moseley found a great similarity between the X-ray spectra and the hydrogen spectrum. He showed that the strongest line of the K-series can be represented by

$$\nu = K(N-\alpha)^2\left(1-\frac{1}{2^2}\right)$$

and the strongest line of the L-series by

$$\nu = K(N-\beta)^2\left(\frac{1}{2^2}-\frac{1}{3^2}\right).$$

This result points to a great similarity between the origin of the X-ray spectra and the hydrogen spectrum. However, while the hydrogen spectrum arises when the first electron is bound by a nucleus, we have in the X-ray spectra to do with the disturbance and reorganization of a complete atom. The observed similarity shows that the energies involved in X-rays are of the same order of magnitude as those involved in the binding of an electron by an N-fold charged nucleus, and that the conditions governing the inner and outer parts of an atom are of the same kind. The fact that the energy required to excite the characteristic X-ray spectra increases roughly with the square of the atomic weight (i.e., as we know now, with the square of the atomic number) has been known for a long time. A year before my attempt to apply the quantum theory to explain the hydrogen spectrum, this result led me to the conviction that the stability of atoms must be connected with the quantum theory.

According to our views, the electrons in an atom are divided into various groups, so that different amounts of energies are required to remove the different electrons. From such a view, Kossel was able to give a simple survey of certain major features of the X-ray spectra. In the excitation of X-ray radiation, according to Kossel, an electron is removed from a shell in the interior of an atom, and the empty place, thus created, is in turn filled by an electron from a shell lying farther out. For example, if an electron is removed from the K-level, the empty place can be filled by an electron from the M-level. However, it can also be filled by an electron from the L-level; when this happens, a vacancy occurs in the L-level which in turn may be filled by an electron from the M-level. From this view, Kossel concluded that there must exist simple additive relations among certain frequencies of the X-ray spectra. In fact, the difference in energy between the M- and K-levels must be composed additively of the energy difference between the L- and K-levels, on the one hand, and between the M- and L-levels, on the other hand. This rule of Kossel has, on the whole, been found valid. However, from this view, it is difficult to see off-hand how an outer electron can know that an electron is missing farther in the interior.

The analogy between the hydrogen spectrum and the X-ray spectra was further clarified by Sommerfeld's investigations. As already mentioned several times, Sommerfeld obtained the following formula for the energy of the hydrogen atom, when relativity is taken into consideration,

$$E = -\frac{N^2 Kh}{n^2}\left\{1 + \left(\frac{\pi N e^2}{hc}\right)^2 \left(-\frac{3}{n^2} + \frac{4}{nk}\right)\right\}.$$

The fine structure of the X-ray spectra is also represented, to a certain degree, by this formula. However, because of the factor N^4, the relativity correction, depending on the atom, becomes up to nearly 10^8 times larger than for hydrogen.

It is very difficult to harmonize the far-reaching analogy between the X-rays and the spectra accompanying the binding of a single electron with the conceptions of a shell structure. Sommerfeld and Vegard have attempted to reach an understanding of this analogy; recently, Sommerfeld has regarded the difficulties as so great that he has assumed that, even in the unexcited state, the electron configurations in different atoms of the same element are different. This is a very questionable idea that does not seem compatible with an understanding of the specific character of the physical and chemical properties of the elements. However, their papers have achieved a complete formal classification of levels in terms of quantum numbers. It has turned out that the levels are multiple. Through the work of Smekal, Coster, and Wentzel, the nature of the complexity is known in detail. Fig. 2*, taken from Coster's last paper, gives an outline of the levels to be expected for niton, based on the spectra of the preceding and subsequent elements. It is seen that it is possible, in a simple way, to assign quantum numbers to the levels. In the first place, only such transitions occur for which $k'-k''=0$, ± 1. However, the transitions are still more restricted. For this reason, Wentzel has introduced a third quantum number, while Coster divides the levels into a- and b-levels. It is seen that there is a group of orbits with the same quantum numbers for each level in our model.

All these levels are absorption levels connected with the absorption edges. In fact, no ordinary absorption lines are found in the X-ray spectra. How are we to picture the absorption process? I believe that we may reach an understanding of the phenomena just by our assumptions, according to which we do not have groups in the atom that are symmetrical at every moment in time. According to the classical theory, absorption and emission depend on the resultant electric moment of the atom. According to the correspondence principle, we can only expect that an electron is removed from its orbit under absorption when the harmonic components of vibration of this electron occur in the resultant electric moment. When forms of motion are assumed in which the electron configuration at every moment in time possesses polygonal or polyhedral symmetry, the harmonic components of the motion of the individual electrons will cancel one another. On the other hand, in our model we assume that all motions have different phases, so that the motion of each electron makes a contribution to the electric moment; in this way, the radiation is able to influence the motion of each individual electron. Our views lead to the further conclusion that an electron cannot be brought into an already completed group by any process involving absorption or emission of radiation. Thus, if an electron is to be removed from a group by absorption, this is only possible if it is removed completely from the atomic

* [See p. [524]; see also diagram I, p. [417].]

assemblage or is taken up into one of the unoccupied outer orbits. If incomplete groups of orbits are present, the electron can also be admitted to such an orbit. Now, since the energy required to transfer the electron to an outer orbit or to orbits in an incomplete group is equal, to a first approximation, to the energy required for a complete removal of the electron, the existing absorption lines will coincide so nearly with the absorption edges that they are difficult to observe.

Now, what happens when an electron is removed from an atom, say, from the innermost group? Each electron, independently of the others, will then, so to speak, seek to become bound more firmly. Thus, if an electron is removed from the K-level of niton, there is then only one electron left in the K-level, and all the other 84 electrons assume the role of the second electron in helium. In this race, it will depend on the transition probabilities which electron will arrive first in the K-level.

On this view, we must expect that, as we proceed through the periodic system, new lines will appear when new electron groups appear. In connection with Kossel's theory, Swinne has already pointed to such a connection between the periodic system and X-ray spectra.

I shall now only add a few words about the most recent work. The necessity, first emphasized by Smekal, for introducing a third quantum number to characterize the levels is easy to comprehend. Even in the optical spectra we had to use three quantum numbers to fix the spatial orientation of an orbit. Here, when an electron is removed, there exist different possibilities for the orientation of the orbits of the other electrons. The problem is not to describe the motion of an electron with the aid of three quantum numbers; for the motion contains only two periods. The third quantum number has to do only with the mutual orientation of the orbits. Coster has measured the X-ray spectra of a large number of elements. It turns out that the K-level is always single. This is easy to understand, since the orbit of an electron can be oriented relative to itself in only a single way. Moreover, it appears that only the uppermost level of the other groups is always single. Thus, for example, for xenon (see diag. II), the levels belonging to the quantum numbers 2_2, 3_3, 4_3 and 5_2 are all single. This seems quite comprehensible; for, when an electron is removed from an uppermost level, that is a simple reversal of the process by which the last electron of the group was bound; hence, when this is done, a uniquely determined result is obtained. We may picture this in the following way: if two boxes are standing between two walls, one above the other, we can remove the upper one in only one way; however, for the lower box, there are two possibilities: we can pull it away from under the upper box either to the right or to the left.

Very recently, Coster has carried out an investigation, not yet published, of the X-ray spectra of the rare earths. He finds that 4_4-orbits must actually be assumed here for the first time. Also, the levels do not vary as much with atomic number here, as for the other elements. The reason is simply that the binding in a 4_4-orbit is not appreciably stronger than the binding in the outermost orbit and, while the 4-quantic group is being developed, the binding in a 4_4-orbit does not become appreciably stronger.

Moreover, Coster has dealt with the lines interpreted by Wentzel as X-ray spark lines. Wentzel imagines that, when these lines appear, two electrons are removed, one after the other. I do not regard this as possible; for, before the removal of the second electron, the first must already have been replaced. The removal of several electrons is conceivable only if they are removed *simultaneously*. For example, when helium is bombarded with α-rays, as has been done by Millikan, the possibility of a double ionization does not depend on the strength of the binding. Rather, the probability that two electrons are removed simultaneously is greater the closer the electrons are to each other. Rosseland* has investigated these matters more closely and, from Millikan's results, has calculated the size of the electron orbits; he found that it agreed with that calculated from the quantum theory. It is clear that a simultaneous removal of two electrons can never occur by absorption; it can be achieved only by impacts. In X-ray spectra, everything is, of course, just as in helium. Hence, Coster has given a new interpretation of Wentzel's lines. According to his view, these lines originate when an electron jumps from a complete into an incomplete group. In the case of such lines, it may happen that an X-ray emission line coincides with an X-ray absorption line. However, Coster's interpretation cannot be regarded as entirely certain.

* [Rosseland, Svein (1884–   ). Norwegian physicist. Became professor of astrophysics at the University of Oslo in 1926.]

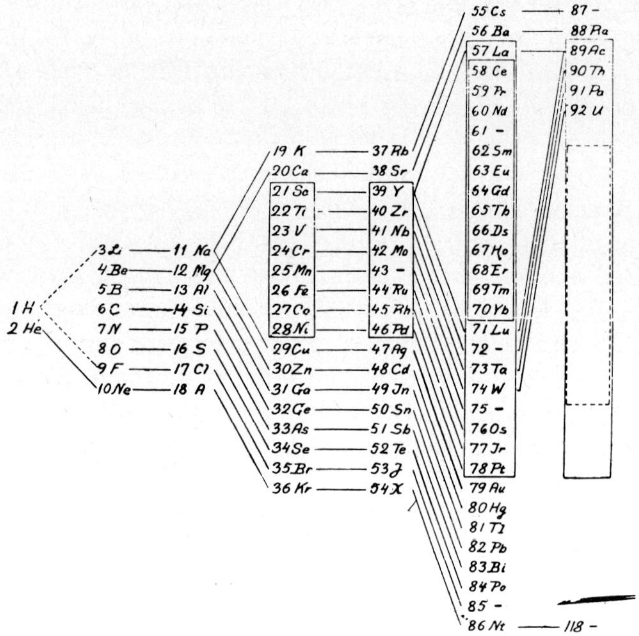
1 H
2 He
3 Li
4 Be
5 B
6 C
7 N
8 O
9 F
10 Ne
11 Na
12 Mg
13 Al
14 Si
15 P
16 S
17 Cl
18 A
19 K
20 Ca
21 Sc
22 Ti
23 V
24 Cr
25 Mn
26 Fe
27 Co
28 Ni
29 Cu
30 Zn
31 Ga
32 Ge
33 As
34 Se
35 Br
36 Kr
37 Rb
38 Sr
39 Y
40 Zr
41 Nb
42 Mo
43 -
44 Ru
45 Rh
46 Pd
47 Ag
48 Cd
49 In
50 Sn
51 Sb
52 Te
53 J
54 X
55 Cs
56 Ba
57 La
58 Ce
59 Pr
60 Nd
61 -
62 Sm
63 Eu
64 Gd
65 Tb
66 Ds
67 Ho
68 Er
69 Tm
70 Yb
71 Lu
72 -
73 Ta
74 W
75 -
76 Os
77 Ir
78 Pt
79 Au
80 Hg
81 Tl
82 Pb
83 Bi
84 Po
85 -
86 Nt
87 -
88 Ra
89 Ac
90 Th
91 Pa
92 U
118 -

Fig. 1

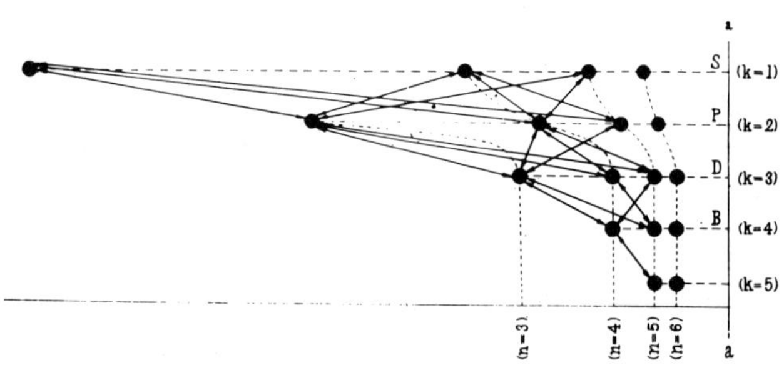
a
S
(k=1)
P
(k=2)
D
(k=3)
B
(k=4)
(k=5)
(n=3)
(n=4)
(n=5)
(n=6)
a

Fig. 2

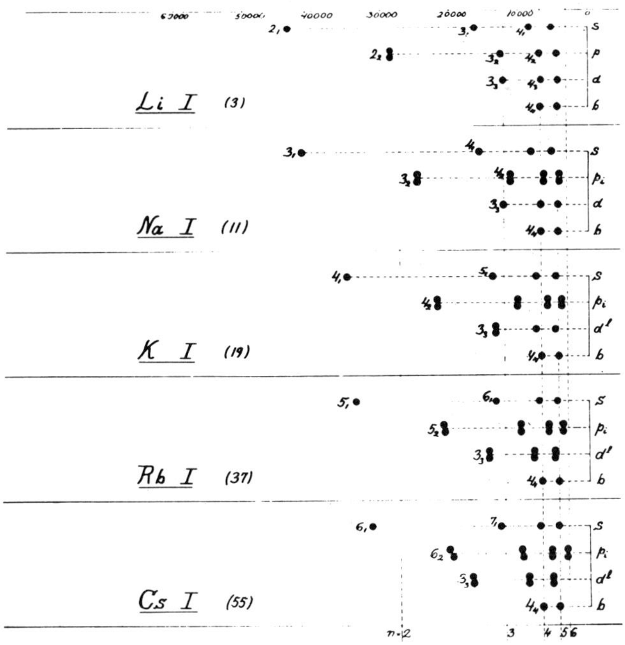

Fig. 3

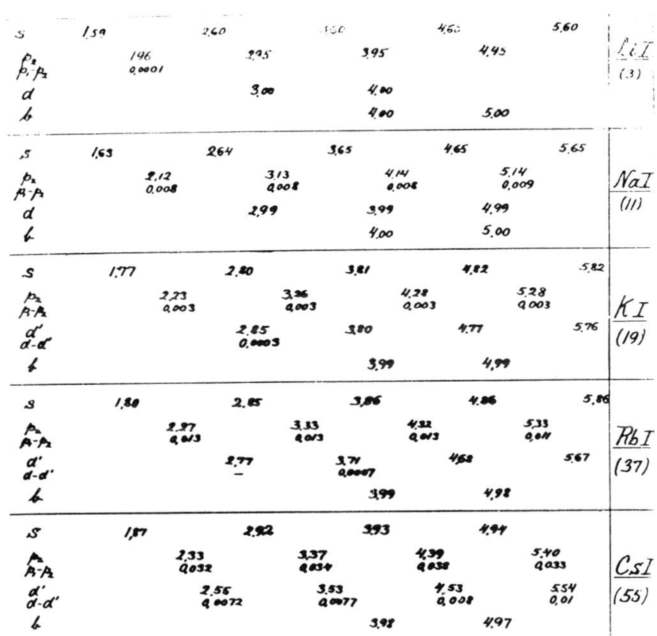

Fig. 4

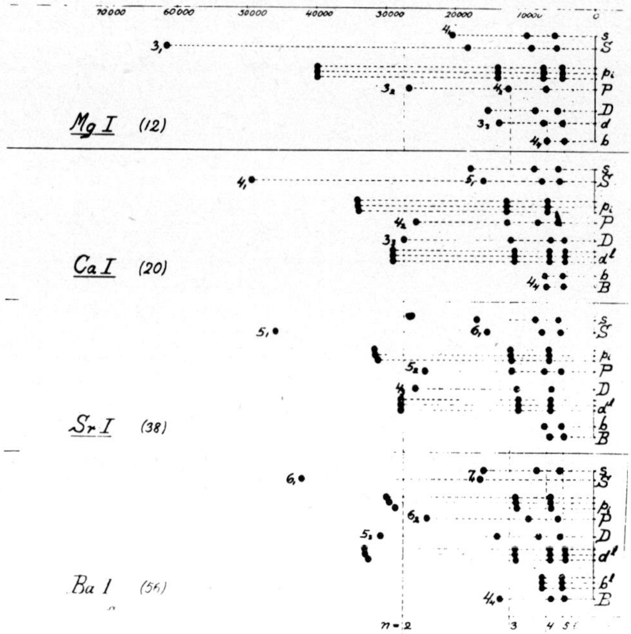

Fig. 5

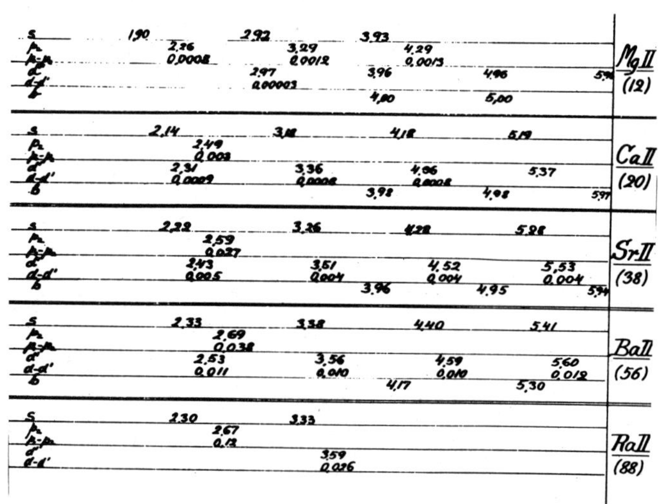

Fig. 6

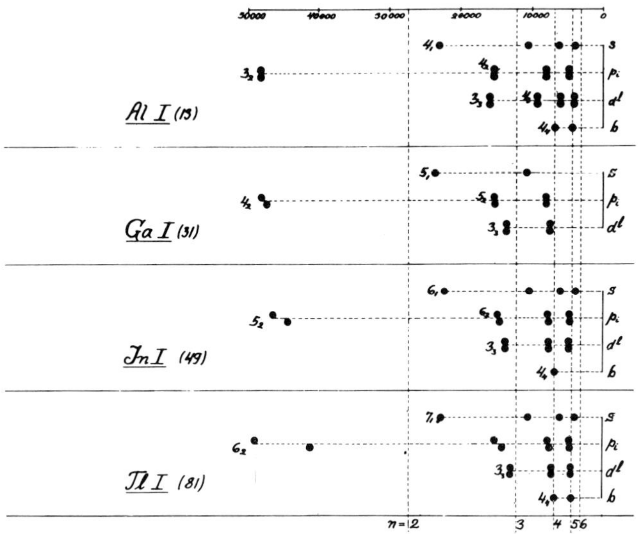

Fig. 7

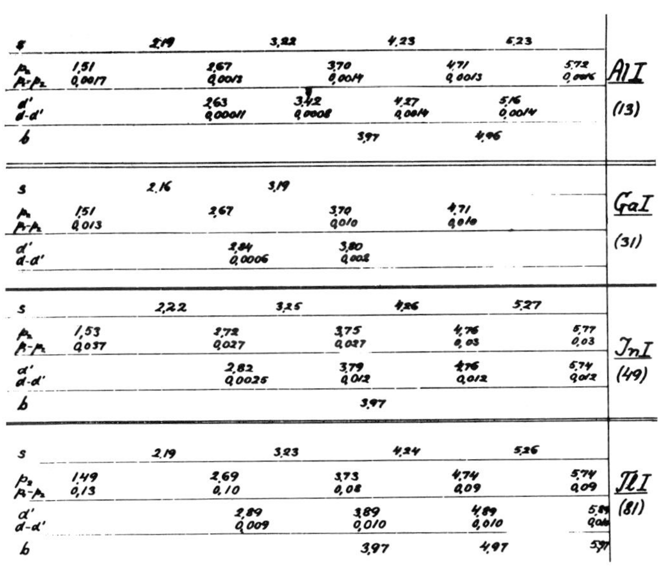

Fig. 8

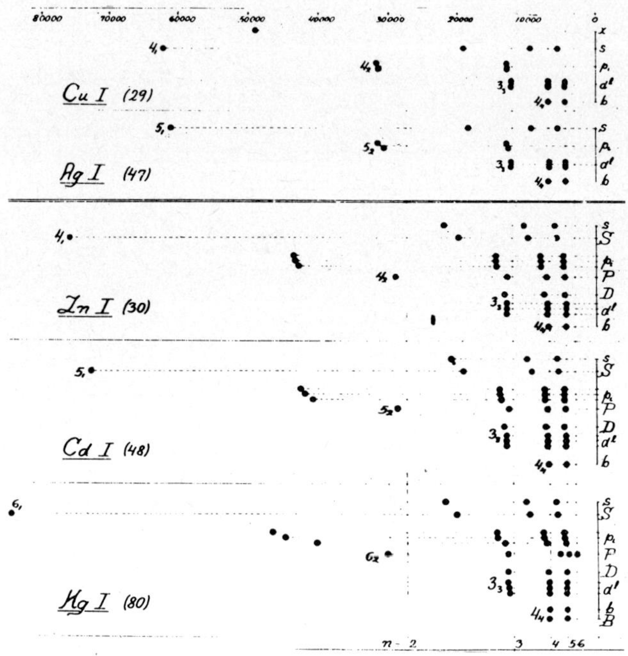
Cu I (29)
Ag I (47)
Zn I (30)
Cd I (48)
Hg I (80)

Fig. 9

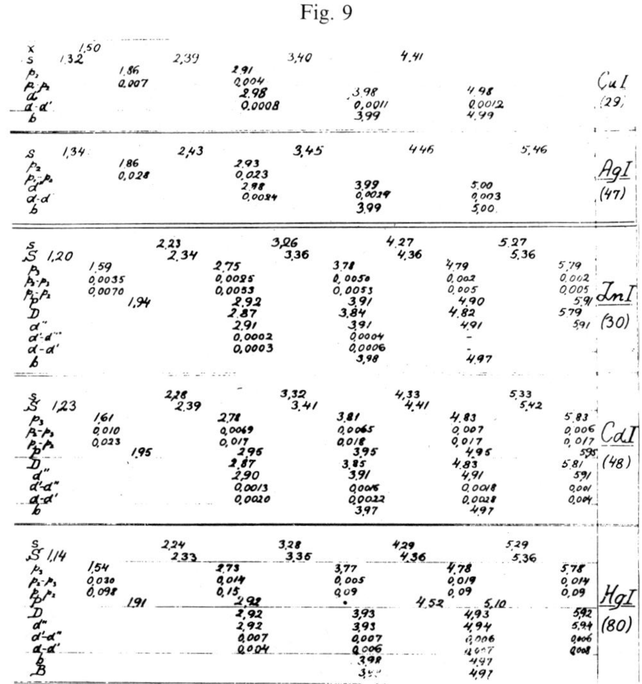
Cu I (29)
Ag I (47)
Zn I (30)
Cd I (48)
Hg I (80)

Fig. 10

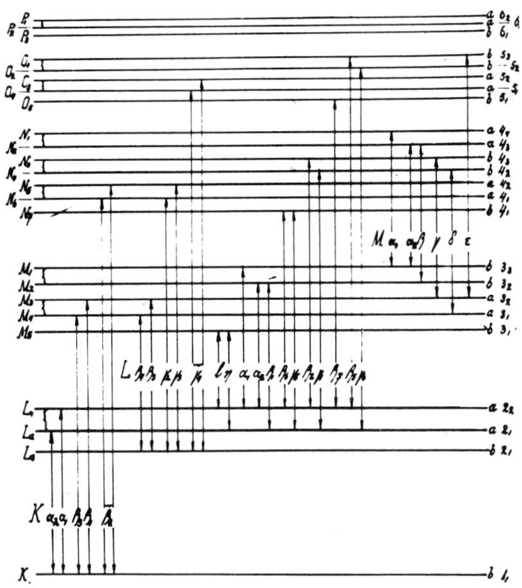

Diagram I.
Niton.

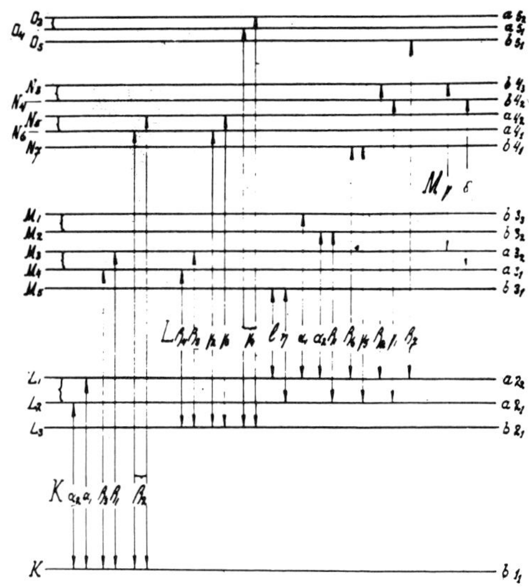

Diagram II.
Xenon.

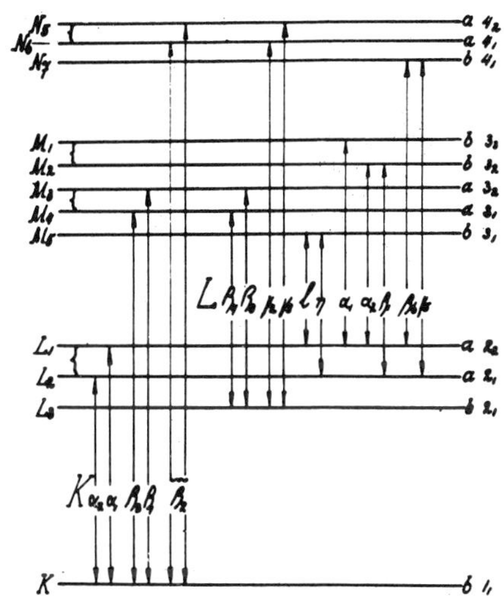

Diagram III
Krypton.

Table I

	1_1
1 H	1
2 He	2

Table II

	1	2_1	2_2
3 Li	2	1	
4 Be	2	2	
5 B	2	(3)	
6 C	2	4	
7 N	2	4	1
8 O	2	4	2
9 F	2	4	3
10 Ne	2	4	4

Table III

	1_1	2_1	2_2	3_1	3_2	3_3
11 Na	2	4	4	1		
12 Mg	2	4	4	2		
13 Al	2	4	4	2	1	
14 Si	2	4	4	(4)		
15 P	2	4	4	4	1	
16 S	2	4	4	4	2	
17 Cl	2	4	4	4	3	
18 A	2	4	4	4	4	

Table IV

	1_1	2_1	2_2	3_1	3_2	3_3	4_1	4_2	4_3	4_4
19 K	2	4	4	4	4		1			
20 Ca	2	4	4	4	4		2			
21 Sc	2	4	4	4	4	1	(2)			
22 Ti	2	4	4	4	4	2	(2)			
. . .										
. . .										
29 Cu	2	4	4	6	6	6	1			
30 Zn	2	4	4	6	6	6	2			
31 Ga	2	4	4	6	6	6	2	1		
32 Ge	2	4	4	6	6	6	(4)			
33 As	2	4	4	6	6	6	4	1		
34 Se	2	4	4	6	6	6	4	2		
35 Br	2	4	4	6	6	6	4	3		
36 Kr	2	4	4	6	6	6	4	4		

Table V

	1_1	2_1	2_2	3_1	3_2	3_3	4_1	4_2	4_3	4_4	5_1	5_2	5_1	5_2	5_3
37 Rb	2	4	4	6	6	6	4	4			1				
38 Sr	2	4	4	6	6	6	4	4			2				
39 Y	2	4	4	6	6	6	4	4	1		(2)				
40 Zr	2	4	4	6	6	6	4	4	2		(2)				
. . .															
. . .															
47 Ag	2	4	4	6	6	6	6	6	6		1				
48 Cd	2	4	4	6	6	6	6	6	6		2				
49 Zn	2	4	4	6	6	6	6	6	6		2	1			
50 Sn	2	4	4	6	6	6	6	6	6		(4)				
51 Sb	2	4	4	6	6	6	6	6	6		4	1			
52 Te	2	4	4	6	6	6	6	6	6		4	2			
53 J	2	4	4	6	6	6	6	6	6		4	3			
54 X	2	4	4	6	6	6	6	6	6		4	4			

Table VI

	1_1	2_1	2_2	3_1	3_2	3_3	4_1	4_2	4_3	4_4	5_1	5_2	5_3	5_4	5_5	6_1	6_2	6_3	6_4	6_5	6_6
55 Cs	2	4	4	6	6	6	6	6	6		4	4				1					
56 Ba	2	4	4	6	6	6	6	6	6		4	4				2					
57 La	2	4	4	6	6	6	6	6	6	.	4	4	1			(2)					
58 Ce	2	4	4	6	6	6	6	6	6	2	4	4	(2)			1					
. . .																					
. . .																					
. . .																					
71 Lu	2	4	4	6	6	6	8	8	8	8	4	4	(2)			1					
72 –	2	4	4	6	6	6	8	8	8	8	4	4	2			(2)					
. . .																					
. . .																					
79 Au	2	4	4	6	6	6	8	8	8	8	6	6	6			1					
80 Hg	2	4	4	6	6	6	8	8	8	8	6	6	6			2					
81 Tl	2	4	4	6	6	6	8	8	8	8	6	6	6			2	1				
82 Pb	2	4	4	6	6	6	8	8	8	8	6	6	6			(4)					
83 Bi	2	4	4	6	6	6	8	8	8	8	6	6	6			4	1				
84 Po	2	4	4	6	6	6	8	8	8	8	6	6	6			4	2				
85 –	2	4	4	6	6	6	8	8	8	8	6	6	6			4	3				
86 Nt	2	4	4	6	6	6	8	8	8	8	6	6	6			4	4				

Table VII

	1_1	2_1	2_2	3_1	3_2	3_3	4_1	4_2	4_3	4_4	5_1	5_2	5_3	5_4	5_5	6_1	6_2	6_3	6_4	6_5	6_6	7_1	7_2	7_3
87 –	2	4	4	6	6	6	8	8	8	8	6	6	6			4	4					1		
88 Ra	2	4	4	6	6	6	8	8	8	8	6	6	6			4	4					2		
89 Ac	2	4	4	6	6	6	8	8	8	8	6	6	6			4	4	1				(2)		
90 Th	2	4	4	6	6	6	8	8	8	8	6	6	6			4	4	2				(2)		
. . .																								
. . .																								
118 –	2	4	4	6	6	6	8	8	8	8	8	8	8	8		6	6	6				4	4	

Table VIII

Element		Number of electrons in n_k-orbits																						
		1_1	2_1	2_2	3_1	3_2	3_3	4_1	4_2	4_3	4_4	5_1	5_2	5	5_4	5_5	6_1	6_2	6_3	6_4	6_5	6_6	7_1	7_2	7_3
Helium	2	2																							
Neon	10	2	4	4																					
Argon	18	2	4	4	4	4	–																		
Krypton	36	2	4	4	6	6	6	4	4	–	–														
Xenon	54	2	4	4	6	6	6	6	6	6	–	4	4	–	–	–									
Niton	86	2	4	4	6	6	6	8	8	8	8	6	6	6	–	–	4	4	–	–	–	–			
?	118	2	4	4	6	6	6	8	8	8	8	8	8	8	8	–	6	6	6	–	–	–	4	4	–

IX. ON THE EXPLANATION OF THE PERIODIC SYSTEM

(OM FORKLARINGEN AF DET PERIODISKE SYSTEM)

ABSTRACT (AUTOREFERAT) OF A LECTURE DELIVERED BEFORE THE SECOND NORDIC PHYSICS CONGRESS, HELD AT UPPSALA, 24–26 AUGUST 1922*

(Edv. Berlings Boktryckeri A.-B, 1922; also Fysisk Tidsskrift **20** (1922) 112)

* [See Introduction, sect. 8.]

TRANSLATION

The peculiar variation in the properties of the elements, which appears when these are arranged according to increasing atomic weight, and which has been brought to light by Mendelejeff and Lothar Meyer, points to the existence of a close relationship among the atomic structures of the different elements. Shortly after the discovery of the electrons as common building stones of all atoms, the possibility of explaining the periodic system was investigated by J. J. Thomson who sought the cause of the periodic variation of the properties of the elements in a periodic variation of the stability of electron configurations with increasing number of electrons. In spite of the many new and fruitful ideas contained in his work, Thomson's theory did not lead to decisive results; for one thing, no sufficient knowledge existed at that time about the distribution of the positive electricity in atoms.

After Rutherford's discovery of the atomic nuclei, we may assume that we have an, in certain respects complete, knowledge of the building stones of atoms, and the problem of explaining the periodic system has become particularly acute after the recognition that the number of electrons in a neutral atom is just equal to the number indicating the place of this atom in the system. As is well known, this assumption, first proposed by van den Broek, has been convincingly supported by the changes in the properties of elements which follow upon the emission of α- and β-rays in radioactive processes, and also by Moseley's discovery of the simple regular manner in which X-ray spectra of the elements change as one goes from one element to the next in the system of elements. As is well known, these latter investigations have succeeded in fixing uniquely the atomic numbers of all elements, and have thus permitted us to recognize accurately the empty places in the periodic system, places which may be filled when new elements are discovered.

However, when attempting a closer account of the properties of the elements on the basis of our picture of atomic structure, we encounter difficulties of a peculiar kind. In fact, it is clear that it is not possible to explain the peculiar stability, demanded by the characteristic properties of the elements, by applying our usual conceptions of mechanics and electrodynamics. Attempts to overcome these difficulties have been made by invoking concepts borrowed from the so-called quantum theory, the foundation of which was created by Planck's investigation of the law of heat radiation and Einstein's investigations of the variation of the specific heat with temperature and of the photoelectric effect.

The application of the quantum theory to the problem of atomic structure rests on two postulates which may be formulated as follows:

I. Among the conceivable kinematically possible relative motions of the particles in an atom there exist certain states, the so-called "stationary states", which are distinguished by a peculiar stability, revealing itself in that every change in the motion of the system must result in a complete transition of the system from the original state to another of these states.

II. Every emission of radiation connected with such a transition consists of a train of simple harmonic electromagnetic waves, the frequency of which is given by the condition $h\nu = E_1 - E_2$, where E_1 and E_2 are the energies of the atom in the two stationary states. Moreover, any absorption process, in which the atom is transferred from one stationary state to another under the influence of electromagnetic radiation, will require irradiation by waves having a frequency given by this same condition.

In direct connection with these postulates, it has been possible to account for the laws holding for the spectra of the elements, not merely in outline but also in many details; especially after it has become possible, through the contributions of a large number of researchers, among whom particularly Sommerfeld and Ehrenhaft may be mentioned, to formulate conditions which permit the determination of the stationary states from among the possible motions of the particles in an atom. Although it has been possible, with the aid of the postulates mentioned, to throw light on the general character of chemical processes, nevertheless, the formulation of methods for determining stationary states has so far not led to decisive results as regards the explanation of the chemical properties of individual elements of the periodic system. In recognition of this circumstance, the work on these problems has, in the last few years, been limited essentially to attempts at accounting for the properties of the elements on the basis of formal assumptions about the stability of atomic structure, delaying questions about the cause of this stability. The best known of these attempts are due to Kossel and Langmuir.

The purpose of this lecture will be to attempt to show that the difficulties, met with in the application of the quantum theory to the problem of the chemical properties of the elements, are only apparent, and further that, with a consistent extension of the ideas underlying the application of the theory to spectra, it seems possible to formulate conceptions about atomic structure which offer a detailed interpretation of the periodic system. This involves partly a closer examination of the conceivable stationary states of atoms, through which it has been possible, for the purpose mentioned, to make fruitful ideas concerning the fixing of the electronic orbits in atoms which had already, more or less clearly, been proposed by Sommerfeld, Vegard, Landé, and others. It also involves an elucidation of a peculiar limitation of the possible transitions between conceivable stationary states of atoms. We seek to achieve this through a connection between the principal ideas of the quantum theory and those of the classical theory which, while emphasizing the essential difference between the two theories, nevertheless permits us to espy, behind the postulates of the quantum theory, an edifice which may be regarded as a natural generalization of the classical conceptions, and which assigns these to a limited boundary domain.

X. THE STRUCTURE OF THE ATOM

(1922)

TEXT AND TRANSLATION*

* [See Introduction, sect. 9.]

LES PRIX NOBEL

EN 1921—1922

OM ATOMERNES BYGNING

Nobelforedrag afholdet i Stockholm

den 11 December 1922

af

NIELS BOHR

STOCKHOLM 1923
IMPRIMERIE ROYALE. P. A. NORSTEDT & FILS
221969

OM ATOMERNES BYGNING.

Nobelforedrag afholdet i Stockholm den 11 December 1922

af

NIELS BOHR.

Højtærede Forsamling,

Naar jeg i Dag, som Følge af den store Hæder, som det Svenske Vetenskapsakademien har vist mig ved at tilkende mig Nobelprisen for Fysik for dette Aar for mine Arbejder over Atomernes Bygning, har den Forpligtelse her at gøre Rede for Resultaterne af disse Arbejder, har jeg ment, at jeg vilde handle i Overensstemmelse med Nobelstiftelsens Traditioner, dersom jeg formede denne Redegørelse som en Oversigt over den Udvikling, der i de senere Aar har fundet Sted indenfor det Omraade af Fysikken, hvorunder de omhandlede Arbejder falder.

Atomteoriens nuværende Standpunkt er karakteriseret derigennem, at vi ikke alene tør betragte Atomernes Eksistens som utvivlsomt godtgjort, men endogsaa tør antage, at vi besidder et indgaaende Kendskab til de enkelte Atomers Byggestene. Det vil ikke være muligt ved denne Lejlighed at give Dem et Overblik over den Udvikling af Videnskaben, der har ført til dette Resultat. Jeg skal blot minde om *Elektronernes* Opdagelse i Slutningen af det forrige Aarhundrede, der gav den direkte Bekræftelse og endelige Afklaring af de Forestillinger om Elektricitetens atomistiske Natur, som langsomt havde udviklet sig siden FARADAY's Opdagelser af de elektrolytiske Grundlove og BERZELIUS' elektrokemiske Teori, og som fandt deres smukkeste Triumf i ARRHENIUS' elektrolytiske Dissociationsteori. Opdagelsen af Elektronerne og Klarlæggelsen af deres Egenskaber var Resultatet af et stort Antal Forskeres Arbejder, blandt hvilke særlig LENARD og J. J. THOMSON kan nævnes. Navnlig den sidstnævnte Forsker har, gennem sine tankerige Forsøg paa at udvikle Forestillinger om Atomernes Bygning

1—221969. *Les prix Nobel en 1921—1922.*

paa Grundlag af Elektronteorien, ydet betydningsfulde Bidrag til Atomteoriens Udvikling. Den foreløbige Slutsten paa Udviklingen af vort Kendskab til Atomernes Byggestene naaedes imidlertid ved Opdagelsen af *Atomkernerne*, som skyldes RUTHERFORD, hvis Arbejder over de omkring Aarhundredskiftet opdagede radioaktive Stoffer paa saa mange Maader har beriget den fysiske og kemiske Videnskab.

Efter vore nuværende Forestillinger er et Atom af et Grundstof opbygget af en Kerne, der har en positiv elektrisk Ladning, og som er Sædet for langt den største Del af Atomets Masse, samt af et Antal Elektroner, der alle har samme negative Ladning og samme Masse, og som bevæger sig i Afstande fra Kernen, der er overordentlig store i Forhold till Kernens og Elektronernes egne Dimensioner. Dette Billede udviser ved første Øjekast en stor Lighed med et Planetsystem, saaledes som vi kender det fra vort Solsystem. Ligesom de simple Lovmæssigheder, der behersker Solsystemets Bevægelser, hænger nøje sammen med den Omstændighed, at de enkelte Himmellegemers Udstrækninger er smaa i Forhold til deres Baner, medfører det tilsvarende Forhold hos Atombygningen en umiddelbar Forstaaelse af et væsentligt Træk ved Naturfænomerne, forsaavidt disse afhænger af Grundstoffernes Egenskaber. Det tillader os nemlig at forstaa, at disse Egenskaber kan skilles i to skarpt adskilte Klasser. Til den første Klasse hører de fleste af Stoffernes sædvanlige fysiske og kemiske Egenskaber, som Tilstandsform, Farve og kemisk Reaktionstilbøjelighed. Disse Egenskaber afhænger af Elektronsystemets Bevægelse og den Maade, hvorpaa denne vil ændres under ydre Paavirkeninger. Paa Grund af Kernens store Masse i Forhold til Elektronernes og ringe Størrelse i Forhold til Elektronbanernes vil Elektronsystemets Bevægelse kun i ringe Grad afhænge af Kernens Masse og med stor Tilnærmelse være bestemt alene ved Kernens samlede elektriske Ladning. Navnlig vil Kernens indre Bygning og den Maade, hvorpaa Elektriciteten og Massen er fordelt paa Kernens enkelte Partikler, have en forsvindende ringe Indflydelse paa Beskaffenheden af det omgivende Elektronsystem. Kernens Bygning paa den anden Side vil være afgørende for den anden Klasse af Stoffernes Egenskaber, som kommer til Syne i Stoffernes Radioaktivitet. I de radioaktive Processer iagttager vi jo en Eksplosion af Kernen, hvorved positive eller negative Partikler, saakaldte α- og β-Partikler, udslynges fra disse med overordentlig store Hastigheder. Vore Forestillinger om Atombygningen giver os derfor en umiddelbar Forklaring af den fuldstændige Mangel paa Sammenhæng mellem de to Klasser af Grundstoffernes Egenskaber, der som bekendt faar sit mest slaaende Udtryk i Eksistensen af Stoffer, som med overordentlig stor Tilnærmelse besidder de samme sædvanlige fysiske og kemiske Egenskaber, selv om Atomvægten ikke er den samme, og de radioaktive Egenskaber er fuldstændig forskellige. Saadanne Stoffer, hvis Eksistens man først har faaet Vidnesbyrd om ved SODDYS og andre Forskeres Undersøgelser over de radioaktive Stoffers kemiske Egenskaber, betegnes med Henblik paa Grundstoffernes Klassifikation efter de sædvanlige fysiske og kemiske Egenskaber som *Isotoper*. Jeg behøver ikke her nærmere at omtale, hvorledes det i de senere Aar har vist sig, at Isotopi ikke alene optræder hos radioaktive Stoffer, men ogsaa hos Grundstoffer af den sædvanlige bestandige Natur, idet et større Antal af disse, der hidtil har været anset for usammensatte, efter ASTONS velkendte Undersøgelser har vist sig at bestaa af en Blanding af Isotoper med forskellige Atomvægte. Spørgsmaalet om Kernens indre Bygning, som ikke mindst disse Undersøgelser har givet stor Interesse, er endnu kun lidet opklaret, selv om en Vej til Udforskning heraf er aabnet ved RUTHERFORDS Undersøgelser over Spaltning af Atomkerner ved Bombardement med α-Straaler, ved hvilke en ny Epoke i Naturvidenskaben tør siges at være indledet, idet det for første Gang er lykkedes kunstigt at forvandle et Grundstof til et andet. I det følgende skal vi imidlertid indskrænke os til at betragte Stoffernes sædvanlige fysiske og kemiske Egenskaber og de Forsøg, der har været gjort paa at forklare disse paa Grundlag af de nævnte Forestillinger om Atombygningen.

Som vel kendt lader Grundstofferne sig, hvad deres fysiske og kemiske Egenskaber angaar, indordne i et saakaldt *naturligt System*, som lader et ejendommeligt Slægtskab mellem de forskellige Stoffer komme til Syne. Som først paapeget af MENDELEJEFF og LOTHAR MEYER, viser Grundstoffernes kemiske og fysiske Egenskaber en udtalt Periodicitet, naar Stofferne ordnes i en Rækkefølge, der i Hovedsagen falder sammen med Rækkefølgen af voksende Atomvægte. En Oversigt over Grundstoffernes naturlige eller periodiske System er givet i Fig. 1, hvor dog Stofferne ikke er ordnede paa den Maade, som benyttes ved sædvanlig Fremstilling af Systemet, men hvor Ordningen er en Modifikation af en Fremstilling, der først er angivet af den danske Kemiker JULIUS THOMSON, som ogsaa paa dette Omraade har gjort en betydelig Indsats i Videnskaben. Paa Figuren er Grundstofferne betegnet ved deres sædvanlige kemiske Tegn, og de forskellige lodrette Kolonner angiver de saakaldte Perioder. Stoffer i paa hinanden følgende Kolonner, der besidder homologe kemiske og fysiske Egenskaber, er forbundne ved Streger. De firkantede Parenteser om visse

Rækker af Grundstoffer i de senere Perioder, hvis Egenskaber udviser typiske Afvigelser fra den simple Periodicitet hos Stofferne i de første Perioder, har en Betydning, til hvilken vi i det følgende skal komme tilbage. Ved Udviklingen af vore Forestillinger om Atombygningen er den Lovmæssighed, som det periodiske System giver Udtryk for, kommet i en overraskende

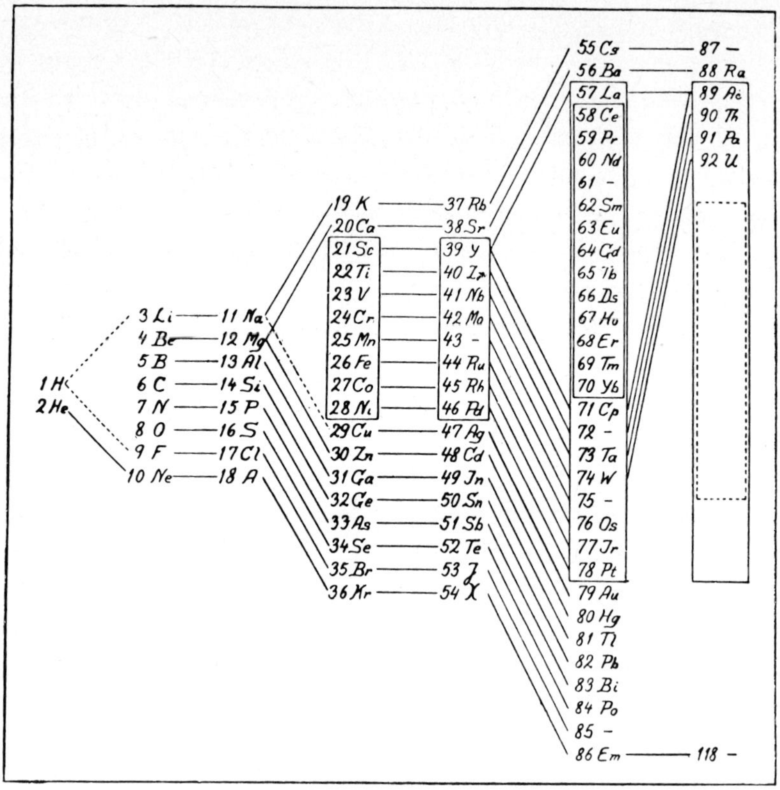

Fig. 1.

simpel Belysning. Saaledes er vi ført til at antage, at det Tal, det saakaldte Atomnummer, der i Fig. 1 er vedføjet de forskellige Grundstoffer, og som angiver vedkommende Stofs Plads i Systemet, netop er lig Antallet af Elektroner, der i det neutrale Atom bevæger sig omkring Kernen. Denne simple Lov blev, omend i ufuldkommen Form, først fremsat af VAN DEN BROEK, efter at den var lagt nær ved Bestemmelser af Antallet af Elektroner i Atomet efter Metoder, udviklede af J. J. THOMSON, samt ved RUTHERFORDS Undersøgelser, der tillod en direkte Maaling af Atomkernens Ladning. Som vi skal se, har den omhandlede Lov faaet overbevisende Støtte paa forskellig Maade, særlig gennem MOSELEYS Undersøgelser over Grundstoffernes Røntgenstraalespektre. Jeg kan maaske ogsaa her minde om, hvorledes den simple Sammenhæng mellem Atomnummer og Kerneladning fører til en umiddelbar Forstaaelse af de Love, som behersker de Ændringer i de radioaktive Stoffers kemiske Egenskaber, der ledsager Udsendelsen af α- eller β-Straaler, og som finder saa enkelt et Udtryk i den saakaldte *radioaktive Forskydningslov*.

Saa snart vi søger at opnaa en nøjere Forbindelse mellem Stoffernes Egenskaber og Atomernes Bygning, støder vi imidlertid paa dybtliggende Vanskeligheder, idet det viser sig, at der, til Trods for den før omtalte Analogi, bestaar en Væsensforskel mellem et Atom og et Planetsystem. Bevægelserne af Legemerne i et Planetsystem vil, selv om de adlyder den almindelige Tyngdelov, ikke være fuldkommen bestemte ved denne Lov alene, men vil i væsentlig Grad være afhængig af Systemets Forhistorie. Saaledes er Aarets Længde ikke bestemt alene ved Solens og Jordens Masser, men tillige ved de Forhold, der har været gældende ved Solsystemets Dannelse, og om hvilke vi ikke i det enkelte har Kendskab. Dersom der en Dag gennem vort Solsystem skulde bevæge sig et fremmed Himmellegeme, der kom Jorden nær, maatte vi endvidere være forberedt paa, at Aarets Længde fra den Dag af kunde blive væsentlig forskellig fra dets nuværende Længde. Ganske anderledes forholder det sig med Atomerne. Grundstoffernes bestemte uforanderlige Egenskaber fordrer nemlig, at et Atoms Tilstand ikke kan undergaa blivende Forandringer ved ydre Paavirkninger. Saa snart Atomet igen overlades til sig selv, maa Atomdelene ordne og bevæge sig paa en Maade, der er fuldkommen bestemt ved Delenes Antal og deres elektriske Ladninger og Masser. Det mest slaaende Vidnesbyrd herom har vi vel i Spektrene, d. v. s. i Beskaffenheden af den Straaling, der under Omstændigheder kan udsendes fra Stofferne, og som ved Hjælp af egnede Apparater kan undersøges med saa overordentlig stor Nøjagtighed. Som vel kendt, er Bølgelængderne for Linierne i Stoffernes Spektre, der i mange Tilfælde kan maales med en større Nøjagtighed end en Del i en Million, indenfor Maalenøjagtigheden under samme ydre Forhold altid de samme, ganske uafhængig af den Behandling Stofferne tidligere har været underkastet. Paa dette Forhold beror jo netop Spektralanalysen, der har været Kemikerne et saa uvurderligt Hjælpemiddel ved Eftersporingen af Grundstofferne, og som endda har belært os om, at der selv paa de fjerneste Himmellegemer befinder sig Stoffer med akkurat de samme Egenskaber som her paa Jorden.

Udfra vort Billede af Atombygningen er det saaledes ikke muligt, saa længe vi alene støtter os til de sædvanlige mekaniske Love, at gøre Rede for den karakteristiske *Stabilitet hos Atomerne*, som en Forklaring af Grundstoffernes Egenskaber kræver. Sagen stiller sig ingenlunde bedre, hvis vi i Betragtningerne inddrager de velkendte elektrodynamiske Love, som det paa Grundlag af de store Opdagelser i første Halvdel af forrige Aarhundrede af ØRSTED og FARADAY lykkedes MAXWELL at opstille. MAXWELLS Teori viste sig ikke alene i Stand til at gøre Rede for de allerede kendte elektriske og magnetiske Fænomener, men fejrede som bekendt sin store Triumf ved Forudsigelsen af de af HERTZ opdagede elektromagnetiske Bølger, der nu benyttes i saa stor Udstrækning i Radiotelegrafi. En Tid syntes det ogsaa, som om Teorien i dens navnlig af LORENTZ og LARMOR udarbejdede Tilpasning til den atomistiske Opfattelse af Elektriciteten var skikket til at danne Grundlag for en Forklaring af Stoffernes Egenskaber i Enkeltheder. Jeg behøver blot at minde om den store Opsigt, det vakte, da LORENTZ, straks efter ZEEMANS Opdagelse af den ejendommelige Forandring, som Spektrallinier undergaar, naar et lysende Stof anbringes i et magnetisk Felt, kunde give en utvungen og simpel Forklaring af Hovedtrækkene af dette Fænomen. LORENTZ antog, at den Straaling, som vi iagttager i en Spektrallinie, udsendes fra en Elektron, der udfører en harmonisk Svingning om en Ligevægtsstilling, paa ganske samme Maade som de elektromagnetiske Bølgesystemer i Radiotelegrafien udsendes paa Grund af de elektriske Svingninger i Antennen, og han paapegede, hvorledes den af ZEEMAN iagttagne Ændring af Spektrallinierne nøje svarede til de Ændringer i den svingende Elektrons Bevægelse, som man maatte vente frembragt af Magnetfeltet. Det skulde dog ikke vise sig muligt paa dette Grundlag at gennemføre en nærmere Forklaring af Grundstoffernes Spektre eller blot af den almindelige Type af de Love, der med stor Nøjagtighed gælder for Bølgebredderne af Linierne i disse Spektre, og som er blevet klarlagt igennem de bekendte Arbejder af BALMER, RYDBERG og RITZ. Efter at vi har faaet Oplysninger om Atomernes Bygning, er disse Vanskeligheder kommet endnu klarere for Dagen, idet vi, saalænge vi holder os til den klassiske elektrodynamiske Teori, ikke engang forstaar, at vi overhovedet faar Spektre bestaaende af skarpe Linier. Ja, denne Teori er overhovedet uforenelig med Antagelsen af Eksistensen af Atomer med den beskrevne Bygning, idet Elektronernes Bevægelse vilde kræve en stadig Udstraaling af Energi fra Atomet, der ikke vilde ophøre, før Elektronerne var faldne ind i Kernen.

En Udvej til at komme over de forskellige nævnte Vanskeligheder har man imidlertid fundet ved at inddrage Betragtninger hentet fra den saakaldte *Kvanteteori*, der betegner et fuldstændigt Brud med de Forestillinger, der hidtil har været benyttet ved Forsøg paa Forklaring af Naturfænomenerne. Den første Kim til denne Teori blev som bekendt lagt af PLANCK i Aaret 1900 gennem hans Undersøgelser over Varmestraalingsloven, der paa Grund af dens Uafhængighed af Stoffernes specielle Egenskaber var særlig egnet til Prøvesten for Anvendeligheden af den klassiske Fysiks Love paa Atomprocesser. PLANCK betragtede Straalingsligevægten mellem et Antal Systemer af samme Beskaffenhed som det, hvorpaa LORENTZ havde baseret sin Teori for Zeemaneffekten, men kunde nu vise, ikke alene at den klassiske elektrodynamiske Teori ikke tillod at gøre Rede for Varmestraalingsfænomenerne, men at en fuldkommen Overensstemmelse med Varmestraalingsloven lod sig opnaa, hvis man — i bestemteste Modstrid til den klassiske Teori — antog, at den svingende Elektrons Energi ikke kunde forandre sig kontinuert, men kun paa en saadan Maade, at Systemets Energi altid forblev lig med et helt Antal saakaldte »Energikvanter«. Størrelsen af et saadant Kvantum fandtes at være proportional med den svingende Partikels Svingningstal, som i Tilknytning til de klassiske Forestillinger antoges at være lig Svingningstallet for den udsendte Straaling. Proportionalitetsfaktoren, den saakaldte PLANCK'ske Konstant, maatte efter Betragtningernes Karakter anses for en ny universel Naturkonstant i Lighed med Lyshastigheden og Elektronens Ladning og Masse.

PLANCKS overraskende Resultat stod til at begynde med fuldstændig isoleret i Naturvidenskaben men fandt faa Aar efter ved EINSTEINS betydningsfulde Indgriben paa dette Omraade en flersidig Anvendelse. For det første gjorde EINSTEIN opmærksom paa, at Fordringen om Begrænsningen af Energiværdierne svarende til Partiklens Svingninger kunde prøves ved Undersøgelser over krystallinske Legemers Varmefylde, idet man ved disse netop har at gøre med lignende Svingninger vel ikke af den enkelte Elektron, men af hele Atomet omkring Ligevægtsstillinger i Krystalgitteret. Den Overensstemmelse med PLANCKS Teori, som EINSTEIN her straks kunde paavise, er som velkendt ved senere Forskeres Arbejde blevet uddybet paa meget betydningsfuld Maade. Endvidere fremhævede EINSTEIN en anden Konsekvens af PLANCKS Resultat, nemlig at Straalingsenergi kun kunde emitteres eller absorberes af den svingende Partikel i saakaldte »Straalingskvanter», hvis Størrelse var lig med Produktet af PLANCKS Konstant og Svingningstallet. I sine Bestræbelser for at give dette Resultat en anskuelig Betydning, førtes EINSTEIN til Opstillingen af den saakaldte »Lyskvantehypothese», efter hvilken Straalingsenergi, i Modsætning til MAXWELLS elektromagnetiske Lysteori, ikke skulde forplante sig som elektromagnetiske Bølger, men som Lysatomer af ringe Størrelse, der hver især netop indeholdt en Energimængde svarende til et Straalingskvantum. Denne Forestilling ledte EINSTEIN til den velkendte Teori for den fotoelektriske Effekt, der kastede helt nyt Lys over dette efter den klassiske Teori fuldstændig uforstaaelige Fænomen, og hvis Forudsigelser i de seneste Aar har faaet saa nøje en experimentel Bekræftelse, at vi i Maalinger over den fotoelektriske Effekt maaske besidder det nøjagtigste Middel til Bestemmelsen af PLANCKS Konstant. Til Trods for dens Værdi som heuristisk Hjælpemiddel er dog Lyskvantehypothesen, der staar fuldkommen fremmed overfor de saakaldte Interferensfænomener, ikke egnet til at bringe en Afklaring af Spørgsmaalet om Straalingens Natur. Vi behøver jo blot at erindre om, at Interferensfænomenerne udgør vort eneste Middel til at undersøge Straalingens Beskaffenhed og til at tillægge det Svingningstal, der er bestemmende for Lyskvanternes Størrelse, nogen nærmere Betydning.

I de paafølgende Aar blev der fra forskellig Side gjort Bestræbelser for at anvende de kvanteteoretiske Synspunkter paa Spørgsmaalet om Atombygningen, idet Hovedvægten snart lagdes paa den ene, snart paa den anden af de af EINSTEIN fremdragne Konsekvenser af PLANCKS Resultat. Som de bekendteste Forsøg i denne Retning, ved hvilke der dog ikke naaedes afklarede Resultater, kan jeg nævne Arbejder af STARK, SOMMERFELD, HASENOHRL; HAAS og NICHOLSON. Fra dette Tidspunkt stammer ogsaa et Arbejde af den danske Kemiker BJERRUM, der, selv om det ikke tager direkte Sigte paa Atombygningen, har været af Betydning for Kvanteteoriens Udvikling. BJERRUM gjorde (1912) opmærksom paa, at den fra Varmebevægelsen hidrørende Rotation af Molekylerne i en Luftart maatte kunne efterspores i Ændringer af visse Absorptionslinier med Temperaturen. Samtidig fremhævede han, at Virkningen ikke skulde bestaa i en kontinuert Udbredning af Linierne, saaledes som man maatte vente efter den klassiske Teori, der ikke paalægger Molekylernes Rotationsbevægelser nogen Indskrænkning, men i Tilknytning til Kvanteteorien forudsagde han, at Linierne maatte være opspaltede i en Række Komponenter, svarende til en Række adskilte Rotationsmuligheder af Molekylerne. Denne Forudsigelse blev faa Aar efter paa smukkeste Maade bekræftet ved Forsøg af den svenske Fysiker

EVA VON BAHR, og Fænomenet maa stadig betragtes som et af de mest slaaende Vidnesbyrd om Kvantetoriens Realitet, selv om fra vort nuværende Standpunkt dets oprindelige Tydning har undergaaet en Modifikation med Hensyn til væsentlige Enkeltheder.

* * *

Spørgsmaalet om Kvanteteoriens nærmere Udformning var imidlertid blevet stillet i et nyt Lys ved RUTHERFORDS Opdagelse af Atomkernen (1911). Som vi allerede har set, gjorde denne Opdagelse det nemlig klart, at de klassiske Forestillinger ikke tilbød noget Grundlag for Forstaaelsen af Atomernes allerværsentligste Egenskaber. Man førtes derfor ind paa at søge en Formulering af Kvanteteoriens Principper, der var egnet til umiddelbart at imødekomme de Fordringer til Stabilitet hos Atombygningen og Beskaffenheden af den fra Atomerne udsendte Straaling, som Stoffernes iagttagne Egenskaber kræver. En saadan Formulering foresloges i 1913 af Foredragsholderen ved Opstilling af to *Postulater*, hvis Indhold kan gengives saaledes:

I. Blandt de tænkelige Bevægelsestilstande i et Atomsystem findes der et Antal saakaldte *stationære Tilstande*, der til Trods for at Partiklernes Bevægelse i disse Tilstande i betydeligt Omfang antages at adlyde de klassiske mekaniske Love, udmærker sig ved en ejendommelig, mekanisk uforklarlig, Stabilitet, der medfører, at enhver blivende Forandring i Systemets Bevægelse maa bestaa i en fuldstændig Overgang fra en stationær Tilstand til en anden stationær Tilstand.

II. Medens der i Modstrid til den klassiske elektromagnetiske Teori ingen Udstraaling finder Sted i de stationære Tilstande selv, vil en *Overgangsproces* mellem to stationære Tilstande kunne ledsages af en Udsendelse af elektromagnetisk Straaling, der vil have samme Beskaffenhed som den, der efter den klassiske Teori vilde udsendes fra en elektrisk Partikel, der udfører en harmonisk Svingning med konstant Svingningstal. Dette Svingningstal v har imidlertid ingen simpel Tilknytning til Bevægelsen af Atomets Partikler, men er givet ved Betingelsen:

$$hv = E' - E'',$$

hvor h er PLANCK'S Konstant og E' og E'' er Værdierne for Atomets Energi i de to stationære Tilstande, der danner Begyndelses- og Slutningstilstand for Straalingsprocessen. Omvendt kan en Bestraaling af

Atomet med elektromagnetiske Bølger af dette Svingningstal give Anledning til en Absorptionsproces, hvorved Atomet føres tilbage fra den sidstnævnte Tilstand til den første Tilstand.

Medens det første Postulat tager Sigte paa Atomernes almindelige Stabilitetsforhold, saaledes som det ytrer sig i Stoffernes kemiske og fysiske Egenskaber, tager det andet Postulat i første Række Sigte paa Eksistensen af Spektre bestaaende af skarpe Linier. Tillige tillbød den i det sidste Postulat indgaaende kvanteteoretiske Betingelse et Udgangspunkt for en Tydning af de i det foregaaende omtalte empiriske Love for Stoffernes Spektre. Den almindeligste af disse Love, det af RITZ opstillede *Kombinationsprincip*, udsiger, at Svingningstallet v for enhver af Linierne i et Stofs Spektrum kan fremstilles ved Formeln

$$\nu = T'' - T',$$

hvor T'' og T' er to saakaldte »Spektralled», tilhørende en Mangfoldighed af saadanne Led karakteristiske for det paagældende Stof.

Efter vore Postulater fortolkes denne Lov umiddelbart ved at antage, at Spektret udsendes ved Overgange mellem et Antal stationære Tilstande, i hvilke den numeriske Værdi af Atomets Energi er lig Værdien for Spektralleddet multipliceret med PLANCK'S Konstant. Denne Tydning af Kombinationsprincippet afviger fra vore sædvanlige elektrodynamiske Forestillinger, ikke alene derigennem at vi antager, at der ikke er nogen simpel Forbindelse mellem Atomets Bevægelse og den udsendte Straaling, men Betragtningernes Afvigelse fra det Grundlag, hvorpaa vor sædvanlige Naturbeskrivelse hviler, kommer maaske klarest frem, naar vi tænker paa, at Fremkomsten af Spektrallinier svarende til Kombinationer af et vist Spektralled med forskellige andre tydes derved, at Beskaffenheden af den Straaling, der udsendes af Atomet, ikke er afhængig alene af Atomets Tilstand ved Straalingsprocessens Begyndelse, men ogsaa af den Tilstand til hvilken Atomet under Processen overføres. I første Øjeblik kunde man maaske vente, at den omhandlede formelle Tydning af Kombinationsprincippet derfor næppe vilde kunne bringes i Forbindelse med Oplysningerne om Atomernes Byggesten, der jo er baseret paa Erfaringer tydet ved Hjælp af de klassiske mekaniske og elektrodynamiske Love. En nærmere Undersøgelse skulde dog aabenbare, at en Forbindelse af nøje Art mellem de forskellige Stoffers Spektre og Atomernes Bygning lod sig opnaa paa Grundlag af Postulaterne.

Af alle Spektre, vi kender, udviser *Brintspektret* de simpleste Forhold; Svingningstallene for dette Spektrums Linier lader sig som bekendt med stor Nøjagtighed fremstille gennem BALMERS Formel:

$$\nu = K\left(\frac{1}{n''^2} - \frac{1}{n'^2}\right),$$

hvor K er en Konstant og n' og n'' er to hele Tal. I Spektret møder vi altsaa en enkelt Række Spektralled af Form K/n^2, der aftager regelmæssigt med stigende Ledmummer n. I Overensstemmelse med Postulaterne skal

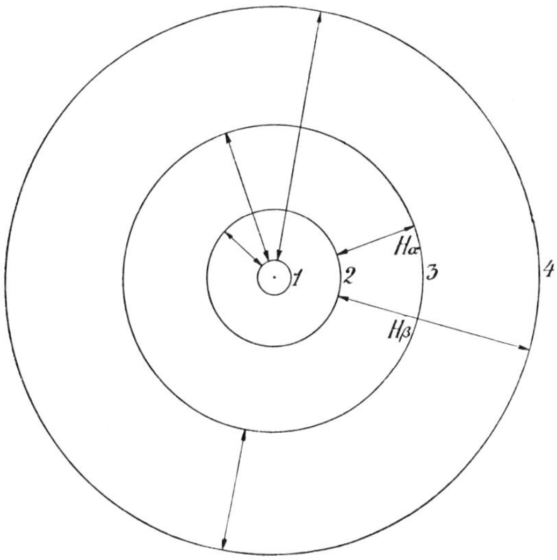

Fig. 2.

vi derfor tænke os, at hver af Brintspektrets Linier udsendes ved en Overgangsproces imellem to blandt en Række af stationære Tilstande af Brintatomet, i hvilke den numeriske Værdi for Atomets Energi er lig med hK/n^2. Ifølge vort Billede af Atombygningen bestaar nu et Brintatom af en positiv Kerne og en Elektron, der — dersom de sædvanlige mekaniske Forestillinger kan anvendes med stor Tilnærmelse vil beskrive en periodisk elliptisk Bane med Kernen i det ene Brændpunkt. Som en enkel Regning viser, er Banens Storakse proportional med det Arbejde, der skal udføres for at fjerne Elektronen fuldstændig fra Kernen, og i Tilknytning til det ovenstaaende skal vi nu antage, at dette Arbejde i de stationære Tilstande netop er lig hK/n^2. Vi kommer saaledes til en Mangfoldighed af stationære Tilstande, for hvilke Elektronbanens Storakse antager en Række diskrete Værdier proportionale med Kvadraterne af de hele Tal. Nedenstaaende Figur 2 illustrerer disse Forhold paa skematisk Maade. For Enkelheds Skyld er Elektronbanerne i de stationære Tilstande fremstillet som Cirkler, medens Teorien i Virkeligheden ikke paalægger Banens Ekscentricitet nogen Indskrænkning, men kun bestemmer Storaksens Længde. Pilene symboliserer de Overgangsprocesser, der svarer til den røde og grønne Brintlinie, hvis Svingningstal er givet ved Balmerformlen, naar vi for n'' sætter 2 og for n' henholdsvis 3 og 4. Endvidere er angivet de Overgangsprocesser, der svarer til de tre første Linier i den af LYMAN i 1914 fundne Række af ultraviolette Linier, hvis Svingningstal er givet ved Formlen, naar man sætter $n'' = 1$, samt den første Linie i den ultrarøde Linierække, der var opdaget nogle Aar i Forvejen af PASCHEN, og som er givet ved Formlen, naar man sætter $n'' = 3$.

Den omhandlede Tydning af Brintspektrets Fremkomst fører paa naturlig Maade til at opfatte dette Spektrum som et Vidnesbyrd om en Proces, hvorved Elektronen »bindes» af Kernen. Medens det største Spektralled med Ledinummer 1 svarer til Bindningsprocessens Slutningsstadium, svarer de smaa Spektralled, der tilhører store Værdier af Lednummeret, til stationære Tilstande, der betegner Begyndelsesstadier af Bindningsprocessen, hvor Elektronens Bane endnu har store Dimensioner, og hvor det Arbejde, der skal til for at trække Elektronen bort fra Kernen, endnu er lille. Slutningsstadiet af Bindingsprocessen kan vi betegne som Atomets »Normaltilstand», og den udmærker sig fremfor de andre stationære Tilstande ved den Egenskab, at Atomets Tilstand efter Postulaterne kun kan ændres ved en Tilførsel af Energi, hvorved Elektronen overføres til en Bane med større Dimensioner svarende til et tidligere Stadium af Bindingsprocessen.

Størrelsen af Elektronbanen i Normaltilstanden beregnet paa Grundlag af den angivne Tydning af Spektret stemmer omtrentlig overens med de Værdier for Størrelsen af Grundstoffernes Atomer, som man ved Hjælp af den kinetiske Luftteori havde beregnet ud fra Luftarternes Egenskaber. Idet vi dog imidlertid som en umiddelbar Følge af den ved Postulaterne krævede Stabilitet af de stationære Tilstande maa antage, at Vekselvirkningen mellem to Atomer under et Sammenstød ikke lader sig fuldstændig beskrive ved Hjælp af de klassiske mekaniske Love, kan en Sammenligning som denne dog ikke forfølges nærmere paa det hidtil beskrevne Grundlag.

En mere indgaaende Tilknytning af Spektret til Atommodellen opnaaedes imidlertid ved at undersøge Bevægelsen i de stationære Tilstande, hvor Lednumrene er store, og hvor Størrelsen af Elektronens Bane samt dens Omløbstal kun ændres forholdsvis lidt, naar vi gaar fra en stationær Tilstand til den næstfølgende. Det kunde nemlig paavises, at Svingningstallet for den Straaling, der udsendes ved Overgange mellem Tilstande, hvis Lednumres Forskel er lille i Forhold til disse Numres egen Størrelse, falder meget nær sammen med Svingningstallet for en af de harmoniske Svingningskomponenter, hvori Elektronbevægelsen kan opløses, og altsaa ogsaa med Svingningstallet for et af Bølgesystemerne i den Straaling, som ifølge de klassiske elektrodynamiske Love vilde udsendes paa Grund af Elektronens Bevægelse. Fordringen, at et saadant Sammenfald finder Sted i den omhandlede Grænse, hvor de stationære Tilstande relativt kun afviger lidt fra hverandre, er ensbetydende med, at Konstanten i Balmerformlen kan udtrykkes ved Relationen

$$K = \frac{2\pi^2 e^4 m}{h^3},$$

hvor e og m betyder henholdsvis Elektronens Ladning og Masse, medens h er PLANCK'S Konstant. Denne Relation har vist sig at være opfyldt indenfor den betydelige Nøjagtighed, hvormed navnlig efter MILLIKANS Undersøgelser Værdierne for e, m og h er kendte.

Dette Resultat betød ikke alene en Paavisning af en Forbindelse mellem Brintspektret og Modellen for Brintatomet, der var saa nøje, som vi overhovedet efter Postulaternes Afvigelse fra de klassiske mekaniske og elektrodynamiske Love kunde haabe, men afgav tillige et Fingerpeg om, hvorledes Kvanteteorien til Trods for denne Afvigelses indgribende Karakter tør opfattes som en naturlig Omformning af den klassiske elektrodynamiske Teoris Grundbegreber. Til dette betydningsfulde Spørgsmaal skal vi i det følgende komme tilbage. Først skal vi omtale, hvorledes Fortolkningen af Brintspektret paa Grundlag af Postulaterne viste sig egnet til paa forskellig Maade at belyse Slægtskabet mellem de forskellige Stoffers Egenskaber.

De i det ovenstaaende omtalte Betragtninger tillader en umiddelbar Anvendelse paa den Proces, hvorved en Elektron bindes af en Kerne med vilkaarlig given Ladning. Beregningens Udførelse fører til den Slutning, at i en stationær Tilstand svarende til en given Værdi af Tallet n, vil Banens Storakse være omvendt proportional med Kerneladningen, medens det Arbejde, som kræves til Fjernelsen af Elektronen fra Kernen, vil være direkte proportionalt med Kerneladningens Kvadrat. Det Spektrum, der udsendes under en Elektrons Binding af en Kerne, hvis Ladning er N Gange større end Brintkernens, kan derfor fremstilles ved Formlen

$$\nu = N^2 K\left(\frac{1}{n''^2} - \frac{1}{n'^2}\right)$$

Dersom i denne Formel N sættes lig 2, faar vi et Spektrum, der indeholder en Linierække i det synlige Spektralomraade, som for adskillige Aar tilbage var observeret i visse Stjernespektre, og som af RYDBERG paa Grund af den nøje Analogi med de ved Balmerformlen fremstillede Linierækker var tilskrevet Brint. Det var aldrig lykkedes at frembringe Linierne i ren Brint, men umiddelbart før Opstillingen af Teorien for Brintspektret var det lykkedes FOWLER at iagttage den omhandlede Linierække ved at sende stærke Udladninger gennem en Blanding af Brint og Helium. Ogsaa denne Forsker antog imidlertid Linierne for Brintlinier, idet man efter de hidtidige Erfaringer ikke var forberedt paa, at to forskellige Stoffer kunde vise Egenskaber, der besad saa nøje en Lighed som den, der udvistes af det omhandlede Spektrum og Brintspektret. Efter Teorien blev det imidlertid klart, at de iagttagne Linier maatte tilhøre et *Heliumspektrum*, der blot ikke saaledes som det sædvanlige Heliumspektrum udsendes fra det neutrale Atom men fra et ioniseret Heliumatom, der jo bestaar af en enkelt Elektron, der bevæger sig omkring en Kerne med dobbelt Ladning. Derved var der bragt et nyt Træk i Slægtskabet mellem Grundstoffernes Egenskaber for Dagen af en Art, som nøje svarer til vore nuværende Forestillinger om Atombygningen, hvorefter et Stofs fysiske og kemiske Egenskaber i første Række er betinget alene af Atomkernens elektriske Ladning.

Kort efter dette Spørgsmaals Opklaring blev Tilstedeværelsen af et almindeligt Slægtskab af lignende Art mellem Stoffernes Egenskaber bragt for Dagen ved MOSELEYS velkendte Undersøgelser over Grundstoffernes karakteristiske *Røntgenspektre*, der var gjort mulig ved LAUES Opdagelse af Røntgenstraalernes Interferens i Krystaller og de derpaa grundede Undersøgelser af W. H. og W. L. BRAGG. Det viste sig nemlig, at Røntgenspektrene for de forskellige Stoffer udviste en langt simplere Bygning og langt større indbyrdes Lighed end Stoffernes optiske Spektre, og navnlig viste det sig, at Spektrene ændrede sig fra Stof til Stof paa en Maade, der nøje svarede til den ovenstaaende Formel for det Spektrum, der udsendes ved en Elektrons Binding af en Kerne, naar i denne Formel N sættes lig vedkommende Stofs Atomnummer. Ja, denne Formel viste sig endda i Stand til med ikke ringe Tilnærmelse at gengive Svingningstallene for de stærkeste Røntgenlinier, naar man for n' og n'' indsatte smaa hele Tal.

Denne Opdagelse var i flere Henseender af stor Betydning. For det første var Slægtskabet mellem Røntgenspektrene af forskellige Stoffer saa enkelt, at det blev muligt at fastlægge Atomnummeret entydigt for alle kendte Stoffer, og derigennem med Sikkerhed at forudsige Atomnumrene for alle saadanne hidtil ukendte Stoffer, som der tilkommer en Plads i det naturlige System. For to karakteristiske Røntgenlinier, der hører til den saakaldte

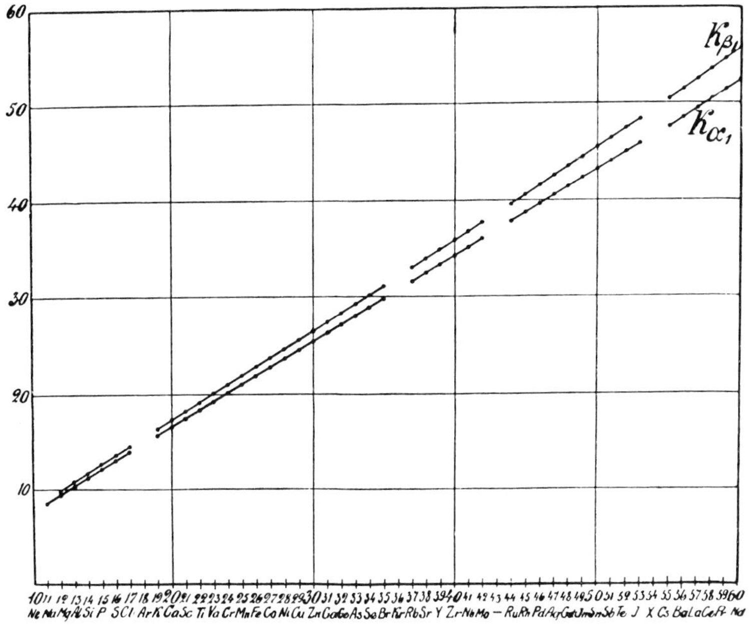

Fig. 3.

K-Gruppe, der besidder den største Gennemtrængningsevne, er paa Fig. 3 Kvadratroden af Forholdet mellem Svingningstallet og Konstanten i Balmerformlen angivet som Funktion af Atomnummeret. Vi ser, hvorledes Punkterne med meget stor Tilnærmelse ligger paa rette Linier, og hvordan dette er betinget af, at man ikke alene tager Hensyn til de allerede kendte Stoffer i det betragtede Omraade af Grundstoffernes System, men tillige lader en Plads staa aaben imellem Molybdæn (42) og Rutenium (44), saaledes som det allerede var angivet i MENDELEJEFFS oprindelige Fremstilling af det naturlige System. Endvidere afgav Lovene for Røntgenspektrene en Bekræftelse paa de almindelige teoretiske Forestillinger, saavel hvad Atombygningens Grundtræk angik, som med Henblik paa de Synspunkter, som var lagt til Grund for Fortolkningen af Spektrene. Ligheden mellem Røntgenspektrene og det Spektrum, som udsendes ved en enkelt Elektrons

Binding af en Atomkerne, beror nemlig simpelthen derpaa, at det ved Røntgenspektrene drejer sig om Overgange mellem stationære Tilstande, ledsaget af Ændringer af Bevægelsen af en Elektron i det indre Omraade af Atomet, hvor Indflydelsen af Kernens Tiltrækning er overvejende sammenlignet med de frastødende Kræfter fra de andre Elektroner.

Slægtskabet mellem Stoffernes andre Egenskaber er ofte af en langt mere indviklet Karakter, hvad der hidrører fra, at det her drejer sig om Processer, der angaar Bevægelsen af Elektroner i Atomets ydre Dele, hvor

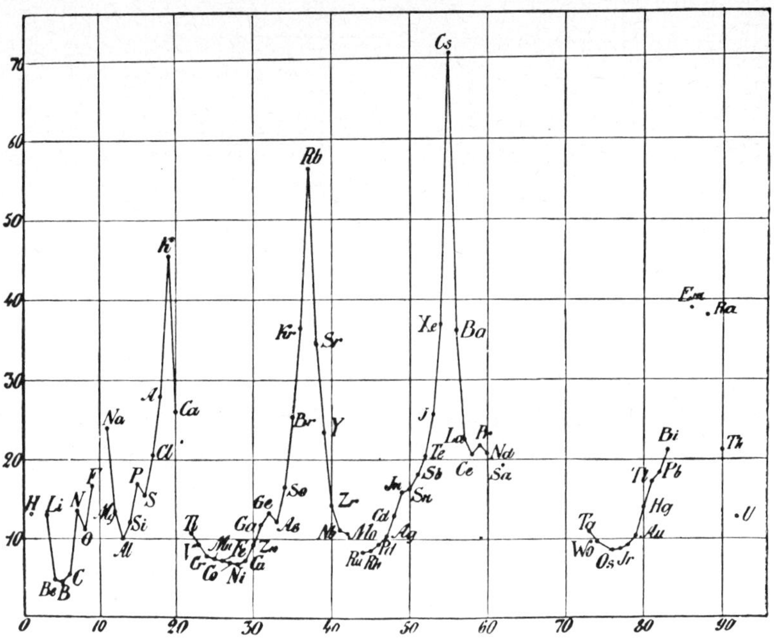

Fig. 4.

de Kræfter, som Elektronerne udøver paa hverandre, er af samme Størrelsesorden som Kernens Tiltrækning, og hvor derfor det nøjere Vekselspil af Elektronerne i Atomet kommer til at spille en afgørende Rolle.

Et karakteristisk Eksempel paa saadanne Forhold gives ved Stoffernes Rumopfyldning. Som bekendt gjorde allerede LOTHAR MEYER opmærksom paa den ejendommelige periodiske Variation som Forholdet mellem Atomvægt og Vægtfylde, det saakaldte Atomvolumen, udviser indenfor det naturlige System. Et Indtryk af denne Variation giver Figur 4, hvor Atomvolumenet er fremstillet som Funktion af Atomnummeret. En større Modsætning mellem denne og den forrige Figur kan næppe tænkes. Medens Røntgenspektrene varierede jævnt med Atomnummeret, viser Atomvolumenerne en udpræget periodisk Ændring, der nøje svarer til den Ændring i Stoffernes kemiske Egenskaber, som kommer til Udtryk i Grundstoffernes naturlige System.

Lignende Forhold genspejler sig i Stoffernes sædvanlige optiske Spektre. Til Trods for de store indbyrdes Forskelle, som disse Spektre udviser, var det dog allerede lykkedes RYDBERG at efterspore et vist almindeligt Slægtskab mellem Brintspektret og de andre Stoffers Spektre. Selv om Spektrallinierne i disse Spektre optræder som Kombinationer af en mere indviklet Mangfoldighed af Spektralled, der ikke simpelt kan tilordnes Rækken af hele Tal, kan dog Spektralleddene ordnes i Rækker, der hver især udviser en stor Lighed med Ledrækken i Brintspektret. Denne Lighed viser sig ved, at det empiriske Udtryk for Leddene i enhver Række med stor Nøjagtighed kan skrives paa Formen $K/(n+\alpha_k)^2$, hvor K er den samme Konstant som den, der optræder i Brintspektret, og som ofte kaldes RYDBERGS Konstant, medens n er Lednummeret og α_k er en Konstant forskellig for de forskellige Rækker. Dette Slægtskabsforhold med Brintspektret fører os umiddelbart til at opfatte de omhandlede Spektre som *det sidste Afsnit af en Proces, hvorved det neutrale Atom dannes ved, at Kernen indfanger og binder Elektronerne en for en.* Det vil nemlig være klart, at den sidst indfangne Elektron under Begyndelsesstadierne af Bindingsprocessen, i hvilke dens Bane endnu er stor i Forhold til de tidligere bundne Elektroners Baner, vil være underkastet Kræfter fra Kernens og disse Elektroners Side, der kun afviger lidt fra de Kræfter, hvormed Elektronen i Brintatomet tiltrækkes af Kernen under dens Bevægelse i Baner af tilsvarende Dimensioner.

Medens de her omhandlede Spektre, for hvilke de RYDBERG'ske Love gælder, udsendes af Stofferne ved elektriske Udladninger under sædvanlige Udladningsbetingelser og ofte benævnes *Buespektre*, udsender Stofferne, naar de underkastes særlig stærke Udladninger, andre Spektre, de saakaldte *Gnistspektre*, hvis Lovmæssigheder det hidtil ikke var lykkedes at udrede.

Kort efter at ovenstaaende Tydning af Buespektrenes Fremkomst var bleven fremsat, fandt FOWLER (1914) imidlertid, at man kunde opstille empiriske Udtryk for Gnistspektrenes Linier, der ganske svarede til de RYDBERG'ske Love, naar undtages den Forskel, at Konstanten K maatte erstattes med en Konstant 4 Gange saa stor. Idet, som vi har set, den Konstant, der optræder i det Spektrum, der udsendes under en Elektrons Binding af en Heliumkerne, netop er lig med 4 K, blev det derfor klart, at Gnistspektre skyldes ioniserede Atomer, og at deres Udsendelse svarede til det *næstsidste Afsnit i det neutrale Atoms Dannelse* ved Elektronernes sukcessive Indfangning og Binding.

Den skildrede Opfattelse af Spektrenes Oprindelse viste sig ogsaa egnet til at bringe en Forklaring af de ejendommelige Love, der behersker Stoffernes *Absorptionsspektre*. Som allerede paavist af KIRCHHOFF og BUNSEN er der en nøje Relation imellem Stoffernes selektive Absorption for Straaling og deres Emissionsspektre, et Forhold paa hvilket Spektralanalysens Anvendelse paa Himmellegemerne er væsentlig begrundet. Det var imidlertid ud fra den klassiske elektromagnetiske Teori uforstaaeligt, hvorfor Stoffer i Dampform viser Absorption for visse af Linierne i Emissionsspektret og ikke for andre. Paa Grundlag af Postulaterne føres vi imidlertid til at antage, at Absorption af Straaling svarende til en Spektrallinie, der udsendes under en Overgang fra en af Atomets stationære Tilstande til en Tilstand med mindre Energi, er betinget af Tilbageførelsen af Atomet under Energioptagelse fra den sidstnævnte Tilstand til den første. Vi forstaar derfor umiddelbart, at en Damp eller Luftart under sædvanlige Omstændigheder kun viser selektiv Absorption for Spektrallinier, der fremkommer ved en Overgang til Normaltilstanden fra en Tilstand svarende til et tidligere Stadium af Bindingsprocessen. Først ved højere Temperaturer eller under Indflydelsen af elektriske Udladninger, hvorved et betydeligt Antal Atomer til Stadighed vil være bragt ud af Normaltilstanden, kan vi i Overensstemmelse med Erfaringen vente Absorption for andre Linier i Emissionsspektret.

En Støtte af den mest direkte Art for vor Tolkning af Spektrene paa Grundlag af Postulaterne har man endvidere faaet ved Forsøg over Frembringelse af Spektrallinier og Ionisation af Atomer ved Sammenstød med fri Elektroner med givne Hastigheder. Det afgørende Fremskridt paa dette Omraade indledtes med de velkendte Undersøgelser af FRANCK og HERTZ (1914). Det viste sig ved disse, at man ved Stød af Elektroner ikke kunde tilføre et Atom en vilkaarlig Energimængde, men kun en saadan, der svarede til en Overførelse fra Atomets Normaltilstand til en anden af de stationære Tilstande, om hvis Eksistens Spektrene belærer os, og hvis Energiindhold bestemmes ud fra Spektralleddenes Størrelse. Endvidere fik man et slaaende Vidnesbyrd om den Uafhængighed, der efter Postulaterne tilkommer de Processer, der giver Anledning til Udsendelsen af Spektrets forskellige Linier, idet det direkte kunde paavises, at Atomer, som saaledes var bragt i en stationær Tilstand med større Energi, var i Stand til at vende tilbage til Normaltilstanden under Udsendelse af en Straaling svarende til en enkelt Spektrallinie. Fortsatte Undersøgelser over Elektronstød, hvori et stort

Antal Fysikere har taget Del, har bragt en i Enkeltheder gaaende Bekræftelse paa de nærmere Antagelser om Seriespektrenes Fremkomst. Navnlig har det kunnet vises, at der for *Ionisation* af Atomer ved Elektronstød kræves en Energi, der netop svarer til det Arbejde, der efter Teorien er nødvendigt for at fjerne den sidst indfangne Elektron fra Atomet, et Arbejde, der umiddelbart bestemmes ved Produktet af PLANCK'S Konstant og det til Normaltilstanden svarende Spektralled, der efter det ovenstaaende netop er Grænseværdien for Svingningstallene af de ved selektiv Absorption forbundne Spektralserier.

* * *

Medens det i umiddelbar Tilknytning til Kvanteteoriens Grundpostulater var muligt at gøre Rede for visse almindelige Træk af Grundstoffernes Egenskaber, krævedes en nøjere Udformning af Teorien for at opnaa en mere indgaaende Fortolkning af disse Egenskaber. Et bredere teoretisk Grundlag er i Løbet af de senere Aar blevet skabt gennem Udviklingen af formelle Metoder, der tillader at fastlægge de stationære Tilstande for Elektronbevægelser af en mere almindelig Type end dem, vi hidtil har betragtet. For en ren periodisk Bevægelse, saaledes som vi møder den ved en simpel harmonisk Oscillator, og i det mindste i første Tilnærmelse ved Bevægelsen af en Elektron omkring en positiv Kerne, vil Mangfoldigheden af stationære Tilstande simpelt kunne tilordnes Rækken af hele Tal. For Bevægelser af den nævnte mere almindelige Type, de saakaldte flerfold periodiske Bevægelser, vil imidlertid de stationære Tilstande udgøre en mere sammensat Mangfoldighed, i hvilken ved de omhandlede formelle Metoder hver Tilstand er karakteriseret ved flere hele Tal, de saakaldte »Kvantetal». I Udviklingen af Teorien har et stort Antal Forskere taget Del, og Indførelsen af flere Kvantetal kan føres tilbage til Arbejder af PLANCK selv. Stødet til det afgørende Fremskridt indenfor Atomforskningen blev dog givet ved den af SOMMERFELD (1915) fremsatte Forklaring af den *Finstruktur*, som Brintlinierne udviser, naar Brintens Spektrum iagttages ved Hjælp af Spektroskoper med stor Opløsningsevne. Denne Finstruktur hidrører derfra, at Bevægelsen af Elektronen i Brintatomet ikke er nøjagtig periodisk. Som Følge af den af Relativitetsteorien krævede Ændring af Elektronens Masse med dens Hastighed vil nemlig Elektronbanen undergaa en ganske langsom Præcession i Baneplanen. Atomets Bevægelse er derfor dobbelt periodisk, og foruden det Tal, som karakteriserer Leddene i Balmerformlen, og som vi betegner som »Hovedkvantetallet», idet det i første Række bestemmer Atomets Energi, vil Fastlæggelsen af de stationære Tilstande kræve endnu et Kvantetal, som vi betegner som »Bikvantetallet».

En Oversigt over Bevægelsen i de saaledes fastlagte stationære Tilstande er givet paa Figur 5, der angiver relativ Størrelse og Form for Elektronens Baner. Enhver Bane er betegnet med Symbolet n_k hvor

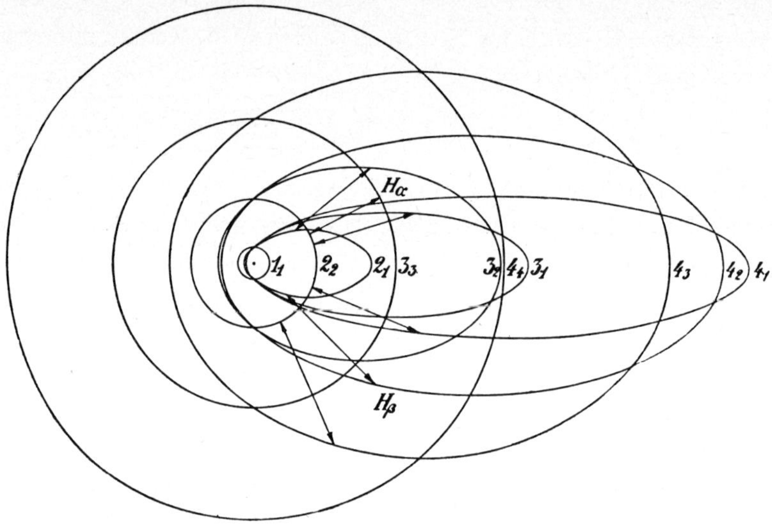

Fig. 5.

n er Hovedkvantetallet, medens k er det nævnte Bikvantetal. Alle Baner med samme Værdi for Hovedkvantetallet har i første Tilnærmelse samme store Akse, medens Baner med samme Værdi for k besidder den samme Længde for Parameteren, d. v. s. den mindste Korde gennem Brændpunktet. Idet Energiværdierne for forskellige Tilstande med samme Værdi for n, men forskellig Værdi for k afviger lidt fra hverandre, faar vi nu for enhver Brintlinie svarende til bestemte Værdier af n' og n'' i Balmerformlen et Antal forskellige Overgangsprocesser, for hvilke Svingningstallene beregnet efter det andet Postulat ikke bliver nøjagtig ens. Som SOMMERFELD kunde paavise, stemmer det saaledes beregnede Komponentbillede for hver Brintlinie indenfor Forsøgsnøjagtigheden overens med Iagttagelserne over Brintliniernes Finstruktur. Paa Figuren betegner Pilene de Overgange, der frembringer Komponenterne af den røde og den grønne Linie i Brintspektret, hvis Svingningstal faas ved at sætte i Balmerformlen n'' lig 2 og n' lig henholdsvis 3 og 4.

Ved Betragtningen af Figuren maa det dog ikke glemmes, at Banernes

Beskrivelse er ufuldstændig, forsaavidt som med den benyttede Maalestok den langsomme Præcession ikke har kunnet bringes til Udtryk paa Tegningen. Denne Præcession er nemlig saa langsom, at Elektronen selv for de Baner, som drejer hurtigst, foretager omtrent 40 000 Omløb, før Perihelet er naaet en Gang rundt. Ikke desto mindre er det denne Præcession, der ene er Grunden til Beskaffenheden af den ved Bikvantetallet karakteriserede Mangfoldighed af stationære Tilstande. Er eksempelvis Brintatomet udsat for smaa ydre Kræfter, der forstyrrer den regelmæssige Præcession, vil Elektronbanen i de stationære Tilstande faa helt andre Former end angivet i Figuren. Samtidig vil Finstrukturen udviskes, men Brintspektret vil stadig bestaa af Linier, der med stor Tilnærmelse er givet ved Balmerformlen, hvad der hænger sammen med, at den tilnærmet periodiske Karakter af Bevægelsen vil bibeholdes. Først naar de forstyrrende Kræfter er saa store, at Banen allerede under et enkelt Omløb væsentlig forstyrres, vil Spektret undergaa betydelige Ændringer. Den Anskuelse, som man ofte ser fremsat, at Indførelsen af to Kvantetal er en nødvendig Betingelse for en Forklaring af Balmerformlen, er derfor en Misforstaaelse af Teoriens Væsen.

SOMMERFELD's Teori viste sig ikke alene at være i Stand til at gøre Rede for Brintliniernes Finstruktur, men ogsaa for Finstrukturen af Linierne i det med Brintspektret analoge Heliumgnistspektrum, hvor Afstanden mellem Liniekomponenterne paa Grund af Elektronens større Hastighed er langt større og har kunnet udmaales med betydelig større Nøjagtighed; ja, det var endogsaa muligt at gøre Rede for visse Træk i Finstrukturen i Røntgenspektrene, ved hvilke det drejede sig om Svingningstaldifferenser, der kunde naa op til mere end en Million Gange Værdien af Svingningstaldifferenserne for Brintliniernes Komponenter.

Kort efter at dette Resultat var vundet, lykkedes det samtidig EPSTEIN og SCHWARZSCHILD (1916) ved tilsvarende Betragtninger i Enkeltheder at gøre Rede for den karakteristiske Forandring, som Brintlinierne undergaar i et elektrisk Felt, og som opdagedes af STARK i Aaret 1913. En Forklaring af væsentlige Træk af Zeemaneffekten paa Brintlinierne blev dernæst udarbejdet samtidigt af SOMMERFELD og af DEBYE (1917). I dette Tilfælde førte Anvendelsen af Postulaterne til den Fordring, at kun visse Orienteringer af Atomet relativt til Magnetfeltet var tilladte, og denne ejendommelige Konsekvens af Kvanteteorien har fornylig faaet den mest direkte Bekræftelse ved de smukke Forsøg af STERN og GERLACH over Afbøjning af Molekularstraaler i et inhomogent magnetisk Felt.

Medens denne Udvikling af Spektralteorien beroede paa Udarbejdelsen af formelle Metoder til Fastlæggelse af stationære Tilstande, lykkedes det Foredragsholderen i den følgende Tid, i Tilknytning til betydningsfulde Arbejder af EHRENFEST og EINSTEIN at belyse Teorien fra et nyt Synspunkt ved at forfølge den allerede ved Brintspektret eftersporede ejendommelige Forbindelse mellem Kvanteteorien og den klassiske elektrodynamiske Teori. Disse Bestræbelser førte til Opstillingen af det saakaldte *Korrespondensprincip*, efter hvilket Forekomsten af de med Udstraaling ledsagede Overgange mellem Atomets stationære Tilstande føres tilbage til de i Atomets Bevægelse optrædende harmoniske Svingningskomponenter, der efter

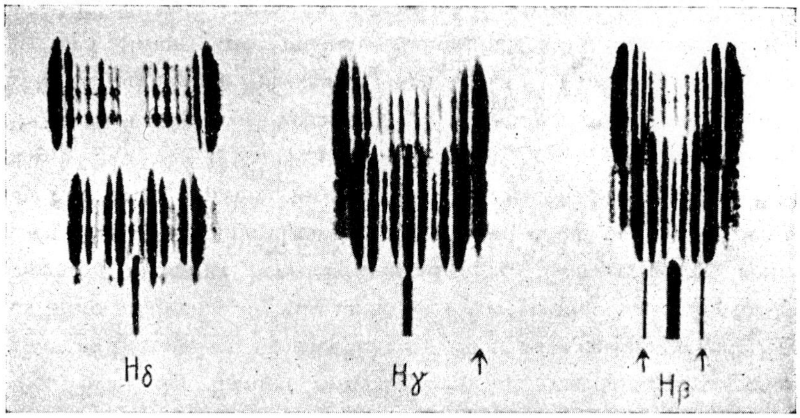

Fig. 6.

den klassiske Teori betinger Beskaffenheden af den paa Grund af Partiklernes Bevægelse udsendte Straaling. Efter Princippet antages saaledes enhver Overgangsproces mellem to stationære Tilstande at være knyttet til en korresponderende harmonisk Svingningskomponent paa en saadan Maade, at Sandsynligheden for Overgangens Fremkomst er afhængig af Svingningens Amplitude, medens Polarisationstilstanden af den ved Overgangen udsendte Straaling er betinget af Svingningens nærmere Beskaffenhed paa tilsvarende Maade som Straalingsintensitet og Polarisationstilstand i det Bølgesystem, der ifølge den klassiske Teori skulde udsendes fra Atomet som Følge af den omhandlede Svingningskomponents Tilstedeværelse, vilde være bestemt ved henholdsvis Svingningens Amplitude og Beskaffenhed.

Ved Hjælp af Korrespondensprincippet har det været muligt at uddybe og videreføre de ovenfor nævnte Resultater. Saaledes lykkedes det at udvikle en fuldstændig kvanteteoretisk Forklaring af Zeemaneffekten paa

Brintlinierne, der til Trods for den væsensforskellige Karakter af de Antagelser, der ligger til Grund for de to Teorier, udviser en dybtgaaende Lighed med den af LORENTZ oprindelig paa Grundlag af den klassiske Teori givne Forklaring. For Starkeffektens Vedkommende, som den klassiske Teori derimod havde staaet fuldstændig raadvild overfor, lykkedes det ved Korrespondensprincippets Hjælp at videreføre den kvanteteoretiske Forklaring til ogsaa at omfatte en Redegørelse for Polarisationen af de forskellige Komponenter, hvori Linierne opspaltes, samt for den ejendommelige Intensitetsfordeling, Komponentbilledet frembyder. Dette sidste Spørgsmaal er blevet nærmere undersøgt af KRAMERS, og de nedenstaaende Fi-

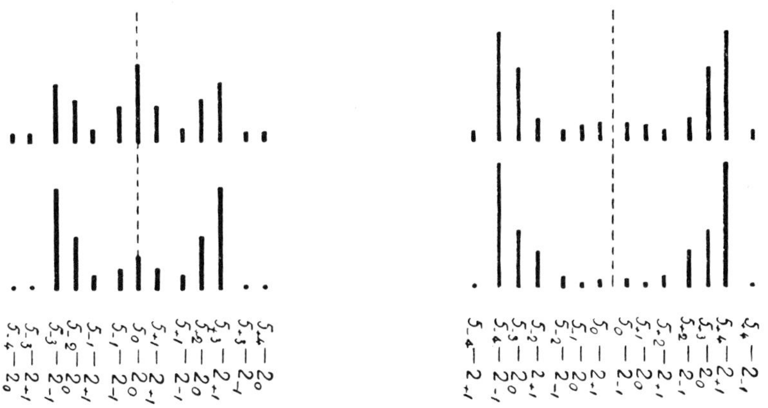

Fig. 7.

gurer vil give et Indtryk af, hvor indgaaende en Redegørelse for Fænomenet det drejer sig om. Figur 6 gengiver en af STARK's velkendte smukke Optagelser af Brintliniernes Opspaltning. Billedet giver et stærkt Indtryk af, hvor righoldigt Fænomenet er, og paa hvilken ejendommelig Maade Intensiteten varierer fra Komponent til Komponent; medens Komponenterne nederst paa Billedet er polariseret vinkelret paa Feltet, er de øverste Komponenter polariseret parallelt med Feltet. I Figur 7 er foroven givet en fra STARK's Afhandlinger stammende skematisk Fremstilling af Forsøgsresultaterne for Linien $H\gamma$, hvis Svingningstal er givet ved Balmerformlen, naar vi sætter $n'' = 2$ og $n' = 5$. De lodrette Linier betegner Opspaltningens Komponenter, idet Billedet til højre gengiver de parallelt polariserede Komponenter, og Billedet til venstre de vinkelret polariserede. Liniernes Afstand fra den punkterede Linie angiver Komponenternes Forskydning fra den uopspaltede Linie, og deres Længder er proportionale med Komponenternes relative Intensitet bedømt ud fra den fotografiske

Plades Sværtning. Forneden er til Sammenligning, efter en Tegning fra KRAMERS' Afhandling, givet en Fremstilling af de teoretiske Resultater. Ligesom i STARK's Fremstilling betegner de lodrette Linier Opspaltningens Komponenter, medens de til Linierne føjede Symboler ($n'_{s'}—n''_{s''}$) angiver de Overgangsprocesser mellem Atomets stationære Tilstande i det elektriske Felt, der giver Anledning til disse Komponenter. Foruden ved Hovedkvantetallet n er de stationære Tilstande karakteriseret ved et Bikvantetal s, der kan være saavel positivt som negativt, og som har en helt anden Betydning end det Kvantetal k, der optræder i Teorien for Brintliniernes relativistiske Finstruktur, og som bestemmer Formen af den roterende Elektronbane i det uforstyrrede Atom. Under Indflydelsen af det elektriske Felt undergaar saavel Banens Form som dens Beliggenhed gennemgribende Ændringer, men visse af Banens Egenskaber forbliver uforandrede, og det er til disses Beskrivelse, at Bikvantetallet s er knyttet. Paa Figuren svarer Komponenternes Beliggenhed til Svingningstallene, beregnet for de forskellige Overgange, og Liniernes Længde er afsat proportional med Sansynligheden for de forskellige Overgangsprocesser, vurderet paa Grundlag af Korrespondensprincippet, som ogsaa fastlægger Polarisationstilstanden af den til Overgangen svarende Straaling. Det vil ses, at Teorien gengiver alle Forsøgsresultaternes Hovedtræk, og vi kan ud fra Korrespondensprincippet sige, at Starkeffekten indtil de mindste Enkeltheder afspejler den Virkning, som det elektriske Felt har paa Elektronbanen i Brintatomet, selv om, i Modsætning til Forholdene ved Zeemaneffekten, Afspejlingen i dette Tilfælde er saa forvrænget, at man næppe direkte vilde kunne genkende Bevægelsen ud fra den klassiske Teoris Opfattelse af elektromagnetisk Straalings Oprindelse.

Ogsaa for Seriespektrene af Grundstoffer med højere Atomnummer, hvis Forklaring i Mellemtiden paa betydningsfuld Maade var videreført af SOMMERFELD ved Indførelsen af flere Kvantetal til Elektronbanernes Beskrivelse, naaedes interessante Resultater ved Korrespondensprincippets Hjælp. Navnlig lykkedes det fuldkommen at gøre Rede for de ejendommelige Regler, der behersker den tilsyneladende lunefulde Fremkomst af Kombinationslinier, og man tør sige, at Kvanteteorien ikke alene har bragt en simpel Fortolkning af Kombinationsprincippet, men at den tillige har bidraget væsentlig til at fjerne den Mystik, der længe hvilede over dette Princips Anvendelser.

De samme Synspunkter har ogsaa vist sig frugtbare for Udforskningen af de saakaldte *Baandspektre*. Disse hidrører ikke som Seriespektrene fra enkelte Atomer, men fra Molekyler, og Spektrenes store Linierigdom er begrundet i den Komplikation i Bevægelsen, som Atomkernernes Svingning relativt til hverandre og Rotationen af Molekylet som Helhed medfører. Den første, der anvendte Postulaterne paa dette Problem, var SCHWARZSCHILD, men Teorien er navnlig blevet udformet af den svenske Fysiker HEURLINGER, der ved sit betydningsfulde Arbejde har kastet et indgaaende Lys over Baandspektrenes Bygning og Oprindelse. De herhen hørende Betragtninger gaar direkte tilbage til den i Begyndelsen af Foredraget omtalte BJERRUMske Teori for Indflydelsen af Molekylrotationen paa Luftarternes ultrarøde Absorptionslinier. Vel tænker vi ikke længere, at Rotationen afspejler sig i Spektret paa den Maade, som den klassiske Elektrodynamik forlanger, men derimod at Liniekomponenterne skyldes Overgangsprocesser mellem stationære Tilstande med forskellig Rotationsbevægelse. At Fænomenet dog derved bevarer sine væsentlige Træk er en typisk Konsekvens af den Lovmæssighed, som Korrespondensprincippet giver Udtryk for.

* * *

De i det foregaaende udviklede Synspunkter vedrørende Spektrenes Fortolkning har tilbudt et Grundlag for en Teori om Bygningen af Grundstoffernes Atomer, der har vist sig egnet til i store Træk at gøre Rede for Stoffernes Egenskaber, saaledes som de fremtræder i Grundstoffernes naturlige System. Denne Teori støtter sig i første Række paa Betragtninger over den Maade, hvorpaa Atomet kan tænkes opbygget ved Elektronernes Indfangning og Binding til Kernen en for en. Som vi har set, afgiver Stoffernes optiske Spektre netop Vidnesbyrd om Forløbet af de sidste Afsnit af denne Opbygningsproces. Et Indblik i Karakteren af de Oplysninger, som den nærmere Undersøgelse af Spektrene i denne Henseende har bragt, faas gennem Figur 8, der giver en skematisk Fremstilling af de stationære Tilstande, der svarer til Udsendelsen af Kaliums Buespektrum. Kurverne angiver Formerne af de Baner, som den sidst indfangne Elektron i Kaliumatomet beskriver i de stationære Tilstande, der kan optræde som Stadier under den Proces, hvorved den 19. Elektron bindes, efter at de 18 første er bundet i deres normale Baner. For ikke at komplicere Figuren er det ikke forsøgt at tegne nogen af disse indre Elektroners Baner, men det Omraade, indenfor hvilket de bevæger sig, er blot angivet ved den punkterede Cirkel. I et Atom med flere Elektroner vil Banerne i Almindelighed have en indviklet Karakter. Paa Grund af den symmetriske Natur af det Kernen omgivende Kraftfelt kan hver enkelt Elektrons Bevægelse dog med Tilnærmelse beskrives som en plan periodisk Bevægelse, paa hvilken der er overlejret en jævn Drejning i Banens Plan. Enhver Elektronbane vil derfor i første Tilnærmelse være dobbelt periodisk og være fastlagt ved Benyttelse af to Kvantetal, i Lighed med de stationære Tilstande i et Brintatom, naar der tages Hensyn til den fra Relativitetseffekten hidrørende Præcession.

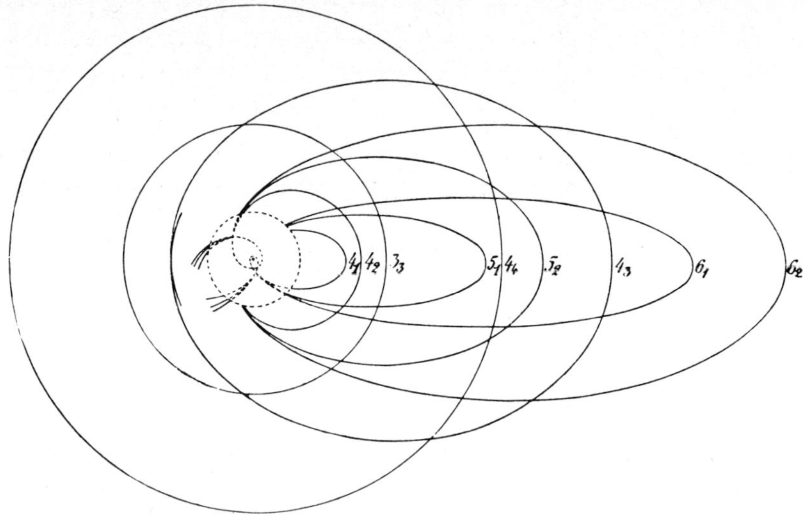

Fig. 8.

Paa samme Maade som i Figur 5 er Elektronbanerne i Figur 8 derfor betegnet med Symbolet n_k, hvor n er Hovedkvantetallet og k Bikvantetallet. Medens for Begyndelsesstadierne, hvor Kvantetallene er store, den sidst indfangne Elektrons Bane forløber helt udenom de tidligere indfangne Elektroners Omraade, forholder det sig anderledes i de sidste Stadier. Saaledes vil i Kaliumatomet Elektronbanerne med Bikvantetallet 2 og 1, saaledes som det er antydet paa Figuren, under deres Omløb trænge ind i det indre Omraade. Paa Grund af denne Omstændighed vil Banerne afvige overordentlig meget fra en simpel Keplerbevægelse, idet de vil bestaa af en Række paa hinanden følgende ydre Baneslynger, der har samme Størrelse og Form, men af hvilke enhver er drejet en betydelig Vinkel i Forhold til den foregaaende. Disse ydre Baneslynger, hvoraf kun en enkelt er tegnet paa Figuren, og som hver især falder nøje sammen med en Del af en Keplerellipse, er som antydet forbundne ved en Række indre Baneslynger af mere indviklet Karakter, og i hvilke Elektronen vil nærme sig

Kernen overordentlig stærkt. Dette gælder navnlig for Banerne med Bikvantetallet 1, der, som en nøjere Undersøgelse viser, endda vil komme Kernen nærmere end nogen af de tidligere bundne Elektroner. Denne Indtrængen i det indre Omraade bevirker, at, tiltrods for at de vedkommende Elektronbaner for største Delen forløber i et Kraftfelt af samme Karakter som det, der omgiver Kernen i Brintatomet, er den Styrke, hvormed Elektronen i en given Bane fastholdes af Atomet, langt større end den, hvormed Elektronen i Brintatomet er bundet i en Bane med samme Hovedkvantetal, ligesom Elektronens Maximalafstand fra Kernen under Omløbet er betydelig mindre end i en saadan Bane. Som vi skal se, er disse Træk ved Elektronbindingen i Atomer med mange Elektroner væsentlige for Forstaaelsen af den ejendommelige periodiske Maade, hvorpaa Stoffernes Egenskaber varierer med Atomnummeret, saaledes som de kommer til Syne i det naturlige System.

I omstaaende Tabel er givet en Oversigt over de Resultater, Foredragsholderen er naaet til angaaende Bygningen af Grundstoffernes Atomer gennem Betragtninger over Elektronernes sukcessive Indfangning og Binding af Atomkernen. De udfor de forskellige Stoffer staaende Tal er Atomnummeret, der angiver det samlede Elektrontal i det neutrale Atom. Tallene i de forskellige Kolonner angiver Antallet af Elektronerne i Baner svarende til de foroven angivne Værdier for Hoved- og Bikvantetal. Efter en almindelig benyttet Betegnelse skal vi for Kortheds Skyld betegne en Bane med Hovedkvantetal n for en n-kvantet Bane. Den først bundne Elektron i hvert Atom bevæger sig i en Bane, der svarer til Normaltilstanden i Brintatomet, og som betegnes ved Kvantebetegnelsen 1_1. I Brintatomet er der jo kun en Elektron; men i de andre Stoffers Atomer antager vi, at ogsaa den næste Elektron vil bindes i en saadan 1-kvantet Bane af Typen 1_1. Som Tabellen angiver, vil de følgende Elektroner bindes i 2-kvantede Baner. Til at begynde med vil Bindingen resultere i en 2_1 Bane, men dernæst vil der bindes Elektroner i 2_2 Baner, indtil vi efter Bindingen af de 10 første Elektroner i Atomet har naaet en afsluttet Konfiguration af de 2-kvantede Baner, i hvilken vi antager, at der er 4 Baner af hver Slags. Denne Konfiguration træffer vi i det neutrale Atom første Gang hos Neon, der danner Afslutningen paa den anden Periode i Grundstoffernes System. Naar vi gaar videre, bliver de følgende Elektroner bundne i 3-kvantede Baner, indtil vi, efter Afslutningen af den tredie Periode i Systemet, hos Elementerne i den 4. Periode for første Gang møder Elektroner i 4-kvantede Baner o. s. v.

	1_1	$2_1 2_2$	$3_1 3_2 3_3$	$4_1 4_2 4_3 4_4$	$5_1 5_2 5_3 5_4 5_5$	$6_1 6_2 6_3 6_4 6_5 6_6$	$7_1 7_2$
1 H	1						
2 He	2						
3 Li	2	1					
4 Be	2	2					
5 B	2	2 (1)					
— —	—	— —					
10 Ne	2	4 4					
11 Na	2	4 4	1				
12 Mg	2	4 4	2				
13 Al	2	4 4	2 1				
— —	—	— —	— —				
18 A	2	4 4	4 4				
19 K	2	4 4	4 4	1			
20 Ca	2	4 4	4 4	2			
21 Sc	2	4 4	4 4 1	(2)			
22 Ti	2	4 4	4 4 2	(2)			
— —	—	— —	— — —	—			
29 Cu	2	4 4	6 6 6	1			
30 Zn	2	4 4	6 6 6	1			
31 Ga	2	4 4	6 6 6	2 1			
— —	—	— —	— — —	— —			
36 Kr	2	4 4	6 6 6	4 4			
37 Rb	2	4 4	6 6 6	4 4	1		
38 Sr	2	4 4	6 6 6	4 4	2		
39 Y	2	4 4	6 6 6	4 4 1	(2)		
40 Zr	2	4 4	6 6 6	4 4 2	(2)		
— —	—	— —	— — —	— — —	—		
47 Ag	2	4 4	6 6 6	6 6 6	1		
48 Cd	2	4 4	6 6 6	6 6 6	2		
49 In	2	4 4	6 6 6	6 6 6	2 1		
— —	—	— —	— — —	— — —	— —		
54 X	2	4 4	6 6 6	6 6 6	4 4		
55 Cs	2	4 4	6 6 6	6 6 6	4 4	1	
56 Ba	2	4 4	6 6 6	6 6 6	4 4	2	
57 La	2	4 4	6 6 6	6 6 6	4 4 1	(2)	
58 Ce	2	4 4	6 6 6	6 6 6 1	4 4 1	(2)	
59 Pr	2	4 4	6 6 6	6 6 6 2	4 4 1	(2)	
— —	—	— —	— — —	— — — —	— — —	—	
71 Cp	2	4 4	6 6 6	8 8 8 8	4 4 1	(2)	
72 —	2	4 4	6 6 6	8 8 8 8	4 4 2	(2)	
— —	—	— —	— — —	— — — —	— — —	—	
79 Au	2	4 4	6 6 6	8 8 8 8	6 6 6	1	
80 Hg	2	4 4	6 6 6	8 8 8 8	6 6 6	2	
81 Tl	2	4 4	6 6 6	8 8 8 8	6 6 6	2 1	
— —	—	— —	— — —	— — — —	— — —	— —	
86 Em	2	4 4	6 6 6	8 8 8 8	6 6 6	4 4	
87 —	2	4 4	6 6 6	8 8 8 8	6 6 6	4 4	1
88 Ra	2	4 4	6 6 6	8 8 8 8	6 6 6	4 4	2
89 Ac	2	4 4	6 6 6	8 8 8 8	6 6 6	4 4 1	(2)
90 Th	2	4 4	6 6 6	8 8 8 8	6 6 6	4 4 2	(2)
— —	—	— —	— — —	— — — —	— — —	— — —	—
118?	2	4 4	6 6 6	8 8 8 8	8 8 8 8	6 6 6	4 4

Dette Billede af Atombygningen gengiver mange Træk, som har været fremdragne i tidligere Forfatteres Arbejder. Saaledes gaar Forsøgene paa at fortolke Slægtskabsforholdene mellem Grundstofferne i det naturlige System ved Antagelse af en Gruppedeling af Elektronerne i Atomet helt tilbage til J. J. THOMSONS Arbejder fra 1904. Senere er dette Synspunkt navnlig blevet udviklet af KOSSEL (1916), der tillige har bragt en saadan Gruppedeling i Forbindelse med de Lovmæssigheder, som Undersøgelserne over Røntgenspektrene har bragt for Dagen. Omtrent paa samme Tid har ogsaa LEWIS og LANGMUIR forsøgt at gøre Rede for Slægtskabsforholdene mellem Grundstoffernes Egenskaber paa Grundlag af en Gruppedeling indenfor Atomet. Disse Forfattere antager imidlertid, at Elektronerne ikke bevæger sig omkring Kernen, men indtager Ligevægtsstillinger. Paa denne Maade kan dog ingen nærmere Forbindelse opnaas mellem Stoffernes Egenskaber og de eksperimentelle Resultater vedrørende Atomernes Byggestene. Statiske Ligevægtskonfigurationer for Elektronerne er nemlig ikke mulige, dersom Kræfterne mellem Atomdelene blot tilnærmelsevis adlyder de Love, der gælder for Tiltrækninger og Frastødninger mellem elektriske Ladninger. Muligheden for en indgaaende Redegørelse for Stoffernes Egenskaber baseret paa de sidstnævnte Love er netop det karakteristiske for det paa Kvanteteorien opbyggede Billede af Atombygningen. Hvad dette Billede angaar, var den Tanke at forbinde Gruppedelingen med en Klassifikation af Elektronbanerne efter stigende Kvantetal lagt nær ved MOSELEY's Opdagelse af Lovene for Røntgenspektrene og ved SOMMERFELD's Arbejder over disse Spektres Finstruktur. Dette har navnlig været fremhævet af VEGARD, der for nogle Aar tilbage i Forbindelse med Undersøgelser over Røntgenspektre har foreslaaet en Gruppeinddeling af Elektronerne i Grundstoffernes Atomer, der paa flere Maader viser Lighed med den, der er angivet i den ovenstaaende Tabel.

Et Grundlag for den nærmere Udformning af det omhandlede Billede af Atombygningen er dog først skabt i den sidste Tid ved Studiet af de Bindingsprocesser af Elektronerne i Atomet, som vi har eksperimentelt Vidnesbyrd om i de optiske Spektre, og hvis karakteristiske Træk navnlig Korrespondensprincippet har formaaet at belyse. En væsentlig Omstændighed er det her, at den Begrænsning i Bindingsprocessernes Forløb, der ytrer sig ved Optræden af flerkvantede Elektronbaner i Atomets Normaltilstand, kan bringes i naturlig Forbindelse med den almindelige Betingelse for Straalingsprocessernes Fremkomst ved Overgang mellem stationære Tilstande, som formuleres ved det nævnte Princip. Et andet væsentligt Træk hos Teorien er den Indflydelse paa Bindingens Styrke og Banernes Dimensioner, som skyldes de senere bundne Elektroners Indtrængen i de tidligere bundne Elektroners Omraade, som vi har set Eksempel paa ved Omtalen af Kaliumspektrets Oprindelse. Denne Omstændighed tør nemlig betragtes som den egentlige Aarsag til den udtalte periodiske Vekslen i Grundstoffernes Egenskaber, idet den medfører, at Atomdimensioner og kemiske Egenskaber af homologe Stoffer i de forskellige Perioder, som for Eksempel i Alkalimetallerne, udviser en langt større Lighed end den, som en direkte Sammenligning mellem den sidst indfangne Elektrons Bane med Banen med samme Kvantetal i Brintatomet vilde lade formode.

Den omhandlede gradvise Stigning i Hovedkvantetallet hos den sidst bundne Elektron i Atomet, som vi møder, naar vi skrider frem i Stoffernes Række, muliggør endvidere en umiddelbar Forstaaelse af de karakteristiske Afvigelser fra simpel Periodicitet, som det naturlige System udviser, og som i Fremstillingen paa Figur 1 er fremhævet ved Indramning af visse Grundstofrækker i Systemets senere Perioder. Den første Gang vi møder en saadan Afvigelse, er i den fjerde Periode, og Aarsagen dertil kan paa simpel Maade illustreres ved Figuren over Banerne af den sidst indfangne Elektron hos Kalium, der jo er det første Stof i denne Periode. Her træffer vi for første Gang i Grundstoffernes Række det Forhold, at Hovedkvantetallet af den sidst indfangne Elektrons Bane i Atomets Normaltilstand er større end i et af de tidligere Stadier af Bindingsprocessen. Normaltilstanden svarer her til en 4_1 Bane, hvor Elektronbindingen paa Grund af dens Indtrængen i det indre Omraade er langt stærkere end hos en 4-kvantet Bane i Brintatomet, ja, endda stærkere end hos en 2-kvantet Bane i dette Atom. Elektronbindingen er derfor mere end dobbelt saa stærk som i den cirkulære 3_3 Bane, som forbliver fuldstændig udenfor det indre Omraade, og hvor Bindingsstyrken kun afviger lidt fra en 3-kvantes Banes Binding i Brintatomet. Dette Forhold vil imidlertid ikke blive ved at gælde, naar vi betragter den 19. Elektrons Binding hos Stoffer af højere Atomnummer, paa Grund af den langt mindre relative Forskel mellem Kraftfeltet udenfor og indenfor de 18 først bundne Elektroners Omraade. Som det fremgaar af en Undersøgelse af Gnistspektret hos Calcium, er allerede her Elektronbindingen i 4_1-Banen kun lidt stærkere end i 3_3 Banen, og saa snart vi kommer til Scandium, maa vi antage, at 3_3 Banen vil repræsentere den 19. Elektrons Bane i Normaltilstanden, idet denne vil svare til en væsentlig stærkere Binding end en 4_1 Bane. Medens Elektrongruppen med 2-kvantede Baner har naaet sin endelige Afslutning ved Slutningen af 2. Periode, kan den Udvikling, som Elektrongruppen med 3-kvantede Baner faar i Løbet af den 3. Periode, derfor kun betegnes som en foreløbig Afslutning, og som angivet i Tabellen vil denne Elektrongruppe undergaa et videre Udviklingstrin under Optagelsen af Elektroner i 3-kvantede Baner hos de indrammede Stoffer i den 4. Periode. Dette bringer nye Forhold med sig, idet Udviklingen af Elektrongruppen med 4-kvantede Baner saa at sige gaar i Staa, indtil den 3-kvantede Elektrongruppe har naaet sin endelige Afslutning. Selvom vi endnu ikke er i Stand til i alle Enkeltheder at gøre Rede for Forløbet af den 3-kvantede Elektrongruppes gradvise Udvikling, kan vi dog sige, at vi ud fra Kvanteteorien umiddelbart forstaar, at der for første Gang i den 4. Periode af Grundstoffernes System optræder paa hinanden følgende Stoffer med Egenskaber, der ligner hverandre saa meget som Egenskaberne hos *Jernmetallerne*, ja endda kan forstaa, hvorfor disse Elementer viser de velkendte paramagnetiske Egenskaber. Uden nærmere Forbindelse med Kvanteteorien har den Tanke at bringe de omhandlede Grundstoffers kemiske og magnetiske Egenskaber i Forbindelse med Udviklingen af en indre Elektrongruppe i Atomet iøvrigt allerede for nogle Aar siden været fremsat af LADENBURG.

Jeg skal ikke gaa ind paa mange flere Enkeltheder, blot sige at de Forhold, vi møder i den 5. Periode, faar en ganske lignende Forklaring som Forholdene i den 4. Periode, idet Egenskaberne af de indrammede Stoffer i denne Periode, saaledes som det fremgaar af Tabellen, beror paa et Trin i Udviklingen af den 4-kvantede Elektrongruppe, der indledes ved Optagelsen af Elektroner i 4_3 Baner. I den 6. Periode derimod møder vi nye Forhold. I denne Periode møder vi ikke alene et Udviklingstrin af Elektrongrupperne med 5- og 6-kvantede Baner, men tillige den endelige Afslutning paa Udviklingen af den 4-kvantede Elektrongruppe, der indledes ved den første Optræden i Atomets Normaltilstand af Elektronbaner af 4_4-Typen. Denne Udvikling kommer paa karakteristisk Maade til Udtryk ved Optræden af den ejendommelige Familie af Grundstoffer, som vi møder i den 6. Periode, og som betegnes som de *sjældne Jordarter*. Disse Stoffer viser som bekendt endnu langt større indbyrdes Slægtskab i deres kemiske Egenskaber end Stofferne i Jernmetallernes Familie, hvad der hidrører fra, at vi har at gøre med Udviklingen af en Elektrongruppe, der ligger dybere i Atomet. Det er af Interesse at bemærke, at Teorien ogsaa paa naturlig Maade kan gøre Rede for, at disse Stoffer, der ligner hinanden paa saa mange Maader, dog viser stor Forskel i deres magnetiske Egenskaber. Den Tanke, at de sjældne Jordarters Forekomst hidrører fra Udviklingen af en indre Elektrongruppe i Atomet, er fremsat fra forskellig Side. Den findes saaledes hos VEGARD, og samtidig med Foredragsholderens Arbejde er denne Tanke i Tilknytning til LANGMUIR's statiske Atommodel blevet nærmere forfulgt af BURY. Medens der imidlertid ikke hidtil kunde gives nogen teoretisk Begrundelse for en saadan Udvikling af en indre Elektrongruppe, ser vi, hvorledes Kvanteteoriens nærmere Udformning giver en saa utvungen Forklaring deraf, at det neppe er nogen Overdrivelse at sige, at dersom de sjældne Jordarters Tilstedeværelse ikke forlængst var fastslaaet ved direkte kemiske Undersøgelser, kunde Forekomsten af en Grundstoffamilie af denne Karakter indenfor den 6. Periode af Grundstoffernes naturlige System have været teoretisk forudsagt.

Naar vi kommer til den 7. Periode af Systemet, møder vi for første Gang 7-kvantede Baner, og vi skal vente indenfor denne Periode at finde Forhold væsentlig lig med dem i den 6. Periode, idet vi foruden det første Trin i Udviklingen af de 7-kvantede Baner maa vente at møde et videre Udviklingstrin i Gruppen med 6- og 5-kvantede Baner. Imidlertid har man ikke kunnet faa nogen direkte Bekræftelse paa disse Forventninger, idet kun faa Stoffer i Begyndelsen af den 7. Periode er bekendte, hvad der maa antages at hænge sammen med den Instabilitet hos Atomkerner med stor Ladning, der ytrer sig i den fremherskende Radioaktivitet hos Stofferne med høje Atomnumre.

* * *

Ved Omtalen af Forestillingerne om Atomstrukturen har vi hidtil lagt Hovedvægten paa Atomernes Dannelse ved sukcessiv Indfangning af Elektroner. Fremstillingen vilde dog være ufuldstændig uden en Henvisning til den Støtte for Teorien, som Undersøgelserne over *Røntgenspektrene* har ydet. Siden Afbrydelsen af MOSELEY's grundlæggende Undersøgelser ved hans altfor tidlige Død er Studiet af disse Spektre paa beundringsværdig Maade blevet videreført af Professor SIEGBAHN i Lund. Paa Grund af det af ham og hans Medarbejdere fremskaffede store Materiale er det lykkedes i den sidste Tid at opnaa en Klassifikation af Røntgenspektrene, der tillader en umiddelbar Fortolkning paa Grundlag af Kvanteteoriens Postulater, og til hvis Tilvejebringelse de ovennævnte Arbejder af SOMMERFELD og KOSSEL har bidraget paa væsentlig Maade.

Først og fremmest har det været muligt, ligesom ved de optiske Spektre, at fremstille Svingningstallet for enhver af Røntgenspektrenes Linier som Forskellen mellem to blandt en Mangfoldighed af Spektralled, karakteristisk for det paagældende Stof. Endvidere er en direkte Tilknytning til Atomteorien opnaaet paa Grundlag af den Antagelse, at ethvert af disse Spektralled multipliceret med PLANCKS Konstant er lig det Arbejde, som maa tilføres Atomet for at fjerne en af de indre Elektroner. I Overensstemmelse med de forudgaaende Betragtninger vedrørende Atomernes Dannelse ved Indfangning af Elektroner bliver der efter Fjernelsen af en indre Elektron i det færdigdannede Atom nemlig Mulighed for Overgangsprocesser, hvorved den manglende Elektron i vedkommende Gruppe erstattes af en Elektron, der tilhører en af Atomets løsere bundne Elektrongrupper, med det Resultat, at en Elektron kommer til at mangle i denne Gruppe. Linierne i Røntgenspektrene kan derfor opfattes som Vidnesbyrd om Stadier i en Proces, hvorved Atomet undergaar en Reorganisation efter en Forstyrrelse frembragt i Atomets Indre. Efter vore Antagelser om Elektronsystemets Stabilitetsforhold maa en saadan Forstyrrelse netop bestaa i den fuldstændige Fjernelse af Elektroner fra Atomet, eller i det mindste i Overførelsen af Elektroner fra deres Baner i Normaltilstanden til Baner med højere Kvantetal end de, der svarer til de afsluttede Elektrongrupper; en Omstændighed, som kommer klart til Syne i den karakteristiske Forskel mellem selektiv Absorption i Røntgenomraadet og Absorptionen i det sædvanlige optiske Omraade.

En nærmere Undersøgelse af den Maade, paa hvilken Spektraltermerne i Røntgenspektrene ændrer sig med Atomnummeret har for nylig gjort det muligt at opnaa en indgaaende Prøve paa et Antal teoretiske Slutninger vedrørende Atomets Bygning. I Fig. 9 er Atomnummeret afsat som Abcisse, medens Ordinaten angiver Kvadratroden af Forholdet mellem Spektralleddenes Værdier, og den RYDBERGSKE Konstant. Betegnelserne K, L, M, N, O, der er vedføjet de enkelte Kurver, refererer til de ejendommelige Diskontinuiteter i Stoffernes selektive Absorption for Røntgenstraaler, der oprindelig blev opdaget af BARKLA, endnu før vi i Interferensen i Krystaller besad et Middel til den nærmere Undersøgelse af Røntgenstraalerne. Skønt Kurverne i Almindelighed forløber jævnt, viser de et Antal Afvigelser fra det jævne Forløb, som navnlig er bragt for Dagen ved Undersøgelser af COSTER, der i nogle Aar har arbejdet i SIEGBAHNS Laboratorium. Disse Afvigelser, hvis Eksistens først blev opdaget efter Offentliggørelsen af den i det foregaaende omhandlede Teori for Atombygningen, svarer nøje til, hvad man efter denne Teori maatte vente. Paa Foden af Figuren er ved de lodrette Streger angivet hvor vi efter Teorien for første Gang skal vente, at der i Atomets Normaltilstand optræder n_k-Baner af den angivne Type, og vi ser, hvorledes det har været muligt at drage de forskellige Kurver saadan, at Forekomsten af ethvert Spektralled netop er knyttet til Tilstedeværelsen i Atomet af en Elektron, der bevæger sig i en Bane af bestemt Type, og til hvis Fjernelse det paagældende Led svarer. At der i Al-

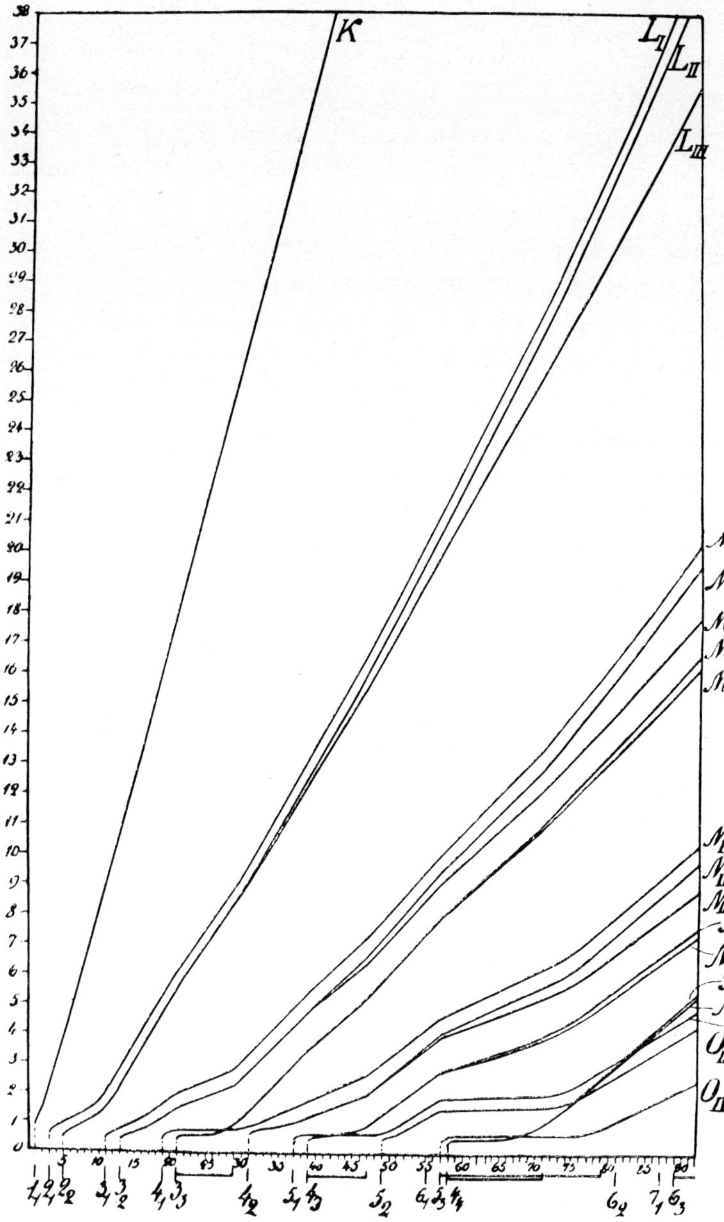

Fig. 9.

mindelighed svarer mere end en Kurve til hver Banetype n_k, er Udtryk for en Komplikation i Spektrene, som det vilde føre for vidt at komme ind paa her, og som maa tilskrives Afvigelserne fra den tidligere beskrevne simple Bevægelsestype hos Elektronbanerne, stammende fra Vekselvirkningen af de forskellige Elektroner indenfor samme Gruppe. De Intervaller indenfor Grundstoffernes System, i hvilke en Udvikling af en indre Elektrongruppe finder Sted som Følge af Optagelsen i Atomet af Elektroner i Baner af given Type, er endvidere paa Figuren angivet ved de vandrette Linier, der udgaar fra de lodrette Streger betegnet ved vedkommende Kvantesymbol. Vi ser hvorledes en saadan Udvikling af en indre Gruppe overalt paa karakteristisk Maade afspejler sig i Kurvernes Forløb. Navnlig tør det ejendommelige Forløb af *N*- og *O*-Kurverne betragtes som afgivende et direkte Vidnesbyrd om det Udviklingsstadium i Elektrongruppen med 4-kvantede Baner, der er Aarsagen til de sjældne Jordarters Optræden. Tiltrods for at det var det typiske og betydningsfulde Træk i MOSELEYS Opdagelse, at det indviklede Slægtskab mellem de fleste andre af Grundstoffernes Egenskaber tilsyneladende ikke afspejlede sig i Røntgenspektrene, erkender vi altsaa nu efter de sidste Aars Fremskridt alligevel Tilstedeværelsen af en nøje Forbindelse mellem Røntgenspektrene og de almindelige Slægtskabsforhold af Grundstofferne indenfor det naturlige System.

I Sammenhæng med dette sidste Spørgsmaal vil jeg gerne have Lov til endnu at nævne et Punkt, hvor røntgenspektroskopiske Undersøgelser har været til Støtte for Teorien. Det drejer sig om Egenskaberne hos det hidtil ukendte Grundstof med Atomnummer 72. Om dette Spørgsmaal har Meningerne været delte med Hensyn til de Slutninger, der kunde drages ud fra Slægtskabsforholdene indenfor det naturlige System, og i mange Fremstillinger af Systemet er der anvist dette Stof en Plads indenfor de sjældne Jordarters Familie. Allerede i JULIUS THOMSENS Fremstilling af det periodiske System var dog paa lignende Maade som i vor Fremstilling i Figur 1 det hypotetiske Elements homologe Stilling i Forhold til Titan og Zirkon angivet. Et saadant Slægtskabsforhold er at betragte som en nødvendig Konsekvens af den her udviklede Teori for Atombygningen, og kommer i Oversigten paa Side 35 derved til Udtryk, at Elektronkonfigurationerne hos Titan og Zirkon viser en tilsvarende Lighed og Forskel som Elektronkonfigurationerne hos Zirkon og Grundstoffet med Atomnummer 72. En tilsvarende Opfattelse var ogsaa blevet fremsat af BURY ud fra hans ovennævnte systematiske Betragtninger over Sammenhængen mellem Grundstoffernes Egenskaber og Gruppeinddelingen af Elektronerne i Atomet. For et halvt Aar siden meddelte imidlertid DAUVILLIER, at han i Røntgenspektret af et Præparat indeholdende sjældne Jordarter havde iagttaget nogle svage Linier, der blev tilskrevet et Stof med Atomnummer 72, som antoges at være identisk med et Element af de sjældne

Jordarters Familie, hvis Tilstedeværelse i det omhandlede Præparat allerede for flere Aar siden var formodet af URBAIN. Denne Slutning vilde imidlertid, hvis den kunde opretholdes, berede Teorien overordentlig store, for ikke at sige uoverstigelige Vanskeligheder, idet den vilde kræve en Ændring i Elektronbindingens Fasthed med Atomnummeret, som syntes uforeneligt med Kvanteteoriens Fordringer. Under disse Omstændigheder har for ganske kort Tid siden Dr. COSTER og Professor HEVESY, der begge for Tiden opholder sig i København, taget Spørgsmaalet op til Prøvelse ved røntgenspektroskopisk Undersøgelse af Præparater fremstillet af zirkonholdige Mineraler, og jeg er i Stand til att meddele, at de nævnte Forskere netop i disse Dage har kunnet konstatere Tilstedeværelsen i de undersøgte Mineraler af betydelige Mængder af et Grundstof med Atomnummer 72, og hvis kemiske Egenskaber viser et nøje Slægtskab med Zirkons og afviger væsentlig fra de sjældne Jordarters.[1]

Idet jeg haaber, at det ved denne Redegørelse er lykkedes mig at give Dem et Overblik over nogle af de vigtigste Resultater, som i de senere Aar er opnaat paa Atomteoriens Omraade, vil jeg gerne, før jeg slutter, tilføje et Par Bemærkninger af almindelig Art vedrørende de Synspunkter, udfra hvilke de vundne Resultater maa bedømmes og da nævnlig Spørgsmaalet, hvorvidt der ved disse Resultater kan være Tale om en Forklaring i Ordets sædvanlige Forstand. Ved en teoretisk Forklaring af Naturfænomener vil man vel i Almindelighed forstaa en Klassifikation af et vist Iagttagelsesomraade ved Hjælp af Analogier hentet fra andre Iagttagelsesomraader, hvor man formentlig har med enklere Foreteelser at gøre; og det meste, man kan forlange af en Teori, er, at denne Klassifikation kan drives saa vidt, at den kan bidrage til en Udvidelse af Iagttagelsesomraadet gennem Forudsigelser af nye Fænomener. Naar vi ser paa Atomteorien, befinder vi os i den ejendommelige Stilling, at der paa den ene Side ikke kan være Tale om Forklaring i den nævnte Betydning, idet det jo her netop drejer sig om Foreteelser, der ifølge Sagens Natur er enklere end ved noget andet Iakttagelsesomraade, hvor vi altid har at gøre med Fænomener, der er betinget af Samvirkningen af et stort Antal Atomer. Vi er derfor nødt til at være beskedne i vore Krav og at nøjes med Forestillinger, der

[1] Ved fortsatte Undersøgelser er det lykkedes COSTER og HEVESY at fremstille det nye Grundstof, for hvilket de har foreslaaet Navnet *Hafnium*, i saa ren Tilstand, at en foreløbig Atomvægtsbestemmelse har kunnet udføres. Den gav omtrent 180, hvilket passer godt ind i Systemet, idet Grundstoffet 71 (Cp) har Atomvægten 174, og Grundstoffet 73 (Ta) har Atomvægten 181,5.

er formelle i den Forstand, at de ikke besidder en Anskuelighed af den Art, som man er vant til at kræve af de Forestillinger, hvormed naturvidenskabelige Teorier opererer. Ikke mindst med Henblik herpaa har jeg søgt at give Dem Indtryk af, at Resultaterne paa den anden Side i det mindste i nogen Grad opfylder de Forventninger, der stilles til en Teori, idet jeg har søgt at vise, hvorledes Atomteoriens Udvikling har bidraget til at klassificere udstrakte Iagttagelsesomraader og ved Forudsigelser anvise Vejen for denne Klassifikations Fuldstændiggørelse. Dog vil det neppe være nødvendigt at understrege, i hvor høj Grad Teorien endnu befinder sig i et Begyndelsesstadium, og hvor mange Grundspørgsmaal der endnu venter deres Besvarelse.

Stockholm 1923. P. A. Norstedt & Söner. 221969

(*Reprinted from* NATURE, *July* 7, 1923.)

The Structure of the Atom.[1]

By Prof. N. BOHR.

THE GENERAL PICTURE OF THE ATOM.

THE present state of atomic theory is characterised by the fact that we not only believe the existence atoms to be proved beyond a doubt, but also we ven believe that we have an intimate knowledge of the nstituents of the individual atoms. I cannot on is occasion give a survey of the scientific developents that have led to this result; I will only recall e discovery of the electron towards the close of the st century, which furnished the direct verification d led to a conclusive formulation of the conception the atomic nature of electricity which had evolved nce the discovery by Faraday of the fundamental ws of electrolysis and Berzelius's electrochemical eory, and had its greatest triumph in the electrolytic ssociation theory of Arrhenius. This discovery of e electron and elucidation of its properties was the sult of the work of a large number of investigators, nong whom Lenard and J. J. Thomson may be rticularly mentioned. The latter especially has ade very important contributions to our subject by s ingenious attempts to develop ideas about atomic nstitution on the basis of the electron theory. The esent state of our knowledge of the elements of omic structure was reached, however, by the disvery of the atomic nucleus, which we owe to Rutherrd, whose work on the radioactive substances scovered towards the close of the last century has uch enriched physical and chemical science.

According to our present conceptions, an atom of element is built up of a nucleus that has a positive ectrical charge and is the seat of by far the greatest rt of the atomic mass, together with a number of ectrons, all having the same negative charge and ass, which move at distances from the nucleus that e very great compared to the dimensions of the cleus or of the electrons themselves. In this picture at once see a striking resemblance to a planetary stem, such as we have in our own solar system. st as the simplicity of the laws that govern the otions of the solar system is intimately connected th the circumstance that the dimensions of the moving bodies are small in relation to the orbits, so the corresponding relations in atomic structure provide us with an explanation of an essential feature of natural phenomena in so far as these depend on the properties of the elements. It makes clear at once that these properties can be divided into two sharply distinguished classes.

To the first class belong most of the ordinary physical and chemical properties of substances, such as their state of aggregation, colour, and chemical reactivity. These properties depend on the motion of the electron system and the way in which this motion changes under the influence of different external actions. On account of the large mass of the nucleus relative to that of the electrons and its smallness in comparison to the electron orbits, the electronic motion will depend only to a very small extent on the nuclear mass, and will be determined to a close approximation solely by the total electrical charge of the nucleus. Especially the inner structure of the nucleus and the way in which the charges and masses are distributed among its separate particles will have a vanishingly small influence on the motion of the electron system surrounding the nucleus. On the other hand, the structure of the nucleus will be responsible for the second class of properties that are shown in the radioactivity of substances. In the radioactive processes we meet with an explosion of the nucleus, whereby positive or negative particles, the so-called α- and β-particles, are expelled with very great velocities.

Our conceptions of atomic structure afford us, therefore, an immediate explanation of the complete lack of interdependence between the two classes of properties, which is most strikingly shown in the existence of substances which have to an extraordinarily close approximation the same ordinary physical and chemical properties, even though the atomic weights are not the same, and the radioactive properties are completely different. Such substances, of the existence of which the first evidence was found in the work of Soddy and other investigators on the chemical properties of the radioactive elements, are called isotopes, with reference to the classification of the elements according to ordinary physical and chemical properties. It is

[1] Lecture delivered at Stockholm, December 11, 1922, on the occasion the receipt of the Nobel prize in physics for the year 1922. English nslation by Dr. Frank C. Hoyt.

not necessary for me to state here how it has been shown in recent years that isotopes are found not only among the radioactive elements, but also among ordinary stable elements; in fact, a large number of the latter that were previously supposed simple have been shown by Aston's well-known investigations to consist of a mixture of isotopes with different atomic weights.

The question of the inner structure of the nucleus is still but little understood, although a method of attack is afforded by Rutherford's experiments on the disintegration of atomic nuclei by bombardment with α-particles. Indeed, these experiments may be said to open up a new epoch in natural philosophy in that for the first time the artificial transformation of one element into another has been accomplished. In what follows, however, we shall confine ourselves to a consideration of the ordinary physical and chemical properties of the elements and the attempts which have been made to explain them on the basis of the concepts just outlined.

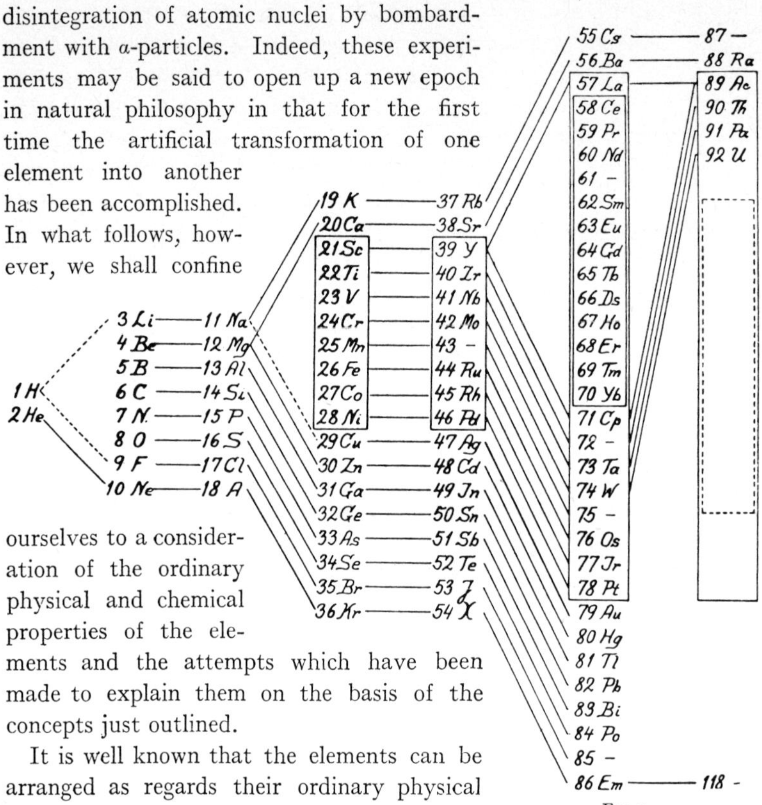

FIG. 1.

It is well known that the elements can be arranged as regards their ordinary physical and chemical properties in a *natural system* which displays most suggestively the peculiar relationships between the different elements. It was recognised for the first time by Mendeléeff and Lothar Meyer that when the elements are arranged in an order which is practically that of their atomic weights, their chemical and physical properties show a pronounced periodicity. A diagrammatic representation of this so-called periodic table is given in Fig. 1, where, however, the elements are not arranged in the ordinary way but in a somewhat modified form of a table first given by Julius Thomsen, who has also made important contributions to science in this domain. In the figure the elements are denoted by their usual chemical symbols, and the different vertical columns indicate the so-called periods. The elements in successive columns which possess homologous
[468] chemical and physical properties are connected with lines. The meaning of the square brackets aroun certain series of elements in the later periods, th properties of which exhibit typical deviations fror the simple periodicity in the first periods, will be dis cussed later.

In the development of the theory of atomic struc ture the characteristic features of the natural syste have found a surprisingly simple interpretation. Thu we are led to assume that the ordinal number an element in the periodic table, the so-called atomi number, is just equal to the number of electror which move about the nucleus in th neutral atom. In an imperfect form this law was first stated by Va den Broek; it was, however, fore shadowed by J. J. Thomson's in vestigations of the number of elec trons in the atom, as well as b Rutherford's measurements of th charge on the atomic nucleus. A we shall see, convincing support fo this law has since been obtained i various ways, especially by Moseley' famous investigations of the X-ra spectra of the elements. We ma perhaps also point out, how th simple connexion between atomi number and nuclear charge offers a explanation of the laws governin the changes in chemical propertie of the elements after expulsion o α- or β-particles, which found simple formulation in the so-calle radioactive displacement law.

Atomic Stability and Electrodynamic Theory.

As soon as we try to trace more intimate connexion between the propertie of the elements and atomic structure, we encounte profound difficulties, in that essential difference between an atom and a planetary system sho themselves here in spite of the analogy we hav mentioned.

The motions of the bodies in a planetary system even though they obey the general law of gravitation will not be completely determined by this law alon but will depend largely on the previous history the system. Thus the length of the year is no determined by the masses of the sun and the eart alone, but depends also on the conditions that existe during the formation of the solar system, of whic we have very little knowledge. Should a sufficientl large foreign body some day traverse our solar system we might among other effects expect that from tha day the length of the year would be different from its present value.

It is quite otherwise in the case of atoms. The definite and unchangeable properties of the elements demand that the state of an atom cannot undergo permanent changes due to external actions. As soon as the atom is left to itself again, its constituent particles must arrange their motions in a manner which is completely determined by the electric charges and masses of the particles. We have the most convincing evidence of this in spectra, that is, in the properties of the radiation emitted from substances in certain circumstances, which can be studied with such great precision. It is well known that the wave-lengths of the spectral lines of a substance, which can in many cases be measured with an accuracy of more than one part in a million, are, in the same external circumstances, always exactly the same within the limit of error of the measurements, and quite independent of the previous treatment of this substance. It is just to this circumstance that we owe the great importance of spectral analysis, which has been such an invaluable aid to the chemist in the search for new elements, and has also shown us that even on the most distant bodies of the universe there occur elements with exactly the same properties as on the earth.

On the basis of our picture of the constitution of the atom it is thus impossible, so long as we restrict ourselves to the ordinary mechanical laws, to account for the characteristic atomic stability which is required for an explanation of the properties of the elements.

The situation is by no means improved if we also take into consideration the well-known electrodynamic laws which Maxwell succeeded in formulating on the basis of the great discoveries of Oersted and Faraday in the first half of the last century. Maxwell's theory has not only shown itself able to account for the already known electric and magnetic phenomena in all their details, but has also celebrated its greatest triumph in the prediction of the electromagnetic waves which were discovered by Hertz, and are now so extensively used in wireless telegraphy.

For a time it seemed as though this theory would also be able to furnish a basis for an explanation of the details of the properties of the elements, after it had been developed, chiefly by Lorentz and Larmor, into a form consistent with the atomistic conception of electricity. I need only remind you of the great interest that was aroused when Lorentz, shortly after the discovery by Zeeman of the characteristic changes that spectral lines undergo when the emitting substance is brought into a magnetic field, could give a natural and simple explanation of the main features of the phenomenon. Lorentz assumed that the radiation which we observe in a spectral line is sent out from an electron executing simple harmonic vibrations about a position of equilibrium in precisely the same manner as the electromagnetic waves in radio-telegraphy are sent out by the electric oscillations in the antennæ. He also pointed out how the alteration observed by Zeeman in the spectral lines corresponded exactly to the alteration in the motion of the vibrating electron which one would expect to be produced by the magnetic field.

It was, however, impossible on this basis to give a closer explanation of the spectra of the elements, or even of the general type of the laws holding with great exactness for the wave-lengths of lines in these spectra, which had been established by Balmer, Rydberg, and Ritz. After we obtained details as to the constitution of the atom, this difficulty became still more manifest; in fact, so long as we confine ourselves to the classical electrodynamic theory we cannot even understand why we obtain spectra consisting of sharp lines at all. This theory can even be said to be incompatible with the assumption of the existence of atoms possessing the structure we have described, in that the motions of the electrons would claim a continuous radiation of energy from the atom, which would cease only when the electrons had fallen into the nucleus.

The Origin of the Quantum Theory.

It has, however, been possible to avoid the various difficulties of the electrodynamic theory by introducing concepts borrowed from the so-called quantum theory, which marks a complete departure from the ideas that have hitherto been used for the explanation of natural phenomena. This theory was originated by Planck, in the year 1900, in his investigations on the law of heat radiation, which, because of its independence of the individual properties of substances, lent itself peculiarly well to a test of the applicability of the laws of classical physics to atomic processes.

Planck considered the equilibrium of radiation between a number of systems with the same properties as those on which Lorentz had based his theory of the Zeeman effect, but he could now show not only that classical physics could not account for the phenomena of heat radiation, but also that a complete agreement with the experimental law could be obtained if—in pronounced contradiction to classical theory—it were assumed that the energy of the vibrating electrons could not change continuously, but only in such a way that the energy of the system always remained equal to a whole number of so-called energy-quanta. The magnitude of this quantum was found to be proportional to the frequency of oscillation of the particle, which, in accordance with classical concepts, was supposed to be also the frequency of the emitted radiation. The proportionality factor had to be regarded as a new universal constant, since termed Planck's constant, similar to the velocity of light and the charge and mass of the electron.

Planck's surprising result stood at first completely isolated in natural science, but with Einstein's significant contributions to this subject a few years after, a great variety of applications was found. In the first place, Einstein pointed out that the condition limiting the amount of vibrational energy of the particles could be tested by investigation of the specific heat of crystalline bodies, since in the case of these we have to do with similar vibrations, not of a single electron, but of whole atoms about positions of equilibrium in the crystal lattice. Einstein was able to show that the experiment confirmed Planck's theory, and through the work of later investigators this agreement has proved quite complete. Furthermore, Einstein emphasised another consequence of Planck's results, namely, that radiant energy could only be emitted or absorbed by the oscillating particle in so-called "quanta of radiation," the magnitude of each of which was equal to Planck's constant multiplied by the frequency.

In his attempts to give an interpretation of this result, Einstein was led to the formulation of the so-called "hypothesis of light-quanta," according to which the radiant energy, in contradiction to Maxwell's electromagnetic theory of light, would not be propagated as electromagnetic waves, but rather as concrete light atoms, each with an energy equal to that of a quantum of radiation. This concept led Einstein to his well-known theory of the photo-electric effect. This phenomenon, which had been entirely unexplainable on the classical theory, was thereby placed in a quite different light, and the predictions of Einstein's theory have received such exact experimental confirmation in recent years, that perhaps the most exact determination of Planck's constant is afforded by measurements on the photo-electric effect. In spite of its heuristic value, however, the hypothesis of light-quanta, which is quite irreconcilable with so-called interference phenomena, is not able to throw light on the nature of radiation. I need only recall that these interference phenomena constitute our only means of investigating the properties of radiation and therefore of assigning any closer meaning to the frequency which in Einstein's theory fixes the magnitude of the light-quantum.

In the following years many efforts were made to apply the concepts of the quantum theory to the question of atomic structure, and the principal emphasis was sometimes placed on one and sometimes on the other of the consequences deduced by Einstein from Planck's result. As the best known of the attempts in this direction, from which, however, no definite results were obtained, I may mention the work of Stark, Sommerfeld, Hasenöhrl, Haas, and Nicholson.

From this period also dates an investigation by Bjerrum on infra-red absorption bands, which, although it had no direct bearing on atomic structure, proved significant for the development of the quantum theory. He directed attention to the fact that the rotation of the molecules in a gas might be investigated by means of the changes in certain absorption lines with temperature. At the same time he emphasised the fact that the effect should not consist of a continuous widening of the lines such as might be expected from classical theory, which imposed no restrictions on the molecular rotations, but in accordance with the quantum theory he predicted that the lines should be split up into a number of components, corresponding to a sequence of distinct possibilities of rotation. This prediction was confirmed a few years later by Eva von Bahr, and the phenomenon may still be regarded as one of the most striking evidences of the reality of the quantum theory, even though from our present point of view the original explanation has undergone a modification in essential details.

The Quantum Theory of Atomic Constitution.

The question of further development of the quantum theory was in the meantime placed in a new light by Rutherford's discovery of the atomic nucleus (1911). As we have already seen, this discovery made it quite clear that by classical conceptions alone it was quite impossible to understand the most essential properties of atoms. One was therefore led to seek for a formulation of the principles of the quantum theory that could immediately account for the stability in atomic structure and the properties of the radiation sent out from atoms, of which the observed properties of substances bear witness. Such a formulation was proposed (1913) by the present lecturer in the form of two postulates, which may be stated as follows:

1. Among the conceivably possible states of motion in an atomic system there exist a number of so-called *stationary states* which, in spite of the fact that the motion of the particles in these states obeys the laws of classical mechanics to a considerable extent, possess a peculiar, mechanically unexplainable stability, of such a sort that every permanent change in the motion of the system must consist in a complete transition from one stationary state to another.

2. While in contradiction to the classical electromagnetic theory no radiation takes place from the atom in the stationary states themselves, a process of transition between two stationary states can be accompanied by the emission of electromagnetic radiation, which will have the same properties as that which would be sent out according to the classical theory from an electrified particle executing an harmonic vibration with constant frequency. This frequency ν has, however, no simple relation to the motion of the particles of the atom, but is given by the relation

$$h\nu = E' - E'',$$

where h is Planck's constant, and E' and E'' are the values of the energy of the atom in the two stationary states that form the initial and final state of the radiation process. Conversely, irradiation of the atom with electromagnetic waves of this frequency can lead to an absorption process, whereby the atom is transformed back from the latter stationary state to the former.

While the first postulate has in view the general :ability of the atom, the second postulate has chiefly ı view the existence of spectra with sharp lines. 'urthermore, the quantum theory condition entering ı the last postulate affords a starting point for the ıterpretation of the laws of series spectra. The ıost general of these laws, the combination principle ınunciated by Ritz, states that the frequency ν for ach of the lines in the spectrum of an element can e represented by the formula

$$\nu = T'' - T',$$

'here T'' and T' are two so-called " spectral terms " elonging to a manifold of such terms characteristic f the substance in question.

According to our postulates, this law finds an nmediate interpretation in the assumption that the pectrum is emitted by transitions between a number f stationary states in which the numerical value of he energy of the atom is equal to the value of the pectral term multiplied by Planck's constant. This xplanation of the combination principle is seen ɔ differ fundamentally from the usual ideas of lectrodynamics, as soon as we consider that there ; no simple relation between the motion of the atom nd the radiation sent out. The departure of our onsiderations from the ordinary ideas of natural hilosophy becomes particularly evident, however, 'hen we observe that the occurrence of two spectral nes, corresponding to combinations of the same pectral term with two other different terms, implies ıat the nature of the radiation sent out from the tom is not determined only by the motion of the atom at the beginning of the radiation process, but also depends on the state to which the atom is transferred by the process.

At first glance one might, therefore, think that it would scarcely be possible to bring our formal explanation of the combination principle into direct relation with our views regarding the constitution of the atom, which, indeed, are based on experimental evidence interpreted on classical mechanics and electrodynamics. A closer investigation, however, should make it clear that a definite relation may be obtained between the spectra of the elements and the structure of their atoms on the basis of the postulates.

The Hydrogen Spectrum.

The simplest spectrum we know is that of hydrogen. The frequencies of its lines may be represented with great accuracy by means of Balmer's formula :

$$\nu = K\left(\frac{1}{n''^2} - \frac{1}{n'^2}\right),$$

where K is a constant and n' and n'' are two integers. In the spectrum we accordingly meet a single series of spectral terms of the form K/n^2, which decrease regularly with increasing term number n. In accordance with the postulates, we shall therefore assume that each of the hydrogen lines is emitted by a transition between two states belonging to a series of stationary states of the hydrogen atom in which the numerical value of the atom's energy is equal to hK/n^2.

Following our picture of atomic structure, a hydrogen atom consists of a positive nucleus and an electron which—so far as ordinary mechanical conceptions are applicable—will with great approximation describe a periodic elliptical orbit with the nucleus at one focus. The major axis of the orbit is inversely proportional to the work necessary completely to remove the electron from the nucleus, and, in accordance with the above, this work in the stationary states is just equal to hK/n^2. We thus arrive at a manifold of stationary states for which the major axis of the electron orbit takes on a series of discrete values proportional to the squares of the whole numbers. The accompanying Fig. 2 shows these relations diagrammatically. For the sake of simplicity the electron orbits in the stationary states are represented by circles, although in reality the theory places no restriction on the eccentricity of the orbit, but only determines the length of the major axis. The arrows represent the transition processes that correspond to the red and green hydrogen lines, $H\alpha$ and $H\beta$, the frequency of which is given by means of the Balmer formula when we put $n''=2$ and $n'=3$ and 4 respectively. The transition processes are also represented which correspond to the first three lines of the series of ultra-violet lines found by Lyman in 1914, of which the frequencies are given by the formula when n is put equal to 1, as well as to the first line of the infra-red series discovered some years previously by Paschen, which are given by the formula if n'' is put equal to 3.

This explanation of the origin of the hydrogen spectrum leads us quite naturally to interpret this spectrum as the manifestation of a process whereby the electron is bound to the nucleus. While the largest spectral term with term number 1 corresponds to the final stage in the binding process, the small spectral terms that have larger values of the term number correspond to stationary states which represent the initial states of the binding process, where the

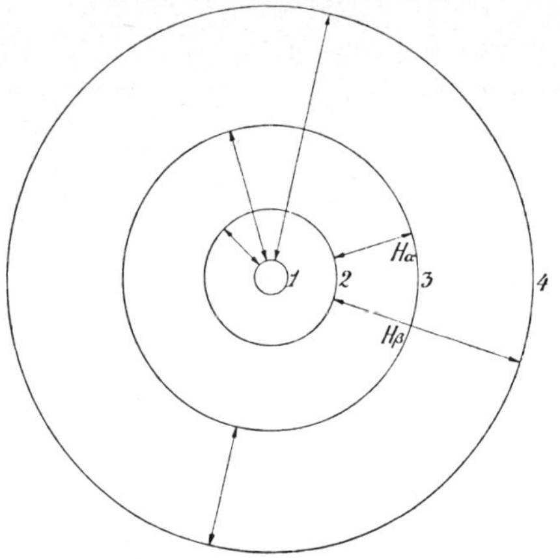

FIG. 2.

electron orbits still have large dimensions, and where the work required to remove an electron from the nucleus is still small. The final stage in the binding process we may designate as the normal state of the atom, and it is distinguished from the other stationary states by the property that, in accordance with the postulates, the state of the atom can only be changed by the addition of energy whereby the electron is transferred to an orbit of larger dimensions corresponding to an earlier stage of the binding process.

The size of the electron orbit in the normal state calculated on the basis of the above interpretation of the spectrum agrees roughly with the value for the dimensions of the atoms of the elements that have been calculated by the kinetic theory of matter from the properties of gases. Since, however, as an immediate consequence of the stability of the stationary states that is claimed by the postulates, we must suppose that the interaction between two atoms during a collision cannot be completely described with the aid of the laws of classical mechanics, such a comparison as this cannot be carried further on the basis of such considerations as those just outlined.

A more intimate connexion between the spectra
[472] and the atomic model has been revealed, however, by an investigation of the motion in those stationar states where the term number is large, and wher the dimensions of the electron orbit and the frequenc of revolution in it vary relatively little when we g from one stationary state to the next following. I was possible to show that the frequency of the radiatio sent out during the transition between two stationar states, the difference of the term numbers of whic is small in comparison to these numbers themselves tended to coincide in frequency with one of th harmonic components into which the electron motio could be resolved, and accordingly also with th frequency of one of the wave trains in the radiatio which would be emitted according to the laws o ordinary electrodynamics.

The condition that such a coincidence should occu in this region where the stationary states differ bu little from one another proves to be that the constan in the Balmer formula can be expressed by means o the relation

$$K=\frac{2\pi^2e^4m}{h^3},$$

where e and m are respectively the charge and mas of the electron, while h is Planck's constant. Thi relation has been shown to hold to within the con siderable accuracy with which, especially through th beautiful investigations of Millikan, the quantitie e, m, and h are known.

This result shows that there exists a connexio between the hydrogen spectrum and the model fo the hydrogen atom which, on the whole, is as clos as we might hope considering the departure of th postulates from the classical mechanical and electro dynamic laws. At the same time, it affords som indication of how we may perceive in the quantun theory, in spite of the fundamental character of thi departure, a natural generalisation of the fundamenta concepts of the classical electrodynamic theory. T this most important question we shall return later but first we will discuss how the interpretation o the hydrogen spectrum on the basis of the postulate has proved suitable in several ways, for elucidating th relation between the properties of the different elements

RELATIONSHIPS BETWEEN THE ELEMENTS.

The discussion above can be applied immediatel to the process whereby an electron is bound to a nucleus with any given charge. The calculation show that, in the stationary state corresponding t a given value of the number n, the size of the orbi will be inversely proportional to the nuclear charge while the work necessary to remove an electron wil be directly proportional to the square of the nuclea charge. The spectrum that is emitted during th binding of an electron by a nucleus with charge N mes that of the hydrogen nucleus can therefore be ·presented by the formula :

$$\nu = \mathrm{N}^2\mathrm{K}\left(\frac{\mathrm{I}}{n''^2} - \frac{\mathrm{I}}{n'^2}\right).$$

If in this formula we put N=2, we get a spectrum hich contains a set of lines in the visible region hich was observed many years ago in the spectrum : certain stars. Rydberg assigned these lines to ydrogen because of the close analogy with the series : lines represented by the Balmer formula. It was ever possible to produce these lines in pure hydrogen, ıt just before the theory for the hydrogen spectrum as put forward, Fowler succeeded in observing the ·ries in question by sending a strong discharge rough a mixture of hydrogen and helium. This vestigator also assumed that the ıes were hydrogen lines, because ıere existed no experimental vidence from which it might be ıferred that two different sub- ances could show properties ·sembling each other so much as ıe spectrum in question and that : hydrogen. After the theory as put forward, it became clear, owever, that the observed lines ıust belong to a spectrum of elium, but that they were not ke the ordinary helium spectrum nitted from the neutral atom. hey came from an ionised helium tom which consists of a single ectron moving about a nucleus ith double charge. In this way ıere was brought to light a new ·ature of the relationship between ıe elements, which corresponds xactly with our present ideas of tomic structure, according to which the physical and ıemical properties of an element depend in the first ıstance only on the electric charge of the atomic ucleus.

Soon after this question was settled the existence f a similar general relationship between the properties f the elements was brought to light by Moseley's ·ell-known investigations on the characteristic X-ray ɔectra of the elements, which was made possible y Laue's discovery of the interference of X-rays in ·ystals and the investigations of W. H. and W. L. ·ragg on this subject. It appeared, in fact, that ıe X-ray spectra of the different elements possessed much simpler structure and a much greater mutual ·semblance than their optical spectra. In particular, appeared that the spectra changed from element to element in a manner that corresponded closely to the formula given above for the spectrum emitted during the binding of an electron to a nucleus, provided N was put equal to the atomic number of the element concerned. This formula was even capable of expressing, with an approximation that could not be without significance, the frequencies of the strongest X-ray lines, if small whole numbers were substituted for n' and n''.

This discovery was of great importance in several respects. In the first place, the relationship between the X-ray spectra of different elements proved so simple that it became possible to fix without ambiguity the atomic number for all known substances, and in this way to predict with certainty the atomic number of all such hitherto unknown elements for which there is a place in the natural system. Fig. 3 shows how the square root of the frequency for two characteristic X-ray lines depends on the atomic number. These lines belong to the group of so-called K-lines, which are the most penetrating of the characteristic rays. With very close approximation the points lie on straight lines, and the fact that they do so is conditioned not only by our taking account of known elements, but also by our leaving an open place between molybdenum (42) and ruthenium (44), just as in Mendeléeff's original scheme of the natural system of the elements.

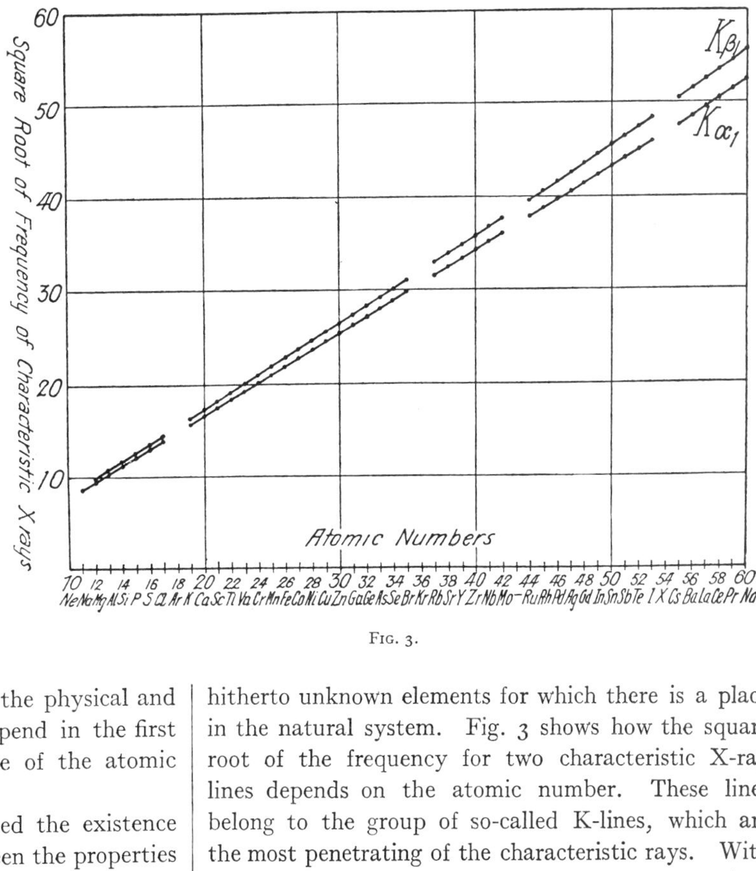

FIG. 3.

Further, the laws of X-ray spectra provide a confirmation of the general theoretical conceptions, both with regard to the constitution of the atom and the ideas that have served as a basis for the interpretation of spectra. Thus the similarity between X-ray spectra and the spectra emitted during the binding of a single electron to a nucleus may be simply interpreted from the fact that the transitions between stationary states with which we are concerned in X-ray spectra are accompanied by changes in the motion of an electron in the inner part of the atom, where the influence of the attraction of the nucleus is very great compared with the repulsive forces of the other electrons.

The relations between other properties of the elements are of a much more complicated character, which originates in the fact that we have to do with processes concerning the motion of the electrons in the outer part of the atom, where the forces that the electrons

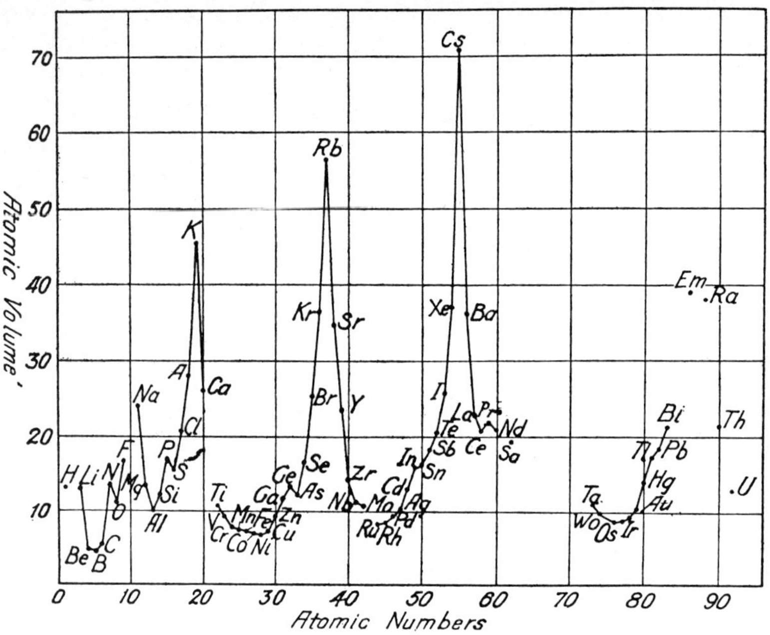

FIG. 4.

exert on one another are of the same order of magnitude as the attraction towards the nucleus, and where, therefore, the details of the interaction of the electrons play an important part. A characteristic example of such a case is afforded by the spatial extension of the atoms of the elements. Lothar Meyer himself directed attention to the characteristic periodic change exhibited by the ratio of the atomic weight to the density, the so-called atomic volume, of the elements in the natural system. An idea of these facts is given by Fig. 4, in which the atomic volume is represented as a function of the atomic number. A greater difference between this and the previous figure could scarcely be imagined. While the X-ray spectra vary uniformly with the atomic number, the atomic volumes show a characteristic periodic change which corresponds exactly to the change in the chemical
[474] properties of the elements.

Ordinary optical spectra behave in an analogou way. In spite of the dissimilarity between thes spectra, Rydberg succeeded in tracing a certai general relationship between the hydrogen spec trum and other spectra. Even though the spectra lines of the elements with higher atomic numbe appear as combinations of a more complicated manifol of spectral terms which is not so simply co-ordinate with a series of whole numbers, still the spectral term can be arranged in series each of which shows a stron similarity to the series of terms in the hydroge spectrum. This similarity appears in the fact tha the terms in each series can, as Rydberg pointed out be very accurately represented by the formula $K/(n+\alpha)^2$ where K is the same constant that occurs in th hydrogen spectrum, often called th Rydberg constant, while n is th term number, and α a constan which is different for the differen series.

This relationship with the hydro gen spectrum leads us immediatel to regard these spectra as the *las step of a process whereby the neutra atom is built up by the capture an binding of electrons to the nucleus* one by one. In fact, it is clear tha the last electron captured, so long a it is in that stage of the bindin process in which its orbit is stil large compared to the orbits of th previously bound electrons, will b subjected to a force from th nucleus and these electrons, tha differs but little from the force wit which the electron in the hydroge atom is attracted towards the nucleu while it is moving in an orbit of correspondin dimensions.

The spectra so far considered, for which Rydberg' laws hold, are excited by means of electric dis charge under ordinary conditions and are often calle arc spectra. The elements emit also another typ of spectrum, the so-called spark spectra, when the are subjected to an extremely powerful discharge Hitherto it was impossible to disentangle the spar spectra in the same way as the arc spectra. Shortl after the above view on the origin of arc spectr was brought forward, however, Fowler found (1914 that an empirical expression for the spark spectrum lines could be established which corresponds exactl to Rydberg's laws with the single difference that th constant K is replaced by a constant four times a large. Since, as we have seen, the constant tha appears in the spectrum sent out during the bindin f an electron to a helium nucleus is exactly equal ɔ 4 K, it becomes evident that spark spectra are due ɔ the ionised atom, and that their emission corresponds ɔ *the last step but one in the formation of the neutral tom* by the successive capture and binding of electrons.

Absorption and Excitation of Spectral Lines.

The interpretation of the origin of the spectra was lso able to explain the characteristic laws that govern bsorption spectra. As Kirchhoff and Bunsen had lready shown, there is a close relation between the elective absorption of substances for radiation and heir emission spectra, and it is on this that the pplication of spectrum analysis to the heavenly odies essentially rests. Yet on the basis of the lassical electromagnetic theory, it is impossible to nderstand why substances in the form of vapour how absorption for certain lines in their emission pectrum and not for others.

On the basis of the postulates given above we are, owever, led to assume that the absorption of radiation orresponding to a spectral line emitted by a transition rom one stationary state of the atom to a state of ess energy is brought about by the return of the atom rom the last-named state to the first. We thus nderstand immediately that in ordinary circumstances gas or vapour can only show selective absorption or spectral lines that are produced by a transition rom a state corresponding to an earlier stage in the inding process to the normal state. Only at higher emperatures or under the influence of electric discharges whereby an appreciable number of atoms are eing constantly disrupted from the normal state, an we expect absorption for other lines in the emission pectrum in agreement with the experiments.

A most direct confirmation for the general interretation of spectra on the basis of the postulates as also been obtained by investigations on the xcitation of spectral lines and ionisation of atoms y means of impact of free electrons with given elocities. A decided advance in this direction was narked by the well-known investigations of Franck nd Hertz (1914). It appeared from their results hat by means of electron impacts it was impossible o impart to an atom an arbitrary amount of energy, ut only such amounts as corresponded to a transfer f the atom from its normal state to another stationary tate of the existence of which the spectra assure us, nd the energy of which can be inferred from the nagnitude of the spectral term.

Further, striking evidence was afforded of the inependence that, according to the postulates, must e attributed to the processes which give rise to the mission of the different spectral lines of an element. Thus it could be shown directly that atoms that were transferred in this manner to a stationary state of greater energy were able to return to the normal state with emission of radiation corresponding to a single spectral line.

Continued investigations on electron impacts, in which a large number of physicists have shared, have also produced a detailed confirmation of the theory concerning the excitation of series spectra. Especially it has been possible to show that for the *ionisation* of an atom by electron impact an amount of energy is necessary that is exactly equal to the work required, according to the theory, to remove the last electron captured from the atom. This work can be determined directly as the product of Planck's constant and the spectral term corresponding to the normal state, which, as mentioned above, is equal to the limiting value of the frequencies of the spectral series connected with selective absorption.

The Quantum Theory of Multiply-Periodic Systems.

While it was thus possible by means of the fundamental postulates of the quantum theory to account directly for certain general features of the properties of the elements, a closer development of the ideas of the quantum theory was necessary in order to account for these properties in further detail. In the course of the last few years a more general theoretical basis has been attained through the development of formal methods that permit the fixation of the stationary states for electron motions of a more general type than those we have hitherto considered. For a simply periodic motion such as we meet in the pure harmonic oscillator, and at least to a first approximation, in the motion of an electron about a positive nucleus, the manifold of stationary states can be simply co-ordinated to a series of whole numbers. For motions of the more general class mentioned above, the so-called *multiply-periodic* motions, however, the stationary states compose a more complex manifold, in which, according to these formal methods, each state is characterised by several whole numbers, the so-called "quantum numbers."

In the development of the theory a large number of physicists have taken part, and the introduction of several quantum numbers can be traced back to the work of Planck himself. But the definite step which gave the impetus to further work was made by Sommerfeld (1915) in his explanation of the fine structure shown by the hydrogen lines when the spectrum is observed with a spectroscope of high resolving power. The occurrence of this fine structure must be ascribed to the circumstance that we have to deal, even in hydrogen, with a motion which is not exactly simply periodic. In fact, as a consequence of the change in the electron's mass with velocity that is claimed by the theory of relativity, the electron orbit will undergo a very slow precession in the orbital plane. The motion will therefore be doubly periodic, and besides a number characterising the term in the Balmer formula, which we shall call the *principal quantum number* because it determines in the main the energy of the atom, the fixation of the stationary states demands another quantum number which we shall call the *subordinate quantum number*.

A survey of the motion in the stationary states thus fixed is given in the diagram (Fig. 5), which reproduces the relative size and form of the electron orbits. Each orbit is designated by a symbol n_k, where n is the principal quantum number and k the subordinate quantum number. All orbits with the same principal quantum number have, to a first approximation, the same major axis, while orbits with the same value of k have the same parameter, *i.e.* the same value for the shortest chord through the focus. Since the energy values for different states with the same value of n but different values of k differ a little from each other, we get for each hydrogen line corresponding to definite values of n' and n'' in the Balmer formula a number of different transition processes, for which the frequencies of the emitted radiation as calculated by the second postulate are

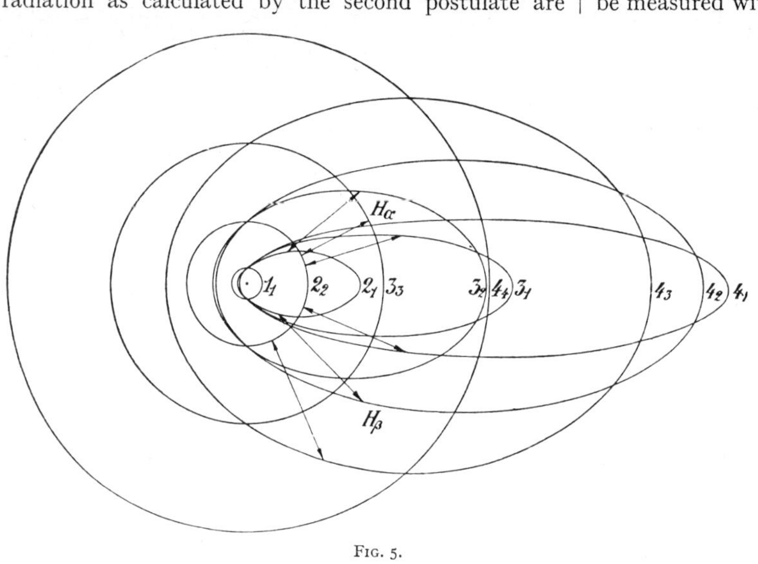

FIG. 5.

not exactly the same. As Sommerfeld was able to show, the components this gives for each hydrogen line agree with the observations on the fine structure of hydrogen lines to within the limits of experimental error. In the figure the arrows designate the processes that give rise to the components of the red and green lines in the hydrogen spectrum, the frequencies of which are obtained by putting $n''=2$ and $n'=3$ or 4 respectively in the Balmer formula.

In considering the figure it must not be forgotten that the description of the orbit is there incomplete, in so much as with the scale used the slow precession does not show at all. In fact, this precession is so slow that even for the orbits that rotate most rapidly the electron performs about 40,000 revolutions before the perihelion has gone round once. Nevertheless, it is this precession alone that is responsible for the multiplicity of the stationary states characterised by the subordinate quantum number. If, for example, the hydrogen atom is subjected to a small disturbing force which perturbs the regular precession, the electron orbit in the stationary states will have a form altogether different from that given in the figure.
[476] This implies that the fine structure will change its character completely, but the hydrogen spectrum wi continue to consist of lines that are given to a clos approximation by the Balmer formula, due to th fact that the approximately periodic character of th motion will be retained. Only when the disturbin forces become so large that even during a sing revolution of the electron the orbit is appreciabl disturbed, will the spectrum undergo essential change The statement often advanced that the introduction o two quantum numbers should be a necessary conditio for the explanation of the Balmer formula must there fore be considered as a misconception of the theory.

Sommerfeld's theory has proved itself able t account not only for the fine structure of the hydroge lines, but also for that of the lines in the helium spar spectrum. Owing to the greater velocity of th electron, the intervals between the components int which a line is split up are here much greater and ca be measured with much greater accuracy. The theor was also able to account for certai features in the fine structure o X-ray spectra, where we meet fre quency differences that may eve reach a value more than a millio times as great as those of the fre quency differences for the com ponents of the hydrogen lines.

Shortly after this result had bee attained, Schwarzschild and Epstei (1916) simultaneously succeeded, b means of similar considerations, i accounting for the characterist changes that the hydrogen line undergo in an electric field, whic had been discovered by Stark i the year 1914. Next, an ex planation of the essential feature of the Zeeman effect for the hydro gen lines was worked out at th same time by Sommerfeld and Deby (1917). In this instance the applica tion of the Postulates involved the consequenc that only certain orientations of the atom relativ to the magnetic field were allowable, and this characte istic consequence of the quantum theory has quit recently received a most direct confirmation in th beautiful researches of Stern and Gerlach on th deflexion of swiftly-moving silver atoms in a non homogenous magnetic field.

THE CORRESPONDENCE PRINCIPLE.

While this development of the theory of spectr was based on the working out of formal method for the fixation of stationary states, the presen lecturer succeeded shortly afterwards in throwin light on the theory from a new view-point, by pursuin further the characteristic connexion between th quantum theory and classical electrodynamics alread traced out in the hydrogen spectrum. In connexio with the important work of Ehrenfest and Einstei these efforts led to the formulation of the so-calle *correspondence principle,* according to which th occurrence of transitions between the stationary state accompanied by emission of radiation is traced bac to the harmonic components into which the motio of the atom may be resolved and which, according to the classical theory, determine the properties of the radiation to which the motion of the particles gives rise.

According to the correspondence principle, it is assumed that every transition process between two stationary states can be co-ordinated with a corresponding harmonic vibration component in such a way that the probability of the occurrence of the transition is dependent on the amplitude of the vibration. The state of polarisation of the radiation emitted during the transition depends on the further characteristics of the vibration, in a manner analogous to that in which on the classical theory the intensity and state of polarisation in the wave system emitted by the atom as a consequence of the presence of this vibration component would be determined respectively by the amplitude and further characteristics of the vibration.

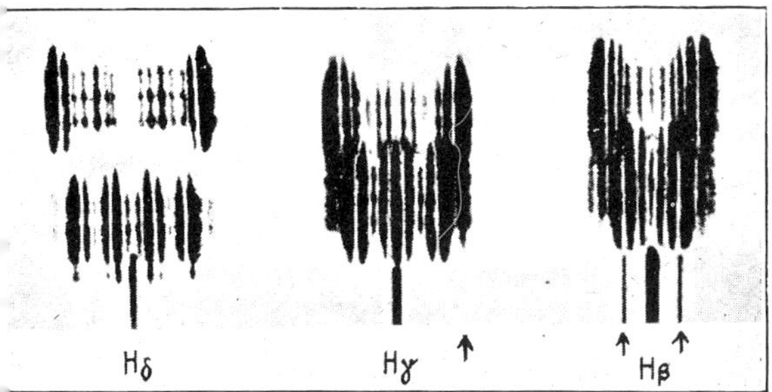

FIG. 6.

With the aid of the correspondence principle it has been possible to confirm and to extend the above-mentioned results. Thus it was possible to develop a complete quantum theory explanation of the Zeeman effect for the hydrogen lines, which, in spite of the essentially different character of the assumptions that underlie the two theories, is very similar throughout to Lorentz's original explanation based on the classical theory. In the case of the Stark effect, where, on the other hand, the classical theory was completely at a loss, the quantum theory explanation could be so extended with the help of the correspondence principle as to account for the polarisation of the different components into which the lines are split, and also for the characteristic intensity distribution exhibited by the components. This last question has been more closely investigated by Kramers, and the accompanying figure will give some impression of how completely it is possible to account for the phenomenon under consideration.

Fig. 6 reproduces one of Stark's well-known photographs of the splitting up of the hydrogen lines. The picture displays very well the varied nature of the phenomenon, and shows in how peculiar a fashion the intensity varies from component to component. The components below are polarised perpendicular to the field, while those above are polarised parallel to the field.

Fig. 7 gives a diagrammatic representation of the experimental and theoretical results for the line Hγ, the frequency of which is given by the Balmer formula with $n''=2$ and $n'=5$. The vertical lines denote the components into which the line is split up, of which the picture on the right gives the components which are polarised parallel to the field and that on the left those that are polarised perpendicular to it. The experimental results are represented in the upper half of the diagram, the distances from the dotted line representing the measured displacements of the components, and the lengths of the lines being proportional to the relative intensity as estimated by Stark from the blackening of the photographic plate. In the lower half is given for comparison a representation of the theoretical results from a drawing in Kramers' paper.

The symbol $(n'_{s'}-n''_{s''})$ attached to the lines gives the transitions between the stationary states of the atom in the electric field by which the components are emitted. Besides the principal quantum integer n, the stationary states are further characterised by a subordinate quantum integer s, which can be negative as well as positive and has a meaning quite different from that of the quantum number k occurring in the relativity theory of the fine structure of the hydrogen lines, which fixed the form of the electron orbit in the undisturbed atom. Under the influence of the electric field both the form of the orbit and its position undergo large changes, but certain properties of the orbit remain unchanged, and the subordinate quantum number s is connected with these. In Fig. 7 the position of the components corresponds to the frequencies calculated for the different transitions, and the lengths of the lines are proportional to the probabilities as calculated on the basis of the correspondence principle, by which also the polarisation of the radiation is determined. It is seen that the theory reproduces completely the main feature of the experimental results, and in the light of the correspondence principle we can say that the Stark effect reflects down to the smallest details the action of the electric field on the orbit of the electron in the hydrogen atom, even though in this case the reflection is so distorted that, in contrast with the case of the Zeeman effect, it would scarcely be possible directly

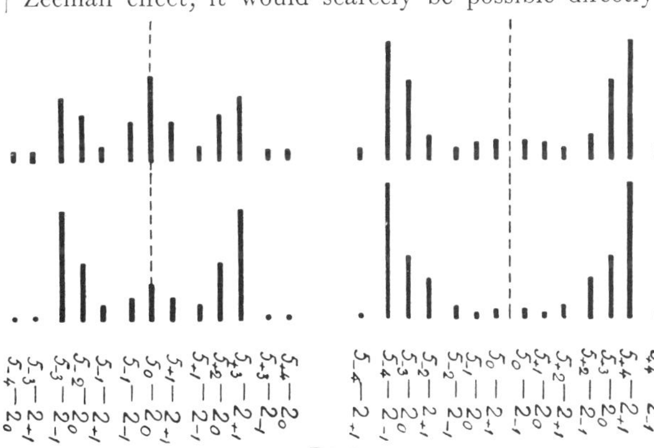

FIG. 7.

to recognise the motion on the basis of the classical ideas of the origin of electromagnetic radiation.

Results of interest were also obtained for the spectra of elements of higher atomic number, the explanation of which in the meantime had made important progress through the work of Sommerfeld, who introduced several quantum numbers for the description of the electron orbits. Indeed, it was possible, with the aid of the correspondence principle, to account completely for the characteristic rules which govern the seemingly capricious occurrence of combination lines, and it is not too much to say that the quantum theory has not only provided a simple interpretation of the combination principle, but has further contributed materially to the clearing up of the mystery that has long rested over the application of this principle.

The same view-points have also proved fruitful in the investigation of the so-called band spectra. These do not originate, as do series spectra, from individual atoms, but from molecules; and the fact that these spectra are so rich in lines is due to the complexity of the motion entailed by the vibrations of the atomic nuclei relative to each other and the rotations of the molecule as a whole. The first to apply the postulates to this problem was Schwarzschild, but the important work of Heurlinger especially has thrown much light on the origin and structure of band spectra. The considerations employed here can be traced back directly to those discussed at the beginning of this lecture in connexion with Bjerrum's theory of the influence of molecular rotation on the infra-red absorption lines of gases. It is true we no longer think that the rotation is reflected in the spectra in the way claimed by classical electrodynamics, but rather that the line components are due to transitions between stationary states which differ as regards rotational motion. That the phenomenon retains its essential features, however, is a typical consequence of the correspondence principle.

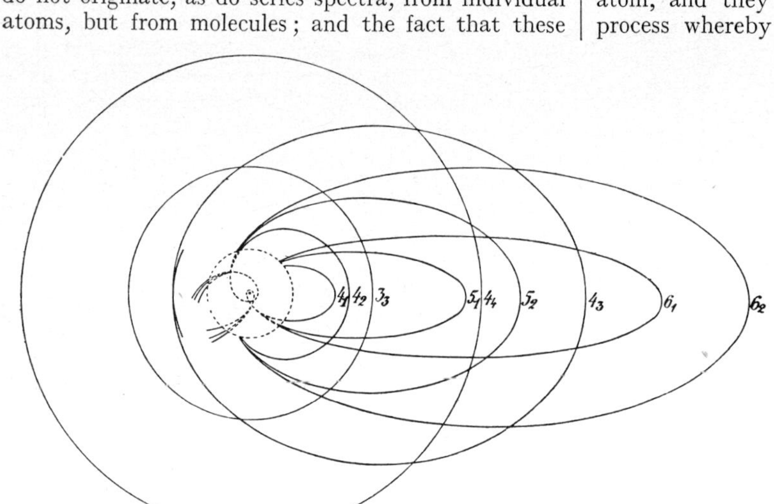

Fig. 8.

The Natural System of the Elements.

The ideas of the origin of spectra outlined in the preceding have furnished the basis for a theory of the structure of the atoms of the elements which has shown itself suitable for a general interpretation of the main features of the properties of the elements, as exhibited in the natural system. This theory is based primarily on considerations of the manner in which the atom can be imagined to be built up by the
[478] capture and binding of electrons to the nucleus, one by one. As we have seen, the optical spectra of elements provide us with evidence on the progress of the last steps in this building up process.

An insight into the kind of information that the closer investigation of the spectra has provided in this respect may be obtained from Fig. 8, which gives a diagrammatic representation of the orbital motion in the stationary states corresponding to the emission of the arc-spectrum of potassium. The curves show the form of the orbits described in the stationary states by the last electron captured in the potassium atom, and they can be considered as stages in the process whereby the 19th electron is bound after the 18 previous electrons have already been bound in their normal orbits. In order not to complicate the figure, no attempt has been made to draw any of the orbits of these inner electrons, but the region in which they move is enclosed by a dotted circle. In an atom with several electrons the orbits will, in general, have a complicated character. Because of the symmetrical nature of the field of force about the nucleus, however, the motion of each single electron can be approximately described as a plane periodic motion on which is superimposed a uniform rotation in the plane of the orbit. The orbit of each electron will therefore be to a first approximation doubly periodic and will be fixed by two quantum numbers, as are the stationary states in a hydrogen atom when the relativity precession is taken into account.

In Fig. 8, as in Fig. 5, the electron orbits are marked with the symbol n_k, where n is the principal quantum number and k the subordinate quantum number. While for the initial states of the binding process, where the quantum numbers are large, the orbit of the last electron captured lies completely outside of those of the previously bound electrons, this is not the case for the last stages. Thus, in the potassium atom, the electron orbits with subordinate quantum numbers 2 and 1 will, as indicated in the figure, penetrate partly into the inner region. Because of this circumstance, the orbits will deviate very greatly from a simple Kepler motion, since they will consist of a series of successive outer loops that have the same size and form, but each of which is turned through an appreciable angle relative to the preceding one. Of these outer loops only one is shown in the figure. Each of them coincides very nearly with a piece of a Kepler ellipse, and they are connected, as indicated by a series of inner loops of a complicated character in which the electron approaches the nucleus closely. This holds especially for the orbit with subordinate quantum number 1, which, as a closer investigation shows, will approach nearer to the nucleus than any of the previously bound electrons.

On account of this penetration into the inner region, the strength with which an electron in such an orbit is bound to the atom will—in spite of the fact that for the most part it moves in a field of force of the ame character as that surrounding the hydrogen ucleus—be much greater than for an electron in a ıydrogen atom that moves in an orbit with the same ›rincipal quantum number, the maximum distance f the electron from the nucleus at the same time ›eing considerably less than in such a hydrogen orbit. \s we shall see, this feature of the binding process n atoms with many electrons is of essential importance n order to understand the characteristic periodic vay in which many properties of the elements as lisplayed in the natural system vary with the atomic umber.

In the accompanying table (Fig. 9) is given a ummary of the results concerning the structure of

	1_1	$2_1 2_2$	$3_1 3_2 3_3$	$4_1 4_2 4_3 4_4$	$5_1 5_2 5_3 5_4 5_5$	$6_1 6_2 6_3 6_4 6_5 6_6$	$7_1 7_2$
1 H	1						
2 He	2						
3 Li	2	1					
4 Be	2	2					
5 B	2	2(1)					
– –	–	– –					
10 Ne	2	4 4					
11 Na	2	4 4	1				
12 Mg	2	4 4	2				
13 Al	2	4 4	2 1				
– –	–	– –	– –				
18 A	2	4 4	4 4				
19 K	2	4 4	4 4	1			
20 Ca	2	4 4	4 4	2			
21 Sc	2	4 4	4 4 1	(2)			
22 Ti	2	4 4	4 4 2	(2)			
– –	–	– –	– – –	–			
29 Cu	2	4 4	6 6 6	1			
30 Zn	2	4 4	6 6 6	2			
31 Ga	2	4 4	6 6 6	2 1			
– –	–	– –	– – –	– –			
36 Kr	2	4 4	6 6 6	4 4			
37 Rb	2	4 4	6 6 6	4 4	1		
38 Sr	2	4 4	6 6 6	4 4	2		
39 Y	2	4 4	6 6 6	4 4 1	(2)		
40 Zr	2	4 4	6 6 6	4 4 2	(2)		
– –	–	– –	– – –	– – –	–		
47 Ag	2	4 4	6 6 6	6 6 6	1		
48 Cd	2	4 4	6 6 6	6 6 6	2		
49 In	2	4 4	6 6 6	6 6 6	2 1		
– –	–	– –	– – –	– – –	– –		
54 X	2	4 4	6 6 6	6 6 6	4 4		
55 Cs	2	4 4	6 6 6	6 6 6	4 4	1	
56 Ba	2	4 4	6 6 6	6 6 6	4 4	2	
57 La	2	4 4	6 6 6	6 6 6	4 4 1	(2)	
58 Ce	2	4 4	6 6 6	6 6 6 1	4 4 1	(2)	
59 Pr	2	4 4	6 6 6	6 6 6 2	4 4 1	(2)	
– –	–	– –	– – –	– – – –	– – –	–	
71 Cp	2	4 4	6 6 6	8 8 8 8	4 4 1	(2)	
72 –	2	4 4	6 6 6	8 8 8 8	4 4 2	(2)	
– –	–	– –	– – –	– – – –	– – –	–	
79 Au	2	4 4	6 6 6	8 8 8 8	6 6 6	1	
80 Hg	2	4 4	6 6 6	8 8 8 8	6 6 6	2	
81 Tl	2	4 4	6 6 6	8 8 8 8	6 6 6	2 1	
– –	–	– –	– – –	– – – –	– – –	– –	
86 Em	2	4 4	6 6 6	8 8 8 8	6 6 6	4 4	
87 –	2	4 4	6 6 6	8 8 8 8	6 6 6	4 4	1
88 Ra	2	4 4	6 6 6	8 8 8 8	6 6 6	4 4	2
89 Ac	2	4 4	6 6 6	8 8 8 8	6 6 6	4 4 1	(2)
90 Th	2	4 4	6 6 6	8 8 8 8	6 6 6	4 4 2	(2)
– –	–	– –	– – –	– – – –	– – –	– – –	–
118 ?	2	4 4	6 6 6	8 8 8 8	8 8 8 8	6 6 6	4 4

FIG. 9.

:he atoms of the elements to which the author has ›een led by a consideration of successive capture and ›inding of electrons to the atomic nucleus. The igures before the different elements are the atomic ıumbers, which give the total number of electrons n the neutral atom. The figures in the different :olumns give the number of electrons in orbits corre-sponding to the values of the principal and subordinate ıuantum numbers standing at the top. In accordance vith ordinary usage we will, for the sake of brevity, designate an orbit with principal quantum number n as an n-quantum orbit. The first electron bound in each atom moves in an orbit that corresponds to the normal state of the hydrogen atom with quantum symbol 1_1. In the hydrogen atom there is of course only one electron; but we must assume that in the atoms of other elements the next electron also will be bound in such a 1-quantum orbit of type 1_1. As the table shows, the following electrons are bound in 2-quantum orbits. To begin with, the binding will result in a 2_1 orbit, but later electrons will be bound in 2_2 orbits, until, after binding the first 10 electrons in the atom, we reach a closed configuration of the 2-quantum orbits in which we assume there are four orbits of each type. This configuration is met for the first time in the neutral neon atom, which forms the conclusion of the second period in the system of the elements. When we proceed in this system, the following electrons are bound in 3-quantum orbits, until, after the conclusion of the third period of the system, we encounter for the first time, in elements of the fourth period, electrons in 4-quantum orbits, and so on.

This picture of atomic structure contains many features that were brought forward by the work of earlier investigators. Thus the attempt to interpret the relations between the elements in the natural system by the assumption of a division of the electrons into groups goes as far back as the work of J. J. Thomson in 1904. Later, this view-point was developed chiefly by Kossel (1916), who, moreover, has connected such a grouping with the laws that investigations of X-ray spectra have brought to light.

Also G. R. Lewis and I. Langmuir have sought to account for the relations between the properties of the elements on the basis of a grouping inside the atom. These investigators, however, assumed that the electrons do not move about the nucleus, but occupy positions of equilibrium. In this way, though, no closer relation can be reached between the properties of the elements and the experimental results concerning the constituents of the atoms. Statical positions of equilibrium for the electrons are in fact not possible in cases in which the forces between the electrons and the nucleus even approximately obey the laws that hold for the attractions and repulsions between electrical charges.

The possibility of an interpretation of the properties of the elements on the basis of these latter laws is quite characteristic for the picture of atomic structure developed by means of the quantum theory. As regards this picture, the idea of connecting the grouping with a classification of electron orbits according to increasing quantum numbers was suggested by Moseley's discovery of the laws of X-ray spectra, and by Sommerfeld's work on the fine structure of these spectra. This has been principally emphasised by Vegard, who some years ago in connexion with investigations of X-ray spectra proposed a grouping of electrons in the atoms of the elements, which in many ways shows a likeness to that which is given in the above table.

A satisfactory basis for the further development of this picture of atomic structure has, however, only recently been created by the study of the binding [479] processes of the electrons in the atom, of which we have experimental evidence in optical spectra, and the characteristic features of which have been elucidated principally by the correspondence principle. It is here an essential circumstance that the restriction on the course of the binding process, which is expressed by the presence of electron orbits with higher quantum numbers in the normal state of the atom, can be naturally connected with the general condition for the occurrence of transitions between stationary states, formulated in that principle.

Another essential feature of the theory is the influence, on the strength of binding and the dimensions of the orbits, of the penetration of the later bound electrons into the region of the earlier bound ones, of which we have seen an example in the discussion of the origin of the potassium spectrum. Indeed, this circumstance may be regarded as the essential cause of the pronounced periodicity in the properties of the elements, in that it implies that the atomic dimensions and chemical properties of homologous substances in the different periods, as, for example, the alkali-metals, show a much greater similarity than that which might be expected from a direct comparison of the orbit of the last electron bound with an orbit of the same quantum number in the hydrogen atom.

The increase of the principal quantum number which we meet when we proceed in the series of the elements, affords also an immediate explanation of the characteristic deviations from simple periodicity which are exhibited by the natural system and are expressed in Fig. 1 by the bracketing of certain series of elements in the later periods. The first time such a deviation is met with is in the 4th period, and the reason for it can be simply illustrated by means of our figure of the orbits of the last electron bound in the atom of potassium, which is the first element in this period. Indeed, in potassium we encounter for the first time in the sequence of the elements a case in which the principal quantum number of the orbit of the last electron bound is, in the normal state of the atom, larger than in one of the earlier stages of the binding process. The normal state corresponds here to a 4_1 orbit, which, because of the penetration into the inner region, corresponds to a much stronger binding of the electron than a 4-quantum orbit in the hydrogen atom. The binding in question is indeed even stronger than for a 2-quantum orbit in the hydrogen atom, and is therefore more than twice as strong as in the circular 3_3 orbit which is situated completely outside the inner region, and for which the strength of the binding differs but little from that for a 3-quantum orbit in hydrogen.

This will not continue to be true, however, when we consider the binding of the 19th electron in substances of higher atomic number, because of the much smaller relative difference between the field of force outside and inside the region of the first eighteen electrons bound. As is shown by the investigation of the spark spectrum of calcium, the binding of the 19th electron in the 4_1 orbit is here but little stronger than in 3_3 orbits, and as soon as we reach scandium, we must assume that the 3_3 orbit will represent the
[480] orbit of the 19th electron in the normal state, since this type of orbit will correspond to a stronger bindin than a 4_1 orbit. While the group of electrons i 2-quantum orbits has been entirely completed at th end of the 2nd period, the development that the grou of 3-quantum orbits undergoes in the course of th 3rd period can therefore only be described as a pro visional completion, and, as shown in the table, th electron group will, in the bracketed elements of th 4th period, undergo a stage of further developmen in which electrons are added to it in 3-quantum orbit

This development brings in new features, in tha the development of the electron group with 4-quantur orbits comes to a standstill, so to speak, until th 3-quantum group has reached its final closed form Although we are not yet in a position to account i all details for the steps in the gradual developmen of the 3-quantum electron group, still we can say tha with the help of the quantum theory we see at onc why it is in the 4th period of the system of the element that there occur for the first time successive element with properties that resemble each other as muc as the properties of the *iron group*; indeed, we ca even understand why these elements show their wel known paramagnetic properties. Without furthe reference to the quantum theory, Ladenburg had o a previous occasion already suggested the idea c relating the chemical and magnetic properties of thes elements with the development of an inner electro group in the atom.

I will not enter into many more details, but onl mention that the peculiarities we meet with in th 5th period are explained in much the same way a those in the 4th period. Thus the properties of th bracketed elements in the 5th period as it appear in the table, depend on a stage in the developmen of the 4-quantum electron group that is initiated b the entrance in the normal state of electrons in 4_3 orbit In the 6th period, however, we meet new feature In this period we encounter not only a stage of th development of the electron groups with 5- an 6-quantum orbits, but also the final completion c the development of the 4-quantum electron grour which is initiated by the entrance for the first tim of electron orbits of the 4_4 type in the normal stat of the atom. This development finds its characteristi expression in the occurrence of the peculiar famil of elements in the 6th period, known as the *rare-earth* These show, as we know, a still greater mutual similarit in their chemical properties than the elements of th iron family. This must be ascribed to the fact tha we have here to do with the development of an electro group that lies deeper in the atom. It is of interes to note that the theory can also naturally accoun for the fact that these elements, which resembl each other in so many ways, still show great difference in their magnetic properties.

The idea that the occurrence of the rare-earth depends on the development of an inner electro group has been put forward from different side Thus it is found in the work of Vegard, and at th same time as my own work, it was proposed by Bur in connexion with considerations of the systemati relation between the chemical properties and th grouping of the electrons inside the atom from th point of view of Langmuir's static atomic model. Whil until now it has not been possible, however, to give any theoretical basis for such a development of an inner group, we see that our extension of the quantum theory provides us with an unforced explanation. Indeed, it is scarcely an exaggeration to say that if the existence of the rare-earths had not been established by direct chemical investigation, the occurrence of a family of elements of this character within the 6th period of the natural system of the elements might have been theoretically predicted.

When we proceed to the 7th period of the system, we meet for the first time with 7-quantum orbits, and we shall expect to find within this period features that are essentially similar to those in the 6th period, in that besides the first stage in the development of the 7-quantum orbits, we must expect to encounter further stages in the development of the group with 6- or 5-quantum orbits. However, it has not been possible directly to confirm this expectation, because only a few elements are known in the beginning of the 7th period. The latter circumstance may be supposed to be intimately connected with the instability of atomic nuclei with large charges, which is expressed in the prevalent radioactivity among elements with high atomic number.

X-ray Spectra and Atomic Constitution.

In the discussion of the conceptions of atomic structure we have hitherto placed the emphasis on the formation of the atom by successive capture of electrons. Our picture would, however, be incomplete without some reference to the confirmation of the theory afforded by the study of X-ray spectra. Since the interruption of Moseley's fundamental researches by his untimely death, the study of these spectra has been continued in a most admirable way by Prof. Siegbahn in Lund. On the basis of the large amount of experimental evidence adduced by him and his collaborators, it has been possible recently to give a classification of X-ray spectra that allows an immediate interpretation on the quantum theory. In the first place it has been possible, just as in the case of the optical spectra, to represent the frequency of each of the X-ray lines as the difference between two out of a manifold of spectral terms characteristic of the element in question. Next, a direct connexion with the atomic theory is obtained by the assumption that each of these spectral terms multiplied by Planck's constant is equal to the work which must be done on the atom to remove one of its inner electrons. In fact, the removal of one of the inner electrons from the completed atom may, in accordance with the above considerations on the formation of atoms by capture of electrons, give rise to transition processes by which the place of the electron removed is taken by an electron belonging to one of the more loosely bound electron groups of the atom, with the result that after the transition an electron will be lacking in this latter group.

The X-ray lines may thus be considered as giving evidence of stages in a process by which the atom undergoes a *reorganisation* after a disturbance in its interior. According to our views on the stability of the electronic configuration such a disturbance must consist in the total removal of electrons from the atom, or at any rate in their transference from normal orbits to orbits of higher quantum numbers than those belonging to completed groups; a circumstance which is clearly illustrated in the characteristic difference between selective absorption in the X-ray region, and that exhibited in the optical region.

The classification of the X-ray spectra, to the achievement of which the above-mentioned work of Sommerfeld and Kossel has contributed materially, has recently made it possible, by means of a closer examination of the manner in which the terms occurring in the X-ray spectra vary with the atomic number, to obtain a very direct test of a number of the theoretical

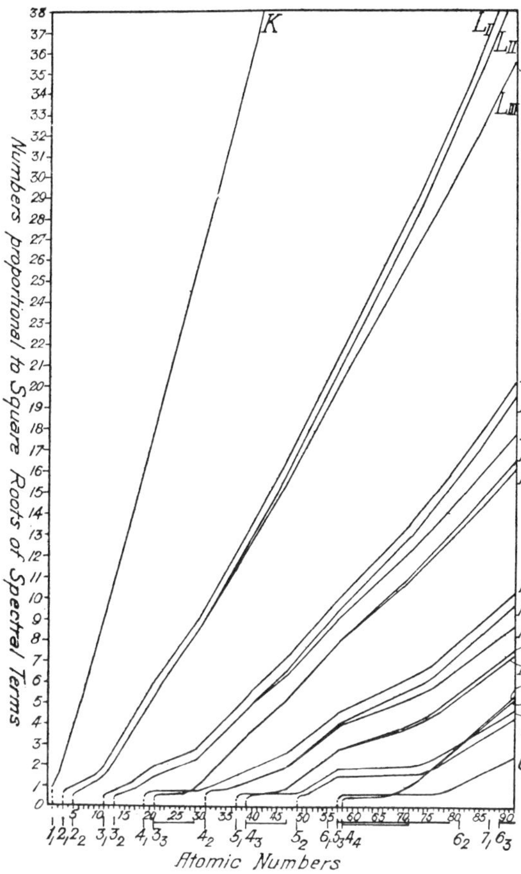

Fig. 10.

conclusions as regards the structure of the atom. In Fig. 10 the abscissæ are the atomic numbers and the ordinates are proportional to the square roots of the spectral terms, while the symbols K, L, M, N, O, for the individual terms refer to the characteristic discontinuities in the selective absorption of the elements for X-rays; these were originally found by Barkla before the discovery of the interference of X-rays in crystals had provided a means for the closer investigation of X-ray spectra. Although the curves generally run very uniformly, they exhibit a number of deviations from uniformity which have been especially brought to light by the recent investigation of Coster, who has for some years worked in Siegbahn's laboratory.

These deviations, the existence of which was not discovered until after the publication of the theory of atomic structure discussed above, correspond exactly to what one might expect from this theory. At the foot of the figure the vertical lines indicate where, according to the theory, we should first expect, in the normal state of the atom, the occurrence of n_k orbits of the type designated. We see how it has been possible to connect the occurrence of every spectral term with the presence of an electron moving in an orbit of a definite type, to the removal of which this term is supposed to correspond. That in general there corresponds more than one curve to each type of orbit n_k is due to a complication in the spectra which would lead us too far afield to enter into here, and may be attributed to the deviation from the previously described simple type of motion of the electron arising from the interaction of the different electrons within the same group.

The intervals in the system of the elements, in which a further development of an inner electron group takes place because of the entrance into the normal atom of electron orbits of a certain type, are designated in the figure by the horizontal lines, which are drawn between the vertical lines to which the quantum symbols are affixed. It is clear that such a development of an inner group is everywhere reflected in the curves. Particularly the course of the N- and O-curves may be regarded as a direct indication of that stage in the development of the electron groups with 4-quantum orbits of which the occurrence of the rare-earths bears witness. Although the apparent complete absence of a reflection in the X-ray spectra of the complicated relationships exhibited by most other properties of the elements was the typical and important feature of Moseley's discovery, we can recognise, nevertheless, in the light of the progress of the last years, an intimate connexion between the X-ray spectra and the general relationships between the elements within the natural system.

Before concluding this lecture I should like to mention one further point in which X-ray investigations have been of importance for the test of the theory. This concerns the properties of the hitherto unknown element with atomic number 72. On this question opinion has been divided in respect to the conclusions that could be drawn from the relationships within the periodic table, and in many representations of the table a place is left open for this element in the rare-earth family. In Julius Thomsen's representation of the natural system, however, this hypothetical element was given a position homologous to titanium and zirconium in much the same way as in our representation in Fig. 1. Such a relationship must be considered as a necessary consequence of the theory of atomic structure developed above, and is expressed in the table (Fig. 9) by the fact that the electron configurations for titanium and zirconium show the same sort of resemblances and differences as the electron configurations for zirconium and the element with atomic number 72. A corresponding view was proposed by Bury on the basis of his above-mentioned systematic considerations of the connexion between the grouping of the electrons in the atom and the properties of the elements.

Recently, however, a communication was published
[482] by Dauvillier announcing the observation of some weak lines in the X-ray spectrum of a preparation containing rare-earths. These were ascribed to an element with atomic number 72 assumed to be identical with an element of the rare-earth family, the existence of which in the preparation used had been presumed by Urbain many years ago. This conclusion would however, if it could be maintained, place extraordinarily great, if not unsurmountable, difficulties in the way of the theory, since it would claim a change in the strength of the binding of the electrons with the atomic number which seems incompatible with the conditions of the quantum theory. In these circumstances Dr. Coster and Prof. Hevesy, who are both for the time working in Copenhagen, took up a short time ago the problem of testing a preparation of zircon-bearing minerals by X-ray spectroscopic analysis. These investigators have been able to establish the existence in the minerals investigated of appreciable quantities of an element with atomic number 72, the chemical properties of which show a great similarity to those of zirconium and a decided difference from those of the rare-earths.[2]

I hope that I have succeeded in giving a summary of some of the most important results that have been attained in recent years in the field of atomic theory and I should like, in concluding, to add a few general remarks concerning the view-point from which these results may be judged, and particularly concerning the question of how far, with these results, it is possible to speak of an explanation, in the ordinary sense of the word. By a theoretical explanation of natural phenomena we understand in general a classification of the observations of a certain domain with the help of analogies pertaining to other domains of observation, where one presumably has to do with simpler phenomena. The most that one can demand of a theory is that this classification can be pushed so far that it can contribute to the development of the field of observation by the prediction of new phenomena.

When we consider the atomic theory, we are, however, in the peculiar position that there can be no question of an explanation in this last sense, since here we have to do with phenomena which from the very nature of the case are simpler than in any other field of observation, where the phenomena are always conditioned by the combined action of a large number of atoms. We are therefore obliged to be modest in our demands and content ourselves with concepts which are formal in the sense that they do not provide a visual picture of the sort one is accustomed to require of the explanations with which natural philosophy deals. Bearing this in mind I have sought to convey the impression that the results, on the other hand fulfil, at least in some degree, the expectations that are entertained of any theory; in fact, I have attempted to show how the development of atomic theory has contributed to the classification of extensive fields of observation, and by its predictions has pointed out the way to the completion of this classification. It is scarcely necessary, however, to emphasise that the theory is yet in a very preliminary stage, and many fundamental questions still await solution.

[2] For the result of the continued work of Coster and Hevesy with the new element, for which they have proposed the name hafnium, the reader may be referred to their letters in NATURE of January 20, February 10 and 24, and April 7.

XI. X-RAY SPECTRA AND THE PERIODIC SYSTEM OF THE ELEMENTS

(WITH D. COSTER)

RÖNTGENSPEKTREN UND PERIODISCHES SYSTEM DER ELEMENTE

(Z. Phys. **12** (1923) 342)

TEXT AND TRANSLATION*

* [See Introduction, sect. 11.]

ZEITSCHRIFT FÜR PHYSIK

HERAUSGEGEBEN VON DER
DEUTSCHEN PHYSIKALISCHEN GESELLSCHAFT
ALS ERGÄNZUNG ZU IHREN „VERHANDLUNGEN"
UNTER DER REDAKTION VON

KARL SCHEEL

Sonderabdruck Band 12, Heft 6.

N. Bohr und D. Coster

Röntgenspektren und periodisches System der Elemente

FRIEDR. VIEWEG & SOHN AKT.-GES. BRAUNSCHWEIG UND JULIUS SPRINGER BERLIN

1923

Röntgenspektren und periodisches System der Elemente.

Von **N. Bohr** und **D. Coster** in Kopenhagen.

Mit vier Abbildungen. (Eingegangen am 2. November 1922.)

In einer Arbeit in dieser Zeitschrift [1]) hat der eine der Verfasser in den Hauptzügen eine Theorie des Atombaus entwickelt, von der aus es möglich scheint, eine Interpretation der optischen sowie der Röntgenspektren zu geben, welche in engster Beziehung zu einer Deutung des periodischen Systems der Elemente steht. Nach dem Erscheinen dieser Arbeit ist das zur Prüfung der theoretischen Vorstellungen geeignete röntgenspektroskopische Versuchsmaterial durch eine Untersuchung des anderen von uns wesentlich vermehrt und in einer kürzlich veröffentlichten Arbeit [2]) in seiner Beziehung zur Theorie diskutiert worden, wodurch mehrere interessante Ergebnisse zutage getreten sind. In der vorliegenden gemeinsamen Arbeit wird nun auf den Vergleich des Versuchsmaterials mit der Theorie etwas genauer eingegangen als es in den erwähnten Arbeiten geschehen ist, indem besonders auf die Beziehung der Röntgenspektren zur Deutung des periodischen Systems Rücksicht genommen wird.

§ 1. Theoretische Vorstellungen vom Atombau. Die erwähnte Theorie beruht auf einer Klassifikation der Elektronenbahnen im Atom mittels eines Vergleichs mit den stationären Zuständen eines Elektrons, das eine Zentralbewegung vollführt. Dabei werden die Bahnen durch das Symbol n_k gekennzeichnet. Hier bedeutet n die sogenante „Hauptquantenzahl", die in dem Grenzfall, wo die Zentralbewegung in eine einfach periodische Keplerbewegung übergeht, für die Bestimmung der Energie allein ausschlaggebend ist, wie es in der Balmerformel des Wasserstoffspektrums zutage tritt. Die „Nebenquantenzahl" k, deren Einfluß auf die Energie von der Abweichung der Bewegung von einer einfach periodischen bedingt ist, bestimmt

[1]) N. Bohr, ZS. f. Phys. **9**, 1, 1922. Die Abhandlung ist inzwischen, mit einigen erläuternden Figuren versehen, zusammen mit zwei früheren Aufsätzen allgemeiner Art als ein Heft der Sammlung Vieweg unter dem Titel „Drei Aufsätze über Spektren und Atombau" erschienen. Es ist die Absicht des Verfassers, in einer Reihe von Aufsätzen in dieser Zeitschrift die Ergebnisse dieser Abhandlung, sowohl was die allgemeine Grundlage wie die Ausführung von Einzelheiten betrifft, näher zu diskutieren.

[2]) D. Coster, Phil. Mag. **43**, 1070, 1922, **44**, 546, 1922. Siehe auch die Leidener Doktordissertation 1922.

in wohlbekannter Weise das Impulsmoment des Elektrons um das Bahnzentrum. Auf der Einführung dieser Quantenzahl beruht bekanntlich die Sommerfeldsche Theorie der Feinstruktur der Wasserstofflinien; diese Feinstruktur wird durch den Umstand verursacht, daß wegen der Relativitätsmodifikation der mechanischen Gesetze schon die Bahn eines einzelnen Elektrons um einen positiven Atomkern nicht streng periodisch ist, sondern beschrieben werden kann als eine Keplerellipse, die eine langsame Drehung in ihrer Ebene ausführt. Eine Übersicht über die Anzahlen von Elektronen im normalen Atom, die nach der Theorie den verschiedenen Typen von n_k-Bahnen angehören, ist in der nachstehenden Tabelle 1 gegeben, welche Vorstellungen vom Atombau zum Ausdruck bringt, die abgesehen von den charakteristischen neuen Zügen in mehrfacher Hinsicht Ähnlichkeiten aufweisen mit den Vorstellungen, die den Arbeiten von Sommerfeld und Vegard über Röntgenspektren zugrunde liegen.

Wie man sieht, sind die Elektronenbahnen in Gruppen angeordnet, die zu demselben Wert der Hauptquantenzahl n gehören. Die Ausbildung dieser Gruppen, wenn man zu Elementen höherer Atomnummer geht, ist mit den Gesetzmäßigkeiten, die im periodischen System auftreten, in Verbindung gebracht. Ein wesentlicher Zug dabei ist, daß die Gruppen wieder in Untergruppen geteilt sind, die verschiedenen Werten der Nebenquantenzahl k entsprechen. Das allmähliche Hinzukommen neuer Typen von Elektronenbahnen mit derselben Hauptquantenzahl wie die schon in vorangehenden Elementen vorhandenen Gruppen von Elektronenbahnen und das dadurch veranlaßte Fortschreiten der stufenweisen Ausbildung dieser Gruppen ist weiter als der Grund der typischen Abweichungen von der einfachen Periodizität im System der Elemente angenommen (Familien der Eisen- und Platinmetalle sowie der seltenen Erden). Eine Übersicht über das periodische System, die diese Züge besonders zum Ausdruck bringt, ist gegeben in Fig. 1 [1]), wo die Elemente, die zur

---

[1]) Vgl. N. Bohr, Drei Aufsätze über Spektren und Atombau, S. 132, Fig. 5, wo eine ausführlichere Erläuterung dieser Figur zu finden ist. Auf die in dem Nachtrag dieser Aufsätze erwähnte Note von Dauvillier, C. R. Mai 1922, über die Beobachtung einiger schwachen Linien des Röntgenspektrums eines Elementes mit Atomnummer 72 in einer Mischung von seltenen Erden ist weder im Schema noch in der Figur Rücksicht genommen, weil diese Beobachtung wohl kaum schon als gesichert angesehen werden kann. Das Vorhandensein eines Elementes zwischen Lu (71), dessen höchste Valenz 3 ist und Ta (73), dessen höchste Valenz 5 ist, mit ähnlichen chemischen Eigenschaften wie Lu würde eine Abweichung von der sonst allgemeinen Regel darstellen, daß die höchste Valenz beim Übergang von einem Elemente zum folgenden sich niemals um mehr als eine Einheit ver-

Tabelle 1. Elektronenbahntypen der Elemente.

N \ n_k		1_1	2_1	2_2	3_1	3_2	3_3	4_1	4_2	4_3	4_4	5_1	5_2	5_3	5_4	5_5	6_1	6_2	6_3	6_4	6_5	6_6	7_1	7_2
1	H	1																						
2	He	2																						
3	Li	2	1																					
4	Be	2	2																					
5	B	2	2	(1)																				
—	—	—	—	—																				
10	Ne	2	4	4																				
11	Na	2	4	4	1																			
12	Mg	2	4	4	2																			
13	Al	2	4	4	2	1																		
—	—	—	—	—	—	—																		
18	A	2	4	4	4	4																		
19	K	2	4	4	4	4		1																
20	Ca	2	4	4	4	4		2																
21	Sc	2	4	4	4	4	1	(2)																
22	Ti	2	4	4	4	4	2	(2)																
—	—	—	—	—	—	—	—	—																
29	Cu	2	4	4	6	6	6	1																
30	Zn	2	4	4	6	6	6	2																
31	Ga	2	4	4	6	6	6	2	1															
—	—	—	—	—	—	—	—	—	—															
36	Kr	2	4	4	6	6	6	4	4															
37	Rb	2	4	4	6	6	6	4	4			1												
38	Sr	2	4	4	6	6	6	4	4			2												
39	Y	2	4	4	6	6	6	4	4	1		(2)												
40	Zr	2	4	4	6	6	6	4	4	2		(2)												
—	—	—	—	—	—	—	—	—	—	—		—												
47	Ag	2	4	4	6	6	6	6	6	6		1												
48	Cd	2	4	4	6	6	6	6	6	6		2												
49	In	2	4	4	6	6	6	6	6	6		2	1											
—	—	—	—	—	—	—	—	—	—	—		—	—											
54	X	2	4	4	6	6	6	6	6	6		4	4											
55	Cs	2	4	4	6	6	6	6	6	6		4	4				1							
56	Ba	2	4	4	6	6	6	6	6	6		4	4				2							
57	La	2.	4	4	6	6	6	6	6	6		4	4	1			(2)							
58	Ce	2	4	4	6	6	6	6	6	6	1	4	4	1			(2)							
59	Pr	2	4	4	6	6	6	6	6	6	2	4	4	1			(2)							
—	—	—	—	—	—	—	—	—	—	—	—	—	—	—			—							
71	Lu	2	4	4	6	6	6	8	8	8	8	4	4	1			(2)							
72	—	2	4	4	6	6	6	8	8	8	8	4	4	2			(2)							
—	—	—	—	—	—	—	—	—	—	—	—	—	—	—			—							
79	Au	2	4	4	6	6	6	8	8	8	8	6	6	6			1							
80	Hg	2	4	4	6	6	6	8	8	8	8	6	6	6			2							
81	Tl	2	4	4	6	6	6	8	8	8	8	6	6	6			2	1						
—	—	—	—	—	—	—	—	—	—	—	—	—	—	—			—	—						
86	Nt	2	4	4	6	6	6	8	8	8	8	6	6	6			4	4						
87	—	2	4	4	6	6	6	8	8	8	8	6	6	6			4	4					1	
88	Ra	2	4	4	6	6	6	8	8	8	8	6	6	6			4	4					2	
89	Ac	2	4	4	6	6	6	8	8	8	8	6	6	6			4	4	1				(2)	
90	Th	2	4	4	6	6	6	8	8	8	8	6	6	6			4	4	2				(2)	
—	—	—	—	—	—	—	—	—	—	—	—	—	—	—			—	—	—				—	
—	—	—	—	—	—	—	—	—	—	—	—	—	—	—	—		—	—	—				—	—
118	?	2	4	4	6	6	6	8	8	8	8	8	8	8	8		6	6	6				4	4

[488]

selben Periode des Systems gehören, in vertikalen Kolonnen angeordnet sind. Elemente in aufeinanderfolgenden Kolonnen, die sowohl in ihren chemischen wie in ihren optischen Eigenschaften als homolog angesehen werden können, sind durch gerade Linien miteinander verbunden. Reihen von aufeinander folgenden Elementen,

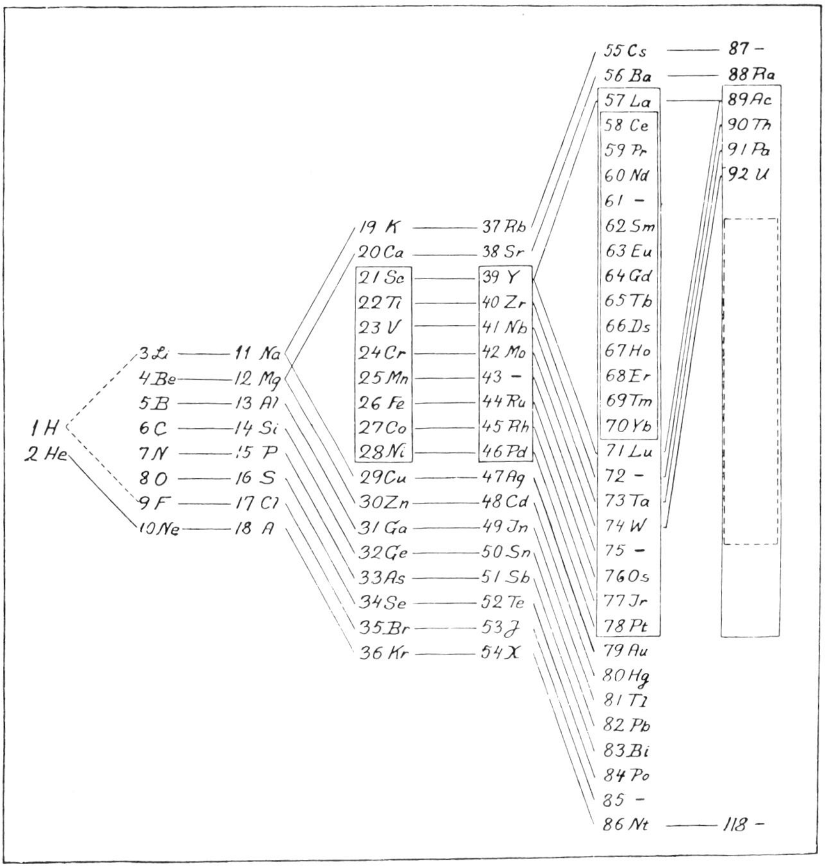

Fig. 1.

die dadurch eine Sonderstellung einnehmen, daß bei ihnen innere Gruppen in Ausbildung begriffen sind, sind in viereckigen Klammern eingeschlossen.

Die Aufstellung der Tabelle 1 beruht, was die Hauptzüge betrifft, auf Betrachtungen allgemeiner Art, die in der oben erwähnten Arbeit

größert; eine Regel, die eine unmittelbare Erklärung erhält aus den allgemein angenommenen Vorstellungen vom Atombau (vgl. hierzu auch C. R. Bury, Journ. Am. Chem. Soc. **43**, 1902, 1921).

dargelegt sind. Bei der Ausarbeitung von Einzelheiten spielt eine eingehende Untersuchung der Serienspektren eine wesentliche Rolle. Obwohl in diesen Spektren sich auch die gradweise Ausbildung innerer Elektronengruppen deutlich abspiegelt, betrifft die direkte Auskunft, die man aus den Serienspektren entnehmen kann, doch in erster Linie das allmähliche Hinzukommen solcher Elektronenbahnen neuer Typen, durch welche die Ausbildungsstufen eingeleitet werden. Dem entsprechend sind in Tabelle 1 wesentlich nur solche Elemente, die dem Anfang der Ausbildungsstufen angehören, angegeben, wobei außerdem überall, wo das Spektralmaterial es nicht zuläßt, eindeutig auf die Zahl der Elektronen in den äußersten Gruppen zu schließen, dies durch das Einklammern der Elektronenzahlen, die den höchstquantigen Bahnen entsprechen, angedeutet ist. Die weitere fortgesetzte Ausbildung der Gruppen ist nur durch ihr Endresultat angedeutet; auch in der Beschreibung der endgültig ausgebildeten Gruppen bleibt aber zurzeit noch viel Unsicheres übrig, besonders was das feinere Wechselspiel der Elektronenbewegungen innerhalb derselben Hauptgruppe und ihrer Untergruppen anbelangt. Ohne hier näher auf diese Frage einzugehen, wollen wir nur darauf hinweisen, daß die in der oben erwähnten Arbeit eingeführten allgemeinen Annahmen über den Charakter dieses Wechselspiels, wie dort näher ausgeführt, auf Grund des Korrespondenzprinzips einen Anhaltspunkt geben für ein theoretisches Verständnis der typischen Gesetze, welche die Emissions- und Absorptionsspektren im Röntgengebiete regeln, und deren Hauptzüge durch die formale Theorie von Kossel klargelegt wurden.

§ 2. Klassifikation der Röntgenspektren. Wie in der oben zitierten Arbeit von Coster dargelegt wird, ist es möglich gewesen, für eine große Zahl der Elemente eine Klassifikation der Röntgenspektren zu erreichen, die darauf beruht, daß die Frequenz ν jeder Linie als Differenz zweier Spektralterme T' und T'' dargestellt werden kann. In Übereinstimmung mit den Prinzipien der Quantentheorie der Linienspektren werden diese Spektralterme — mit h multipliziert — als Energieniveaus eines Atoms gedeutet, das durch Entfernung eines Elektrons aus einer inneren Gruppe aus seinem Normalzustand gebracht ist

Über das charakteristische Röntgenspektrum, wie es für das Edelgas Niton (86) zu erwarten ist, gibt Fig. 2 eine Übersicht. In der üblichen Weise sind die Spektrallinien durch vertikale Pfeile angedeutet, während die horizontalen Geraden die Niveaus darstellen. Nur solche Linien, welche für mehrere Elemente in der Umgebung dieses Edel-

[490]

gases wirklich gemessen wurden, sind eingetragen. Die Figur gibt zugleich eine Übersicht über das Spektrum der Elemente in der Nähe der Edelgase Xenon und Krypton. So sind die Niveaus, welche beim Fortschreiten zu Elementen niedriger Atomnummer zwischen Niton

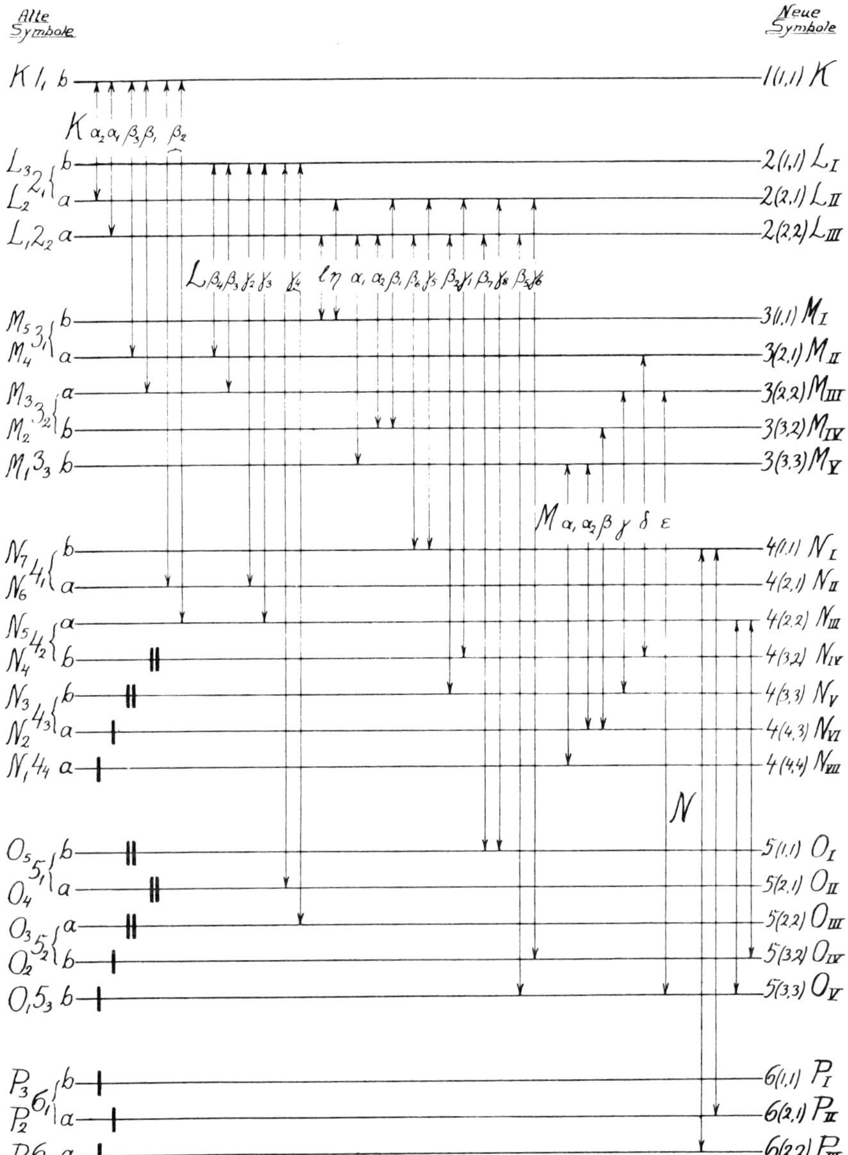

Fig. 2.

(86) und Xenon (54) ausfallen, in der Figur mit einem vertikalen Strich angegeben, und die Niveaus, die zwischen Xenon (54) und Krypton (36) außerdem noch verschwinden, mit zwei Strichen bezeichnet worden.

Was zunächst die Symbole betrifft, mit denen die Niveaus bezeichnet sind, ist, wie in der Figur angegeben, gegenüber den vorigen Abhandlungen eine Änderung vorgenommen. Erstens sind die Unterniveaus, die zur selben Niveaugruppe gehören, in der Reihenfolge mit römischen Ziffern bezeichnet, in der sie allmählich bei steigender Atomnummer in den Röntgenspektren der Elemente auftreten. Weiter ist an jedes Niveau ein Zahlensymbol der Form n (k_1, k_2) geknüpft. Auf die Beziehung dieser Symbole zur Theorie des Atombaus werden wir später näher eingehen. Hier soll nur darauf hingewiesen werden, daß die Niveaus, die in Fig. 2 für die Darstellung der Spektren jedes der drei Edelgase benutzt worden sind, alle durch solche Werte von n und den k charakterisiert sind, die in der obigen Tabelle 1 als Werte der Quantenzahlen n und k für die Elektronenbahnen in den Atomen dieser Elemente vorkommen. Ferner bemerken wir, daß das Auftreten von Linien, die Kombinationen zwischen den Niveaus entsprechen, dadurch geregelt wird, daß die Zahl k_1 bei einem Übergangsprozeß sich immer um eine Einheit ändert, während k_2 sich entweder um eine Einheit ändert oder konstant bleibt. Diese Regeln sind äquivalent mit den Regeln, die in den früheren Arbeiten von Coster mittels der alten Symbole ausgedrückt wurden, und sie entsprechen formell genau den Regeln, die gleichzeitig von Wentzel[1]) aufgestellt wurden. Die für k_1 und k_2 eingeführten Zahlenwerte sind gleich den Werten der von diesem Verfasser zur Klassifikation der Niveaus benutzten Zahlen n und m. Ihre theoretische Bedeutung werden wir aber in etwas anderer Weise auffassen.

§ 3. Die Ermittlung der Energieniveaus aus dem Versuchsmaterial. Die Resultate der jetzt vorliegenden Messungen der Röntgenspektren erlauben uns, die Energieniveaus für eine große Anzahl von Elementen mit beträchtlicher Genauigkeit zu berechnen. Im folgenden geben wir eine Übersicht über die dabei herangezogenen experimentellen Untersuchungen[2]), während die durch

---

[1]) G. Wentzel, ZS. f. Phys. 6, 84, 1921.

[2]) Eine Tabelle über die Spektralterme der Röntgenspektren findet sich auch in Sommerfelds Buch: Atombau und Spektrallinien, 3. Aufl., S. 630. Durch die neuen Versuchsergebnisse ist diese Tabelle jedoch in vieler Hinsicht änderungsbedürftig.

In einer eben erschienenen Arbeit von Dauvillier (Journal de Physique et le Radium, Juli 1922) ist ebenfalls eine Übersicht über das Versuchsmaterial

[492]

die Rydbergsche Konstante R dividierten Werte der Spektralterme T in Tabelle 2 und die Wurzelwerte derselben in Tabelle 3 zusammengestellt sind.

Übersicht über die benutzten Arbeiten.

K-Absorption [1]	Mg (12) — Cr (24)	Fricke, Phys. Rev. **16**, 202, 1920.
„ [2]	Mn (25) — U (92)	Duane u. Shimizu, Phys. Rev. **14**, 522, 1919.
		Duane und Kang-Fuh-Hu, Phys. Rev. **14**, 516, 1919.
		Duane, Fricke, Stenström, Proc. Nat. Ac. Sc. **6**, 607, 1920.
„ [1]	Cd (48) — U (92)	Siegbahn u. Jönsson, Phys. ZS. **20**, 251, 1919.
L-Absorption [2] [3]	W (74) — Pb (82)	Duane und Patterson, Proc. Nat. Ac. Sc. **6**, 509, 1920.
„ [1]	G (55) — Eu (63)	Coster, Phil. Mag. **43**, 1070, 1922 und **44**, 546, 1922. *
„ [1]	Sb (51) — Ba (56)	Lindsay, C. R. **175**, 150, 1922.
„ [1]	Ag (47)	Coster, Phil. Mag. **43**, 1070, 1922.
M-Absorption [1]	Th (90) u. U (92)	Stenström, Dissertation Lund 1919.
„ [1]	Bi (83), Th (90) u. U (92)	Coster, Phys. Rev. **19**, 20, 1922.
K-Linien [1]	Na (11) — Zn (30)	Hjalmar, ZS. f. Phys. **1**, 439, 1920 und **7**, 341, 1921.
„ [1]	P (15) — Zn (30)	Siegbahn u. Dolejsek, ZS. f. Phys. **10**, 159, 1922.
„ [2]	Mo (42) u. Rh (45)	Duane, Phys. Rev. **14**, 373, 1919.
L-Linien [1]	Cu (30) — Ta (73)	Hjalmar, ZS. f. Phys. **3**, 262, 1920 und **7**, 341, 1921.
„ [1]	W (74)	Siegbahn, Phys. ZS. **20**, 533, 1919.
„ [1]	Rb (37) — Ba (56)	Coster, Phil. Mag. **43**, 1070, 1922.
„ [1]	La (57) — Lu (71)	Coster, Ebenda **44**, 546, 1922.
„ [1]	Ta (73) — U (92)	Coster, ZS. f. Phys. **6**, 186, 1921.
M-Linien [1]	Dy (66) — U (92)	Stenström, Dissertation Lund 1919.

Mit Ausnahme der mit einem Sternchen bezeichneten Niveaus sind alle Werte der Tabelle 2 aus den oben zitierten experimentellen Daten berechnet unter Benutzung von Additionsbeziehungen, welche man unmittelbar aus Fig. 2 ablesen kann. In einer früheren Arbeit von Coster[4]) wurde schon darauf hingewiesen, daß Duane und Pattersons Werte für die L-Kanten besser mit den Ergebnissen des Lunder Institutes übereinstimmen, wenn man die von ihnen gegebenen

sowie eine graphische Darstellung der Abhängigkeit der Niveauwerte der Elemente von der Atomnummer gegeben. Die meisten der unten besprochenen und uns speziell interessierenden Eigentümlichkeiten dieser Abhängigkeit treten aber in Dauvilliers Darstellung wegen der Unvollständigkeit der von ihm benutzten experimentellen Daten nicht zutage.

[1]) Gemessen in Siegbahns Laboratorium.

[2]) Gemessen in Duanes Laboratorium.

[3]) Um eine X-Einheit korrigiert.

[4]) D. Coster, ZS. f. Phys. **6**, 185, 1921.

24*

* [Instead of G(55) read Cs(55).]

Tabelle 2.

	K	L_{I}	L_{II}	L_{III}	M_{I}	M_{II}	M_{III}	M_{IV}	M_{V}
92 U	8477,0	1603,5	1543,1	1264,3	408,9	382,1	317,2	274,0	261,
90 Th	8073,5	1509,7	1451,5	1200,6	381,6	354,4	298,0	256,6	244,
83 Bi	6646,7	1207,9	1159,4	990,0	295,9	273,6	234,0	199,4	191,
82 Pb	6463,0	1169,3	1121,9	960,5	283,8	262,3	226,0	190,5	183,
81 Tl	6289,0	1132,4	1084,2	933,2	273,9	253,8	219,2	184,8	176,
80 Hg	6115,9	1094,6	1048,6	906,1	—	—	—	—	—
79 Au	5940,4	1060,2	1014,4	878,5	252,9	235,1	202,8	169,3	163,
78 Pt	5764,0	1026,8	978,7	852,0	243,4	227,3	198,0	162,3	156,
74 W	5113,8	890,6	850,1	751,6	207,6	188,9	167,4	137,8	133,
73 Ta	—	860,3*	821,0	728,6	200,1	181,0*	160,7*	132,8	128,
71 Lu	—	801,1*	763,1*	682,0*	184,5*	167,0*	149,4*	121,7*	117,
70 Yb	—	773,4*	735,7*	659,6*	177,5*	161,1*	144,7*	116,8*	113,
68 Er	—	718,1*	682,8*	616,0*	163,8*	147,3*	132,2*	107,3*	104,
67 Ho	4115,9	692,9*	657,4*	595,0*	157,3*	142,4*	129,0*	103,0*	100,
66 Dy	3972,5	666,9*	632,3*	574,0*	150,9*	136,1*	123,7*	98,3*	95,
65 Tb	—	642,4*	608,2*	553,8*	145,0*	130,9*	119,4*	94,1*	91,
64 Gd	3711,9	617,4*	584,3*	533,7*	138,6*	124,6*	114,2*	89,8*	87,
63 Eu	3583,4	593,6*	561,3	514,1	132,9	119,5*	109,6*	85,8	83,
62 Sm	3457,0	570,7*	539,0	495,0	127,1	114,3*	105,3*	81,9	79,
60 Nd	3214,2	525,8*	495,9	457,9	116,6	104,4*	96,4*	74,3	72,
59 Pr	3093,3	504,2*	474,7	439,4	111,4	99,0*	92,1*	70,1	68,
58 Ce	2972,2	483,0*	454,4	421,9	105,9	94,2*	87,8*	66,7	65,
56 Ba	2756,4	442,3	414,7	386,7	95,4	84,9	79,4	58,8	57,
55 Cs	2649,1	421,8	395,0	369,5	89,6	79,3	74,4	54,8	53,
53 J	2448,3	382,6	357,6	336,0	—	69,0	64,8	46,8	46,
52 Te	2345,0	364,1	339,6	320,1	74,5	64,4	60,5	43,2	42,
51 Sb	2241,7	346,1	321,9	304,3	69,3	59,9	56,4	39,5	38,
47 Ag	1878,9	279,5	260,0	247,3	53,3	43,5	41,2	27,9	27,
45 Rh	1709,1	253,4	231,4	220,9	45,8	39,7	38,5	22,5	22,
42 Mo	1473,1	211,9	193,9	186,2	37,6	30,5	29,7	17,4	17,
41 Nb	1401,3	—	181,4	174,4	35,1	—	—	15,0	14,
40 Zr	1325,8	—	169,6	163,5	31,4	—	—	13,0	
29 Cu	661,1	—	71,3	69,8	—	5,2		0	
28 Ni	612,0	—	62,6	61,3	—	3,3		—	
27 Co	568,9	—	59,7	58,5	—	5,3		—	
26 Fe	523,8	—	53,2	52,2	—	4,0		—	
25 Mn	482,4	—	48,7	47,9	—	4,2		—	
24 Cr	441,1	—	43,0	42,3	—	3,5		—	
23 Va	402,3	—	38,2	37,6	—	2,6		—	
22 Ti	365,4	—	32,6	32,2	—	2,2		—	
21 Sc	331,2	—	30,3	30,0	—	2,7		—	
20 Ca	297,5	—	25,9	25,6	—	2,0		—	
19 K	265,3	—	21,4	21,2	—	0,9		—	
17 Cl	207,8	—	14,8	14,7	—	0,4		—	
16 S	181,8	—	11,8		—	0,3		—	
15 P	158,3	—	9,9		—	0,8		—	
13 Al	114,7	—	5,2		—	0		—	
12 Mg	95,8	—	3,5		—	—		—	

* Interpoliert oder aus einem interpolierten Wert berechnet.

Niveauwerte (T/R).

N_I	N_{II}	N_{III}	N_{IV}	N_V	N_{VI}	N_{VII}	O_I	$O_{II,III}$	$O_{IV,V}$	
06,6	95,7	77,1	56,3	53,6	28,4	27,6	26,2	15,4	5,8	92 U
97,8	—	—	51,2	48,7	24,8	24,1	—	—	5,7	90 Th
71,0	58,7	50,3	35,7	33,7	13,6	13,0	—	11,0	2,0	83 Bi
66,0	55,4	49,3	32,2	30,5	10,8	10,3	10,3	6,4	0,8	82 Pb
63,7	53,6	44,9	30,6	29,0	10,0	9,6	10,6	7,4	1,7	81 Tl
—	—	—	—	—	—	—	—	—	—	80 Hg
58,0	49,1	42,8	26,4	25,0	6,4		7,8	8,3	0,8	79 Au
52,5	48,7	42,3	24,5	23,2	5,2		7,1	8,6	0,4	78 Pt
43,6	35,6	30,6	18,7	17,8	2,5		5,2	2,9	(0)	74 W
41,7	33,0*	30,0*	18,0	17,2	2,5		5,3	2,6*	—	73 Ta
37,7*	30,9*	27,2*	16,2*	15,5*	1,3*		4,9*	2,5*	—	71 Lu
36,6*	29,9*	26,4*	15,2*	14,6*	1,0*		—	2,5*	—	70 Yb
33,2*	26,9*	23,4*	13,4*	12,8*	0,3*		4,1*	2,4*	—	68 Er
32,1*	26,6*	23,5*	12,7*	12,2*	0,1*		—	2,4*	—	67 Ho
30,8*	25,3*	22,4*	11,9*	11,4*	(—0,2)*		2,9*	2,5*	—	66 Dy
29,3*	24,1*	21,8*	11,5*	11,1*	—		3,5*	2,4*	—	65 Tb
28,3*	22,2*	20,2*	11,0*	10,6*	—		3,5*	2,4*	—	64 Gd
26,9	21,9*	19,7*	10,4	10,1	—		2,9	—	—	63 Eu
25,8	20,2*	18,8*	10,1	9,8	—		3,0	2,3*	—	62 Sm
23,8	18,6*	17,2*	9,6	9,3	—		3,2	2,5*	—	60 Nd
22,5	18,0*	16,7*	8,8	8,5	—		2,6	2,2*	—	59 Pr
21,7	17,0*	15,8*	8,7	8,5	—		3,1	2,2*	—	58 Ce
18,8	15,3	14,4	7,1	6,9	—		2,1	2,4	—	56 Ba
17,3	13,6	12,6	6,1	5,9	—		2,1	1,7	—	55 Cs
—	—	—	4,4	4,2	—		—	—	—	53 J
12,7	8,8		3,4	3,3	—		—	0,4	—	52 Te
11,1	7,2		2,3	2,2	—		(0)	0,1	—	51 Sb
7,5	3,3		0,6		—		—	—	—	47 Ag
5,3	5,0		(— 0,3)		—		—	—	—	45 Rh
4,7	3,0		0,5		—		—	—	—	42 Mo
4,0	—		—		—		—	—	—	41 Nb
3,5	—		—		—		—	—	—	40 Zr
—	—		—		—		—	—	—	29 Cu
—	—		—		—		—	—	—	28 Ni
—	—		—		—		—	—	—	27 Co
—	—		—		—		—	—	—	26 Fe
—	—		—		—		—	—	—	25 Mn
—	—		—		—		—	—	—	24 Cr
—	—		—		—		—	—	—	23 Va
—	—		—		—		—	—	—	22 Ti
—	—		—		—		—	—	—	21 Sc
—	—		—		—		—	—	—	20 Ca
—	—		—		—		—	—	—	19 K
—	—		—		—		—	—	—	17 Cl
—	—		—		—		—	—	—	16 S
—	—		—		—		—	—	—	15 P
—	—		—		—		—	—	—	13 Al
—	—		—		—		—	—	—	12 Mg

Tabelle 3.

	K	L_I	L_{II}	L_{III}	M_I	M_{II}	M_{III}	M_{IV}	M_V
92 U	92,07	40,04	39,28	35,56	20,22	19,55	17,81	16,55	16,1
90 Th	89,85	38,85	38,10	34,65	19,53	18,83	17,26	16,02	15,6
83 Bi	81,53	34,75	34,05	31,46	17,20	16,54	15,32	14,12	13,8
82 Pb	80,39	34,19	33,50	30,99	16,85	16,20	15,08	13,80	13,5
81 Tl	79,31	33,65	32,92	30,55	16,55	15,93	14,81	13,60	13,3
80 Hg	78,20	33,08	32,38	30,10	—	—	—	—	—
79 Au	77,07	32,54	31,85	29,64	15,90	15,33	14,25	13,01	12,7
78 Pt	75,92	32,01	31,28	29,19	15,60	15,08	13,99	12,74	12,5
74 W	71,51	29,84	29,16	27,41	14,41	13,74	12,93	11,74	11,5
73 Ta	—	29,33*	28,65	26,99	14,15	13,45*	12,67*	11,52	11,3
71 Lu	—	28,30*	27,62*	26,12*	13,58*	12,92*	12,22*	11,03*	10,8
70 Yb	—	27,81*	27,12*	25,68*	13,32*	12,69*	12,03*	10,83*	10,6
68 Er	—	26,80*	26,15*	24,82*	12,80*	12,14*	11,50*	10,36*	10,2
67 Ho	64,16	26,32*	25,64*	24,39*	12,54*	11,93*	11,36*	10,15*	9,9
66 Dy	63,03	25,82*	25,15*	23,96*	12,28*	11,67*	11,12*	9,91*	9,7
65 Tb	—	25,35*	24,66*	23,57*	12,04*	11,44*	10,93*	9,70*	9,5
64 Gd	60,93	24,85*	24,17*	23,10*	11,77*	11,16*	10,69*	9,48*	9,3
63 Eu	59,86	24,36*	23,69	22,67	11,53	10,93*	10,47*	9,26	9,1
62 Sm	58,80	23,89*	23,22	22,25	11,27	10,68*	10,25*	9,05	8,9
60 Nd	56,69	22,93*	22,27	21,40	10,79	10,21*	9,81*	8,62	8,5
59 Pr	55,62	22,45*	21,79	20,96	10,55	9,95*	9,60*	8,37	8,2
58 Ce	54,52	21,98*	21,32	20,54	10,29	9,71*	9,37*	8,17	8,0
56 Ba	52,50	21,03	20,36	19,66	9,77	9,21	8,91	7,67	7,6
55 Cs	51,47	20,54	19,87	19,22	9,47	8,91	8,63	7,40	7,3
53 J	49,48	19,56	18,91	18,33	—	8,31	8,05	6,84	6,7
52 Te	48,43	19,08	18,43	17,89	8,63	8,02	7,78	6,57	6,5
51 Sb	47,35	18,60	17,94	17,44	8,32	7,74	7,51	6,29	6,2
47 Ag	43,35	16,72	16,12	15,73	7,30	6,60	6,42	5,28	5,2
45 Rh	41,34	15,90	15,20	14,86	6,76	6,30	6,16	4,74	4,7
42 Mo	38,38	14,54	13,92	13,64	6,13	5,52	5,40	4,17	4,1
41 Nb	37,43	—	13,47	13,21	5,92	—	—	3,87	3,8
40 Zr	36,41	—	13,02	12,79	5,60	—	—	3,61	
29 Cu	25,71	—	8,44	8,36	—	2,28		0	
28 Ni	24,74	—	7,91	7,83	—	1,82		—	
27 Co	23,85	—	7,73	7,65	—	2,30		—	
26 Fe	22,89	—	7,29	7,23	—	2,00		—	
25 Mn	21,96	—	6,98	6,92	—	2,05		—	
24 Cr	21,00	—	6,56	6,50	—	1,87		—	
23 Va	20,06	—	6,18	6,13	—	1,61		—	
22 Ti	19,11	—	5,71	5,66	—	1,48		—	
21 Sc	18,20	—	5,51	5,48	—	1,64		—	
20 Ca	17,25	—	5,09	5,06	—	1,42		—	
19 K	16,29	—	4,63	4,61	—	0,95		—	
17 Cl	14,42	—	3,85	3,84	—	0,63		—	
16 S	13,48	—	3,44		—	0,55		—	
15 P	12,58	—	3,15		—	0,89		—	
13 Al	10,71	—	2,28		—	0		—	
12 Mg	9,79	—	1,87		—	—		—	

* Interpoliert oder aus einem interpolierten Wert berechnet.

Wurzeln der Niveauwerte $(\sqrt{T/R})$.

N_I	N_{II}	N_{III}	N_{IV}	N_V	N_{VI}	N_{VII}	O_I	$O_{II,III}$	$O_{IV,V}$	
10,32	9,78	8,78	7,50	7,32	5,33	5,25	5,12	3,92	2,41	92 U
9,89	—	—	7,16	6,98	4,98	4,91	—	—	2,39	90 Th
8,43	7,66	7,09	5,98	5,81	3,69	3,61	—	3,39	1,42	83 Bi
8,13	7,44	7,02	5,68	5,52	3,29	3,21	3,21	2,79	0,89	82 Pb
7,98	7,32	6,70	5,54	5,39	3,16	3,10	3,26	2,72	1,30	81 Tl
—	—	—	—	—	—		—	—	—	80 Hg
7,62	7,01	6,54	5,14	5,00	2,53		2,79	2,88	0,89	79 Au
7,25	6,98	6,50	4,95	4,82	2,28		2,67	2,93	0,63	78 Pt
6,60	5,97	5,53	4,32	4,22	1,58		2,28	1,70	0	74 W
6,46	5,75*	5,48*	4,24	4,15	1,58		2,30	1,61*	—	73 Ta
6,14*	5,56*	5,22*	4,03*	3,94*	1,14*		2,21*	1,58*	—	71 Lu
6,05*	5,47*	5,14*	3,90*	3,82*	1,0*		—	1,58*	—	70 Yb
5,76*	5,18*	4,84*	3,66*	3,58*	0,5*		2,03*	1,55*	—	68 Er
5,67*	5,16*	4,84*	3,56*	3,49*	—		—	1,55*	—	67 Ho
5,55*	5,03*	4,73*	3,45*	3,38*	—		1,70*	1,58*	—	66 Dy
5,41*	4,91*	4,67*	3,39*	3,33*	—		1,87*	1,55*	—	65 Tb
5,32*	4,71*	4,49*	3,32*	3,25*	—		1,87*	1,55*	—	64 Gd
5,19	4,68*	4,44*	3,23	3,18	—		1,70	—	—	63 Eu
5,08	4,49*	4,34*	3,18	3,13	—		1,73	1,52*	—	62 Sm
4,88	4,31*	4,15*	3,10	3,05	—		1,79	1,58*	—	60 Nd
4,74	4,24*	4,09*	2,97	2,92	—		1,61	1,48*	—	59 Pr
4,66	4,12*	3,98*	2,95	2,92	—		1,76	1,48*	—	58 Ce
4,34	3,91	3,80	2,67	2,63	—		1,45	1,55	—	56 Ba
4,16	3,69	3,55	2,47	2,43	—		1,45	1,30	—	55 Cs
—	—	—	2,10	2,05	—		—	—	—	53 J
3,56	2,97		1,84	1,82	—		—	0,63	—	52 Te
3,33	2,68		1,52	1,48	—		0	0,32	—	51 Sb
2,74	1,82		0,77		—		—	—	—	47 Ag
2,30	2,24		—		—		—	—	—	45 Rh
2,17	1,58		—		—		—	—	—	42 Mo
2,00	—		—		—		—	—	—	41 Nb
1,83	—		—		—		—	—	—	40 Zr
—	—		—		—		—	—	—	29 Cu
—	—		—		—		—	—	—	28 Ni
—	—		—		—		—	—	—	27 Co
—	—		—		—		—	—	—	26 Fe
—	—		—		—		—	—	—	25 Mn
—	—		—		—		—	—	—	24 Cr
—	—		—		—		—	—	—	23 Va
—	—		—		—		—	—	—	22 Ti
—	—		—		—		—	—	—	21 Sc
—	—		—		—		—	—	—	20 Ca
—	—		—		—		—	—	—	19 K
—	—		—		—		—	—	—	17 Cl
—	—		—		—		—	—	—	16 S
—	—		—		—		—	—	—	15 P
—	—		—		—		—	—	—	13 Al
—	—		—		—		—	—	—	12 Mg

Wellenlängen für die Kanten alle um eine Einheit verkleinert. Um besser vergleichbare Werte zu bekommen, haben wir diese Korrektion an den Duane- und Pattersonschen L-Kantenwellenlängen vorgenommen.

Für W (74) fällt die Kante L_{III}[1]) noch nahezu mit der Linie L_{β_5} zusammen. Ebenso die Kante L_{II} mit der Linie L_{γ_6}. Für das vorangehende Element Ta (73) konnten also ohne Willkür statt der Kanten L_{III} und L_{II} die Linien L_{β_5} und L_{γ_6} eingesetzt werden. Ebenso wurden für die Elemente Zr (40) und Nb (41) für dieselben Kanten die Werte der Linien L_{β_2} und L_{γ_1} benutzt.

Für die Elemente zwischen Ba (56) und W (74) sind die L-Kanten nur teilweise gemessen; für die unbekannten Werte wurde versucht, eine rationelle Interpolation zu erhalten. So sind für die Elemente zwischen Eu (63) und Ta (73) die Werte für die M_V-Niveaus durch graphische Interpolation ermittelt worden. Für die Elemente zwischen Ba (56) und W (74) wurden die Werte für die M_{III}-Niveaus auf ähnliche Weise bestimmt. Die anderen Niveaus der besprochenen Elemente sind mit Hilfe von experimentell gefundenen Daten aus dem interpolierten M_{III}- oder M_V-Werte berechnet. Wo für die Berechnung eines Niveaus ein interpolierter Wert benutzt worden ist, wurde dieses mit einem Sternchen in der Tabelle bezeichnet. Obwohl es natürlich immer möglich sein kann, daß durch die Interpolation einige spezielle Eigenarten der Änderungen der Niveauwerte mit der Atomnummer verloren gegangen sind, so scheint es uns doch sehr unwahrscheinlich, daß weitere experimentelle Messungen die in dieser Arbeit gezogenen Schlüsse in Gefahr bringen können.

In diesem Zusammenhang möge darauf hingewiesen werden, daß der in Tabelle 2 angegebene Wert für L_I des W, die den Messungen von Duane und Patterson unter Anbringung der besprochenen Korrektion entnommen sind, ziemlich genau ist, obwohl im allgemeinen der experimentelle Fehler bei der direkten Bestimmung dieser Kante wegen ihrer Schwäche und Unschärfe ziemlich groß sein mag. Man kann nämlich den Wert für L_I noch aus der viel schärferen und genauer bekannten Kante L_{III} auf drei verschiedene Weisen berechnen, und zwar unter Zuhilfenahme von einer der folgenden Additionsbeziehungen:

$$L_{\beta_2} + (L_I - L_{III}) = L_{\beta_3} + M_\gamma$$
$$L_I - L_{III} = L_{\beta_9}{}^{2)} - L_{\alpha_1}$$
$$L_I - L_{III} = L_{\beta_{10}}{}^{2)} - L_{\alpha_2}.$$

[1]) Früher L_1 genannt. Die Kanten sind jetzt auch in Übereinstimmung mit den zugehörigen Niveaus anders bezeichnet.

[2]) Siehe D. Coster, ZS. f. Phys. **6**, 185, 1921.

In allen diesen drei Fällen bekommt man für Wolfram T/R-Werte, welche innerhalb einer Unsicherheit von $\pm 0{,}6$ mit dem direkt gemessenen Werte übereinstimmen. Hieraus kann man schließen, daß der Fehler in dem T/R-Werte für die $L_{\rm I}$-Kante von W sicher weniger als eine Einheit beträgt. Dies ist von Interesse für die Interpolation der $M_{\rm III}$-Kurve zwischen W und Ba.

Weiter mag noch bemerkt werden, daß, wo für die Berechnung der Niveaus ziemlich kurze Wellenlängen benutzt wurden, die Genauigkeit kleiner ist als dort, wo größere Wellenlängen gebraucht werden konnten. So entspricht ein Fehler von 2 X-E in der Wellenlänge der $L_{\rm I}$-Kante einem Fehler des T/R-Wertes von weniger als $1/_2$ bei Ba, aber von etwa 2 und 3 bei W und bzw. Bi. Dies bewirkt z. B., daß in die niedrigsten Niveauwerte für die schwereren Elemente ziemlich große Fehler eingehen können[1]); während hingegen in der Nähe des Ba auch T/R-Werte der niedrigsten Niveaus noch verhältnismäßig genau bekannt sind.

In dieser Verbindung mag noch auf eine gewisse Zweideutigkeit hingewiesen werden, welche den aus der Röntgenspektroskopie stammenden Daten auch bei den genauesten Messungen anhaften. Die Röntgenabsorptionsspektren werden in der Regel mit Hilfe eines Stoffes in festem Zustand, und zwar oft in chemischer Verbindung, aufgenommen. Nun hat aber Bergengren[2]) gezeigt, daß die Zustandsform dabei von wesentlicher Bedeutung ist, indem ein merklicher Unterschied der K-Kanten des Phosphors beobachtet wurde, wenn verschiedene Modifikationen dieses Elementes benutzt wurden. Weiter hat Lindh[3]) meßbare Unterschiede in den Absorptionsspektren gefunden, die mit verschiedenen chemischen Verbindungen desselben Stoffes aufgenommen wurden. So hat er für die K-Kante des Chlors und Schwefels Unterschiede gefunden, die 0,5 bis 0,8 Einheiten in den T/R-Werten betragen, wenn Chlor- und Schwefelverbindungen benutzt wurden, in denen diese Stoffe einerseits mit positiver, andererseits mit negativer Wertigkeit auftreten.

Eine noch ernstere Unbestimmtheit in den Röntgendaten rührt von der sogenannten Feinstruktur der Absorptionskanten her, die sich in dem Auftreten einer zweiten schwächeren Diskontinuität

[1]) Anmerkung bei der Korrektur: Neuerdings ist es Hjalmar durch neue Messungen in der M-Serie (C. R. **175**, 878, 1922) gelungen, die O- und M-Niveaus der schwersten Elemente mit größerer Genauigkeit zu bestimmen. Seine Ergebnisse konnten nicht mehr in diese Arbeit aufgenommen werden, sie sind aber mit unseren Schlußfolgerungen in dem besten Einklang.

[2]) J. Bergengren, ZS. f. Phys. **3**, 247, 1920.

[3]) A. E. Lindh, ZS. f. Phys. **6**, 303, 1922; C. R. **175**, 25, 1922.

meistens auf der harten Seite der Hauptdiskontinuität äußert. Für den Unterschied der T/R-Werte dieser Diskontinuitäten gibt Fricke [1]) für die K-Kante der leichteren Elemente Beträge von ein bis zwei Einheiten an. Bekanntlich hat Kossel [2]) eine Erklärung für diese Kantenstruktur vorgeschlagen, nach welcher sie kleinen Verschiedenheiten der Ablösungsarbeit eines Elektrons in einer inneren Gruppe entsprechen sollte, je nachdem dieses Elektron völlig vom Atom entfernt wird oder in eine außerhalb der Elektronenhülle verlaufende, den stationären Zuständen der Serienspektren entsprechende Quantenbahn übergeführt wird. Die Beziehung der verschiedenen hier besprochenen Umstände, welche eine Unbestimmtheit der Röntgendaten verursachen, zu den in dieser Arbeit zugrunde gelegten Vorstellungen des Atombaues werden wir am Schluß des § 6 kurz berühren. Hier möchten wir nur darauf hinweisen, daß wir es unter diesen Umständen mit eben derartigen individuellen Eigenartigkeiten der Niveauwerte zu tun haben, von deren Berücksichtigung wir bei der oben erwähnten Interpolation von vornherein absehen mußten.

§ 4. Theoretische Abschätzung der Energieniveaus. Was die Verwertung der Theorie für die nähere Deutung der Niveaus der Röntgenspektren betrifft, so handelt es sich in erster Linie darum, die Arbeit zu berechnen, die zur Entfernung eines Elektrons aus einer inneren Gruppe des Atoms aufgewendet werden muß. Zu diesem Zweck müssen wir natürlich erst die Bewegung der Elektronen im normalen Zustand des Atoms betrachten. Hierbei ist zu beachten, daß es ein wesentlicher Zug der Theorie ist, daß die Elektronen der verschiedenen Gruppen nicht in räumlich getrennten Gebieten des Atoms sich bewegen, sondern daß die Elektronen in den Gruppen mit größeren Hauptquantenzahlen (äußeren Gruppen), jedenfalls was die Bahnen in gewissen Untergruppen betrifft, während ihrer Bewegung in die Gebiete der Elektronenbahnen niedriger Hauptquantenzahlen (inneren Gruppen) eindringen. Dieser Umstand bringt mit sich, daß die totale Anziehungskraft zum Mittelpunkt des Atoms, die auf ein Elektron wirkt, für die verschiedenen Teile der Bahn, relativ zur Coulombschen Anziehungskraft durch den Kern selbst starken Änderungen unterworfen sein kann. Für die Beurteilung der Verhältnisse ist aber der Umstand von Wichtigkeit, daß im allgemeinen größere Teile der Zentralbahn, die man als dessen äußere Schlingen bezeichnen kann, nahe mit einer Keplerellipse verglichen werden können. Die Dimensionen und die Energie dieser elliptischen Schlingen weichen

* 1) R. Fricke, Phys. Rev. **16**, 202, 1920.

2) W. Kossel, Verh. d. D. Phys. Ges. **18**, 339, 1916.

* [Instead of R. Fricke read H. Fricke.]

dabei aber jedoch im allgemeinen wesentlich ab von den Dimensionen und der Energie einer durch dieselbe Hauptquantenzahl wie die tatsächliche Elektronenbahn charakterisierenden Keplerellipse, die in einem Coulombschen Felde von derselben Stärke wie diejenige, in welchem sich die betreffenden Bahnschlingen befinden, von einem Elektron durchlaufen würde. Energie und Dimension einer gequantelten Keplerellipse sind bekanntlich durch die einfachen Formeln

$$W = Rh\frac{N^2}{n^2}, \qquad 2a = \frac{e^2}{Rh}\frac{n^2}{N} \tag{1}$$

gegeben, wo W die zur Entfernung erforderliche Arbeit, a die große Halbachse der Ellipse, R die Rydbergsche Konstante, Ne die Kernladung und n die Hauptquantenzahl bezeichnen. Für die in Frage stehenden Bahnschlingen werden wir nun nicht nur mit einer „effektiven Kernladungszahl" N^*, sondern auch mit einer „effektiven Quantenzahl" n^* zu rechnen haben, definiert durch die Ausdrücke

$$W = Rh\frac{N^{*2}}{n^{*2}}, \qquad 2a = \frac{e^2}{Rh}\frac{n^{*2}}{N^*}. \tag{2}$$

Dieser Umstand ist besonders für die Beurteilung der Bindungsstärke der Elektronen der äußeren Gruppen entscheidend, indem hier die effektiven Quantenzahlen der elliptischen Schlingen beträchtlich kleiner sein können als die Hauptquantenzahlen und wesentlich verschieden für Bahnen, die zum gleichen n, aber verschiedenen Werten von k gehören. So beruht die typische Periodizität der chemischen Eigenschaften und der optischen Spektren darauf, daß für die äußersten Elektronenbahnen die effektiven Quantenzahlen, im Gegensatz zu den Hauptquantenzahlen, sich nur wenig ändern, wenn man von einem Element zum homologen Elemente der nächsten Periode im System der Elemente übergeht. Andererseits beruht der auffallende Mangel von Periodizität in wesentlichen Zügen der Röntgenspektren darauf, daß es sich hier in der Hauptsache um das Verhalten der innersten Elektronen im Atom handelt, wo wir es mit Elektronengruppen zu tun haben, die bereits vollständig ausgebildet sind und sich bei allen folgenden Elementen in derselben Weise wiederholen. Die effektive Quantenzahl der Bahnschlingen weicht in einem solchen Falle nur wenig ab von der Hauptquantenzahl der Bahn, und diese geringe Abweichung ist, wie eine einfache Rechnung lehrt, für einen bestimmten Bahntypus in erster Annäherung der Atomnummer N umgekehrt proportional. Das Eintauchen der Elektronenbahn in das Gebiet innerhalb liegender Gruppen macht sich daher, was die Stärke der Bindung in ihrer Abhängigkeit von der Kernladungszahl betrifft, in ganz ähnlicher

Weise geltend wie die direkte Abschirmung eines Teils der Kernladung durch die inneren Elektronen während des Durchlaufens der äußeren Bahnschlingen. Es ist schon deshalb um so weniger Grund, die zwei Wirkungen voneinunder zu unterscheiden, als noch eine Wirkung auf die Ablösungsarbeit hinzukommt, die schwer zu berechnen ist, und die in derselben Weise sich mit der Kernladung ändert. Wir denken hier an die Verminderung der Ablösungsarbeit, die mit der Verstärkung der Bindung der in derselben Gruppe übrigbleibenden Elektronen bei der Entfernung des Elektrons zusammenhängt. Den Einfluß aller dieser Wirkungen können wir also durch die einfache Formel

$$W = Rh \frac{(N - \alpha_{n,k})^2}{n^2} \tag{3}$$

ausdrücken, wo $\alpha_{n,k}$ eine Konstante für jede Bahntype n_k ist, die man der üblichen Sprechweise gemäß als „Abschirmungszahl“ bezeichnen kann.

Bei der Bestimmung der Ablösungsarbeit eines Elektrons aus einer inneren Gruppe ist es jedoch wesentlich, den Einfluß der Anwesenheit der äußeren Elektronengruppen noch in Betracht zu ziehen. Während der Entfernung des Elektrons üben diese Elektronengruppen eine Abschirmungswirkung aus, die am einfachsten in Rechnung gestellt werden kann, dadurch, daß wir von dem durch (3) gegebenen Ausdruck die Zunahme der Bindungsstärke sämtlicher äußerer Elektronen abziehen, welche durch die Entfernung des betreffenden Elektrons veranlaßt wird. Die gesamte Bindungsenergie der äußeren Elektronen mag folgendermaßen geschrieben werden:

$$B = Rh \sum_{p > n} A_p \frac{N_p^{*2}}{p^{*2}}, \tag{4}$$

wo A_p die Anzahl der Elektronen in der Gruppe mit der Hauptquantenzahl p bedeutet, während N_p^* und p^* als mittlere effektive Kernladungszahl bzw. mittlere effektive Quantenzahl der Gruppe bezeichnet werden können. Wenn wir der Einfachheit halber uns auf die Betrachtung solcher Elemente beschränken, bei denen sich keine Gruppe in einer Stufe geradweiser Entwicklung befindet, können wir bei einer angenäherten Abschätzung der Größe B einfach $\frac{A_p}{p^{*2}}$ gleich 2 setzen. Bei den vollständig ausgebildeten inneren Gruppen ist ja eben $A_p = 2p^2$ und gleichzeitig p^* nur wenig von p verschieden. Bei den äußeren nur vorläufig abgeschlossenen Gruppen, die an und für sich wegen der kleineren Werte der N_p^* einen geringeren Einfluß haben, ist p^* wesentlich kleiner als p, aber, wie eine einfache Abschätzung lehrt, eben in solchem Maße, daß wir p^{*2} in grober Annäherung auch hier stets $p^{*2} = {}^1/_2\, A_p$ setzen können. Was nun die Änderung von B bei der Entfernung eines Elektrons aus der Gruppe mit der Hauptquantenzahl n betrifft, können wir also mit beträchtlicher Annäherung schreiben:

$$\Delta B = Rh \sum_{p>n} 2\,\Delta N_p^{*2} = Rh\, 4 \sum_{p>n} N_p^*. \tag{5}$$

Um diesen Ausdruck weiter abzuschätzen, müssen wir die Weise betrachten, in welcher N_p^* für wachsendes p abnimmt. In erster Annäherung setzen wir

$$N_p^* = N - \sum_{q<p} A_q, \tag{6}$$

wodurch wir zugleich erreichen, daß die Summe in (5) sich einfach ausführen läßt. Aus der typischen symmetrischen Weise, in welcher bei den betrachteten Elementen nach der Tabelle 1 die Elektronenzahl in den Gruppen bei wachsender Hauptquantenzahl erst steigt und dann wieder sinkt, sehen wir nämlich, daß diese Summe stets annähernd gleich wird dem halben Produkt von N_n^*, wie es durch (6) für $p = n$ definiert ist, und der Anzahl der Glieder der Summe, d. h. der Anzahl der äußeren Elektronengruppen. Wenn der Ausdruck (6) genau richtig wäre, würde in der Tat, wie man leicht einsieht, diese Behauptung im Falle des K-Niveaus ($n = 1$) streng gelten bei den Elementen, die zwei Stellen nach den Edelgasen stehen, und für welche die äußerste Gruppe zwei Elektronen enthält; im Falle der L-Niveaus ($n = 2$) bei den Edelgasen selbst; im Falle der M-Niveaus ($n = 3$) für die Elemente, die acht Stellen vor den Edelgasen stehen und bei denen eine Gruppe mit 18 Elektronen eben im Abschluß begriffen ist. Mit Annäherung bekommen wir also

$$\Delta B = Rh \,.\, 2\, N_n^* \,.\, \beta_{N,n},$$

wo $\beta_{N,n}$ eine Größe ist, die kontinuierlich mit fortschreitender Atomnummer anwächst und die sich annähernd um eine Einheit vermehrt, wenn wir von einem Element zum homologen Element in der nächsten Periode übergehen, in solcher Weise, daß $\beta_{N,n}$ niemals viel von der Anzahl der äußeren Gruppen abweicht. Annähernd können wir jetzt den Einfluß der äußeren Elektronen auf die Ablösungsarbeit eines Elektrons aus einer inneren Gruppe berücksichtigen, indem wir statt (3) schreiben:

$$W = Rh \frac{(N - \alpha_{n,k} - n^2 \beta_{N,n})^2}{n^2}.$$

Während die Größe $\alpha_{n,k}$, die wir als „innere Abschirmungszahl" bezeichnen können, nicht von der Atomnummer abhängt, ändert sich die Größe $n^2\beta_{N,n}$, die „äußere Abschirmungszahl", in charakteristischer

Weise mit der Atomnummer. Nicht nur wird diese letztere Abschirmungszahl, wie es aus dem Obigen folgt, sich innerhalb einer Periode des Systems der Elemente um n^2 Einheiten vermehren, sondern wir müssen auch erwarten, daß die Einzelheiten des allmählichen Hinzukommens und der gradweisen Ausbildung der äußeren Gruppen, wovon bei der obigen Rechnung abgesehen wurde, sich abspiegeln werden in der Weise, in welcher die äußere Abschirmungszahl im einzelnen mit wachsender Atomnummer sich ändert. Die Ausbildung einer Elektronengruppe muß sich natürlich desto deutlicher zeigen, je tiefer sie in dem Atominneren liegt und wir müssen erwarten, daß das Anwachsen der äußeren Abschirmungszahlen besonders rasch sein wird in Gebieten des periodischen Systems, wo eine tiefliegende Gruppe sich in einer Stufe gradweiser Ausbildung befindet.

Wir wollen hier nicht näher auf die Berechnung der Größen $\alpha_{n,k}$ und $\beta_{N,n,k}$ eingehen, die sich schon beim jetzigen Stand der Theorie wesentlich verfeinern ließe, da es uns bei dieser Gelegenheit mehr darauf ankommt, das Versuchsmaterial mit den Hauptzügen der Theorie zu vergleichen. Bei diesem Vergleich kommt jedoch noch ein wesentlicher Umstand in Frage, von dem wir der Einfachheit halber bisher abgesehen haben. Wir denken hier an die Relativitätskorrektion, die bei der Beschreibung der Bewegung der inneren Elektronen bekanntlich keineswegs zu vernachlässigen ist. Diese wird dadurch berücksichtigt, daß wir anstatt der einfachen Formel (1) die Sommerfeldsche Formel

$$W = Rh\,\frac{N^2}{n^2} + Rh\left(\frac{2\pi e^2}{hc}\right)^2 \frac{N^4}{n^4}\left(\frac{n}{k} - \frac{3}{4}\right) + \cdots \tag{7}$$

zugrunde legen, die von (1) durch das Hinzukommen der sogenannten Relativitätsglieder abweicht, von denen wir nur das erste hingeschrieben haben, dessen Berücksichtigung für viele Anwendungen genügt.

Allen besprochenen Korrektionen mag formell dadurch Rechnung getragen werden, daß wir für die Ablösungsarbeit eines Elektrons mit dem Bahntypus n_k schreiben:

$$W = Rh\,\frac{(N-\gamma)^2}{n^2} + Rh\,\frac{(N-\delta)^4}{n^4}\left(\frac{2\pi e^2}{hc}\right)^2\left(\frac{n}{k} - \frac{3}{4}\right) + \cdots \tag{8}$$

Die im Hauptgliede vorkommende Größe γ bezeichnen wir als „Gesamtabschirmungszahl"; für die betrachteten innersten Elektronengruppen ist diese angenähert durch die Formel

$$\gamma = \alpha_{n,k} + n^2 \beta_{N,n} \tag{9}$$

gegeben. Die im zweiten Gliede vorkommende Größe $N-\delta$ soll einfach eine effektive Kernladungszahl bedeuten, die für eine angenäherte

Berechnung der Relativitätskorrektion des betreffenden Bahntypus als maßgebend betrachtet werden kann. Bei einer Abschätzung von δ ist zu bedenken, daß die Abschirmungswirkung sowohl der inneren wie der äußeren Elektronengruppen sich in wesentlich verschiedener Weise in den zwei Gliedern der Formel (8) äußern wird. So ist zu erwarten, daß im allgemeinen δ beträchtlich kleiner sein wird als die im Hauptgliede auftretende Abschirmungszahl.

Wenn wir nicht, wie bisher, nur die innersten Gruppen des Atoms betrachten, sondern uns den weiter außen befindlichen Gruppen zuwenden, werden die Verhältnisse viel verwickelter und bei der Berechnung der Ablösungsarbeit ist es wie beim optischen Spektrum von wesentlicher Bedeutung, die Abweichung der Bahn von einer Keplerellipse und die daraus folgende Erniedrigung der effektiven Quantenzahlen gesondert zu berücksichtigen. Im besonderen treffen wir auf Verhältnisse, die von den bisher besprochenen abweichen, wenn es sich um die Ablösungsarbeit von Elektronen in solchen Gruppen handelt, die sich in einer Stufe von gradweiser Ausbildung befinden. Obwohl die quantitative Berechnung der Niveaus in solchen Fällen beim jetzigen Stand der Theorie noch sehr unvollständig sein muß, kann jedoch eine Anzahl von unmittelbaren Folgerungen aus der Theorie über die Änderung der Niveauwerte mit der Atomnummer angegeben werden, auf die wir aber erst später bei dem Vergleich des Versuchsmaterials mit der Theorie näher eingehen werden.

§ 5. Allgemeiner Vergleich der Klassifikation der empirischen Röntgenniveaus mit der Theorie des Atombaues. Der Übersicht halber und um den Vergleich mit den theoretischen Vorstellungen über den Atombau zu erleichtern, sind in der untenstehenden Fig. 3 Kurven eingetragen, die für jedes Niveau den Verlauf der in Tabelle 2 berechneten $\sqrt{T/R}$-Werte als Funktion von N darstellen. Wo eine besonders große Unsicherheit über die Niveauwerte herrscht, ist der Verlauf der sonst voll ausgezogenen Niveaukurven nur gestrichelt angedeutet. Jede Kurve ist in eben derselben Weise wie die Niveaus in Fig. 2 mit Symbolen $n(k_1 k_2)$ bezeichnet.

Erstens muß nun darauf hingewiesen werden, daß man die Klassifikation der zu jedem Element gehörigen Niveauschar mit eben denselben Werten der Zahl n und der Zahlen k hat durchführen können, welche in Tabelle 1 für die Klassifikation aller im Atom anwesenden Elektronenbahnen mit den Quantenzahlen n und k benutzt wurden. Dies gilt nicht nur für die Elemente mit abgeschlossenen Elektronengruppen im Atom, wie die in § 2 besonders betrachteten Edelgase, sondern das Versuchsmaterial gestattet für eine größere Anzahl von Niveaus,

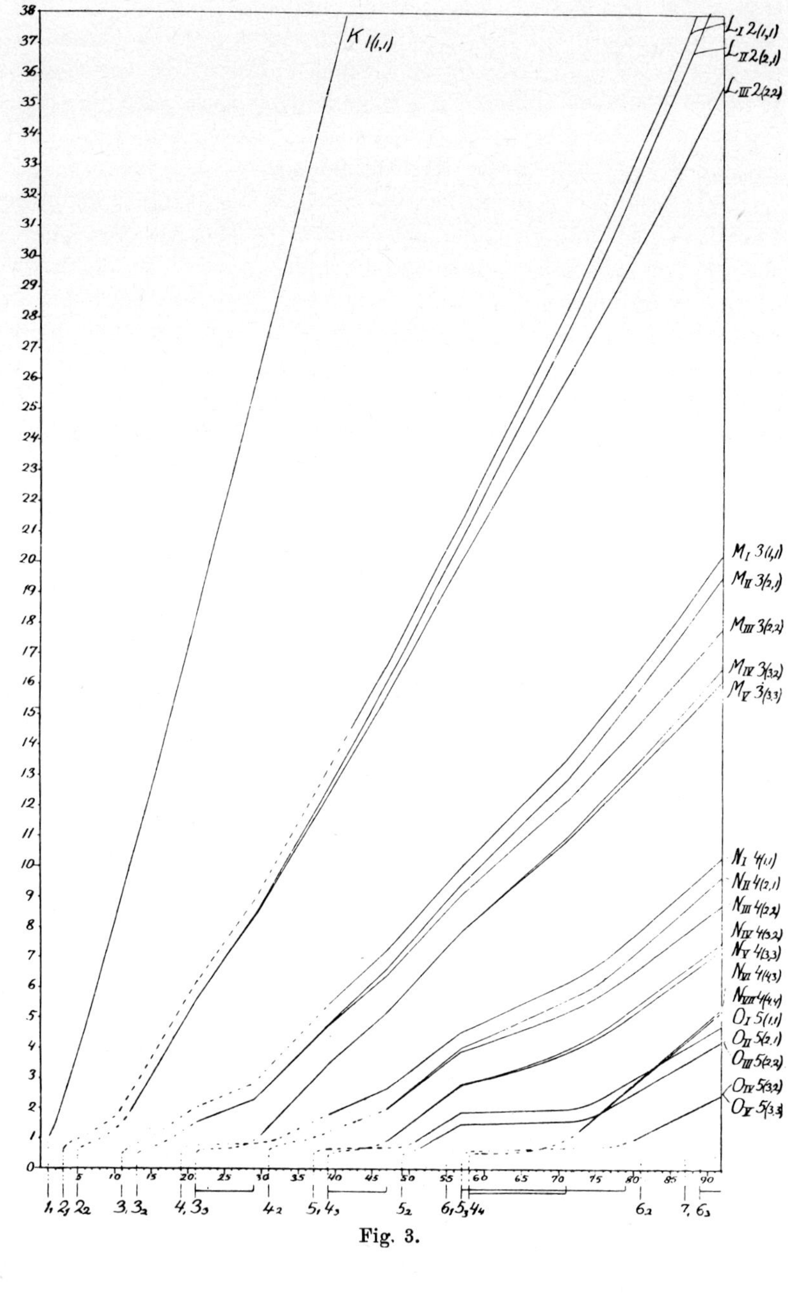
K 1(1,1)
L_I 2(1,1)
L_{II} 2(2,1)
L_{III} 2(2,2)
M_I 3(1,1)
M_{II} 3(2,1)
M_{III} 3(2,2)
M_{IV} 3(3,2)
M_V 3(3,3)
N_I 4(1,1)
N_{II} 4(2,1)
N_{III} 4(2,2)
N_{IV} 4(3,2)
N_V 4(3,3)
N_{VI} 4(4,3)
N_{VII} 4(4,4)
O_I 5(1,1)
O_{II} 5(2,1)
O_{III} 5(2,2)
O_{IV} 5(3,2)
O_V 5(3,3)
1_1 2_1 2_2 3_1 3_2 4_1 3_3 4_2 5_1 4_3 5_2 6_1 5_3 4_4 6_2 7_1 6_3

Fig. 3.

den Elementen nachzuspüren, bei welchen diese sich in den Röntgenspektren zuerst kennlich machen, und es hat sich dabei gezeigt, daß ein Niveau, das zu bestimmten Werten von n und k gehört, soweit es das Versuchsmaterial zu entscheiden erlaubt[1]), eben dort zum ersten Male erscheint, wo nach der Theorie im normalen Atom eine Bahn auftritt, die denselben Werten von n und k entspricht. Diese Stellen sind unten an der Fig. 3 durch kleine vertikale Striche angegeben.

Ein Blick auf die Figur bringt nun weiter sogleich den wichtigen Umstand zutage, daß in dem sonst regelmäßigen Verlauf der Kurven an gewissen Stellen eigentümliche Unregelmäßigkeiten auftreten, und wie ein Vergleich mit der Fig. 1 zeigt, gilt dieses für eben die Stellen des periodischen Systems, wo eine innere Gruppe in einer Stufe von gradweiser Ausbildung sich befindet. Diese Stellen sind unten an der Fig. 3 außerdem noch durch horizontale Linien, die bei den früher erwähnten vertikalen Strichen anfangen, angegeben. Dabei ist aber zu bemerken, daß eine solche Linie nicht einfach das gradweise Hinzukommen von Elektronen in n_k-Bahnen andeutet, sondern allgemeiner der ganzen Ausdehnung der $k$ten Stufe in der Ausbildung der $n$ten Elektronengruppe im Atom entspricht, bei welcher Stufe sich auch die Anzahl der Elektronenbahnen mit demselben Wert von n, aber kleineren Werten von k vermehrt.

Ehe wir aber auf weitere Einzelheiten in dem Kurvenverlauf eingehen, müssen wir einige allgemeine Betrachtungen über die sogenannten Niveaudubletts anstellen. Diese sind bekanntlich von zweierlei Art, die üblicher Weise als „Abschirmungsdublette“ und „Relativitätsdublette“ bezeichnet werden. In der Fig. 3 sind die ersteren kenntlich durch den parallelen Verlauf der Bildkurven gewisser Niveaupaare, wie: (L_{I} L_{II}); (M_{I} M_{II}), (M_{III} M_{IV}); (N_{I} N_{II}), (N_{III} N_{IV}), (N_{V} N_{VI}); (O_{I} O_{II}), (O_{III} O_{IV}). Die zweite Art von Dubletten sind dagegen kenntlich durch das immer raschere Anwachsen der Abstände der Bildkurven gewisser Niveaupaare, wie: (L_{II} L_{III}); (M_{II} M_{III}), M_{IV} M_{V}); (N_{II} N_{III}), (N_{IV} N_{V}), (N_{VI} N_{VII}); (O_{II} O_{III}), (O_{IV} O_{V}), die anfänglich so nahe beisammen liegen, daß sie bei den Messungen der Linien nicht zu trennen sind, während wir jedoch aus den Kombinationsregeln schließen dürfen, daß sie bei steigender Atomnummer zu gleicher Zeit zuerst im Atom auftreten. Während die Größe der ersten Dubletts formal dargestellt werden kann durch einen Unterschied zwischen zwei Werten des ersten Gliedes in der Formel (7), die einem Unterschied in der inneren Abschirmungszahl $\alpha_{n,k}$ ent-

[1]) D. Coster, Phil. Mag. **43**, 1070, 1922.

sprechen, kann die Größe der zweiten Dublette angenähert dargestellt werden durch den Unterschied zwischen zwei Werten des zweiten Gliedes dieser Formel, die dadurch zustande kommen, daß man für k die zwei verschiedenen Werte einsetzt, die bei der Klassifikation der Niveaus mit k_2 bezeichnet wurden.

Für das Auftreten dieser Dublette, die bekanntlich bei der formalen Klassifikation der Röntgenspektren von großer Bedeutung gewesen sind, kann beim jetzigen Stand der Theorie keine vollständige Erklärung gegeben werden. Es scheint nichtsdestoweniger möglich, die erfahrungsmäßige Struktur der Niveaugruppen in enge Verbindung zu bringen mit der bei der Deutung des periodischen Systems angenommenen Struktur der Elektronengruppen im Atom. Beim ersten Anblick möchte man vielleicht erwarten, daß die Zahl der zu einem bestimmten Wert von n gehörigen Niveaus gleich sein sollte der Anzahl der im normalen Atom auftretenden Untergruppen derjenigen Elektronengruppe, die zur selben Hauptquantenzahl gehört, indem jedes Niveau der Entfernung eines Elektrons aus jeder dieser Untergruppen entsprechen sollte. Wir müssen aber bedenken, daß die Röntgenniveaus wesentlich bestimmt sind durch die Zustände eines Atoms, aus dem ein Elektron aus einer inneren Gruppe entfernt ist, und wir müssen uns die Komplexität vor Augen halten, die durch diese Entfernung eintreten kann. Denken wir uns einen Augenblick, daß auch für das angeregte Atom die verschiedenen Untergruppen als unabhängig voneinander betrachtet werden können in dem Sinne, daß die Entfernung eines Elektrons aus einer Untergruppe keine Änderung wesentlicher Züge der Struktur der anderen Untergruppen mit sich bringt. In einem solchen Falle dürfen wir eine Anzahl von Niveaus erwarten, die sich aber nicht allein durch ihr Relativitätsglied, sondern immer auch durch ihre Abschirmungszahl unterscheiden müßten, so daß wir uns nicht von dem Auftreten der typischen Relativitätsdublette Rechenschaft geben könnten. Wenn wir nun aber das Niveaubild betrachten, das entstehen würde, sobald alle Niveaus weggelassen würden, bei deren Klassifikation ungleiche Werte von k_1 und k_2 benutzt worden sind, so sehen wir, daß es eben alle die Züge wiedergibt, die eine solche einfache Auffassung des Ursprungs der Röntgenniveaus erwarten lassen würde. Nicht nur sehen wir, daß die Größe der Abstände aller übrigbleibenden Niveaupaare sich als eine Summe einer Abschirmungs- und einer Relativitätskorrektion ausdrücken läßt, wie sie der Formel (8) entspricht, wenn wir bei gleichem n die Werte der Nebenquantenzahl k ändern, sondern auch die Regeln für das Auftreten von beobachteten Röntgenlinien, welche Kombinationen der betreffenden Niveaus entsprechen, sind eben diejenigen, welche eine einfache Anwendung des Korrespondenzprinzips auf eine Zentralbewegung ergibt, indem sich hier k immer um eine Einheit ändert. Deshalb wollen wir die Niveaus, bei denen $k_1 = k_2$ gesetzt wurde, der Kürze halber als „normale" Niveaus bezeichnen.

Was das Vorhandensein der anderen Niveaus, die mit ungleichen Werten von k_1 und k_2 versehen sind, betrifft, so ist zunächst zu sagen, daß die zur Klassifikation dieser „anormalen" Niveaus benutzten Werte von k_1 und k_2 einfach zum Ausdruck bringen sollen, daß jedes dieser Niveaus rein formell zu zwei normalen Niveaus in enger quantitativer Beziehung steht. So tritt jedes dieser Niveaus im Atom zugleich mit dem normalen Niveau auf, für das k_1 denselben Wert hat, mit dem es auch die Abschirmungszahl gemeinsam hat. Andererseits ist es mit dem normalen Niveau, für das k_2 denselben Wert hat, dadurch verknüpft, daß es mit ihm das Relativitätsglied gemeinsam hat, soweit man von Verschiedenheiten der in diesem vorkommenden effektiven Kernladungszahl absieht. Dieser Tatbestand führt dazu, den Ursprung dieser anormalen Niveaus in einer engen Wechselwirkung der Elektronenbewegung zweier benachbarter Untergruppen des angeregten Atoms zu suchen, die eine wesentliche Änderung des harmonischen Wechselspiels der Elektronenbewegung innerhalb beider Untergruppen bewirkt, wenn ein Elektron aus einer dieser Untergruppen entfernt wird. Durch die Theorie ist eine solche Auffassung insofern nahegelegt, als angenommen wird, daß die Ausbildungsstufe einer Elektronengruppe eben wesentlich durch das Wechselspiel der Elektronenbewegung innerhalb der verschiedenen Untergruppen bedingt ist [1]). Andererseits ist noch ganz unklar, wie durch eine nähere Ausführung dieser Betrachtung im Einzelnen eine Erklärung für das gesonderte Auftreten der Abschirmungs- und Relativitätsdublette gegeben werden kann. Ebenso steht eine Erklärung aus für die empirischen Kombinationsregeln, soweit sie die anormalen Niveaus betreffen, obwohl sie auf das Auftreten einer zweiten Nebenquantenzahl hindeuten, in ähnlicher Weise, wie sie bei den optischen Serienspektren zur Geltung kommt, wo die Komplexität der Energieniveaus sowie die Kombinationsregeln eine einfache Erklärung auf Grund des Korrespondenzprinzips zu gestatten scheinen. Die obenstehende Betrachtung verfolgt nur den Zweck, zu zeigen, daß, obwohl eine vollständige Erklärung der Komplexstruktur der Niveaus beim jetzigen Stand der Theorie nicht gegeben werden kann, es dennoch möglich scheint, den experimentellen Befund

[1]) Vgl. N. Bohr, ZS. f. Phys. 9, 1, 1922.

in ungezwungener Weise mit den charakteristischen Zügen der Theorie in Einklang zu bringen.

§ 6. **Theoretische Deutung der näheren Abhängigkeit der Energieniveaus von der Atomnummer.** Wenn wir zu einem genaueren Vergleich der aus dem Versuchsmaterial ermittelten Niveauwerte mit der Theorie übergehen, werden wir erst kurz die Niveaus betrachten, die mit vollständig ausgebildeten Gruppen im Atom zusammenhängen und für welche wir annehmen können, daß die Formel (8) für die Ablösungsarbeit eines Elektrons einer inneren Gruppe eine beträchtliche Genauigkeit besitzt. Bei der Benutzung dieser Formel zur Berechnung der im Hauptgliede eingehenden Abschirmungszahl wird es, jedenfalls bei höheren Kernladungen, von Bedeutung sein, welchen Wert wir der im Relativitätsglied eingehenden Abschirmungszahl δ beilegen. Für die K-Niveaus, wo eine einfache Abschätzung lehrt, daß δ nie mehr als einige Einheiten beträgt, führt dieses nicht zu besonderen Schwierigkeiten. Bei den Elementen höchster Atomnummer, wo der Wert, den man der Abschirmungszahl δ beilegt, von wesentlicher Bedeutung ist, kommt überdies die Unsicherheit in der Abschätzung dieser Größe gegenüber der Unsicherheit in den Messungsergebnissen nicht in Betracht. Bei den anderen Niveaus können wir der Einfachheit halber so vorgehen, daß wir — unter Hinweis auf die Überlegungen im vorigen Paragraphen — die aus den Relativitätsdubletten direkt berechneten Werte von δ benutzen. Dieses ist um so mehr berechtigt, als diese Werte annähernd denjenigen Werten gleich sind, die man aus dem Einfluß der Relativitätsmodifikation auf die Bewegung der Elektronen in Bahnen vom n_k-Typus theoretisch erwarten dürfte. Die hierbei eingeführte Unsicherheit ist besonders gering bei denjenigen Niveaus, die zirkulären Bahnen entsprechen, weil hier die absoluten Werte der Relativitätskorrektionen am kleinsten sind. Auf diese Weise haben wir aus der Tabelle 2 die in die untenstehende Tabelle eingetragenen Werte für die in das Hauptglied der Formel (8) eingehende Gesamtabschirmungszahl γ berechnet.

Element	K	L_{III}		M_{V}		N_{VII}	
	γ	γ	$^1/_4\,\gamma$	γ	$^1/_9\,\gamma$	γ	$^1/_{16}\,\gamma$
Ar (18)	2,7	(9,5)	(2,4)	—	—	—	—
Kr (36)	3,7	13,5	3,4	(28,2)	(3,1)	—	—
X (54)	4,7	16,8	4,2	33,5	3,7	—	—
Nt (86)	5,5—6,0	21,7	5,4	43,0	4,8	(65)	(4,1)

Um diese Werte von γ mit der theoretischen Formel (9) vergleichen zu können, sind für die L-, M- und N-Niveaus zugleich die durch n^2 dividierten Abschirmungszahlen angegeben. Wie wir sehen, bekommen wir dadurch für alle betrachteten Niveaus Werte, die im großen und ganzen um 1 anwachsen, wenn wir von einer Periode im System der Elemente zur nächsten übergehen. Diese Übereinstimmung ist besonders gut für die K-Niveaus, wo der theoretischen Formel nach der Art der Abschätzung auch die größte Genauigkeit beizulegen ist. Dieser Übereinstimmung kommt eine wesentliche Bedeutung zu, weil sie, ganz abgesehen von der noch offenen Frage der Deutung der Relativitätsdublette, eine direkte Stütze liefert für die Richtigkeit der bei der Klassifikation der Elektronenbahnen im Atom benutzten Quantenzahlen. Wenn z. B., wie in mehreren neueren Arbeiten vorgeschlagen, der Nebenquantenzahl k anstatt ganzzahlige Werte halbzahlige Werte beizulegen wären, würden die Verhältnisse sich ganz anders gestalten. Infolge des viel größeren Wertes des Relativitätsgliedes für die K-Niveaus, welche nach der Formel der Annahme $k = {}^1/_2$ entspricht, würde diese Annahme bedeuten, daß die Abschirmungszahl anstatt der theoretisch geforderten Zunahme um eine Einheit von Edelgas zu Edelgas, von Krypton bis Xenon, eine Zunahme um etwa 5 Einheiten, und von Xenon bis Niton sogar eine Zunahme um mehr als 20 Einheiten aufwiese.

Aus der Tabelle finden wir weiter, wenn wir in der Formel (9) $\beta_{N,n}$ gleich der Anzahl der außerhalb liegenden Gruppen setzen, Werte für die innere Abschirmungszahl $\alpha_{n,k}$, nämlich $\alpha_{1,1} = 0{,}7$, $\alpha_{2,2} \sim 5{,}5$, $\alpha_{3,3} \sim 16$, ($\alpha_{4,4} \sim 33$), die von der nach der Theorie zu erwartenden Größenordnung sind. Wie schon in § 5 erwähnt, ist es jedoch nicht die Absicht, in dieser Arbeit näher auf eine genauere Berechnung der Abschirmungszahlen einzugehen. Wir wollen nur an der Hand der Figur aufmerksam machen auf die Unregelmäßigkeiten, die in den Kurven von Niveaus auftreten, die schon vollständig abgeschlossenen Gruppen zugehören, und die als Eigentümlichkeiten in der Änderung der äußeren Abschirmungszahlen gedeutet werden müssen. Wie aus der Figur ersichtlich, treten derartige Unregelmäßigkeiten besonders stark bei den M-Niveaus auf, erst in dem Gebiete der Palladiumfamilie und später noch deutlicher im Gebiete der seltenen Erden, in welchen Gebieten die Neigung der Niveaukurven in Übereinstimmung mit der theoretischen Erwartung wesentlich kleiner ist als in den unmittelbar angrenzenden Gebieten. Auch in den L-Niveaus sind entsprechende Unregelmäßigkeiten der Änderung der Niveauwerte mit der Atomnummer in denselben Gebieten sicher nachweisbar. Sie sind aber zu gering, als daß sie bei dem benutzten Maßstabe in den Kurven der Fig. 3 deutlich zum Vorschein kommen könnten. In der Nähe der Eisenfamilie aber sind aus der Figur die Niveauänderungen in den L-Kurven deutlich zu erkennen. In der Nähe dieser Familie scheint auch eine kleine Unregelmäßigkeit in dem Verlauf der K-Kurve vorhanden zu sein.

Wenden wir uns nun dem Verhalten der Niveaukurven zu, die Elektronengruppen entsprechen, die selbst in Ausbildung begriffen sind. Hier läßt das Versuchsmaterial eine Anzahl von Zügen neuer Art zutage treten, die einen eingehenden Vergleich mit der Theorie zulassen. Die ersten Stellen im periodischen System, wo wir solche Verhältnisse treffen, sind die Umgebungen der Eisen- und der Palladiumfamilie, wo wir in dem Verlauf der M- bzw. N-Kurven charakteristische Eigentümlichkeiten erkennen. Wegen der relativ großen Unsicherheit des bezüglichen experimentellen Materials wollen wir uns jedoch hier nicht länger bei diesen aufhalten, weil wir bei der Familie der seltenen Erden noch ausgeprägteren Eigentümlichkeiten der Kurven begegnen. Um den Kurvenverlauf in diesem Gebiet, wo die experimentellen Ergebnisse auch am meisten gesichert erscheinen, deutlicher hervortreten zu lassen, sind in Fig. 4 die Kurven der uns hier interessierenden Niveaus in größerem Maßstab gezeichnet.

Das Auftreten der Familie der seltenen Erden ist nach der Theorie verursacht durch das Hinzukommen von Elektronen in 4_4-Bahnen und durch die damit verbundene Entwicklung der vierquantigen Elektronengruppe, während welcher die Anzahl der Elektronen dieser Gruppe allmählich von 18 bis 32 anwächst. Die Anwesenheit der 4_4-Bahnen äußert sich im Röntgenspektrum durch das Auftreten der Linien M_α und M_β. Bisher ist das M-Spektrum unterhalb Dy (66) noch nicht untersucht. Für dieses Element sind die erwähnten Linien von Stenström gefunden worden, und wir können aus den experimentellen Daten schließen, daß die Bindungsenergie der 4_4-Bahnen für Dy (66) und die nächstfolgenden Elemente von etwa derselben Größe ist wie die der sechsquantigen Valenzelektronen und wesentlich kleiner als die der 5_1- und 5_2-Bahnen, welche im O_{I}-Niveau sowie im Niveaupaar O_{II}, O_{III} in Erscheinung tritt. Dies ist eben das, was nach der Theorie zu erwarten wäre, denn solange die vierquantige Gruppe in Ausbildung begriffen ist, erfolgt das Hinzukommen von Elektronen in 4_4-Bahnen sozusagen unter Wettstreit mit den Valenzelektronen in den sechsquantigen Bahnen. Von dem Augenblick aber, wo die vierquantige Gruppe zu ihrem endgültigen

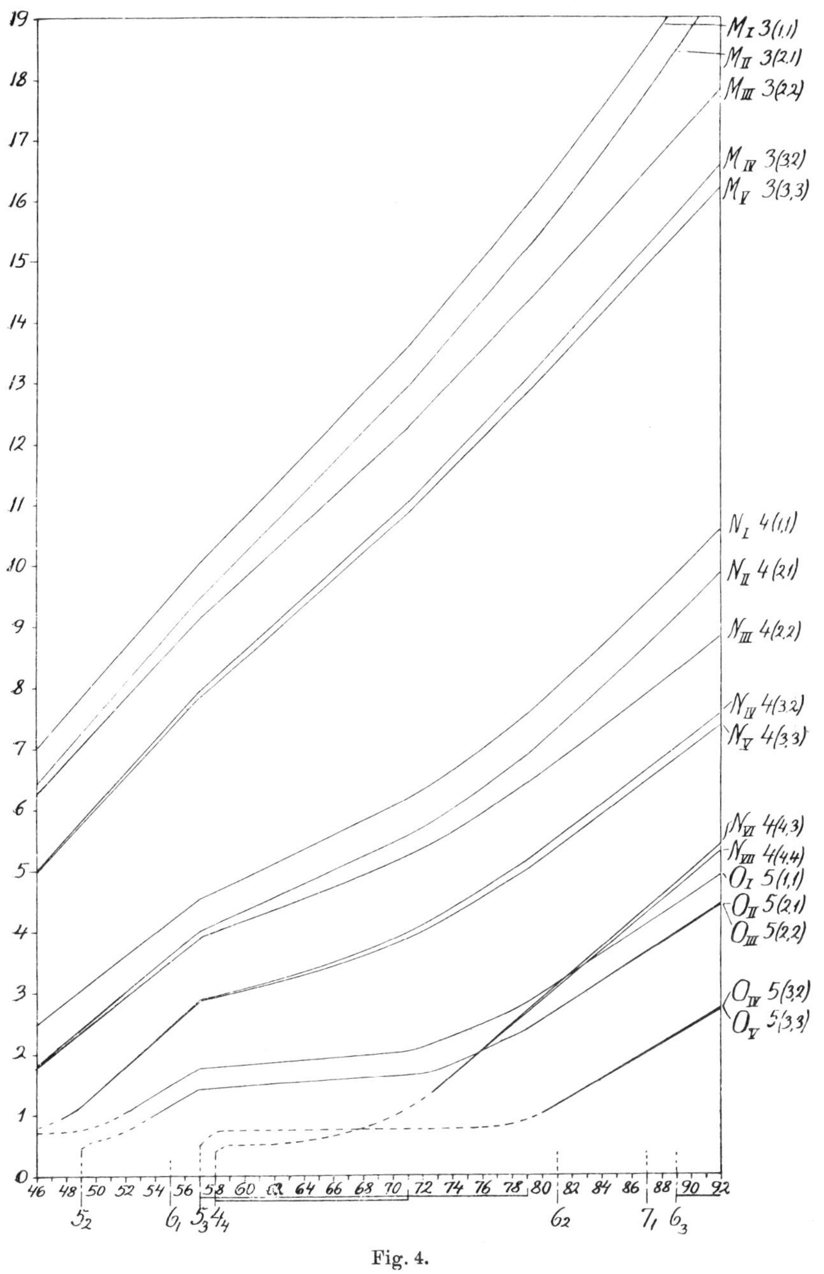
19
18
17
16
15
14
13
12
11
10
9
8
7
6
5
4
3
2
1
0
46 48 50 52 54 56 58 60 62 64 66 68 70 72 74 76 78 80 82 84 86 88 90 92
M_I 3(1,1)
M_{II} 3(2,1)
M_{III} 3(2,2)
M_{IV} 3(3,2)
M_V 3(3,3)
N_I 4(1,1)
N_{II} 4(2,1)
N_{III} 4(2,2)
N_{IV} 4(3,2)
N_V 4(3,3)
N_{VI} 4(4,3)
N_{VII} 4(4,4)
O_I 5(1,1)
O_{II} 5(2,1)
O_{III} 5(2,2)
O_{IV} 5(3,2)
O_V 5(3,3)
5_2
6_1
5_3
4_4
6_2
7_1
6_3

Fig. 4.

Abschluß gekommen ist, machen sich gänzlich andere Verhältnisse geltend, und wir müssen erwarten, daß die Bindungsenergie der 4_4-Bahnen mit wachsender Atomnummer schnell und regelmäßig ansteigt. Dies entspricht ganz dem experimentellen Befund, daß nach dem Abschluß der Familie der seltenen Erden die Kurven der Niveaupaare N_{VI}, N_{VII} schnell ansteigen, und, wie wir in der Fig. 4 sehen, zwischen W und Bi die Kurven der O_I-, O_{II}- und O_{III}-Niveaus überkreuzen.

Obwohl, wie erwähnt, das vorhandene Versuchsmaterial nicht direkt den Beginn des Auftretens der 4_4-Bahnen durch das Nachspüren nach dem ersten Erscheinen der Niveaupaare (N_{VI}, N_{VII}) zu prüfen gestattet, tritt jedoch das Hinzukommen von Elektronenbahnen dieses Typus deutlich zutage im Verlauf der Kurven der anderen vierquantigen Niveaus sowie der Niveaus mit der Hauptquantenzahl 5. Diese Kurven zeigen alle, wie aus der Fig. 4 zu ersehen ist, einen ausgesprochenen Knick in der Nähe von Ce (58), wo nach der Theorie, wie in Tabelle 1 auf S. 344 angegeben, eine 4_4-Bahn im normalen Atom zum erstenmal vorhanden ist. Diese Knicke müssen als herrührend von einem plötzlich beginnenden Anwachsen der inneren Abschirmungszahl in dem Ausdruck der diesen Niveaus entsprechenden Ablösungsarbeiten gedeutet werden. Dieses Anwachsen ist dadurch bedingt, daß, wie es theoretisch zu erwarten ist, die 4_4-Bahnen schon bei ihrem ersten Auftreten im Normalzustand des Atoms in dem Gebiete der bereits vorhandenen Elektronengruppen mit vierquantigen Bahnen liegen und also innerhalb des Gebietes, in welchem die Elektronen mit fünfquantigen Bahnen sich während des größten Teiles ihres Umlaufes bewegen. In dieser Verbindung ist daran zu erinnern, daß es für die nähere Beurteilung des oben besprochenen Gleichgewichtes der Bindungsstärke der 4_4-Elektronen und derjenigen der Valenzelektronen entscheidend ist, einerseits, daß die Stärke der Bindung der 4_4-Elektronen durch den äußeren Abschirmungseffekt der fünfquantigen Elektronengruppe außerordentlich abgeschwächt ist, andererseits, daß zufolge des großen Unterschiedes zwischen Hauptquantenzahl und effektiver Quantenzahl die Bindung der Valenzelektronen vielmal stärker ist, als einer sechsquantigen Keplerellipse entsprechen würde, die sich ganz außerhalb der inneren Elektronengruppe befände.

Wegen der Unvollständigkeit der Messungen herrscht für die Elemente in der Umgebung von (72), die den Abschluß der Familie der seltenen Erden bilden, einige Unsicherheit über den Verlauf der Kurven der Niveaus $N_I, N_{II}, N_{III}, N_{IV}, N_V$. Das vorhandene Versuchsmaterial scheint jedoch deutlich zu zeigen, daß diese Kurven wie in der Figur angedeutet eine mehr kontinuierliche Änderung ihrer Neigungen aufweisen als in der Umgebung von Ce (58). Dies dürfte zusammenhängen mit der in diesem Gebiet stattfindenden Entwicklungsstufe der fünfquantigen Elektronengruppe, die für das Auftreten der Platinmetalle verantwortlich ist, und die sich in der Bindung der vierquantigen Niveaus bemerkbar machen durch eine Beeinflussung der äußeren Abschirmungszahl.

Was zuletzt die O-Niveaus betrifft, begegnen wir neuen Zügen von großem theoretischen Interesse, indem wir es hier mit einer inneren Elektronengruppe zu tun haben, deren Bahnen teilweise außerhalb einer Gruppe verlaufen, die sich in einer Stufe gradweiser Ausbildung befindet. Obwohl noch eine beträchtliche Unsicherheit des Versuchsmaterials vorhanden ist, scheint es jedoch festzustehen, daß die T/R-Werte dieser Niveaus, wie es dem Verlauf der Kurven in der Figur entspricht, sich innerhalb der Familie der seltenen Erden außerordentlich wenig ändern. Dies ist gerade, was man nach der Theorie erwarten sollte, da die äußeren Schlingen der fünfquantigen Bahnen außerhalb des Gebietes der vierquantigen Elektronenbahnen liegen und daher die für die Ablösungsarbeit der O-Niveaus maßgebende effektive Kernladungszahl sowie die effektive Quantenzahl innerhalb der seltenen Erden nahezu konstant bleiben müssen.

Es wird von größtem Interesse sein zu versuchen, den Verlauf der Röntgenniveaus zwischen Xenon und Niton mit möglichst großer Meßgenauigkeit zu bestimmen, weil wir hoffen können, daß man daraus vieles über den näheren Verlauf der Ausbildung der vier- und fünfquantigen Elektronengruppen erfahren wird.

§ 7. Schlußbemerkungen. Bekanntlich war die Entdeckung der einfachen Verbindung zwischen dem Röntgenspektrum eines Elementes und der Atomnummer das typische Merkmal bei den grundlegenden Untersuchungen von Moseley und im besonderen war der scheinbar völlige Mangel von näheren Beziehungen dieser Spektren zu jenen Gesetzmäßigkeiten in den anderen Eigenschaften der Elemente, die im periodischen System zum Ausdruck kommen, auffallend. Durch die erhöhte Meßgenauigkeit, die wir besonders den Untersuchungen von Siegbahn verdanken, und das erweiterte Meßgebiet ist jedoch inzwischen eine Anzahl von Erscheinungen zutage getreten, die, wie wir gesehen haben, in engster Beziehung zu wesentlichen Zügen der Theorie gebracht werden können, die auch für die Deutung des periodischen Systems herangezogen sind.

Hierbei muß jedoch betont werden, daß bei den bisher besprochenen Gesetzmäßigkeiten der Röntgenspektren keine Rede war von einer Periodizität in dem Sinne, wie sie in den optischen und chemischen Eigenschaften der Elemente zutage tritt. Bei näherer Überlegung müssen wir aber auch bei den Röntgenspektren Verhältnisse erwarten, die eben mit jenen Zügen des Atombaus zusammenhängen, die direkt für die Periodizität der genannten Eigenschaften maßgebend sind. Was in den gewöhnlichen Gesetzmäßigkeiten im periodischen System zum Ausdruck kommt, sind ja die Abschließungsprozesse, die sich fortgesetzt in den äußersten Elektronengruppen im Atom abspielen, und die mit sich bringen, daß bei wachsender Atomnummer im Normalzustand des Atoms immer wieder Elektronen in Quantenbahnen neuer Typen gebunden werden. Bei der Frage, in welcher Weise sich diese Verhältnisse in den Röntgenspektren äußern können, kommen wir auf einen Zug der Theorie, den wir bisher noch nicht besonders erwähnt haben. Für die Deutung der allgemeinen Gesetze der Emission und Absorption von Röntgenstrahlen ist es gerade wesentlich, daß die inneren Gruppen im allgemeinen einen abgeschlossenen Charakter besitzen. Dies hat nämlich zur Folge, daß weder ein von außen hinzukommendes Elektron unter Emission von Strahlung in einer Bahn vom selben Typus gebunden werden kann, wie die Elektronenbahnen bereits abgeschlossener Gruppen, noch ein Elektron aus einer inneren Gruppe unter Absorption von Strahlung in eine Bahn dieser Typen gelangen kann.

Eben diese Betrachtung, die eine unmittelbare Erklärung des charakteristischen Unterschiedes der Gesetze bietet, welche die Absorption einerseits im Röntgengebiet, andererseits bei den optischen Spektren regeln, gibt zugleich einen Fingerzeig, daß wir bei der Absorption von Röntgenspektren Sonderverhältnisse erwarten dürfen, wenn es sich um Elemente handelt, bei denen eine Gruppe sich in einer Stufe gradweiser Ausbildung befindet[1]). Wir müssen nämlich erwarten, daß ein Elektron aus einer inneren Gruppe durch Absorption von Strahlung direkt in eine solche unabgeschlossene Gruppe gelangen kann. Dies wird sich äußern in dem Auftreten von Röntgenlinien, die sowohl in Emission wie in Absorption erscheinen, und die sich nicht einordnen lassen in die besprochene Klassifikation der Röntgenspektren, die dem Diagramm in Fig. 2 entspricht, und die strenggenommen nur für Elemente mit abgeschlossenen Gruppen Gültigkeit besitzt[2]). Solche Linien scheinen in der Tat bei den

[1]) Vgl. N. Bohr, ZS. f. Physik **6**, 1, 1922.

[2]) Diese Klassifikation bezieht sich außerdem nur auf die Spektren, die dadurch angeregt werden, daß nur ein einziges Elektron aus einer inneren Gruppe des Atoms entfernt wird. Wie kürzlich von Wentzel hervorgehoben wurde

Elementen der Eisenfamilie sowohl im Emissionsspektrum als langwellige Satelliten zu Diagrammlinien wie im Absorptionsspektrum als langwellige Komponenten der Kantenfeinstruktur aufzutreten [1]). Weiter sind auch im Gebiet der seltenen Erden mehrfach Linien beobachtet, die auf der langwelligen Seite von Diagrammlinien liegen und die einen ähnlichen Ursprung zu haben scheinen. Die nähere Untersuchung dieser Linien, besonders was ihre Absorptionsverhältnisse betrifft, ist in Angriff genommen und dürfte wichtige Auskunft liefern über die Weise, in welcher die allmähliche Komplettierung der vierquantigen Elektronengruppe im Gebiet der seltenen Erden vor sich geht.

Die letzten Betrachtungen führen uns unmittelbar dazu eine allgemeine nähere Verbindung von röntgenspektroskopischen Daten mit den periodischen Eigenschaften der optischen und chemischen Eigenschaften der Elemente zu erwarten. In der charakteristischen Änderung dieser Eigenschaften mit der Atomnummer spiegelt sich ja, wie erwähnt, eben der im allgemeinen unabgeschlossene Charakter der am losesten gebundenen Elektronengruppe im Atom ab. Wir dürfen daher annehmen, daß ein Elektron von einer inneren Gruppe durch einen Absorptionsprozeß nicht nur von dem Atom völlig entfernt werden oder in eine Quantenbahn eines anderen Typus als die im Atom schon vorhandenen Elektronenbahnen übergeführt werden kann, sondern daß es im allgemeinen auch direkt in die äußerste Elektronengruppe gelangen kann. Um die Bindungsenergie abzuschätzen, die einer solchen Elektronenbewegung entsprechen wird, müssen wir bedenken, daß die Verhältnisse, unter denen die Bewegung der Elektronen in den äußeren Gebieten des durch die Entfernung eines Elektrons aus einer inneren Gruppe angeregten Atoms stattfindet, sehr nahe mit

(Ann. d. Phys. **66**, 437, 1921), lassen sich einige Linien, die in dem Diagramm in Fig. 2 keinen Platz finden und die als kurzwellige Satelliten von Diagrammlinien auftreten, Atomen zuordnen, bei denen mehr als ein Elektron aus ihren normalen Bahnen in inneren Gruppen entfernt sind. Im Gegensatz zu den gewöhnlichen Röntgenspektren, die als Spektren erster Art bezeichnet werden können, mag es zweckmäßig sein, ein Spektrum, das durch die Entfernung von p-Elektronen aus inneren Gruppen angeregt wird, ein Spektrum $p$ter Art zu nennen. In Verbindung mit den im Text stehenden Betrachtungen muß jedoch hervorgehoben werden, daß bei den Erscheinungen der Absorption von Röntgenstrahlen die Spektren höherer Art kaum eine Rolle spielen können, weil bei den experimentellen Verhältnissen die Anzahl von Atomen, die durch Entfernung von inneren Elektronen angeregt sind, wohl immer als verschwindend betrachtet werden kann gegenüber der Anzahl von Atomen, deren inneres Gebäude sich in seinem Normalzustand befindet. (Vgl. D. Coster, Phil. Mag. **44**, 546, 1922 und S. Rosseland, Phil. Mag., erscheint demnächst.)

[1]) Vgl. D. Coster, Phil. Mag. **44**, 546, 1922.

denjenigen zusammenfallen, welche unter normalen Verhältnissen die Bewegung der äußersten Elektronen im Atom des nächstfolgenden Elementes beherrschen.

Außer den Gesetzmäßigkeiten allgemeiner Art in der Änderung der Niveauwerte mit der Atomnummer, deren Verbindung mit dem periodischen System in den vorigen Paragraphen besprochen wurde, müssen wir überhaupt erwarten, daß die fortgesetzten Untersuchungen eine Fülle von individuellen Einzelheiten in den röntgenspektroskopischen Daten enthüllen werden. Viele Andeutungen solcher Erscheinungen sind schon im Versuchsmaterial vorhanden; wir wollen jedoch darauf bei dieser Gelegenheit nicht näher eingehen, besonders weil wir, wie schon in § 3 betont, bei der Ermittlung der Niveauwerte der Tabellen 2 und 3 von ihrer Abhängigkeit vom physikalischen und chemischen Zustand, in dem das Element sich befindet und der hier von entscheidender Bedeutung ist, abgesehen haben.

Kopenhagen, Oktober 1922.

Universitetets Institut for teoretisk Fysik.

TRANSLATION

X-Ray Spectra and the Periodic System of the Elements.

By N. Bohr and D. Coster in Copenhagen.

With four figures. (Received 2 November 1922.)

In a paper in this journal* one of the authors has developed the main features of a theory of atomic structure, on the basis of which it seems possible to interpret both the optical and the X-ray spectra with the closest possible bearing on the interpretation of the periodic system of the elements. After the appearance of this paper, the experimental X-ray spectroscopic material, suited to test the theoretical conceptions, has been substantially extended by an investigation of the other one of us, and its bearing on the theory has been discussed in recent publications**, which have brought to light several interesting results. In the present joint paper, the comparison of the experimental data with the theory will be discussed somewhat more fully than in the papers mentioned, special regard being given to the bearing of X-ray spectra on the interpretation of the periodic system.

* N. Bohr, Z. Phys. **9** (1922) 1. In the meantime, this paper, with some explanatory figures added, together with two previous articles of a general nature, have appeared as a paperbound volume of "Sammlung Vieweg" with the title "Drei Aufsätze über Spektren und Atombau". It is the intention of the author to discuss more fully the results of this paper, the general fundamentals as well as the exposition of details, in a series of articles in this journal.

** D. Coster, Phil. Mag. **43** (1922) 1070; **44** (1922) 546. See also the Doctor's Dissertation, Leiden, 1922.

§ 1. Theoretical conceptions of atomic structure.

The theory mentioned depends upon classifying the electronic orbits in an atom by comparing them with the stationary states of an electron executing a central motion. In this procedure the orbits are characterized by the symbols n_k. Here n represents the "principal quantum number" which, in the limit in which the central motion goes over into a simple periodic Kepler motion, is alone effective in determining the energy, as shown by the Balmer formula for the hydrogen spectrum. The "subordinate quantum number" ["Nebenquantenzahl"] k, whose influence upon the energy depends on the deviation of the motion from being simple periodic, determines in the well-known manner the angular momentum of the electron about the center of the orbit. As is well known, Sommerfeld's theory of the fine structure of the hydrogen lines depends on the introduction of this quantum number; this fine structure is caused by the circumstance that, because of the relativistic modification of the laws of mechanics, even the orbit of a single electron around a positive nucleus is not strictly periodic, but can be described as a Kepler ellipse carrying out a slow rotation in its plane. A survey of the number of electrons in the normal atom, belonging, according to the theory, to the different types of n_k-orbits, is given below in table 1; it expresses conceptions of atomic structure which, apart from characteristic new features, in many respects resemble the conceptions on which papers on X-ray spectra by Sommerfeld and Vegard are based.

As is seen, the electron orbits are arranged in groups corresponding to a given value of the principal quantum number n. The development of these groups, as one goes to elements of higher atomic number, is connected with the regularities appearing in the periodic system. An essential feature of this process is that these groups, in turn, are divided into subgroups corresponding to different values of the subordinate quantum number k. The gradual addition of new types of electronic orbits, with the same principal quantum number as groups of electron orbits already present in preceding elements, and the resulting advance of the stepwise formation of these groups, are further assumed to cause typical depar-

Table 1. Types of electronic orbits in the elements.

N \ n_k	1_1	2_1 2_2	3_1 3_2 3_3	4_1 4_2 4_3 4_4	5_1 5_2 5_3 5_4 5_5	6_1 6_2 6_3 6_4 6_5 6_6	7_1 7_2
1 H	1						
2 He	2						
3 Li	2	1					
4 Be	2	2					
5 B	2	2 (1)					
— —	—	— —					
10 Ne	2	4 4					
11 Na	2	4 4	1				
12 Mg	2	4 4	2				
13 Al	2	4 4	2 1				
— —	—	— —	— —				
18 A	2	4 4	4 4				
19 K	2	4 4	4 4	1			
20 Ca	2	4 4	4 4	2			
21 Sc	2	4 4	4 4 1	(2)			
22 Ti	2	4 4	4 4 2	(2)			
— —	—	— —	— — —	—			
29 Cu	2	4 4	6 6 6	1			
30 Zn	2	4 4	6 6 6	2			
31 Ga	2	4 4	6 6 6	2 1			
— —	—	— —	— — —	— —			
36 Kr	2	4 4	6 6 6	4 4			
37 Rb	2	4 4	6 6 6	4 4	1		
38 Sr	2	4 4	6 6 6	4 4	2		
39 Y	2	4 4	6 6 6	4 4 1	(2)		
40 Zr	2	4 4	6 6 6	4 4 2	(2)		
— —	—	— —	— — —	— — —	—		
47 Ag	2	4 4	6 6 6	6 6 6	1		
48 Cd	2	4 4	6 6 6	6 6 6	2		
49 In	2	4 4	6 6 6	6 6 6	2 1		
— —	—	— —	— — —	— — —	— —		
54 X	2	4 4	6 6 6	6 6 6	4 4		
55 Cs	2	4 4	6 6 6	6 6 6	4 4	1	
56 Ba	2	4 4	6 6 6	6 6 6	4 4	2	
57 La	2	4 4	6 6 6	6 6 6	4 4 1	(2)	
58 Ce	2	4 4	6 6 6	6 6 6 1	4 4 1	(2)	
59 Pr	2	4 4	6 6 6	6 6 6 2	4 4 1	(2)	
— —	—	— —	— — —	— — — —	— — —	—	
71 Lu	2	4 4	6 6 6	8 8 8 8	4 4 1	(2)	
72 —	2	4 4	6 6 6	8 8 8 8	4 4 2	(2)	
— —	—	— —	— — —	— — — —	— — —	—	
79 Au	2	4 4	6 6 6	8 8 8 8	6 6 6	1	
80 Hg	2	4 4	6 6 6	8 8 8 8	6 6 6	2	
81 Tl	2	4 4	6 6 6	8 8 8 8	6 6 6	2 1	
— —	—	— —	— — —	— — — —	— — —	— —	
86 Nt	2	4 4	6 6 6	8 8 8 8	6 6 6	4 4	
87 —	2	4 4	6 6 6	8 8 8 8	6 6 6	4 4	1
88 Ra	2	4 4	6 6 6	8 8 8 8	6 6 6	4 4	2
89 Ac	2	4 4	6 6 6	8 8 8 8	6 6 6	4 4 1	(2)
90 Th	2	4 4	6 6 6	8 8 8 8	6 6 6	4 4 2	(2)
— —	—	— —	— — —	— — — —	— — —	— — —	—
— —	—	— —	— — —	— — — —	— — — —	— — —	— —
118 ?	2	4 4	6 6 6	8 8 8 8	8 8 8 8	6 6 6	4 4

tures from a simple periodicity in the system of the elements (families of iron and platinum metals, as well as the rare earths). A survey of the periodic system which particularly expresses these features is given in fig. 1*, in which elements

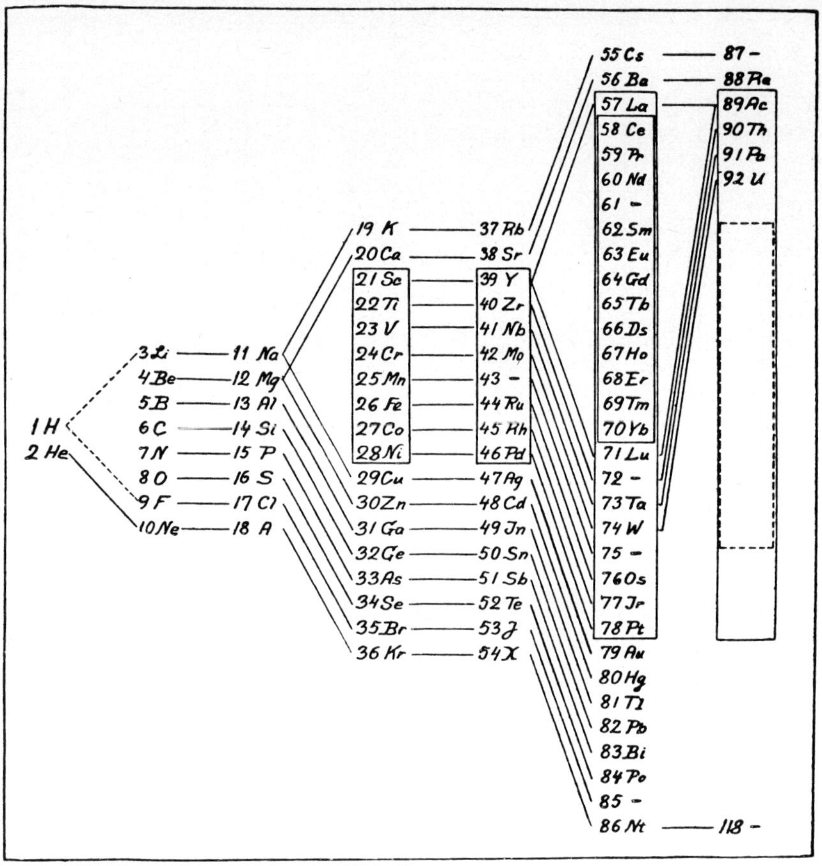

Fig. 1.

* Cf. N. Bohr, Drei Aufsätze über Spektren und Atombau, p. 132, fig. 5, where a more detailed explanation of this figure is to be found. Neither in the table, nor in the figure, has regard been given to the note of Dauvillier, C. R. May 1922, mentioned in the addendum to these articles, and dealing with the observation of some weak lines of an element with atomic number 72 in the X-ray spectrum of a mixture of rare earths, for this observation can hardly be considered ensured. The presence of an element between Lu (71), whose highest valency is 3, and Ta (73), whose highest valency is 5, having similar chemical properties as Lu, would represent a departure from the otherwise general rule that, as one goes from one element to the next, the highest valency never increases by more than one unit; a rule which receives its immediate explanation by the generally assumed conceptions of atomic structure (in this connection, see also C. R. Bury, Journ. Am. Chem. Soc. **43** (1921) 1902).

belonging to the same period of the system are placed in vertical columns. Elements in successive columns which can be regarded as homologous, in their chemical as well as in their optical properties, are connected by straight lines. Columns of successive elements, which are exceptional in that inner groups are being developed in them, are enclosed in square brackets.

As far as the principal features are concerned, the setting up of table 1 depends on considerations of a general nature which are set forth in the above-mentioned paper. In working out the details, a thorough investigation of the series spectra plays an essential role. Although the gradual formation of inner electron groups is also plainly reflected in these spectra, the direct information that can be obtained from the series spectra deals, after all, primarily with the gradual addition of such electron orbits of a new type with which the process of formation takes its beginning. Accordingly, in table 1 only such elements are given, on the whole, which belong to the beginning of the processes of formation; moreover, wherever the spectral data do not allow a unique conclusion as to the number of electrons in the outermost group, this is indicated by placing in a parenthesis the number of electrons in the orbits corresponding to the highest number of quanta. The further development of the groups is indicated only by the end result; even in the description of the finally completed groups, there remains at present much uncertainty, especially as regards the finer interplay among the electrons within a main group and those in its subgroups. Without entering further into this question here, we shall only point out that the general assumptions about the nature of this interplay, made in the paper mentioned above and discussed in greater detail there, give a foothold, on the basis of the correspondence principle, for a theoretical understanding of the typical laws governing the emission and absorption spectra in the X-ray region, the main features of which were elucidated by the formal theory of Kossel.

§ 2. Classification of X-ray spectra.

As explained in the above-mentioned paper by Coster, it has been possible, for a large number of elements, to achieve a classification of X-ray spectra, based on the circumstance that the frequency ν of each line can be represented as the difference between two spectral terms T' and T''. In accordance with the principles of the quantum theory of line spectra, these spectral terms—multiplied by h—are interpreted as energy levels of an atom which has been brought out of its normal state by the removal of an electron from one of its inner groups.

Fig. 2 gives a survey of the characteristic X-ray spectrum to be expected for the noble gas niton (86). The spectral lines are indicated in the usual manner by vertical arrows, while the horizontal straight lines represent the energy levels. Only such lines have been drawn which have actually been measured for several elements in the neighborhood of this noble gas. The figure also gives a survey of the spectra of the elements in the neighborhood of the noble gases xenon and

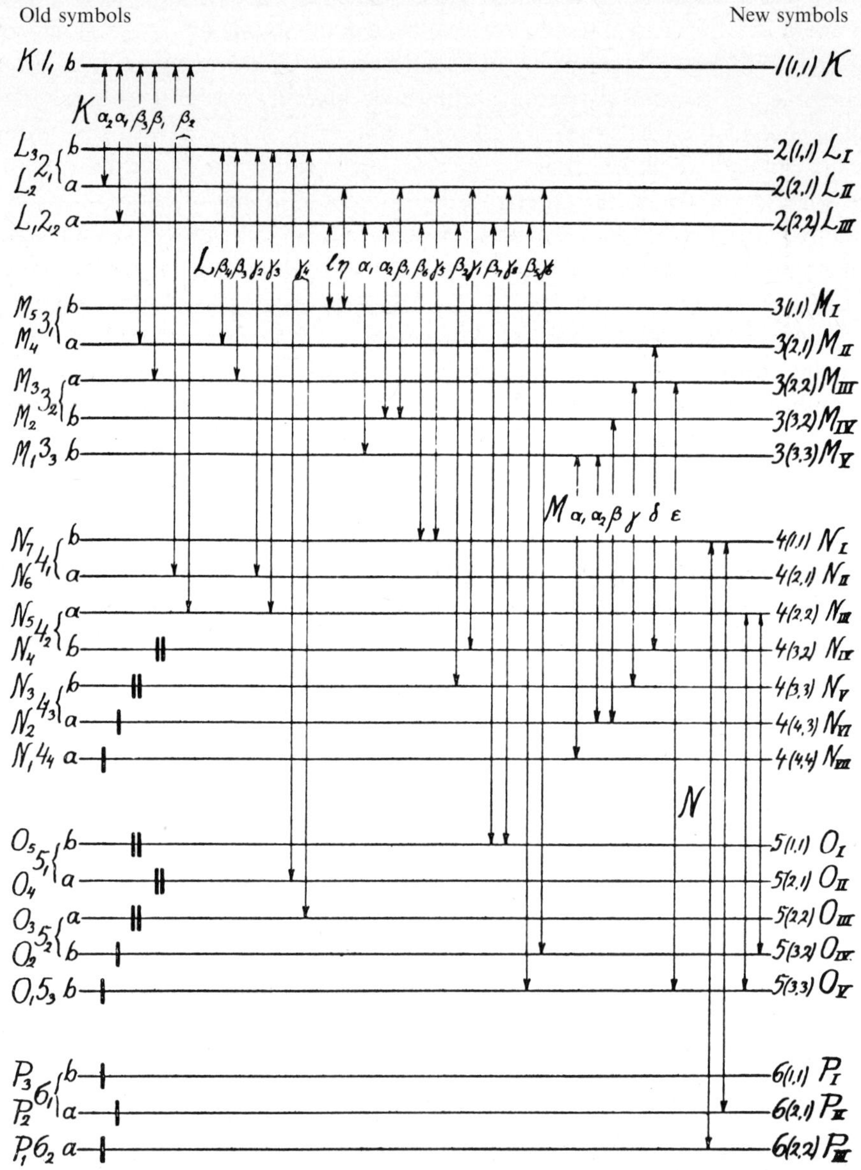

Fig. 2.

krypton. Thus, levels which fall out, as one proceeds to elements of lower atomic number between niton (86) and xenon (54), are indicated in the figure by a vertical bar, and the additional levels disappearing between xenon (54) and krypton (36) are marked with two vertical bars.

Now, with regard to the symbols marking the levels shown in the figure, a change has been made from those used in the previous papers. In the first place, sublevels belonging to one and the same main group of levels are marked by Roman numbers in the order in which they gradually appear with increasing atomic number in the X-ray spectra of elements. Further, to each level is assigned a number symbol of the form $n(k_1, k_2)$. We shall later discuss the relation of these symbols to the theory of atomic structure. Here we shall merely point out that the levels used in fig. 2 to represent the spectra of each of the three noble gases are all characterized by the same values of n and k which occur in table 1 as those of the quantum numbers n and k of the electron orbits present in the atoms of these elements. We remark further that the appearance of lines corresponding to combinations between the levels follows the rule that in a transition process the number k_1 always changes by one unit, while k_2 either changes by one unit or remains unaltered. These rules are equivalent to those expressed in the old symbols in the earlier papers of Coster, and they correspond formally quite accurately to the rules stated at the same time by Wentzel*. The numerical values introduced for k_1 and k_2 are equal to those of the numbers n and m used by the latter author to classify the levels. However, we shall conceive their theoretical significance in a somewhat different manner.

§ 3. Determination of energy levels from experimental data.

The results of the measurements of X-ray spectra available at present allow us to calculate the energy levels of a large number of elements with considerable accuracy. Below we give a survey of the experimental investigations employed for this purpose**, while the spectral terms T, divided by Rydberg's constant R, are listed in table 2, and their square roots are listed in table 3.

* G. Wentzel, Z. Phys. **6** (1921) 84.

** A table of X-ray spectral terms is also found in Sommerfeld's book: Atombau und Spektrallinien, 3rd ed., p. 630. However, this table has in many respects been made obsolete by new experimental results.

In a paper by Dauvillier (Journal de Physique et le Radium, July 1922), which has just appeared, a survey of the experimental material is also given as well as a graphical representation of the dependence of the energy values of the elements upon their atomic numbers. However, because of the incompleteness of the experimental data used by Dauvillier, most of the peculiarities discussed below, in which we are particularly interested, are not revealed in his representation.

Table

	K	L_I	L_{II}	L_{III}	M_I	M_{II}	M_{III}	M_{IV}	M_V
92 U	8477,0	1603,5	1543,1	1264,3	408,9	382,1	317,2	274,0	261,
90 Th	8073,5	1509,7	1451,5	1200,6	381,6	354,4	298,0	256,6	244,
83 Bi	6646,7	1207,9	1159,4	990,0	295,9	273,6	234,0	199,4	191,
82 Pb	6463,0	1169,3	1121,9	960,5	283,8	262,3	226,0	190,5	183,
81 Tl	6289,0	1132,4	1084,2	933,2	273,9	253,8	219,2	184,8	176,
80 Hg	6115,9	1094,6	1048,6	906,1	—	—	—	—	—
79 Au	5940,4	1060,2	1014,4	878,5	252,9	235,1	202,8	169,3	163,
78 Pt	5764,0	1026,8	978,7	852,0	243,4	227,3	198,0	162,3	156,
74 W	5113,8	890,6	850,1	751,6	207,6	188,9	167,4	137,8	133,
73 Ta	—	860,3*	821,0	728,6	200,1	181,0*	160,7*	132,8	128,
71 Lu	—	801,1*	763,1*	682,0*	184,5*	167,0*	149,4*	121,7*	117,
70 Yb	—	773,4*	735,7*	659,6*	177,5*	161,1*	144,7*	116,8*	113,
68 Er	—	718,1*	682,8*	616,0*	163,8*	147,3*	132,2*	107,3*	104,
67 Ho	4115,9	692,9*	657,4*	595,0*	157,3*	142,4*	129,0*	103,0*	100,
66 Dy	3972,5	666,9*	632,3*	574,0*	150,9*	136,1*	123,7*	98,3*	95,
65 Tb	—	642,4*	608,2*	553,8*	145,0*	130,9*	119,4*	94,1*	91,
64 Gd	3711,9	617,4*	584,3*	533,7*	138,6*	124,6*	114,2*	89,8*	87,
63 Eu	3583,4	593,6*	561,3	514,1	132,9	119,5*	109,6*	85,8	83,
62 Sm	3457,0	570,7*	539,0	495,0	127,1	114,3*	105,3*	81,9	79,
60 Nd	3214,2	525,8*	495,9	457,9	116,6	104,4*	96,4*	74,3	72,
59 Pr	3093,3	504,2*	474,7	439,4	111,4	99,0*	92,1*	70,1	68,
58 Ce	2972,2	483,0*	454,4	421,9	105,9	94,2*	87,8*	66,7	65,
56 Ba	2756,4	442,3	414,7	386,7	95,4	84,9	79,4	58,8	57,
55 Cs	2649,1	421,8	395,0	369,5	89,6	79,3	74,4	54,8	53,
53 J	2448,3	382,6	357,6	336,0	—	69,0	64,8	46,8	46,
52 Te	2345,0	364,1	339,6	320,1	74,5	64,4	60,5	43,2	42,
51 Sb	2241,7	346,1	321,9	304,3	69,3	59,9	56,4	39,5	38,
47 Ag	1878,9	279,5	260,0	247,3	53,3	43,5	41,2	27,9	27,
45 Rh	1709,1	253,4	231,4	220,9	45,8	39,7	38,5	22,5	22,
42 Mo	1473,1	211,9	193,9	186,2	37,6	30,5	29,7	17,4	17,
41 Nb	1401,3	—	181,4	174,4	35,1	—	—	15,0	14,
40 Zr	1325,8	—	169,6	163,5	31,4	—	—	13,0	
29 Cu	661,1	—	71,3	69,8	—	5,2		0	
28 Ni	612,0	—	62,6	61,3	—	3,3		—	
27 Co	568,9	—	59,7	58,5	—	5,3		—	
26 Fe	523,8	—	53,2	52,2	—	4,0		—	
25 Mn	482,4	—	48,7	47,9	—	4,2		—	
24 Cr	441,1	—	43,0	42,3	—	3,5		—	
23 Va	402,3	—	38,2	37,6	—	2,6		—	
22 Ti	365,4	—	32,6	32,2	—	2,2		—	
21 Sc	331,2	—	30,3	30,0	—	2,7		—	
20 Ca	297,5	—	25,9	25,6	—	2,0		—	
19 K	265,3	—	21,4	21,2	—	0,9		—	
17 Cl	207,8	—	14,8	14,7	—	0,4		—	
16 S	181,8	—	11,8		—	0,3		—	
15 P	158,3	—	9,9		—	0,8		—	
13 Al	114,7	—	5,2		—	0		—	
12 Mg	95,8	—	3,5		—	—		—	

* Interpolated or computed from an interpolated value.

lues of levels (T/R).

N_I	N_{II}	N_{III}	N_{IV}	N_V	N_{VI}	N_{VII}	O_I	$O_{II,III}$	$O_{IV,V}$	
6,6	95,7	77,1	56,3	53,6	28,4	27,6	26,2	15,4	5,8	92 U
7,8	—	—	51,2	48,7	24,8	24,1	—	—	5,7	90 Th
1,0	58,7	50,3	35,7	33,7	13,6	13,0	—	11,0	2,0	83 Bi
6,0	55,4	49,3	32,2	30,5	10,8	10,3	10,3	6,4	0,8	82 Pb
3,7	53,6	44,9	30,6	29,0	10,0	9,6	10,6	7,4	1,7	81 Tl
—	—	—	—	—	—	—	—	—	—	80 Hg
8,0	49,1	42,8	26,4	25,0	6,4		7,8	8,3	0,8	79 Au
2,5	48,7	42,3	24,5	23,2	5,2		7,1	8,6	0,4	78 Pt
43,6	35,6	30,6	18,7	17,8	2,5		5,2	2,9	(0)	74 W
41,7	33,0*	30,0*	18,0	17,2	2,5		5,3	2,6*	—	73 Ta
37,7*	30,9*	27,2*	16,2*	15,5*	1,3*		4,9*	2,5*	—	71 Lu
36,6*	29,9*	26,4*	15,2*	14,6*	1,0*		—	2,5*	—	70 Yb
33,2*	26,9*	23,4*	13,4*	12,8*	0,3*		4,1*	2,4*	—	68 Er
32,1*	26,6*	23,5*	12,7*	12,2*	0,1*		—	2,4*	—	67 Ho
30,8*	25,3*	22,4*	11,9*	11,4*	(—0,2)*		2,9*	2,5*	—	66 Dy
29,3*	24,1*	21,8*	11,5*	11,1*	—		3,5*	2,4*	—	65 Tb
28,3*	22,2*	20,2*	11,0*	10,6*	—		3,5*	2,4*	—	64 Gd
26,9	21,9*	19,7*	10,4	10,1	—		2,9	—	—	63 Eu
25,8	20,2*	18,8*	10,1	9,8	—		3,0	2,3*	—	62 Sm
23,8	18,6*	17,2*	9,6	9,3	—		3,2	2,5*	—	60 Nd
22,5	18,0*	16,7*	8,8	8,5	—		2,6	2,2*	—	59 Pr
21,7	17,0*	15,8*	8,7	8,5	—		3,1	2,2*	—	58 Ce
18,8	15,3	14,4	7,1	6,9	—		2,1	2,4	—	56 Ba
17,3	13,6	12,6	6,1	5,9	—		2,1	1,7	--	55 Cs
—	—	—	4,4	4,2	—		—	—	—	53 J
12,7	8,8		3,4	3,3	—		—	0,4	—	52 Te
11,1	7,2		2,3	2,2	—		(0)	0,1	—	51 Sb
7,5	3,3		0,6		—		—	—	—	47 Ag
5,3	5,0		(— 0,3)		—		—	—	—	45 Rh
4,7	3,0		0,5		—		—	—	—	42 Mo
4,0	—		—		—		—	—	—	41 Nb
3,5	—		—		—		—	—	—	40 Zr
—	—		—		—		—	—	—	29 Cu
—	—		—		—		—	—	—	28 Ni
—	—		—		—		—	—	—	27 Co
—	—		—		—		—	—	—	26 Fe
—	—		—		—		—	—	—	25 Mn
—	—		—		—		—	—	—	24 Cr
—	—		—		—		—	—	—	23 Va
—	—		—		—		—	—	—	22 Ti
—	—		—		—		—	—	—	21 Sc
—	—		—		—		—	—	—	20 Ca
—	—		—		—		—	—	—	19 K
—	—		—		—		—	—	—	17 Cl
—	—		—		—		—	—	—	16 S
—	—		—		—		—	—	—	15 P
—	—		—		—		—	—	—	13 Al
—	—		—		—		—	—	—	12 Mg

Table

	K	L_{I}	L_{II}	L_{III}	M_{I}	M_{II}	M_{III}	M_{IV}	M_{V}
92 U	92,07	40,04	39,28	35,56	20,22	19,55	17,81	16,55	16,1[illegible]
90 Th	89,85	38,85	38,10	34,65	19,53	18,83	17,26	16,02	15,6[illegible]
83 Bi	81,53	34,75	34,05	31,46	17,20	16,54	15,32	14,12	13,8[illegible]
82 Pb	80,39	34,19	33,50	30,99	16,85	16,20	15,08	13,80	13,5[illegible]
81 Tl	79,31	33,65	32,92	30,55	16,55	15,93	14,81	13,60	13,3[illegible]
80 Hg	78,20	33,08	32,38	30,10	—	—	—	—	—
79 Au	77,07	32,54	31,85	29,64	15,90	15,33	14,25	13,01	12,7[illegible]
78 Pt	75,92	32,01	31,28	29,19	15,60	15,08	13,99	12,74	12,5[illegible]
74 W	71,51	29,84	29,16	27,41	14,41	13,74	12,93	11,74	11,5[illegible]
73 Ta	—	29,33*	28,65	26,99	14,15	13,45*	12,67*	11,52	11,3[illegible]
71 Lu	—	28,30*	27,62*	26,12*	13,58*	12,92*	12,22*	11,03*	10,8[illegible]
70 Yb	—	27,81*	27,12*	25,68*	13,32*	12,69*	12,03*	10,83*	10,64
68 Er	—	26,80*	26,15*	24,82*	12,80*	12,14*	11,50*	10,36*	10,2[illegible]
67 Ho	64,16	26,32*	25,64*	24,39*	12,54*	11,93*	11,36*	10,15*	9,9[illegible]
66 Dy	63,03	25,82*	25,15*	23,96*	12,28*	11,67*	11,12*	9,91*	9,7[illegible]
65 Tb	—	25,35*	24,66*	23,57*	12,04*	11,44*	10,93*	9,70*	9,57
64 Gd	60,93	24,85*	24,17*	23,10*	11,77*	11,16*	10,69*	9,48*	9,35
63 Eu	59,86	24,36*	23,69	22,67	11,53	10,93*	10,47*	9,26	9,14
62 Sm	58,80	23,89*	23,22	22,25	11,27	10,68*	10,25*	9,05	8,93
60 Nd	56,69	22,93*	22,27	21,40	10,79	10,21*	9,81*	8,62	8,53
59 Pr	55,62	22,45*	21,79	20,96	10,55	9,95*	9,60*	8,37	8,28
58 Ce	54,52	21,98*	21,32	20,54	10,29	9,71*	9,37*	8,17	8,09
56 Ba	52,50	21,03	20,36	19,66	9,77	9,21	8,91	7,67	7,60
55 Cs	51,47	20,54	19,87	19,22	9,47	8,91	8,63	7,40	7,34
53 J	49,48	19,56	18,91	18,33	—	8,31	8,05	6,84	6,78
52 Te	48,43	19,08	18,43	17,89	8,63	8,02	7,78	6,57	6,52
51 Sb	47,35	18,60	17,94	17,44	8,32	7,74	7,51	6,29	6,23
47 Ag	43,35	16,72	16,12	15,73	7,30	6,60	6,42	5,28	5,24
45 Rh	41,34	15,90	15,20	14,86	6,76	6,30	6,16	4,74	4,71
42 Mo	38,38	14,54	13,92	13,64	6,13	5,52	5,40	4,17	4,15
41 Nb	37,43	—	13,47	13,21	5,92	—	—	3,87	3,86
40 Zr	36,41	—	13,02	12,79	5,60	—	—	3,61	
29 Cu	25,71	—	8,44	8,36	—	2,28		0	
28 Ni	24,74	—	7,91	7,83	—	1,82		—	
27 Co	23,85	—	7,73	7,65	—	2,30		—	
26 Fe	22,89	—	7,29	7,23	—	2,00		—	
25 Mn	21,96	—	6,98	6,92	—	2,05		—	
24 Cr	21,00	—	6,56	6,50	—	1,87		—	
23 Va	20,06	—	6,18	6,13	—	1,61		—	
22 Ti	19,11	—	5,71	5,66	—	1,48		—	
21 Sc	18,20	—	5,51	5,48	-	1,64		—	
20 Ca	17,25	—	5,09	5,06	—	1,42		—	
19 K	16,29	—	4,63	4,61	—	0,95		—	
17 Cl	14,42	—	3,85	3,84	—	0,63		—	
16 S	13,48	—	3,44		—	0,55		—	
15 P	12,58	—	3,15		—	0,89		—	
13 Al	10,71	—	2,28		—	0		—	
12 Mg	9,79	—	1,87		—	—		—	

* Interpolated or computed from an interpolated value.

Square root of values of levels ($\sqrt{T/R}$)

N_I	N_{II}	N_{III}	N_{IV}	N_V	N_{VI}	N_{VII}	O_I	$O_{II,III}$	$O_{IV,V}$	
,32	9,78	8,78	7,50	7,32	5,33	5,25	5,12	3,92	2,41	92 U
,89	—	—	7,16	6,98	4,98	4,91	—	—	2,39	90 Th
3,43	7,66	7,09	5,98	5,81	3,69	3,61	—	3,39	1,42	83 Bi
3,13	7,44	7,02	5,68	5,52	3,29	3,21	3,21	2,79	0,89	82 Pb
7,98	7,32	6,70	5,54	5,39	3,16	3,10	3,26	2,72	1,30	81 Tl
—	—	—	—	—	—		—	—	—	80 Hg
7,62	7,01	6,54	5,14	5,00	2,53		2,79	2,88	0,89	79 Au
7,25	6,98	6,50	4,95	4,82	2,28		2,67	2,93	0,63	78 Pt
6,60	5,97	5,53	4,32	4,22	1,58		2,28	1,70	0	74 W
6,46	5,75*	5,48*	4,24	4,15	1,58		2,30	1,61*	—	73 Ta
6,14*	5,56*	5,22*	4,03*	3,94*	1,14*		2,21*	1,58*	—	71 Lu
6,05*	5,47*	5,14*	3,90*	3,82*	1,0*		—	1,58*	—	70 Yb
5,76*	5,18*	4,84*	3,66*	3,58*	0,5*		2,03*	1,55*	—	68 Er
5,67*	5,16*	4,84*	3,56*	3,49*	—		—	1,55*	—	67 Ho
5,55*	5,03*	4,73*	3,45*	3,38*	—		1,70*	1,58*	—	66 Dy
5,41*	4,91*	4,67*	3,39*	3,33*	—		1,87*	1,55*	—	65 Tb
5,32*	4,71*	4,49*	3,32*	3,25*	—		1,87*	1,55*	—	64 Gd
5,19	4,68*	4,44*	3,23	3,18	—		1,70	—	—	63 Eu
5,08	4,49*	4,34*	3,18	3,13	—		1,73	1,52*	—	62 Sm
4,88	4,31*	4,15*	3,10	3,05	—		1,79	1,58*	—	60 Nd
4,74	4,24*	4,09*	2,97	2,92	—		1,61	1,48*	—	59 Pr
4,66	4,12*	3,98*	2,95	2,92	—		1,76	1,48*	—	58 Ce
4,34	3,91	3,80	2,67	2,63	—		1,45	1,55	—	56 Ba
4,16	3,69	3,55	2,47	2,43	—		1,45	1,30	—	55 Cs
—	—	—	2,10	2,05	—		—	—	—	53 J
3,56	2,97		1,84	1,82	—		—	0,63	—	52 Te
3,33	2,68		1,52	1,48	—		0	0,32	—	51 Sb
2,74	1,82		0,77		—		—	—	—	47 Ag
2,30	2,24		—		—		—	—	—	45 Rh
2,17	1,58		—		—		—	—	—	42 Mo
2,00	—		—		—		—	—	—	41 Nb
1,83	—		—		—		—	—	—	40 Zr
—	—		—		—		—	—	—	29 Cu
—	—		—		—		—	—	—	28 Ni
—	—		—		—		—	—	—	27 Co
—	—		—		—		—	—	—	26 Fe
—	—		—		—		—	—	—	25 Mn
—	—		—		—		—	—	—	24 Cr
—	—		—		—		—	—	—	23 Va
—	—		—		—		—	—	—	22 Ti
—	—		—		—		—	—	—	21 Sc
—	—		—		—		—	—	—	20 Ca
—	—		—		—		—	—	—	19 K
—	—		—		—		—	—	—	17 Cl
—	—		—		—		—	—	—	16 S
—	—		—		—		—	—	—	15 P
—	—		—		—		—	—	—	13 Al
—	—		—		—		—	—	—	12 Mg

Summary of papers used.

K-absorption [1])	Mg(12)–Cr(24)	Fricke, Phys. Rev. **16** (1920) 202.
K-absorption [2])	Mn(25)–U(92)	Duane and Shimizu, Phys. Rev. **14** (1919) 522.
		Duane and Kang-Fuh-Hu, Phys. Rev. **14** (1919) 516.
		Duane, Fricke, Stenström, Proc. Nat. Ac. Sc. **6** (1920) 607.
K-absorption [1])	Cd(48)–U(92)	Siegbahn and Jönsson, Phys. Zeitschr. **20** (1919) 251.
L-absorption [2, 3])	W(74)–Pb(82)	Duane and Patterson, Proc. Nat. Ac. Sc. **6** (1920) 509.
L-absorption [1])	Cs(55)–Eu(63)	Coster, Phil. Mag. **43** (1922) 1070; **44** (1922) 546.
L-absorption [1])	Sb(51)–Ba(56)	Lindsay, C. R. **175** (1922) 150.
L-absorption [1])	Ag(47)	Coster, Phil. Mag. **43** (1922) 1070.
M-absorption [1])	Th(90), U(92)	Stenström, Dissertation, Lund, 1919.
M-absorption [1])	Bi(83), Th(90), U(92)	Coster, Phys. Rev. **19** (1922) 20.
K-lines [1])	Na(11)–Zn(30)	Hjalmar, Z. Phys. **1** (1920) 439; **7** (1921) 341.
K-lines [1])	P(15)–Zn(30)	Siegbahn and Dolejsek, Z. Phys. **10** (1922) 159.
K-lines [2])	Mo(42), Rh(45)	Duane, Phys. Rev. **14** (1919) 373.
L-lines [1])	Cu(30)–Ta(73)	Hjalmar, Z. Phys. **3** (1920) 262; **7** (1921) 341.
L-lines [1])	W(74)	Siegbahn, Phys. Zeitschr. **20** (1919) 533.
L-lines [1])	Rb(37)–Ba(56)	Coster, Phil. Mag. **43** (1922) 1070.
L-lines [1])	La(57)–Lu(71)	Coster, Phil. Mag. **44** (1922) 546.
L-lines [1])	Ta(73)–U(92)	Coster, Z. Phys. **6** (1921) 186.
M-lines [1])	Dy(66)–U(92)	Stenström, Dissertation, Lund, 1919.

[1]) Measured in Siegbahn's laboratory.
[2]) Measured in Duane's laboratory.
[3]) Corrected by one X-unit.

With the exception of levels marked by an asterisk, all values in table 2 are calculated from the above-mentioned experimental data, with the use of addition

* D. Coster, Z. Phys. **6** (1921) 185.

relations that can be taken directly from fig. 2. In a previous paper by Coster*, it was pointed out that Duane and Patterson's values for the L-edges agree better with the results from the Lund Institute if the values for all edges given by them are reduced by one unit. In order to obtain more comparable values, we have made this correction to all of Duane and Patterson's L-edge wavelengths.

For W(74) the edge L_{III}** still coincides nearly with the line L_{β_5}. Likewise the edge L_{II} coincides with the line L_{γ_6}. For the preceding element Ta(73), the lines L_{β_5} and L_{γ_6} can therefore without arbitrariness be substituted for the edges L_{III} and L_{II}. Also, for the elements Zr(40) and Nb(41), the values of the lines L_{β_2} and L_{γ_1} were used for these same edges.

For the elements between Ba(56) and W(74), the L-edges are only partly measured; a rational interpolation for the unknown values was attempted. Thus, the values of the M_V-levels for the elements between Eu(63) and Ta(73) were determined by a graphical interpolation. The values of the M_{III}-levels for the elements between Ba(56) and W(74) were determined in a similar manner. The other levels of the elements mentioned were calculated, with the aid of the experimentally obtained data, from the interpolated M_{III}- and M_V-values. Whenever an interpolated value was used in the calculation of a level, this level is marked with an asterisk in the table. Although it is always possible, of course, that certain special kinds of variations with atomic number have been lost by the interpolation, we regard it as highly improbable that further experimental measurements could endanger the conclusions drawn in the present paper.

In this connection, it may be pointed out that the value listed in table 2 for L_I of W, which has been taken from the measurements of Duane and Patterson after application of the correction mentioned, is rather accurate, although the experimental error in the direct determination of this edge can be fairly large because of its weakness and diffuseness. In fact, it is possible to calculate the value of L_I from the much sharper and more accurately known edge L_{III} in three different ways, namely with the aid of one of the following addition relations:

$$L_{\beta_2}+(L_I-L_{III}) = L_{\beta_3}+M_{\gamma},$$

$$L_I-L_{III} = L^*_{\beta_9}-L_{\alpha_1},$$

$$L_I-L_{III} = L^*_{\beta_{10}}.$$

In all of these three cases one gets T/R values for tungsten which agree within an inaccuracy of ± 0.6 with the directly measured value. It may be concluded from

* See D. Coster, Z. Phys. **6** (1921) 185.

** Formerly called L_I. In conformity with the corresponding levels, the edges are now also denoted differently.

this that the error in the T/R value for the L_I edge of W is certainly less than one unit. This is of interest for the interpolation of the M_{III} curve between W and Ba.

Further, it may be remarked that, where rather short wavelengths were used in the calculation of levels, the accuracy is lower than where larger wavelengths could be used. Thus, an error of two X-units in the wavelength of the L_I edge corresponds to an error in the T/R value of less than $\frac{1}{2}$ for Ba, but of about 2 and 3 for W and Bi, respectively. This means, for example, that rather large errors can occur in the values of the lowest levels for the heavier elements*, while, on the other hand, in the neighborhood of Ba the T/R values for the lowest levels are still comparatively accurately known.

In this connection, mention may be made of a certain ambiguity still adhering to data derived from X-ray spectroscopy, even for the most accurate measurements. As a rule, the X-ray absorption spectra are obtained with elements in the solid state, and indeed often in chemical compounds. However, Bergengren** has shown that the condition of state is of essential significance; indeed, an appreciable difference in the K edge of phosphorus was observed when different modifications of this element were used. Moreover, Lindh*** has found measurable differences in the absorption spectra of a given element observed in different chemical compounds. Thus, for the K edges of chlorine and sulphur, he has found differences amounting to from 0.5 to 0.8 units in the T/R values, according to whether chlorine and sulphur compounds were used in which these elements occur with positive or negative valency.

A still more serious uncertainty in X-ray data arises from the so-called fine structure of the absorption edges, revealing itself in the appearance of a second weaker discontinuity, usually on the hard side of the principal discontinuity. For the difference in the T/R values of these continuities Fricke† gives, for the K edge of the lighter elements, amounts of one to two units. As is well known, Kossel†† has proposed an explanation of this structure of the edges, according to which it should correspond to small differences in the work required to remove an electron from an inner group, depending on whether the electron is completely removed from the atom or is transferred to a quantum orbit lying outside the electron sheet ["Elektronenhülle"] and corresponding to a stationary state of

* *Note added in proof*: Recently Hjalmar, by new measurements in the M-series (C. R. **175** (1922) 878), has succeeded in determining the O- and M-levels of the heaviest elements with greater accuracy. His results could not be included in the present paper; however, they are in very good agreement with our conclusions.

** J. Bergengren, Z. Phys. **3** (1920) 247.

*** A. E. Lindh, Z. Phys. **6** (1922) 303; C. R. **175** (1922) 25.

† H. Fricke, Phys. Rev. **16** (1920) 202.

†† W. Kossel, Verh. d. D. Phys. Ges. **18** (1916) 339.

the series spectrum. The bearing of these various circumstances, causing an uncertainty in the X-ray data, on the conceptions of atomic structure, taken as basic in this paper, will be briefly touched on at the end of § 6. Here we shall merely point out that under these circumstances we have to do with just such individual peculiarities of energy values which we had to neglect at the outset in the above-mentioned interpolation.

§ 4. Theoretical estimation of energy levels.

With regard to the utilization of the theory to a detailed interpretation of the levels of X-ray spectra, the problem is first of all to calculate the work required to remove an electron from an inner group in an atom. For this purpose we must, of course, first consider the motion of the electrons in the normal state of the atom. In this connection, it must be realized that it is an essential feature of the theory that the electrons in the various groups do not move in spatially separated regions of the atom, but that the electrons in groups with higher principal quantum numbers (outer groups), at any rate those in certain subgroups, during their motion penetrate into the region of electron orbits of lower principal quantum numbers (inner groups). This circumstance involves that the total attractive force toward the midpoint of the atom, acting on an electron in the different parts of its orbit, can be subject to great changes relative to the Coulomb attraction from the nucleus itself. However, for estimating these conditions, the circumstance is important that, in general, major parts of the central orbit, which may be called its outer loops, can be compared closely to a Kepler ellipse. However, the dimensions and energy of these elliptic loops generally deviate essentially from the dimensions and energy of a Kepler ellipse characterized by the same principal quantum number as the actual orbit described by an electron in a Coulomb field of the same strength as that in which the orbital loops in question are located. As is well known, the energy and dimension of a quantized Kepler ellipse are given by the simple formulas

$$W = RhN^2/n^2, \qquad 2a = e^2n^2/RhN, \tag{1}$$

where W is the work required to remove the electron, a is half the major axis of the ellipse, R is Rydberg's constant, Ne the nuclear charge, and n the principal quantum number. For the orbital loops in question we must not only count on an "effective nuclear charge number" N^*, but also on an "effective quantum number" n^*, defined by the expressions

$$W = RhN^{*2}/n^{*2}, \qquad 2a = e^2n^{*2}/RhN^*. \tag{2}$$

This circumstance is especially decisive for estimating the strength of binding of electrons in the outer groups, since there the effective quantum numbers of the elliptic loops can be considerably smaller than the principal quantum numbers, and essentially different for orbits belonging to the same n but to different values of k. Thus, the *typical periodicity of the chemical properties and of the optical spectra* depends on the circumstance that, for the *outermost electron orbits*, the effective quantum numbers, in contrast to the principal quantum numbers, vary only little as one goes from an element to the homologous element in the next period in the system of elements. On the other hand, the *striking lack of periodicity of the essential features of X-ray spectra* depends on the circumstance that we are here primarily concerned with the conditions of the *innermost electrons* in the atom, which move in groups that are already completely formed and that repeat themselves unchanged in all subsequent elements. In such cases, the effective quantum number of the orbital loops deviates only little from the principal quantum number of the orbit and, as shown by a simple calculation, this small deviation is, for a given type of orbit, to a first approximation inversely proportional to the atomic number N. Hence, as regards the dependence of the strength of binding upon the nuclear charge, the penetration of the electron orbit into the region of the inner groups has an effect quite similar to a direct screening of part of the nuclear charge by the inner electrons when the electron in question is passing through its outer loops. It is hardly necessary to distinguish between these two effects, as there is yet another influence upon the work needed to remove an electron, an effect which is difficult to calculate, but which varies in the same manner with the nuclear charge. We refer here to the decrease in the work of removal connected with the increase in the binding of electrons left behind in the same group resulting from the removal of the electron in question. Hence, we can express the influence of all these effects by the simple formula

$$W = Rh(N-\alpha_{n,k})^2/n^2, \tag{3}$$

where $\alpha_{n,k}$ is a constant for each type of orbit n_k; according to the usual terminology, we shall call it a "screening constant".

However, in determining the work required to remove an electron from an inner group, it is essential to also take the presence of the outer electron groups into account. During the removal of the electron, these electron groups exert a screening effect which can most simply be taken into account by subtracting from the expression given by (3) the increase in the strength of binding of all outer electrons resulting from the removal of the electron in question. The total binding energy of the outer electrons may be written as follows:

$$B = Rh\sum_{p>n} A_p N_p^{*2}/p^{*2}, \tag{4}$$

where A_p is the number of electrons in the group with principal quantum number p, while N_p^* and p^* may be called the average effective nuclear charge and the average effective quantum number, respectively, of the group. If, for simplicity, we limit ourselves to consider such elements in which no group is in a process of gradual development we can, in an approximate estimate of the quantity B, simply put A_p/p^{*2} equal to 2. In fact, in the completely formed inner groups, A_p is just equal to $2p^2$, and, at the same time, p^* is not very different from p. For the outer, only temporarily closed, groups, which in themselves have a smaller influence because of the lower values of the N_p^*, p^* is substantially smaller than p, but, as shown by a simple estimate, just to such an extent that, to a rough approximation, we can also here always put $p^{*2} = \frac{1}{2} A_p$. Now, for the change in B resulting from the removal of an electron with principal quantum number n, we can therefore write with considerable approximation

$$\Delta B = Rh \sum_{p>n} 2\Delta N_p^{*2} = Rh4 \sum_{p>n} N_p^*. \tag{5}$$

In order to estimate this expression more closely, we must consider the manner in which N_p^* decreases with increasing p. To a first approximation, we put

$$N_p^* = N - \sum_{q<p} A_q, \tag{6}$$

by which we also achieve that the summation in (5) can easily be carried out. From the typical symmetrical manner in which, according to table 1, for the elements considered, the number of electrons in the groups first increases with increasing principal quantum number and then again decreases, we see that this sum is, in fact, always approximately equal to half the product of N_n^*, as defined in (6) for $p = n$, times the number of terms in the sum, i.e., the number of outer electron groups. If expression (6) were accurately valid, this assertion would, as easily seen, actually hold strictly in case of the K levels ($n = 1$) for elements located two places after the noble gases, for which the outermost group contains two electrons; in case of L levels ($n = 2$), it would hold strictly for the noble gases themselves, and in case of M levels ($n = 3$) for elements located 8 places before the noble gases, in which a group of 18 electrons is just being completed. Hence, we obtain approximately

$$\Delta B = Rh2N_n^*\beta_{N,n},$$

where $\beta_{N,n}$ is a quantity which increases continually with increasing atomic number, and increases approximately by one unit as one goes from an element to the homologous element in the next period, in such a way that $\beta_{N,n}$ never differs much from the number of outer groups. We can now account approximately for the influence of the outer electrons on the work required to remove an electron from an inner group by writing, instead of (3),

$$W = Rh(N-\alpha_{n,k}-n^2\beta_{N,n})^2/n^2.$$

While the quantity $\alpha_{n,k}$, which we may call the "inner screening constant", does not depend on the atomic number, the quantity $n^2\beta_{N,n}$, the "outer screening constant", varies in a characteristic manner with the atomic number. Not only will this latter screening constant, as follows from the above, increase by n^2 units within a period of the system of elements, but we must also expect that the details of the gradual appearance and formation of the outer groups, which have been ignored in the above calculation, will be reflected in the precise manner in which the outer screening constant varies with increasing atomic number. The formation of an electron group must, of course, reveal itself the more plainly the deeper it lies in the interior of the atom, and we must expect that the increase in the outer screening constant will be particularly rapid at places in the periodic system where a deep-lying group is in a process of gradual formation.

We shall not here enter more closely into the computation of the quantities $\alpha_{n,k}$ and $\beta_{N,n,k}$, which even at the present state of the theory could be substantially refined, since on this occasion we are more concerned with comparing the experimental material with the main features of the theory. However, in this comparison, attention must be given to another essential circumstance which, for simplicity, we have so far ignored. We have in mind the relativistic correction which, as is well known, is not at all negligible in the description of the motion of the inner electrons. It is taken into account by using as a basis, instead of the simple formula (1), the Sommerfeld formula

$$W = Rh\frac{N^2}{n^2} + Rh\left(\frac{2\pi e^2}{hc}\right)^2 \frac{N^4}{n^4}\left(\frac{n}{k}-\frac{3}{4}\right)+\dots, \tag{7}$$

which differs from (1) by the addition of the so-called relativistic terms, of which we have written only the first one, the consideration of which suffices for many applications.

All the corrections mentioned can be formally taken into account by writing the work required to remove an electron from an orbit of type n_k as follows:

$$W = Rh\frac{(N-\gamma)^2}{n^2} + Rh\frac{(N-\delta)^4}{n^4}\left(\frac{2\pi e^2}{hc}\right)^2\left(\frac{n}{k}-\frac{3}{4}\right)+\dots. \tag{8}$$

The quantity γ appearing in the main term we call the "total screening constant"; for the innermost electron groups considered, it is given approximately by the formula

$$\gamma = \alpha_{n,k}+n^2\beta_{N,n}. \tag{9}$$

The quantity $N-\delta$ appearing in the second term simply signifies an effective number of nuclear charges that can be considered appropriate for calculating approximately the relativistic correction for the orbital type in question. In estimating δ, it must be considered that the effect of screening, as well of the inner as of the outer electron groups, reveals itself in essentially different ways in the two terms of formula (8). Thus, it is to be expected that δ in general will be considerably smaller than the screening constant appearing in the main term.

If we do not, as heretofore, limit our attention to the innermost groups in the atom, but turn to the groups located farther out, the conditions become much more complicated and, in calculating the work of removing an electron, it is, as for the optical spectra, of essential importance to consider separately the deviation of the orbit from a Kepler ellipse and the consequent lowering of the effective quantum number. We meet conditions differing from those so far discussed especially when dealing with the work of removing electrons from groups which are in the process of gradual formation. Although the quantitative calculation of levels in such cases must remain very incomplete at the present state of the theory, it is nevertheless possible to state a number of immediate consequences of the theory concerning the variation of the values of the levels with atomic number; however, we shall delay the discussion of these until later when we compare the experimental material with the theory.

§ 5. General comparison of the classification of empirical X-ray levels with the theory of atomic structure.

To provide a comprehensive survey and to facilitate a comparison with the theoretical conceptions of atomic structure, curves have been drawn in fig. 3 representing the $\sqrt{T/R}$ values, calculated in table 2, as functions of N. Where there is a particularly great uncertainty in the values of the levels, the otherwise full level curves are merely dashed. Each curve is marked with the symbols $n(k_1, k_2)$ in just the same manner as the levels in fig. 2.

Now, it must be pointed out first that it has been possible to classify the multitude of levels belonging to each element with just the same values of the numbers n and k used in table 1 to classify all electron orbits present in the atoms. This is true not only for elements with completed electron groups in their atoms, such as the noble gases considered especially in § 2; but for a considerable number of levels the experimental material allows us to trace the elements in which these levels first appear in the X-ray spectra, and it has been found besides that a level belonging to definite values of n and k, as far as this can be decided from the experimental data*, appears for the first time just where, according to the theory,

* D. Coster, Phil. Mag. **43** (1922) 1070.

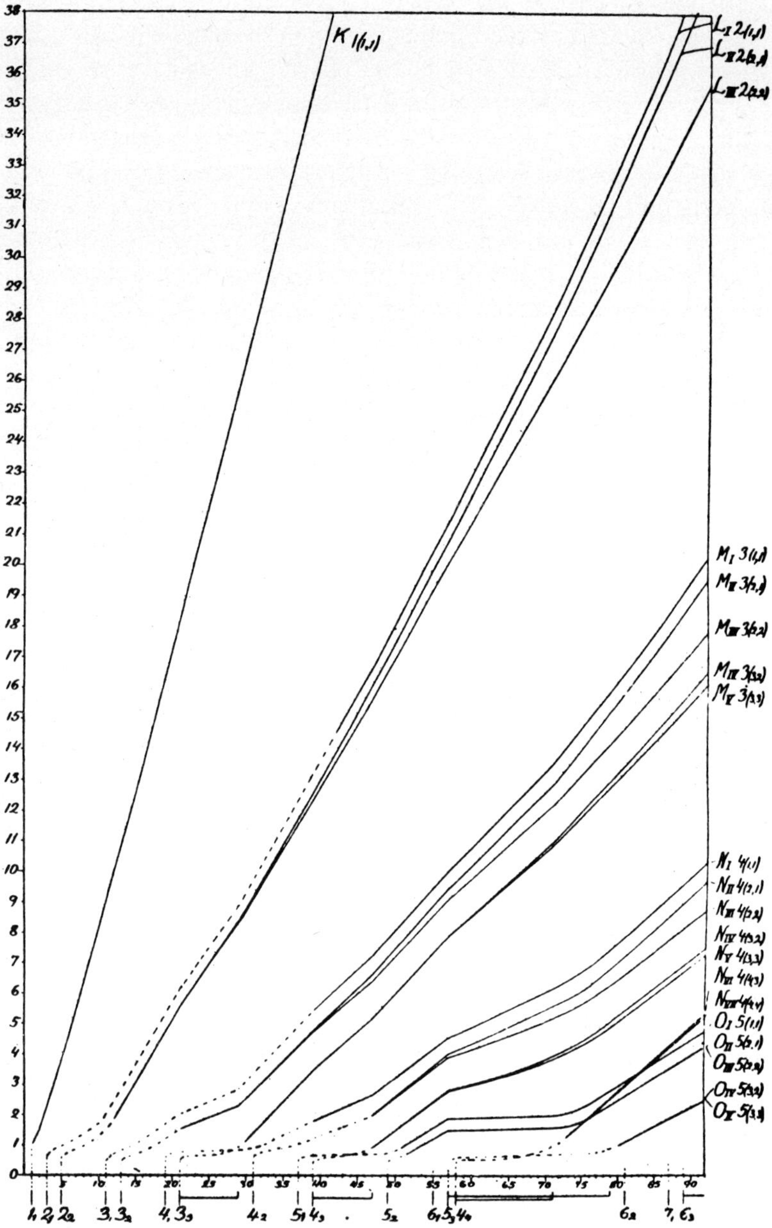
K 1(1,1)
L_I 2(1,1)
L_II 2(2,1)
L_III 2(2,2)
M_I 3(1,1)
M_II 3(2,1)
M_III 3(2,2)
M_IV 3(3,2)
M_V 3(3,3)
N_I 4(1,1)
N_II 4(2,1)
N_III 4(2,2)
N_IV 4(3,2)
N_V 4(3,3)
N_VI 4(4,3)
N_VII 4(4,4)
O_I 5(1,1)
O_II 5(2,1)
O_III 5(2,2)
O_IV 5(3,2)
O_V 5(3,3)
1₁ 2₁ 2₂ 3₁ 3₂ 4₁ 3₃ 4₂ 5₁ 4₃ 5₂ 6₁ 5₃ 4₄ 6₂ 7₁ 6₃
Fig. 3.

an orbit corresponding to the same values of n and k occurs in the normal atom. These places are indicated by short vertical bars at the bottom of fig. 3.

A glance at the figure further reveals the important circumstance that, on the otherwise regular curves, peculiar irregularities appear at certain places, and a comparison with fig. 1 shows that this occurs just at those places in the periodic system where an inner group is in the process of being gradually formed. These places are indicated at the bottom of fig. 3 by horizontal lines beginning at the previously mentioned bars. However, in this connection, it must be mentioned that such a line does not merely indicate the gradual addition of electrons in n_k-orbits, but more generally corresponds to the entire development of the k'th step in the formation of the n'th electron group in the atom, in which process also the number of electrons with the same value of n but smaller values of k is increased.

However, before entering into further details of the course of the curves, we must make some general remarks about the so-called level doublets. As is well known, these are of two kinds, commonly designated as "screening doublets" and "relativity doublets", respectively. In fig. 3 the former are recognizable by the parallel course of the curves belonging to certain pairs of levels, such as (L_IL_{II}), (M_IM_{II}), $(M_{III}M_{IV})$, (N_IN_{II}), $(N_{III}N_{IV})$, (N_VN_{VI}), (O_IO_{II}), and $(O_{III}O_{IV})$. On the other hand, the second kind of doublets can be recognized by the ever greater increase in the separation between the curves corresponding to certain pairs of levels such as $(L_{II}L_{III})$, $(M_{II}M_{III})$, $(M_{IV}M_V)$, $(N_{II}N_{III})$, $(N_{IV}N_V)$, $(N_{VI}N_{VII})$, $(O_{II}O_{III})$, and $(O_{IV}O_V)$, which to begin with lie so close together that they cannot be separated by measurements of the lines. However, from the rules of combination, we may conclude that, as the atomic number increases, they begin to appear in the atom at the same time. While the separation of the former doublets can be formally represented by the difference between two values of the first term in formula (7), corresponding to a difference in the inner screening constant $\alpha_{n,k}$, the separation of the latter doublets can approximately be represented as the difference between two values of the second term in this formula, obtained by substituting for k the two different values that in the classification of the terms were denoted by k_2.

No complete explanation can be given, at the present state of the theory, for the appearance of these doublets which, as well known, have been of great importance for the formal classification of X-ray spectra. Nevertheless, it seems possible to connect the empirical structure of the group of levels closely with the structure of the electron groups in atoms assumed in the interpretation of the periodic system. At a first glance, one might perhaps expect that the number of levels belonging to a definite value of n would be equal to the number of subgroups present in the normal atom belonging to the electron group having the same principal quantum number, such that each level would correspond to the removal of an electron from each of these subgroups. However, we must realize that X-ray levels are essentially determined by the states of an atom from which an electron is removed from an inner group, and we must keep in mind the complexity which may arise from this removal. Let us for a moment assume that, also for an excited atom, the different subgroups can be regarded as independent of one another, in the sense that the removal of an electron from a subgroup does not produce any change in essential features of the structure of the other subgroups. In such a case, we should expect a number of levels differing not solely in their relativistic term but also by their screening constants, so that we could not account for the appearance of the typical relativity doublets. However, if we now consider the pattern of levels that would result if all levels were omitted which are classified by unequal values of k_1 and k_2, then we see that this would reproduce just all those features to be expected from such a simple conception of the origin of the X-ray levels. We see not only that the separations of all the remaining pairs of levels can be expressed as a sum of a screening term and a relativity correction corresponding to formula (8), when, for constant n, we vary the value of the subordinate quantum number k; but aiso that the rules for the appearance of the observed X-ray lines, corresponding to combinations of the levels in question, are just those given by a simple application of the correspondence principle to a central motion, since k now always changes by one unit. Therefore, we shall for brevity call levels for which k_1 and k_2 have been taken to be equal "normal levels".

With regard to the existence of the other levels, having different values of k_1 and k_2, it may first of all be said that the values of k_1 and k_2 used to classify these "abnormal" levels simply are intended to express that each of these levels, purely formally, is quantitively closely related to two normal levels. Thus, each of these levels appears in the atom together with a normal level for which k_1 has the same value, and with which it also has the screening constant in common. On the other hand, it is related to a normal level having the same value of k_2, in that both have the same relativity term, provided the difference in the effective nuclear charge occurring in this term be ignored. This state of affairs leads one to seek the origin of these abnormal levels in a close interplay between the electronic motions in two neighboring subgroups of the excited atom, involving an essential change in the harmonious interplay of the electronic motions within *both* subgroups when an electron is removed from *one* of these subgroups. Such a conception is suggested by the theory, inasmuch as it is assumed that the process of forming an electron group depends essentially just upon the interplay of electronic motions within different subgroups*. On the other hand, it is still quite obscure

* Cf. N. Bohr, Z. Phys. **9** (1922) 1.

how a detailed explanation of the separate appearance of screening and relativity doublets can be given by an elaboration of this view. Likewise, an explanation of the empirical combination rules, as far as the abnormal levels are concerned, is lacking, although they hint at the appearance of a second subordinate quantum number, much as it appears in the optical series spectra where the complexity of the energy levels, as well as the combination rules, seem to permit a simple explanation on the basis of the correspondence principle. This consideration only intends to show that, although a complete explanation of the complex structure of the levels cannot be given at the present state of the theory, it seems nevertheless possible to harmonize the experimental findings in a natural manner with characteristic features of the theory.

§ 6. Theoretical interpretation of the detailed dependence of energy levels upon the atomic number.

Proceeding to a closer comparison with the theory of the values of levels determined from the experimental data, we shall first briefly consider the levels associated with completely formed groups in the atom, for which we can assume that formula (8), for the work required to remove an electron from an inner group, possesses a considerable accuracy. When applying this formula to calculate the screening constant γ appearing in the main term, it becomes important, at least for higher nuclear charges, what value we ascribe to the screening constant δ occurring in the relativity term. For K levels, for which a simple estimate shows that δ never amounts to more than a few units, this does not lead to particular difficulties. Moreover, for the elements of highest atomic number, for which the value ascribed to the screening constant δ is of essential significance, the uncertainty in estimating this quantity is not important as compared with the uncertainty in the experimental results. For other levels, we can, for simplicity, proceed in such a way that—referring to the considerations of the previous section—we use the values of δ calculated directly from the relativity doublets. This is so much the more justified, as these values are approximately equal to those that might be expected theoretically from the influence of the relativistic modification of the motion of electrons in orbits of type n_k. The uncertainty introduced hereby is particularly small for levels corresponding to circular orbits, since for these the absolute values of the relativistic corrections are the smallest. In this manner we have calculated from table 2 the values of the total screening constant γ occurring in the main term of formula (8); these are listed in the table below.

Element	K	L_{III}		M_V		N_{VII}	
	γ	γ	$\frac{1}{4}\gamma$	γ	$\frac{1}{9}\gamma$	γ	$\frac{1}{16}\gamma$
A (18)	2.7	(9.5)	(2.4)	–	–	–	–
Kr (36)	3.7	13.5	3.4	(28.2)	(3.1)	–	–
X (54)	4.7	16.8	4.2	33.5	3.7	–	–
Nt (86)	5.5–6.0	21.7	5.4	43.0	4.8	(65)	(4.1)

In order to be able to compare these values of the total screening constant γ with the theoretical formula (9), we have, for the L, M, and N levels, also listed the screening constant divided by n^2. As we see, we thereby obtain values for all the levels considered which, on the whole, increase by one unit on going from one period in the system of elements to the next. This agreement is particularly good for the K levels where the theoretical formula, from the nature of the estimation, must also be regarded as most accurate. This agreement is of essential significance in that, quite apart from the still open question of the interpretation of the relativity doublets, it provides a direct support for the correctness of the quantum numbers used in the classification of the electron orbits in the atoms. If, for example, as proposed in several recent papers, the subordinate quantum number k should be assigned half integral rather than integral values, matters would appear quite differently. Because of the much larger value of the relativity term for the K levels corresponding, according to the formula, to the assumption $k = \frac{1}{2}$, this assumption would mean that the screening constant, instead of the increase of one unit from noble gas to noble gas claimed by the theory, would show an increase of about 5 units from krypton to xenon, and even an increase of more than 20 units from xenon to niton.

Moreover, when in formula (9) we put $\beta_{N,n}$ equal to the number of outer groups, we find from the table values for the inner screening constant $\alpha_{n,k}$, namely $\alpha_{11} = 0.7$, $\alpha_{22} \sim 5.5$, $\alpha_{33} \sim 16$, ($\alpha_{44} \sim 33$), which are of the order of magnitude to be expected according to the theory. However, as already mentioned in § 5, it is not our intention to enter more closely into an accurate calculation of screening constants in the present paper. With the aid of the figures, we shall only point out the irregularities that appear in the curves at levels belonging to completely closed groups, which must be interpreted as peculiarities in the variation of the outer screening constants. As seen in the figure, such irregularities appear particularly strongly in the M levels, first in the region of the palladium family and later, still more plainly, in the region of the rare earths, in which region the slope of the level curves, in agreement with the theoretical expectation, is substantially smaller than in the immediately adjoining regions. Also for the L levels, similar irregularities in the variation of the values of levels with atomic number can be detected with certainty in these regions. However, because of the scale used, they are too small to be plainly visible in the curves of fig. 3. In the neighborhood of the iron family, however, the change in the slope of the L curves can be plainly detected. In the neighborhood of this family, there also seems to be a small irregularity in the course of the K curve.

Let us now turn to the condition of level curves corresponding to electron groups which are themselves in the process of being formed. Here, the experimental material brings a number of new features to light, allowing a thorough comparison with the theory. The first places in the periodic system where such conditions are met are in the vicinity of the iron and palladium families, where we discern characteristic peculiarities in the M and N curves, respectively. However, because of the relatively great uncertainty of the experimental material in question, we shall not dwell any longer on these phenomena, especially since we meet still more pronounced peculiarities of the curves at the *family of rare earths*. In order to let the course of the curves in this region, where the experimental results also seem most reliable, appear more plainly, the curves of the levels of interest here are drawn on a larger scale in fig. 4.

According to the theory, the appearance of the family of rare earths is caused by the addition of electrons in 4_4-orbits, and by the consequent development of the 4-quantic electron group, during which the number of electrons in this group gradually increases from 18 to 32. The presence of the 4_4-orbits manifests itself in the X-ray spectrum by the appearance of the lines M_α and M_β. So far, the M spectrum has not been investigated below Dy (66). For this element, the lines mentioned have been found by Stenström, and we can conclude from the experimental data that the binding energy of the 4_4-orbits for Dy (66) and the closest following elements is of about the same magnitude as that of the 6-quantic valence electrons, and substantially smaller than that of the 5_1- and 5_2-orbits, a fact that reveals itself in the O_I level, as well as in the pair of levels O_{II} and O_{III}. This is just what should be expected from the theory; for, as long as the 4-quantic group is in the process of being formed, the addition of electrons to 4_4-orbits occurs, so to speak, in competition with the valence electrons in the 6-quantic orbits. However, from the moment when the 4-quantic group has been completed, quite different circumstances will assert themselves, and we must expect that the binding energy of the 4_4-orbits will increase rapidly and regularly with increasing atomic number. This corresponds reasonably with the experimental finding that, after the completion of the family of rare earths, the curves of the pair of levels N_{VI} and N_{VII} rise steeply and, as may be seen in fig. 4, cross the curves of the O_I, O_{II} and O_{III} levels between W and Bi.

Although, as mentioned, the available experimental material does not allow a

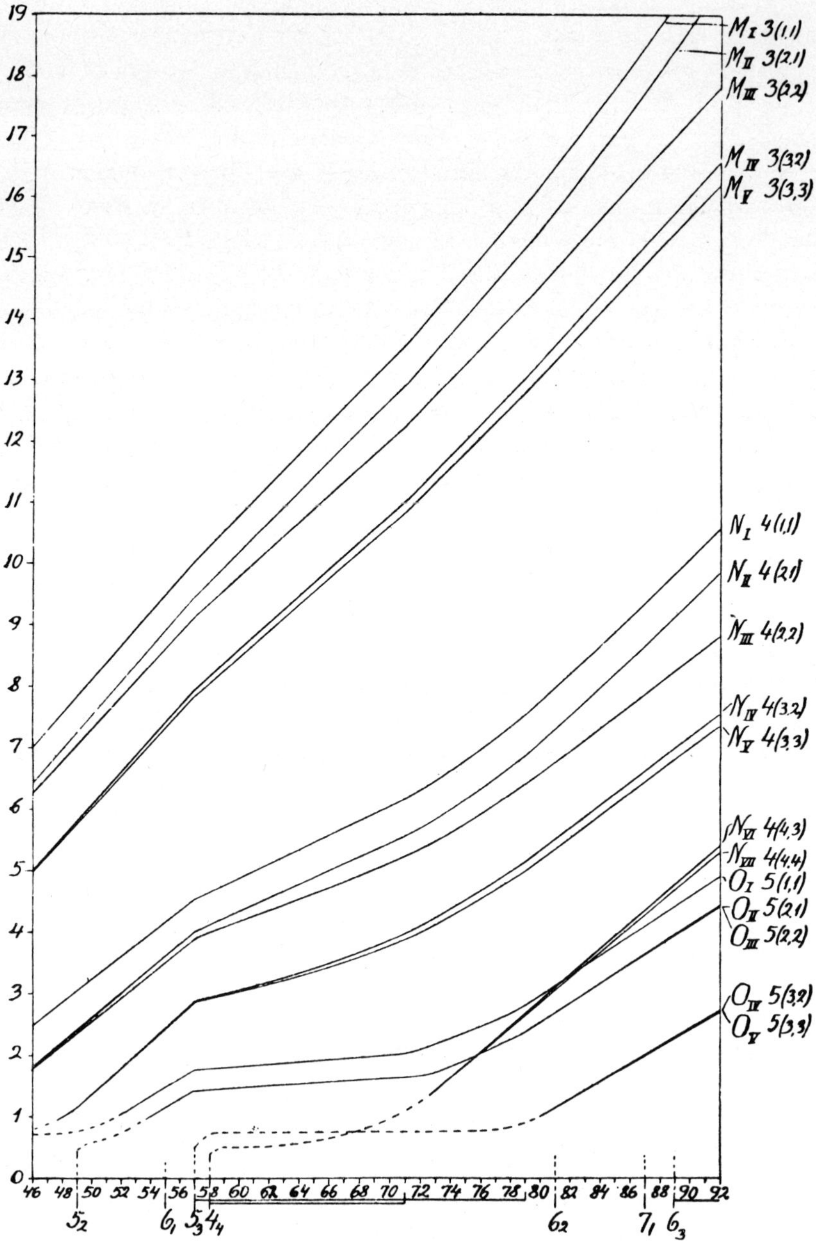

19
18
17
16
15
14
13
12
11
10
9
8
7
6
5
4
3
2
1
0
46 48 50 52 54 56 58 60 62 64 66 68 70 72 74 76 78 80 82 84 86 88 90 92
5_2 6_1 5_3 4_4 6_2 7_1 6_3
M_I 3(1,1)
M_{II} 3(2,1)
M_{III} 3(2,2)
M_{IV} 3(3,2)
M_V 3(3,3)
N_I 4(1,1)
N_{II} 4(2,1)
N_{III} 4(2,2)
N_{IV} 4(3,2)
N_V 4(3,3)
N_{VI} 4(4,3)
N_{VII} 4(4,4)
O_I 5(1,1)
O_{II} 5(2,1)
O_{III} 5(2,2)
O_{IV} 5(3,2)
O_V 5(3,3)

Fig. 4.

direct verification of the onset of the presence of 4_4-orbits, by revealing the first appearance of the pair of levels N_{VI} and N_{VII}, the addition of electron orbits of this type reveals itself plainly in the course of the curves of the other 4-quantic levels, as well as in those of levels with principal quantum number 5. These curves all show a pronounced sharp bend in the neighborhood of Ce (58) where, according to the theory as given in table 1 on p. [521], a 4_4-orbit is present in the atom for the first time. These bends must be interpreted as arising from an abruptly beginning increase in the inner screening constant in the expression for the energy required for removal of an electron from these levels. This increase depends on the circumstance that, as to be expected theoretically, the 4_4-orbits, even when they first appear in the normal state of the atom, lie in the region of already present electron groups with 4-quantic orbits, and therefore within the region in which electrons in 5-quantic orbits move during the greatest part of their revolution. In this connection, it must be remembered that, for the closer estimation of the above-mentioned balance between the strength of binding of the 4_4-electrons and that of the valence electrons, it is decisive, on the one hand, that the strength of binding of the 4_4-electrons is enormously weakened by the outer screening effect of the 5-quantic electron group and, on the other hand, that, because of the great difference between principal quantum number and effective quantum number, the binding of the valence electrons is many times stronger than that which would correspond to a 6-quantic Kepler ellipse lying entirely outside the inner electron group.

Because of the incompleteness of the measurements for the elements in the vicinity of 72, which form the end of the family of rare earths, there exists here some uncertainty about the course of the curves of the levels N_I, N_{II}, N_{III}, N_{IV}, and N_V. However, the available experimental material seems to show clearly that, as indicated in the figure, these curves exhibit here a more continuous change in slope than in the neighborhood of Ce (58). This may be connected with the process of development of the 5-quantic electron group taking place in this region; this is responsible for the appearance of the platinum metals, and reveals itself in the 4-quantic levels by an influence upon the outer screening constant.

Finally, as regards the O levels, we encounter new features of great theoretical interest, in that we here have to do with an inner electron group whose orbits lie partly outside a group in a process of gradual formation. Although there still is considerable uncertainty in the experimental data, it seems nevertheless certain that the T/R values of these levels, as indicated by the course of the curves in the figure, change exceedingly little within the family of the rare earths. This is just what should be expected from the theory, since the outer loops of the 5-quantic orbits lie outside the region of the 4-quantic electron orbits, so that the effective nuclear charge, decisive for the work of removal of an electron from the O levels, as well as the effective quantum number, must remain nearly constant within the rare earths.

It would be of great interest to attempt to determine the course of the X-ray levels between xenon and niton with the greatest possible accuracy, since we might hope thereby to learn much about the detailed process of development of the 4- and 5-quantic electron groups.

§7. Concluding remarks.

As is well known, the discovery of a simple connection between the X-ray spectrum of an element and its atomic number was the typical earmark of Moseley's fundamental investigations. Especially striking was the seemingly complete lack of any close relation of these spectra to the regularities of those other properties of the elements expressed in the periodic system. However, through the increased accuracy of measurements, which we owe largely to Siegbahn's investigations, and the extended range of measurements, a number of phenomena have in the meantime come to light which, as we have seen, can be related closely to essential features of the theory that are also invoked to interpret the periodic system.

However, it must be emphasized here that, among the regularities of the X-ray spectra so far discussed, no mention was made of any periodicity of the kind revealed in the optical and chemical properties of the elements. However, on closer consideration, we must also in the case of the X-ray spectra expect traits connected with just those features of atomic structure which are directly responsible for the periodicity of the properties mentioned. In fact, what are expressed in the usual regularities of the periodic system are the processes of closure taking place continually in the outermost electron groups in the atom which, with increasing atomic number, cause the electrons to be continually bound in new types of quantum orbits in the normal state of the atom. With regard to the manner in which these circumstances may reveal themselves in the X-ray spectra, we come to a feature of the theory which we have so far not mentioned in particular. For the interpretation of the general laws of emission and absorption of X-rays, it is quite essential that the inner groups in general have a closed character. In fact, this involves that an electron coming from the outside cannot be bound, under emission of radiation, in an orbit of the same type as the electron orbits in an already closed group; nor can an electron from an inner group, under absorption of radiation, be admitted to an orbit of these types.

This very consideration, which offers an immediate explanation of the characteristic difference in the laws holding for absorption in the X-ray region, on the one hand, and for the optical spectra, on the other hand, suggests at the same time that, for the absorption of X-ray spectra, we must expect special circumstances in the case of elements in which a group is in the process of being gradually formed*. In fact, we must expect that an electron can be transferred directly, by absorption of radiation, to such an uncompleted group. This will reveal itself in the appearance, both in emission and in absorption, of X-ray lines which cannot be subsumed under the classification of X-ray spectra corresponding to the diagram in fig. 2 which, strictly speaking, is valid only for elements with completed groups**. In fact, such lines seem to occur for the elements of the iron family, in the emission spectrum as longwave satellites to lines in the diagram as well as in the absorption spectrum as longwave components of the fine structure of the edges***. Moreover, also in the region of the rare earths, several lines have been observed which lie on the longwave side of lines in the diagram and seem to have a similar origin. A closer investigation of these lines, especially as to their behavior in absorption, has begun and should yield important information about the manner in which the gradual conclusion of the 4-quantic electron group takes place in the region of the rare earths.

These last considerations lead us immediately to expect a general close connection between the X-ray spectroscopic data and the periodic nature of the optical and chemical properties of the elements. Indeed, the characteristic variation of these properties with the atomic number is, as mentioned, simply a reflection of the generally non-closed character of the loosest bound electron group in the atom. We may therefore assume that an electron from an inner group can not only be entirely removed from the atom by an absorption process, or transferred into a quantum orbit of another kind than those already present in the atom, but that in general it can also land directly in the outermost electron group. To estimate the

* Cf. N. Bohr, Z. Phys. **6** (1922) 1.

** Moreover, this classification applies only to spectra excited by removal of a single electron from an inner group in an atom. As recently pointed out by Wentzel (Ann. d. Phys. **66** (1921) 437), certain lines which do not find a place in the diagram in fig. 2, but which appear as shortwave satellites to lines in the diagram, can be ascribed to atoms from which more than one electron have been removed from their normal orbits in inner groups. In contrast to the usual X-ray spectra, which may be called spectra of the first kind, it may be useful to call a spectrum excited by the removal of p electrons from an inner group a spectrum of the p'th kind. However, in connection with the considerations in the text, it must be emphasized that, among the phenomena of X-ray absorption, spectra of higher kind can hardly play any appreciable role, since, under the experimental conditions, the number of atoms excited by removal of inner electrons is probably always negligible compared to the number of atoms whose inner structure is in the normal state. (Cf. D. Coster, Phil. Mag. **44** (1922) 546; and S. Rosseland, to appear shortly.)

*** Cf. D. Coster, Phil. Mag. **44** (1922) 546.

binding energy corresponding to such an electronic motion, we must bear in mind that the circumstances under which the motion of the electrons takes place in the outer regions of an atom excited by the removal of an electron from an inner group will coincide very nearly with those normally governing the motion of the outermost electrons in the succeeding element.

Besides regularities of a general nature in the variation of the values of the levels with atomic number, whose connection with the periodic system was discussed in the previous sections, we must expect in general that continued investigations will unravel a multitude of individual details of X-ray spectroscopic data. Many indications of such phenomena are already present in the experimental material; however, we shall not enter further into these matters on this occasion, especially since, as already emphasized in § 3, in deriving the values of levels in tables 2 and 3, we have neglected their dependence on the physical and chemical states in which the element exists, something that would be of decisive importance in this connection.

Copenhagen, October 1922.
Institute for Theoretical Physics of the University.

XII. LINE SPECTRA AND ATOMIC STRUCTURE

LINIENSPEKTREN UND ATOMBAU
(Ann. d. Phys. **71** (1923) 228)

TEXT AND TRANSLATION*

* [See Introduction, sect. 12.]

16. *Linienspektren und Atombau; von Niels Bohr.*

§ 1. Allgemeine Gesichtspunkte.

Seit der Entdeckung der für die Elemente charakteristischen Linienspektren war es die allgemeine Überzeugung der Physiker, daß man in diesen Spektren ein wertvolles Mittel zur Erforschung des Atombaues besaß. Die Hoffnung, auf diesem Wege vordringen zu können, wurde vor allem gestützt durch die Entdeckung der eigentümlichen Gesetze, welche die Schwingungszahlen der Linien mancher Spektren befolgen, und dies mit einer in der Physik sonst unerreichten Genauigkeit. Diese Sachlage ist ja der Anlaß für die Bemühungen gewesen, die auf die Ausmessung der Spektren sowie auf die Entwirrung der Spektralgesetze verwandt worden sind. Hiervon bekommt man wohl den stärksten Eindruck durch das große Werk, mit dem Kayser[1]), der selbst an der Bewältigung dieser beiden Aufgaben einen hervorragenden Anteil hat, die physikalische Literatur bereichert hat und das stets ein unentbehrliches Hilfsmittel für die experimentelle sowie für die theoretische spektroskopische Forschung gewesen ist.

Die Bestrebungen, das empirische spektroskopische Material für die Probleme des Atombaues theoretisch zu verwerten, wurden besonders verlockend durch die allmähliche Erkenntnis, daß eine sinngemäße Deutung der Spektralgesetze auf Grund der klassischen Mechanik und Elektrodynamik nicht durchführbar war.[2]) Diese Schwierigkeit trat noch schärfer hervor

1) H. Kayser, Handbuch der Spektroskopie, Bd. I—VI. Leipzig 1900—1912.

2) In dem Buche von H. Konen, Das Leuchten der Gase und Dämpfe, Braunschweig 1913, das eine Übersicht über die Resultate der empirischen Spektroskopie enthält, findet sich am Schluß eine interessante Zusammenstellung der Deutungsversuche der Spektralgesetze auf Grund der Vorstellungen der klassischen Mechanik und Elektrodynamik.

durch den großen Fortschritt unserer Kenntnisse vom Atombau, den die Untersuchungen über die elektrischen Entladungen in Gasen und die radioaktiven Phänomene gebracht haben. Das Ergebnis dieser Untersuchungen war, daß die Atome aus elektrischen Teilchen, und zwar negativen Elektronen und positiven Kernen, aufgebaut sind, deren Dimensionen außerordentlich klein sind verglichen mit der Größe des ganzen Atoms, wie sie sich aus der Raumerfüllung der Stoffe ergibt. So nehmen wir auf Grund der Rutherfordschen Vorstellung des *Kernatoms* an, daß ein Atom eines Elementes aus einem zentralen positiven Kern besteht, der der Sitz des ganz überwiegenden Teiles der Atommasse ist, und den Elektronen umkreisen, deren Anzahl im neutralen Atom eben gleich ist der sogenannten Atomnummer, welche die Stelle des Elementes im periodischen System bezeichnet. Solange wir auf die Prinzipien der gewöhnlichen Mechanik und Elektrodynamik bauen, ist diese Vorstellung aber nicht vereinbar mit der Forderung der Stabilität der Atome, die in den spezifischen Eigenschaften der Elemente zum Ausdruck kommt, und von der eben die Linienspektren nicht am wenigsten ein schlagendes Zeugnis ablegen.

Inzwischen hatten aber die Physiker in dem sogenannten Wärmestrahlungsgesetz bereits einen Prüfstein für die Anwendbarkeit der klassischen Gesetze auf die Emission und Absorption von Strahlung durch Atome gefunden, der unabhängig von speziellen Voraussetzungen über deren Bau ist. Die theoretischen Untersuchungen über das Wärmestrahlungsgesetz hatten nun nicht nur das Scheitern der klassischen Elektrodynamik mit Sicherheit ergeben, sondern sogar auch einen Hinweis gegeben, in welcher Richtung die notwendigen Änderungen der klassischen Gesetze zu suchen waren. So wurde durch die wohlbekannte Plancksche Herleitung dieses Gesetzes der Anstoß zur Entwicklung der sogenannten *Quantentheorie* gegeben.

Der Versuch, die auf dem Gebiete des Atombaues gewonnenen Resultate mit den Ergebnissen der Spektroskopie in Verbindung zu bringen, führte nun zu einer Formulierung der Prinzipien der Quantentheorie, die direkt darauf hinzielte, von der Stabilität der Atome sowie vom Charakter der Gesetzmäßigkeiten der Linienspektren Rechenschaft zu geben.[1]) Diese Formulierung beruht auf der Aufstellung der folgenden Postulate:

I. Unter den denkbaren Bewegungszuständen eines Atomsystems gibt es eine Anzahl von sogenannten *stationären Zuständen*, in denen zwar die Bewegung der Teilchen in großem Umfang die klassischen mechanischen Gesetze erfüllt, die sich aber durch eine eigentümliche, mechanisch unerklärbare Stabilität auszeichnen, die zur Folge hat, daß jede bleibende Veränderung in der Bewegung des Systems in einem vollständigen Übergang von einem stationären Zustand zu einem anderen bestehen muß.

II. Während, im Widerspruch zur klassischen Elektrodynamik, in den stationären Zuständen selbst keine Ausstrahlung vom Atom stattfindet, kann ein Übergang zwischen zwei stationären Zuständen von der Emission elektromagnetischer Strahlung begleitet sein, welche dieselbe Beschaffenheit hat wie diejenige, die nach der klassischen Theorie von einem elektrischen Teilchen ausgesandt wird, das eine harmonische Schwingung mit konstanter Schwingungszahl vollführt. Diese Schwingungszahl ν hat indessen keinen einfachen Zusammenhang mit der Bewegung der Teilchen des Atoms, sondern ist durch die sogenannte *Frequenzbedingung*

$$\nu = \frac{E' - E''}{h} \tag{1}$$

gegeben, wo h die Plancksche Konstante und E' und E'' die Energiewerte des Atoms in dem Anfangs- und Endzustand des Ausstrahlungsprozesses bedeuten. Umgekehrt kann eine Bestrahlung des Atoms mit einem Wellenzug dieser Schwingungszahl zu einem Absorptionsprozeß Anlaß geben, bei dem das Atom von dem zweitgenannten stationären Zustand in den ersteren zurückgeführt wird.

Während das erste Postulat auf die allgemeinen Stabilitätseigenschaften der Atome hinzielt, faßt das zweite Postulat in erster Linie die Existenz von aus scharfen Linien bestehenden

1) Vgl. N. Bohr, Phil. Mag. (6) **26**. S. 1, 476, 857. 1913. In deutscher Übersetzung: Abhandlungen über Atombau, Braunschweig 1921 (im folgenden als „Abh. über Atombau" zitiert) S. 1, 26, 52.

Spektren ins Auge. Die Bedingung (1), deren Verwandtschaft mit den in der Einsteinschen Theorie des lichtelektrischen Effektes benutzten Annahmen ersichtlich ist, gab nun eine Grundlage für eine Deutung der empirischen Spektralgesetze.

Gemäß dem allgemeinsten dieser Gesetze, dem von Ritz formulierten *Kombinationsprinzip*, kann die Schwingungszahl ν jeder Linie eines Spektrums durch die Formel

$$\nu = T' - T'' \tag{2}$$

dargestellt werden, worin T' und T'' zwei sogenannte Spektralterme bedeuten, die einer für das betreffende Spektrum charakteristischen Mannigfaltigkeit von Spektraltermen angehören. Dieses Prinzip, das den Gesetzen, welche die Schwingungszahlen mechanischer Systeme regeln, ganz fremdartig gegenübersteht, läßt auf Grund der Postulate eine einfache Interpretation zu, indem wir nämlich die Termwerte T mit Rücksicht auf die Bedingung (1) direkt gleichsetzen können den durch h dividierten Energiewerten E in den verschiedenen stationären Zuständen des das Spektrum emittierenden Atoms.

Der vollständige Bruch der Theorie mit den Vorstellungen der klassischen Elektrodynamik tritt in dieser Deutung des Kombinationsprinzips besonders auffallend zutage. In der Tat enthält diese Deutung die Annahme, daß die Beschaffenheit der emittierten Strahlung selbst bei gleichem Bewegungszustand des Atoms beim Beginn des Strahlungsprozesses je nach dem Zustand des Atoms am Ende des Prozesses eine ganz verschiedene sein kann. Nichtsdestoweniger werden wir sehen, daß es in gewissem Sinne möglich ist, die Theorie als eine sinngemäße Verallgemeinerung der klassischen Strahlungstheorie anzusehen. So können wir das Vorkommen eines jeden der verschiedenen mit Ausstrahlung verbundenen Übergangsprozesse auf eine gewisse der in der Atombewegung auftretenden harmonischen Schwingungskomponenten zurückführen, die nach der klassischen Theorie die in der emittierten Strahlung auftretenden Wellenzüge bestimmen würden. Diesen Zusammenhang zwischen Spektrum und Atombewegung bezeichnen wir als *Korrespondenzprinzip.* In bezug auf die Deutung der Spektralgesetze hat dieses Prinzip insbesondere ermöglicht, eine einfache Erklärung zu geben für die schein-

bare Launenhaftigkeit im Auftreten der durch das Kombinationsprinzip vorausgesagten Linien.

§ 2. Quantentheorie der Periodizitätssysteme.

Die Formulierung des Korrespondenzprinzips stützt sich auf eine quantentheoretische Untersuchung von Atomsystemen, bei denen die Bewegung der Teilchen in eine Mannigfaltigkeit von harmonischen Komponenten mit diskreten Schwingungszahlen aufgelöst werden kann, d. h. daß die Verschiebung jedes Teilchens in einer gegebenen Richtung als Funktion der Zeit sich durch einen Ausdruck der folgenden Form darstellen läßt:

$$(3) \qquad \xi = \sum C_{\tau_1 \ldots \tau_r} \cdot \cos\left[2\pi(\tau_1 \omega_1 + \ldots + \tau_r \omega_r) t + \gamma_{\tau_1 \ldots \tau_r}\right],$$

worin $\tau_1 \ldots \tau_r$ positive oder negative ganze Zahlen und die Größen $\omega_1 \ldots \omega_r$ die Grundfrequenzen bedeuten. Ihre Anzahl r, die als der Periodizitätsgrad der Bewegung bezeichnet werden kann, ist eindeutig durch die Bedingung festgelegt, daß zwischen ihnen keine Verbindung der Form

$$m_1 \omega_1 + \ldots + m_r \omega_r = 0$$

bestehen soll, worin $m_1 \ldots m_r$ positive oder negative ganze Zahlen bedeuten. Im allgemeinen ist dabei die Summation in (3) über alle positiven und negativen Werte der Zahlen $\tau_1 \ldots \tau_r$ zu erstrecken.

Die stationären Zustände solcher r-fach periodischen Systeme sind nun durch r *Zustandsbedingungen* der Form

$$(4) \qquad J_k = n_k h \qquad (k = 1, \ldots r)$$

gegeben, wo h die Plancksche Konstante, $n_1 \ldots n_r$ ganze positive Zahlen, die sogenannten *Quantenzahlen* bedeuten, während $J_1 \ldots J_r$ eine Anzahl von Größen bedeuten, die mechanische Eigenschaften der Bewegung charakterisieren, und welche die Bedingung

$$(5) \qquad \delta E = \sum_k \omega_k \delta J_k$$

erfüllen, wenn E die totale Energie des Systems bedeutet, und wo das Differentialzeichen δ sich auf den Unterschied der Werte einer Größe in zwei benachbarten Bewegungszuständen bezieht. Durch diese Bedingung werden die Größen J nur bis auf additive Konstanten festgelegt. Diese sind aber vollständig bestimmt durch die weitere Bedingung

$$\sum_k \omega_k J_k = \overline{\sum_i p_i \dot{q}_i}, \tag{6}$$

wo $q_1 \ldots q_s$ beliebig gewählte, unser System von s Freiheitsgraden beschreibende generalisierte Ortskoordinaten und $p_1 \ldots p_s$ die zugehörigen Impulskoordinaten bedeuten, und wo der Strich auf der rechten Seite bedeutet, daß der Mittelwert zu bilden ist über Zeiten, die lang sind verglichen mit den Grundperioden der Bewegung.

Gemäß den Bedingungen (1) und (4) bekommen wir nun für die Schwingungszahl ν der Strahlung, die bei einem Übergang zwischen zwei durch die Werte $n_1' \ldots n_r'$ und $n_1'' \ldots n_r''$ der Quantenzahlen charakterisierten stationären Zustände emittiert wird:

$$\nu = \frac{1}{h}(E' - E'') = \frac{1}{h}\int_{''}^{'} \sum \omega_k \, d J_k = \overline{\sum (n_k' - n_k'') \, \omega_k}, \tag{7}$$

wo der Mittelwert auf der rechten Seite der Gleichung (7) über die Bewegungszustände zu nehmen ist, die in dem r-dimensionalen J-Raum durch eine gerade Verbindungslinie zwischen den den stationären Zuständen entsprechenden Punkten $(J_1' \ldots J_r')$ und $(J_1'' \ldots J_r'')$ dargestellt werden.

In der Grenze, wo die Werte der Quantenzahlen groß sind, verglichen mit deren Differenzen $n_k' - n_k''$, können wir bei der Mittelwertbildung die Größen ω als Konstante betrachten und bekommen die asymptotische Gleichung

$$\nu \sim \sum_k \omega_k (n_k' - n_k''). \tag{8}$$

In dieser Grenze fällt also die Schwingungszahl der Strahlung asymptotisch zusammen mit der Schwingungszahl derjenigen harmonischen Komponente in der Bewegung, für welche die Gleichungen gelten

$$\tau_k = n_k' - n_k''. \tag{9}$$

Um in der Grenze großer Quantenzahlen einen Anschluß der statistischen Resultate der Quantentheorie an die klassische Theorie zu erreichen, werden wir also dazu geführt, das Vorkommen von Übergängen zwischen zwei stationären Zuständen als durch das Auftreten einer durch (9) definierten harmonischen Komponente in der Bewegung, oder genauer im elektrischen Moment des Systems, bedingt anzusehen. Die Überlegung, daß es sich nur um ein statistisches Resultat und nicht etwa um ein allmähliches Verschwinden des fundamentalen Unterschiedes der Quantentheorie und der klassischen Theorie im Gebiet großer Quantenzahlen handelt, führt uns weiter dazu, den in diesem Grenzgebiet nachgewiesenen Zusammenhang zwischen dem Vorkommen von Strahlungsübergängen und dem Auftreten von harmonischen Bewegungskomponenten im Atom, die sich nach (9) den Übergängen zuordnen lassen, als ein allgemeines quantentheoretisches Gesetz anzusehen, das bei beliebigen Werten der Quantenzahlen Geltung hat. Dieses Gesetz stellt den Inhalt des oben erwähnten *Korrespondenzprinzips* dar, und wir bezeichnen die zugeordnete Bewegungskomponente als die mit dem Übergange korrespondierende harmonische Schwingung.

Bei Problemen statistischer Natur ist außer der Festlegung der stationären Zustände eines Systems auch die Kenntnis der Gewichte erforderlich, die den einzelnen stationären Zuständen bei der Berechnung der Wahrscheinlichkeit einer Verteilung der Energie auf eine große Anzahl von Atomen zuzuschreiben ist. Eine Grundlage zur Behandlung dieses Problems ist in einer von Ehrenfest aus dem zweiten Hauptsatz der Wärmetheorie abgeleiteten allgemeinen Bedingung enthalten, die verlangt, daß bei einer kontinuierlichen Transformation eines Systems die statistischen Gewichte der einzelnen stationären Zustände sich nicht ändern, solange der Periodizitätsgrad unverändert bleibt. Für Systeme, für welche der Periodizitätsgrad der Anzahl s der Freiheitsgraden gleich ist, führt diese Bedingung mit sich, daß allen stationären Zuständen dasselbe Gewicht h^s beizulegen ist. Für die sogenannten „entarteten" Systeme dagegen, wo der Periodizitätsgrad kleiner ist als die Zahl der Freiheitsgrade, ist das Gewicht der einzelnen durch (5) definierte Zustände im allgemeinen verschieden. Dieses Gewicht kann aber einfach ermittelt werden durch einen Vergleich mit einem durch eine kleine Transformation des ursprünglichen Systems hervorgebrachten nichtentarteten System, indem das Gewicht der ursprünglichen stationären Zustände gleich der Summe der Gewichte der im Grenzfall der Entartung zusammenfallenden Zustände des transformierten Systems ist.[1])

Das einfachste Beispiel der Anwendung der besprochenen Prinzipien der Quantentheorie ist nun eben dasjenige System, das in der Planckschen Wärmestrahlungstheorie betrachtet wurde, nämlich ein elektrisch geladenes Teilchen, das mit einer charakteristischen Eigenfrequenz rein harmonische Schwingungen um eine Gleichgewichtslage vollführt. In diesem Fall ist das elektrische Moment des Systems einfach gegeben durch die Formel

$$\xi = C \cos(2\pi\omega_0 t + \nu), \tag{10}$$

worin ω_0 die konstante Eigenfrequenz bedeutet. Aus den Gleichungen (5) und (6) folgt nun unmittelbar, daß hier die Bedingung (4) gleichbedeutend ist mit der wohlbekannten Planckschen Formel

$$E_n = n h \omega_0, \tag{11}$$

wenn mit E_n die Energie im n^{ten} stationären Zustand bezeichnet wird. Wenn wir nun gemäß (1) die Frequenz der bei einem Übergang zwischen zwei stationären Zuständen mit den Quantenzahlen n' und n'' emittierten Strahlung berechnen, bekommen wir

$$\nu = (n' - n'')\,\omega_0. \tag{12}$$

Im Hinblick auf (10) fordert nun das Korrespondenzprinzip gemäß (9), daß unter Emission von Strahlung nur solche Übergänge vorkommen können, bei denen $n' - n'' = 1$ ist. Wie in der ursprünglichen Planckschen Theorie angenommen, kommen wir also zu dem einfachen Resultat, daß die Beschaffenheit der emittierten Strahlung in diesem Fall dieselbe ist, wie es nach der klassischen Theorie zu erwarten wäre.

---

1) Eine ausführliche Darstellung der Theorie ist vom Verfasser in einer zusammenfassenden Abhandlung „On the quantum theory of line spectra", Schriften der Kopenhagener Akademie 1918, gegeben, von der kürzlich eine deutsche Übersetzung erschienen ist (Über die Quantentheorie der Linienspektren, Braunschweig 1922). Dort finden sich auch ausführliche Hinweise auf die ältere Literatur. Vgl. auch N. Bohr, Zeitschr. f. Phys. **13.** S. 117. 1923, wo eine kürzere Zusammenfassung des jetzigen Standes der Theorie gegeben ist.

Im Falle eines linearen Oszillators ist ferner nach dem obigen das statistische Gewicht der verschiedenen stationären Zustände, wie es auch in der ursprünglichen Planckschen Theorie angenommen wurde, das gleiche. Im Falle eines isotropen Oszillators mit zwei oder drei Freiheitsgraden ist das statistische Gewicht mit $(n+1)$ bzw. $(n+1)(n+2)/2$ proportional.

§ 3. Das Wasserstoffspektrum.

Ein fast ebenso einfaches System wie der Plancksche Oszillator findet sich nun in der Natur im *Wasserstoffatom* verwirklicht und gibt uns einen einfachen Ausgangspunkt für die Erklärung der Spektralgesetze. Nach der Rutherfordschen Vorstellung des Kernatoms besteht dieses aus einem positiv geladenen Kern, um den sich ein einziges Elektron bewegt. Wenn wir auf die Bewegung eines solchen Systems die Gesetze der Mechanik anwenden, so finden wir, daß die Bewegung der Teilchen mit außerordentlich hoher Annäherung eine rein periodische ist, indem diese sich in Ellipsenbahnen mit einem gemeinsamen Brennpunkt bewegen, der mit dem Schwerpunkt des Systems zusammenfällt. In diesem Fall läßt sich also das elektrische Moment des Atoms durch die Formel

$$\xi = \sum_{\tau} C_{\tau} \cos (2\pi \tau \omega t + \gamma_{\tau}) \tag{13}$$

darstellen, wo ω die Umlaufszahl der Teilchen bedeutet. Diese Umlaufszahl ω sowie die große Achse $2a$ der Bahn des Elektrons sind durch die folgenden einfachen Gleichungen bestimmt, die den wohlbekannten Keplerschen Gesetzen Ausdruck geben:

$$\omega = \sqrt{\frac{2 W^3 (M+m)}{\pi^2 e^4 N^2 m M}}, \quad 2a = \frac{N e^2 M}{W(M+m)}, \tag{14}$$

worin Ne und e die elektrischen Ladungen des Kerns und des Elektrons, M und m deren Masse bedeuten, während W gleich ist der Arbeit, die erforderlich ist, um die Teilchen in unendlich großen Abstand voneinander zu bringen. Mittels einer einfachen Rechnung gemäß (5) und (6) folgt nun aus (14), daß

$$W = \frac{2\pi^2 N^2 e^4 m M}{J^2 (M+m)}, \tag{15}$$

wo $J = \overline{\sum_i p_i \dot{q}_i} / \omega$ die Größe ist, die gemäß den Zustandsgleichungen (4) in den stationären Zuständen des Atoms durch die Bedingung

$$J = n h \tag{16}$$

festgelegt ist, in der n eine ganze Zahl bedeutet. Für die Energie und Bahndimensionen in den stationären Zuständen bekommen wir also

$$E_n = -W_n = -\frac{1}{n^2}\frac{2\pi^2 e^4 N^2 m M}{h^2 (M+m)}, \quad 2a_n = n^2 \frac{h^2}{2\pi^2 e^2 N m} \tag{17}$$

und für die Schwingungszahlen ν der vom System während eines Überganges emittierten Strahlung

$$\nu = K \frac{N^2}{1 + \frac{m}{M}} \left(\frac{1}{n''^2} - \frac{1}{n'^2}\right), \tag{18}$$

wo

$$K = \frac{2\pi^2 e^4 m}{h^3}. \tag{19}$$

Der Ausdruck (18) hat nun eben dieselbe Form wie die Balmersche Formel für das *Wasserstoffspektrum*, die bekanntlich das einfachste und das zuerst erkannte unter den empirischen Spektralgesetzen ist.

Bevor wir auf die nähere Diskussion dieser Übereinstimmung eingehen, wollen wir kurz darauf hinweisen, daß die Möglichkeit des Auftretens von Spektrallinien, die jeder Kombination der Werte der beiden Zahlen n' und n'' entsprechen, nach dem Korrespondenzprinzip darauf beruht, daß hier im elektrischen Moment des Atoms gemäß (13) Oberschwingungen vorkommen, die jedem Werte der Zahl τ entsprechen. Wir bekommen somit eine einfache Erklärung der eigentümlichen unbegrenzten Kombinationsmöglichkeit der in der Balmerformel auftretenden Spektralterme, die bekanntlich für Ritz eben der Anstoß zur Formulierung des allgemeinen Kombinationsprinzips war.

Was ferner die quantitative Wiedergabe der empirischen Werte für die Schwingungszahlen der Spektrallinien betrifft, zeigte es sich, daß innerhalb der Meßgenauigkeit der durch (19) gegebene Ausdruck für die Konstante K mit der empirischen Konstante der Balmerformel übereinstimmt. Dies entspricht ganz den theoretischen Vorstellungen, weil für Wasserstoff $N = 1$ ist, und das Verhältnis m/M zu klein ist, um sich bei diesem Vergleich bemerkbar zu machen. Für die Bahndimensionen in den stationären Zuständen bekommt man weiter nach (17) eine Reihe von Werten für die große Achse der Elektronenbahn, die mit wachsendem n stark ansteigen. Für den Zustand $n = 1$, wo die Energie am kleinsten ist, und der als *Normalzustand* des Atoms bezeichnet werden kann, bekommen wir einen Wert für die große Achse der Bahn (ca. 10^{-8} cm), der von derselben Größenordnung ist, wie die Werte der Atomdimensionen, die sich aus der Raumerfüllung der Stoffe ergeben.

Eine weitere Stütze dieser Deutung des Wasserstoffspektrums[1]) bekam man durch die Erkenntnis, daß einige Spektrallinien, die wegen der Verwandtschaft ihrer formelmäßigen Darstellung zur Balmerformel früher dem Wasserstoff zugeschrieben wurden, in Wirklichkeit dem *Spektrum des einfach ionisierten Heliumatoms* angehören. Ein solches Atom besteht ja nach der Theorie aus einem Kern mit der Ladung $2e$, um den sich ein einziges Elektron bewegt. Es zeigte sich nun, daß die erwähnten Linien durch die Formel (18) wiedergegeben werden konnten, wenn $N = 2$ gesetzt wird. Ja, es war sogar durch einen Vergleich mit den Wasserstofflinien möglich, die kleine von dem Massenverhältnis m/M herrührende Korrektion als vorhanden nachzuweisen.

Bisher haben sich Spektren eines ionisierten Atoms, das aus einem höher geladenen Kern und einem einzigen Elektron besteht, nicht nachweisen lassen. Eine Stütze für die allgemeine Richtigkeit der Formel (18) hat man aber aus den von Moseley entdeckten Gesetzen der *Röntgenspektren* bekommen. Hier handelt es sich zwar nicht um ein Atom, das nur ein Elektron enthält, sondern um einen Prozeß, bei dem die Bewegung eines der inneren Elektronen im Atom sich ändert; für große Werte der Kernladungszahl N kann hier jedoch in erster Näherung von der Anwesenheit der anderen Elektronen abgesehen werden. Dieser Umstand ist von größter Bedeutung, weil er zuläßt, aus den Röntgenspektren die Kernladungszahl eindeutig festzulegen.

Während wir uns bisher auf die Betrachtung einfach

1) Abh. über Atombau, S. 8, 70.

periodischer Systeme stützen konnten, gibt schon die Erklärung des *Einflusses äußerer elektrischer und magnetischer Felder auf das Wasserstoffspektrum* Anlaß, Bewegungen von höherem Periodizitätsgrad zu betrachten.[1]) So beruht die Aufspaltung der Wasserstofflinien in Anwesenheit äußerer Felder darauf, daß diese Felder Störungen der einfachen Keplerbewegung des ungestörten Atoms verursachen. Dabei treten neue der Intensität der äußeren Felder proportionale Grundschwingungszahlen in der Bewegung auf, und wir bekommen durch Anwendung der Zustandsbedingungen (4) eine Mannigfaltigkeit von stationären Zuständen, die außer durch die bei der rein periodischen Bewegung des ungestörten Atoms vorhandene Quantenzahl, die wir als „Hauptquantenzahl" bezeichnen können, durch neu hinzukommende „Nebenquantenzahlen" charakterisiert sind. Für jeden Wert der Hauptquantenzahl bekommen wir daher nicht mehr einen einzigen Energiewert, sondern eine Anzahl von Energiewerten, die verschiedenen Werten der Nebenquantenzahlen entsprechen, und deren Differenzen der Intensität der äußeren Felder proportional sind. Gemäß der Frequenzbedingung (1) wird also jede Linie des Spektrums in Anwesenheit des Feldes in mehrere Komponenten aufgelöst, deren Abstände ebenfalls der Feldintensität proportional sind. In dieser Weise ist es in der Tat möglich gewesen, eine Erklärung der beim Stark- und Zeemaneffekt der Wasserstofflinien beobachteten Aufspaltungsbilder zu erreichen. Diese Effekte bieten zugleich eine besonders lehrreiche Anwendungsmöglichkeit des Korrespondenzprinzips, indem dieses Prinzip eine direkte Erklärung der Beschränkung der Kombinationsmöglichkeiten der aufgespaltenen Terme sowie der charakteristischen Polarisation der Komponenten, ja sogar ein Verständnis der

1) Eine kurze Darstellung der Theorie der Einwirkung äußerer Felder auf das Wasserstoffspektrum, so wie sie sich im Lichte des Korrespondenzprinzips gestaltet, findet sich in einem Vortrag des Verfassers in der Deutschen Physikalischen Gesellschaft in Berlin, der in Zeitschr. f. Phys. **2**. S. 423. 1920 abgedruckt ist. Hinsichtlich des Anteiles der verschiedenen Forscher an der Entwicklung der Theorie sei auf diesen Vortrag verwiesen und auf einen demnächst in den Proceedings of the London physical Society erscheinenden Vortrag des Verfassers, wo eine mehr eingehende Behandlung mit ausführlichen Literaturangaben zu finden ist.

beobachteten Intensitätsverteilung im Aufspaltungsbild liefert. Die Deutung der in Rede stehenden Effekte, die man in dieser Weise erzielt, läßt einerseits die grundsätzliche Verschiedenheit, andererseits zugleich auch die formale Ähnlichkeit mit der Behandlungsweise des Problems auf Grund der klassischen Theorie erkennen. So weist die Deutung des Zeemaneffektes eine weitgehende Analogie auf mit der ursprünglichen Lorentzschen Theorie, deren grundsätzliche Bedeutung bekanntlich darin lag, daß sie in überzeugender Weise den Ursprung der Spektren auf Elektronenbewegungen zurückführte.

Eine Betrachtung der stationären Zustände des Wasserstoffatoms in äußeren Feldern gestattet weiter das statistische Gewicht der durch (16) angegebenen stationären Zustände des ungestörten Atoms zu bestimmen. Bei dieser Bestimmung ist jedoch zu beachten, daß mehrere der Zustände des nicht entarteten Systems, das unter dem Einfluß passend gewählter äußerer Kräfte hervorgebracht werden kann, mechanische Singularitäten aufweisen, die das Hineinfallen des Elektrons in den Kern mit sich bringen würden und deshalb auszuschließen sind. Die Untersuchung ergibt, daß das statistische Gewicht des n-ten Zustandes gleich $n(n+1)\,h^3$ ist.

Bis jetzt haben wir davon abgesehen, daß die Linien des Wasserstoffspektrums bei Beobachtung mit großer Dispersion die sogenannte *Feinstruktur* aufweisen. Diese beruht darauf, daß schon die Bewegung des ungestörten Wasserstoffatoms strenggenommen keine rein periodische ist. Wegen der kleinen Änderung der Masse des Elektrons mit seiner Geschwindigkeit ist nämlich die Elektronenbahn keine geschlossene Ellipse, sondern das Perihel vollführt eine langsame Drehung, die eine hinzukommende Grundschwingungszahl von sehr kleiner Größe und somit gemäß den Zustandbedingungen (4) das Auftreten einer neuen Quantenzahl in der Mannigfaltigkeit der stationären Zustände mit sich bringt. Ein charakteristischer Zug der Theorie der Feinstruktur besteht darin, daß die Exzentrizität der Ellipsenbahn durch die hinzukommende Quantenbedingung festgelegt ist. Es ist aber ein Mißverständnis, wenn angenommen wird, daß eine derartige Festlegung der Ellipsenbahnen schon für die Deutung der Balmerformel erforderlich ist. Die große Instabilität der beobachteten Feinstruktur gegenüber äußeren

[562]

Feldern, deren Intensität noch klein ist verglichen mit Feldintensitäten derjenigen Größenordnung, wie sie in Versuchen über den **Stark**- und **Zeeman**effekt zur Anwendung kommen, zeigt eben in deutlichster Weise, daß die in der Theorie der Feinstruktur infolge der Periheldrehung hinzukommende Quantenbedingung schon durch den Einfluß solcher kleiner Felder hinfällig wird. Andererseits kann man erst dann erwarten, daß die Gültigkeit der Balmerformel wesentlich beeinträchtigt wird, wenn die äußeren Kräfte solche Intensität besitzen, daß sie imstande sind, die annähernd periodische Keplerbewegung des Atoms stark zu verändern. Die Theorie der Feinstruktur, die nicht nur für die Wasserstofflinien Geltung hat, sondern auch beim damit analogen Spektrum des ionisierten Heliumatoms sowie bei den Röntgenspektren eine glänzende Bestätigung fand, rührt bekanntlich von **Sommerfeld**[1]) her und gab den Anstoß zur Weiterentwicklung der quantentheoretischen Deutung der Spektralgesetze durch Heranziehung von Systemen mit höherem Periodizitätsgrad.

§ 4. Allgemeine Spektralgesetze der Elemente höherer Atomnummer.

Bei den Spektren der Elemente mit höherer Atomnummer treffen wir kompliziertere Verhältnisse als bei Wasserstoff. Jedoch gelang es bekanntlich, für manche Spektren empirische Spektralgesetze aufzustellen, die eine eigentümliche Analogie mit der formalen Darstellung des Wasserstoffspektrums besitzen. So können die Schwingungszahlen der Linien dieser Spektren, wenn wir zunächst von der Komplexstruktur (Dubletts, Tripletts usw.) der einzelnen Linien absehen, mit großer Annäherung durch die Formel dargestellt werden:

$$\nu = \frac{K}{(n'' + \alpha_k'')^2} - \frac{K}{(n' + \alpha_k')^2}, \tag{20}$$

wo n' und n'' ganze Zahlen sind, während α_k' und α_k'' zwei Glieder einer Folge von Konstanten bezeichnen, die für das betrachtete Spektrum charakteristisch sind. K ist die sogenannte **Rydberg**sche Konstante, die für alle unter gewöhnlichen Entladungsbedingungen hervorgebrachten Spektren, die sogenannten Bogenspektren, dieselbe ist und mit der Konstante der Balmerformel zusammenfällt. Ebenso wie das Wasserstoffspektrum

1) A. **Sommerfeld**, Münch. Ber. 1915.

[563]

wird ein Spektrum mit einem derartigen Bau ein *Serienspektrum* genannt, da sich die Linien in Reihen, den sogenannten Spektralserien, anordnen lassen, innerhalb welcher die Schwingungszahlen gegen bestimmte Grenzwerte konvergieren. Eine solche Serie wird durch die Formel (20) dargestellt, wenn man, bei der Benutzung von zwei bestimmten Konstanten α_k'' und α_k' die Zahl n'' unverändert läßt, während n' eine Folge von wachsenden Werten annimmt.

Der mehr komplizierte Bau der in Rede stehenden Spektren, verglichen mit dem des Wasserstoffspektrums, rührt einfach davon her, daß wir es hier mit Atomen zu tun haben, die mehr als ein Elektron enthalten, und wir daher mehr komplizierte Bewegungsformen erwarten müssen. In Anlehnung an die Deutung des Wasserstoffspektrums bekommt man aber zugleich ein Verständnis für das allgemeine Auftreten der Rydbergschen Konstante auf Grund der Annahme, daß wir bei der Emission der Bogenspektren mit Übergängen zwischen stationären Zuständen zu tun haben, in denen sich ein Elektron jedenfalls auf dem größten Teil seiner Bahn in einem Abstand vom Kern bewegt, der groß ist verglichen mit dem Abstand der anderen Elektronen vom Kern. Die Kraft, die dieses Elektron vom Rest des Atoms erfährt, wird dann offenbar annähernd gleich sein der Anziehung des Elektrons vom Kern im Wasserstoffatom in entsprechenden Abständen.

Diese Deutung des Ursprungs der Bogenspektren[1]) bekam bald eine überzeugende Stütze durch den bedeutungsvollen Befund von Fowler[2]), daß die sogenannten Funkenspektren, die von den Elementen in besonders kräftigen elektrischen Entladungen emittiert werden, in manchen Fällen durch eine Formel dargestellt werden können, die sich nur dadurch von der Formel (20) für die Bogenspektren unterscheidet, daß die Konstante K durch $4K$ ersetzt wird. Dies entspricht genau der theoretischen Erwartung, wenn angenommen wird, daß die Funkenspektren nicht von neutralen, sondern von einfach ionisierten Atomen emittiert werden. In diesem Fall wird ja ein Elektron, das sich in größerem Abstand vom Kern als die übrigen bewegt, annähernd derselben Kraft unterworfen sein

1) Abh. über Atombau, S. 11, 115.

2) A. Fowler, Trans. Roy. Soc. (A) **214**. S. 225. 1914.

wie das Elektron im ionisierten Heliumatom. Kürzlich wurde diese Auffassung in sehr interessanter Weise durch die Entdeckung von Paschen[1]) ergänzt, wonach im Funkenspektrum des Aluminiums außer einem Serienspektrum der Fowlerschen Art ein Serienspektrum vorhanden ist, in dessen formelmäßiger Darstellung die Konstante K durch $9K$ ersetzt ist, und das demnach dem zweifach ionisierten Aluminiumatom zugeschrieben werden muß. Allgemein werden wir ein Serienspektrum, welches von einem Atom herrührt, das $p-1$ Elektronen verloren hat, während sich noch ein p-tes Elektron in größerem Abstand vom Kern bewegt als die übrigen Elektronen, ein *Serienspektrum p-ter Art* nennen.

Während die bisherigen Betrachtungen sich direkt auf die Deutung des einfachen Wasserstoffspektrums stützen, zeigt andererseits die größere Mannigfaltigkeit der Terme, die in den Spektren anderer Elemente auftreten, daß wir hier mit einer mehr verwickelten Bewegung des äußeren Elektrons als im Wasserstoffatom zu tun haben. Die Frage, die wir hier treffen, ist zunächst eine sehr schwierige, da das Problem der Bewegung in einem Atom mit mehreren Elektronen, wenn wir uns auf den Boden der gewöhnlichen Mechanik stellen, auf außerordentlich verwickelte Lösungen führt. Insbesondere muß hervorgehoben werden, daß die Bahnkonfiguration der inneren Elektronen im Atom sich nach den klassischen mechanischen Gesetzen als unstabil gegen die störenden Einflüsse des äußeren Elektrons erweisen würde. Nichtsdestoweniger ist es möglich gewesen, jedenfalls in formaler Hinsicht, eine eingehende Deutung der Gesetzmäßigkeiten der Serienspektren zu erzielen mittels der Einführung von einfachen Annahmen über den Charakter der Bewegung des äußeren Elektrons. So ist es, wenn wir von der Komplexstruktur absehen, gelungen, die Spektralterme zu klassifizieren und die Kombinationsregeln zu deuten durch die Annahme, daß die Abweichung dieser Bewegung von einer Keplerbahn in erster Annäherung in einer Präzession der großen Achse in der Bahnebene besteht. In der Tat führt, wie Sommerfeld[2]) zuerst nachwies, diese An-

1) F. Paschen, vgl. S. 142. Für die Kenntnis dieser Arbeit bin ich Hrn. Paschen zu Dank verpflichtet.

2) A. Sommerfeld, Münch. Ber. 1916.

nahme, nach der die Bewegung des Elektrons einer Zentralbewegung sehr ähnlich ist, zu einer Mannigfaltigkeit stationärer Zustände, die von derselben Art ist wie die Mannigfaltigkeit der in den Serienspektren auftretenden Terme. Zugleich liefert sie unter Heranziehung des Korrespondenzprinzips eine einfache Erklärung für die eigentümliche, für die Serienspektren charakteristische Begrenzung der Kombinationsmöglichkeiten dieser Terme.[1])

Eine Zentralbewegung kann immer aufgefaßt werden als eine rein periodische ebene Bewegung, der eine gleichförmige Rotation in ihrer Bahnebene überlagert ist. Durch diese überlagerte Rotation wird jede in der periodischen Bewegung anwesende elliptische harmonische Schwingungskomponente in zwei zirkulare harmonische Rotationen aufgelöst, deren Schwingungszahlen gleich sind der ursprünglichen Schwingungszahl, vermehrt oder vermindert um die Umlaufszahl der Rotation. Bezeichnen wir die Umlaufszahl der periodischen Bahn mit ω und diejenige der überlagerten Rotation mit σ, so kann also das elektrische Moment des Atoms, soweit es vom äußeren Elektron abhängt, durch die Formel dargestellt werden

$$(21) \qquad \xi = \sum_{\tau} C_{\tau, \pm 1} \cos\left[2\pi(\tau\omega \pm \sigma)t + \gamma_{\tau, \pm 1}\right]$$

Gemäß (4) werden die zugehörigen stationären Zustände durch zwei Bedingungen charakterisiert, die wir schreiben können

$$(22) \qquad J = nh, \qquad P = kh,$$

wo n und k ganze Zahlen sind. Während die erste dieser Bedingungen im Grenzfall, wo die Zentralbewegung in eine einfache Keplerbewegung übergeht, mit der Bedingung (16) zusammenfällt, welche die stationären Zustände des einfachen Wasserstoffatoms festlegt, kann die zweite Bedingung als eine infolge der Rotation der Bahn hinzukommende Quantenbedingung angesehen werden. Die in dieser Bedingung auftretende Größe P ist einfach gleich dem mit 2π multiplizierten Impulsmoment des Elektrons in seiner Bahn. Andererseits ist die

1) Eine kurze Darstellung der Theorie der Serienspektren auf Grund des Korrespondenzprinzips ist in dem S. 239 erwähnten Vortrag des Verfassers (Zeitschr. f. Phys. 2. 423. 1920) gegeben. Eine ausführlichere Diskussion mit näheren Literaturhinweisen findet sich im dritten Teil der S. 235 zitierten Arbeit „Über die Quantentheorie der Linienspektren".

kinematische Bedeutung der Größe J für eine beliebige Zentralbahn durch die Gleichung

$$R = J - P = \int p_r \, dr = \int m \left(\frac{dr}{dt}\right)^2 dt \tag{23}$$

festgelegt, wo r der Radiusvektor des Elektrons und $p_r = m \frac{dr}{dt}$ der radiale Impuls ist, und wo das Integral über eine vollständige Periode der radialen Bewegung zu nehmen ist. Der Ausdruck auf der rechten Seite der Gleichung (23) ist derselbe, der in die bekannten Wilson-Sommerfeldschen Bedingungen für die stationären Zustände einer Zentralbewegung eingeht. Die hier benutzte Form der Bedingungen (22) ist durch die Wahl der Grundfrequenzen ω und σ geboten, indem entsprechend (5) der Energieunterschied zwischen zwei benachbarten Bewegungszuständen durch die Gleichung

$$\delta E = \omega \, \delta J + \sigma \, \delta P \tag{24}$$

gegeben ist. Aus dieser Gleichung erhellt, daß die Energie, falls σ klein gegenüber ω ist, in erster Linie nur von J abhängt. Wir werden deshalb, wie beim Wasserstoff, n immer als die Hauptquantenzahl und k als Nebenquantenzahl bezeichnen. In dem genannten Grenzfall, wo die Abweichung von einer Keplerbewegung nur sehr gering ist, werden ferner die Bahndimensionen angenähert durch die Gleichungen

$$2a = \frac{n^2 h^2}{2 \pi^2 N e^2 m}, \quad 2p = \frac{k^2 h^2}{2 \pi^2 N e^2 m} \tag{25}$$

bestimmt, wo $2a$ die große Achse und $2p$ den Parameter (die kleinste Sehne durch den Brennpunkt) der nahezu elliptischen rotierenden Bahn bezeichnen.

Aus der Relation (23) folgt allgemein, daß P niemals größer als J sein kann. Für jeden Wert n der Hauptquantenzahl sollen wir daher Spektralterme erwarten, die den Werten $k = 1, 2 \ldots n$ entsprechen. Aus dem Korrespondenzprinzip folgt weiter, daß ein Übergang von einem Zustande n', k' nach einem Zustande n'', k'' durch die Anwesenheit einer harmonischen Schwingungskomponente der Schwingungszahl

$$(n' - n'')\,\omega + (k' - k'')\,\sigma$$

in der Bewegung bedingt ist. Gemäß (21) bedeutet diese Zuordnung aber, daß nur solche Übergänge stattfinden können, bei denen

$$k' - k'' = 1\,. \tag{26}$$

Dieses Resultat gibt nicht nur eine Erklärung der bei den Serienspektren beobachteten eigentümlichen Begrenzung der Kombinationsmöglichkeiten der Spektralterme, sondern liefert zugleich eine überzeugende Stütze für die von Sommerfeld vorgeschlagene Zuordnung der möglichen k-Werte zu den empirischen Termreihen. So müssen wir den s-Termen den Wert $k = 1$, den p-Termen $k = 2$, den d-Termen $k = 3$, den f-Termen $k = 4$ zuordnen und höhere k-Werte den Termreihen, die besonders in den Funkenspektren auftreten, und die sich sehr nahe den Termreihen des Wasserstoffs anschließen. Während die Festlegung der Hauptquantenzahl n bei solchen Termreihen offenbar zu keinen Schwierigkeiten Anlaß gibt, hat die Frage nach der Festlegung dieser Quantenzahl bei den Termreihen, die wesentlich von der Wasserstoffreihe abweichen, neue Verhältnisse an den Tag gebracht, auf die wir in den zwei letzten Paragraphen zurückkommen werden.

Die erwähnte, durch Gleichung (26) bedingte Begrenzung der Kombinationsmöglichkeiten hat zur Folge, daß im Gegensatz zu dem, was beim einfachen Wasserstoffspektrum gilt, bei den Serienspektren der anderen Elemente eigentliche Kombinationslinien nicht zu erwarten sind in dem Sinne, daß keine Linie auftritt, deren Schwingungszahl als die Summe oder Differenz der Schwingungszahlen zweier anderer Linien des Spektrums dargestellt werden kann. Bekanntlich ist aber eine große Zahl von derartigen Linien in der Tat beobachtet, was seinerzeit als eine der stärksten Stützen des Kombinationsprinzips angesehen wurde. Der Ursprung solcher Linien muß aber, wie schon die eigentümliche Abhängigkeit ihres Auftretens von den Anregungsbedingungen andeutet, auf die Anwesenheit von störenden äußeren elektrischen Feldern in der Entladung zurückgeführt werden. Durch den Einfluß solcher Kräfte wird nämlich die einfache Zentralbewegung des Serienelektrons in solcher Weise gestört, daß außer den durch (21) angegebenen harmonischen Schwingungskomponenten auch Komponenten mit den Schwingungszahlen $\tau\omega$ und $\tau\omega \pm 2\sigma$ auftreten, deren Amplituden proportional den äußeren Kräften sind. Dies bringt mit sich, daß im gestörten Atom außer den Übergängen, für die k sich nur um eine Einheit ändert, Übergänge zu erwarten sind, bei denen k unverändert bleibt, oder sich um zwei Einheiten ändert, und die formal den Summen- und Differenztönen in der Akustik analog sind.

Das Auftreten solcher „neuer" Linien ist insbesondere das wesentliche Ergebnis der Versuche über den Starkeffekt der Serienspektren, wo es sich um die *Einwirkung homogener äußerer elektrischer Felder* bekannter Intensität handelt. In vollständiger Übereinstimmung mit der Theorie zeigt es sich, daß die Intensität dieser Linien bei gleicher äußerer Feldstärke um so größer ist, je weniger die betreffenden Kombinationsterme von den Wasserstofftermen mit derselben Hauptquantenzahl abweichen. Diese Abweichung gibt nämlich gemäß (24) ein unmittelbares Maß für die Größe der Umlaufszahl σ und also für die Abweichung der Bahn von einer rein periodischen und bedingt ja überhaupt den grundsätzlichen Unterschied zwischen dem Starkeffekt des Wasserstoffspektrums und dem der anderen Spektren. Insbesondere treten bei den letzteren keine neue Grundschwingungszahlen in der Bewegung auf, deren Größe der Intensität der äußeren Kräfte proportional ist, und wir bekommen deshalb nur Aufspaltungen der Linien, bei denen der Komponentenabstand mit einer höheren Potenz der Feldstärke als der ersten anwächst.

Im Gegensatz hierzu ist die *Wirkung eines homogenen magnetischen Feldes* auf die Serienspektren, soweit wir von der Komplexstruktur absehen, ganz analog der Wirkung auf das Wasserstoffspektrum. Nach einem Theorem von Larmor ist bekanntlich nach der klassischen Elektrodynamik zu erwarten, daß die Änderung der Elektronenbewegung eines Kernatoms in einem äußeren Magnetfeld einfach in einer Überlagerung einer gleichförmigen Rotation besteht, deren Umlaufszahl durch die Gleichung

$$\varrho_{\mathrm{H}} = \frac{e\,H}{4\,\pi\,m\,c} \tag{27}$$

dargestellt wird, wo H die Intensität des Magnetfeldes, e und m wie oben die Ladung und Masse des Elektrons und c die Lichtgeschwindigkeit bedeutet. Durch eine Betrachtung des Einflusses dieser Rotation auf die Auflösung der Elektronenbewegung in harmonische Komponenten läßt sich unter Heranziehung des Korrespondenzprinzips einfach schließen, daß jede Linie voraussichtlich in ein normales Lorentzsches Triplett aufgespalten wird. Diese Erwartung ist bekanntlich erfüllt, solange wir es mit Spektren zu tun haben, deren Serienlinien keine Komplexstruktur aufweisen.

§ 5. Komplexstruktur der Serienlinien.

Bisher haben wir ganz vom feineren Bau abgesehen, den die Linien vieler Serienspektren zeigen. Eine formale quantentheoretische Analyse dieser „Komplexstrukturen" mittels Einführung einer dritten Quantenzahl, die als innere Quantenzahl bezeichnet wurde, ist zuerst von Sommerfeld[1]) versucht worden. Um ein näheres Verständnis der Verhältnisse bei der Komplexstruktur zu erzielen, besonders was die empirischen Kombinationsregeln betrifft, die hier Geltung haben, müssen wir in Anlehnung an das Korrespondenzprinzip vor allem nach der Art der Elektronenbewegung fragen. Wie im vorigen Paragraph dargelegt ist, zeigt der charakteristische Bau der Spektren, daß die Bewegung des Serienelektrons in erster Annäherung eine Zentralbewegung ist. Die Komplexstruktur zeigt aber, daß in der Bewegung strenggenommen mehr als zwei Grundschwingungszahlen auftreten, was zur Folge hat, daß die Mannigfaltigkeit der stationären Zustände mit mehr als zwei Quantenzahlen klassifiziert werden muß. Es liegt nun nahe zu versuchen, bei der Beschreibung der Elektronenbewegung wenigstens eine axiale Symmetrie beizubehalten, was dazu führt, anzunehmen, daß die Abweichung der Bewegung von einer Zentralbahn in erster Näherung in der Überlagerung einer *gleichförmigen Präzession der Bahnebene* besteht. Durch diese Präzession wird die Auflösung der Elektronenbewegung in ihre harmonischen Komponenten in der Weise verändert, daß an Stelle jeder früheren Komponente nunmehr drei Komponenten treten, von welchen die eine in einer linearen Schwingung parallel der Präzessionsachse mit unveränderter Schwingungszahl besteht, während die zwei anderen zirkulare Rotationen in einer Ebene senkrecht zu dieser Achse sind, deren Schwingungszahlen gleich sind der ursprünglichen Schwingungszahl, vermehrt oder vermindert um die Präzessionsgeschwindigkeit. Nennen wir diese ϱ, so bekommen wir also für die Komponenten des elektrischen

1) A. Sommerfeld, Ann. d. Phys. 63. S. 221. 1920.

[570]

Momentes, parallel und senkrecht zur Präzessionsachse,

$$\xi = \sum_\tau C_{\tau,\,\pm 1} \cos[2\pi(\tau\omega \pm \sigma)t + \gamma_{\tau,\,\pm 1}] \tag{28a}$$

und

$$\eta = \sum_\tau C_{\tau,\,\pm 1,\,\pm 1} \cos[2\pi(\tau\omega \pm \sigma \pm \varrho)t + \gamma_{\tau,\,\pm 1,\,\pm 1}]. \tag{28b}$$

Die drei Zustandsbedingungen, die gemäß (4) die stationären Zustände einer derartigen Bewegung festlegen, können wir nun schreiben

$$J = nh; \qquad P = kh; \qquad Q = jh, \tag{29}$$

wo n, k und j ganze Zahlen sind. Während J und P Größen sind, die im Grenzfall der reinen Zentralbewegung mit den in (22) auftretenden Größen übereinstimmen, dürfen wir die in der hinzukommenden Quantenbedingung auftretende Größe Q als das mit 2π multiplizierte gesamte resultierende Impulsmoment des Atoms ansehen. In der Tat erfüllen diese Größen die zu (5) und (24) analoge Gleichung

$$\delta E = \omega\delta J + \sigma\delta P + \varrho\delta Q \tag{30}$$

für die Energiedifferenz zweier benachbarter Bewegungen eines Systems, das aus einem Elektron und einem starren mit einem Drehimpuls behafteten Körper besteht, welch letzterer ein um die Impulsachse symmetrisches Kraftfeld erzeugt.

Während wir im vorigen Paragraph einen engen Vergleich mit der Quantentheorie der Periodizitätssysteme erreichen konnten, indem wir jedenfalls formal uns auf die Bewegung des äußeren Elektrons beschränken konnten, ist die Sachlage im jetzigen Fall etwas verwickelter und gibt Anlaß zu einer grundsätzlichen Unsicherheit, was die Deutung der hinzukommenden Quantenbedingung betrifft, schon weil wir mechanisch nicht streng begründen können, daß das Impulsmoment des Atomrestes während der Bewegung des äußeren Elektrons konstant verbleibt. Zunächst nehmen wir aber an, daß die Bewegung des Atoms, soweit es das äußere Elektron betrifft, streng mit der des erwähnten vielfach periodischen Systems verglichen werden kann, und schreiben für das Impulsmoment des Atomrestes $ih/2\pi$, wo i ebenso wie j als eine ganze Zahl angesehen wird. Die letztere Quantenzahl, die die relative Orientierung der äußeren Bahn zum Atomreste festlegt, muß offenbar der Bedingung

(31) $$|i-k| \leqq j \leqq i+k$$
genügen.

Das mit Hilfe der Zustandbedingung (29) und der Frequenzbedingung (1) ermittelte Komplexstrukturbild steht, wie aus (30) erhellt, in innigem Zusammenhang mit der Präzessionsbewegung der Bahn des äußeren Elektrons. Weiter ist nach dem Korrespondenzprinzip, gemäß (29) und (30), das Vorkommen eines Übergangs von einem Zustand n', k', j', nach einem Zustand n'', k'', j'', durch das Auftreten einer harmonischen Komponente der Schwingungszahl $(n''-n')\omega+$
* $(k''-k')\sigma+(j''-j')\varrho$ bedingt. Gemäß (27) und (28) bedeutet dies, daß außer der Bedingung (26), die möglichen Übergänge durch die weitere Bedingung

(32) $$j'-j''=0 \quad \text{oder} \quad j'-j''=\pm 1$$
begrenzt sind.

Dieses Resultat bietet einen Ausgangspunkt dar für ein Verständnis der schon von Rydberg hervorgehobenen charakteristischen Begrenzung der Kombinationsmöglichkeiten zwischen den Spektraltermen, die in den beobachteten Strukturbildern zutage tritt. Zugleich liefert es zusammen mit der Beziehung (31) auch in gewissem Umfang ein Hilfsmittel für die Festlegung der Werte der Quantenzahlen j und i. Im Gegensatz zu dem, was für die Quantenzahl k gilt, begegnen wir jedoch hier tiefliegenden Schwierigkeiten, sobald wir versuchen die beobachteten Komplexstrukturbilder zu deuten. Hierzu kommt, daß diese Strukturbilder oft einen höheren Grad von Komplexität aufweisen als nach der hier besprochenen einfachen Theorie zu erwarten wäre. In der empirischen spektroskopischen Klassifikation kommt diese höhere Komplexität darin zum Ausdruck, daß in manchen Fällen zwischen *verschiedenen Arten von Spektraltermen* unterschieden werden muß, die z. B. bei den Spektren der Erdalkalien als Einfach- und Tripletterme bezeichnet werden. Gemäß den obigen Kombinationsregeln wird man dazu geführt, in diesen Fällen mehreren Termen dieselben Werte der Quantenzahlen n, k und j zuzuschreiben. Wenn wir an den einfachen Annahmen über den Charakter der Bewegung des äußeren Elektrons festhalten, kommen wir daher zu der Vorstellung, daß wir es bei den verschiedenen Arten der Terme mit verschiedenen Konfigurationen der Elektronenbahnen im Atom-

* [Instead of (27) und (28) read (28a) und (28b).]

reste zu tun haben, die verschiedenen Werten ihres Gesamtimpulses $ih/2\pi$ entsprechen. Diese Annahme ist auch dadurch gestützt, daß Übergänge, die Kombinationen zwischen verschiedenen Termarten entsprechen, im allgemeinen viel weniger wahrscheinlich sind als Übergänge, die Kombinationen zwischen gleichartigen Termen entsprechen. Wir müssen nämlich bedenken, daß ein Übergang, bei dem eine Konfigurationsänderung der Elektronenbahnen des Atomrestes stattfindet, nicht mit einer harmonischen Komponente der Hauptschwingungen korrespondiert, die durch (27), (28) gegeben sind, sondern mit einer harmonischen Komponente kleinerer Amplitude, die durch das feinere Wechselspiel des äußeren Elektrons und der Elektronen des Atomrestes bedingt ist. Abgesehen von der Art dieses Wechselspiels müssen wir an der Kombinationsregel, welche die Änderung der Quantenzahl j bei den möglichen Übergängen begrenzt, auch bei den Übergängen zwischen ungleichartigen Termen festhalten. Korrespondenzmäßig folgt dies schon aus der allgemeinen Achsensymmetrie, die für die Elektronenbewegung in einem Kernatom zu erwarten ist. Es ist von großer prinzipieller Bedeutung, daß diese Regel auch aus einer Betrachtung der Erhaltung des Drehimpulses bei den Strahlungsprozessen gefolgert werden kann.[1]) *

Was die Wirkung äußerer Felder auf die Komplexstruktur der Serienlinien betrifft, so stoßen wir hier auf sehr interessante Fragen, die noch ihrer endgültigen Lösung harren. Bekanntlich treffen wir bei der Wirkung *magnetischer Felder* auf Serienlinien mit einem komplizierten Bau auf die reizvolle Erscheinung der *anomalen Zeemaneffekte.* Diese muß gemäß dem Korrespondenzprinzip auf ein Versagen des früher erwähnten Larmorschen Theorems zurückgeführt werden. Aus der von Paschen und Back[2]) entdeckten Erscheinung der allmählichen Umwandlung des Aufspaltungsbildes aller Komplexstrukturkompo-

1) Die Beziehung einer solchen Betrachtung — die unabhängig von Rubinowicz und vom Verfasser vorgeschlagen wurde, — zu den Prinzipien der Quantentheorie ist näher diskutiert in einer Abhandlung des Verfassers (Zeitschr. f. Phys. **6.** S. 1. 1921), wo auch Literaturhinweise gegeben sind.

2) F. Paschen und E. Back, Ann. d. Phys. **39.** S. 897. 1912 und Physica **1.** S. 261. 1921.

* [Instead of (27), (28) read (28a) und (28b).]

nenten einer und derselben Serienlinie in ein normales Lorentzsches Triplett bei wachsender Feldintensität können wir insbesondere entnehmen, daß das Magnetfeld außer seiner Wirkung auf die Bewegung des Atoms als Ganzes auch wesentlich in das Wechselspiel der inneren und äußeren Elektronen eingreift, auf dem die Komplexstruktur beruht.[1]) Dies kommt auch klar darin zum Ausdruck, daß das Magnetfeld neue Linienkomponenten der Komplexstruktur hervorruft, die im feldlosen Bild nicht vorhanden sind. Diese neuen Komponenten sind in der Tat vollständig anolog den neuen Serien, die, wie oben erwähnt, im elektrischen Felde auftreten. Im Falle der Einwirkung eines Magnetfeldes können wir aber nicht mit Hilfe der klassischen Elektrodynamik in Einzelheiten den Einfluß der äußeren Kräfte auf die Elektronenbewegung berechnen. Vielmehr lehrt uns das Versagen des Larmorschen Theorems, daß wir hier vor ganz neuartigen Erscheinungen stehen, bei denen wir hoffen müssen, daß das empirische spektroskopische Material uns Hinweise geben wird, welche Gesetze hier an Stelle derjenigen der klassischen Theorie treten.

In der Tat hat die quantentheoretische Analyse des Zeemaneffektes schon Gesetzmäßigkeiten zutage gebracht, die eine einfache formale Deutung zulassen. So hat Landé[2]) nachweisen können, daß das Aufspaltungsbild bei schwachen Feldern mittels der Frequenzbedingung wiedergegeben werden kann, wenn man annimmt, daß jeder Spektralterm unter Einwirkung des Feldes in einer Reihe von äquidistanten Termwerten aufgespalten wird. Auf Grund des Korrespondenzprinzips läßt sich dieses Resultat so deuten, daß die Änderung der Bewegung des Atoms als Ganzes bei schwachen Feldern in einer überlagerten Präzessionsbewegung um eine durch den Kern gelegte Achse parallel dem Feld mit einer von der Orientierung der Atoms relativ zum Feld unabhängigen Umlaufszahl besteht. Diese Deutung wird auch dadurch gestützt, daß Sommerfeld und Heisenberg[3]) mittels einer Unter-

1) Vgl. N. Bohr, Zeitschr. f. Phys. **9.** S. 27. 1922.

2) A. Landé, Zeitschr. f. Phys. **5.** S. 231. 1921.

3) A. Sommerfeld u. W. Heisenberg, Zeitschr. f. Phys. **11.** S. 131. 1922.

suchung der harmonischen Komponenten, die nach dieser Auffassung in der Bewegung des Atoms im Felde auftreten sollten, gezeigt haben, daß man unter Zugrundelegung des Korrespondenzprinzips ein allgemeines Verständnis bekommen kann von den relativen Intensitäten der beobachteten Komponenten des anomalen Zeemaneffektes. Im Gegensatz zu dem, was nach der klassischen Theorie zu erwarten wäre, sind aber die von Landé ermittelten Werte für die Umlaufszahl der überlagerten Präzession nicht durch die Gleichung (27) gegeben, sondern sind für die verschiedenen stationären Zustände des ungestörten Atoms charakteristisch. Dies bedeutet, daß die Verbindung zwischen magnetischem Moment und Impulsmoment des Atoms eine kompliziertere ist als die von klassischer Theorie geforderte. Ferner ergibt die Landésche Analyse, daß der Wert der Komponente des Impulsmomentes des Atoms parallel zum Feld durch $h/2\pi$ dividiert nicht immer gleich einer ganzen Zahl sein dürfte, wie es bei der Deutung des Zeemaneffektes der Wasserstofflinien angenommen wird. Während dies nur bei Spektren mit gewissen Typen der Komplexstruktur gelten sollte, gelangt Landé für andere Typen tatsächlich zu Werten dieser Größen, die sich um $^1/_2$ von einer ganzen Zahl unterscheiden. Dieser Befund hat die Berechtigung, die Bewegung in den stationären Zuständen vom Atom mit mehreren Elektronen in unmittelbarer Anlehnung an die Quantentheorie der Periodizitätssysteme festzulegen, in Frage gestellt. Auf die Änderungen, die die im obigen beschriebene einfache Deutung der Komplexstruktur bei näherer Betrachtung erleiden dürfte, werden wir im letzten Paragraphen zurückkommen.

Die Wirkung äußerer *elektrischer Felder* auf die Komplexstruktur der Serienlinien ist von wesentlich anderer Art als die Wirkung magnetischer Felder. Aus dem angenäherten Zentralbahncharakter der Bewegung des Serienelektrons folgt nämlich, wie schon oben erwähnt, daß die Bewegung in einem äußeren homogenen elektrischen Feld keine durch neue, der Feldintensität proportionale Grundschwingungszahlen charakterisierten harmonischen Komponenten enthalten wird; insbesondere wird keine Drehung der Bahnebene mit einer dem Feld proportionalen Präzessionsgeschwindigkeit auftreten. Das elektrische Feld bekommt daher keine Möglichkeit, direkt in den der

Komplexstruktur zugrunde liegenden Mechanismus einzugreifen. Dies bietet eine einfache Erklärung dar für die Erfahrung, daß selbst enge Dubletts und Tripletts nicht von einem elektrischen Feld beeinflußt werden, auch wenn das elektrische Feld eine Intensität hat, die zu einer Aufspaltung der Wasserstofflinien Anlaß geben würde, die groß ist gegenüber dem Komponentenabstand der betreffenden Strukturen. Dies steht alles im klarsten Gegensatz zum Verhalten der Komplexstruktur in äußeren Magnetfeldern. Überhaupt kann man sagen, daß, während Magnetfelder nur die Komplexstruktur angreifen und keinen Einfluß haben auf den allgemeinen Bau der Serienspektren, elektrische Felder erst dann die Komplexstruktur beeinflussen, wenn sie imstande sind, den allgemeinen Zentralbahncharakter der Bewegung zu stören. Von elektrischen Feldern können wir also nur einen Einfluß auf die Linienstruktur erwarten, wenn einer der die Linien bedingenden Spektralterme sehr wenig vom entsprechenden Term des Wasserstoffspektrums abweicht.

§ 6. Näherer Zusammenhang zwischen Serienspektren und Atombau.

Wenn wir versuchen auf Grund der im vorhergehenden besprochenen quantentheoretischen Deutung der allgemeinen empirischen Spektralgesetze nähere Vorstellungen über das Zustandekommen der beobachteten *Spektren der einzelnen Elemente*[1]) zu entwickeln, begegnen wir einem Problem, dessen Lösung noch immer grundsätzliche Schwierigkeiten entgegenstehen. Im Gegensatze zum Falle, wo das Atom nur ein

1) In Kaysers Handbuch ist die in mehreren Beziehungen immer noch vollständigste Übersicht über das spektroskopische Material zu finden. Außerdem sind in letzter Zeit sowohl von Fowler (Report on Series in line spectra, Physical Society of London, 1922) wie von Paschen u. Götze (Seriengesetze der Linienspektren, Berlin 1922) zusammenfallende Darstellungen veröffentlicht, in denen besonders Rücksicht auf die Gesetzmäßigkeiten der Serienspektren genommen sind, und die für die theoretische Diskussion deshalb sehr wertvoll sind, weil die Spektralterme selber angegeben sind. Das im folgenden benutzte Material ist aus diesen Büchern entnommen. Um den Lesern die Übersicht zu erleichtern, ist in den Tafeln am Schluß dieser Abhandlung von einer Anzahl von Spektren eine schematische graphische Darstellung gegeben, die sich den theoretischen Erörterungen anschließt.

Elektron besitzt, führt, wie erwähnt, die Anwendung der mechanischen Gesetze auf die Bewegung in Atomen mit mehreren Elektronen im allgemeinen auf Bewegungsformen, welche nicht die Periodizitätseigenschaften aufweisen, die für eine Klassifikation von stationären Zuständen mittels Quantenzahlen erforderlich scheinen. Bei dieser Sachlage scheint zunächst kein anderer Ausweg vorhanden zu sein als anzunehmen, daß die klassischen mechanischen Gesetze nicht nur bei der Erklärung der Stabilität der Atome versagen, sondern schon, wenn es sich darum handelt, die Bewegung in Atomen mit mehreren Elektronen in feineren Einzelheiten zu beschreiben. Die empirischen Spektralgesetze, die im Lichte des Korrespondenzprinzips uns über die allgemeine Art der Bewegung in den stationären Zuständen, die für das Spektrum maßgebend sind, Aufschluß geben, bieten uns jedoch in Verbindung mit den allgemeinen gesicherten Zügen des Atombaus einen Ausgangspunkt für die Beseitigung dieser Schwierigkeit. So legen diese Gesetze, sowie die zentrale Symmetrie des Kernatoms vor allem die Annahme nahe, daß die Bahn eines jeden Elektrons im Atom in erster Annäherung als eine Zentralbahn angesehen werden kann, die sich in der beschriebenen Weise durch zwei Quantenzahlen n und k charakterisieren läßt. Im folgenden wird eine solche Bahn kurz als eine n_k-Bahn bezeichnet. Die nähere Deutung eines jeden Spektrums ist aber andererseits untrennbar verknüpft mit der Entwicklung von detaillierten Vorstellungen über den Bau der betreffenden Atome, besonders was die Festlegung der Werte der Quantenzahlen n und k für die Bahn jedes Elektrons im Normalzustand des Atoms sowie auch die Festlegung der Werte der weiteren Quantenzahlen betrifft, welche die relative Orientierung dieser Bahnen bestimmen. Streng genommen läßt die Bedeutung dieser Quantenzahlen sich nicht in Anlehnung an die Theorie der Periodizitätssysteme, auf Grundlage der klassischen mechanischen Gesetze genau feststellen. Die allgemeine Stabilität des Atomgebäudes führt vielmehr zu der Anschauung, daß es in jedem Atom mit mehreren Elektronen Eigenschaften sowohl der Bewegungen der einzelnen Elektronen wie des Wechselspiels zwischen diesen gibt, die einen mechanisch unerklärbar invarianten Charakter besitzen, und die eben durch die Quantenzahlen einen sinngemäßen Ausdruck bekommen. Diese Anschauung kann als ein formales *Postulat der Invarianz und Permanenz der Quantenzahlen* bezeichnet werden.[1])

Diese Betrachtungsweise tritt besonders in den Vordergrund, wenn wir versuchen zu einem Bilde des Atombaus zu gelangen durch *Verfolgung des Aufbauprozesses der Atome durch sukzessives Einfangen der Elektronen.* Über den Verlauf der Stufen dieses Prozesses geben uns ja eben die Serienspektren Auskunft. So können wir das Wasserstoffspektrum als Zeugnis eines Prozesses betrachten, bei dem das Atom gebildet wird durch Bindung des Elektrons durch den Kern unter Emission von Strahlung; und allgemein erfahren wir aus einem Spektrum, das durch die Formel (18) gegeben ist, den Verlauf der ersten Stufe des Aufbaues eines Atoms mit der Kernladungszahl N, die in der Einfangung des ersten Elektrons besteht. Nach den obigen Betrachtungen über den Ursprung der Serienspektren erfahren wir in analoger Weise aus dem Bogenspektrum eines Elementes den Verlauf eines Bindungsprozesses, bei dem der Aufbau des neutralen Atoms unter Einfangung des letzten Elektrons beendet wird. Im allgemeinen erblicken wir im Serienspektrum p-ter Art eines Elementes den Verlauf der p-letzten Stufe dieses Aufbaues. Um die Gesetzmäßigkeiten der Spektren für diesen Zweck zu verwerten, nehmen wir nach dem oben genannten Postulat an, daß während eines Einfangungsprozesses die Quantenzahlen, die die Bewegung jedes einzelnen der früher gebundenen Elektronen im Normalzustand der Aufbaustufen bezeichnen, unter dem störenden Einfluß des hinzukommenden Elektrons unverändert bleiben. Durch die Aufnahme weiterer Elektronen können höchstens diejenigen Quantenzahlen sich ändern, die die relative Orientierung der Bahnen solcher Elektronen betreffen, die, was die Werte von n und k betrifft, mit dem hinzukommenden Elektron äquivalent sind.

Unsere bisherige Kenntnis der Serienspektren höherer Ordnung ist zwar bei den meisten Elementen zu unvollständig, um den Aufbau in allen Stufen verfolgen zu können. Diese Lücke läßt sich aber in bedeutendem Umfange ausfüllen durch

1) Vgl. N. Bohr, Zeitschr. f. Phys. **13**. S. 135. 1923.

[578]

die Betrachtung einerseits der mechanischen Gleichartigkeit des Problems der Elektronenbewegung in Atomen mit verschiedener Kernladungszahl, aber gleicher Anzahl von Elektronen, andererseits der allmählichen Änderung der relativen Größe der verschiedenen Quantenbahnen, die durch das Anwachsen der Kernladungszahl hervorgebracht wird. Der letzterwähnte Umstand betrifft besonders die durch die Hauptquantenzahl n charakterisierten Eigenschaften der Bahn. Hier treffen wir in den verschiedenen Gebieten des Atoms auf wesentlich verschiedene Verhältnisse. Im innersten Gebiet, wo die Wechselwirkung zwischen den verschiedenen Elektronen nur gering ist verglichen mit der Anziehung durch den Kern, wird die Elektronbahn nahezu einer Keplerbewegung entsprechen, und die Dimensionen der Bahn sowie die Energie, die erforderlich ist, um das Elektron vom Atom zu entfernen, werden annähernd gleich sein den Dimensionen und dem Energiewert eines stationären Zustandes derselben Hauptquantenzahl von einem Atom mit derselben Kernladung, das nur ein einziges Elektron enthält. In äußeren Teilen des Atoms dagegen wird die Anziehung des Kernes zum großen Teil durch die Abstoßung der inneren Elektronen aufgehoben. Besonders gilt dieses für die Bahnen in den stationären Zuständen, die den letzten Stufen des Aufbaues des neutralen Atoms angehören. Hier begegnen wir zwei wesentlich verschiedenen Fällen, je nach dem das äußere Elektron während seiner ganzen Bewegung in größerem Abstand vom Kern wie die inneren Elektronen bleibt oder während eines Teiles seines Umlaufs in die Gebiete der früher gebundenen Elektronenbahnen eindringt.

Im ersten Fall wird sich die Bewegung, wenn es sich um die p-letzte Stufe des Bindungsprozesses handelt, nur wenig von einer Keplerbewegung unterscheiden, die sich im Felde einer punktförmigen Ladung der Größe $p\,e$ bewegt und die Bindungsenergie wird annähernd gleich sein derjenigen eines stationären Zustandes einer solchen Keplerbewegung mit derselben Hauptquantenzahl. In einem solchen Fall werden wir von einer *Bahn erster Art* sprechen.

Im zweiten Fall, wo wir von einer *Bahn zweiter Art* sprechen werden, kann die Zentralbahn, die das Elektron beschreibt, als aus einer Reihe aufeinander folgender innerer und äußerer Schlingen bestehend beschrieben werden, von denen die äußeren Bahnschlingen nahezu mit einem Teil einer um eine Punktladung pe beschriebenen Keplerbewegung zusammenfallen, während die inneren Bahnschlingen im allgemeinen sehr stark von einer Keplerbewegung abweichen. Die Dimensionen der äußeren Bahnschlingen weichen indessen im allgemeinen wesentlich ab von denen einer Keplerbewegung mit derselben Hauptquantenzahl um einen Kern mit Ladung pe. Dies sieht man leicht ein, wenn man beachtet, daß das äußere Elektron auf den inneren Schlingen seiner Bahn in einem gegebenen Abstand vom Kerne nahezu derselben resultierenden Anziehung nach dem Atomzentrum unterworfen ist wie eines der inneren Elektronen. Infolge dieses Umstandes muß der Wert des radialen Impulses des Elektrons auf jedem Punkt der inneren Schlinge immer größer sein als der Wert dieses Impulses im selben Abstand vom Kerne für ein Elektron, das sich mit demselben Impulsmoment auf einer ganz im inneren Gebiet verlaufenden Bahn bewegt. Gemäß der Formel (23) ist deshalb die Größe $R = J - P$, und nach (22) also auch die Differenz zwischen den Quantenzahlen n und k, immer größer für eine Bahn zweiter Art als für irgendeine innere Bahn mit demselben Wert von k. Schon der Beitrag zum Werte des Integrals (23), der der inneren Bahnschlinge entspricht, ist sogar immer größer als $m - k$, wo m der größte vorkommende Wert von n bei den inneren Bahnen mit demselben k ist. Der Beitrag zum genannten Integral, der von der äußeren Schlinge herrührt, ist deshalb immer kleiner als $n - m$.

Dieses Resultat zeigt zunächst, daß für eine Bahn der zweiten Art die Hauptquantenzahl n immer größer sein wird als m. Weiter liefert die Überlegung eine Möglichkeit, in manchen Fällen eine einfache Abschätzung der Dimensionen der äußeren Bahnschlinge, sowie der zur Abtrennung des Elektrons vom Atom erforderlichen Energie zu bekommen. Diese sind ja einfach bestimmt durch den Wert J^* der Größe J, die zu der reinen Keplerbewegung gehört, von der die äußere Bahnschlinge ein Teil bildet. Dieser Wert ist nach (23) gleich der Summe von P und des über die vervollständigte Ellipsenbahn genommenen Integrals R. Dieses Integral wird nun in vielen Fällen nicht viel abweichen vom Beitrag, der der äußeren Schlinge allein entspricht, der selber nicht viel vom oben genannten Wert $n - m$ abweicht. In solchen Fällen wird der gesuchte J^*-Wert also annähernd durch $(n - m + k)\,h$ gegeben. Allgemein definieren wir als *effektive Quantenzahl* eine n-Bahn zweiter Art, die der p-letzten Bindungsstufe im Aufbau des Atoms entspricht die Zahl n^*, die in den Formeln (17) für n einzusetzen ist (wobei $N = p$ gesetzt ist), um die Abtrennungsenergie und Bahndimension für das betreffende Elektron zu erhalten. Ein sehr wichtiges Hilfsmittel bei der Untersuchung ist nun, daß diese effektive Quantenzahl sich offenbar direkt aus den empirischen Spektraltermen entnehmen läßt. *

Der Gesichtspunkt des sukzessiven Aufbaus hat zu einem allgemeinen Bilde von der Konstitution der Atome geführt, das an der Hand der Deutung der empirischen Spektralgesetze eine ungezwungene Interpretation des eigentümlichen Wechsels der Eigenschaften der Elemente mit der Atomnummer zuläßt, wie sie im *periodischen System der Elemente* zum Ausdruck kommt.[1]) Die Hauptzüge dieses Bildes sind in der beigefügten Tabelle, die die Klassifikation der Elektronenbahnen im normalen Zustand des Atoms für eine Anzahl von Elementen angibt, zum Ausdruck gebracht. Die Zahlen bedeuten die Anzahl der Elektronen in dem Normalzustand des Atoms, die sich in Bahnen bewegen, welche durch das obenstehende Symbol n_k gekennzeichnet sind. Die Periodizität in den Eigenschaften der Elemente hängt damit zusammen, daß mit fortschreitender Atomnummer eine ständige Ausbildung und Abschließung der verschiedenen Elektronengruppen stattfindet. Diese Ausbildung

1) Die Hauptzüge der Theorie sind in einem Vortrag des Verfassers, der in Zeitschr. f. Phys. **9**. S. 1. 1922 abgedruckt ist, entwickelt. Dort findet sich auch eine eingehende Besprechung von älteren Arbeiten anderer Verfasser über denselben Gegenstand. In der vorliegenden Abhandlung sind die Betrachtungen über den Zusammenhang zwischen optischen Spektren und Atombau mehr im einzelnen ausgeführt. Eine nähere Diskussion des Beobachtungsmaterials der Röntgenspektren im Verhältnis zur behandelten Theorie des Atombaus ist ferner in einer kürzlich erschienenen Arbeit von D. Coster und dem Verfasser (Zeitschr. f. Phys. **12**. S. 342. 1923) gegeben.

* [Instead of eine read einer.]

Tabelle.

N	n_k	1_1	2_1	2_2	3_1	3_2	3_3	4_1	4_2	4_3	4_4	5_1	5_2	5_3	5_4	5_5	6_1	6_2	6_3	6_4	6_5	6_6	7_1	7_2
1	H	1																						
2	He	2																						
3	Li	2	1																					
4	Be	2	2																					
5	B	2	2	(1)																				
—	—	—	—	—																				
10	Ne	2	4	4																				
11	Na	2	4	4	1																			
12	Mg	2	4	4	2																			
13	Al	2	4	4	2	1																		
—	—	—	—	—	—	—																		
18	A	2	4	4	4	4																		
19	K	2	4	4	4	4		1																
20	Ca	2	4	4	4	4		2																
21	Sc	2	4	4	4	4	1	(2)																
22	Ti	2	4	4	4	4	2	(2)																
—	—	—	—	—	—	—	—	—																
29	Cu	2	4	4	6	6	6	1																
30	Zn	2	4	4	6	6	6	2																
31	Ga	2	4	4	6	6	6	2	1															
—	—	—	—	—	—	—	—	—	—															
36	Kr	2	4	4	6	6	6	4	4															
37	Rb	2	4	4	6	6	6	4	4			1												
38	Sr	2	4	4	6	6	6	4	4			2												
39	Y	2	4	4	6	6	6	4	4	1		(2)												
40	Zr	2	4	4	6	6	6	4	4	2		(2)												
—	—	—	—	—	—	—	—	—	—	—		—												
47	Ag	2	4	4	6	6	6	6	6	6		1												
48	Cd	2	4	4	6	6	6	6	6	6		2												
49	In	2	4	4	6	6	6	6	6	6		2	1											
—	—	—	—	—	—	—	—	—	—	—		—	—											
54	X	2	4	4	6	6	6	6	6	6		4	4											
55	Cs	2	4	4	6	6	6	6	6	6		4	4				1							
56	Ba	2	4	4	6	6	6	6	6	6		4	4				2							
57	La	2	4	4	6	6	6	6	6	6		4	4	1			(2)							
58	Ce	2	4	4	6	6	6	6	6	6	1	4	4	1			(2)							
59	Pr	2	4	4	6	6	6	6	6	6	2	4	4	1			(2)							
—	—	—	—	—	—	—	—	—	—	—	—	—	—	—			—							
71	Cp	2	4	4	6	6	6	8	8	8	8	4	4	1			(2)							
72	Hf	2	4	4	6	6	6	8	8	8	8	4	4	2			(2)							
—	—	—	—	—	—	—	—	—	—	—	—	—	—	—			—							
79	Au	2	4	4	6	6	6	8	8	8	8	6	6	6			1							
80	Hg	2	4	4	6	6	6	8	8	8	8	6	6	6			2							
81	Tl	2	4	4	6	6	6	8	8	8	8	6	6	6			2	1						
—	—	—	—	—	—	—	—	—	—	—	—	—	—	—			—	—						
86	Em	2	4	4	6	6	6	8	8	8	8	6	6	6			4	4						
87	—	2	4	4	6	6	6	8	8	8	8	6	6	6			4	4					1	
88	Ra	2	4	4	6	6	6	8	8	8	8	6	6	6			4	4					2	
89	Ac	2	4	4	6	6	6	8	8	8	8	6	6	6			4	4	1				(2)	
90	Th	2	4	4	6	6	6	8	8	8	8	6	6	6			4	4	2				(2)	
—	—	—	—	—	—	—	—	—	—	—	—	—	—	—			—	—	—				—	
—	—	—	—	—	—	—	—	—	—	—	—	—	—	—	—		—	—	—				—	—
118	?	2	4	4	6	6	6	8	8	8	8	8	8	8	8		6	6	6				4	4

findet im allgemeinen statt in der äußersten Elektronengruppe, die dem größten vorkommenden Wert der Hauptquantenzahl entspricht und tritt besonders deutlich in den optischen Spektren zutage. Zuweilen findet aber auch, wie aus der Tabelle ersichtlich, eine weitere Entwicklung einer der inneren Gruppen statt, was ein Verständnis liefert für die Abweichungen von einfacher Periodizität in der Änderung der Eigenschaften der Elemente mit wachsender Atomnummer. Diese Verhältnisse sind in der nachstehenden Übersicht des periodischen Systems zum Ausdruck gebracht, in der die vertikalen Reihen die Perioden bezeichnen, während Elemente mit homologen Eigenschaften durch Striche verbunden sind. Die Elementenreihen in eckigen Klammern sind solche, für welche innere Gruppen in Ausbildung begriffen sind. Eine derartige Ausbildung äußert sich in der nahen Verwandtschaft in manchen Eigenschaften der betreffenden Elemente, was davon herrührt, daß während der Ausbildung der inneren Gruppen die Entwicklung der äußersten Gruppen vorübergehend keine Fortschritte macht. Zur selben Zeit kommt diese Ausbildung jedoch deutlich zum Ausdruck in den charakteristischen Unterschieden der Serienspektren zwischen solchen Elementen innerhalb und außerhalb der Klammern, die sonst in chemischer Hinsicht sehr ähnliche Eigenschaften besitzen und deshalb in den üblichen Darstellungen des periodischen Systems homologe Stellungen einnehmen. Obwohl in der letzten Zeit versprechende Fortschritte in diesen Fragen gemacht wurden, sind die einzelnen Stufen der Ausbildung nicht derart geklärt, daß wir den Versuch gewagt haben, sie in der Tabelle näher anzugeben.

Ein besonders lehrreiches Beispiel der innigen Verbindung zwischen dem in Rede stehenden Bilde des Atombaues und den spektralen Gesetzmäßigkeiten finden wir in den *Bogenspektren der Alkalimetalle,* die bekanntlich einen besonders einfachen Bau besitzen. Dies hängt damit zusammen, daß für diese Elemente, die ja am Anfang der Perioden des Systems stehen, bei der Einfangung des letzten Elektrons die Elektronenbahnen im Atomrest eine abgeschlossene Konfiguration bilden von der Art wie sie im neutralen Atom des vorausgehenden Edelgases auftritt. Die Frage, der wir zunächst begegnen, ist nun, welcher der beiden oben besprochenen Arten von Bahnen die verschiedenen Terme dieser Spektren zugehören. Hierfür ist in erster Linie ausschlaggebend der Parameter der Keplerbahn, welche die wirkliche Bahn jedenfalls in ihren äußeren Schlingen approximiert. Wenn nämlich dieser Parameter, der bei beiden Bahnarten gemäß (25) nur vom Wert der Nebenquantenzahl

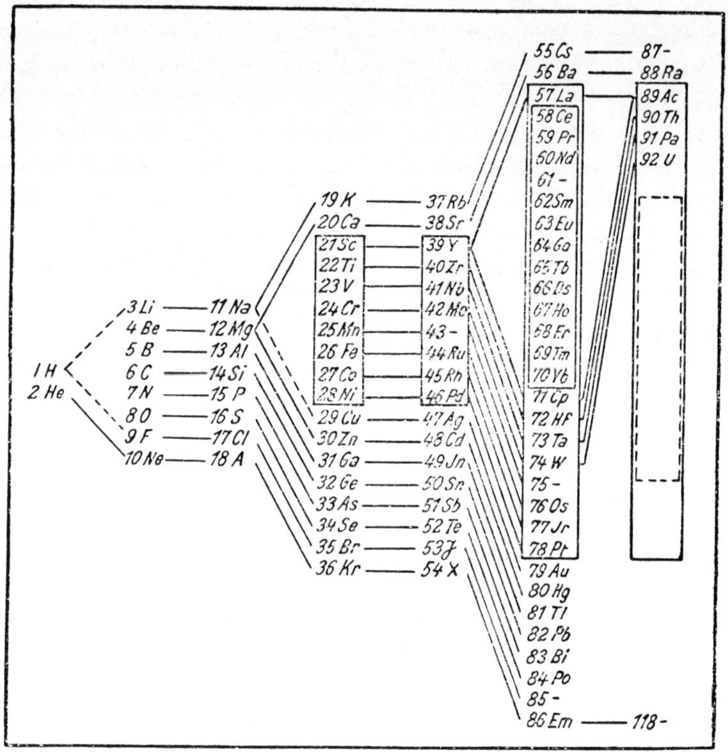

Fig. 1.

abhängt, wesentlich größer ist als die Dimensionen der Bahnen des Atomrestes, können wir annehmen, daß wir es mit einer Bahn erster Art zu tun haben, die gänzlich außerhalb des Atomrestes verläuft. Bei den Spektren aller Alkalien ist dies der Fall bei Bahnen, für die $k \geqq 3$ ist. In der Tat weichen schon die d- und f-Terme dieser Spektren nur wenig von Wasserstofftermen ab und zwar in dem Sinne einer stärkeren Anziehungskraft als im Wasserstoffatom, so wie es zu erwarten wäre wegen der elektrischen Polarisation, die das äußere

[584]

Elektron im Atomreste induziert und deren Vorkommen den dielektrischen Erscheinungen völlig analog ist. Bei den s- und p-Termen dagegen treffen wir im allgemeinen andere Verhältnisse.

Bei Lithium bleiben noch die Bahnen für $k = 2$ in großen Abständen von den Bahnen des Atomrestes, wie es sich auch in der kleinen Abweichung der p-Terme des Lithiumspektrums von den Wasserstofftermen kundgibt. Bei den Bahnen mit $k = 1$ treffen wir im Lithium auf besondere Verhältnisse, weil die Bahnen nicht ausgesprochen zu der einen oder anderen Klasse gezählt werden können. Der auf Grund der Formel für die Keplerbewegung berechnete Parameter der Bahnen ist zwar noch wesentlich größer als die Dimensionen des inneren Systems, aber die von der induzierten Polarisation des Atomrestes herrührenden Kräfte, die aus der Abweichung der p-Terme von den Wasserstofftermen geschätzt werden können, ergeben sich so groß, daß eine zirkulare 1_1-Bahn in das Gebiet der inneren 1_1-Bahnen gezogen würde. Der abgeschlossene Charakter der Bahnenkonfiguration des Atomrestes hängt aber nach unseren Vorstellungen eben damit zusammen, daß die Aufnahme eines dritten Elektrons in eine mit den zwei ersten äquivalente Bahn als Endresultat eines von Strahlung begleiteten Bindungsprozesses nicht vorkommen kann. Was die n_1-Bahnen mit $n > 1$ betrifft, so kommen sie während ihres Umlaufes, selbst wenn sie nicht in das Gebiet der 1_1-Bahnen eindringen, diesen Bahnen so nahe, daß wir unmittelbar verstehen, daß sie infolge der innigen Wechselwirkung einer wesentlich stärkeren Bindung des Elektrons entsprechen als die Bahnen im Wasserstoffatom mit gleicher Hauptquantenzahl.

Im Falle der anderen Alkalimetalle liefern die stationären Zustände mit $k = 1$, wie eine einfache Abschätzung der Bahndimensionen zeigt, ausgesprochene Beispiele für Bahnen zweiter Art. Weiter zeigt ein Vergleich der Bewegung in der inneren Bahnschlinge mit den Bahnen der früher gebundenen Elektronen, daß das Serienelektron in den s-Bahnen während seines Umlaufes dem Kern sogar wesentlich näher kommt als die beiden in zirkularen 1_1-Bahnen laufenden innersten Elektronen. Bei den Bahnen mit $k = 2$ liegen die Verhältnisse mehr analog dem Fall der n_1-Bahnen im Lithium. So dürfen wir schließen, daß wegen der von der Polarisation des Atomrestes herrührenden Kräfte ein Elektron in einer 2_2-Bahn sich nicht im äußeren Gebiet des Atoms bewegen kann, obgleich die entsprechende Bahn im Wasserstoffatom größere Dimensionen als der Atomrest haben würde. Für $n > 2$ kommen die n_2-Bahnen, schon als Wasserstoffbahnen abgeschätzt, in Abstände vom Kern von derselben Größenordnung wie die Elektronen des Atomrestes, und wir müssen annehmen, daß sie in das innere Gebiet eindringen, obwohl sie immer in einem größeren Abstand vom Kern bleiben als die zirkularen 1_1-Bahnen der beiden innersten Elektronen. Dieser Sachverhalt zeigt sich auch deutlich in der Größe der entsprechenden Terme. Ihre effektiven Quantenzahlen weichen nämlich von den Wasserstofftermen nicht nur in einem viel höheren Maße ab als die der d- und f-Terme, sondern auch in einem Sinn, der, falls die Bahnen ganz im äußeren Gebiet verliefen, einer schwächeren Anziehungskraft als im Wasserstoffatom entsprechen würden, was dem Charakter der Atommodelle widerspricht.

Was nun die Werte der s- und p-Terme betrifft, so werden nach unserer Definition die effektiven Quantenzahlen der entsprechenden Bahnen eben mit den Nennern in der Rydbergschen Formel (20) zusammenfallen, und es ist zunächst besonders befriedigend, daß wir durch unsere Vorstellungen über die Bahnformen eine unmittelbare Erklärung dafür bekommen, daß die Größen α_k für die Reihen der s- und p-Terme so nahe konstant sind. Innerhalb jeder dieser Reihen wird nämlich die Elektronenbewegung auf den inneren Schlingen der zugehörigen Bahnen mit außerordentlich großer Annäherung dieselbe sein und die Differenz der Hauptquantenzahl und der effektiven Quantenzahl wird daher innerhalb jeder Reihe nahe konstant sein. Setzen wir für die Zahl n in der Rydbergschen Formel die Hauptquantenzahl ein, so wird die Konstante α für alle Spektren negativ. Für Bahnen zweiter Art sollen wir, nach der Überlegung auf S. 259 erwarten, daß der numerische Wert von α nie viel von der Zahl $m - k$ abweichen kann. Diese Erwartung ist bei den Alkalispektren auch erfüllt, indem, wie es aus den schematischen Darstellungen auf Tafel II ersichtlich ist, die effektiven Quantenzahlen, die dem ersten

Gliede der s-Reihe und p-Reihe entsprechen, beim Übergang zum nächsten Elemente nur langsam anwachsen, ohne die Werte 2 bzw. 3 zu überschreiten. Bei den Alkalimetallen entspricht, wie in der Tabelle angegeben, das erste Glied der s-Reihe dem Normalzustand des Atoms, und der Umstand, daß trotz des beständigen Anwachsens der Hauptquantenzahl die numerischen Werte dieser Glieder verhältnismäßig nur wenig verschieden sind, liefert ein charakteristisches Beispiel für die *Erklärung des periodischen Wechsels in den Eigenschaften der Elemente mit steigender Atomnummer* auf Grund unseres Bildes der Entwicklung und Abschließung von Elektrongruppen mit höherer Quantenzahl im Atom.[1])

1) Zur selben Zeit, als die hier besprochenen allgemeinen Züge der Vorstellungen vom Atombau vom Verfasser entwickelt wurden (vgl. Nature, März 24, 1921), wurde unabhängig von E. Schrödinger (Zeitschr. f. Phys. **4.** 347. 1921) hervorgehoben, daß wir es bei den s-Termen der Alkalien mit stationären Zuständen zu tun haben, wo das Serienelektron während seines Umlaufes in das innere Gebiet hineindringt, und es wurde von diesem Forscher versucht, mit Hilfe der einfachen Annahme, daß die Wirkung der inneren Elektronen mit der einer gleichmäßig mit Ladung belegten Kugelschale verglichen werden kann, eine Abschätzung der Werte der s-Terme beim Natriumspektrum zu erzielen, wobei angenommen wurde, daß der erste s-Term einer 2_1-Bahn entspricht. Obgleich diese Betrachtung ein einfaches Verständnis der großen Abweichungen der s-Terme von den Wasserstofftermen lieferte, konnte aber in dieser Weise eine quantitative Berechnung nicht erreicht werden, denn eine solche verlangt in erster Linie eine nähere Klassifikation der Elektronenbahnen im Atomreste, um eine Grundlage zu erlangen für die Festsetzung der den betreffenden Spektraltermen entsprechenden Hauptquantenzahl. Die Rechnungen von Schrödinger sind neuerdings weitergeführt von Th. van Urk Zeitschr. f. Phys. **13.** 268. 1923); dieser glaubt schon aus einer Betrachtung der zur Ermittelung der s-Terme geforderten Dimensionen der Kugelschale schließen zu können, daß der erste s-Term im Natriumspektrum einer 3_1-Bahn entspricht. Andererseits hat E. Fues (Zeitschr. f. Phys. **11.** 364. 1922; **12.** 1. 1922; **13.** 211. 1923) unter Zugrundelegung der oben angeführten Werte der Quantenzahlen ein zentralsymmetrisches Kraftfeld zu ermitteln versucht, das die beobachteten Terme des Natriumspektrums quantentheoretisch direkt zu berechnen erlaubt. Obgleich hierdurch in mehrerer Hinsicht interessante Resultate erreicht wurden, besonders was die Verwandtschaft mit den analogen Funkenspektren der folgenden Elemente betrifft, kann man doch nicht hoffen, die Einzelheiten der Spektren zu erklären mittels einer derartigen Annahme, die von dem dynamischen Charakter des

Ähnliche Verhältnisse wie bei den Alkalispektren, treffen wir bei den *Funkenspektren der alkalischen Erden*, die von den einwertigen Ionen dieser Elemente herrühren. Die Bahnen im Atomreste bilden ja auch hier eine Edelgaskonfiguration; nur ist die Gesamtladung doppelt so groß als bei den Alkalien. Wie bei den Bogenspektren dieser Elemente, dringt auch hier das äußere Elektron in die Bahnen, die s- und p-Termen entsprechen, in den Atomrest ein. Die Werte dieser Terme zeigen ein Verhältnis zu den Werten der entsprechenden Terme bei den Alkalien, das eben derart ist, wie es nach der

Atomgebäudes absieht. Außer dem oben erwähnten polarisierenden Einflusse eines äußeren Elektrons auf das innere System spielt in manchen Fällen dieser Charakter eine wesentliche Rolle, auch dadurch, daß die Abtrennungsarbeit eines Elektrons oft in hohem Maße von dem Einfluß auf die Bewegung der anderen Elektronen abhängt, die seine Entfernung aus dem Atome mit sich bringt. Dieser Umstand zeigt sich besonders in Fällen, wo im Atom bereits Elektronenbahnen mit derselben oder größerer Hauptquantenzahl wie diejenigen der Bahn des betreffenden Elektrons vorhanden sind.

Besonders lehrreiche Beispiele von Verhältnissen der letzterwähnten Art begegnen wir bei den Röntgenspektren. Was diesen Punkt betrifft, sei auf die in der Fußnote auf S. 259 erwähnte Arbeit von Coster und dem Verfasser hingewiesen. Es kann in dieser Verbindung erwähnt werden, daß die dort angeführten, auf dem großen Einfluß der Relativitätskorrektion auf die Elektronenbahnen im Atominneren beruhenden Argumente (a. a. O. S. 367) für die Ganzzahligkeit der Werte der Quantenzahlen sich noch verschärfen lassen, und daß ähnliche Verhältnisse auch bei den Serienspektren von wesentlicher Bedeutung sind. So läßt sich mittels einer einfachen Rechnung zeigen, daß jede n_k-Bahn in einem Kernatom in den Kern stürzen würde, wenn $N/k = hc/2\pi e^2 = 137$. Für die schwereren Elemente sieht man also, daß das Eindringen des Serienelektrons in das Atominnere mit sich führt, daß Bahnen mit $k = 1/2$ nicht existieren können. Schon für $k = 1$ kommt für diese Elemente das Elektron in Abstände vom Kernmittelpunkt, die von derselben Größenordnung wie die aus radioaktiven Daten ermittelten Werte der Kerndimensionen sind. Wie S. Rosseland bemerkt hat (Nature, März 17, 1923), bietet schon dieser Umstand einen Fingerzeig für ein Verständnis von der Begrenzung der Atomnummer der vorkommenden Elemente dar. Weiter möchte derselbe Umstand auch eine Erklärung darbieten für die beobachteten Differenzen in der Wellenlänge gewisser Linien in den Spektren von Bleiisotopen, die, obwohl sehr klein, doch vielmals größer sind als die Unterschiede, die voraussichtlich gemäß (18) von der Verschiedenheit in der Kernmasse der beiden Isotope herrühren könnten (vgl. Nature, Juni 10, 1922).

Änderung der Bahndimensionen, die die größere Kernladung bedingt, zu erwarten wäre. Was die d-Terme betrifft, bringt diese Änderung einen neuen Zug mit sich, indem die Bahnen mit $k = 3$ bei den höheren Homologen der betreffenden Elementenfamilie nicht mehr ganz im Außenraum verlaufen. Nur im Funkenspektrum des Magnesiums treffen wir Verhältnisse, die denen bei den Alkalispektren analog sind, indem die d-Terme hier nur sehr wenig von den Termen des Heliumfunkenspektrums abweicht. Beim Calcium und den höheren Homologen weichen aber, wie die schematische Darstellung auf Tafel IV zeigt, die d-Terme außerordentlich stark von diesen Termen ab, was bedeutet, daß die Bahn wenigstens teilweise im inneren Gebiet liegt. Ja, aus der Größe des ersten Gliedes der d-Reihe folgt sogar, daß die entsprechende Bahn ganz in diesem Gebiete verläuft, was auch durch eine Abschätzung der Bahndimensionen verständlich gemacht wird. Bei Ca, Sr, Ba entspricht dieses Glied bzw. einer 3_3-, 4_3- und 5_3-Bahn. Die den b-Termen entsprechenden Bahnen verlaufen dagegen völlig im Gebiete außerhalb des Atomrestes.

Ein *Serienspektrum dritter Art* ist bis jetzt nur für Aluminium bekannt. Hier sind die Verhältnisse, wie von Paschen hervorgehoben, ganz denen bei dem Bogenspektrum des Na und bei dem Funkenspektrum des Mg analog, indem die Werte entsprechender Terme in diesen Spektren eben solche Verhältnisse zeigen wie es nach der allmählichen Änderung der relativen Größen der Bahndimensionen zu erwarten wäre. Eine Untersuchung der Serienspektren dritter Art der anderen Elemente, deren Atomnummer um drei Einheiten größer ist, als die der Edelgase (Sc, Y, La, Ac), dürfte sehr interessante Verhältnisse zutage bringen. Wir müssen nämlich hier erwarten, nicht nur daß die d-Terme, wie bei den Funkenspektren der alkalischen Erden, stark von den Wasserstofftermen abweichen, sondern daß die Abtrennungsenergie, die dem ersten Gliede dieser Reihe entspricht, sogar größer wird als die Abtrennungsenergie, die dem größten s-Term entspricht, der bei den bisher besprochenen Spektren immer zum Normalzustand des Atoms gehört. Wie in der Übersichtstabelle auf S. 260 angedeutet, bringt dieser Umstand mit sich, daß eine neue Entwicklungsstufe derjenigen Elektronengruppe, die in dem vorhergehenden

Edelgase zum vorläufigen Abschluß gebracht war, eingeleitet wird. Hierbei ist zu bemerken, daß die Klammern, die in der Tabelle bei den betreffenden Elementen um die Zahl der Elektronen in der Gruppe mit höchster Hauptquantenzahl angebracht sind, anzeigen sollen, daß die Spektren dieser Elemente nicht genügend bekannt sind, um uns zu erlauben, die Quantenzahlen der letztgebundenen Elektronen im normalen Atom mit Sicherheit festzulegen.

Solange diese Entwicklung dauert, treffen wir bei steigender Atomnummer Elemente, deren Spektren neuartige Züge aufweisen; sobald sie aber beendigt ist, treffen wir wieder Bahnkonfigurationen im Atomreste, die einen abgeschlossenen Charakter besitzen, obgleich diese Konfiguration nicht im neutralen Atom des vorangehenden Elementes auftritt. Wir verstehen daher, daß die *Bogenspektren der Elemente Kupfer und Silber*, die im periodischen System nach der Familie der Eisen- und Palladiummetalle stehen, einen ähnlichen Bau wie die Alkalispektren besitzen (vgl. Tafel III). Bei den betreffenden Spektren wird, ebenso wie bei den Alkalispektren, das äußere Elektron in Bahnen, die den s-Termen entsprechen, in das Atominnere eindringen, und wie in der Übersichtstabelle auf S. 260 angegeben, entspricht wieder hier der Normalzustand einer Bahn, für welche die Hauptquantenzahl der Nummer der Periode im System gleich ist. Dagegen zeigen die Werte der p-Terme, daß bei den Kupfer- und Silberatomen die Bahnen mit $k = 2$ ganz im Gebiete außerhalb des Atomrestes verlaufen, was mit sich bringt, daß der erste p-Term einer 2_2-Bahn entspricht. Das wird auch verständlich, da nach der Theorie die einwertigen Cu^+- und Ag^+-Ionen einen wesentlich kleineren Raum einnehmen, als die positiven Alkaliionen, ein Umstand, der übrigens schon in den größeren Werten der s-Terme in Cu und Ag, verglichen mit den entsprechenden Termen bei den Alkalimetallen zum Ausdruck kommt. Die Bahnen, die den d- und f-Termen entsprechen ($k = 3,4$) verlaufen natürlich bei Kupfer und Silber, wie bei den Alkalien, ganz im Außengebiet.[1])

1) Es mag von Interesse sein zu bemerken, daß ein im Kupferspektrum nach H. M. Randall (Ann. d. Phys. 33. S. 741. 1910) auftretender, außer den gewöhnlichen Termreihen stehender Spektralterm, der nach

Die *Funkenspektren der Elemente Zink* und *Cadmium*, die direkt nach *Cu* und *Ag* folgen, sind nur wenig bekannt. Doch dürfte es schon möglich sein, aus den entwirrten Termkombinationen zu schließen, daß bei den p-Termen das äußere Elektron in den Atomrest eindringt, was man auch nach einer theoretischen Abschätzung der Bahndimensionen erwarten dürfte. Wie bei den Alkalispektren und den Funkenspektren der Erden sollten dann hier die Hauptquantenzahlen für den größten s-Term und p-Term gleich sein, und zwar gleich der Nummer der Periode im System der Elemente.

§ 7. Nähere Diskussion der Komplexstruktur der Serienspektren.

Im vorhergehenden Paragraphen haben wir die Hauptzüge des Bildes des Atombaus, auf das man durch die Deutung der Spektralgesetze geführt wird, kurz auseinandergesetzt. Die Betrachtungen über spezielle Spektren, die wir daran knüpften, betreffen jedoch nur die angenäherte Beschreibung

---

den Kombinationsregeln und dem Zeemaneffekt sich den d-Termen anschließt, wie in der Tafel angegeben, sich als von einer innerhalb des Atomrestes verlaufenden 3_3-Bahn herrührend deuten lassen dürfte. In der Tat weist eine einfache Schätzung der Bahndimensionen des Atomrestes auf die dynamische Möglichkeit einer 3_3-Elektronenbahn hin, die zugleich außerhalb des Gebietes der symmetrischen Untergruppe der sechs 3_3-Bahnen liegt, und innerhalb der Grenzen des Gebietes, wo die Elektronen der zwei anderen Untergruppen der dreiquantigen Hauptgruppe mit je 6 Elektronen in 3_1- und 3_2Bahnen sich bewegen. Diese Deutung wird dadurch gestützt, daß im Kupferspektrum noch einige Linien auftreten, die, wie Hr. Werner gefunden hat, als Kombinationen bekannter Spektralterme mit einem zweiten d-Term dargestellt werden können, der mit dem erwähnten Term hinsichtlich Kombinationsmöglichkeiten und Zeemaneffekt in gewöhnlicher Weise ein Dublett bildet. Wie aus der Tafel III ersichtlich, ist hier jedoch, entgegen dem gewöhnlichen Verhalten der Dublettspektren der d_1-Term größer als der d_2-Term. Dieser Umstand dürfte aber eben damit zusammenhängen, daß bei der angenommenen Lage der Bahn die Termgröße in besonders hohem Maße von der Änderung abhängt, welche die Bindungsstärke der anderen dreiquantigen Bahnen bei der Entfernung des Elektrons erfährt.

Anm. bei der Korrektur: Eine nähere Untersuchung des Kupferspektrums von Hrn. Nissen und Hrn. Werner hat inzwischen ergeben, daß noch mehrere außerhalb der gewöhnlichen Termreihen stehenden Terme vorhanden zu sein scheinen. Es muß daher mit der Möglichkeit gerechnet werden, daß wir im Kupferspektrum mit einer komplizierteren Erscheinung als oben angenommen zu tun haben.

der Bahn jedes einzelnen Elektrons und ihre Klassifikation mit Hilfe der Quantenzahlen n und k, die auf dem angenäherten Zentralbahncharakter beruht, und wir haben bisher die Fragen der Einzelheiten der näheren Orientierung der Bahnebenen der verschiedenen Elektronbahnen außer Acht gelassen. Diese Orientierung bedingt in erster Linie die Art des feineren Wechselspiels zwischen den Bewegungen der verschiedenen Elektronen, das wieder als verantwortlich angesehen werden muß für die bei steigender Atomnummer immer wiederkehrenden Abschließungen gewisser Elektronengruppen, die sich in in der Abneigung gegen die Aufnahme neu hinzukommender Elektronen in Bahnen mit demselben Wert der Quantenzahlen n und k äußern. Wie wir sahen, waren diese Abschließungen ein wesentlicher Zug bei der Erklärung des periodischen Wechsels der Eigenschaften der Elemente, und ein lehrreiches Beispiel treffen wir eben beim Helium, das am Ende der ersten Periode des Systems steht.

Das *Bogenspektrum des Heliums* zeigt einen wesentlich verschiedenen Bau von den im vorigen Paragraphen betrachteten Spektren. Im Gegensatz zu den Spektren der Elemente, die im Anfang der Perioden des Systems stehen, besteht das Heliumspektrum aus zwei Kombinationsspektren des Rydbergschen Typus, dem sogenannten „Orthoheliumspektrum" und dem „Parheliumspektrum", die in dem Sinne scharf getrennt sind, daß keine gegenseitige Kombinationslinien der zwei Termgruppen auftreten. Nach dem Korrespondenzprinzip muß dieser Umstand so gedeutet werden, daß wir es in diesen Spektren mit Bewegungen der Elektronen zu tun haben, die nicht zu einer zusammenhängenden Klasse gehören, innerhalb welcher die Bewegung des Serienelektrons eine Auflösung in harmonische Komponenten zuläßt von der Art, die einer Zentralbahn entspricht, deren Ebene eine gleichmäßige Präzession vollführt. Bei den beiden Heliumspektren läßt sich daher eine rationale Klassifikation der stationären Zustände mit Hilfe dreier Quantenzahlen n, k, j auf den bisherigen Grundlagen nicht durchführen.[1])

1) Der erste Versuch einer theoretischen Deutung des Heliumspektrums ist von Landé gegeben (Phys. Ztschr. 20. S. 228. 1919 u. 21, S. 114. 1920). In direkter Anlehnung an die Theorie der Periodizitätssysteme beschreibt dieser Verfasser die verschiedenen stationären Zu-

Bei dieser Sachlage müssen wir uns damit begnügen, einige Vermutungen allgemeiner Natur über den Ursprung der in Rede stehenden Spektren zu erwähnen, die nur den Zweck verfolgen, einen Anschluß an die folgenden Betrachtungen herzustellen.

Im *Orthoheliumspektrum* begegnen wir, von den engen Dubletts abgesehen, einem Spektrum desselben Charakters wie das eines Elektrons, das sich in einem Zentralfelde bewegt, mit der Besonderheit, daß ein Zustand fehlt, dem die Hauptquantenzahl 1 entspricht, sowie es aus der Existenz des von Franck und seinen Mitarbeitern durch Elektronenstoßversuche entdeckten metastabilen Orthoheliumzustandes hervorgeht. Ein Spektrum von diesem Bau wäre in der Tat zu erwarten, wenn

---

stände durch drei Quantenzahlen und nimmt für die Orthoheliumzustände ($j = k + 1$) (komplanare Bahnen) und für die Parheliumzustände $j = k$ (gekreuzte Bahnebenen) an. Obwohl man nicht hoffen kann, durch eine derartige Behandlung zu einer befriedigenden Erklärung der Spektren zu gelangen, ist die Landésche Theorie schon dadurch von großer Bedeutung, daß sie den ersten Versuch einer kinematischen Deutung der Komplexstruktur eines Spektrums darstellt, in welcher als dritte den Bedingungen (22) hinzuzufügende Quantenbedingung eine Festlegung des resultierenden Impulsmementes des Atoms auftritt.

Eine nähere Untersuchung der mechanischen Eigenschaften des Heliumatoms, die durch die gegenseitigen Störungen der beiden Bahnen bedingt sind, ist von Kramers und vom Verfasser unternommen, die zur Erkenntnis gewisser besonders einfacher Bewegungstypen geführt hat, zu denen der unten bei der Besprechung des Orthoheliumspektrums erwähnte Typus gehört. Die Resultate dieser Untersuchung, die vom Verfasser bei Gelegenheit einer in Göttingen im Juni 1922 gehaltenen Vortragsreihe mitgeteilt wurden, sind wegen ihres unabgeschlossenen Charakters nicht publiziert worden.

Wie mir von Hrn. Born freundlichst mitgeteilt wurde, hat er neuerdings gemeinsam mit Hrn. Heisenberg das mechanische Problem der gegenseitigen Störungen der Elektronen im Heliumatom einer umfassenden Untersuchung unterzogen, wobei sich herausgestellt hat, daß zwar formal eine konsequente Behandlung im Anschluß an die Quantentheorie der Periodizitätssysteme durchführbar ist, daß aber in dieser Weise nicht nur keine Erklärung des Nichtvorkommens von Kombinationen zwischen Orthohelium- und Parheliumtermen, sondern auch nicht eine angenäherte Übereinstimmung mit den empirischen Werten dieser Terme erreicht werden kann. Diese Untersuchung dürfte daher besonders geeignet sein, ein Zeugnis zu liefern für das prinzipielle Versagen der mechanischen Gesetze bei der Beschreibung der feineren Einzelheiten der Bewegung von Systemen mit mehreren Elektronen.

* [Instead of Impulsmementes read Impulsmomentes.]

die Bewegung des äußeren und inneren Elektrons in derselben Ebene vor sich ginge, denn unter dieser Annahme gibt es eine mechanisch mögliche Bewegungsform, wo die gegenseitigen Störungen besonders einfachen Charakter besitzen, in dem die Exzentrizität der inneren Bahn und die Lage ihrer großen Achse relativ zum Radius-Vektor des äußeren Elektrons angenähert unverändert bleibt. Die stationären Zustände lassen sich in diesem Fall durch eine sinngemäße Anpassung der Theorie einfacher Zentralbewegungen festlegen, und die Bahn des äußeren Elektrons kann außer durch die Hauptquantenzahl n durch eine Nebenquantenzahl k' beschrieben werden, die unter der Annahme, daß beide Elektronen den gleichen Umlaufsinn haben, um eine Einheit kleiner ist als der durch $h/2\pi$ dividierte Gesamtdrehimpuls des Atoms. Die Annahme, daß die Aussendung des Orthoheliumspektrums an einen solchen speziellen Bewegungstypus geknüpft wäre, würde in direkter Anlehnung an das Korrespondenzprinzip ein Verständnis der allgemeinen Kombinationsmöglichkeiten der Orthoheliumterme ermöglichen, und zugleich eine Erklärung dafür darbieten, daß kein Orthoheliumzustand mit der Hauptquantenzahl 1 existiert. Ein Übergang von einem höherquantigen Orthoheliumzustand in einen Zustand, wo beide Elektronen sich in äquivalenten einquantigen Bahnen bewegen und der deshalb einem anderen Bewegungstypus entsprechen müßte, würde nämlich mit keiner harmonischen Komponente im elektrischen Moment des Atoms korrespondieren. Andrerseits stehen der Deutung des Orthoheliumspektrums noch große Schwierigkeiten entgegen, weil es sich als unmöglich erwiesen hat, mittels einer Berechnung auf Grund der klassischen mechanischen Gesetze unter der Annahme des genannten Bewegungstypus von der Größe der beobachteten Orthoheliumterme genauer Rechenschaft zu geben.

Der Umstand, daß der Normalzustand des Atoms, wie aus Lymans Untersuchungen hervorgeht, den stationären Zuständen des *Parheliumspektrums* angehört, dürfte andrerseits die Annahme nahelegen, daß wir es hier mit Bewegungsformen zu tun haben, denen sich eine Konfiguration, in der beide Elektronen sich in äquivalenten Bahnen bewegen, in natürlicher Weise anschließt. Dieses dürfte mit sich bringen, daß in den stationären Zuständen des Parheliums die Elektronenbahnen nicht in derselben Ebene verlaufen, und daß sowohl ihre Gestalt wie ihre Lage sich während der Bewegung immer ändern. Fragen wir nun nach der Bedeutung der in der Klassifikation des Parheliumspektrums auftretenden Nebenquantenzahl, so haben wir es kaum mit einem genügenden Anschluß an die Theorie der Zentralbewegungen zu tun, als daß es gerechtfertigt wäre, sie mit der in dieser Theorie auftretenden Quantenzahl k zu identifizieren; vielmehr bietet sich die Annahme dar, daß die zu quantisierende Größe der Gesamtdrehimpuls des Atoms, oder — wenn der Impulssatz in der gewöhnlichen mechanischen Fassung versagen sollte — eine mit dem Drehimpuls formal analoge Invariante ist, deren Existenz aus der allgemeinen Symmetrie des Kernatoms hervorgehen dürfte. Was speziell den Normalzustand betrifft, so sei bemerkt, daß es in Anlehnung an die Theorie der Periodizitätssysteme die einfachste Annahme wäre, den Gesamtdrehimpuls des Atoms in diesem Zustand gleich $h/2\pi$ zu setzen.[1]) Diese Annahme werden wir in die folgenden formalen Betrachtungen über die Komplexstruktur der Serienspektren einführen, indem wegen des normalen Zeemaneffektes der Heliumlinien hier kaum eine Begründung für eine Abweichung in einem bestimmten Sinne von den formalen Gesetzen der genannten Theorie gegeben werden könnte.

Die *Spektren der schwereren Edelgase* zeigen einen sehr komplizierten Bau, die eine große Verwickeltheit der Gesamtheit der stationären Zustände fordert. Wir dürfen aber annehmen, daß wir es im Normalzustand der Atome der Edelgase mit Elektronenkonfigurationen mit ausgeprägter Symmetrie zu tun haben. In der Tat erblicken wir eben hierin eine Erklärung des abgeschlossenen Charakters dieser Elektronengruppen, denn die Aufnahme weiterer Elektronen, die diese Symmetrie

1) Daß man bei dieser Annahme unter Zugrundelegung des einfachst möglichen mechanischen Bewegungstypus jedoch keine Übereinstimmung mit dem dem Normalzustande des Heliums entsprechenden Termwerte erreichen kann, hat neuerdings Kramers (Zeitschr. f. Phys. **13.** S. 312. 1923) gezeigt. (Vgl. auch J. van Vleck, Phil. Mag. **44.** S. 842. 1922.) In Kramers Abhandlung werden auch die grundsätzlichen Schwierigkeiten, die wegen der Instabilität der Elektronbewegung einer mechanischen Behandlung des Problems entgegenstehen, ausführlich besprochen.

vernichten würde, würde in der Tat nicht diejenige Analogie mit einem klassischen Strahlungsprozeß aufweisen, die bei den möglichen Übergängen zwischen den stationären Zuständen einfach oder vielfach periodischer Systeme nachgewiesen werden konnte. Im besonderen scheint es notwendig anzunehmen, daß die Symmetrie der Elektronengruppen, für die die Hauptquantenzahl größer als 1 ist, derart ist, daß diese Gruppen nicht zum Gesamtimpuls des Atoms beitragen, und daß der Wert dieses Impulses also für alle Edelgase gleichzusetzen wäre. Bekanntlich zeigen die Edelgase keinen Paramagnetismus, woraus oft der Schluß gezogen wird, daß der Gesamtimpuls im Normalzustand verschwindet. Nach der Quantentheorie ist jedoch dieser Schluß nicht zwingend, denn für ein Atom, das einen endlichen Wert des Gesamtimpulses und des resultierenden magnetischen Momentes besitzt, fordert ja diese Theorie, daß bei Anwesenheit des Magnetfeldes die Impulsachse nur ganz bestimmte Lagen relativ zum Felde einnehmen kann, eine Forderung, die neuerdings eine so glänzende Bestätigung gefunden hat durch die schönen Versuche von Gerlach und Stern[1]) über die Ablenkung von Silberatomen in einem inhomogenen Magnetfelde. Unabhängig von dem Werte, der dem magnetischen Momente der einzelnen Atome zuzuschreiben ist, ist der Mangel von Paramagnetismus der Edelgase daher erklärbar, wenn man annimmt, daß die Atome dieser Elemente in einem magnetischen Felde sich mit ihrer Impulsachse nur senkrecht zur Feldrichtung einstellen können. Diese Annahme lehnt sich, wie wir sehen werden, natürlich an die Resultate der quantentheoretischen Analyse der anomalen Zeemaneffekte, der auf die Edelgase folgenden Elementenfamilien an.

In der *Komplexstruktur der Alkalispektren* treffen wir einen Fall, wo eine einfache Anwendung der Betrachtungen in § 5 zunächst zu erwarten wäre, indem wir annehmen dürfen, daß hier im Gegensatz zu Helium jedenfalls in erster Annäherung die Bahnkonfiguration des Atomrestes unter dem Einfluß des Serienelektrons nur relativ kleine Änderungen erleiden kann. Die genannte Erwartung erweist sich nun tatsächlich insofern

1) W. Gerlach u. O. Stern, Zeitschr. f. Phys. 9. S. 349. 1922.

erfüllt, als von den Kombinationsregeln der Terme und näherungsweise auch von den allgemeinen Intensitätsverhältnissen der Komponenten Rechenschaft abgelegt werden kann, wenn man annimmt, daß die j-Werte der Termdublette für $k \geqq 2$ gleich $k + d$ und $k + d - 1$ sind, während in dem einfachen s-Terme ($k = 1$) wir j gleich $1 + d$ setzen, wo d eine zunächst unbestimmte, von k unabhängige Größe ist. Andererseits ist es aber nicht möglich, einer derartigen Mannigfaltigkeit von j-Werten Rechnung zu tragen, so lange wir, wie in § 5, uns streng auf den Boden der Quantentheorie der Periodizitätssysteme stellen. In der Tat liefern uns die Bedingungen (29) immer eine ungerade Multiplizität der Termkomponenten, wenn wir annehmen, daß die Größen i und j ganze Zahlen sind. Um zu Dubletts zu gelangen, müßten wir daher gewisse j-Werte ausschließen, wofür aus Betrachtungen, wie denen in § 5, keine einfache Begründung abgeleitet werden kann.

Bei dieser Sachlage ist es nun befriedigend, daß die von den empirischen Spektren geforderten Stabilitätseigenschaften von Atomen mit mehreren Elektronen schon Konsequenzen allgemeiner Art mit sich führen, die mit wesentlichen Voraussetzungen der Quantentheorie der Periodizitätssysteme die statistischen Anwendungen betreffend nicht in Einklang zu bringen sind. So liegt es im Wesen dieser Theorie, daß das Gewicht eines stationären Zustandes eines Systems, das aus mehreren zusammengekoppelten Teilsystemen besteht, durch das Produkt der Gewichte derjenigen stationären Zustände der Teilsysteme gegeben ist, die kontinuierlich erhalten werden, wenn die Koppelungskräfte zum Verschwinden gebracht werden. Diese Forderung läßt sich nun aber bei der Koppelung des Serienelektrons an den Atomrest nicht mit der Erfahrung in Einklang bringen, wenn wir an den Quantisierungsregeln des § 5 festhalten.

Bei der Bestimmung des statistischen Gewichts eines stationären Zustandes eines Kernatoms in der Abwesenheit von äußeren Feldern ist zunächst zu berücksichtigen, daß das Atom ein entartetes System bildet, schon deshalb, weil die Lage der Impulsachse im Raum willkürlich ist; in der Anwesenheit eines äußeren Kraftfeldes wird diese Willkür aufgehoben und in der quantentheoretischen Analyse des Zeemaneffektes besitzen wir tatsächlich ein direktes Mittel, die Anzahl von stationären Zuständen, in die sich jeder stationäre Zustand des ungestörten Atoms in Anwesenheit des Magnetfeldes aufspaltet, zu bestimmen. So ergeben für die Alkalispektren die Landéschen Untersuchungen, daß die Anzahl der Termwerte im Magnetfeld, zu der die Zustände, in denen das Serienelektron eine n_k-Bahn beschreibt, Anlaß geben, im ganzen gleich $4k - 2$ ist. Einerseits müssen wir nun in unserem Fall, wie erwähnt, annehmen, daß der Atomrest nur eine Stellung im Magnetfelde annehmen kann. Andererseits kann eine n_k-Bahn in einem reinen Zentralsystem im Felde $2k$ Stellungen einnehmen, wenn man die Stellung ausschließt, wo das Impulsmoment senkrecht zur Feldrichtung steht, und die im Grenzfall, wo die Zentralbahn in eine Keplerbahn übergeht, als singulär anzusehen ist. Das Produkt dieser beiden Zahlen wird nun für große k nur etwa halb so groß wie die Anzahl der beobachteten Zustände des Alkaliatoms im Magnetfelde. Wir müssen deshalb schließen, daß die Koppelung zwischen Atomrest und Serienelektron nicht einfach als analog mit einer Koppelung zwischen zwei reinen Periodizitätssysteme aufgefaßt werden kann, in der einfachen Weise, die bei den Betrachtungen in § 5 vorausgesetzt war. Wir werden vielmehr zu der Ansicht geleitet, daß wegen der mechanisch unbeschreibbaren Stabilitätseigenschaften der Atome die Koppelung des Serienelektrons an den Atomrest einen *Zwang* mitführt, der nicht mit der Wirkung äußerer Kraftfelder analog ist, und der mit sich bringt, daß dem Atomrest anstatt der einzigen Einstellungsmöglichkeit in einem konstanten äußeren Feld zwei verschiedene Stellungen im Atom aufgezwungen werden, während andererseits demselben Zwang zufolge das äußere Elektron anstatt der $2k$ Einstellungsmöglichkeiten in einem äußeren Feld im Atomverband nur $2k - 1$ Einstellungsmöglichkeiten vertritt.

Unter diesen Umständen müssen wir darauf vorbereitet sein, daß die Beschreibung der Koppelung zwischen Serienelektron und Atomrest nicht in direkter Anlehnung an die Quantisierungsregeln für Periodizitätssysteme sich durchführen läßt. Einen bedeutungsvollen Fingerzeig dafür, in welcher Richtung eine Durchbrechung dieser Regeln zu suchen ist, hat nun eben wieder die Landésche Analyse des anomalen Zeemaneffektes gegeben, indem diese, wie erwähnt, unter Zugrundelegung möglichst einfacher Annahmen zu Werten der Drehimpulskomponente des Atoms in der Feldrichtung führt, die durch $\mu \frac{h}{2\pi}$ dargestellt werden können, wo μ nicht immer eine ganze Zahl ist, sondern in gewissen Fällen, so wie eben bei den Alkalispektren, „halbzahlig" ist.

Dieser Befund legt die Vermutung nahe, daß auch die Orientierung der Bahn des Serienelektrons relativ zum Atomreste nicht durch ganzzahlige j-Werte beschrieben werden kann und speziell dürfte aus Symmetriegründen die formale Annahme $d = \frac{1}{2}$ sich darbieten.[1]) Vergleichen wir weiter die nach dieser Annahme festgelegten j-Werte mit den von Landé aus dem Zeemaneffekt formal ermittelten möglichen Werten der Drehimpulskomponenten parallel dem äußeren Magnetfelde, so finden wir, daß diese Werte eben allen möglichen halbzahligen μ-Werten entsprechen mit Ausschließung von solchen Orientierungen des Atoms, wo der Gesamtimpuls

1) In seinem kühnen Versuch, mittels sinnreich gewählter Annahmen die Einzelheiten der anomalen Zeemaneffekte in Anlehnung an Landés und Sommerfelds Arbeiten zu erklären, wurde zuerst von Heisenberg (Zeitschr. für Phys. 8. S. 273. 1921) die Annahme gemacht, daß die stationären Zustände des Atoms nicht direkt mit Hilfe von Bedingungen der Form (29) mit ganzzahligen Quantenzahlen beschrieben werden können. In Heisenbergs Untersuchung sowie in den späteren Arbeiten von Landé wird jedoch angenommen, daß sogar die Werte der Quantenzahl k halbzahlig anzusetzen sind. Im Hinblick auf die Weise, in welcher eine Erklärung des allgemeinen Baues der Serienspektren in Anlehnung an die Quantentheorie der Periodizitätssysteme sich erzielen ließ, dürfte aber diese Annahme, deren Konsequenzen den Erfahrungen über die Spektren an verschiedenen Punkten zu widersprechen scheinen (vgl. z. B. Fußnote S. 266), eine kaum begründbare Abweichung von dieser Theorie bezeichnen. Wie oben darzulegen versucht ist, liegen die Verhältnisse anders, was die Festlegung der Werte der Quantenzahl j betrifft. Hier sind ja wesentliche Voraussetzungen der Quantentheorie reiner Periodizitätssysteme schon durchbrochen durch die allgemeinen Konsequenzen, die die Koppelung zwischen Serienelektron und Atomrest mit sich bringt. Überhaupt gibt die Heranziehung von halbzahligen j-Werten bei der im Texte versuchten formalen Darstellung der Komplexstrukturen nur Ausdruck für die Ansicht, daß in diesen Fragen beim jetzigen Stand der Theorie kein direkter Anschluß an die gesicherten Züge der Quantentheorie der Periodizitätssysteme erreicht werden kann.

der Feldrichtung parallel wäre. Dieser Umstand kommt u. a. darin zum Ausdruck, daß bei den Alkalispektren die s-Terme einfach sind, und dem j-Werte $j = 1\frac{1}{2}$ entsprechen. Ein Zustand $j = \frac{1}{2}$ würde nämlich, unserer formalen Ausschlußregel gemäß, keine Einstellungsmöglichkeit im Magnetfeld besitzen, und würde deshalb kein statistisches Gewicht besitzen. Es mag noch in dieser Verbindung bemerkt werden, daß der betreffende Befund genau dem Verhalten der Edelgaskonfigurationen im Magnetfeld analog ist. Wenn wir für diese wie oben $j = 1$ annehmen, führt die Ausschließung von Parallelstellungen ja eben zu der Konsequenz, daß das Atom sich mit seiner Impulsachse nur senkrecht zur Richtung des Feldes einstellen kann. Übrigens muß hervorgehoben werden, daß wir darauf vorbereitet sein müssen, daß die dynamischen und magnetischen Eigenschaften der Edelgaskonfigurationen im Falle, wo sie als Atomrest auftreten, infolge des Zwanges wesentlich andere sein können, als wenn sie ein freies Atom bilden.

Ganz ähnliche Betrachtungen wie für die Alkalispektren gelten sowohl für die Spektren des Cu und Ag wie für die Funkenspektren der Erdalkalien, die denselben Typus von Komplexstruktur aufweisen. Dem Wesen der Sache nach hängt die Komplexstruktur ja nur von den Werten der Quantenzahlen i, k, j ab, und die Verhältnisse, die für die Festlegung der Hauptquantenzahl n maßgebend waren, kommen hier nicht in Frage. Es handelt sich hier um ein Beispiel von dem sogenannten *spektroskopischen Verschiebungssatz*, der zuerst von Sommerfeld und Kossel[1]) formuliert wurde, und der wertvolle Dienste geleistet hat bei der Arbeit der Entwirrung der Funkenspektren mancher Elemente.[2])

Gehen wir zu den *Bogenspektren der Erdalkalien* über, so treffen wir Verhältnisse, die sich ungezwungen in den oben angedeuteten Rahmen einfügen lassen. In der Mannigfaltigkeit der Terme der betrachteten Spektren tritt bekanntlich ein *System von Triplettermen* auf, und wir dürfen annehmen, um nach (27) dem empirischen Kombinationsregeln formal gerecht zu werden, daß die j-Werte durch $k+t$, $k+t-1$ und $k+t-2$

---

1) A. Sommerfeld u. W. Kossel, Verh. d. deutschen phys. Ges. 21. S. 240. 1919; vgl. auch N. Bohr, Quantentheorie der Linienspektren.

2) Vgl. E. Fues, Ann. d. Phys. 63. S. 1. 1920.

gegeben sind, während für die entsprechenden s-Terme, die einfach sind, j gleich $1 + t$ ist. Für die Annahme $t = 1$ läßt sich nun eine analoge Begründung geben wie für die obige Annahme $d = \frac{1}{2}$ bei den Alkalispektren. In der Tat hat Landés Analyse des Zeemaneffektes der Erdalkalispektren gezeigt, daß wir die Drehimpulskomponenten parallel dem Magnetfelde immer als ganzzahlige Vielfache von $h/2\pi$ ansehen müssen. Diese Analyse hat weiter für die Tripletsysteme als Gesamtzahl der Stellungen eines n_k-Zustandes im Magnetfeld $3(2k-1)$ ergeben. Analog zum obigen kann dieser Befund dahin gedeutet werden, daß die Koppelung des Serienelektrons an den Atomrest einerseits bewirkt, daß letzterem anstatt der zwei Einstellungsmöglichkeiten in einem äußeren Feld, die dem Normalzustand der Alkaliatome zukommen, drei Einstellungsmöglichkeiten zuzuschreiben sind, während sie andererseits bewirkt, daß die Bewegung des äußeren Elektrons anstatt $2k$ nur $2k - 1$ Zustände vertritt. Hiermit dürfte auch eng der Umstand zusammenhängen, daß, wenn die j-Werte in der erwähnten Weise gewählt werden, bei der Einstellung des Atoms im Magnetfelde alle Parallelstellungen der Impulsachse relativ zur Feldrichtung genau so wie bei den Alkalispektren auszuschließen sind. Es muß aber bemerkt werden, daß wir bei der Frage, wie die Einfachheit der s-Terme zu erklären sei, bei den alkalischen Erden nicht ganz so einfache Umstände wie bei den Alkalien treffen, da diese Ausschließungsregel für die Einstellungen im Magnetfelde offenbar eine Reihe von s-Zuständen, für die $j = 1$ ist, keineswegs verbietet. Es ist daher befriedigend, daß die Ausschließung der betreffenden s-Zustände mit der Forderung in Übereinstimmung ist, daß der allgemeine Ausdruck für die gesamte Anzahl der Zustände in einem Magnetfeld, die einer n_k-Bahn des Serienelektrons entsprechen, auch für $k = 1$ gültig bleiben soll; eine Forderung, für die man, wie Hr. Pauli in einer bald erscheinenden Arbeit zeigen wird, durch eine Betrachtung der Umwandlung des Zeemaneffektes für wachsende Feldstärke eine einfache Deutung erhalten kann.

Bei den Spektren der Erdalkalien tritt bekanntlich außer dem Tripletsystem ein *System von Einfachtermen* auf, für welche wir, um den Regeln der empirischen Gesetzmäßigkeiten der

Kombinationen mit dem Triblettermen gerecht zu werden, nach obiger Wahl der j-Werte der letzteren Terme eindeutig auf die Annahme $j = k$ geführt werden. Unabhängig von einer näheren Interpretation der den Einfachtermen entsprechenden Atombewegung werden wir weiter zu einem unmittelbaren Verständnis des empirischen Befundes geführt, nach welchem die Kombinationsmöglichkeiten mit dem Triplettermen außer durch die Bedingung (32) noch dadurch beschränkt sind, daß die Kombination der S-Terme ($j = 1$) mit dem p_3-Term ($j = 1$) nicht vorkommt. In einem äußeren Magnetfeld steht nämlich die Impulsachse des Atoms für die betreffenden Zustände immer senkrecht zur magnetischen Kraft, und es tritt daher offenbar keine harmonische Schwingungskomponente parallel dem Felde auf, die mit einem Übergang $S - p_3$ korrespondieren würde. Eine Betrachtung des entsprechenden Falles des Ausbleibens gewisser Zeemaneffektkomponenten findet sich in der schon genannten Arbeit über die Intensitäten im anomalen Zeemaneffekt, die Sommerfeld und Heisenberg auf Grund des Korrespondenzprinzips gegeben haben (vgl. S. 252). Bei diesen Verfassern, sowie in den früheren Abhandlungen von Landé wird aber aus dem Nichtauftreten der Kombinationen $S - p_3$ der Schluß gezogen, daß das Atom in den beiden betreffenden stationären Zuständen keinen resultierenden Impuls besitzt. Wir sehen aber, daß dieser Schluß nicht zwingend ist, sondern daß die Beschränkung der Einstellungsmöglichkeiten des Atoms im Magnetfeld mit sich bringt, daß die betreffende Kombinationslinie schon bei Anwesenheit schwächster äußerer Felder nicht vorkommen kann.

Wie bereits in § 5 erwähnt, bedeutet das Auftreten der verschiedenen Typen von Termen, daß Bahnkonfigurationen des Atomrestes mit verschiedenen Werten ihres Gesamtdrehimpulses auftreten. Was das nähere Verständnis dieses Umstandes betrifft, ist es ein bedeutungsvoller Fingerzeig, daß unsere auf die Ausschließung der Parallelstellungen im Magnetfelde beruhende Begründung für die Einfachheit des s-Zustandes bei den Alkaliatomen ihre Gültigkeit verliert, wenn die betreffende Bahnkonfiguration als Atomrest auftritt, falls nur der Gesamtimpuls des Atoms einem Werte $j \geqq 1$ entspricht. Unabhängig von diesen Argumenten kann man schon aus einer Betrachtung der Termwerte selbst eine interessante Stütze erreichen für die Annahme, daß das Auftreten der Einfachterme mit einer durchgreifenden Umlagerung der Bahnkonfigurationen im Atomreste zusammenhängt.

Um den beobachteten Termwerten gerecht zu werden, müssen wir bei den Erdalkalien für die Einfachterme bei allen Termreihen entsprechende Werte der Hauptquantenzahl annehmen, wie bei den Funkenspektren desselben Elementes. Bei den Triplettermen gilt zwar ähnliches für die p-, d- und f-Terme. Dem größten s-Term dagegen müssen wir aber einen Wert der Hauptquantenzahl zuordnen, der um eine Einheit größer ist als die Nummer der Periode im System der Elemente. Zu einem Verständnisse dieses Resultats gelangen wir durch die Überlegung, daß das Auftreten mehrerer Elektronenbahnen mit denselben Werten der Quantenzahlen n und k im selben Gebiet des Atoms nur möglich ist, wenn sie ein harmonisches Wechselspiel bilden können. Ähnlich wie bei der Deutung des metastabilen Heliumzustandes ist ein solches Wechselspiel jedoch im allgemeinen nicht als Endresultat eines Strahlungsprozesses zu erreichen. Wenn aber die Bahnkonfiguration des Atomrestes zur selben Zeit eine wesentliche Umlagerung erfährt, wird eine Möglichkeit geöffnet für die Aufnahme eines weiteren Elektrons in eine schon vorhandene Elektrongruppe.

Beim jetzigen Stand der Theorie ist es schwierig, sich bestimmte Vorstellungen zu machen von der Bahnkonfiguration des Atomrestes in den den Einfachtermen entsprechenden stationären Zuständen, die uns in Anlehnung an die Beobachtungen über die Funkenspektren erlauben könnten, die dynamischen Eigenschaften und speziell den Energiewert des Atomrestes zu bestimmen. Eben der Umstand, daß das Vorkommen dieser Konfiguration nach unserer Anschauung durch die Koppelung eines weiteren Elektrons an das Atom bedingt ist, dürfte es aber nicht nur begreiflich machen, daß ein Zustand, in welchem ein harmonisches Wechselspiel der beiden Valenzelektronen hergestellt ist, der Mannigfaltigkeit der Einfachterme angehört, sondern zugleich ein Verständnis dafür ergeben, daß die Energie des Atoms, wie die Spektren verlangen, für die Einfachterme sich mit wachsender Hauptquantenzahl demselben Werte nähert wie für die Tripletterme.

Bekanntlich zeigen die Bogenspektren der Elemente Zn, Cd, die auf Cu und Ag folgen, sowie das Hg-Spektrum denselben charakteristischen Bau wie die Bogenspektren der Erdalkalien. In der Tat müssen wir ja auch annehmen, daß die Verhältnisse, soweit es die Quantenzahlen k und j betrifft, genau dieselben sind. Andererseits dürfen wir annehmen, daß auf Grund der relativ kleineren Dimensionen des Atomrestes die Bahnen, die $k \geqq 2$ entsprechen, ganz außerhalb derselben verlaufen, und daß wie beim Bogenspektrum des Cu und Ag die Werte der Hauptquantenzahl für die Reihe der p- und P-Terme mit zwei, für die Reihe der d- und D-Terme mit drei anfangen. Diese Annahme, die mit der Größe der Terme in Übereinstimmung ist, bekommt eine anscheinend zwingende Stütze durch eine von Hansen, Takamine und Werner in Kopenhagen ausgeführte eingehende Untersuchung über den Starkeffekt des Quecksilberspektrums[1]), die gezeigt hat, daß nicht nur für die d-Terme, sondern auch für die höheren Glieder der p-Terme die Bahn des Serienelektrons nur sehr wenig von einer Keplerellipse abweichen kann.

* Die Spektren der Elemente Al, Ca, In und Tl zeigen wieder eine Komplexstruktur von derselben Art wie die der Alkalibogenspektren, wie es auch zu erwarten wäre wegen der angenommenen Gleichheit des Normalzustandes der Erdalkalien und der Edelgase sowohl was das resultierende Impulsmoment wie die Einstellungsmöglichkeit der Atome im Magnetfeld betrifft. Zugleich zeigen diese Spektren verglichen mit den Alkalispektren typische Unterschiede, die genau den theoretischen Erwartungen entsprechen. Erstens ist die relativ geringe Größe des ersten s-Termes dadurch bedingt, daß die Hauptquantenzahl hier um 1 größer ist als bei dem Alkalimetalle in derselben Periode des Systems der Elemente, was wieder damit zusammenhängt, daß wir in dem Normalzustand der vorangehenden Elemente eine vorläufig abgeschlossene Elektronbahnkonfiguration von zwei n_1-Bahnen begegnen, wo n eben die Nummer der Periode im System der Elemente ist. Infolge

1) Über diese Untersuchung, die mehrere interessante Resultate betreffend die Beeinflussung komplizierter Spektren durch elektrische Felder ergeben hat, wird bald in den Schriften der Kopenhagener Akademie ein ausführlicher Bericht erscheinen.

* [Instead of Ca read Ga.]

dieses Umstandes wird im Normalzustand der betreffenden Elemente die Bahn des zuletzt eingefangenen Elektrons durch $k = 2$ gekennzeichnet sein. Wegen der großen Dimensionen des Atomrestes bei den betreffenden Elementen dringen die Bahnen $k = 2$ nicht nur in das Atominnere hinein, sondern für den ersten p-Term, der dem Normalzustand des Atoms entspricht, liegt die Elektronenbahn sogar ganz innerhalb des Gebietes der früher gebundenen Elektronen. Die Hauptquantenzahl dieser Bahn ist der Nummer der Periode im System der Elemente gleich.

Bei Aluminium bieten die Paschenschen Messungen des Serienspektrums dritter Art eine Grundlage für eine einfache Abschätzung der Dimensionen des Atomrestes. Aus dem Wert des größten s-Terms dieses Spektrums lesen wir direkt ab, daß das Elektron, das sich im zweifach geladenen Ion in einer 3_1-Bahn bewegt, während seines Umlaufs einen Abstand vom Kern erreicht, der nahezu dreimal so groß ist, wie der Radius der Elektronenbahn im normalen Wasserstoffatom. Im einfach geladenen Aluminiumion, wo zwei 3_1-Bahnen vorhanden sind, werden die Bahndimensionen der betreffenden Elektronen noch wesentlich größer. Dieses bringt nicht nur mit sich, daß eine Bahn mit $k = 2$ nicht ganz im äußeren Gebiet liegen kann, sondern sogar, daß das Elektron einer 3_2-Bahn niemals so weit vom Kerne kommt wie die Valenzelektronen in 3_1-Bahnen, was andererseits wieder ein weiteres Anwachsen der Bahndimensionen dieses Elektrons im Normalzustand des Atoms bedingt. In solchen Fällen ist es klar, daß man nicht erwarten darf, daß man eine angenäherte Abschätzung der Termgrößen mittels eines Vergleiches mit einer Elektronenbewegung in einem festen Zentralfeld erreichen kann, so wie es in gewissem Umfang möglich gewesen ist bei den Spektren, wo der Atomrest eine Edelgaskonfiguration bildet. Die besonders große Dimension des Atomrestes bei Aluminium und die geringe Symmetrie seiner Bahnkonfigurationen ermöglichen weiter ein Verständnis für die eigentümliche große Abweichung von den Wasserstofftermen, die die d-Terme des Bogenspektrums zeigen.

Bei der betrachteten Elementengruppe treffen wir einen neuen Zug im Aufbau der Atome, indem wir hier zum erstenmal nach dem Auftreten einer Edelgaskonfiguration Zeugen sind vom Auftreten von Bahnen im Normalzustand des Atoms, für die $k = 2$ ist und für die die Hauptquantenzahl der Nummer der Periode im System der Elemente gleich ist. Dieses bedingt die Einleitung einer neuen Stufe in der Entwicklung der äußeren Elektronengruppe im Atom und bringt mit sich, daß wir neue Verhältnisse erwarten müssen bei den Spektren, die die folgenden Elemente zeigen. Obwohl hier versprechende Anfänge gemacht sind, sind weder die empirischen noch die theoretischen Verhältnisse genügend geklärt, damit wir bei dieser Gelegenheit darauf eingehen können.

Wie schon im vorhergehenden Paragraphen erwähnt, begegnen wir bei den in dem Schema auf S. 262 eingerahmten Elementenreihen Verhältnissen, die durch eine vorübergehende *Entwicklungsstufe einer inneren Elektronengruppe* bedingt sind. Bei diesen Elementen treffen wir, wie es theoretisch zu erwarten wäre, ausgesprochene Abweichungen vom oben erwähnten spektroskopischen Verschiebungssatze. Wie es zuerst durch die eingehende Untersuchung von Catalán über das Manganspektrum klargelegt wurde, begegnen wir hier sogar ganz neuen Typen der Komplexstruktur der Serienlinien, den sogenannten Multiplets.[1]) Das Auftreten dieser Multiplets ist bedingt durch größere Werte des Gesamtimpulses des Atomrestes, als wir bei den Dublet- und Tripletspektren begegnet haben, was sich auch in den magnetischen Eigenschaften der Ionen dieser Elementenreihen in charakteristischer Weise äußert. Auch in anderer Weise bewirken die Verhältnisse, die die Ausbildung der Elektronengruppen begleiten, daß die Mannigfaltigkeit der stationären Zustände, mit denen wir bei der Deutung der Spektren zu rechnen haben, eine viel größere Verwickeltheit aufweist, als wir nach der einfachen in §§ 4 und 5 dargelegten schematischen Deutung der Spektralgesetze erwarten sollten.

Wir werden hier nicht näher auf diese komplizierteren Fragen eingehen, bei deren Untersuchung die Unvollständig-

---

1) M. Catalán, Trans. Roy. Soc. A. 223. S. 127—173. 1922. Vgl. in dieser Verbindung auch die Untersuchung von H. Gieseler über das Chromspektrum. (Ann. d. Phys. 69. S. 147. 1922.) Für die theoretische Diskussion der betreffenden Typen von Spektren siehe weiter A. Sommerfeld, Ann. d. Phys. 70. S. 32. 1923, sowie G. Wentzel, Phys. Zeitschr. 24. S. 104. 1923.

keit der theoretischen Grundlagen, die sich schon bei der Behandlung der einfacheren in diesem letzten Paragraphen besprochenen Fragen so stark geltend gemacht hat, uns immer deutlicher entgegentritt. Diese Unvollständigkeit hervorzuheben mußte dem Wesen der Sachen nach eine der Hauptaufgaben dieses Berichtes bilden, deren Absicht es ja in erster Linie war, einen Eindruck von dem jetzigen Stand der Bestrebungen zu geben, die darauf hinzielen, die Gesetzmäßigkeiten der Spektren zur Entwicklung von Vorstellungen über den Atombau zu verwerten, um so die Hoffnung zu erfüllen, die vom Anfang der spektroskopischen Wissenschaft ihren Pflegern vor Augen geschwebt hat.

Bei dem Abschluß dieser Arbeit werde ich gern meinen Mitarbeitern herzlich danken, besonders den Herren H. A. Kramers und W. Pauli, die mich bei der Abfassung des Manuskriptes und der Diskussion vieler Einzelheiten der Theorie tatkräftig unterstützt haben, sowie Hrn. S. Werner, der mir bei der Diskussion des empirischen Materials und der Ausführung der graphischen Darstellungen der Spektren wertvolle Hilfe geleistet hat.

(Eingegangen 15. März 1923.)

Erklärung der Tafeln.

Die schwarzen Punkte vertreten einige der größten Glieder der Reihen von empirischen Spektraltermen, deren übliche spektroskopischen Bezeichnungen auf der linken Seite angeführt sind. Der Abstand der Punkte bis zur Begrenzungslinie auf der rechten Seite der Figur ist den numerischen Termwerten, für die eine Skala in Wellenzahlen unten an den Tafeln angebracht ist, proportional. Die jeder Tafel oben beigefügte Skala gibt die Werte der effektiven Quantenzahl n^* an, die für ein Spektrum p-ter Art (vgl. S. 243) mit dem Termwerte T (in Wellenzahlen ausgedrückt) eindeutig durch die Formel $T = Rp^2/n^{*2}$ verknüpft ist, wo R die Rydbergsche Zahl, deren Wert durch den Ausdruck $R = 109\,737\,(1 - 5{,}4 \cdot 10^{-4}/A)$ gegeben ist, wo A das Atomgewicht des Elementes relativ zum Wasserstoff bedeutet.

Die den Punkten beigefügten Zahlen geben die Werte der Hauptquantenzahl der den Termen entsprechenden Bahnen des Serienelektrons an. Diese Zahl ist geeignet, die Laufzahl innerhalb der einzelnen Termreihen zu vertreten. Die Nebenquantenzahlen k und j der Elektronenbahnen, deren Werte den einzelnen Termreihen zugeordnet sind, ergeben in erster Linie die Beschränkung in den Kombinationsmöglichkeiten der Terme bei Abwesenheit von äußeren Kräften, die dadurch geregelt wird, daß k sich nur um eine Einheit ändern kann, während j sich entweder um eine Einheit ändert oder unverändert bleibt. Dazu tritt noch die weitere Einschränkung, daß eine Kombination, bei der die j-Werte beide gleich 1 sind, nicht auftreten kann. Es muß besonders betont werden, daß die Theorie hinsichtlich der Bedeutung und Festlegung der Absolutwerte der Quantenzahl j keinen abgeklärten Charakter besitzt, was in den Tafeln durch die angebrachten Klammern zum Ausdruck gebracht ist.

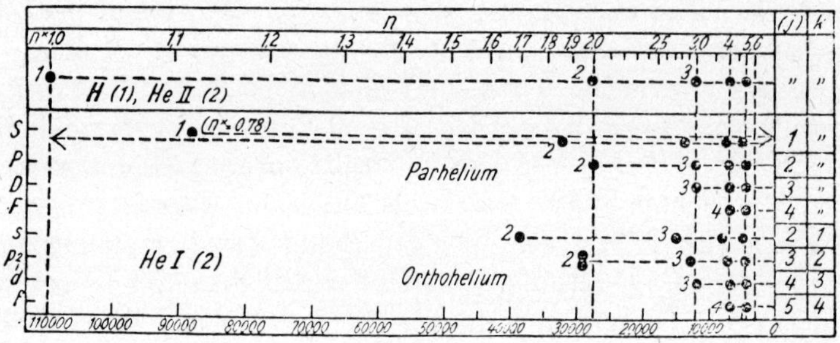

Tafel I.

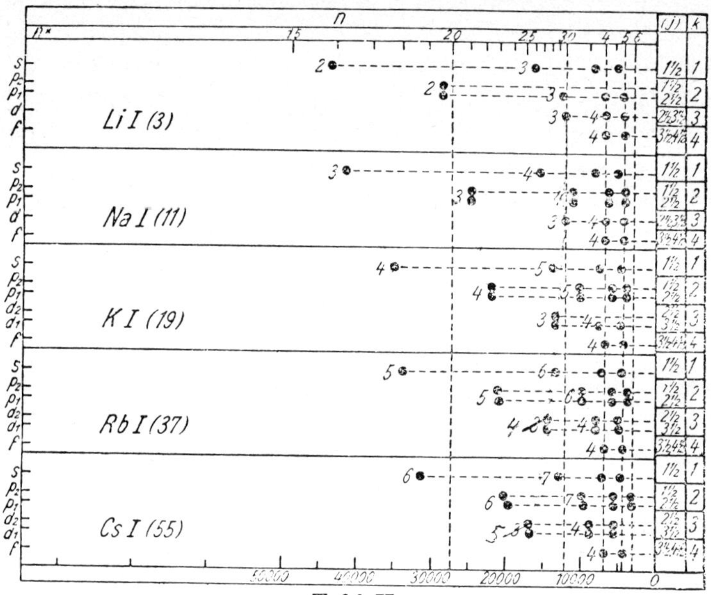

Tafel II.

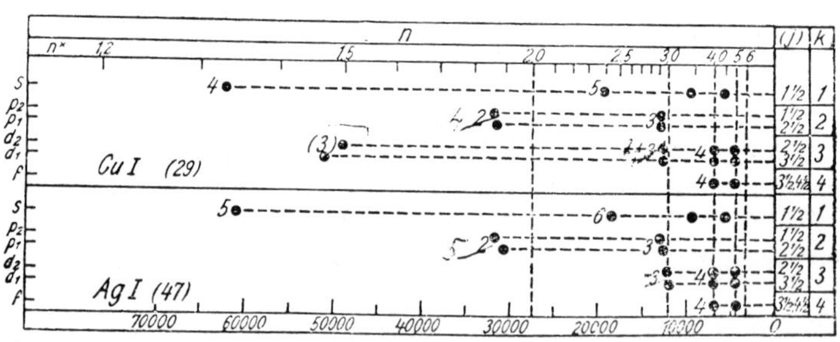

Tafel III.

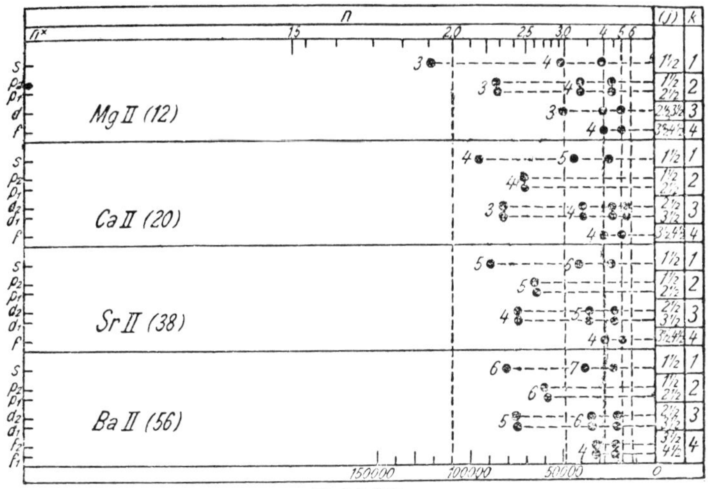

Tafel IV.

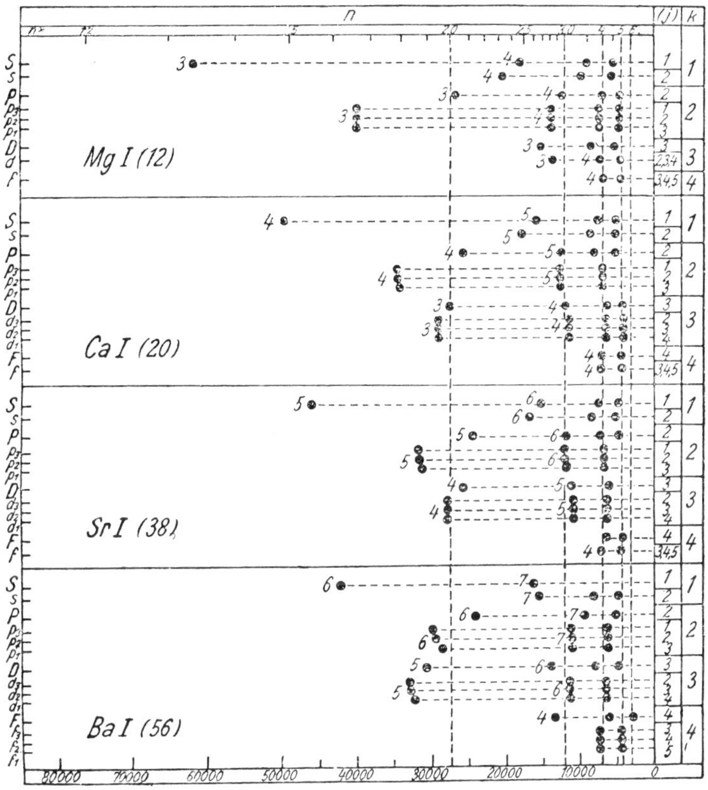

Tafel V.

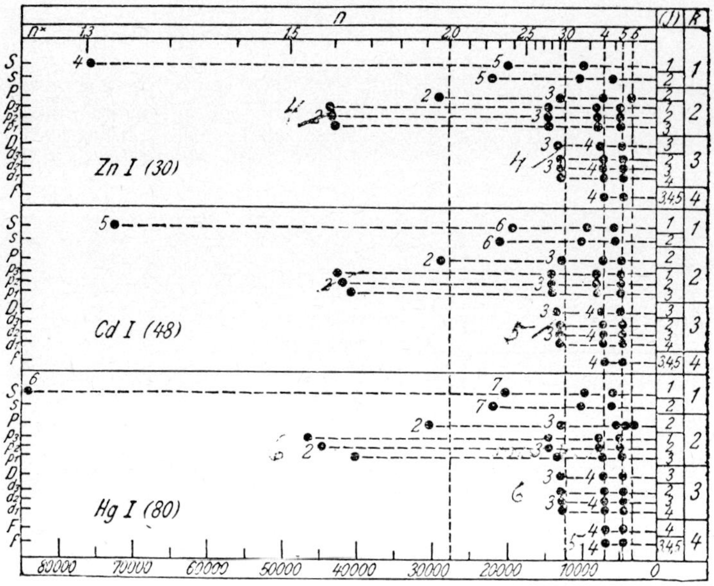

Tafel VI.

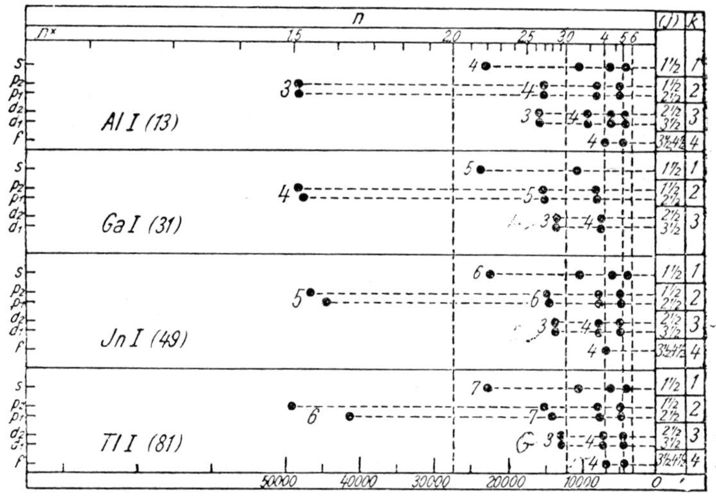

Tafel VII.

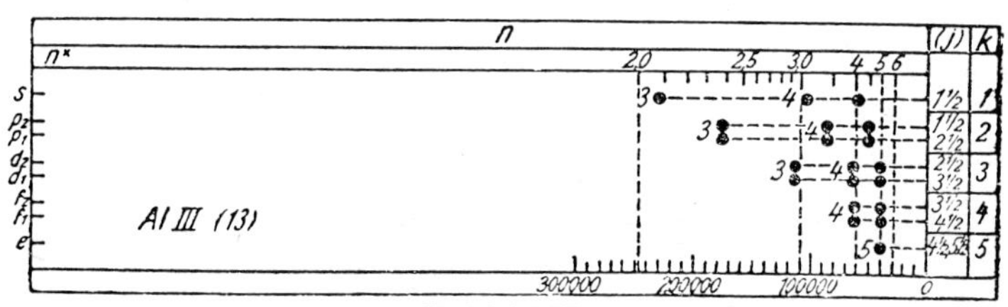

Tafel VIII.

Druck von Metzger & Wittig in Leipzig.

TRANSLATION

Line Spectra and Atomic Structure

§ 1. General points of view.

Since the discovery of the line spectra characteristic of the elements, it has been a general conviction of physicists that we possess in these spectra a valuable means of exploring atomic structure. The hope of being able to advance in this way was supported above all by the discovery of the peculiar laws obeyed by the frequencies of the lines of many spectra; obeyed, in fact, with an accuracy not otherwise attained in physics. This circumstance has formed the inducement to the efforts made to measure the spectra and disentangle the spectral laws. The strongest impression of this effort is probably given by the great work with which Kayser*, who himself has had a prominent share in accomplishing both of these tasks, has enriched the physical literature, a work which has always been an indispensable source for theoretical as well as for experimental spectroscopic research.

The efforts to make theoretical use of the empirical spectroscopic material for the problem of atomic structure became especially tempting after the gradual recognition that a meaningful interpretation of the spectral laws cannot be

* H. Kayser, *Handbuch der Spektroskopie*, Vols. I–VI (Leipzig 1900–1912).

carried out on the basis of classical mechanics and electrodynamics*. This difficulty stood out even more sharply after the great progress in our knowledge of atomic structure brought about by investigations of electric discharges in gases and of radioactive phenomena. The result of these investigations was that atoms are built up of electric particles, i.e., negative electrons and positive nuclei, whose dimensions are extremely small compared to the size of an entire atom, as given by the atomic volumes of the elements. Thus, on the basis of Rutherford's conception of the *nuclear atom*, we assume that an atom of an element consists of a central positive nucleus, which is the site of the preponderate part of the mass of the atom, around which there revolve a number of electrons equal, in the neutral atom, to the so-called atomic number giving the place of the element in the periodic system. However, as long as we rely on the principles of ordinary mechanics and electrodynamics, this conception is not compatible with the demand that the atoms possess a stability revealed in the specific properties of the elements, and about which not least the line spectra bear such striking witness.

However, in the meantime physicists had already, in the so-called law of heat radiation, found a touch-stone for the applicability of the classical laws to the emission and absorption of radiation by atoms, independent of any special assumptions about their structure. The theoretical investigations of the law of heat radiation had not only revealed with certainty the failure of classical electrodynamics, but had even given a hint as to the direction in which the necessary changes in the classical laws were to be sought. Thus, by Planck's well-known derivation of this law, the impetus was given to the development of the so-called *quantum theory*.

Now, the attempt to connect the results won in the field of atomic structure with the data of spectroscopy led to a formulation of the principles of the quantum theory which aimed directly at accounting for the stability of atoms, as well as for the character of the regularities of line spectra**. This formulation depends on setting up the following postulates:

I. Among the conceivable states of motion of an atomic system, there exist a number of so-called *stationary states*, in which the motion of the particles to a great extent follows the classical mechanical laws, but which are distinguished by a peculiar stability that cannot be explained mechanically and that brings about

* At the end of the book by H. Konen, *Das Leuchten der Gase und Dämpfe* (Braunschweig 1913), which contains a survey of empirical spectroscopy, is an interesting compilation of the attempts made to interpret the spectral laws on the basis of the conceptions of classical mechanics and electrodynamics.

** Cf. N. Bohr, Phil. Mag. (6) **26** (1913) pp. 1, 476, 857. In German translation: *Abhandlungen über Atombau* (Braunschweig 1921) pp. 1, 26, 52. (In the following quoted as "*Abh. über Atombau*".)

that every lasting change in the motion of a system must consist in a complete transition from one stationary state to another.

II. While, contrary to classical electrodynamics, no radiation is emitted in the stationary states themselves, a transition between two stationary states may be accompanied by emission of electromagnetic radiation of the same nature as that emitted, according to the classical theory, by an electrical particle executing a harmonic oscillation with constant frequency. However, this frequency ν is not related in a simple way to the motion of the particles in the atom, but is given by the so-called *frequency relation*

$$\nu = \frac{E' - E''}{h}, \tag{1}$$

where h is Planck's constant, and E' and E'' are the energies of the atom in the initial and final states of the emission process. Conversely, irradiation of the atom by a train of waves with this frequency may give rise to a process of absorption in which the atom is returned from the latter to the former stationary state.

While the first postulate aims at the general stability of atoms, the second postulate primarily keeps in view the existence of spectra consisting of sharp lines. Now, condition (1), whose relationship to the assumptions used in Einstein's theory of the photoelectric effect is evident, provided a basis for the interpretation of the empirical spectral laws. According to the most general of these laws, the *combination principle* formulated by Ritz, the frequency ν of any line in a spectrum can be represented by the formula

$$\nu = T' - T'', \tag{2}$$

where T' and T'' are two so-called spectral terms belonging to a set of terms characteristic of the spectrum in question. On the basis of the postulates stated, this principle, quite alien to the laws governing the frequencies of mechanical systems, provides a simple interpretation, since, according to condition (1), we can put the term values T directly equal to the energy values E, divided by h, of the different stationary states of the atom emitting the spectrum.

The complete break of the theory with the concepts of classical electrodynamics appears particularly strikingly in this interpretation of the combination principle. In fact, this interpretation involves the assumption that the nature of the emitted radiation, even for the same state of motion of the atom at the beginning of the radiation process, can be quite different, all depending on the state of the atom at the end of the process. Nevertheless, we shall see that it is possible, in a certain sense, to regard the theory as a natural generalization of the classical radiation theory. Thus, we can correlate the occurrence of each of the different transitions accompanying emission with a certain harmonic oscillation component present in the atomic motion, a component which, according to the classical theory, would determine the wave trains appearing in the emitted radiation. We call this connection between spectrum and atomic motion the *correspondence principle*. As regards the interpretation of the spectral laws, this principle has, in particular, made possible a simple explanation of the apparent capriciousness with which lines predicted by the combination principle actually occur.

§ 2. Quantum theory of periodicity systems.

The formulation of the correspondence principle rests on a quantum-theoretical examination of atomic systems in which the motion of the particles can be resolved into a set of harmonic components with discrete frequencies, i.e., in which the displacement in a given direction of each particle, as a function of time, can be represented by an expression of the following form

$$\xi = \sum C_{\tau_1 \ldots \tau_r} \cos\left[2\pi(\tau_1\omega_1 + \ldots + \tau_r\omega_r)t + \gamma_{\tau_1 \ldots \tau_r}\right], \tag{3}$$

where $\tau_1 \ldots \tau_r$ are positive or negative integers, and where $\omega_1 \ldots \omega_r$ are the fundamental frequencies. Their number, which may be called the degree of periodicity of the motion, is uniquely determined by the condition that no relation of the form

$$m_1\omega_1 + \ldots + m_r\omega_r = 0,$$

where $m_1 \ldots m_r$ are positive or negative integers shall exist among them. In general, the summation in (3) is to be extended over all positive and negative values of $\tau_1 \ldots \tau_r$.

Now, the stationary states of such r-fold periodic systems are given by r *state conditions* of the form

$$J_k = n_k h, \quad (k = 1, \ldots r), \tag{4}$$

where h is Planck's constant, $n_1 \ldots n_r$ are positive integers, the so-called *quantum numbers*, while $J_1 \ldots J_r$ are a number of quantities which characterize the mechanical properties of the motion and obey the condition

$$\delta E = \sum_k \omega_k \delta J_k, \tag{5}$$

where E is the total energy of the system, and where the differential sign δ refers to the difference between the values of a quantity in two neighboring states of motion. The quantities J are fixed by this condition only to within additive constants. However, these are completely determined by the further condition

$$\sum_k \omega_k J_k = \sum_i \overline{p_i \dot{q}_i}, \tag{6}$$

where $q_1 \dots q_s$ are arbitrarily chosen generalized spatial coordinates describing our system of s degrees of freedom and $p_1 \dots p_s$ are the corresponding momenta; the bar on the right side indicates that an average is to be taken over a time long compared to the fundamental periods of the motion.

According to conditions (1) and (4), we now obtain for the frequency ν of the radiation emitted in a transition between two stationary states characterized by the quantum numbers $n'_1 \dots n'_r$ and $n''_1 \dots n''_r$, respectively,

$$\nu = \frac{1}{h}(E'-E'') = \frac{1}{h}\int_{''}^{'} \sum \omega_k \mathrm{d}J_k = \sum \overline{(n'_k - n''_k)\omega_k}, \tag{7}$$

where the averaging on the right side of eq. (7) is to be carried out over the states of motion which are represented, in the r-dimensional J-space, by a straight line joining the points $(J'_1 \dots J'_r)$ and $(J''_1 \dots J''_r)$ corresponding to the stationary states.

In the limit where the values of the quantum numbers are large compared to the differences $n'_k - n'_k$, we can regard the quantities ω as constants in the averaging, and we then obtain the asymptotic equation

$$\nu \sim \sum_k \omega_k (n'_k - n''_k). \tag{8}$$

Thus, in this limit the frequency of the radiation approaches asymptotically the frequency of the harmonic component in the motion for which the equations

$$\tau_k = n'_k - n''_k \tag{9}$$

hold. In order to obtain a connection between the statistical results of the quantum theory and the classical theory in the limit of large quantum numbers, we are thus led to regard the occurrence of a transition between two stationary states as dependent on the presence of a harmonic component defined by (9) in the motion, or rather in the electrical moment, of the system. The consideration that we are dealing merely with a statistical result, and not, say, with a gradual disappearance of the fundamental difference between the quantum theory and the classical theory in the region of large quantum numbers, leads us further to regard the connection, proven in this limiting region, between the occurrence of radiative transitions and the existence in the motion of the atom of harmonic components correlated according to (9) with the transitions, as a general quantum-theoretical law, valid for arbitrary values of the quantum numbers. This law represents the content of the above-mentioned *correspondence principle*, and we denote the associated component of the motion as the harmonic oscillation corresponding to the transition.

In addition to fixing the stationary states, we need, in problems of a statistical nature, a knowledge of the weights that must be assigned to the individual stationary states to allow us to calculate the probability of a distribution of energy over a large number of atoms. A basis for the treatment of this problem is contained in a general condition, derived by Ehrenfest from the second law of thermodynamics, stating that, in a continuous transformation of a system, the statistical weights of the individual stationary states do not change, as long as the degree of periodicity remains the same. For systems for which the degree of periodicity is equal to the number s of the degrees of freedom, this condition involves that all stationary states must be assigned the same weight h^s. On the other hand, for so-called "degenerate" systems, for which the degree of periodicity is smaller than the number of the degrees of freedom, the weights of the individual states defined by (5) are in general different. However, the weight can be simply determined by comparison with a non-degenerate system, obtained by a small transformation of the original system; for the weight of the original stationary state is equal to the sum of the weights of those states of the transformed system which merge in the limit of degeneracy*.

Now, the simplest example of the application of these principles of the quantum theory is just the system that was considered in Planck's theory of heat radiation, namely an electrically charged particle executing simple harmonic vibrations about an equilibrium position with a characteristic proper frequency. In this case the electrical moment of the system is simply given by the formula

$$\xi = C \cos (2\pi\omega_0 t + \nu), \tag{10}$$

where ω_0 is the constant proper frequency. Now, it follows directly from eqs. (5) and (6) that condition (4) is in this case equivalent to Planck's well-known formula

$$E_n = nh\omega_0, \tag{11}$$

where the energy in the n'th state is denoted by E_n. If we now, according to (1), calculate the frequency of the radiation emitted during a transition between two stationary states with quantum numbers n' and n'', we obtain

$$\nu = (n' - n'')\omega_0. \tag{12}$$

With regard to (10), the correspondence principle now claims, according to (9), that only such transitions can occur with emission of radiation for which $n' - n'' = 1$. Thus, as assumed in the original theory of Planck, we arrive at the

* A detailed exposition of the theory has been given by the author in a comprehensive survey paper "*On the quantum theory of line spectra*", Memoirs of the Copenhagen Academy, 1918, of which a German translation has recently appeared (*Über die Quantentheorie der Linienspektren*, Braunschweig 1922). Detailed references to the older literature are also found there. Cf. also N. Bohr, Z. Phys. **13** (1923) 117, where a briefer summary of the present status of the theory is given.

simple result that the nature of the emitted radiation in this case is the same as that to be expected in the classical theory.

Moreover, according to the foregoing, the statistical weights of the different stationary states of a linear oscillator are equal, as was also assumed in the original theory of Planck. In case of isotropic oscillators with two or three degrees of freedom, the statistical weights are proportional to $(n+1)$ and $(n+1)\,(n+2)/2$, respectively.

§ 3. The hydrogen spectrum.

Now, a system almost as simple as the Planck oscillator is found realized in nature in the *hydrogen atom*, and this gives us a simple starting point for the explanation of the spectral laws. According to Rutherford's conception of the nuclear atom, this consists of a positively charged nucleus around which a single electron moves. Applying the laws of mechanics to the motion of such a system, we find that the particles to an extraordinarily high approximation carry out a purely periodic motion, i.e., that they move in elliptical orbits with a common focus coinciding with the center of mass of the system. In this case, the electric moment of the atom can therefore be represented by the formula

$$\xi = \sum_{\tau} C_{\tau} \cos\,(2\pi\tau\omega t + \gamma_{\tau}), \tag{13}$$

where ω is the frequency of revolution of the particles. This frequency ω, as well as the major axis $2a$ of the orbit of the electron, are determined by the following simple equations expressing the well-known Kepler laws:

$$\omega = \sqrt{\frac{2W^3(M+m)}{\pi^2 e^4 N^2 mM}}, \qquad 2a = \frac{Ne^2 M}{W(M+m)}, \tag{14}$$

where Ne and e are the electrical charges of the nucleus and the electron, and M and m their masses, while W is the work required to remove the particles to an infinite distance from each other. From (14) it now follows, by a simple calculation in accordance with (5) and (6), that

$$W = 2\pi^2 N^2 e^4 mM/J^2(M+m), \tag{15}$$

where $J = \sum_i p_i \dot{q}_i/\omega$ is the quantity which, according to state equations (4), in the stationary states of the atom is determined by the condition

$$J = nh, \tag{16}$$

where n is an integer. Hence, we obtain for the energy and the orbital dimension in the stationary states

$$E_n = -W_n = -\frac{1}{n^2}\frac{2\pi^2 e^4 N^2 mM}{h^2(M+m)}, \qquad 2a_n = n^2\frac{h^2}{2\pi^2 e^2 Nm}, \tag{17}$$

and for the frequency of the radiation emitted by the system during a transition

$$\nu = K\frac{N^2}{1+\dfrac{m}{M}}\left(\frac{1}{n''^2}-\frac{1}{n'^2}\right), \tag{18}$$

where

$$K = 2\pi^2 e^4 m/h^3. \tag{19}$$

Now, the expression (18) has just the same form as the Balmer formula for the *hydrogen spectrum*, which, as is well known, is the simplest, and first recognized, of the empirical spectral laws.

Before entering upon a closer discussion of this agreement, we shall briefly point out that, according to the correspondence principle, the possibility that spectral lines may appear corresponding to any combination of values of the two numbers n' and n'' depends upon the circumstance that, in the electrical moment of this atom, there occur overtones, according to (13), corresponding to any value of the number τ. Thus, we obtain a simple explanation of the peculiar unlimited ability of the terms occurring in the Balmer formula to combine; as is well known, this was the impetus that led Ritz to formulate the general combination principle.

Further, as regards the quantitative rendering of the empirical values of the frequencies of the spectral lines, it was found that, within the accuracy of the measurements, the constant K, given by (19), agrees with the empirical constant in the Balmer formula. This is in perfect accord with the theoretical conceptions, since for hydrogen $N = 1$, and the ratio m/M is too small to be noticeable in this comparison. Moreover, for the orbital dimensions one obtains from (17) a number of values for the major axis of the orbit of the electron which increase greatly with increasing n. For the state $n = 1$, in which the energy is a minimum, and which may be called the *normal state* of the atom, we obtain a value for the major axis of the orbit (ca. 10^{-8} cm) of the same order of magnitude as the value of the atomic dimensions given by the atomic volumes of the elements.

A further support for this interpretation of the hydrogen spectrum* was obtained when it was recognized that certain spectral lines which, because of the relationship of the formula representing them with the Balmer formula, had

* *Abh. über Atombau*, pp. 8, 70.

previously been ascribed to hydrogen actually belong to the *spectrum of the singly ionized helium atom*. In fact, according to the theory, this atom consists of a nucleus with charge $2e$, around which a single electron moves. It now turned out that the lines mentioned can be represented by formula (18) when N is put equal to 2. It was even possible, by a comparison with the hydrogen lines, to verify the existence of the small correction arising from the ratio m/M between the masses.

So far, no evidence has been obtained of spectra of ionized atoms consisting of a more highly charged nucleus and a single electron. However, support for the correctness of formula (18) has been obtained from the laws of *X-ray spectra* discovered by Moseley. It is true that we are here dealing not with atoms having a single electron, but rather with processes in which the motion of one of the inner electrons undergoes a change; nevertheless, for large values of the nuclear charge N, it is possible to a first approximation to neglect the presence of the other electrons. This circumstance is of the greatest importance; for it allows us to determine the atomic number uniquely from the X-ray spectrum.

While we could hitherto depend on a consideration of simple periodic systems, already the explanation of *the influence of external electrical and magnetic fields upon the hydrogen spectrum* gives us occasion to consider motions of a higher degree of periodicity*. Thus, the splitting of the hydrogen lines in the presence of external fields depends on the fact that these fields perturb the simple Kepler motion of the unperturbed atom. As a result, new fundamental frequencies, proportional to the intensity of the external fields, appear in the motion. By applying the state conditions (4), we obtain a set of stationary states which, besides being characterized by the quantum number of the purely periodic unperturbed motion, which we may call the "principal quantum number", are characterized by additional "subordinate quantum numbers" ["Nebenquantenzahlen"]. Hence, for each value of the principal quantum number we no longer obtain a single value of the energy, but rather a number of energy values corresponding to different values of the subordinate quantum numbers, the differences between the energy values being proportional to the intensity of the external fields. According to the frequency condition (1), in the presence of the field each line in the spectrum is therefore resolved into several components, the separations of which are also proportional to the field. In this way it has actually been possible

* A brief exposition of the theory of the influence of external fields upon the hydrogen spectrum, in the light of the correspondence principle, was given by the author in a lecture delivered before the German Physical Society in Berlin and printed in Z. Phys. **2** (1920) 423. With regard to the contributions of different researchers to the development of this theory, reference is made to this lecture as well as to a lecture by the author which will appear shortly in Proceedings of the Physical Society of London. Here a more detailed treatment is given, with comprehensive literature references.

to explain the splittings of the hydrogen lines observed in the Stark and Zeeman effects. These effects likewise offer an especially instructive possibility of applying the correspondence principle, since this principle provides a direct explanation of the limitation of the possibilities of combinations among the increased number of terms, as well as of the polarizations of the spectral components; it even provides an understanding of the observed intensity distribution in the splitting pattern. The interpretation of the effects achieved in this way reveals, on the one hand, the fundamental difference and, on the other hand, the formal similarity with the treatment of the problem on the basis of the classical theory. Thus, the interpretation of the Zeeman effect shows an extensive analogy with the original theory of Lorentz, the fundamental significance of which, as is well known, was that, in a convincing manner, it traced the origin of spectra back to the motion of electrons.

Further, a consideration of the stationary states of the hydrogen atom in external fields permits us to determine the statistical weights of the stationary states of the unperturbed atom given by (16). However, in this determination it must be taken into consideration that several of the states of the non-degenerate system, which can be obtained by the influence of suitably chosen external forces, will show mechanical singularities which would result in the electron falling into the nucleus; these states must therefore by excluded. An investigation shows that the statistical weight in the n'th state is equal to $n(n+1)h^3$.

So far, we have ignored the fact that the lines of the hydrogen spectrum, when observed with high dispersion, show a *fine structure*. This arises from the fact that, strictly speaking, even the motion of the unperturbed hydrogen atom is not purely periodic. In fact, because of the small variation of its mass with velocity, the orbit of the electron is not a closed ellipse, but its perihelion turns slowly, resulting in an additional fundamental frequency of very small magnitude and, hence, according to the state conditions (4), introducing a new quantum number in the multitude of stationary states. A characteristic feature of the theory of the fine structure is that the eccentricity of the elliptical orbit becomes fixed by the additional quantum condition. However, it is a misconception to assume that such a fixing of the elliptical orbits is necessary already for the interpretation of the Balmer formula. In fact, the great instability of the observed fine structure under the influence of external fields of intensities that are still small compared to field strengths of the magnitude applied in experiments on the Stark and Zeeman effects, shows most clearly that the additional quantum condition, arising from the turning of the perihelion in the theory of the fine structure, becomes untenable, even under the influence of such small fields. On the other hand, one must expect that the validity of the Balmer formula is only appreciably affected when the external forces have such an intensity that they are able to strongly alter the approximate Kepler motion of the atom. The theory of the fine structure, which holds not only for the hydrogen lines, but which was brilliantly confirmed also for the analogous spectrum of the ionized helium atom, as well as for the X-ray spectra, is, as is well known, due to Sommerfeld*, and gave the impetus to his extension of the quantum-theoretical interpretation of the spectral laws by taking into consideration systems of a higher degree of periodicity.

§ 4. General spectal laws for elements of higher atomic number.

In the spectra of elements of higher atomic number we encounter more complicated conditions than with hydrogen. As is well known, it became possible, nevertheless, to set up empirical spectral laws for many spectra showing a peculiar analogy to the formal representation of the hydrogen spectrum. Thus, if for the time being we ignore the complex structure of the individual lines (doublets, triplets, etc.), the frequencies of the lines of these spectra can be represented to a high approximation by the formula

$$\nu = \frac{K}{(n'' + \alpha''_k)^2} - \frac{K}{(n' + \alpha'_k)^2}, \tag{20}$$

where n' and n'' are integers, while α'_k and α''_k are two members of a series of constants characteristic of the spectrum considered. K is the so-called Rydberg constant, which is the same for all spectra produced under the usual discharge conditions, the so-called arc spectra, and is equal to the constant in the Balmer formula. Just as the hydrogen spectrum, a spectrum with such a structure is called a *series spectrum*, since the lines can be arranged in rows, the so-called spectral series, within which the frequency converges toward a definite limit. Such a series is represented by formula (20) if, for two definite values of α'_k and α''_k, the number n'' is left unchanged, while n' assumes a succession of increasing values.

The more complicated structure of these spectra, as compared with the hydrogen spectrum, simply arises from the fact that we are here dealing with atoms containing more than one electron, and that we must therefore expect more complicated forms of motion. However, with reference to the interpretation of the hydrogen spectrum, we obtain immediately an understanding of the general occurrence of the Rydberg constant if we assume that, in the emission of arc spectra, we are dealing with transitions between states in which one electron moves at a distance from the nucleus which, at least in the greater part of its orbit, is large compared to the distance of the other electrons from the nucleus.

* A. Sommerfeld, Münch. Ber. (1915).

The force which this electron experiences from the rest of the atom will then obviously be nearly the same as the attraction by the nucleus of the electron in the hydrogen atom at a similar distance.

This interpretation of the origin of arc spectra* soon received convincing support by the finding of Fowler** that the so-called spark spectra, emitted by elements in especially strong electric discharges, in many cases can be represented by a formula differing from formula (20) holding for arc spectra only in that the constant K is replaced by $4K$. This agrees precisely with the theoretical expectation when it is assumed that the spark spectra are emitted not by neutral atoms but by singly ionized atoms. In fact, in this case an electron, moving at a greater distance from the nucleus than the other electrons, will be subjected to nearly the same force as the electron in the ionized helium atom. A short time ago this conception was supplemented in a very interesting manner by the discovery by Paschen*** that in the spark spectrum of aluminum there is present, in addition to a series spectrum of the Fowler kind, another series spectrum which can be represented by a formula in which the constant K is replaced by $9K$, and which therefore must be assigned to a doubly ionized aluminum atom. In general, we shall call a series spectrum originating in an atom which has lost $p-1$ electrons, while a p'th electron moves at a greater distance from the nucleus than the remaining electrons, a *series spectrum of the p th kind.*

While the previous considerations depended directly on the interpretation of the simple hydrogen spectrum, the greater multitude of terms appearing in the spectra of the other elements show, on the other hand, that in them we have to do with a more complicated motion of the outer electron than in the hydrogen atom. The question that we meet here is first of all very difficult, since the problem of the motion in an atom with several electrons, when treated on the basis of ordinary mechanics, leads to extraordinarily complicated solutions. It must especially be emphasized that, according to the classical mechanical laws, the orbital configurations of the inner electrons in the atom would prove to be unstable under the perturbing influence of the outer electron. Nevertheless, by introducing simple assumptions about the character of the motion of the outer electron, it has been possible, at least formally, to attain a detailed interpretation of the regularities of the series spectra. Thus, ignoring the complex structure, it has been possible to classify the spectral terms, and interpret the rules of combination, on the basis of the assumption that the deviation of this motion from a Kepler orbit to a first approximation consists in a precession of

* *Abh. über Atombau*, pp. 11, 115.

** A. Fowler, Trans. Roy. Soc. (A) **214** (1914) 225.

*** F. Paschen, cf. p. 142 ⟨of Ann. d. Phys. **71** (1923)⟩. I am indebted to Mr. Paschen for my knowledge of this work.

the major axis in the orbital plane. Actually, as first shown by Sommerfeld*, this assumption, according to which the motion of the electron is very similar to a central motion, leads to a manifold set of stationary states of the same kind as the multitude of terms occurring in the spectra. At the same time, when the correspondence principle is taken into account, it provides a simple explanation of the peculiar limitation of the possible combinations among these terms characteristic of the series spectra**.

A central motion can always be regarded as a purely periodic plane motion on which there is superposed a uniform rotation in the orbital plane. By this superposed rotation, each elliptical harmonic component of oscillation present in the periodic motion becomes resolved into two circular harmonic rotations, whose frequencies are equal to the original frequency increased or diminished by the frequency of the superposed rotation. Thus, if the frequency of revolution in the periodic orbit is denoted by ω and that of the superposed rotation by σ, the electric moment of the atom, in so far as it depends on the outer electron, can be represented by the formula

$$\xi = \sum_{\tau} C_{\tau\pm 1} \cos\left[2\pi(\tau\omega \pm \sigma)t + \gamma_{\tau,\pm 1}\right]. \tag{21}$$

According to (4), the corresponding stationary states are characterized by two conditions which can be written

$$J = nh, \qquad P = kh, \tag{22}$$

where n and k are integers. While the first of these conditions, in the limit where the central motion goes over into a Kepler motion, coincides with condition (16) which fixes the stationary states of the simple hydrogen atom, the second condition may be regarded as an additional quantum condition resulting from the rotation of the orbit. The quantity P appearing in this condition is simply equal to the angular momentum of the electron in its orbit, multiplied by 2π. On the other hand, the kinematical interpretation of the quantity J for an arbitrary central orbit is determined by the equation

$$R = J - P = \int p_r \mathrm{d}r = \int m \left(\frac{\mathrm{d}r}{\mathrm{d}t}\right)^2 \mathrm{d}t, \tag{23}$$

where r is the radius vector of the electron and $p_r = m\,\mathrm{d}r/\mathrm{d}t$ is the radial momentum; the integral is taken over a complete period of the radial motion. The

* A. Sommerfeld, Münch. Ber. 1916.

** A brief exposition of the theory of series spectra on the basis of the correspondence principle is given in the lecture by the author mentioned on p. 239 (Z. Phys. **2** (1920) 423). A more detailed discussion with fuller literature references is found in the third part of the work quoted on p. 235, "*On the Quantum Theory of Line Spectra*".

expression on the right side of eq. (23) is the same as that entering in the well-known Wilson–Sommerfeld conditions for the stationary states of a central motion. The form of conditions (22) used here is proposed in view of the choice of the fundamental frequencies ω and σ, since, according to (5), the energy difference between two neighboring states is given by the equation

$$\delta E = \omega\delta J + \sigma\delta P. \tag{24}$$

It is clear from this equation that if σ is small compared to ω the energy will depend primarily on J. Just as in the case of hydrogen, we shall therefore always call n the principal quantum number and k the subordinate quantum number. Moreover, in the limiting case mentioned, in which the deviation from a Kepler motion is very small, the orbital dimensions are determined approximately by the equations

$$2a = \frac{n^2h^2}{2\pi^2Ne^2m}, \qquad 2p = \frac{k^2h^2}{2\pi^2Ne^2m}, \tag{25}$$

where $2a$ denotes the major axis and $2p$ the parameter (the smallest chord through the focus) of the nearly elliptical rotating orbit.

From (23) it follows generally that P can never be larger than J. For every value n of the principal quantum number we shall therefore expect spectral terms corresponding to the values $k = 1, 2 \ldots n$. Further, it follows from the correspondence principle that a transition from a state n', k' to a state n'', k'' depends on the presence of a harmonic oscillation component of frequency

$$(n'-n'')\omega + (k'-k'')\sigma$$

in the motion. However, according to (21), this correspondence means that only such transitions can occur for which

$$k' - k'' = 1. \tag{26}$$

This result not only explains the peculiar limitation of possible combinations of spectral terms observed in the series spectra, but it also supports convincingly the assignment of the possible k-values to the empirical term series proposed by Sommerfeld. Thus, we must assign the value $k = 1$ to the s-terms, the value $k = 2$ to the p-terms, $k = 3$ to the d-terms, $k = 4$ to the f-terms, and higher values of k to the series of terms which appear especially in spark spectra and which coincide closely with the term series of hydrogen. While the assignment of the principal quantum number n clearly presents no difficulty for such series of terms, the problem of determining this quantum number for series of terms deviating considerably from the hydrogen series has revealed new circumstances, to which we shall return in the last two sections.

As a consequence of this limitation of possible combinations, implied by eq. (26), in contrast to what holds for the simple hydrogen spectrum, no true combination lines are to be expected in the series spectra of the other elements. This means that no line should occur, the frequency of which can be represented as the sum of or the difference between the frequencies of two other lines in the spectrum. However, as is well known, a large number of such lines are actually observed, and this was formerly regarded as one of the strongest supports for the combination principle; but the occurrence of these lines, as indicated already by the peculiar dependence of their appearance on the conditions of excitation, must be ascribed to the presence of perturbing external electric fields in the discharge. In fact, through the influence of such forces, the simple central motion of the series electron is perturbed in such a way that, besides the harmonic components of oscillation given by (21), there will occur also components with frequencies $\tau\omega$ and $\tau\omega \pm 2\sigma$, with amplitudes proportional to the external forces. This has the consequence that in the perturbed atom, in addition to transitions in which k changes by one unit, we must expect transitions in which k remains unchanged or changes by two units, i.e., frequencies formally analogous to the sum and difference tones in acoustics.

In particular, the appearance of such "new" lines is the essential result of experiments on the Stark effect of series spectra, in which we are dealing with the *influence of homogeneous external electric fields* of known intensity. In complete agreement with the theory, it is found that the intensity of these lines, for equal external field strength, is the greater the less the combining terms in question deviate from hydrogen terms corresponding to the same principal quantum number. In fact, according to (24), this deviation gives directly a measure of the magnitude of the frequency of rotation σ, and therefore of the deviation of the orbit from a purely periodic one, and in general causes the fundamental difference between the Stark effect of the hydrogen spectrum and of the other spectra. In particular, in the latter spectra, no new fundamental frequencies occur in the motion of a magnitude proportional to the intensity of the external forces, and we therefore obtain only such splittings of the lines in which the separation of the components increases with a higher power of the field strength than the first.

Contrary to this, the *influence of a homogeneous magnetic field* upon series spectra, in so far as we can ignore the complex structure, is quite analogous to its effect upon the hydrogen spectrum. As is well known, according to a theorem of Larmor, it is to be expected from classical electrodynamics that the change in the electronic motion of a nuclear atom in an external magnetic field will consist simply in the superposition of a uniform rotation with a frequency given by the equation

$$\zeta_H = eH/4\pi mc, \tag{27}$$

where H is the intensity of the magnetic field, e and m, as above, are the charge and mass of an electron, and c is the velocity of light. By considering the influence of this rotation on the resolution of the electronic motion into harmonic components, it is simply concluded from the correspondence principle that each line will be split into a normal Lorentz triplet. As is well known, this expectation is fulfilled, as long as we are dealing with spectra whose series lines show no complex structure.

§ 5. Complex structure of series lines.

Heretofore, we have ignored completely the finer structure shown by the lines of many series spectra. A formal quantum-theoretical analysis of these "complex structures" by introducing a third quantum number, called the inner quantum number, was first attempted by Sommerfeld*. In order to reach a closer understanding of the phenomena of complex structure, especially of the empirical combination rules holding, we must, relying on the correspondence principle, first of all ask about the nature of the electronic motion. As explained in the previous section, the characteristic structure of the spectra shows that the motion of the series electron to a first approximation is a central motion. However, the complex structure shows that strictly speaking there occur more than two fundamental frequencies in the motion, with the consequence that more than two quantum numbers are required to classify the multitude of stationary states. Now, in attempting to describe the electronic motion, it is natural to retain at least an axial symmetry; this leads to the assumption that the deviation of the motion from a central orbit consists in the superposition of a *uniform precession of the orbital plane*. By this precession, the resolution of the electronic motion into its harmonic components is altered in that, instead of each of the previous components, there now occur three components, of which one consists of a linear vibration parallel to the axis of precession with unchanged frequency, while the other two are circular rotations in a plane perpendicular to this axis; their frequencies are equal to the original frequency increased or diminished by the frequency of precession. If this is denoted by ρ, we obtain for the components of the electric moment parallel and perpendicular to the axis of precession

$$\xi = \sum_{\tau} C_{\tau,\pm 1} \cos \left[2\pi(\tau\omega \pm \sigma)t + \gamma_{\tau,\pm 1}\right] \tag{28a}$$

* A. Sommerfeld, Ann. d. Phys. **63** (1920) 221.

and

$$\eta = \sum_{\tau} C_{\tau, \pm 1, \pm 1} \cos \left[2\pi(\tau\omega \pm \sigma \pm \rho)t + \gamma_{\tau, \pm 1, \pm 1}\right]. \tag{28b}$$

We can now write the three conditions of state which, according to (4), determine the stationary states of such a motion

$$J = nh, \qquad P = kh, \qquad Q = jh, \tag{29}$$

where n, k and j are integers. While J and P are quantities which, in the limiting case of a pure central motion, coincide with the quantities appearing in (22), we may regard the quantity Q appearing in the additional quantum condition as the total resultant angular momentum of the atom, multiplied by 2π. Actually, these quantities satisfy an equation which is analogous to (5) and (24), i.e.,

$$\delta E = \omega\delta J + \sigma\delta P + \rho\delta Q \tag{30}$$

for the energy difference between two neighboring motions of a system consisting of an electron and a rigid body having an angular momentum, and producing a field of force symmetrical with respect to the axis of the angular momentum.

While, in the previous section, we could make a close comparison with the quantum theory of periodicity systems, in that, at least formally, we could confine our attention to the motion of the outer electron, the situation is somewhat more complicated in the present case, giving rise to a fundamental uncertainty as regards the interpretation of the additional quantum condition; for we cannot even prove by rigorous mechanics that the angular momentum of the atomic core will remain constant during the motion of the outer electron. However, to begin with, we shall assume that the motion of the atom, as far as the outer electron is concerned, can be strictly compared with the multiply periodic system mentioned, and we write for the angular momentum of the atomic core $ih/2\pi$, where i, just as j, is assumed to be an integer. The latter quantum number, which fixes the orientation of the outer orbit relative to the atomic core, must obviously satisfy the condition

$$|i-k| \leqq j \leqq i+k. \tag{31}$$

As seen from (30), the pattern of complex structure, obtained from state conditions (29) and the frequency condition (1), is intimately related to the precession movement of the orbit of the outer electron. Further, according to the correspondence principle and to (29) and (30), the occurrence of a transition from a state $n'k'j'$ to a state $n''k''j''$ depends on the presence of a harmonic component of frequency $(n'-n'')\omega + (k'-k'')\sigma + (j'-j'')\rho$. According to (28a) and (28b), this means that, besides by condition (26), the possible transitions are limited by the further condition

$$j'-j'' = 0 \quad \text{or} \quad j'-j'' = \pm 1. \tag{32}$$

This result offers a starting point for an understanding of the characteristic limitation of the possible combinations among the spectral terms appearing in the observed structural patterns, a limitation already emphasized by Rydberg. At the same time, together with condition (31), it provides to a certain extent a means of determining the values of the quantum numbers j and i. However, contrary to what holds for the quantum number k, we here encounter profound difficulties, as soon as we attempt to interpret the observed complex structure patterns. Furthermore, these structure patterns often exhibit a higher degree of complexity than is to be expected from the simple theory mentioned here. In the empirical classification of spectra, this greater complexity reveals itself in that in many cases we must distinguish between *different kinds of spectral terms*; for example, in the spectra of the alkaline earths, these are called singlet and triplet terms. According to the above combination rules, one is led in such cases to assign the same values of the quantum numbers n, k and j to several different terms. Hence, if we retain the simple assumptions about the nature of the motion of the outer electron, we are led to the conception that, in the different kinds of terms, we have to do with different configurations of the electron orbits in the atomic core, corresponding to different values of its total angular momentum $ih/2\pi$. This assumption is also supported by the fact that transitions involving combinations between different kinds of terms are generally far less probable than transitions corresponding to combinations between terms of the same kind. For we must bear in mind that a transition in which a change occurs in the configurations of the electronic orbits in the atomic core does not correspond to a harmonic component of the fundamental vibrations, given by (28a) and (28b), but rather to a harmonic component of smaller amplitude, depending on the subtler interplay between the outer electron and the electrons in the atomic core. Apart from the nature of this interplay, we must retain the combination rule limiting the change of j in the possible transitions also for transitions between terms of different kinds. According to the correspondence principle, this follows already from the general axial symmetry to be expected for the electronic motion in a nuclear atom. It is of great fundamental significance that this rule can also be derived from a consideration of the conservation of angular momentum in radiation processes*.

* The bearing of such a consideration on the quantum theory—proposed independently by Rubinowicz and the author—is discussed in greater detail in a paper by the author (Z. Phys. **6** (1921) 1) in which literature references are also given.

With regard to the effect of external fields upon the complex structure of series lines, we meet with very interesting problems, the definitive solution of which is still outstanding. As is well known, in the influence of *magnetic fields* upon series lines of complicated structure, we encounter the alluring phenomenon of the *anomalous Zeeman effect*. According to the correspondence principle, this must be ascribed to a failure of the Larmor theorem previously mentioned. In particular, from the phenomenon discovered by Paschen and Back* that the splitting pattern of the complex structure components of a single series line is gradually transformed with increasing field strength into a normal Lorentz triplet, we can conclude that the magnetic field, besides its influence upon the motion of the atom as a whole, also affects the interplay of the inner and outer electrons, upon which the complex structure depends**. This is also clearly indicated by the fact that the magnetic field causes new line components of the complex structure to appear which are not present in the field-free pattern. In fact, these new components are quite analogous to the new series appearing, as mentioned above, in an electric field. However, in the case of the effect of a magnetic field, we cannot calculate the detailed influence of the external forces on the electronic motion by means of classical electrodynamics. Rather, the failure of the Larmor theorem teaches us that we here face quite novel phenomena; with respect to these, we must hope that the empirical spectroscopic data may give us an indication of what laws must take the place of those of the classical theory.

Actually, the quantum-theoretical analysis of the Zeeman effect has already revealed regularities which permit a formal interpretation. Thus, Landé*** has been able to show that the splitting pattern in weak fields can be represented by means of the frequency condition, when it is assumed that each of the spectral terms is split into a series of equidistant term values under the influence of the field. On the basis of the correspondence principle, this result can be interpreted by assuming that the change in the motion of the atom as a whole consists in a superposed precession around an axis through the nucleus, and parallel to the field, with a frequency of precession independent of the orientation of the atom relative to the field. This interpretation is supported by the fact that Sommerfeld and Heisenberg†, by means of an investigation of the harmonic components which, according to this view, should appear in the motion of the atom in the field, have shown that one can obtain a general understanding of the relative intensities of the observed components in the Zeeman effect on the basis

* F. Paschen and E. Back, Ann. d. Phys. **39** (1921) 897 and Physica **1** (1921) 261.

** Cf. N. Bohr, Z. Phys. **9** (1922) 27.

*** A. Landé, Z. Phys. **5** (1921) 231.

† A. Sommerfeld and W. Heisenberg, Z. Phys. **11** (1922) 131.

of the correspondence principle. However, contrary to what was to be expected from the classical theory, the values of the frequencies of rotation of the superposed precession, calculated by Landé, are not given by (27), but are characteristic of the different stationary states of the unperturbed atom. This means that the connection between the magnetic moment and the angular momentum of the atom is more complicated than demanded by the classical theory. Furthermore, it follows from Landé's analysis that the value of the component of the angular momentum of the atom parallel to the field, divided by 2π, may not always be equal to an integer, as assumed in the interpretation of the Zeeman effect of the hydrogen lines. While this is supposed to hold only for spectra with certain kinds of complex structure, Landé actually obtained values for this quantity for other types differing from an integer by $\frac{1}{2}$. This finding has made questionable the justification of fixing the motion in the stationary states of an atom with several electrons in direct analogy with the quantum theory of periodicity systems. In the last section we shall return to the changes which the simple interpretation of the complex structure described above may undergo upon closer consideration.

The effect of external *electric fields* on the complex structure of series lines is of an essentially different nature from the effect of magnetic fields. In fact, from the approximate central character of the orbit of the series electron, it follows, as already mentioned above, that the motion in an external homogeneous electric field will contain no components characterized by new fundamental frequencies proportional to the field intensity; in particular, there will occur no rotation of the orbital plane with a frequency of precession proportional to the field intensity. Thus, it is not possible for the electric field to affect the mechanism responsible for the complex structure. This offers a simple explanation of the observed fact that even narrow doublets and triplets are unaffected by an electric field, even when this is strong enough to produce a splitting of hydrogen lines large compared to the distance between components of the structures in question. All this contrasts sharply with the behavior of the complex structure in external magnetic fields. In general, one may say that, while magnetic fields only affect the complex structure and do not influence the general structure of series spectra, electric fields only influence the complex structure when they are able to disturb the general central character of the motion. Thus, we can only expect an influence of electric fields on the line structure when one of the spectral terms, on which the lines depend, deviates very little from the corresponding term of the hydrogen spectrum.

§ 6. Closer connection between series spectra and atomic structure.

When, on the basis of the quantum-theoretical interpretation of the general empirical spectral laws mentioned in the preceding, we attempt to form more definite conceptions of the origin of the observed *spectra of the individual elements**, we meet a problem, the solution of which is still beset with fundamental difficulties. Contrary to the case in which the atom contains only a single electron, the application of the laws of mechanics to the motion in atoms with several electrons leads, as mentioned, in general to types of motion which do not exhibit the periodicity properties that seem necessary for a classification of their stationary states in terms of quantum numbers. Under these circumstances, there seems to be no other way out than to assume that the classical mechanical laws fail not only to explain the stability of atoms but, for atoms with several electrons, even fail to describe the motions in the atoms in finer details. However, the empirical spectral laws, which, in the light of the correspondence principle, give us information about the general nature of the motion in the stationary states responsible for the spectra, in connection with the general established features of atomic structure, offer us nevertheless a starting point for overcoming this difficulty. Thus, these laws, and the central symmetry of the nuclear atom, suggest above all that to a first approximation the orbit of every electron in an atom may be regarded as a central orbit which, in the manner described, is characterized by two quantum numbers n and k. In what follows, such an orbit will, for brevity, be denoted as an n_k-orbit. The closer interpretation of any spectrum is, however, inseparably connected with the development of detailed conceptions of the structure of the atom in question, especially as regards the fixing of the values of the quantum numbers n and k for the orbit of each electron in the normal state of the atom, and also as regards the fixing of the values of the additional quantum numbers determining the orientation of these orbits. Strictly speaking, on the basis of classical mechanics, the meaning of these quantum numbers cannot be accurately determined from the theory of periodicity systems. Rather, the general stability of atomic structures leads to the view that, in every atom with several electrons, there are properties of the motion of each electron, and of the interplay

* In Kayser's handbook may be found a compilation of the general empirical spectroscopic data which in many respects is still the most complete. Moreover, comprehensive expositions have recently been published as well by Fowler (*Report on Series in Line Spectra*, Physical Society of London 1922) as by Paschen and Götze (*Seriengesetze der Linienspektren*, Berlin 1922). In these, special attention has been paid to the regularities in the series spectra, and they are very valuable for the theoretical discussion, because they give the spectral terms themselves. The data used in the following have been taken from these books. In order to make the review easier for the reader, there are given, in tables at the end of this paper, graphical representations of a number of spectra referred to in the theoretical discussion.

among the electrons, which possess an invariant character that cannot be explained mechanically, but that finds its meaningful expression just in the quantum numbers. This view may be called a formal *postulate of the invariance and permanence of quantum numbers**.

This way of looking at the matter is especially emphasized when we attempt to form a picture of atomic structure by *following the process of the building up of atoms by the successive capture of electrons*. In fact, just the series spectra provide information about the course of the steps of this process. Thus, we can regard the hydrogen spectrum as evidence of a process in which the atom is formed by the binding of an electron by the nucleus under emission of radiation; and in general we learn, from a spectrum given by (18), the course of the first step in the building up of an atom with nuclear charge N, consisting in the capture of the first electron. According to these considerations about the origin of series spectra, we learn from the arc spectrum of an element, in an analogous manner, the course of a binding process by which the neutral atom is completely formed by the capture of the last electron. In general, in a series spectrum of the p'th kind of an element, we witness the course of the p'th last step in its formation. In order to utilize the regularities in the spectra for this purpose, we assume, in accordance with the postulate mentioned, that during a capture process the quantum numbers, characterizing the motions of each of the previously bound electrons in the normal state of the building-up stages, will remain unchanged under the perturbing influence of the electron being captured. By the capture of additional electrons, at most those quantum numbers can change which have to do with the relative orientation of the orbits of such electrons which, as regards the values of n and k, are equivalent with the electron being captured.

Our present knowledge of series spectra of higher order is for most elements too incomplete to enable us to follow all steps in the building-up process. However, this gap we can to a considerable extent fill by considering, on the one hand, the mechanical similarity of the problem of the electronic motion in atoms with different nuclear charge but with the same number of electrons and, on the other hand, the gradual change in the relative size of the different quantum orbits produced by increasing the nuclear charge. The last circumstance concerns especially the properties of the orbit characterized by the principal quantum number n. Here we meet essentially different conditions in the different regions of the atom. In the innermost region, where the interaction among the different electrons is small compared to the attraction by the nucleus, the electronic orbit will correspond closely to a Kepler orbit, and the orbital dimensions, and the energy necessary to remove the electron from the atom, will be approximately

* Cf. N. Bohr, Z. Phys. **13** (1923) 135.

equal to the dimensions and the energy in a stationary state with the same principal quantum number of an atom with the same nuclear charge, but containing only a single electron. On the other hand, in the outer parts of the atom the attraction from the nucleus will to a great extent be cancelled by the repulsion from the inner electrons. This holds especially for orbits of the stationary states belonging to the last steps in the formation of the neutral atom. Here we meet two essentially different cases, according to whether the outer electron during its entire motion remains at a greater distance from the nucleus than the inner electrons, or whether during a part of its revolution it penetrates into the region of the orbits of the previously bound electrons.

In the former case, the motion, if we are dealing with the p'th last step in the binding process, will only differ little from a Kepler motion in a field of point charge pe, and the energy of binding will be nearly equal to that of a stationary state of such a Kepler motion with the same principal quantum number. In this case we shall speak of an *orbit of the first kind.*

In the latter case, in which we shall speak of an *orbit of the second kind*, the central orbit traced by the electron can be described as consisting of successive inner and outer loops, of which the outer loops coincide nearly with part of a Kepler motion around a point charge pe, while the inner loops in general deviate very much from a Kepler motion. However, the dimensions of the outer loops deviate in general essentially from those of a Kepler motion with the same principal quantum number around a nucleus with charge pe. This is easily seen when it is considered that the outer electron, on the inner loops of its orbit at a given distance from the nucleus, is subjected to almost the same resultant attraction toward the center of the atom as is one of the inner electrons. Because of this circumstance, the value of the radial momentum of the electron must at each point of its inner loop always be larger than the value of this momentum at the same distance from the nucleus of an electron moving with the same angular momentum in an orbit lying entirely in the inner region. Hence, according to (23), the quantity $R = J-P$, and according to (22) therefore also the difference between the quantum numbers n and k, will always be larger for an orbit of the second kind than for any inner orbit with the same value of k. The contribution to the value of the integral (23), corresponding just to the inner loop, is always even larger than $m-k$, where m is the largest value of n found among the inner orbits with the same value of k. The contribution from the outer loop to the integral mentioned is therefore always smaller than $n-m$.

This result shows above all that, for an orbit of the second kind, the principal quantum number n will always be larger than m. Furthermore, this consideration offers a possibility of obtaining in many cases a simple estimate of the dimensions of the outer orbital loop, as well as of the energy required to remove the electron from the atom. In fact, this energy is determined simply by the value J^* of the quantity J associated with a pure Kepler motion of which the outer loop forms a part. According to (23), this value is equal to the sum of P and the integral R extended over the completed elliptical orbit. Now, in many cases, this integral will not differ much from the contribution corresponding to the outer loop alone, which itself does not deviate much from the value $n-m$ mentioned above. In such cases the desired J^*-value will therefore be given approximately by $(n-m+k)h$. In general, we define, as *effective quantum number* of an n-orbit of the second kind, corresponding to the p'th last step of binding in the formation of the atom, the number n^* which must be inserted instead of n in formula (17) (where N is put equal to p), in order to obtain the ionization energy and the orbital dimension of the electron in question. It is now a very important aid to the investigation that this effective quantum number obviously can be obtained directly from the empirical spectral terms.

The point of view of the successive building-up has led to a general picture of the constitution of atoms which, with the aid of the interpretation of the empirical spectral laws, offers a natural explanation of the peculiar alternation of the properties of the elements with atomic number expressed in the *periodic system of the elements**. The main features of this picture are represented in the accompanying table, giving the classification of electron orbits in the normal state of the atom for a number of elements. The numbers given indicate the number of electrons in the normal state of the atom moving in orbits characterized by the n_k at the top. The periodicity in the properties of the elements depends on the fact that with increasing atomic number a continual formation and closure of the different groups of electrons take place. This formation generally occurs in the outermost electron group, having the largest value of the principal quantum number, and it reveals itself most clearly in the optical spectra. However, as seen in the table, occasionally a further development of one of the inner groups occurs. This provides an understanding of the departure from simple periodicity in the change of the properties of the elements with increasing atomic number. These circumstances are expressed in figure 1 in an arrangement of the periodic system in which the vertical columns indicate the periods, while elements with homologous properties are connected by lines. The columns of elements enclosed in square brackets are those in which inner groups are being built up. Such a

* The main features of the theory have been developed by the author in a lecture reprinted in Z. Phys. **9** (1922) 1. A detailed discussion of older works by other authors on the same subject can also be found there. In the present paper, the considerations concerning the connection between optical spectra and atomic structure are given in greater detail. Moreover, a detailed discussion of the relation of the observed data on X-rays to this theory of atomic structure is given in a recent paper by D. Coster and the author (Z. Phys. **12** (1923) 342).

[634]

TABLE.

N \ n_k		1_1	2_1	2_2	3_1	3_2	3_3	4_1	4_2	4_3	4_4	5_1	5_2	5_3	5_4	5_5	6_1	6_2	6_3	6_4	6_5	6_6	7_1	7_2
1	H	1																						
2	He	2																						
3	Li	2	1																					
4	Be	2	2																					
5	B	2	2	(1)																				
—	—	—	—	—																				
10	Ne	2	4	4																				
11	Na	2	4	4	1																			
12	Mg	2	4	4	2																			
13	Al	2	4	4	2	1																		
—	—	—	—	—	—	—																		
18	A	2	4	4	4	4																		
19	K	2	4	4	4	4		1																
20	Ca	2	4	4	4	4		2																
21	Sc	2	4	4	4	4	1	(2)																
22	Ti	2	4	4	4	4	2	(2)																
—	—	—	—	—	—	—	—	—																
29	Cu	2	4	4	6	6	6	1																
30	Zn	2	4	4	6	6	6	2																
31	Ga	2	4	4	6	6	6	2	1															
—	—	—	—	—	—	—	—	—	—															
36	Kr	2	4	4	6	6	6	4	4															
37	Rb	2	4	4	6	6	6	4	4			1												
38	Sr	2	4	4	6	6	6	4	4			2												
39	Y	2	4	4	6	6	6	4	4	1		(2)												
40	Zr	2	4	4	6	6	6	4	4	2		(2)												
—	—	—	—	—	—	—	—	—	—	—		—												
47	Ag	2	4	4	6	6	6	6	6	6		1												
48	Cd	2	4	4	6	6	6	6	6	6		2												
49	In	2	4	4	6	6	6	6	6	6		2	1											
—	—	—	—	—	—	—	—	—	—	—		—	—											
54	X	2	4	4	6	6	6	6	6	6		4	4											
55	Cs	2	4	4	6	6	6	6	6	6		4	4				1							
56	Ba	2	4	4	6	6	6	6	6	6		4	4				2							
57	La	2	4	4	6	6	6	6	6	6		4	4	1			(2)							
58	Ce	2	4	4	6	6	6	6	6	6	1	4	4	1			(2)							
59	Pr	2	4	4	6	6	6	6	6	6	2	4	4	1			(2)							
—	—	—	—	—	—	—		—	—	—	—	—	—	—			—							
71	Cp	2	4	4	6	6	6	8	8	8	8	4	4	1			(2)							
72	Hf	2	4	4	6	6	6	8	8	8	8	4	4	2			(2)							
—	—	—	—	—	—	—	—	—	—	—	—	—	—	—			—							
79	Au	2	4	4	6	6	6	8	8	8	8	6	6	6			1							
80	Hg	2	4	4	6	6	6	8	8	8	8	6	6	6			2							
81	Tl	2	4	4	6	6	6	8	8	8	8	6	6	6			2	1						
—	—	—	—	—	—	—	—	—	—	—	—	—	—	—			—	—						
86	Em	2	4	4	6	6	6	8	8	8	8	6	6	6			4	4						
87	—	2	4	4	6	6	6	8	8	8	8	6	6	6			4	4					1	
88	Ra	2	4	4	6	6	6	8	8	8	8	6	6	6			4	4					2	
89	Ac	2	4	4	6	6	6	8	8	8	8	6	6	6			4	4	1				(2)	
90	Th	2	4	4	6	6	6	8	8	8	8	6	6	6			4	4	2				(2)	
—	—	—	—	—	—	—	—	—	—	—	—	—	—	—			—	—	—				—	
—	—	—	—	—	—	—	—	—	—	—	—	—	—	—	—		—	—	—				—	—
118	?	2	4	4	6	6	6	8	8	8	8	8	8	8	8		6	6	6				4	4

development reveals itself in the close relationship in many of the properties of the elements in question, arising from the fact that, during the building up of the inner groups, the development of the outer group temporarily does not proceed any farther. Nevertheless, this development is at the same time clearly expressed in the characteristic differences between the series spectra of such elements inside

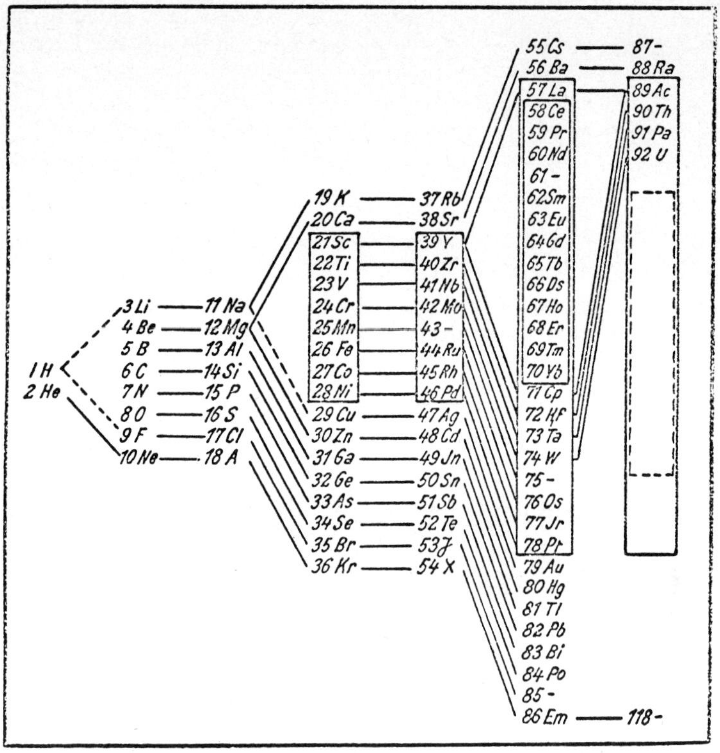

Fig. 1.

and outside the brackets, which otherwise have very similar chemical properties and therefore occupy homologous places in the usual representations of the periodic system. Although very promising advances have recently been made with regard to these questions, the individual steps in the building-up process have not been clarified to such an extent that we have presumed to give them in detail in the table.

An especially instructive example of the intimate connection between the picture of atomic structure under discussion and the spectral regularities is found in the *arc spectra of the alkali metals* which, as is well known, have a particularly simple structure. This is connected with the fact that for these elements, which stand at the beginning of the periods of the system, the electronic orbits in the atomic core at the capture of the last electron form a completed configuration of the same nature as that found in the neutral atom of the preceding noble gas. Now, the question which we meet at the outset is which of the abovementioned two kinds of orbits corresponds to the various terms of these spectra. Decisive for this problem is above all the parameter of the Kepler orbit approximating the real orbit, at least in its outer loops. In fact, when this parameter, which for both kinds of orbit, according to (25), depends only on the value of the subordinate quantum number k, is essentially larger than the dimensions of the orbits of the atomic core, we have to do with an orbit of the first kind lying entirely outside the atomic core. In the spectra of all the alkalis this is the case for orbits for which $k \geqq 3$. Actually, already the d- and f-terms of these spectra deviate only little from the hydrogen terms, and that in the direction of a stronger attractive force than in the hydrogen atom, as indeed was to be expected because of the electric polarization induced by the outer electron in the atomic core, an effect which is perfectly analogous to the dielectric phenomena. On the other hand, we find in general a different behavior of the s- and p-terms.

In lithium, the orbits with $k = 2$ still remain at large distances from the orbits of the atomic core, as is revealed also in the small deviation from the hydrogen terms of the p-terms of the lithium spectrum. For orbits with $k = 1$ in lithium, we meet special circumstances, because the orbits cannot be assigned definitely to one or the other class. The parameter, calculated from the formula for the Kepler motion, is still considerably larger than the dimensions of the inner system, it is true, but the forces arising from the polarization induced in the atomic core, as estimated from the deviation of the p-terms from the hydrogen terms, turn out to be so large that a circular 1_1-orbit would be pulled into the region of the inner 1_1-orbits. However, according to our conceptions, the completed character of the orbital configurations of the atomic core depends just upon the circumstance that the addition of a third electron to an orbit equivalent to that of the first two cannot occur as the end result of a binding process accompanied by radiation. As regards the n_1-orbits with $n > 1$, even if they do not penetrate into the region of the 1_1-orbits, they come during their revolution so close to these orbits that it is immediately clear that, because of the intimate interaction, they correspond to an essentially stronger binding of the electron than hydrogen orbits with the same principal quantum number.

In case of the other alkali metals, the stationary states with $k = 1$, as a simple estimate of the orbital dimensions shows, provide pronounced examples of orbits of the second kind. Moreover, a comparison of the motion in the inner loop with the orbits of the previously bound electrons shows that the series electron in the s-orbits during its revolution comes even considerably closer to the nucleus than the two innermost electrons moving in 1_1-orbits. For the orbits with $k = 2$ the circumstances are more nearly analogous to the case of the n_1-orbits in lithium. We may therefore conclude that, because of the forces arising from the polarization of the atomic core, an electron in a 2_2-orbit cannot move in the outer region of the atom, although the corresponding orbit in hydrogen would have larger dimensions than the atomic core. For $n > 2$, the n_2-orbits, even when estimated as hydrogen orbits, come to distances from the nucleus of the same order of magnitude as the electrons in the atomic core, and we must assume that they penetrate into the inner region, although they always remain at a larger distance from the nucleus than the circular 1_1-orbits of the two innermost electrons. These circumstances are also revealed clearly in the magnitude of the corresponding terms. In fact, their effective quantum numbers deviate from those of the hydrogen terms not only to a much higher degree than those of the d- and f-terms, but also in a direction which, if the orbits lay entirely in the outer region, would correspond to a weaker force of attraction than in hydrogen, and this would contradict the character of the atomic models.

Now, as regards the s- and p-terms, the effective quantum numbers of the corresponding orbits, according to our definition, will just coincide with the denominators in the Rydberg formula (20), and it is above all particularly satisfying that our conceptions yield an immediate explanation of the fact that the quantities α_k are so nearly constant for the series of s- and p-terms. In fact, within each of these series, the electronic motion in the inner loops of the corresponding orbits will to an extraordinarily high approximation be the same, and the difference between the principal quantum number and the effective quantum number will therefore be nearly constant within each series. If the principal quantum number is inserted for the number n in Rydberg's formula, the constant α will be negative for all spectra. According to the considerations on p. [634], we shall expect that, for orbits of the second kind, the numerical value of α will never deviate much from the number $m-k$. This expectation is indeed fulfilled for the alkali spectra, since, as may be seen from the graphical representations in table II, the effective quantum numbers corresponding to the first members of the s- and p-series increase only slowly as one goes to the next element, without exceeding the values 2 and 3, respectively. As indicated in the table, for the alkali metals, the first member of the s-series corresponds to the normal state of the atom, and the circumstance that, despite the constant increase in the principal quantum number, the numerical values of these members differ only rather little, furnishes a characteristic example of the *explanation of the periodic alternation in the properties of the elements with increasing atomic number*, based on our picture of the development and closure in the atom of electron groups with higher quantum number*.

We meet results similar to those in the alkali spectra in the *spark spectra of the alkaline earths*, which arise from the singly ionized atoms of these elements. In

* At the same time as the general features of the views of atomic structure mentioned here were developed by the author (cf. Nature, March 24, 1921), Schrödinger (Z. Phys. **4** (1921) 347) independently emphasized that in the s-terms of the alkalis we have to do with stationary states in which the series electron during its revolution penetrates into the inner region. Aided by the simple assumption that the effect of the inner electrons can be compared with that of a uniformly charged spherical shell, he attempted to obtain an estimate of the values of the s-terms in the sodium spectrum, assuming that the first s-term corresponds to a 2_1-orbit. Although this view provided a simple explanation of the large deviation of the s-terms from the hydrogen terms, no quantitative calculation could be obtained in this manner; for, in order to attain a basis for fixing the principal quantum number of the spectral terms in question, such a calculation requires first of all a closer classification of the electronic orbits in the atomic core. Schrödinger's calculations have recently been extended by Th. van Urk (Z. Phys. **13** (1923) 268); he believes he is able to conclude, merely from a consideration of the dimension of the spherical shell needed to evaluate the s-terms, that the first s-term in the sodium spectrum corresponds to a 3_1-orbit. On the other hand, E. Fues (Z. Phys. **11** (1922) 364; **12** (1922) 1; **13 (1923) 211)** has attempted, on the basis of the values of the quantum numbers given above, to determine a centrally symmetrical force field which would permit a direct quantum-theoretical calculation of the observed terms in the sodium spectrum. Although several interesting results were obtained in this way, especially as regards the relationship with the analogous spark spectra of the subsequent elements, one cannot hope to explain the details of the spectra by means of an assumption of this kind which neglects the dynamic character of atomic structure. Besides the above-mentioned polarizing effect of the outer electron upon the inner system, this character plays in many cases an essential role; also in that the ionization potential of an electron often depends to a high degree upon the influence on the motion of the other electrons, resulting from its removal from the atom. This circumstance occurs especially in cases in which electron orbits are already present having the same principal quantum number as the electron in question or a larger principal quantum number.

Especially instructive examples of conditions of this kind are met with in X-ray spectra. As regards this point, reference is made to the paper by Coster and the author mentioned in the footnote on p. [634]. In this connection, it may be mentioned that the arguments given there (p. [542]) for integral values of the quantum numbers, based on the great influence of the relativistic correction upon electron orbits in the interior of the atom, can be made still more cogent, and that similar circumstances are also of essential significance for series spectra. In fact, it is possible to show by a simple calculation that every electron in an n_k-orbit in a nuclear atom would fall into the nucleus if $N/k = hc/2\pi e^2 = 137$. Thus, it is seen that, for the heavier elements, the penetration of the series electron into the interior of the atom would lead to the result that orbits with $k = \frac{1}{2}$ cannot exist. Even for $k = 1$, the electron in these elements comes to distances from the nucleus of the same order of magnitude as the value of the nuclear dimensions obtained from radioactive data. As remarked by Rosseland (Nature, March 17, 1923), this circumstance alone offers a hint toward an understanding of the limitation in the atomic number of existing elements. Furthermore, this circumstance may also offer an explanation of the observed differences in the wavelengths of certain lines in the spectra of lead isotopes which, although very small, are nevertheless many times larger than the differences which, according to (18), presumably might arise from the difference in the nuclear masses of the two isotopes (cf. Nature, June 10, 1922).

fact, here the orbits in the atomic core also form a noble gas configuration; only the total charge is twice as high as in the alkalis. As in the arc spectra of the latter elements, the outer electron in the orbits corresponding to s- and p-terms penetrates into the atomic core. The values of these terms show a relation to the values of the corresponding alkali terms, which is just what should be expected from the change in the orbital dimensions resulting from the larger nuclear charge. As regards the d-terms, this change leads to a new feature in that, in the higher homologues of the family of elements in question, orbits with $k = 3$ no longer lie entirely in the outer region. Only in the spark spectrum of magnesium do we meet conditions analogous to those in the alkali spectra, in that the d-terms here deviate only very little from the terms of the helium spark spectrum. However, as shown by the schematic representation in table IV, for calcium and the higher homologues the d-terms deviate extraordinarily much from these terms, which means that the orbits, at least partly, lie in the inner region. Indeed, from the magnitude of the first member of the d-series, it even follows that the corresponding orbit lies completely in this region, which can also be made clear from an estimate of the orbital dimensions. In Ca, Sr and Ba, this term corresponds to a 3_3-, 4_3- and 5_3-orbit, respectively. On the other hand, orbits corresponding to b-terms lie entirely in the region outside the atomic core.

A *series spectrum of the third kind* is hitherto known only for aluminum. As emphasized by Paschen, the conditions are here quite analogous to those of the arc spectrum of Na and the spark spectrum of Mg, in that the values of the corresponding terms show just such conditions as should be expected from the gradual change in the relative magnitudes of the orbital dimensions. An investigation of the series spectra of the third kind of the other elements (Sc, Y, La and Ac), whose atomic numbers are three units larger than those of the noble gases, would undoubtedly reveal very interesting conditions. Indeed, we must expect here not only, as in the spark spectra of the alkaline earths, that the d-terms would deviate greatly from the hydrogen terms, but that the ionization potential corresponding to the first member of this series would be even higher than the ionization potential corresponding to the largest s-term, which in all the spectra discussed so far belongs to the normal state of the atom. As indicated in the summary table on p. [635], this circumstance involves the beginning of a new step in the development of that electron group which had reached a temporary completion in the preceding noble gas. It should be remarked here that the brackets, which in the table are placed around the number of electrons of the group with highest principal quantum number of the element in question, signify that the spectra of these elements are not sufficiently known to allow us to fix with certainty the quantum numbers of the last electrons bound in the normal atom.

As long as this development lasts, we meet, with increasing atomic number, elements whose spectra exhibit novel features; however, as soon as it has ended, we meet again an orbital configuration in the atomic core having a closed character, even if this configuration does not occur in the normal atom of the preceding element. Thus, we understand that the *arc spectra of the elements copper and silver*, which follow after the families of the iron and palladium metals in the periodic system, have a similar structure as the alkali spectra (cf. table III). In the spectra in question, just as in the alkali spectra, the outer electron, moving in an orbit corresponding to an s-term, will penetrate into the interior of the atom, and, as stated in the summary table on p. [635], the normal state will again correspond to an orbit for which the principal quantum number is equal to the number of the period in the system. On the other hand, the value of the p-terms shows that, in the copper and silver atoms, orbits with $k = 2$ will lie entirely in the region outside the atomic core, with the result that the first p-term corresponds to a 2_2-orbit. This is also clear, since, according to the theory, the monovalent Cu^+ and Ag^+ ions occupy an essentially smaller volume than the positive alkali ions, a circumstance which, by the way, is also revealed in the higher values of the s-terms in Cu and Ag as compared with the corresponding terms in the alkali metals. Of course, the orbits corresponding to the d- and f-terms ($k = 3$ or 4) lie for copper and silver, as for the alkalis, entirely in the outer region*.

* According to H. M. Randall (Ann. d. Phys. **33** (1910) 741), a spectral term lying outside the usual term series occurs in the spectrum of copper; according to the rules of combination and the Zeeman effect, it conforms to the d-terms, as stated in the table. It may be of interest to remark that this term can be interpreted as corresponding to a 3_3-orbit lying outside the atomic core. In fact, a simple estimate of the orbital dimensions of the atomic core points to the dynamical possibility of a 3_3-orbit lying outside the symmetrical subgroup of the six 3_3-orbits, but at the same time inside the limits of the region within which the two other subgroups of the main 3-quantic group, with six electrons each in 3_1- and 3_2-orbits, move. This interpretation is supported by the occurrence in the copper spectrum of still other lines, as found by Mr. Werner, which can be represented as combinations of known spectral terms with a second d-term which, as regards possible combinations and Zeeman effect, forms a doublet in the usual manner with the term mentioned. However, as may be seen in table III, the d_1-term is here larger than the d_2-term, contrary to the usual condition in doublet spectra. This may be connected, however, with the circumstance that, with the assumed position of the orbit, the magnitude of the term will depend to a particularly high degree upon the change suffered by the strength of binding of the other 3-quantic orbits on the removal of the electron.

Note added in proof: A closer investigation of the copper spectrum by Mr. Nissen and Mr. Werner has shown, in the meantime, that there appear to be still more terms lying outside the usual series of terms. We must therefore count on the possibility that in the copper spectrum we have to do with a more complicated phenomenon than assumed above.

The *spark spectra of the elements zinc and cadmium*, which follow directly after Cu and Ag, are only poorly known. Nevertheless, it may be possible to conclude, from the disentangled term combinations, that in the p-terms the outer electron penetrates into the atomic core, which might also be expected from a theoretical estimate of the orbital dimensions. As in the case of the alkali spectra, and the spark spectra of the alkaline earths, the principal quantum number should then be the same for the largest s- and p-terms and, indeed, equal to the number of the period in the system of elements.

§7. Closer discussion of the complex structure of series spectra.

In the preceding sections we have briefly explained the picture of atomic structure to which we are led by the interpretation of the spectral laws. However, the considerations which we have given to particular spectra have dealt only with the approximate description of the orbit of each electron and its classification with the aid of the quantum numbers n and k, based on its approximative central character; we have hitherto neglected the questions of the more precise orientation of the planes of the different electron orbits. This orïentation determines first of all the nature of the finer interplay among the motions of the different electrons, and this in turn must be regarded as responsible for the recurring closure of certain electron groups with increasing atomic number, a closure expressing itself in a disinclination of these groups to admit any additional electrons in orbits with the same values of the quantum numbers n and k. As we saw, these closures were an essential feature of the explanation of the periodic alternation of the properties of the elements; we meet an instructive example of this in helium which stands at the end of the first period of the system.

The *arc spectrum of helium* shows an essentially different structure from the spectra considered in the preceding sections. Contrary to the spectra of the elements situated at the beginning of the periods of the system, the helium spectrum consists of two combination spectra of the Rydberg type, the "orthohelium spectrum" and the "parhelium spectrum", which are sharply separated in the sense that no mutual combination lines exist between the two groups of terms. According to the correspondence principle, this circumstance must be interpreted to mean that in these two spectra we have to do with electronic motions which do not belong to a connected class, within which the motion of the series electron can be resolved into harmonic components of a kind corresponding to a central orbit whose plane executes a uniform precession. Hence, rational classification of the stationary states involved in the two helium spectra cannot be carried out in terms of three quantum numbers n, k and j, on the basis hitherto applied*. Under these circumstances, we must restrict ourselves to mention some conjectures of a general character about the origin of the spectra in question having only the purpose of serving as a connection with the following considerations.

In the *orthohelium* spectrum, apart from the narrow doublets, we meet a spectrum of the same character as that of an electron moving in a central orbit, with the peculiarity that the state corresponding to the principal quantum number 1 is missing; this follows from the existence of the metastable orthohelium state discovered by Franck and his co-workers by experiments on electron impacts. A spectrum of this kind is, in fact, to be expected if the motions of the outer and inner electrons take place in the same plane; for, when this is assumed, there exists a mechanically possible form of motion in which the mutual perturbations possess a particularly simple character, in that the eccentricity of the inner orbit, and the position of its major axis relative to the radius vector of the outer electron, remain approximately constant. In this case, the stationary states can be fixed by a suitable adaptation of the theory of simple central motions, and the orbit of the outer electron can be described by a principal quantum number n and a subordinate quantum number k', which, on the assumption that both electrons revolve in the same direction, is one unit smaller than the total angular momentum of the atom, divided by $h/2\pi$. By direct application of the correspondence principle, the assumption that the emission of the orthohelium spectrum is connected with

* The first attempt at a theorctical interpretation of the helium spectrum was made by Landé (Phys. Zeitschr. **20** (1919) 2 8, and **21** (1920) 114). With the theory of periodicity systems as a direct model, he describes the different stationary states by three quantum numbers and assumes for the orthohelium states $j = k+1$ (coplanar orbits) and for the parhelium states $j = k$ (crossed orbits). Although one cannot hope to achieve a satisfactory explanation of the spectra by such a treatment, Landé's theory is of great importance, if for no other reason in that it represents a first attempt at a kinematical interpretation of the complex structure of a spectrum, in which the third quantum condition to be added to (22) appears as a fixation of the resultant angular momentum of the atom.

A more detailed investigation of the mechanical properties of the helium atom, depending on the mutual perturbation of the two orbits, has been undertaken by Kramers and the author; it led to the recognition of certain especially simple types of motion, to which that mentioned below in the discussion of the orthohelium spectrum belongs. Because of their incomplete character, the results of this investigation, which were reported by the author on the occasion of a series of lectures delivered in Göttingen in June 1922, have not been published.

As has been kindly communicated to me by Mr. Born, he has recently, together with Mr. Heisenberg, carried out an extensive investigation of the mechanical problem of the mutual perturbations of the electrons in the helium atom. It has turned out that a consistent treatment, modeled after the quantum theory of periodicity systems, can actually be carried out formally; however, in this way, not only is no explanation obtained of the non-occurrence of combinations between ortho- and parhelium terms, but not even an approximate value of these terms can be obtained. This investigation may therefore be particularly well suited to provide evidence of the fundamental failure of the laws of mechanics to describe the finer details of the motion of systems with several electrons.

this type of motion makes possible an understanding of the general possibilities of combinations among the orthohelium terms, and also offers an explanation of the absence of an orthohelium state with principal quantum number 1. In fact, a transition from an orthohelium state with higher quantum number to a state in which both electrons move in equivalent 1-quantic orbits, and which therefore belongs to a different type of motion, would not correspond to any harmonic component in the electric moment of the atom. On the other hand, the interpretation of the orthohelium spectrum still faces great difficulties; for it has proved impossible, on the assumption of the type of motion mentioned, to account more closely for the magnitude of the observed orthohelium terms by means of calculations based on the laws of classical mechanics.

The circumstance that, as follows from the investigations of Lyman, the normal state of the atom belongs to the stationary states of the *parhelium spectrum* suggests that we here have to do with forms of motion to which a configuration of both electrons moving in equivalent orbits would naturally belong. This would probably imply that, in the stationary states of parhelium, the electronic orbits do not lie in the same plane, and that their shape, as well as their position, continually change during the motion. If we now ask about the meaning of the subordinate quantum number occurring in the classification of the parhelium spectrum, we hardly have a sufficient connection with the theory of central motions to justify identifying it with the quantum number k appearing in this theory; rather, the assumption suggests itself that the quantity to be quantized is the total angular momentum of the atom or—if the law of the conservation of momentum should fail in its ordinary mechanical formulation—an invariant formally analogous to the angular momentum which might result from the general symmetry of the nuclear atom. In particular, with regard to the normal state it may be remarked that, from the point of view of the theory of periodicity systems, the simplest assumption would be to put the total angular momentum of the atom in this state equal to $h/2\pi$*. We shall introduce this assumption in the following formal considerations of the complex structure of the series spectra; for, because of the normal Zeeman effect of the helium lines, we have hardly any justification for deviating in any definite direction from the formal laws of the theory mentioned.

The *spectra of the heavier noble gases* have a very complicated structure, requiring a great intricacy of the totality of their stationary states. However, we

* That it is not possible, on this assumption and on the basis of the simplest possible type of mechanical motion, to achieve agreement with the value of the term corresponding to the normal state of helium has recently been shown by Kramers (Z. Phys. **13** (1923) 312). (Cf. also J. van Vleck, Phil. Mag. **44** (1922) 842.) In the paper by Kramers, the fundamental difficulties confronting a mechanical treatment of the problem, because of the instability of the electron motion, are also thoroughly discussed.

[644]

may assume that, in the normal states of the atoms of the noble gases, we are dealing with electron configurations of pronounced symmetry. Actually, we see just in this symmetry the explanation of the closed character of these electron groups; for the admission of additional electrons, which would destroy this symmetry, would, in fact, not show that analogy with a classical radiation process which could be established for the possible transitions between stationary states of simply or multiply periodic systems. In particular, it seems necessary to assume that the symmetry of the electron groups for which the principal quantum number is larger than 1 is such that these groups do not contribute to the total angular momentum of the atom, and that the value of this momentum therefore must be set equal for all the noble gases. As is well known, the noble gases show no paramagnetism, from which it is often concluded that their total angular momenta vanish in the normal state. However, according to the quantum theory, this conclusion is not compelling; for this theory claims, for an atom possessing a finite value of the total angular momentum and the resulting magnetic moment, that in the presence of a magnetic field the axis of the angular momentum can assume only definite orientations relative to the field, a claim that recently has received such brilliant confirmation by the beautiful experiments of Gerlach and Stern* on the deflection of silver atoms in an inhomogeneous magnetic field. Hence, independently of the value of the magnetic moment which must be ascribed to the individual atoms, the lack of paramagnetism in the noble gases can be explained by assuming that the atoms of these elements can only orient themselves in a magnetic field with the axis of the angular momentum at right angles to the direction of the field. As we shall see, this assumption is supported by the quantitative analysis of the anomalous Zeeman effect of the families of elements following after the noble gases.

In the *complex structure of the alkali spectra* we meet a case in which, to begin with, a simple application of the considerations of § 5 would be expected, since we may assume that here, in contrast to helium, the orbital configurations of the atomic core, at least to a first approximation, can only suffer relatively small changes under the influence of the series electron. This expectation is now actually fulfilled, in so far as it is possible to account for the combination rules of the terms, and approximately also for the general intensities of the components, by assuming that for $k > 2$ the j-values for the doublet terms are equal to $k+d$ and $k+d-1$, and that for the singlet s-term ($k = 1$) j is equal to $1+d$, where d is a quantity that is independent of k, but is to begin with undetermined. On the other hand, as long as we restrict ourselves strictly to the basis of the quantum theory of periodicity systems, as in § 5, it is not possible to account for such a variety of values for j. In fact, the conditions (29) always yield an odd multiple

* W. Gerlach and O. Stern, Z. Phys. **9** (1922) 349.

of term components, if we assume that the quantities i and j are integers. Hence, in order to obtain doublets, we should have to exclude certain j-values, for which no simple reason can be given from considerations such as those in § 5.

Under these circumstances it is now satisfying that the stability properties of atoms with several electrons, required by the empirical spectra, involve consequences of a general kind which cannot be reconciled with essential postulates regarding statistical applications of the quantum theory of periodicity systems. Thus, it is an essential claim of this theory that the weight of a stationary state of a system consisting of two coupled component systems is given by the product of the weights of those states of the component systems which are obtained continuously when the force of coupling is made to vanish. However, as long as we retain the quantization rules of § 5, this claim cannot be reconciled with experience in the case of the coupling of the series electron to the atomic core.

In determining the statistical weight of a stationary state of a nuclear atom, we must above all bear in mind that the atom forms a degenerate system, simply because the direction of the angular momentum in space is arbitrary; in the presence of a magnetic field this arbitrariness is removed, and in the quantum-theoretical analysis of the Zeeman effect we actually possess a direct means of determining the number of states into which each stationary state of the unperturbed atom is split in the presence of the magnetic field. Thus, Landé's investigations give for the alkali spectra that in a magnetic field the number of term values, arising from states in which the series electron describes an n_k-orbit, totals $4k-2$. On the one hand, we must in our case assume, as mentioned, that the atomic core can take on only one position in the magnetic field. On the other hand, an n_k-orbit in a pure central system can take on $2k$ positions in the field, if one excludes the position in which the angular momentum is perpendicular to the field and which, in the limiting case when the central orbit becomes a Kepler orbit, must be regarded as singular. Now, for large k the product of these two numbers is only about half as large as the number of observed states of an alkali atom in a magnetic field. We must therefore conclude that the coupling between the atomic core and the series electron cannot be regarded as simply analogous to a coupling between two pure periodicity systems, in the simple manner assumed in § 5. Rather, we are led to the view that, because of stability properties of the atom which cannot be described mechanically, the coupling of the series electron to the atomic core is subject to a *constraint* ["Zwang"] which is not analogous to the effect of an external field, but which forces the atomic core to assume two different positions in the atom, instead of the single orientation possible in a constant external field, while at the same time, as a result of the same constraint, the outer electron, instead of $2k$ possible orientations in an external field, can only assume $2k-1$ orientations in the atomic assemblage.

Under these circumstances, we must be prepared for the possibility that the coupling between the series electron and the atomic core cannot be described directly after the model of the quantization rules for periodicity systems. Now, an important hint as to the direction in which we must look for the breaking of these rules has again been given just by Landé's analysis of the anomalous Zeeman effect. As mentioned, on the basis of the simplest possible assumptions, this analysis leads to values of the components of the angular momentum in the direction of the field which can be represented by $\mu h/2\pi$, where μ is not always a whole number, but in certain cases, such as for the alkali spectra, is a "half integer".

This finding makes it natural to surmise that the orientation of the orbit of the series electron relative to the atomic core cannot be described either by integer values of j; in particular, for reasons of symmetry, the formal assumption $d = \frac{1}{2}$ might offer itself*. Further, if we compare the values of j fixed according to this assumption with the possible values of the component of angular momentum parallel to the external magnetic field, evaluated formally by Landé from the Zeeman effect, we find that these values correspond to all possible half integral values of μ, with the exclusion of such orientations of the atom in which the total angular momentum would be parallel to the field. This circumstance is expressed, e.g., in the fact that in the alkali spectra the s-terms are single and correspond to the value $1\frac{1}{2}$ of j. In fact, according to our formal exclusion rule, a state $j = \frac{1}{2}$ would have no possible orientation in the magnetic field, and would therefore possess no statistical weight. Further, it may be remarked in this connection that this finding is quite analogous to the behavior of the noble gas configurations in a magnetic field. In fact, if, as above, we assume $j = 1$ for these, the exclusion of the parallel orientation leads just to the consequence that the

* In his bold attempt to explain, by ingeniously chosen assumptions, the details of the anomalous Zeeman effect, as revealed in Landé's and Sommerfeld's papers, Heisenberg (Z. Phys. **8** (1921) 273) was the first to assume that the stationary states of an atom cannot be described directly by conditions of the form (29) with integral quantum numbers. In fact, in Heisenberg's investigation, as well as in the later papers of Landé, it is assumed that even the values of the quantum number k must be taken as half integers. However, in view of the manner in which the general structure of the series spectra could be explained with the help of the quantum theory of periodicity systems, this assumption, which at several points has consequences that seem to contradict our experience about spectra (cf. the footnote on p. [639]), must be regarded as a departure from this theory which can hardly be substantiated. As we attempted to show above, the situation is different as regards the fixing of the quantum number j. In fact, in this case, essential postulates of the quantum theory of pure periodicity systems are already broken by the general consequences resulting from the coupling between series electron and atomic core. After all, the introduction in the text of half-integral j-values in the formal representation of complex structure merely expresses the view that, at the present state of the theory, it is impossible in these problems to achieve a direct connection with proven features of the quantum theory of periodicity systems.

atom can only orient itself with its axis of angular momentum at right angles to the direction of the field. Moreover, it must be emphasized that we must be prepared to find that the dynamical and magnetic properties of the noble gas configurations, when these appear as atomic cores, may, as a result of the constraint, be essentially different from when they form free atoms.

Quite similar considerations as hold for the alkali spectra are valid for the spectra of Cu and Ag as well as for the spark spectra of the alkaline earths, which show the same type of complex structure. In the nature of the matter, the complex structure depends only on the values of the quantum numbers i, k and j, and the circumstances determining the fixing of the principal quantum number are irrelevant in this connection. We are here dealing with a case of the so-called *spectroscopic displacement law*, which was first formulated by Sommerfeld and Kossel* and which has rendered valuable service in the disentanglement of the spark spectra of many elements**.

Passing to the *arc spectra of the alkaline earths*, we meet circumstances which can be fitted naturally into the framework indicated above. Among the multiplicity of terms of the spectra considered, there appears, as is well known, a *system of triplet terms*; in order to conform formally, according to (27), to the empirical combination rules, we may assume that the values of j are given by $k+t$, $k+t-1$, and $k+t-2$, while for the corresponding singlet s-terms j is equal to $1+t$. Now, it is possible to give a reason for the assumption that $t = 1$ analogous to the assumption made above that $d = \frac{1}{2}$ for the alkali spectra. Landé's analysis of the Zeeman effect of the alkaline earth spectra has actually shown that we must assume that the components of the angular momentum parallel to the magnetic field are always integral multiples of $h/2\pi$. Further, for the triplet system, this analysis has given $3(2k-1)$ for the total number of orientations of an n_k-system in a magnetic field. In analogy to what was said above, this finding can be interpreted to mean that the coupling of the series electron to the atomic core involves, on the one hand, that the core, instead of two possible orientations in an external field, as in the case of the normal state of the alkali atoms, must be assigned three possible orientations; on the other hand, the coupling also involves that the motion of the outer electron, instead of $2k$, has only $2k-1$ possible states. Closely connected herewith may also be the circumstance that, when the j-values are chosen in this way, the atom orients itself in a magnetic field such that all parallel orientations of the axis of the angular momentum relative to the field are to be excluded, as in the case of the alkali spectra. It must be remarked, however, that,

* A. Sommerfeld and W. Kossel, Verh. d. Deutsch. Phys. Ges. **21** (1919) 240; cf. also N. Bohr: *Quantum Theory of Line Spectra.*

** Cf. E. Fues, Ann. d. Phys. **63** (1920) 1.

in the problem of explaining the singlet nature of the s-terms, we do not meet quite such simple circumstances for the alkaline earths as for the alkalis. Indeed, this exclusion rule for the orientation in a magnetic field clearly does not at all exclude a number of terms for which $j = 1$. It is satisfying, therefore, that the exclusion of these s-states agrees with the demand that the general expression for the total number of states in a magnetic field, corresponding to an n_k-orbit of the series electron, must remain valid also for $k = 1$; a demand which, as Mr. Pauli will show in a paper to appear soon, can be given a simple interpretation from a consideration of the transformation of the Zeeman effect with increasing field strength.

As is well known, in the spectra of the alkaline earths there occurs, besides the triplet terms, a *system of singlet terms*. Taking into account the rules for the empirical regularities of the triplet terms, and after the above choice of j-values, we are led uniquely to the assumption that $j = k$ for these. Independently of a detailed interpretation of the atomic motion corresponding to the singlet terms, we are further led to an immediate understanding of the empirical finding that the possible combinations with the triplet terms are restricted, besides by conditions (32), in such a way that combinations of S-terms ($j = 1$) with the p_3-term ($j = 1$) do not occur. In fact, in an external magnetic field, the axis of the angular momentum of the atom for the states in question is always perpendicular to the field; hence, there are obviously no harmonic components of oscillation parallel to the field corresponding to a transition S–p_3. A consideration of the absence of certain components of the Zeeman effect, based on the correspondence principle, is found in the paper on intensities in the anomalous Zeeman effect by Sommerfeld and Heisenberg mentioned on p. [629]. These authors draw the conclusion, also found in an earlier paper by Landé, that, from the non-appearance of the S–p_3 combinations, the atom in both stationary states in question has zero resultant angular momentum. However, we see that this conclusion is not convincing, but that the limitation of possible orientations of the atom in a magnetic field implies that the combination line in question cannot occur in the presence of even the weakest external fields.

As mentioned already in § 5, the appearance of different types of terms means that orbital configurations occur in the atomic core with different values of its total angular momentum. As regards a deeper understanding of this circumstance, it is a significant hint that our reason for the singlet nature of the s-state of the alkali atoms, based on the exclusion of parallel orientations in a magnetic field, loses its validity when the orbital configuration in question appears as an atomic core, if only the total angular momentum of the atom corresponds to a value $j \geqq 1$. Independently of these considerations, a consideration of the term values themselves yields interesting support for the assumption that the appearance of the singlet terms is connected with a comprehensive rearrangement of the orbital configurations in the atomic core.

In order to account for the observed term values, we must, for the singlet terms of all alkaline earth series, assume corresponding values of the principal quantum number as for the spark spectra of the same element. It is true that in case of the triplet terms, something similar holds for the p-, d-, and f-terms. On the other hand, for the largest s-term we must assign a value of the principal quantum number which is one unit larger than the number of the period in the system of the elements. We understand this result when we reflect that the appearance of several electron orbits with the same quantum numbers n and k in the same region of an atom is possible only if they can develop a harmonious interplay. However, as indicated by the interpretation of the metastable helium state, such interplay is generally not possible as an end result of a radiation process. When the orbital configuration of the atomic core suffers an essential rearrangement at the same time, however, a possibility is opened up for the admission of an additional electron into an already present electron group.

At the present state of the theory, it is difficult to form definite conceptions of the orbital configuration in the atomic core in the stationary states corresponding to the singlet terms, which might allow us, from observations of the spark spectra, to determine the dynamic properties, and in particular the energy value of the atomic core. The very circumstance that, in our view, the appearance of this configuration depends on the coupling of an additional electron to the atom, might not only explain that a state in which a harmonious interplay of the two valence electrons has been established belongs to the multiplicity of singlet terms, but might at the same time explain that, as demanded by the spectra, the energy of the atom for the singlet terms with increasing principal quantum number approaches the same value as for the triplet terms.

As is well known, the arc spectra of the elements Zn and Cd, which follow after Cu and Ag, and also the Hg spectrum, show the same characteristic structure as the arc spectra of the alkaline earths. In fact, we must also assume that the circumstances, as regards the quantum numbers k and j, are exactly the same. On the other hand, because of the relatively smaller dimensions of the atomic core, we may assume that orbits for which $k \geqq 2$ will lie entirely outside it, and that, as for the arc spectra of Cu and Ag, the values of the principal quantum number for the series of p- and P-terms will begin with 2, and for the series of d- and D-terms with 3. This assumption, which agrees with the magnitude of the terms, has apparently been conclusively supported by a thorough investigation carried out in Copenhagen by Hansen, Takamine, and Werner of the Stark effect in the mercury spectrum*; it has shown that not only for the d-terms, but also for the

* A detailed report of this investigation, which has given several interesting results of the influence of electric fields on complicated spectra, will appear soon in the Memoirs of the Copenhagen Academy.

higher members of the p-terms, the orbit of the series electron can only deviate very little from a Kepler ellipse.

The spectra of the elements Al, Ga, In, and Tl, again show a complex structure of the same kind as that of alkali arc spectra, as was indeed to be expected because of the assumed similarity of the normal states of the alkaline earths and the noble gases, as regards the resulting angular momentum as well as the possible orientations in a magnetic field. At the same time, when compared with the alkali spectra, these spectra exhibit typical differences, corresponding exactly to the theoretical expectations. In the first place, the relatively small magnitude of the first s-term results from the principal quantum number being one unit larger than for the alkali metal in the same period of the system of the elements; this, in turn, is connected with the fact that, in the normal state of the preceding elements, there is a temporarily closed electronic orbital configuration consisting of two n_1-orbits, where n is just the number of the period in the system of elements. Because of this circumstance, the orbit of the last electron to be captured in the normal state of the element in question will be characterized by $k = 2$. Because of the large dimensions of the atomic core of the elements in question, the orbits with $k = 2$ do not only penetrate into the interior of the atom but, for the first p-term, which corresponds to the normal state of the atom, the electronic orbit even lies completely within the region of the previously bound electrons. The principal quantum number of this orbit is equal to the number of the period in the system of elements.

For aluminum, Paschen's measurements of the series spectrum of the third kind offer a basis for a simple estimate of the dimensions of the atomic core. From the value of the largest s-term, we see directly that the electron, moving in a 3_1-orbit in the doubly charged ion, during its revolution reaches a distance from the nucleus that is nearly three times as large as the radius of the electronic orbit in the normal hydrogen atom. In the singly charged aluminum ion, in which two 3_1-orbits are present, the orbital dimensions of these electrons are considerably larger still. This involves not only that an orbit with $k = 2$ cannot lie entirely in the outer region, but even that an electron in a 3_2-orbit never gets as far from the nucleus as the valence electrons in 3_1-orbits, which again results in a further increase in the orbital dimensions of this electron in the normal state of the atom. It is clear that in such cases one cannot expect to obtain an approximate estimate of the term values by comparing with the motion of an electron in a fixed central field, as was to a certain extent possible for spectra for which the atomic core forms a noble gas configuration. The particularly large dimensions of the atomic core in aluminum, and the low symmetry of its orbital configurations, further make possible an understanding of the peculiarly large deviations from the hydrogen terms exhibited by the d-terms of the arc spectrum.

In the group of elements considered, we meet a new feature of atomic structure, in that we here for the first time after the appearance of a noble gas configuration witness the appearance of orbits in the normal state of an atom for which $k = 2$, and for which the principal quantum number is equal to the number of the period in the system of elements. This leads to the beginning of a new step in the development of the outer electron group in the atom, and involves that we must expect new traits in the spectra shown by the succeeding elements. Although promising beginnings have been made, neither the empirical nor the theoretical circumstances are sufficiently clarified to warrant going into these matters on this occasion.

As already mentioned in previous sections, in case of the series of elements bracketed in the figure on p. [636], we meet conditions resulting from a temporary *step in the development of an inner electron group*. As expected theoretically, we meet in these elements pronounced deviations from the spectroscopic displacement law mentioned above. As was first made clear by a thorough investigation of the manganese spectrum by Catalán, we encounter here even quite new types of complex structure of series lines, the so-called multiplets*. The appearance of these multiplets depends on the total angular momentum of the atomic core having larger values than in the doublet and triplet spectra; this is also evidenced in a characteristic way in the magnetic properties of the ions of these series of elements. Also in other ways the conditions accompanying the development of the electron groups cause the multiplicity of stationary states that must be considered in order to interpret the spectra to exhibit a much greater intricacy than should be expected from the simple schematic interpretation of the spectral laws given in §4 and §5.

Here we shall not enter further into these more complicated questions, for the investigation of which the incompleteness of the theoretical foundations, so strongly apparent even in the treatment of the simpler problems discussed in the last sections, confronts us ever more plainly. To emphasize this incompleteness had, in the nature of the matter, to be one of the principal tasks of this report, whose aim first of all has been to give an expression of the present state of our efforts to apply the regularities in the spectra to develop conceptions of atomic structure, and thus to fulfill the hope that from the beginning of the science of spectroscopy has always hovered before the eyes of its devotees.

* M. Catalán, Trans. Roy. Soc. A. **223** (1922) 127–173. In this connection, see also the investigation of the chromium spectrum by H. Gieseler (Ann. d. Phys. **69 (1922) 147**). For a theoretical discussion of these types of spectra, see also A. Sommerfeld, Ann. d. Phys. **70** (1923) 32, as well as G. Wentzel, Phys. Zeitschr. **24** (1923) 104.

[652]

In concluding this paper, I wish to express my sincere gratitude to my co-workers, especially to Messrs. H. A. Kramers and W. Pauli, who have helped me energetically with the writing of the manuscript and the discussion of many details of the theory, as well as to Mr. S. Werner who has rendered me valuable help with the discussion of the empirical material and the preparation of the graphical representations of the spectra.

(Received 15 March 1923)

Explanation of tables

The black points represent some of the largest members of the series of empirical spectral terms, the usual spectroscopic notations of which are given on the left side. The distances of the points from the border line at the right of the figure are proportional to the numerical term values; a scale for these, in wave numbers, is given at the bottom of each table. The scale added at the top of each table gives the values of the effective quantum number n^*, which for a spectrum of the p'th kind (cf. p. [622]) is uniquely connected with the term value T (expressed in wave numbers) by the formula $T = Rp^2/n^{*2}$, where R is Rydberg's constant, given by the expression $R = 109737(1 - 5.4 \times 10^{-4}/A)$, where A is the atomic weight of the element relative to hydrogen.

The numbers appended to the points give the values of the principal quantum numbers of the orbits of the series electron corresponding to the terms. This number is suited to replace the running number ("Laufzahl") within an individual term series. The subordinate quantum numbers k and j, values of which are associated with the individual term series, determine first of all the possible combinations of terms in the absence of external forces; the rules are that k can only change by one unit, whereas j either changes by one unit or remains unchanged. The further limitation must be added, namely that a combination cannot occur between terms for both of which j has the value 1. It must particularly be emphasized that the theory does not possess a clarified character as regards the meaning of the fixing of the absolute values of the quantum number j; this is indicated at the top of the tables by placing this quantum number in parentheses.

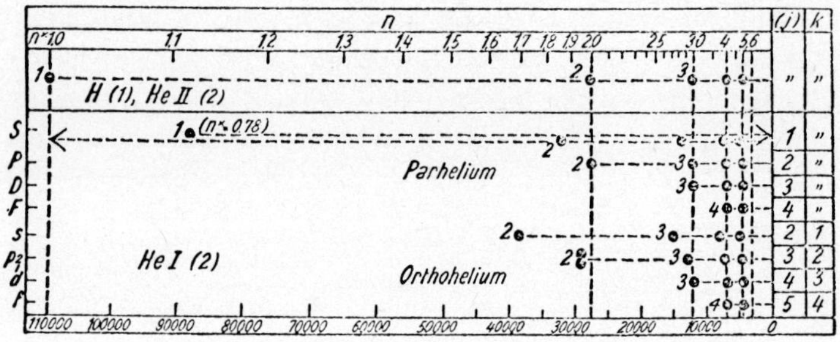

Table I.

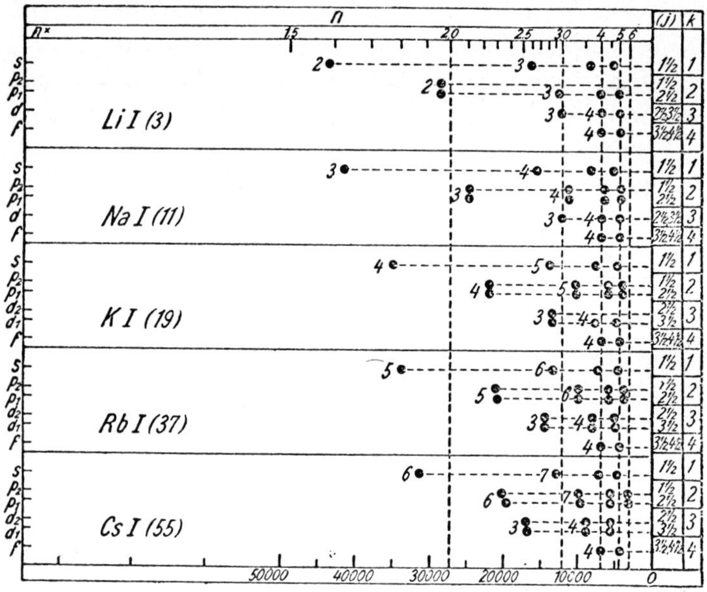

Table II.

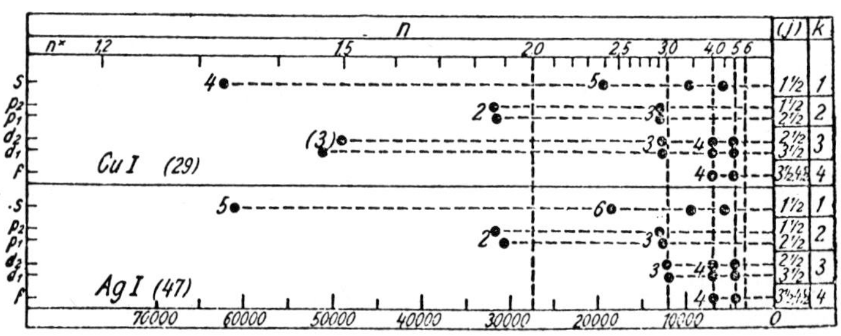

Table III.

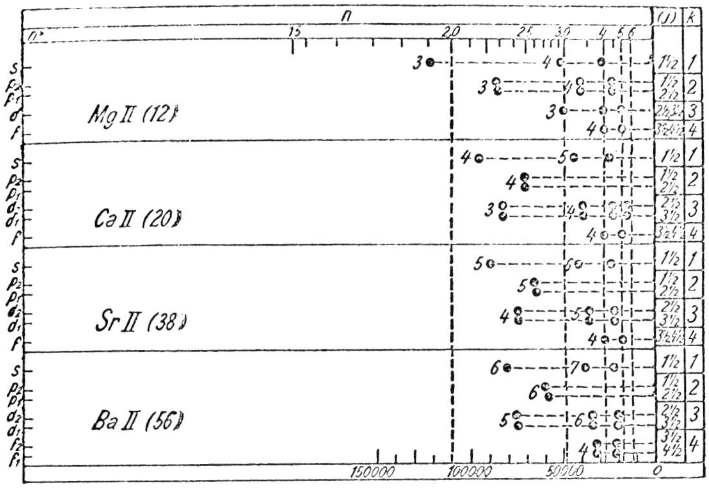

Table IV.

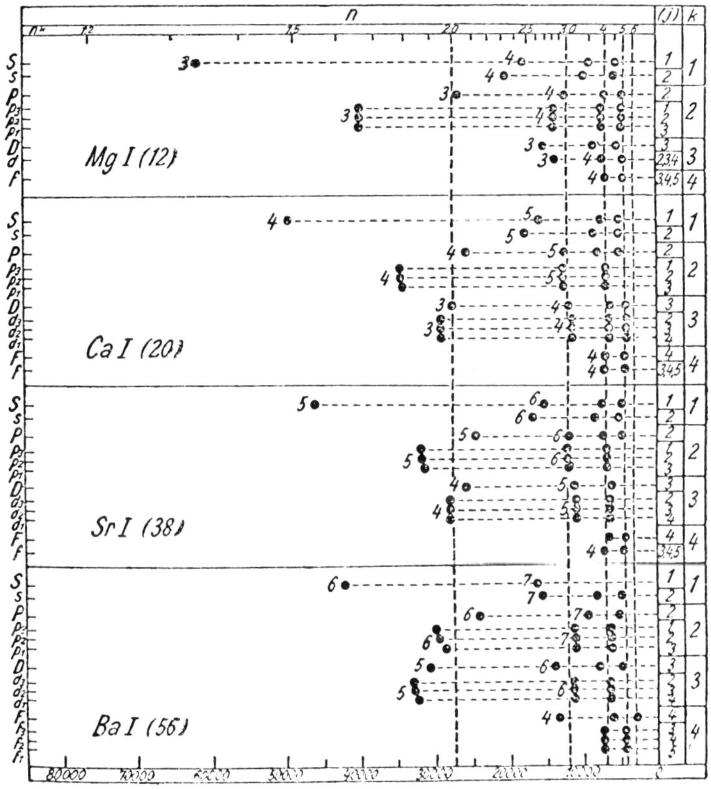

Table V.

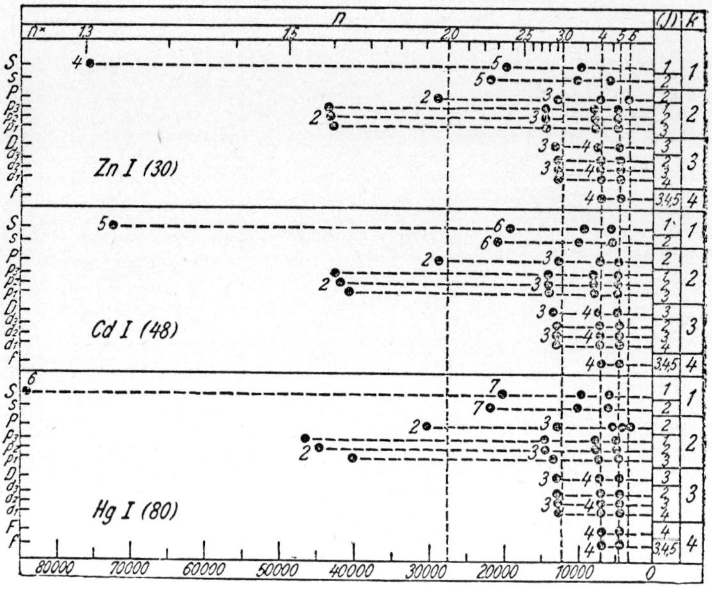

Table VI.

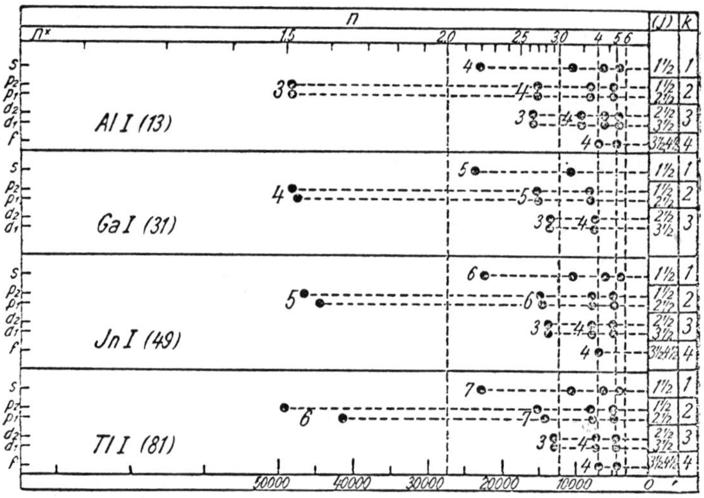

Table VII.

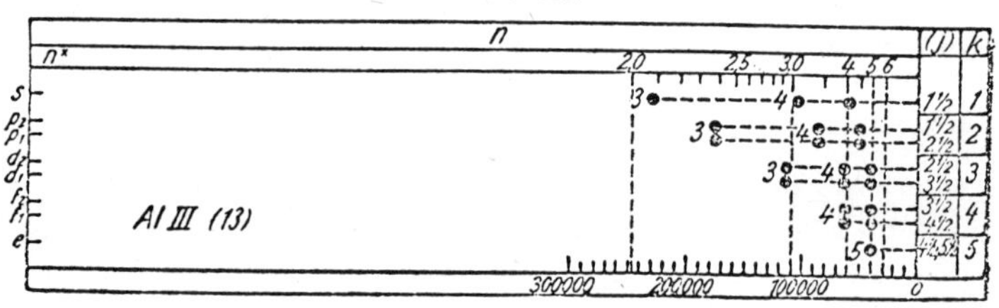

Table VIII.

XIII. ATOM

(Encyclopaedia Britannica 13 (1926) 262*)

* [See Introduction, sect. 14.]

ATOM (*see* 2.870), *Structural Units.*—Through the experimental discoveries of the second half of the 19th century it became gradually clear that the atoms of the elements, far from being indivisible entities, had to be thought of as aggregates built up of separate particles. Thus from experiments on electrical discharges in rarified gases and especially from a closer study of the so-called cathode rays, one was led to recognise the existence of small negatively charged particles the mass of which was found to be about 2,000 times as small as the mass of the lightest atom, the hydrogen atom. These small particles, which may be regarded as atoms of negative electricity are now, following Johnstone Stoney, generally called electrons. Through the investigations of J. J. Thomson and others convincing evidence was obtained that these electrons are a constituent of every atom. On this basis a number of the general properties of matter, especially as regards the interaction between matter and radiation, receive a probable explanation.

In fact the assumption that electrons are vibrating around positions of stable equilibrium in the atom offered a simple picture of the origin of spectral lines which allowed the phenomena of selective absorption and dispersion to be accounted for in a natural way. Even the characteristic effect of magnetic fields on spectral lines discovered by Zeeman could, as was shown by Lorentz, be simply understood on this assumption. The origin of the forces which kept the electrons in their positions remained for a time unknown, as well as the way in which the positive electrification was distributed within the atom. From experiments on the passage through matter of the high speed particles expelled from radioactive substances, however, Rutherford was in 1911 led to the so-called nuclear model of the atom. According to this the positive electricity is concentrated within a nucleus of dimensions very small compared with the total space occupied by an atom. This nucleus is also responsible for practically the whole of the atomic mass.

True Properties of the Elements.—The nuclear theory of the atom has afforded a new insight into the origin of the properties of the elements. These properties can be divided into two sharply distinguished classes. To the first class belong most of the ordinary physical and chemical properties. These depend on the constitution of the electron cluster round the nucleus and on the way in which it is influenced by external agencies. This, however, will depend on the attractive force due to the nucleus which keeps the cluster together. On account of the small size of the nucleus compared with the distance apart of the electrons in the cluster, this force will to a high approximation be determined solely by the total electric charge of the nucleus. The mass of the nucleus and the way in which the charges and masses are distributed among the particles making up the nucleus itself will only have an exceedingly small influence on the behaviour of the electronic cluster.

To the second class belong such properties as the radioactivity of the substance. These are determined by the actual internal structure of the nucleus. In the radioactive processes we witness, in fact, explosions of the nucleus in which positive or negative particles, the so-called α and β particles, are expelled with very great velocities. The complete independence of the two classes of properties is most strikingly shown by the existence of substances which are indistinguishable from one another by any of the ordinary physical and chemical tests, but of which the atomic weights are not the same, and whose radioactive properties are completely different. Any group of two or more such substances are called isotopes (*q.v.*), since they occupy the same position in the classification of the elements according to ordinary physical and chemical properties. The first evidence of their existence was found in the work of Soddy and other investigators on the chemical properties of the radioactive elements. It has been shown that isotopes are found not only among the radioactive elements, but that many of the ordinary stable elements consist of isotopes, for a large number of the latter that were previously supposed to consist of atoms all alike have been shown by Aston's investigations to be a mixture of isotopes with different atomic weights. Moreover the atomic weights of these isotopes are whole numbers, and it is because the so-called chemically pure substances are really mixtures of isotopes, that the atomic weights are not integers.

Inner Structure.—The inner structure of the nucleus is still but little understood, although a method of attack is afforded by Rutherford's experiments on the disintegration of atomic nuclei by bombardment with α particles. Indeed, these experiments may be said to have started a new epoch in natural philosophy in that for the first time the artificial transformation of one element into another has been accomplished (*see* TRANSMUTATION OF ELEMENTS). In what follows, however, we shall confine ourselves to a consideration of the ordinary physical and chemical properties of the elements and the attempts which have been made to explain them on the basis of the concepts just outlined.

THE RELATIONSHIPS BETWEEN THE ELEMENTS

Periodicity of Elements.—It was recognised by Mendelejeff that when the elements are arranged in an order which is practically that of their atomic weights, their chemical and physical properties show a pronounced periodicity. A diagrammatic representation of this so-called periodic table is given in Table I., which represents in a slightly modified form an arrangement first proposed by Julius Thomsen. In the table the elements are denoted by their usual chemical symbols, and the different vertical columns indicate the so-called periods. The elements in successive columns which possess homologous chemical and physical properties are connected by lines. The meaning of the square brackets around certain series of elements in the later periods, the properties of which exhibit typical deviations from the simple periodicity in the first periods, will be mentioned below.

Radiation.—The discovery of the relationship between the elements was primarily based on a study of their chemical prop-

TABLE I.

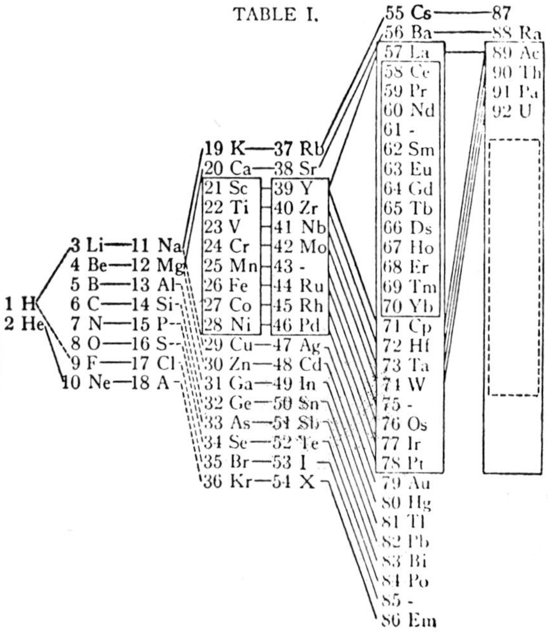

erties. Later it was recognised that this relationship appears also very clearly in the constitution of the radiation which the elements emit or absorb in suitable circumstances. In 1883 Balmer showed that the spectrum of hydrogen, the first element in the table, could be expressed by an extremely simple mathematical law. This so called Balmer formula states that the frequencies ν of the lines in the spectrum are given to a close approximation by

$$\nu = R\left(\frac{1}{(n'')^2} - \frac{1}{(n')^2}\right) \tag{1}$$

where R is a constant, and where n' and n'' are whole numbers. If n'' is put equal to 2 and n' is given successively the values 3, 4, . . . the formula gives the frequencies of the series of lines in the visible part of the hydrogen spectrum. If n'' is put equal to 1 and n' equal to 2, 3, 4, . . . a series of ultra-violet lines is obtained which was discovered by Lyman in 1914. To $n''=3$, 4, . . . correspond series of infra-red hydrogen lines which also have been observed.

Rydberg in his famous investigation of line spectra more than 30 years ago was able to analyse in a similar way many spectra of other elements. Just as in the case of hydrogen he found that the frequencies of a line-spectrum (such as that of sodium) could be represented by a formula of the type

$$\nu = T'' - T' \tag{2}$$

where T'', T' can be approximately represented by

$$T = \frac{R}{(n - \alpha_\kappa)^2} \tag{3}$$

α_κ is a constant for any one series, but takes different values α_1, α_2 . . . for the different series, while n takes a set of successive integral values. R is constant throughout for all spectra, and is the same constant as that appearing in (1); it is generally called the "Rydberg number." In many spectra the terms of most series are multiple, *i.e.*, the terms which we consider as forming a series do actually form two, three or more series corresponding to two, three or more slightly different values of α_κ. Rydberg also discovered that the spectra of elements occupying homologous positions in the periodic table were very similar to each other, a similarity which is especially pronounced as regards the multiplicity of the terms.

Moseley's Discovery.—The study of X-ray spectra made possible by the work of Laue and Bragg brought out relations of a still simpler kind between different elements. Thus Moseley (*q.v.*) in 1913 made the fundamental discovery that the X-ray spectra of all elements show a striking similarity in their structure, and that the frequencies of corresponding lines depend in a very simple way on the ordinal number of the element in the periodic table. Moreover the structure of these spectra was very like that of the hydrogen spectrum. The frequency of one of the strongest X-ray lines for the various elements could for instance be given approximately by

$$\nu = N^2 R\left(\frac{1}{1^2} - \frac{1}{2^2}\right) \tag{4}$$

and that of another line by

$$\nu = N^2 R\left(\frac{1}{2^2} - \frac{1}{3^2}\right) \tag{5}$$

where R is again the Rydberg constant and N the ordinal number of the element in the periodic table. The extreme simplicity of these formulae enabled Moseley to settle any previous uncertainty as to the order of the elements in the periodic table, and also to state definitely the empty places in the table to be filled up by elements not yet discovered.

Atomic Numbers.—In the nuclear model of the atom, the ordinal number of an element in the periodic table receives an extraordinarily simple interpretation. In fact, if the numerical value of the charge on an electron is taken as unity, this ordinal number, which is often called the "atomic number" (*q.v.*), can simply be identified with the magnitude of the nuclear charge. This law which was foreshadowed by J. J. Thomson's investigations of the number of electrons in the atom as well as by Rutherford's original estimate of the charge on the atomic nucleus, was first suggested by van den Broek. It has since been established by refined measurements of the nuclear charge, and it has proved itself an unerring guide in the study of the relationship between the physical and chemical properties of the elements. This law also offers an immediate explanation of the simple rules governing the changes in the chemical properties of radioactive elements following the expulsion of α or β particles.

The Quantum Theory

The discovery of the electron and of the nucleus was based on experiments, the interpretation of which rested on applications of the classical laws of electrodynamics. As soon, however, as an attempt is made to apply these laws to the interaction of the particles within the atom, in order to account for the physical and chemical properties of the elements, we are confronted with serious difficulties. Consider the case of an atom containing one electron: it is evident that an electrodynamical system consisting of a positive nucleus and a single electron will not exhibit the peculiar stability of an actual atom. Even if the electron might be assumed to describe an elliptical orbit with the nucleus in one of the foci, there would be nothing to fix the dimensions of the orbit, so that the magnitude of the atom would be an undetermined quantity. Moreover, according to the classical theory the revolving electron would continually radiate energy in the form of electromagnetic waves of changing frequency and the electron would finally fall into the nucleus. In short, all the promising results of the classical electronic theory of matter would seem at first sight to have become illusory. It has nevertheless been possible to develop a coherent atomic theory based on this picture of the atom by the introduction of the concepts which formed the basis of the famous theory of temperature radiation developed by Planck in 1900.

This theory marked a complete departure from the ideas which had hitherto been applied to the explanation of natural phenomena, in that it ascribed to the atomic processes a certain element of discontinuity of a kind quite foreign to the laws of classical physics. One of its outstanding features is the appearance in the formulation of physical laws of a new universal constant, the so-called Planck's constant, which has the dimensions of energy multiplied by time, and which is often called the "elementary quantum of action." We shall not enter upon the form which the quantum theory exhibited in Planck's original investigations, or on the important theories developed by Einstein in 1905, in which the fertility of Planck's ideas in explaining various physical phenomena was shown in an ingenious way. We shall proceed at once to explain the form in which it has been possible to apply the quantum theory to the problem of atomic constitution. This rests upon the following two postulates:—

1. An atomic system is stable only in a certain set of states, the "stationary states," which in general corresponds to a discrete sequence of values of the energy of the atom. Every change in this energy is associated with a complete "transition" of the atom from one stationary state to another.

2. The power of the atom to absorb and emit radiation is governed by the law that the radiation associated with a transition must be monochromatic and of frequency ν such that

$$h\nu = E_1 - E_2 \tag{6}$$

where h is Planck's constant and E_1 and E_2 are the energies in the two stationary states concerned.

The first of these postulates aims at a definition of the inherent stability of atomic structures, manifested so clearly in a great number of chemical and physical phenomena. The second postulate, which is closely related to Einstein's law of the photo-electric-effect, offers a basis for the interpretation of line spectra; it explains directly the fundamental spectral law expressed by relation (2). We see in fact that the spectral terms appearing in this relation can be identified with the energy values of the stationary states divided by h. This view of the origin of spectra has been found to agree with the experimental results obtained in the excitation of radiation. This is shown especially in the discovery of Franck and Hertz relating to impacts between free electrons and atoms. They found that an energy transfer from the electron to the atom can take place only in amounts which correspond with the energy differences of the stationary states as computed from the spectral terms.

The Hydrogen Spectrum.—From the Balmer formula (1) and the quantum theory postulates, it follows that the hydrogen atom has a single sequence of stationary states, the numerical value of the energy in the nth state being Rh/n^2. Applying this result to the nuclear model of the hydrogen atom, we may assume that this expression represents the work necessary to remove the electron from the nth state to an infinite distance from the nucleus. If the interaction of the atomic particles is to be explained upon the laws of classical mechanics, the electron in any one of the stationary states must move in an elliptical orbit about the nucleus as focus, with a major axis whose length is proportional to n^2. The state for which n is equal to 1 may be considered as the normal state of the atom, the energy then being a minimum. For this state the major axis is found to be approximately 10^{-8} *centimetres*. It is satisfactory that this is of the same order of magnitude as the atomic dimensions derived from experiments of various kinds. It is clear, however, from the nature of the postulates, that such a mechanical picture of the stationary states can have only a symbolic character. This is, perhaps most clearly manifested by the fact that the frequencies of the orbital revolution in these pictures have no direct connection with the frequencies of the radiation emitted by the atom. Nevertheless, the attempts at visualising the stationary states by mechanical pictures have brought to light a far-reaching analogy between the quantum theory and the classical theory. This analogy was traced by examining the radiation processes in the limit where successive stationary states differ comparatively little from each other. Here it was found that the frequencies associated with the transition from any state to the next succeeding one tend to coincide with the frequencies of revolution in these states, if the Rydberg constant appearing in the Balmer formula (1) is given by the following expression:

$$R = \frac{2\pi^2 e^4 m}{h^3} \tag{7}$$

where e and m are the charge and mass of the electron and h is Planck's constant. This relation is actually found to be fulfilled within the limits of the experimental errors involved in the measurements of e, m and h, and may be considered to establish a definite relation between the spectrum and the atomic model of hydrogen.

Correspondence Principle.—The considerations just mentioned constitute an example of the application of the so-called "correspondence principle" which has played an important part in the development of the theory. This principle gives expression to the endeavour, in the laws of the atom, to trace the analogy with classical electrodynamics as far as the peculiar character of the quantum theory postulates permits. On this line much work has been done in the last few years, and quite recently in the hands of Heisenberg has resulted in the formulation of a rational quantum kinematics and mechanics. In this theory the concepts of the classical theories are from the outset transcribed in a way appropriate to the fundamental postulates and every direct reference to mechanical pictures is discarded. Heisenberg's theory constitutes a bold departure from the classical way of describing natural phenomena but may count as a merit that it deals only with quantities open to direct observation. This theory has already given rise to various interesting and important results, and it has in particular allowed the Balmer formula to be derived without any arbitrary assumptions as to the nature of the stationary states. However, the methods of quantum mechanics have not yet been applied to the problem of the constitution of atoms containing several electrons, and in what follows we are reduced to a discussion of results which have been derived by using mechanical pictures of the stationary states. Although in this way a rigorous quantitative treatment is not obtainable it has nevertheless been possible, with the guidance of the correspondence principle, to obtain a general insight into the problem of atomic constitution.

The Spectra of Elements of Higher Atomic Number

The hydrogen spectrum may be considered as evidence of a step-like process in which an electron is captured and bound increasingly strongly in the field surrounding the nucleus, the stages of this process being the stationary states of the atom. Simple arguments lead to the conclusion that the stages corresponding to the binding of an electron by a nucleus of any given charge will be represented by a similar sequence of stationary states and that the energy W_n necessary to remove the electron from the nth state will be given by the expression:

$$W_n = N^2 \frac{Rh}{n^2} \tag{8}$$

where N is the atomic number of the elements under consideration. These states may be visualised as mechanical orbits of the electron in which the major axis is N times as small as the major axis in the corresponding orbit in the hydrogen atom. The spectrum associated with the binding process under consideration is represented by the formula:

$$\nu = N^2 R\left(\frac{1}{(n'')^2} - \frac{1}{(n')^2}\right) \tag{9}$$

For $N=2$, this formula actually represents the spectrum which is emitted by a singly ionised helium atom, *i.e.*, a helium atom, which has lost one of its electrons. Spectra of this type have not yet been observed for values of N larger than 2, but it will be seen that formula (9) includes the approximate formulae (4) and (5) representing the frequencies of the strongest lines in the X-ray spectra of the elements. This may be understood if we assume that an X-ray spectrum is associated which changes in the state of binding of one of the electrons in the inner region of the atom, where, at least when the atomic number is large, the force on the electron due to the nucleus will far outweigh the forces due to the other electrons, and where consequently the presence of these electrons will have a comparatively small influence on the strength of the binding.

Influence of Electrons.—In general the mutual influence of the electrons is very considerable. Consider the stages by which an electron is captured by an atom of which the nucleus already has s electrons circulating round it. In the initial stages of this process while the orbits may be supposed to have dimensions which are large compared with the orbital dimensions of the electrons previously bound, the repulsive forces from these latter electrons may be assumed to neutralise s units of the nuclear charge, and the resultant force will be approximately the same as when an electron is circulating round a nucleus of atomic number N-S. In the later stages, when the dimensions of the orbit of the new electron are smaller, the other electrons can no longer be considered to act as a single central charge, and their repulsion cannot be easily determined. Thus the conditions become more complicated, and the stationary states can no longer be treated by picturing the motion of the new electron as following a Keplerian ellipse. It has been found, however, that many features of the resulting spectra would be explained by assuming the added electron to move in a plane central orbit consisting of a sequence of quasi-elliptic loops. In contrast to a Keplerian orbit, however, the single loops are not closed but the successive maximum radii will be placed at constant angular intervals on a circle with the nucleus at the centre. For such central orbits it is possible, as was first shown by Sommerfeld, to select from the continuous multitude of possible orbits a set of orbits which may be taken as representing stationary states in the sense of the quantum theory. These states are labelled with two integral numbers; the one, denoted by n, corresponds to the integer appearing in the Balmer formula and is called the principal quantum number. The other, denoted by k, may be called the subordinate quantum number. For any given value of n, the number k can take the values $1, 2, 3 \ldots n$, corresponding to a set of orbits with increasing minimum distance from the nucleus. For a given value of k increasing values of n correspond to orbits which exhibit an increasing maximum distance from the nucleus, but which are similar in size and shape in the region where the electron comes nearest to the nucleus.

For the work necessary to remove an electron in an n_k orbit completely from the nucleus, the theory leads to the following approximate expression

$$W_{n,k} = (N-s)^2 \frac{Rh}{(n-\alpha_k)^2} \qquad (10)$$

where α_k depends only on the subordinate quantum number k, and approaches zero for increasing k.

If s is equal to N-1, we see that the $W_{n,k}$ when divided by h coincide exactly with Rydberg's expressions (3) for the spectral terms of the ordinary series spectra of the elements. These spectra may therefore be considered as evidence of processes, representing the last stage in the formation of a neutral atom, in which a nucleus of charge Ne, which holds already N-1 electrons bound in its field, is capturing an Nth electron. In recent years it has been found that many elements under suitable conditions besides their ordinary spectra also emit spectra for which the terms can be represented by

$$T = p^2 \frac{R}{(n-\alpha_k)^2} \qquad (11)$$

where p may take the integral values 2, 3, 4. . . . Comparing (11) with formula (10) we see that these spectra must be ascribed to atoms, which after having lost p electrons are rebinding an electron in the field of the remaining atomic ion.

This interpretation of series spectra allows also the rules governing the possible combinations of spectral terms to be explained. In fact, it has been found that only those lines appear in the spectrum for which the k-values of the spectral terms involved differ by one unit. From an investigation of the constitution of the radiation which on classical electrodynamics would be emitted from an electron performing a central motion, this rule can be shown to be a simple consequence of the correspondence principle.

Multiplet Structure.—The multiplet structure exhibited by the terms of most series spectra makes it necessary to assume that the motion of the electron involved in the emission of these spectra is somewhat more complicated than the simple central motion described above. An analysis based on the correspondence principle indicates that this motion may be described as a central motion on which is superposed a uniform precession of the orbital plane round an invariable axis in space. For a time, however, it seemed very difficult to obtain any closer connection between the observed structures and the above hypothesis of the constitution of the atom. In particular the remarkable analogy between the finer structures of the optical spectra the X ray spectra, which had been brought out by the experiments, was very puzzling. The study of the strange anomalies exhibited by the effect of a magnetic field on the components of the optical multiplets has, however, quite recently led to the view that the electron itself carries besides its electric charge, also a magnetic moment which may be associated with a swift rotation round an axis through its centre. This new assumption allows not only the anomalous Zeeman effect to be accounted for, but affords at the same time a natural explanation for the empirical rules governing the dependency of the widths of the multiplet structures on the atomic number.

Atomic Constitution and the Periodic Table

Soon after the discovery of the electron it was recognised that the relationships between the physical and chemical properties of the elements expressed in the periodic table point towards a group structure of the electronic distribution in the atom. Fundamental work on these lines was done by J. J. Thomson in 1904. After the discovery of the nucleus and the simple interpretation of the atomic number given above, his work has been followed up with great success especially by Kossel and Lewis.

Valency Properties.—It is suggested that the electrons within the atom possess a tendency to form stable groups, each containing a definite number of electrons, which in the neutral state of the atom surround the centre of the atom like successive shells or layers. An explanation of the simple valency properties holding for the second and third period of the periodic table was for instance obtained by assuming that there was a tendency to form completed shells each containing eight electrons. The single valency of sodium and the double valency of magnesium are ascribed to the facility with which the neutral atoms of these elements can loose one or two electrons respectively, as the atomic ions remaining would then contain completed shells only. On the other hand the double negative valency of sulphur and the single negative valency of chlorine are ascribed to the tendency of their outermost shells to take up two or one additional electron respectively in order to form a complete shell of eight electrons, like that contained in the neutral atom of the inactive gas argon.

Statical Arrangement of Electrons.—Attempts have been made to associate the existence of such groups with statical configurations of electrons possessing a high degree of symmetry. The presence of groups of eight electrons for instance has been explained as an arrangement of electrons at the corners of a cube. However suggestive these ideas have been in affording pictures of the constitution of chemical compounds, they do not allow a direct connection with other properties of the atom to be established; the main difficulty being that stable statical arrangements of the electrons are incompatible with the nuclear theory of the atom. In the meantime, however, it has been possible to connect the group structure of the electronic cluster in the atom with the quantum-theory interpretation of spectra. Thus the constitution of the neutral atom in its normal state can be investigated by imagining a process by which N electrons one after one are captured and bound in the field of force surrounding a nucleus of charge Ne.

To each step there corresponds a multitude of stages, *i.e.*, stationary states, in which the electron is more and more firmly bound to the atom. The final state, in which binding is strongest, corresponds to the normal state of the atomic ion. A definite connection between the spectra and the group structure was now established by assuming that in the normal atom only a limited number of electrons can be bound in states visualised as orbits characterised by definite values of the quantum numbers n and k. The electrons bound in orbits corresponding to a given value of n are said to form an n-quantum group, which in its finally completed stage will contain n subgroups, corresponding to the possible values 1, 2 . . . n which k may take. For a sufficiently large nuclear charge, the strength with which the electrons in the different subgroups belonging to one and the same group are bound will be nearly equal. In the gradual building up of the groups in atoms with increasing nuclear charge, it is, however, to be noted that when an n_k orbit appears for the first time in the neutral atom, the strength of the binding will depend very considerably on the value of k.

This is due to the circumstance that this quantum number fixes the closest distance to which the electron may approach the nucleus. The screening of the nuclear charge by the other electrons in the atom may therefore be very different for orbits corresponding to different values of k, and the effect on the strength of the binding can be so large that an orbit characterised by certain values of n and k may correspond to a stronger binding than an orbit for which n is smaller but k larger. This offers a natural explanation of one feature of the periodic table, namely that the periods grow gradually larger, while there appear sequences of elements which differ comparatively little in their chemical and physical properties. Such a sequence marks a stage in the development of an n-quantum group, which consists in the addition of a subgroup corresponding to a value of k which was previously not yet represented in that group, and which takes place after the building up of a group corresponding to a higher value of n has already begun. In fact, during the addition of the subgroup a temporary standstill will occur in the development of the latter group, the constitution of which will primarily determine the chemical affinity of the atom, since it contains the most loosely bound electrons.

In the accompanying table (Table II.) is given a summary of the structure of the normal state of the neutral atoms of the elements. The figures before the different elements are the

TABLE II.

	1_1	2_1	2_2	3_1	3_2	3_3	4_1	4_2	4_3	4_4	5_1	5_2	5_3	5_4	5_5	6_1	6_2	6_3	6_4	6_5	6_6	7_1	7_2
1 H	1																						
2 He	2																						
3 Li	2	1																					
4 Be	2	2																					
5 B	2	2	1																				
⋮																							
10 Ne	2	2	6																				
11 Na	2	2	6	1																			
12 Mg	2	2	6	2																			
13 Ae	2	2	6	2	1																		
18 A	2	2	6	2	6																		
19 K	2	2	6	2	6		1																
20 Ca	2	2	6	2	6		2																
21 Sc	2	2	6	2	6	1	2																
22 Ti	2	2	6	2	6	2	2																
⋮																							
29 Cu	2	2	6	2	6	10	1																
30 Zn	2	2	6	2	6	10	2																
31 Ga	2	2	6	2	6	10	2	1															
⋮																							
36 Kr	2	2	6	2	6	10	2	6															
37 Rb	2	2	6	2	6	10	2	6			1												
38 Sr	2	2	6	2	6	10	2	6			2												
39 Y	2	2	6	2	6	10	2	6	1		2												
40 Zr	2	2	6	2	6	10	2	6	2		2												
47 Ag	2	2	6	2	6	10	2	6	10		1												
48 Cd	2	2	6	2	6	10	2	6	10		2												
49 In	2	2	6	2	6	10	2	6	10		2	1											
⋮																							
54 X	2	2	6	2	6	10	2	6	10		2	6											
55 Cs	2	2	6	2	6	10	2	6	10		2	6				1							
56 Ba	2	2	6	2	6	10	2	6	10		2	6				2							
57 La	2	2	6	2	6	10	2	6	10		2	6	1			2							
58 Ce	2	2	6	2	6	10	2	6	10	1	2	6	1			2							
59 Pr	2	2	6	2	6	10	2	6	10	2	2	6	1			2							
⋮																							
71 Cp	2	2	6	2	6	10	2	6	10	14	2	6	1			2							
72 Hf	2	2	6	2	6	10	2	6	10	14	2	6	2			2							
⋮																							
79 Au	2	2	6	2	6	10	2	6	10	14	2	6	10			1							
80 Hg	2	2	6	2	6	10	2	6	10	14	2	6	10			2							
81 Tl	2	2	6	2	6	10	2	6	10	14	2	6	10			2	1						
⋮																							
86 Em	2	2	6	2	6	10	2	6	10	14	2	6	10			2	6						
87 —	2	2	6	2	6	10	2	6	10	14	2	6	10			2	6					1	
88 Ra	2	2	6	2	6	10	2	6	10	14	2	6	10			2	6					2	
89 Ac	2	2	6	2	6	10	2	6	10	14	2	6	10			2	6					2	1

atomic numbers, which give the total number of electrons in the neutral atom. The figures in the different columns give number of electrons in orbits corresponding to values of the principal and subordinate quantum numbers standing at the top. A comparison with the periodic table (Table I.) will show that those elements, which in chemical respect are homologous will have the same number of electrons in the electronic groups most loosely bound, containing the so-called valence-electrons. The atoms of elements which in Table I. are enclosed in brackets possess electronic configurations in which a subgroup is being added to a group, whose principal number is less than the group containing the typical valence-electrons. An especially conspicuous example of such a completion of an inner group is offered by the elements forming the family of the rare earths. Here we witness the addition of the fourth subgroup to the 4-quantum group, which begins first in Ce (58) while the addition of the third subgroup was already finished in Ag (47).

Table II. is in general agreement not only with the optical spectral evidence but also with that in the region of X-rays. As mentioned earlier, we see in X-ray spectra a change in the binding of an electron in the interior of the atom. This takes place when, for instance, by the impact of a swiftly moving particle on the atom, an electron is removed from one of the electronic groups, and its place is taken by an electron belonging to a group for which the binding energy is smaller. As an example it may be stated that the strong X-ray whose frequency is approximately represented by formula (4) is emitted when an electron has been removed from the 1-quantum group, and one of the 2_2 electrons performs a transition so as to occupy the empty place. The line represented approximately by formula (5) originates from a transition by which a 3_3 electron takes the place left open upon the removal of a 2_2 electron.

The question how many electrons there are in the various groups and subgroups has been subject to much discussion in the last few years. Table II. is the temporary result of this discussion and seems to give an adequate description of the spectral as well as the chemical evidence. It is clear that a full theoretical treatment of the problem cannot be obtained from considerations based only on the simple picture of central orbits. Such a treatment will essentially involve an examination of those features of the binding of the electrons, which appear in the multiplet structure of spectral lines. Indeed it is very probable that the idea that the electron itself has magnetic properties may give the clue to the interpretation of the empirical rules governing the number of electrons in the group structure of the atom.

In this article we have tried to give an idea of the fertility of the use of mechanical pictures visualising stationary states in the analysis of the properties of the elements. Notwithstanding the valuable suggestiveness of such an hypothesis it must once more be emphasised that such pictures do not allow the atomic phenomena to be interpreted quite rationally within the frame of the postulates of the quantum theory. For the adequate description of the properties of atoms it seems impossible to rely directly upon any concept of classical electrodynamics. On the contrary it appears imperative that any such concept should from the outset be transcribed in accordance with the methods of the new quantum mechanics, alluded to above. (*See* also ATOMIC WEIGHTS; CHEMISTRY; GASES, ELECTRICAL PROPERTIES OF; ISOTOPES; QUANTUM THEORY.)

BIBLIOGRAPHY.—E. N. de C. Andrade, *The Structure of the Atom;* G. Birtwistle, *The Quantum Theory of the Atom;* N. Bohr, *The Theory of Spectra and Atomic Constitution;* I. D. Main Smith, *Chemistry and Atomic Structure;* A. Sommerfield, *Atomic Structure and Spectral Lines.*

(N. Bo.)

PART II

SELECTED CORRESPONDENCE

1920–1924

INTRODUCTION

In the Introduction to Part I, passages from many letters having a bearing on the papers and manuscripts presented in the present volume have been quoted in English. The complete texts of all of these letters are given here in the original language. The few Danish letters are accompanied by an English translation. The letters are grouped according to the correspondents (in alphabetical order), each group being preceded by a list of the letters it contains.

M. BORN

LETTER FROM M. BORN TO NIELS BOHR (4 Mar. 23)

Göttingen, d, 4. März 1923.

Lieber Herr Bohr,

Vor kurzem hat Heisenberg an Pauli geschrieben, dass er mit mir zusammen eine Rechnung über die Terme des angeregten Heliumatoms ausgeführt hat. Da Pauli noch nicht geantwortet hat, so möchte ich Sie selber anfragen, was Sie dazu meinen. Unser Resultat ist ganz negativ: Wenn man mit den Methoden der Störungstheorie *alle* möglichen Bahntypen des angeregten He aufsucht, so finden sich (ausser dem p-Term) keine Energieniveaus, die mit den beobachteten Termen übereinstimmen. Es müssen also selbst hier, wo das eine Elektron weit entfernt ist, Abweichungen von den mechanischen Gesetzen oder den Quantenregeln vorliegen. Ich möchte Sie nun nicht damit quälen, alle Möglichkeiten zu erwägen, wie man dieser Katastrophe entgehen könnte. Sondern ich möchte nur von Ihnen wissen, ob Sie oder Kramers oder Pauli ähnliche Rechnungen gemacht haben, und wenn das nicht der Fall ist, ob Sie es für richtig halten, dass wir das Resultat kurz publizieren. Die Rechnung ist ja eine genaue Fortsetzung von Ueberlegungen, die Sie und Pauli (mit mir zusammen) über das Ortho-He angestellt haben. Es lohnt nicht, sie ausführlich darzustellen; aber vielleicht kann man andern Leuten Mühe sparen, wenn man das Resultat kurz mitteilt. Bei dieser Gelegenheit möchte ich Ihnen auch erzählen, dass ich einen Weg gefunden habe, um die Dipolmomente der Gase HCl, HBr, HJ zu schätzen. Man kann nämlich aus Dipolen ein einfaches reguläres Gitter aufbauen, dessen $\frac{1}{8}$-Zelle ich hier neben

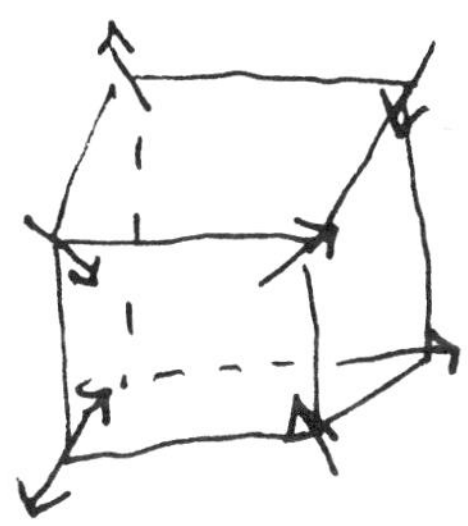

hinzeichne; sie besteht aus 2 Tetraedern, die einen Würfel bilden, und das eine Tetraeder trägt in den Ecken Dipole, die nach dem Centrum zeigen, das andere solche, die vom Centrum wegweisen. Wenn man nun die elektrostatische Kohäsion dieses Gitters berechnet und annimmt, dass die Abstossungsenergie proportional r^{-9} ist (es kommt auf den Exponenten nicht sehr an), so kann man die Sublimationswärme S durch Dipolmoment p und Gitterkonstante δ so ausdrücken, $S = \alpha \cdot p^2/\delta^3$, wo α eine numerische Konstante ist. Berechnet man aus den gemessenen Werten von S die Werte von p, findet man ganz vernünftige

Zahlen, etwa halb so gross, als dem maximalen Moment (Ladungen in den Kernen vereinigt, d. h. fehlende Deformation) entspricht, aber etwas grösser, als nach den Messungen von Frivold* u. Hassel**. Jedenfalls kann man danach sagen, dass ein grosser Teil der Kohäsion dieser Körper auf der Dipolanziehung beruht (vielleicht das Ganze).

Wenn Sie nicht Zeit haben, mir zu schreiben, so beauftragen Sie bitte Pauli. Ich verreise übermorgen und bin bis zum 15. März in Berlin-Grunewald (p.A. Herrn H. Hecht, Königsallee 35), dann in Breslau (Kaiser Wilhelmplatz 13) bis zum 25. März. Mit besten Grüssen, auch von meiner Frau, an Ihre Gattin und die Kollegen

Ihr sehr ergebener
M. Born.

Lieber Bohr!

Ich benutze die Gelegenheit viele herzliche Grüsse an Sie und Ihre Frau beizufügen. Wir haben uns sehr gefreut Familie Nörrlund hier zu sehen. Hoffentlich hindert Sie Ihre Amerikareise nicht uns mit Ihrer Frau hier zu besuchen. Mit Buch Andersen*** bin ich in Briefwechsel. Ich möchte an der Deutung unserer Resultate über die H_2 Dissoziation festhalten und glaube auch zu wissen woher Buch-Andersens negat. Resultat kommt.

Herzlichst Ihr J. Franck.

* Frivold, Olaf Edvin (1885–1944). Norwegian physicist.

** Hassel, Odd, 1897– . Norwegian physical chemist. Became docent at the University of Oslo in 1926, and later became professor there. Received the Nobel Prize in 1969 for his work on the structure and conformations of molecules.

*** Andersen, Emil Buch (1892–1937). Danish chemist and physicist. Became professor of physics at the University of Aarhus in 1933, rektor in 1936.

LETTER FROM NIELS BOHR TO M. BORN (9 Apr. 23)

9. April 23.

Lieber Herr Born,

Ich weiss nicht, wie ich entschuldigen soll, dass ich Ihren freundlichen Brief nicht früher beantwortet habe. Mein Absicht war zu schreiben, sobald ich die Frage mit Kramers und Pauli besprochen hatte. Wegen Arbeit mit unaufschiebbaren Pflichten kann* wir aber in diesen Wochen niemals dazu die Frage gründlich zu diskutieren. Nachdem ich nun eben einen Uebersichtsartikel über die Spektralfragen für das Kayserheft abgeschlossen habe, finde ich mit grössten Bedauern, dass ich Sie so lange ohne Antwort gelassen habe. Ich hoffe nicht, dass Sie sich in Ihren Plänen dadurch haben stören lassen, und kann Ihnen nur empfehlen, Ihre Berechnungen ganz ohne Rücksicht auf uns zu veröffentlichen. Die Frage des Heliumspektrums ist ja eine, die überhaupt noch weit von Ihrer Lösung entfernt ist. Ich habe einige allgemeinen Ansichten in dem genannten Uebersichtsartikel angedeutet und dort auch eine Veröffentlichung von Kramers' und meinen Rechnungen angekündigt, die wir im Laufe der nächsten Woche abzuschicken hoffen, und von der wir Ihnen eine Kopie des Manuskriptes senden werden. Bei der Arbeit mit dem erwähnten Uebersichtsartikel habe ich überhaupt die tiefliegenden Schwierigkeiten betont die wir überall treffen, sobald wir mit Systeme mit mehreren Elektronen zu tun haben. Eine nähere Ausführung der dort gegebenen Andeutungen beabsichtige ich im zweiten Teil meiner Arbeit über die Quantentheorie des Atombaus in der Zs. für Phys. zu geben.

Mit herzlichen Grüssen an Sie selbst und Frau Born und alle gemeinsamen Göttinger Freunde von meiner Frau und

Ihr sehr ergebener

* [Read: kamen.]

LETTER FROM M. BORN TO NIELS BOHR (17 Apr. 23)

Institut für theoretische Physik
der Universität

Göttingen, den 17. April 1923.

Lieber, verehrter Herr Bohr,

Für Ihren liebenswürdigen Brief danke ich Ihnen herzlich. Obwohl Sie schreiben, wir sollten unsere Arbeit über das Helium ohne Rücksicht auf Sie und Ihre Mitarbeiter drucken lassen, so haben Heisenberg und ich doch beschlossen, sie Ihnen erst vorzulegen. Ein Grund dafür ist, dass Pauli mehrmals in Briefen an Heisenberg die Meinung geäussert hat, wir müssten uns verrechnet haben; wir halten das für recht unwahrscheinlich, möchten aber doch gern Paulis Zustimmung gewinnen. Sie werden gewiss so gut sein, ihm das beiliegende Manuskript zu zeigen; er kann ja dann auch für baldige Rücksendung (an mich) sorgen. Vielleicht finden Sie selbst Zeit einen Blick auf die Arbeit zu werfen. Ich möchte auf diesem, Ihrem engsten Arbeitsgebiete nichts publizieren, das von Ihnen nicht gebilligt werden könnte.

Auf Ihre in Aussicht gestellten Publikationen bin ich sehr gespannt. Auch ich denke viel darüber nach, woher es kommt, dass bei mehreren Elektronen die Mechanik zu versagen scheint. Heisenberg schickte mir einen Brief von Pauli, woraus hervorgeht, dass Sie und Pauli die Möglichkeit halber Quantenzahlen ernstlich in Erwägung ziehen. Ich bin sehr neugierig, wie sich das weiter entwickelt. An wirkliche "halbe" Quanten glaube ich nicht, sondern meine, dass diese indirekt irgendwie vorgetäuscht werden; wohl aber teile ich Heisenbergs Ansicht, dass die Zeeman-Terme sich so verhalten, als ob halbe Quanten da wären.

Franck ist in Holland bei Gustav Hertz; wir erwarten ihn Ende der Woche zurück.

Mit herzlichen Grüssen an Pauli und besten Empfehlungen an Ihre Frau, auch von der meinen,

Ihr ergebener
M. Born.

[672]

LETTER FROM NIELS BOHR TO M. BORN (2 May 23)

2. Mai 23.

Lieber Herr Born,

Ich hoffe, dass es Ihnen keine Unbequemlichkeiten veranlasst hat, dass wir die schöne Arbeit von Ihnen und Heisenberg so lange behalten haben. Wie Ihnen Pauli schreibt, hat er und Kramers sich aber eingehend mit der Sache beschäftigt mit dem Resultat, dass wir mit allem einverstanden sind. Auch hoffe ich, dass Sie nichts dagegen haben, dass ich mir erlaubt habe, in einem Bericht über die Probleme der Deutung der Spektren, den ich für das Kayser-Jubiläumsheft verfasst habe, auf Ihre Arbeit hinzuweisen. Das Resultat ist ja sehr wichtig als ein Zeugnis für die Unzulänglichkeit der jetzigen Grundlage der Quantentheorie, soweit es Systeme mit mehreren Elektronen betrifft. Ich sehe jedoch nicht diese Schwierigkeit ganz im selben Licht, wie Sie es andeuten, denn ich glaube, wie schon in meiner letzten Arbeit in der Zeitschrift für Physik angedeutet, dass eine einheitliche Auffassung der Quantentheorie gegeben werden kann, in die sich das Versagen der Mechanik für stationäre Zustände mit mehreren Elektronen sich ungezwungen einfügt. In welcher Weise die Bedingungen für die Festlegung der stationären Zustände dabei zu formulieren sind, ist wohl das Hauptproblem beim jetzigen Stand der Theorie. Ich habe in dem erwähnten Bericht dieses Problem im Zusammenhang mit der Erfahrung besprochen und habe versucht zu zeigen, dass die allgemeinen Schlüsse über die Hauptzüge des Atombaus von diesem Problem in gewisser Hinsicht unabhängig sind, dass aber für solche Probleme wie wir ihnen bei der Komplexstruktur der Serienspektren und deren Zeemaneffekt begegnen, die jetzige Grundlage der Quantentheorie derart versagt, dass es kaum möglich scheint, ohne begriffliche Neuerungen eine rationelle Verwertung der Ganzzahligkeit der Quantenzahlen zu erreichen. Im Gegensatz hierzu scheint schon aus Symmetriegründen eine Normierung mit halbzahligen Quantenzahlen in gewissem Umfang zweckmässig, obwohl ich natürlich ganz Ihre Ansicht teile, dass der Gebrauch halbzahliger Quantenzahlen überhaupt nur ein Ausdruck ist für die Unvollständigkeit der theoretischen Grundlage.

Was die alte Arbeit über das Heliumproblem von Kramers und mir anbelangt, haben wir uns geschämt, dass sie nicht vollständiger durchgeführt wurde, da die Untersuchung auch der nicht komplanaren Bahnen prinzipiell ebensogut nach unserer Methode hätte durchgeführt werden können. Aus Kurzsicht haben wir uns mit der Ansicht begnügt, dass wir es hier wegen der grösseren Komplikation dieser Bahnen nicht mit harmonischen Störungen zu tun haben. Wir sind aber sehr froh, nun aus Ihrer Arbeit gelernt zu haben, dass dies doch der Fall ist, was ja so zwangläufig aus der benutzten schönen Methode folgt. Bei dieser Sachlage denken wir vorläufig kaum daran, unsere Rechnungen zu publizieren.

Mit herzlichen Grüssen,
Ihr sehr ergebener,

D. COSTER

Letter from Niels Bohr to D. Coster (3 July 22)
Letter from D. Coster to Niels Bohr (16 July 22)
Letter from Niels Bohr to D. Coster (5 Aug. 22)
Letter from D. Coster to Niels Bohr (7 Dec. 24)
Letter from Niels Bohr to D. Coster (10 Dec. 24)

LETTER FROM NIELS BOHR TO D. COSTER (3 July 22)

3. Juli 22.

Kære Dr. Coster,

Jeg takker mange Gange for Deres Dissertation, som jeg fandt nu, da jeg kom tilbage fra mit Ophold i Göttingen. Jeg vilde gerne benytte Lejligheden til at sende Dem mine bedste Lykønskninger til Fuldendelsen af Deres betydningsfulde Undersøgelser og de bedste Ønsker for Deres fremtidige Arbejde. Vi glæder os alle her meget til Deres Ophold i København næste Aar, og jeg haaber, at De ogsaa selv skal faa Fornøjelse deraf. Arbejdet her paa Spektrografen skrider frem, og jeg haaber, at det altsammen skal være i Orden, til De kommer. Til Efteraaret kommer Professor Franck her for at holde Foredrag i nogle Uger, og den unge Pauli, som De sikkert har hørt om, kommer her for at arbejde med teoretiske Spørgsmaal et halvt Aar, saa vi faar nok alle en livlig Tid.

Med mange Hilsner til Dem og Deres Hustru og de bedste Ønsker for en god Sommerferie, før De tager fat paa Arbejdet igen,

Deres hengivne,

P.S. Ved min Tilbagekomst fra Göttingen fandt jeg en Note af Rutherford i "Nature" af 17. Juli*, som De sikkert har set, of hvori der omtales nogle Noter af Dauvillier og Urbain angaaende Identificeringen af Grundstoffet med Atomnummer 72 som Celtium. Jeg skulde være overmaade taknemmelig, om De vilde skrive til mig om Deres Syn paa Paalideligheden af Dauvilliers Identifikation af vedkommende Røntgenlinier. Spørgsmaalet er jo af største Interesse, idet, som De ved, Forestillingerne om Atombygningen synes at maatte kræve væsentlig andre kemiske Egenskaber af et Stof med Atomnummer 72 end de, der udvises af de sjældne Jordarter. Tilsyneladende ligger Spørgsmaalet temmelig klart, men man maa dog naturligvis altid være forberedt paa Komplikationer, som i dette Tilfælde maaske kunde hidrøre fra, at vi har at gøre med en samtidig Udvikling af to indre Elektrongrupper.

* [This should be Juni.]

Translation

3 July 1922

Dear Dr. Coster,

Many thanks for your dissertation which I found just now on my return from my stay in Göttingen. I should like to make use of this occasion to send you my warmest congratulations on the completion of your important investigations and my best wishes for your future work. We are all looking forward to your stay in Copenhagen next year, and I hope that you will derive pleasure from it. The work on the spectrograph is progressing, and I hope that everything will be ready when you come. Professor Franck will come in the autumn to lecture for some weeks, and the young Pauli, whom you undoubtedly have heard about, will come here to work on theoretical problems, so we shall probably all have a lively time.

With kind regards to yourself and your wife and best wishes for a good summer vacation, before you resume your work.

Sincerely yours,
N. Bohr.

P.S. On my return from Göttingen, I found a note by Rutherford in "Nature" of July 17* which you probably have seen; it mentions some notes by Dauvillier and Urbain concerning the identification of the element with atomic number 72 as celtium. I should be extremely grateful to you if you would write to me about your opinion of the reliability of Dauvillier's identification of the X-ray lines in question. The problem is of the greatest interest, since, as you know, the conceptions of atomic structure seem to demand essentially different chemical properties of an element with atomic number 72 than those exhibited by the rare earths. The question is apparently rather clear, but one must, of course, always be prepared for complications. They may arise from the circumstance that we have to do with a simultaneous development of two inner electron groups.

* [This should be June 17.]

LETTER FROM D. COSTER TO NIELS BOHR (16 July 22)

Haag, 16 Juli 1922

Lieber Herr Bohr,

Heute traf ich Herrn Siegbahn hier in Haag. Er ist in Paris gewesen, hat mit Dauvillier gesprochen und seine Platten gesehen. Er sagte mir, dass er absolut kein Zutrauen in Dauvilliers Arbeiten bezüglich Element 72 hat. Er könnte die Linien $L\alpha_1$ und $L\beta_2$ von 72 (die einzige Linien welche D. für diesen Element behauptet gefunden zu haben) *gar nicht* sehen. Als er D. hierauf aufmerksam machte (namentlich dass er kein Spur von diesen Linien sah) sagte Herr Dauvillier: Ja, dass ist sehr wohl möglich, dass kommt daher, dass es heute kein klares Wetter ist!! Herr Siegbahn hat mir autorisiert Ihnen als seine Meinung dieses zu schreiben: "Wenn D. behaupten möchte dass in seinem Praeparat etwa 0,01 prozent (*ein hundertel prozent*) vom Elemente 72 anwesend war*, so muss man ihm zugeben, dass die Platten welche er aufnahm nicht dagegen sprechen—aber *ebensowenig dafür*!" Die Platten von D. sind namentlich wie Herr Siegbahn mir sagte im allgemeinen sehr schlecht, die Linien sind sehr breit, die Platten sind sehr überexponiert und die Dispersion ist sehr schlecht. Es ist wohl kaum nötig zu sagen dass ich Herrn Siegbahn immer bewundert habe, für die Weise, auf der er visuell eine Platte schnell und richtig zu interpretieren weiss.

Die unwissenschaftliche Arbeitsmethode von Dauvillier ist mir schon lange ein Aerger gewesen. Ich glaube man muss sehr bedauern dass Rutherford auf D's Anlass diesen Brief in Nature geschrieben hat. Sie können glaube ich Rutherford auch sehr gut Siegbahn's Meinung über die ganze Sache mitteilen, falls Sie dies für gewünscht halten. Ich möchte in diesem Zusammenhang noch besonders betonen, dass, wenn wirklich an der von Dauvillier angegebenen Stelle Linien zu finden wären, dies in sichselber noch absolut kein Beweis für die Anwesenheit von 72 sein sollte. Dafür wäre eine genaue Untersuchung und Diskussion unentbehrlich. (z.B. es muss experimentell festgestellt werden, dass es keine höhere Ordnungen von K-Linien sein können u.s.w.).

Ich will versuchen in einem Nachtrag meiner letzten Arbeit die Unhaltbarkeit von D's Arbeit an dieser Arbeit selber zu demonstrieren. In diesem Augenblick weiss ich aber nicht ob dies möglich sein wird.

Mit freundlichen Grüssen auch für Ihre Frau

Ihr D. Coster.

* Ueber die vermutliche Prozentzahl des Elementes 72 in seinem Praeparate spricht D. sich in seiner Arbeit nicht aus.

LETTER FROM NIELS BOHR TO D. COSTER (5 Aug. 22)

5. august 22.

Kære Coster,

Jeg har lige sendt Deres Afhandling videre til Francis. Jeg synes udmærket om Tilføjelserne, og det gjorde Kramers ogsaa, der lige er kommet hjem fra Finland. Jeg tog Henvisningen i Deres Afhandling til Brevet til "Nature" ud, fordi jeg neppe synes at kunne raade Dem til at sende noget saadant Brev. Jeg var overordentlig glad for Deres venlige og interessante Breve og er ligesom De overbevist om, at man neppe tør tillægge Dauvilliers Resultat nogen større Betydning. Jeg synes imidlertid, at det er meget vanskeligt for Dem direkte at drage hans Resultat offentlig i Tvivl udfra Betragtninger over Dauvilliers andre eksperimentelle Undersøgelser, uden at De selv har været i Stand til at undersøge Dauvilliers Plader eller Urbains Præparat.

Jeg har med samme Post skrevet til Rutherford og rent privat meddelt ham Deres og Siegbahns Tvivl om Paalideligheden af Dauvilliers Resultater. Saa snart Rutherford svarer mig angaaende Sagen, skal jeg skrive til Dem.

Med mange Hilsener til Dem og Deres Hustru fra os alle,
Deres hengivne

Translation

5 August 1922.

Dear Coster,

I have sent your paper on to Francis. I like the additions very much, and so does Kramers who has just returned from Finland. I deleted the reference to the letter to "Nature" from your paper, because I could hardly advise you to send such a letter. I was extremely pleased by your kind and interesting letters, and, like you, I am convinced that one can hardly ascribe much significance to Dauvillier's result. However, I think that it is very difficult for you to directly and publicly express doubt about his result based on the consideration of Dauvillier's other experimental investigations, without having yourself been in a position to examine Dauvillier's plates or Urbain's sample.

By the same mail I have written to Rutherford and quite privately informed him of yours and Siegbahn's doubt about the reliability of Dauvillier's result. As soon as Rutherford replies to me about the matter, I shall write to you.

With kind regards to yourself and your wife.

Sincerely yours,
N. Bohr.

LETTER FROM D. COSTER TO NIELS BOHR (7 Dec. 24)

Haarlem, 7 Dez. 1924

Kære Professor Bohr,

Sie werden es mir hoffentlich verzeihen, dass nur ein einziges dänisches Wort in diesem Brief steht, aber es kostet mir jetzt leider zuviel Mühe mich etwas ausführlicher in dänisch auszudrucken. Ich will dann erst Ihnen und Ihrer lieben Frau ganz herzlich danken für Ihren Glückwunsch mit dem Sie mir sehr viel Freude gemacht haben. Ich glaube es ist eine dankbare Aufgabe zu versuchen von der Physik in Groningen etwas mehr zu machen als es augenblicklich ist. An der anderen Seite bringt es natürlich grosse Schwierigkeiten mit sich, dass das Laboratorium so sehr vernachlässigt worden ist, zumal da jetzt von der holländischen Regierung fast kein Geld für wissenschaftliche Arbeit zu kriegen ist.

Ein Vorteil von Groningen ist es, dass es etwas näher zu Kopenhagen liegt als Haarlem und ich bin Ihnen sehr dankbar für Ihre freundliche Einladung ein einziges Mal nach Kopenhagen zu kommen. Ich bedaure sehr dass ich momentan keine Gelegenheit habe mit Ihnen oder Kramers eine Sache zu besprechen, die mich sehr interessiert, nämlich die Stonersche Arbeit über die Gruppeneinteilung der Elektrone. Ich glaube man soll die Resultate dieser Arbeit für sich betrachten und abstrahieren von der wirklich nicht sehr einwandfreien Motivierung. Kramers schreibt mir, dass Sie abweisend gegenüber diese Arbeit stehen und ich bedaure sehr dass ich Ihre Argumente nicht kenne.

Ich habe mir die Stonerschen Quantenzahlen n, k und j in dieser Weise interpretiert: wenn ein Elektron aus einer kompletten Gruppe weggenommen wird, kann man es formal so deuten: Statt ein Elektron wegzunehmen kann man ein *positives* Elektron so in die Gruppe hineinbringen, dass damit das erste Elektron kompensiert wird. Die n, k und j von Stoner sind die Quantenzahlen dieses virtuellen positiven Elektrons, das dann bei der Aussendung des Röntgenspektrums von der eine Bahn in die andere übergeht. Es scheint mir, dass es wirklich möglich ist eine Elektronengruppe so aufzubauen dass die Stonersche Untergruppen herauskommen.

Experimentell könnte man die Frage vielleicht in folgender Weise entscheiden. Nehmen wir das Röntgenlinienpaar

	L_{III}	$N_V - L_{II}$	N_{IV}
nkj	$2_{2,2}$	$4_{3,3}$ $2_{2,1}$	$4_{3,2}$

Nach Ihrer Auffassung hängt das Auftreten dieser beiden Linien zusammen mit dem Anwesendsein von Elektronen in der 4_3-Gruppe. Stoner behauptet: die erste Linie hängt mit dem Anwesendsein eines $4_{3,3}$ Elektrons, die zweite Linie mit einem $4_{3,2}$ Elektron zusammen. Nun scheint es mir fast eine notwendige Konsequenz zu sein der Stonerschen Auffassung dass es im Element Yttrium *wohl* ein $4_{3,2}$ Elektron *nicht* aber ein $4_{3,3}$ Elektron gäbe. Hier müsste man also nach Stoner erwarten, dass *wohl* die Linie $L_{II}N_{IV}$ *nicht* aber $L_{III}N_V$ auftreten sollte. Es gibt noch einige Stellen im periodischen System wo man ähnliche Umstände hat. Das bis jetzt vorliegende experimentelle Material erlaubt keine sichere Entscheidung. Ich hoffe aber bald zu versuchen hier nähere Auskunft zu bekommen.

Ich hoffe sehr, dass ich Ihnen nicht zu ausführlich über diese Sache geschrieben habe. Natürlich brauchen Sie mir gar nicht zu antworten, ich höre wohl gelegentlich von Kramers, wenn Sie etwas positives in der Sache sehen würden.

Sehr viele herzliche Grüsse für Sie und Ihre liebe Frau

von Ihrem D. Coster.

LETTER FROM NIELS BOHR TO D. COSTER (10 Dec. 24)

10. December 24.

Kære Coster,

Tak for Deres Brev. Selvfølgelig maa De aldrig tænke paa at skrive paa dansk, men det er mig naturligvis en stor Lettelse med Hensyn til vor, som jeg haaber, altid nøje Forbindelse at vide, at De kan forstaa et dansk Brev. Jeg var meget interesseret i, hvad De skrev om Stoners Arbejde; jeg har fra først af forstaaet den formelle Skønhed og Simpelhed i hans Klassifikation af Niveauerne, jeg har blot i Diskussioner her betonet, at det fra kvanteteoretiske Synspunkter ikke kan dreje sig om en endelig Løsning af Spørgsmaalet, idet vi endnu ikke besidder Muligheden for paa rationel Maade at forbinde Niveauernes Klassifikation med en kvanteteoretisk Analyse af Elektronbaner. Ganske uafhængig heraf er det af Dem fremhævede Punkt om Niveauernes første Optræden naturligvis af største Betydning at faa opklaret. Man vil jo deraf faa en ny vigtig Oplysning om Gruppernes Opbygning, selv om man maaske næppe kan benytte den umiddelbart til at afgøre, om en Betragtning som Stoners er at foretrække for en saadan Opfattelse af Niveauernes Komplexitet, der er antydet i vort fælles Arbejde i Zeitschrift für Physik. Overhovedet kan man efter min Mening næppe stærkt nok betone den tvetydige Karakter af enhver Sammenligning mellem Eksperimenter og Teori, saalænge den sidste ikke tillader en rationel Forklaring af de experimentelle Hovedresultater vedrørende Dubletternes Karakter. Jeg forsøger i disse Dage at skrive en lille Note om den almindelige Karakter af de Vanskeligheder, vi for Øjeblikket møder indenfor Atomteorien, og skal, saa snart jeg er færdig, sende Dem et Eksemplar af Manuskriptet.

Med mange venlige Hilsener fra os alle,

Deres hengivne

Translation

10 December 1924.

Dear Coster,

Thank you for your letter. Of course, you must never think of writing in Danish; but for the sake of our, as I hope, always close association, it is naturally a great convenience to me to know that you are able to understand a Danish letter. I was very interested in what you wrote about Stoner's paper; I have from the first understood the formal beauty and simplicity of his classification of the levels; however, in the discussions here I have merely emphasized that, from quantum-theoretical points of view, it cannot mean a final solution of the problem, since we do not yet possess any possibility of connecting the classification of the levels in a rational manner with a quantum-theoretical analysis of electron orbits. Quite apart from this, it is, of course, of the greatest significance to get cleared up the point emphasized by you about the first appearance of the levels. In fact, from that one would get important new information about the building-up of the groups, even if one perhaps could hardly use it to decide directly whether a consideration such as Stoner's is preferable to the view of the complexity of the levels indicated in our joint paper in Zeitschrift für Physik. Above all, in my opinion one can hardly emphasize strongly enough the ambiguous character of any comparison between experiments and theory, as long as the latter does not permit a rational explanation of the main experimental results concerning the character of the doublets. I am attempting at present to write a little note about the general nature of the difficulties we meet at the moment within the atomic theory and, as soon as I have finished, I shall send you a copy of the manuscript.

With kindest regards from all of us,

Sincerely yours,
N. Bohr.

P. EHRENFEST

Letter from P. Ehrenfest to Mrs. Niels Bohr (Nov. 22)

LETTER FROM P. EHRENFEST TO MRS. NIELS BOHR (Nov. 22)

NICHT ANTWORTEN!!

Liebe, Liebe Frau Bohr!

Das war eine grosse, reine Freude als ich erfuhr dass Einstein u. Niels die Nobelpreise für 1921 u. 1922 erhalten haben.

Ich wünsche Euch aus ganzem Herzen Glück und vor allem gute Gesundheit für Euch beide, Eure Kinderln und alle die Euch nahestehen.—Es ist so schön an Euer *helles* Haus zu denken!!

Mit innigen Grüssen
Ihr
P. Ehrenfest.

Mein College u. Freund Prof. Kuenen ist, wie Sie wohl gehört haben werden Ende September plötzlich gestorben.—In Folge dessen musste ich alle Absprachen in Amerika absagen und mit der Amerikareise ist auch der Ausflug zu Euch aufgegeben!–

A. EINSTEIN

Letter from Niels Bohr to A. Einstein (11 Nov. 22)
Letter from A. Einstein to Niels Bohr (11 Jan. 23)

LETTER FROM NIELS BOHR TO A. EINSTEIN (11 Nov. 22)

11. November 22.

Lieber Professor Einstein,

Ich möchte gerne meine herzlichsten Glückwünsche aussprechen zu der Erteilung des Nobelpreises. Die äussere Anerkennung kann ja nichts für Sie bedeuten, aber die zugehörigen Geldmittel dürften vielleicht eine Erleichterung in Ihren Arbeitsverhältnissen bewirken.

Für mich war es die grösste Ehre und Freude, die ich durch äussere Umstände wohl bekommen konnte, dass ich zur selben Zeit wie Sie bei der Preisverteilung in Betracht kommen sollte. Ich weiss, wie wenig ich dies verdient habe, aber ich wollte gern sagen, dass ich es als ein grosses Glück gefühlt habe, dass—ganz unabhängig von Ihrem grossen Einsatz in der menschlichen Gedankenwelt—der von Ihnen gegebene grundlegende Beitrag auf dem spezielleren Gebiet, wo ich arbeite, ebenso wie die Beiträge von Rutherford und Planck auch rein äusserlich anerkannt werden sollte, bevor ich zu einer solchen Ehre in Betracht gezogen wurde.

Mit den herzlichsten Grüssen an Sie und Ihre Frau von meiner Frau und mir.

Ihr ergebener

LETTER FROM A. EINSTEIN TO NIELS BOHR (11 Jan. 23)

Nippon Yusen Kaiska
S.S. "Haruna Maru."

Nahe Singapore 11. Januar 1923.

Lieber oder vielmehr geliebter Bohr!

Ihr herzlicher Brief traf mich kurz vor der Abreise in Japan. Ich kann ohne Übertreibung sagen, dass er mich ebenso freute wie der Nobelpreis. Besonders reizend finde ich Ihre Angst, Sie könnten den Preis vor mir bekommen—das ist ächt bohrisch. Ihre neuen Untersuchungen über das Atom haben mich auf der Reise begleitet und meine Liebe zu Ihrem Geist noch vergrössert. Ich glaube nun endlich den Zusammenhang zwischen Elektrizität und Gravitation begriffen zu haben. Eddington ist der Sache näher gekommen als Weyl.

Die Reise ist herrlich. Von Japan und den Japanern bin ich entzückt und bin sicher, dass Sie es auch wären. Ausserdem ist so eine Seereise eine herrliche Existenz für einen Grübler—wie ein Kloster. Dazu die schmeichelnde Wärme nah am Aequator Warmes Wasser träufelt nachlässig vom Himmel herunter und verbreitet Ruhe und pflanzenhaftes Dämmern—das Briefchen zeugt davon.

Herzliche Grüsse. Auf frohes Wiedersehen spätestens in Stockholm.

Ihr Sie verehrender
A. Einstein.

P. S. EPSTEIN

Letter from P. S. Epstein to Niels Bohr (23 Dec. 22)

LETTER FROM P. S. EPSTEIN TO NIELS BOHR (23 Dec. 22)

CALIFORNIA INSTITUTE OF TECHNOLOGY
PASADENA

December 23rd 22.

Hochgeehrter Herr Professor!

Erlauben Sie mir, Ihnen meine herzlichen Glückwünsche auszusprechen zu der höchsten Auszeichnung, welche ein Physiker erwerben kann. Die engeren Fachgenossen, denen es vergönnt war an der Errichtung Ihres stolzen Gebäudes der Atomistik, wenn auch in noch so bescheidenem Masse, mitzuwirken, haben ja die grandiose Bedeutung Ihrer Arbeiten längst erkannt, aber es ist erfreulich, dass heutzutage noch weitere Kreise sich nicht länger ihrer Beachtung verschliessen.

In meinem Verkehr mit Dr. Nielsen, ebenso wie bei der kurzen Begegnung mit Prof. Hansen hatte ich das Vergnügen einiges über die Arbeiten und Pläne des Institutes für Theoretische Physik zu erfahren. Ich selbst bin zeitweilig durch andere Untersuchungen von der Quantentheorie abgelenkt. Nur wird es Sie, vielleicht interessieren, dass ich die Ionisierungsspannung des symmetrischen Heliummodells vor einiger Zeit berechnet habe. Mein Ergebnis ist leider zu niedrig: 21 Volt. Ich hatte gewisse Zweifel, denn die Berechnung nach meiner Methode gestaltete sich ganz überraschend kurz. Allein vor ein Paar Wochen hatte ich Gelegenheit mit Prof. Brown darüber zu sprechen, und er meinte, dass wenn die Störungen erster und zweiter Ordnung praktisch das gleiche liefern (wie es in meinem Falle ist) die Wahrscheinlichkeit, ein richtiges Resultat zu haben erheblich ist. Später kam die Arbeit von Van Vleck, welcher nach einer ganz anderen Methode das gleiche Resultat findet. Meine Theorie ist, dass die Symmetrie des Atomes eine unvollständige ist. Ich glaube, dass es meine erlauben würde auch den allgemeinen asymmetrischen Fall zu behandeln, weiss aber nicht ob es sich lohnt. Bei der Berechnung des Spectrums hatte ich nämlich keinen Erfolg. Ein Term (etwas wie 29200, ich habe hier kein Material) stimmt ganz gut, während die höheren, von denen man eine bessere Genauigkeit erwarten sollte nicht stimmen. Es wäre mir sehr interessant zu erfahren, ob die Resultate der Herren Kramers und Pauli mit den meinigen übereinstimmen. Von einer Veröffentlichung sehe ich vorläufig ab, da ich erst einige andere Sachen, die aus den Bedürfnissen unseres Instituts erwachsen sind, erledigen möchte.

Ich bitte, den Herren Pauli, Coster und Hansen meine herzlichen Grüsse zu übergeben. Momentan bin ich im fahrenden Eisenbahnzuge (daher meine Handschrift) auf meinem Wege zur Amerikanischen Naturforscherversammlung in

Boston, wo ich viele interessante Bekanntschaften erhoffe. Mit meinem Wirkungskreise in Californien bin ich ausserordentlich zufrieden, besonders weil die Berührung mit bedeutenden Experimentalphysikern sehr anregend ist.

Mit vielen Grüssen
Ihr Sie aufrichtig verehrender verstreuter
Paul S. Epstein.

K. FAJANS

Letter from K. Fajans to Niels Bohr (25 June 21)

LETTER FROM K. FAJANS TO NIELS BOHR (25 June 21)

München, 25. VI. 21.

Sehr geehrter Herr Kollege!

Für die so freundliche Ueberlassung Ihres mir von dem Verlag Vieweg zugesandten Buches danke ich Ihnen verbindlichst. Es freut mich sehr Ihre so bedeutsamen Arbeiten leicht zugänglich zu besitzen, Sie werden noch lange Zeit eine Quelle für Anregungen bilden.

Mit besonders grossen Erwartungen erfüllt uns Chemiker die in Ihrem Nature-Brief erfolgte Ankündigung der Enträtselung des periodischen Systems. Ich glaube, dass wenn wir einmal über die Konfiguration der Elektronen bei einer grösseren Zahl von Elementen orientiert sein werden, viele Fragen der Chemie in einem ganz neuen Licht erscheinen werden. Denn schon die einfache Annahme, dass alle Ionen der Alkalihalogenide dieselbe Struktur haben (bis auf Li^+), die jetzt durch Ihre Angabe der Achterschale bei allen Edelgasen wohl als sichergestellt gelten kann, hat zu vielen überraschenden Aufklärungen geführt. Ich hoffe Ihnen bald eine diesbezügliche Abhandlung übersenden zu können.

Mit den freundlichsten Grüssen
auch für Herrn v. Hevesy
Ihr ergebener
K. Fajans.

J. FRANCK

LETTER FROM NIELS BOHR TO J. FRANCK (16 Sept. 21)

16th of Sept. 21.

Lieber Franck,

Sie wissen nicht, wie froh Sie mich mit Ihrem vorigen freundlichen Brief machten; denn ich war natürlich sehr unruhig, dass die Veränderung der Pläne Ihnen allzuviel Unbequemlichkeit bringen möchte. In der Zwischenzeit habe ich der Ausarbeitung des Berichtes über meine Anschauungen über Atombau meine ganze Zeit gewidmet. Wie ich hervorsehen konnte, bin ich aber noch nicht damit durch, obwohl es mit der Arbeit selbst gut gegangen ist. Es scheint wirklich, dass man eine Fülle von Tatsachen erklären kann. Während der Arbeit hat sich auch meine Anschauung wesentlich geklärt, und manches scheint noch einfacher zu sein, als ich hoffen durfte. Ich habe eben einen Brief an "Nature" geschickt, der in einem wesentlichen Punkt eine Berichtigung und Verbesserung meines vorigen Briefes darstellt. Eine Kopie des Manuskripts hiervon, den Sie vielleicht auch Born zeigen möchten, schliesse ich in diesem Briefe ein.

Ich möchte gern die Gelegenheit benutzen um Ihnen zu fragen, ob in Ihrem Laboratorium Versuche über die Entscheidung der in unseren früheren Briefen diskutierten Frage des Zustandekommens von strahlungslosen Übergängen bei Zusammenstössen zwischen Atomen angestellt werden, oder ob Sie vielleicht eine solche Untersuchung geplant haben. Im hiesigen Laboratorium haben wir nämlich davon gesprochen, dass es wünschenswert wäre, eine systematische Untersuchung über diesen Gegenstand anzustellen; und wenn eine solche Untersuchung in Ihrem Laboratorium nicht in Gang oder beabsichtigt ist, möchte Herr Udden hier gern versuchen, eine solche anzustellen.

Noch eine Sache habe ich auf dem Herzen, über die ich gerne Ihren Rat haben möchte. Es handelt sich um eine Übersetzung meiner Arbeiten in den Schriften der Kopenhagener Akademie. Schon vor Jahren hat Vieweg mir den Vorschlag gemacht, eine solche herauszugeben. Ich habe aber die Sache aufgeschoben, ohne dazu Stellung zu nehmen, bis die mangelnden Teile meiner Arbeit erschienen wären. Nun habe ich aber die Absicht, den Bericht über Atombau an der Stelle des geplanten dritten und vierten Teils treten zu lassen, und ich habe daher Vieweg geschrieben, um zu fragen ob er bei seinem Vorschlag bleiben will, und eine Übersetzung der zwei ersten Teile zusammen mit der neuen Arbeit, die ebenfalls im englischen Sprache in den Schriften der hiesigen Akademie erscheinen wird, herausgeben. Dieselbe konnte da fast gleichzeitig in englisch und deutsch erscheinen. Ich will Ihnen nun gern fragen, ob Sie oder vielleicht Born irgend einen jüngeren Physiker kennen sollten, der die Herstellung einer Übersetzung der betreffenden Arbeiten übernehmen wollte. Es ist natürlich die Absicht ihn ordentlich für seine Mühe zu entschädigen, und Dr. Kramers und Dr. Klein im hiesigen

Laboratorium sowie ich selbst werden das Manuskript des Inhalts wegen durchsehen. Es wäre natürlich schön, wenn man irgend jemand finden könnte, der sich für den Inhalt der Abhandlungen interessieren würde. Schon vor längerer Zeit habe ich einen freundlichen Vorschlag, meine älteren englischen Arbeiten zu übersetzen, von einen jungen Göttinger Physiker bekommen, dessen Brief ich, da ich denselben nicht zur Hand haben, Ihnen in einer der nächsten Tage schicken werde. Ich musste aber damals den Vorschlag ablehnen, da eine solche Übersetzung schon in Gang war.

Mit den besten Grüssen von meiner Frau und mir an Sie und Ihre Familie und alle gemeinsamer Göttinger Freunden.

Ihr

LETTER FROM NIELS BOHR TO J. FRANCK (15 July 22)

15. Juli 22.

Lieber Franck,

Mein ganzer Aufenthalt in Göttingen war für mich ein wunderschönes und lehrreiches Erlebnis, und ich kann nicht sagen, wie glücklich ich mich fühlte für alle die Freundschaft, die mir von allen Seiten erwiesen wurde. Besonders kann ich gar nicht beschreiben, wie eng ich mich an Ihnen geknüpft fühlte, und wie dankbar ich bin für alle Ihre Hilfe und Teilnahme im Kleinen und im Grossen. Auch war es mir eine grosse Freude Ihre Frau und Kinder kennen zu lernen. Meine Frau, der es wie auch dem Kinde sehr gut geht, war traurig, diesmal nicht mit nach Göttingen gehen zu können, sie freut sich aber sehr, im Herbst Ihre Frau kennen zu lernen.

Ich schreibe gleichzeitig an Courant, bitte Sie aber meine Grüsse und Dank an die anderen Kollegen und Freunde auszusprechen. Ich hatte so grosse Freude alle näher kennen zu lernen. An Born werde ich später schreiben, wenn wir etwas mehr über meine Pläne über die Veröffentlichung der Vorlesungen wissen. Ich habe noch nicht näheres darüber gedacht, da ich seit meinem Zurückkunft nach Kopenhagen mit verschiedenen Sachen sehr viel zu tun gehabt habe. Ich gehe jetzt Morgen früh mit meiner Frau auf dem Lande um Ferie zusammen mit den Kindern zu halten und werde alles in Ruhe überlegen.

Mit vielen Grüssen von uns beiden an Sie und Ihre Frau und Kinder und alle gemeinsamen Göttinger Freunde.

Ihr,

P.S. Eben ist die Nummer der Zeitschrift für Physik gekommen mit der schönen Abhandlung von Cario. Beim Durchblicken war es mir aber schwierig zu verstehen, wie er seine quantitative Schlüsse begründet. Ich denke nicht so sehr an die Rechnungsart im Anfang der Arbeit, die kaum ganz streng ist. Diese Rechnung scheint ganz einfach und streng durchgeführt werden zu können, wenn man nur die Wahrscheinlichkeit für spontane Übergänge mit der Wahrscheinlichkeit für Stösse zweiter Art vergleicht. Sei die erste durch $A\mathrm{d}t$ gegeben und die letzte durch $\alpha B\mathrm{d}t$, wo α eine Ausbeutekoeffizient und B die Stosszahl ist, hat man ja einfach, dass das Verhältnis der Zahlen der Atome, die strahlen und die umgekehrte Stösse erleiden, gleich $A/\alpha B = X$ ist, und dass folglich der Bruchteil der angeregten Atome, die zur Strahlung Anlass geben, gleich $X/(1+X) = A/(A+\alpha B)$ ist. Dieser Ausdruck wiedergibt eben die von Stern und Volmer berechnete Abhängigkeit der Resonanzstrahlung vom Druck und stimmt qualitativ sehr gut mit den Experimenten überein. Meine Einwendung bezieht sich aber gar nicht auf die Strenge der Rechnung, sondern was mir schwierig zu verstehen bleibt, ist die Genauigkeit, mit welcher Cario den Wert der Konstante A schätzt. Sie ist in Carios Schreibweise virtuel gleich der Grösse $1/\tau$, und in seiner Abhandlung wird ohne nähere Erläuterung $\tau = 10^{-8}$ gesetzt. (In der Abhandlung steht $\tau = 10^{-3}$ sec; dies ist wohl aber ein Druckfehler). Wie dieses mit der Genauigkeit, welche seine Schlüsse fordern, zu rechtfertigen ist, sehe ich nicht ein. Über die Grösse A wissen wir ja nur sehr wenig, obwohl man vielleicht eine grobe Schätzung durch Anwendung des Korrespondenzprinzips erreichen kann. Im allgemeinen wurde eine solche Anwendung fordern, dass A mit der Frekvenz rasch zunehmen sollte. So möchte man aus Wiens Untersuchungen über die Leuchtdauer der Wasserstofflinien darauf vorbereitet sein, dass der Wert der Grösse A für die betreffende Quecksilberlinie wesentlich grösser als 10^{-8} wäre. Dies würde aber natürlich gar nicht mit dem Experimenten übereinstimmen, und die Sache liegt ja auch so, dass die betreffende Linie auch keine gewöhnliche Serienlinie ist, deren Ursprung in einer der einfachen harmonischen Komponenten einer Zentralbewegung zu suchen ist. Es handelt sich ja um eine Kombination zwischen zwei verschiedenen Termsystemen, die einer wesentlich verschiedenen Bewegungstype entspricht. Aus dem Korrespondenzprinzip ist es daher zu verstehen, dass der betreffende Übergang relativ unwahrscheinlich ist, wie es ja auch aus Fürtbauers Dispersionsuntersuchungen hervorgehen scheint. Wenn ich nicht einen Fehlschluss gemacht habe, dürfte es schwierig sein aus den Woodschen Versuchen nähere quantitative Schlüsse über den Wert der Ausbeutekoeffizient für die umgekehrten Stösse zu ziehen, sondern es scheint, dass man aus diesen Versuchen zunächst nur auf eine obere Grenze für die Wahrscheinlichkeitskoeffizient A schliessen kann. Ich schreibe alles dieses, nur weil die Diskussion für das ganze Untersuchungsgebiet so bedeutungsvoll ist. Ich bin aber ganz darauf vorbereitet, dass ich ein grober

Irrtum vielleicht gemacht habe, und ich sollte sehr dankbar sein, wenn Sie mir mit ein Paar Worte Ihre Ansicht mitteilen wollten.

P.S. 2. Das einzige sicheres was ich noch über meine Vorlesungen in Göttingen weiss, ist, dass mehrere der mitgeteilten Resultate schon falsch sind. Ein erster Punkt ist die Konstitution des Elementes 72, das wie von Urbain und Dauvillier gezeigt im Gegensatz zu der Erwartung sich doch als eine seltene Erde erwiesen hat. Hierauf, wie auch auf die Abhandlung von Götze habe ich in einer Nachschrift eines eben bei Vieweg erscheinenden Heftes hingewiesen. Ein anderer Punkt ist das Quecksilberspektrum und die Spektren von verwandtem Typus, wo die auf den gezeigneten Lichtbildern angegebenen Quantenzahlen nicht alle aufrecht zu erhalten sind. Andererseits glaube ich aber jetzt zu einer ziemlich sicheren Deutung des Spektrums gelangen zu sein.

LETTER FROM J. FRANCK TO NIELS BOHR (29 July 22)

Zweite experimentelle Abteilung
des Physikalischen Instituts
der Universität.

Göttingen, den 29. Juli 1922.
Bunsenstr. 9.

Herrn Professor Dr. Niels Bohr, Kopenhagen,
Blegdansvey 15.

Lieber Bohr!

Vielen Dank für Ihren lieben und herzlichen Brief, über den wir uns geradeso wie über den an Courant geschickten, äusserst gefreut haben. Ich hätte Ihnen gewiss schon geantwortet und versucht auszudrücken, was für uns alle die Zeit, die Sie uns geschenkt haben bedeutet, wenn ich nicht durch vielerlei Umstände gehindert worden wäre. Unter anderem musste ich leider nach Berlin fahren, um mit Pohl zusammen an der Trauerfeier für unseren verstorbenen alten Chef, Geheimrat Rubens in Berlin, teilzunehmen. Es ist traurig, dass er so früh sterben musste. (57 Jahre). Er litt an pernit. Leukämie. Lassen Sie mich jetzt daher sagen, dass wir jede Minute Ihres Aufenthaltes hier als ein grosses Geschenk betrachtet haben und unsere Hoffnung darauf eingestellt haben, dass Sie bald einmal wieder herkommen und dann auch Ihre liebe Frau mitbringen werden. Es freut uns sehr zu hören, dass alles gut gegangen ist, und dass Sie jetzt hoffentlich im Kreise Ihrer lieben Familie einen ruhigen Ferienaufenthalt geniessen. Was die

Reise von meiner Frau und mir nach Kopenhagen betrifft, so freuen wir uns sehr darauf. Ich möchte aber noch einmal betonen, dass wir sie jederzeit gern auf später verschieben, wenn sie mit Ihren Dispositionen irgendwie kollidieren würde.

Lassen Sie mich noch einige Worte über physikalische Fragen anfügen. Sie haben natürlich ganz recht mit Ihrer Kritik des ersten Teiles der Cario'schen Arbeit. Klassisch würde der vierte Teil der dort angegebenen mittleren Lebensdauer, d.h. ca. $2{,}5 \times 10^{-9}$ sec. herauskommen. Das ist uns natürlich vollkommen klar gewesen, leider habe ich es beim Durchsehen der Cario'schen Arbeit übersehen. Dass im übrigen die Rechnung etwas ungeschickt ist, müssen Sie unserer Experimental-Theorie zugute halten. Ihr Hauptargument gegen die Berechtigung der Cario'schen Rechnung, nämlich die Seltenheit des Überganges $1{,}5\ S-2p_2$ haben wir erst durch Sie bei Ihrem hiesigen Aufenthalt kennen gelernt, als die Korrektur der Arbeit schon gelesen war. Ich muss bekennen, dass dieser Gesichtspunkt, der natürlich der absolut richtige ist, uns damals dank mangelnder Einsicht nicht bekannt sein konnte. Hoffentlich findet sich bald Gelegenheit auf Ihren Einwand hinzuweisen und die Sache richtigzustellen. Wir arbeiten nämlich zur Zeit über diese Frage weiter, insbesondere habe ich mir überlegt, dass eine Möglichkeit besteht, experimentell festzustellen, wie beim Übergang von Quantenenergie von einem Atom auf das andere der Überschuss an Energie sich auf die Translationsenergie der beiden zusammenstossenden Atome verteilt. Ich hoffe nämlich, dass sich das zum mindestens qualitativ durch den Doppler-Effekt nachweisen lässt. Besonders interessant dürfte es sein, mit polarisiertem Licht anzuregen und dann zuzusehen, ob auch für die Translationsgeschwindigkeit eine Vorzugsrichtung bestehen bleibt. Würden Sie das für möglich erachten? Ich hatte in den letzten Wochen mich mit einer Idee gequält, den Übergang von quantenhaften Absorptionsprozessen zu Dispersionsprozessen dadurch mir begreiflich zu machen, dass ich annahm, dass die Lebensdauer des Anregungszustandes, wenn ich so sagen darf, von der Mitte der Spektrallinien nach beiden Seiten hin äusserst schnell abnähme, bis man in kleiner Entfernung von der Spektrallinie zur mittleren Lebensdauer Null und damit zum Übergang zur klassischen Behandlungsweise käme. Einstein, dem ich davon erzählte, stimmte dieser Anschauung zu, indem er mir sagte, dass er sie schon seit über einem Jahr verträte. Wahrscheinlich ist die Anschauung, falls Sie sie für richtighalten, Ihnen ebenfalls schon lange geläufig. Bei solchen Ideen liegts immer so: Entweder sie sind falsch, oder Bohr und Einstein kennen sie schon lange. Sympathisch ist mir bei der Anschauung insbesonders, dass man für den Zerfall angeregter Atome auf diese Weise das Kausalitätsprinzip in seine Rechte einsetzt, da jedem angeregten Atom bei seiner Bildung durch die mehr oder minder genaue Definition der Quantenbahn die Zeit seines Zerfalls von vornherein vorgeschrieben ist. Alle Experimente, die ich bisher mir zum Nachweis eines solchen Verhaltens ausgedacht habe, scheitern daran, dass der Dopplereffekt so viel grösser ist, wie die Strahlungsdämpfung. Man muss eben weiter sinnieren.

Nun habe ich Sie aber, wie ich fürchte, zu lange mit allen möglichen physikalischen Fragen belästigt. Ich erwarte keine Antwort darauf, da ich hoffe, Mitte September Sie das Eine oder das Andere mündlich fragen zu können.

Nehmen Sie viele herzliche Grüsse von meiner Frau und mir, auch an Ihre Frau Gemahlin.

Ihr Ihnen herzlich ergebener
J. Franck.

LETTER FROM J. FRANCK TO NIELS BOHR (23 Dec. 22)

Zweite experimentelle Abteilung
des Physikalischen Instituts
der Universität.

Göttingen, den 23. Dezember 1922.
Bunsenstr. 9.

Herrn Prof. Dr. Niels Bohr, Kopenhagen

Lieber Bohr!

Ich hatte lange die Absicht, Ihnen ausführlich zu schreiben um Ihnen, abgesehen von dem kümmerlichen telegraphischen Glückwunsch, noch einmal in einem Briefe zu sagen, wie herzlich wir uns hier gefreut haben, dass Sie endlich Ihren wirklich wohlverdienten Nobelpreis bekommen haben. Es war hier eine allgemein grosse Begeisterung darüber, dass Sie und Einstein ihn nunmehr beide haben. Man sieht, dass es manchmal auch wirklich vernünftige Entscheidungen von Kommissionen gibt. Ich glaube, ich persönlich habe Ihnen überhaupt nicht nach unserer Rückkehr von Dänemark und Schweden geschrieben, sondern das hat nur meine Frau getan. Es war hier aber soviel zu tun, dass ich niemals die Muse fand, Ihnen ordentlich zu schreiben. Nunmehr ergibt sich eine dreifache Gelegenheit. Einmal, dass ich Ihnen und Ihrer Frau ein frohes Weihnachtsfest und gutes neues Jahr wünschen möchte, zweitens, dass Rosseland gerade hier ist, was uns viel Freude macht und uns die Kopenhagener Erinnerungen noch lebendiger macht und drittens, dass endlich die Arbeit von Hund fertig ist, die Sie damals in einem Zustande gesehen haben, den wir lange verlassen haben. Ich möchte mir erlauben, Ihnen diese Arbeit zuzusenden und Sie um Ihre Kritik zu bitten. Wir sind nämlich jetzt sicher geworden, dass die Elektronen im Argon aber auch im Krypton wirklich die Atome ohne Ablenkung durchqueren können und meinen, dass das sogar eine Forderung der Quantentheorie ist, sobald die Elektronen kleine Geschwindigkeiten haben. Ich will nicht in diesem Briefe die Meinung länger begründen, da Sie ja leicht selbst oder durch Kramers die mit gleicher Post gesandte Arbeit von Hund durchsehen oder durchsehen lassen werden. Wesentlich ist nach meiner Meinung, dass dieser Prozess stattfindet, sobald ein Elektron bei seinem Eintritt ins Atom quantenmässig mehr Energie ausstrahlen müsste, als es besitzt. Das Problem ist sicher ganz verwandt mit dem was Kramers beim kontinuierlichen Röntgenspektrum behandelt hat. Ein freies Elektron befindet sich auf einer geradlinigen Quantenbahn; durch einen Zusammenstoss kann ein Übergang in eine andere Quantenbahn angeregt werden. Dabei finden Ablenkungen statt. Für die Elektronen jedoch, die auf ihrer Quantenbahn bleiben, existiert der Zusammenstoss überhaupt nicht. Das Elektron bleibt auf seiner geraden alten Quantenbahn. Es scheint dies im ersten Moment wohl etwas wild, ich glaube aber, dass es sich sinngemäss an Ihre Auffassung anschliesst. Ich hoffe, dass die Rechnungen von Herrn Hund richtig sein werden, für Sie wird es wahrscheinlich nur nötig sein, den zweiten Teil der Arbeit anzuschauen, da im ersten Teil nur gezeigt wird, dass man mit Bahnen, bei denen das Elektron modulo 2π um den Kern läuft, nicht die Versuche wiedergeben kann. Ich freue mich sehr durch Rosseland gehört zu haben, dass die Kramersche Arbeit im Druck ist. Wenn wie ich auch nicht genau weiss, wie Kramers gerechnet hat, so ist doch in dieser Arbeit vermutlich mancherlei Verwandtes vorhanden, und da ich von Kramers und Ihnen von seiner Arbeit hörte, hat mir diese Idee zur Deutung der freien Weglänge, mit der Sie hoffentlich einverstanden sind, doch recht nahe gelegt. Es scheint überhaupt so zu sein, dass jedes Zusammensein mit Ihnen und Ihren Herren bei mir wie die Düngung eines sandigen Ackerlandes wirkt. Zum Experimentieren bin ich leider in dieser Zeit garnicht gekommen. Immerhin können wir über Stösse zweiter Art durch Versuche von Schülern und auch aus den alten Versuchen von Cario jetzt schon aussagen, dass tatsächlich beim Zusammenstoss eines angeregten Atoms mit einem zweiten auf das, das erste seine Quantenenergie überträgt, die zur Quantenanregung überschüssige Energiedifferenz auf Translationsenergie verteilt wird. Nur so lassen sich die Absorptionsverhältnisse der angeregten Absorptionslinien verstehen, aber wir hoffen, das deutlicher beweisen zu können. Ausserdem glauben wir durch kritische Sichtung der Anregungsverhältnisse des H_2 Moleküls und auch durch einige Versuche eines Doktoranden schliessen zu müssen, dass die Dissoziationsarbeit des Wasserstoffmoleküls wohl noch unter 70 Kilokalorien ist anstelle der 84 Kilokalorien, die Langmuir gemessen hat und die wir auch früher aus Elektronenstossmessung (durch die falsche Heliumaichung*) zu bestätigen glaubten. Es ist richtig, dass nunmehr, wo Ihr erstes Wasserstoffmolekülmodell nicht mehr zu Recht besteht, *ein* experimenteller Einwand gegen dasselbe wegfallen würde.

Sonst kann ich von unseren Arbeiten hier nicht sehr viel Neues erzählen. Ich habe mit viel Freude gehört, wie schön es bei Ihnen weitergeht, wenn auch Rosseland keine Einzelheiten erzählt hat. Das Element 72 wird ja ein wunderschöner Erfolg von Koster und Hevesy.

Born hat jetzt ziemlich viel mit Heisenberg gerechnet und sie glauben ein rechtes Stück weitergekommen zu sein. Er wird Ihnen selbst schreiben und ich könnte auch nur sehr schlecht seine Resultate wiedergeben.

Zu Haus geht es jetzt bei uns ganz gut, nur ist leider meine Frau noch nicht völlig wieder von ihrer starken nervösen Abspannung kuriert, aber sie ist doch schon wesentlich frischer geworden.

Ich hoffe, dass Pauli auf der Rückreise hier in Göttingen Station macht und uns noch viel Schönes von Ihnen, Ihrer Familie und Ihrer Tätigkeit erzählt. Ich hoffe, Sie haben auch in Stockholm schöne Tage gehabt, obgleich wir mit Bedauern hörten, dass Ihr Aufenthalt dort durch Schwierigkeiten im Haushalt leider abgekürzt wurde. Haben wir eine Hoffnung, Sie hier im Frühjahr oder im Sommer mit Ihrer Frau begrüssen zu dürfen? Das wäre uns allen eine sehr grosse Freude.

Nun aber Schluss! Hoffentlich sehen Sie diesen langen Brief und die Übersendung der Hund'schen Arbeit nicht als Belästigung an. Ich wünsche Ihnen und Ihrer Frau Gemahlin recht schöne Weihnachtsferien und sende auch viele Grüsse von meiner Frau. Auch bitte ich Sie sehr, die Herren Hevesy, Kramers, Koster und auch die übrigen Herren im Institut herzlichst zu grüssen.

Ihr Ihnen herzlich ergebener

J. Franck.

Courant ist gerade bei mir und lässt herzlich grüssen desgl. Rosseland.

* [Read Heliumeichung.]

H. KONEN

LETTER FROM H. KONEN TO NIELS BOHR (16 Dec. 22)

PHYSIKALISCHES INSTITUT
DER UNIVERSITÄT BONN

Bonn, den 16. Dez. 1922.

Hochverehrter Professor Bohr!

Am 16. März feiert Herr H. Kayser seinen 70. Geburtstag. Aus die diesem Anlass geben Redaction und Verlag der Annalen der Physik ein Gratulationsheft heraus, das wesentlich spectroscopische Beiträge enthält und zu dem die Herren Zeeman, Julius, Siegbahn, Runge, Paschen, Eder u.a. bereits feste Zusagen gemacht haben.

Mir ist die Sammlung und Ordnung der Beiträge übertragen.

So erlaube ich mir die Anfrage, ob Sie bereit sein würden, einen wenn auch ganz kurzen Aufsatz für das Heft zu stiften. Verlag und Redaction gewährleisten den Druck innerhalb 4 Wochen, so dass es früh genug wäre, wenn das Manuscript vor Ende Januar in meinen Händen wäre. Es bedarf nicht erst der Begründung, wie dankbar wir alle Ihnen für einen Beitrag sein würden, und wie sehr er dem Heft zur Zierde gereichen und zur Ehrung für Herrn Kayser beitragen würde.

Ich benutze die Gelegenheit meiner Freude und Genugtuung Ausdruck zu geben, dass der diesjährige Nobelpreis einem Manne zugefallen ist, dessen Verdienste um die Spectroscopie, wenn ich meine eigenen Worte aus 1913 brauchen darf, nur verglichen werden können mit der Leistung Darwins um die Biologie. Wie ich denke, habe ich in dem Vorwort zu meinem kleinen bescheidenen Buche von damals gesagt, ehe Ihre ersten Arbeiten erschienen waren, so dass ich nicht in den Verdacht kommen kann, Ihre ungeheure Leistung erst a posterieri erkannt zu haben. Um so mehr wäre es mir eine persönliche Freude und Genugtuung, in dem geplanten Heft auch einen Beitrag von Ihnen aufweisen zu können.

In dem ich Ihnen im Voraus danke und damit gute Wünsche für die bevorstehende Jahreszeit verbinde bin ich mit den besten Empfehlungen

Ihr hochachtungsvoll ergebener
H. Konen.

LETTER FROM NIELS BOHR TO H. KONEN (13 Mar. 23)

13. März 23.

Lieber Professor Konen,

Ich bedaure überaus, dass ich nicht in der versprochenen Zeit das Manuskript meines Beitrages zum Kayserhefte schicken konnte. Die Ursache der Verspätung liegt nicht nur darin, dass ich mit anderer Arbeit sehr überladen war, sondern in erster Linie darin, dass ich unter der Arbeit mit dem Artikel selbst auf wissenschaftliche Schwierigkeiten stiess, denen ich viel Zeit geben musste, um zu versuchen meine Gedanken zu klären. Ich schliesse die ersten vier Paragraphen des Manuskriptes ein und hoffe Ihnen die letzten zwei, die eben reingeschrieben werden, an einem der erstkommenden Tage zu schicken. Sie werden sehen, dass der Artikel leider viel länger geworden ist, als der ursprüngliche Plan war, und ich fürchte, dass der Umfang allzu gross für das Kayser-Heft ist, wenn dieses nicht schon gar gedruckt ist. Wenn das eine oder das andere der Fall ist, so möchte ich Sie bitten, es für mich womöglichst so zu ordnen, dass mein Artikel im Anfang des nächstfolgenden Heftes der Annalen erscheint, so dass er jedenfalls im fertigen Bande im Anschluss an die anderen Artikeln des Kayserheftes stehen konnte. Eine solche Ordnung wäre deshalb wünschenswert, weil ich mich bestrebt habe, meine Arbeit, was den Stil anbelangt, eben als ein Jubiläumsartikel erscheinen zu lassen. Abgesehen davon dass die Beschäftigung mit den Problemen zu meinem Bedauern den ursprünglichen Plan so sehr gestört hat, hat die Arbeit mir viel Freude gegeben, da sie mir den Anlass gab, manche alte und neue Sachen nochmals zu überlegen und aus einheitlichen Gesichtspunkt zu beleuchten. Besonders gilt dies die letzten zwei Paragraphen, wo die Fragen sich in mancher Hinsicht in Gährung befinden.

Ich schliesse einen Brief an Professor Kayser ein, der einen Glückwunsch enthält, und worin ich die Verspätung meines Artikels bedaure. Ich wollte ihn nicht gern direkt an Kayser schicken, weil ich nicht wusste, ob er schon mit dem Plan des Jubiläumsheftes bekannt ist.

Mit vielen herzlichen Grüssen, und mit der Bitte, mir all die Beschwerden zu entschuldigen, die ich Ihnen besorge,

Ihr sehr ergebener

LETTER FROM NIELS BOHR TO H. KONEN (4 Apr. 23)

4. April 23.

Sehr verehrter Herr Kollege,

Ich schicke Ihnen hiermit den Rest des Manuskriptes meiner Arbeit. Originalzeichnungen zu einigen erläuternden Figuren, woraus in der Arbeit (Fussnote S.27 und später) hingewiesen wird, werden im Laufe der nächsten Tage, sobald sie in reproduzierbaren Stand ausgeführt sind, an Sie abgeschickt.

Ich spreche nochmals mein Bedauern aus, dass die Arbeit mit der Abhandlung so sehr in die Länge gezogen hat, und hoffe dass dadurch die Pläne mit dem Kayserheft nicht allzusehr gestört sind. Wie in meinem vorigen Brief erwähnt, war die Verzögerung durch die Beschäftigung mit der Sache verursacht, die allmählich eine vollständige Aenderung des ursprünglich geplanten kurzen Artikels mit sich brachte. Ich sah nämlich ein, wie schwierig es war, eine befriedigende formale Darstellung der Theorie der Serienspektren zu geben, ohne an der Hand einer Diskussion der tieferen Probleme des Atombaus. So wie die Sache jetzt steht bin ich sehr dankbar für die Anregung zur Abfassung der Arbeit, die Ihre Aufforderung mir gegeben hat, aber ich bin zur selben Zeit sehr beschämt durch das Gefühl, dass die Arbeit, ihren Umfang nach, vielleicht Ihrem Zweck nicht angepasst ist. Ich bitte Sie deshalb mir aufrichtig zu sagen, was Sie am liebsten mit dem Artikel tun möchten. Was mich angeht, ist nichts dagegen, die Arbeit zu teilen und z.B. den schon eingesandten Teil im Kayserheft zu veröffentlichen und den Rest der Arbeit im folgenden Hefte der Annalen zu publizieren. Der Einwand gegen ein solches Verfahren wäre nur der, dass dabei das Jubiläumsheft einen für ihn beabsichtigten Artikel nur teilweise enthalten sollte. Was diesen Punkt angeht, werden Sie natürlich die Sache am besten übersehen; ich möchte nur sagen, dass wenn der Abdruck des ganzen Artikels im Jubiläumsheft zu etwaigen ökonomischen Schwierigkeiten Anlass geben würde, würden wir vollkommen bereit sein vom hiesigen Institut zu den Druckkosten beizutragen.

Ich bekam vor einigen Tagen einen sehr liebenswürdigen und interessanten Brief von Professor Kayser, für den ich Sie bitte ihn herzlich zu danken.

Mit freundlichem Gruss
Ihr ergebener

LETTER FROM NIELS BOHR TO H. KONEN (6 Apr. 23)

6. April 23.

Lieber Professor Konen,

Ich bekam gestern Ihren Brief und die Korrektur des ersten Teils meiner Arbeit. Die Korrektur habe ich eben an die Druckerei zurückgeschickt; leider war ich gezwungen, einige kleinere Änderungen einzuführen, die als Folge der Arbeit mit dem zweiten Teil notwendig wurden. In der Zwischenzeit haben Sie sicherlich den zweiten Teil des Manuskriptes mit den letzten Paragraphen empfangen. Wie Sie auf den mitfolgenden Brief gesehen haben, bedaure ich ausserordentlich alle die Beschwerden, die ich Ihnen verursache, und nach Ihrem letzten freundlichen Brief drückt mich dieses Gefühl noch schwerer. Mein ganzes Benehmen in der Sache wird sicherlich eigentümlich vorkommen, aber ohne zu versuchen mich zu entschuldigen, kann ich Ihnen nur versichern, dass ich erst jetzt beim Abschluss des Artikels darüber klar geworden bin, wie wenig er vielleicht seinem Zwecke angepasst ist. Unter den Eindruck der Rätseln, die immer auftauchen beim Versuche, seine Gedanken über die behandelten Probleme zu klären, hat die Form der Darstellung sich allmählich verändert und verbreitert; und im letzten Monate habe ich alle andere Arbeit und Pflichten aufschieben müssen, um überhaupt den Artikel zu einem Abschluss zu bringen. Ich bitte Sie deshalb ernstlich, meinen Vorschlag die eventuelle Teilung der Arbeit betreffend zu überlegen und den für Sie bequemsten Weg zu wählen. Dass ich auf den an die Druckerei gesandten Korrekturbogen angezeichnet habe, dass nach Verabredung mit Ihnen auch die Fortsetzung in unmittelbarem Anschluss an den Anfang gedruckt werden sollte, war nur um nicht durch etwaige Missverständnisse Ihre eventuelle Pläne zu verhindern.

Mit freundlichem Gruss
Ihr sehr ergebener

W. KOSSEL

LETTER FROM NIELS BOHR TO W. KOSSEL (19 July 21)

19. Juli 21.

Lieber Professor Kossel,

Ich danke Ihnen vielmals für die freundliche Zusendung Ihres schönen Aufsatzes über die Atomfrage, mit dem es mir eine grosse Freude war Bekanntschaft zu machen. Vielleicht haben Sie von Sommerfeld gehört, das ich mich durch Überanstrengung nicht wohl gefühlt habe. Ich habe aber jetzt wieder meine Arbeit aufgenommen und hoffe in den nächsten Wochen einen Bericht über meine Anschauungen über den Atombau, über den ich in meiner Note zu "Nature" kurz berichtet habe, fertig zu stellen. Ich brauche Ihnen nicht zu sagen, wie sehr ich in dieser Arbeit mich auf Ihre Ideen stütze. Wenn ich mich lange zurückhaltig gegenüber Ihrer Valenztheorie verhalten habe, war es, weil ich grosse Schwierigkeiten zu erblicken dachte, um sie in Einzelheiten mit dem Bild des Atombaus zu bringen. Jetzt glaube ich aber einzusehen, wie es möglich ist die Valenztheorie in schönster Harmonie mit den der Anwendung der Quantentheorie zugrundeliegenden Prinzipien hervortreten zu lassen, und ich freue mich sehr Ihnen bald meinen ausführlichen Bericht senden zu können.

Ich denke oft an unseren interessanten Gesprächen bei dem schönen Sommerabend in München in 1914, und wie sehr Sie zu der Klärung der da zur Sprache gekommenen Problemen durch Ihre anregenden Arbeiten beitragen haben. Hiermit komme ich zu der eigentlichen Veranlassung dieses Briefes. Auf dem Umschlag Ihres Büchleins sah ich nämlich, dass Sie zum Professor in Kiel ernannt worden sind, und ich werde gern die Gelegenheit benutzen Sie meine Glückwünsche zu senden, und Sie zu sagen dass ich hoffe, dass die kleine Entfernung zwischen Kiel und Kopenhagen mit sich bringen wird, dass wir bald Gelegenheit haben werden unsere alten Gespräche aufzunehmen.

Mit freundlichem Gruss,
Ihr ergebener

LETTER FROM W. KOSSEL TO NIELS BOHR (15 Aug. 21)

Abtlg. f. Physik
des physikalischen Instituts
der Universität Kiel.

Kiel, den 15. August 1921.

Lieber Herr Professor!

Ich danke Ihnen aufs Beste für Ihren freundlichen Brief, der mir eine grosse Freude war. Auch mir ist jener Münchner Sommerabend in schöner Erinnerung, —ich habe oft daran zurückgedacht. Damals hatte ich gerade Ihre ersten Arbeiten kennen gelernt und begann ihre Tragweite zu verstehen. Auch sah ich, welche Bedeutung sie für verschiedene Fragen haben mussten, die mir am Herzen lagen, und freute mich sehr, mit Ihnen darüber sprechen zu können. Heute sind wir nun Alle aufs Äusserste gespannt auf die erstaunlichen Fortschritte, die Sie in Ihrem Briefe an die "Nature" angekündigt haben. Ich brauche nicht zu sagen, wie besonders wichtig sie mir sein müssen, ich habe Ihren Brief ganz genau studiert und das Referat für die "Physikalischen Berichte" erstattet. Ich freue mich deshalb ganz besonders, dass wir nun bald Ihre vollständige Darstellung darüber erwarten dürfen und es ist mir eine grosse Freude, dass Sie die Schlüsse, die ich aus dem chemischen Verhalten und Anderem zu ziehen suchte, von Ihrem allgemeinen Standpuncte aus brauchbar finden.

Ich hoffe sehr, dass Sie mit Ihrem Gesundheits-Zustand nun wieder ganz zufrieden sind!—Wenn Sie wieder einmal nach Deutschland kommen, wäre es mir natürlich eine grosse Freude, falls Sie den Weg über Kiel nähmen und zu einem Aufenthalt hier Zeit hätten.

Meinen besten Dank möchte ich Ihnen sagen für die deutsche Uebersetzung Ihrer Arbeiten, die mir vom Verlage in Ihrem Auftrage zugesandt wurde. Sie ist mir von grossem Wert, da ich die ersten Ihrer Arbeiten nicht besass. Es ist sehr erfreulich, dass nun jeder Ihre Arbeiten so im Original in die Hand bekommen kann, um alle die daraus fliessenden Anregungen zu geniessen. Hoffentlich sind Sie mit der Uebersetzung zufrieden,—ich glaube sagen zu dürfen, dass sie sehr sorgfältig ausgeführt ist.

Mit freundlichem Grusse
Ihr sehr ergebener
W. Kossel.

R. LADENBURG

Letter from R. Ladenburg to Niels Bohr (18 June 20)
Letter from Niels Bohr to R. Ladenburg (16 July 20)*
Letter from R. Ladenburg to Niels Bohr (27 Aug. 20)
Letter from Niels Bohr to R. Ladenburg (29 Sept. 20)
Letter from R. Ladenburg to Niels Bohr (12 Nov. 22)
Letter from Niels Bohr to R. Ladenburg (23 Dec. 22)

* [Translation of complete letter given in Introduction, pp. [4]–[5].]

LETTER FROM R. LADENBURG TO NIELS BOHR (18 June 20)

Breslau, 18. 6. 1920.
Menzelstr. 87

Sehr geehrter Herr Professor.

Haben Sie vielen Dank für Ihren freundlichen Brief und für die grosse Liebenswürdigkeit mit Ihre wertvollen Arbeiten zu schicken. Sie werden mir stets eine Erinnerung sein an die ausserordentlichen Stunden, die ich in Berlin durch Ihre Anwesenheit erleben durfte. Die entgegenkommende Offenheit, mit der Sie dort Ihre Ansicht zu allen aufgeworfenen Fragen äusserten, ermutigt mich heute wiederum eine Frage an Sie zu richten

Ich hatte in Halle auf dem Bunsenkongress über "Atombau und periodisches System" zu berichten; in der Diskussion kam die Rede auf den Unterschied zwischen Ihrem Versuch (aus dem Jahre 1913) der Elektronenanordnung der Elemente und der von Kossel 1916 vorgeschlagenen Anordnung. Darf ich nun anfragen, was Sie von dem Kosselschen Gedanken halten, dass die Halogene F, Cl, Br, J in ihren einfachen Verbindungen (etwa mit den Alkalien) durch Aufnahme eines Elektrons, die Alkalien ebenso durch Abgabe eine(s) Elektrons die besonders *stabile* Anordnung der Edelgase mit 8 Elektronen in der Aussenschale annehmen, und entsprechend die heteropolaren Verbindungen überhaupt durch Aufnahme bzw. Abgabe von Elektronen die Edelgasanordnung erstreben. Die Bornsche Berechnung der "Elektronenaffinität" der Halogene und des S scheint mir doch eine starke Stütze dieser Vorstellung. Freilich scheint die Grundlage heute noch nicht gesichert, da man die Stabilitätsfrage noch nicht lösen kann.

Oder ziehen Sie Ihre damalige Auffassung der 2 Elektronen in der Aussenschale der Edelgase auch heute noch vor?

Mit besten Grüssen
Ihr sehr ergebener
R. Ladenburg.

LETTER FROM NIELS BOHR TO R. LADENBURG (16 July 20)

UNIVERSITETETS INSTITUT
FOR
TEORETISK FYSIK

BLEGDAMSVEJ 15, KØBENHAVN Ø.
DEN 16. JULI 1920.

Sehr geehrter Professor Ladenburg,

In Antwort auf Ihren freundlichen Brief muss ich damit anfangen zu sagen, dass ich sehr bedaure, dass ich, von den Umständen gezwungen, erst jetzt Gelegenheit habe ihn zu beantworten. Was Ihre Frage über meine Ansichten über die Konstitution der Atome der Elemente anbelangt, muss ich gestehen, dass ich nicht irgend welche Auffassung als schon genügend gesichert ansehe, um einen bestimmten Standpunkt einnehmen zu können. Die Betrachtungen in meinen ersten Abhandlungen sind, was diesen Punkt betrifft, jedenfalls nur als eine versuchsweise angestellte Orientierung anzusehen. Die Schwierigkeiten des Problems liegen besonders in der rationellen Verwertung der verschiedenen vorgeschlagenen Elektronkonfigurationen auf die Erklärung der chemischen Eigenschaften der betreffenden Elemente. Dies hängt ja nämlich nicht allein zusammen mit dem geometrischen Charakter der Konfiguration, sondern in erster Linie mit den Stabilitätsverhältnisse der Konfigurationen. Ausser der Schwierigkeiten, die, wie es von mancher Seite hervorgehoben ist, die Annahme von Elektronringe mit sich bringt für eine Erklärung der Eigenschaften der Krystalle, Bandenspektra, Ionisationspotentiale u.s.w., scheint es auch, dass eine Annahme von Ringe schon wegen der ungenügende Stabilität zu verlassen ist, und dass wir gezwungen sind mit viel verwickelter Bewegungen der Elektronen im Atom zu rechnen. Bevor aber irgend ein bestimmter Fall von einer solchen Bewegung näher untersucht worden ist, besonders was die Stabilitätsverhältnisse anbelangt, scheint es aber schwierig eine endgültige Stellung zu der Verwertung solcher Bewegungen in der Frage nach den chemischen Eigenschaften einzunehmen.

Was Ihre spezielle Frage nach der Konstitution der Atome der Edelgase anbelangt, glaube ich, dass derartige Betrachtungen wie die Kossel'schen vielleicht nicht so schwerwiegende Argumente beibringen, wie man beim ersten Blick geneigt wäre zu glauben, und es scheint mir, dass diese Frage erst nach einer viel genaueren quantitativen Untersuchung, die bis jetzt noch fehlt, erledigt werden kann.

Indem ich Sie um Entschuldigung für die wenig detaillierte Form dieser Bemerkungen bitte, die aber doch, wie ich hoffe, die Tendenz meiner Auffassung hervorbringen, verbleibe ich

mit freundlichen Grüssen
Ihr sehr ergebener

LETTER FROM R. LADENBURG TO NIELS BOHR (27 Aug. 20)

Breslau, 27. 8. 1920.

Sehr geehrter Herr Professor Bohr.

Haben Sie besten Dank für Ihren freundlichen Brief. Ich bin ganz Ihrer Ansicht, dass die Stabilitätsbedingungen das Entscheidende sind. Aber solange man über die Bewegungen der Elektronen im einzelnen nichts Sicheres aussagen kann, meine ich doch, dass man auf Grund des gewaltigen Beobachtungsmaterials über die chemischen und physikalischen Eigenschaften der Elemente an Hand des periodischen Systems schon einige provisorische Schlüsse ziehen und Vermutungen äussern kann, sozusagen Arbeitshypothesen. In diesem Sinne bitte ich die Ueberlegungen aufzufassen, die ich in dem zugleich übersandten Vortrag über Atombau und Period. System zusammengestellt habe. Ich finde vorläufig die chemischen Gründe für die Stabilität der Edelgasanordnung überzeugend. Daneben scheint mir in den grossen Perioden noch eine Zwischenanordnung besonders stabil, die der Elektronenzahl des Ni, Pd, Pt entspricht und für deren spezifische Elektronenzahl sich durch Abzählen die Zahl 10 ergibt. Näheres hierüber enthält der Vortrag.

Nun möchte ich mir noch eine Frage erlauben. Ich habe noch Ende 1913 Versuche gemacht, ob man die elektrische Aufspaltung an den D-linien des Natriumdampfes in Absorption nachweisen könne und gefunden, dass der Effekt jedenfalls kleiner als 0,1 Å pro 46000 Volt/cm ist (vgl. die gleichzeitig übersandte Notiz). Allerdings ist an diesem Serienglied (1,5s–2,p) kein grosser Effekt zu erwarten. Dasselbe gilt von der Hg-linie 2536, an der ja Paschen u Gerlach mit noch viel empfindlicheren Mitteln keinen Effekt fanden. Man müsste höhere Serienglieder untersuchen—oder eine andre Serie. Aber sowohl die D linien wie 2536 werden doch durch Druck sehr leicht verbreitet, was nach Debye und Holtsmark* ebenfalls auf dem elektrischen Effekt beruht. Andrerseits kann man fragen, ob überhaupt ein "inverser Starkeffekt" an Absorptionslinien, speziell an höheren Gliedern der Serie beobachtbar sein kann. Die Ausgangsbahn der Absorption, z.B. 3/2, s, wird wohl durch ein äusseres kaum beeinflussbar sein, wegen zu starker Wirkung des Feldes des eignen Atoms, der übrigen Elektronen etc—und die Endbahn wird vielleicht gar nicht erreicht? Ist das Ihre Meinung, wenn Sie in Ihrer ersten Arbeit über Atombau (S.18) aus Woods Versuchen, der über 50 Glieder der H.S. des Na beobachtete, folgern, dass diese Absorption nicht von einem "vollständigen" Übergang zwischen 2 stationären Zuständen begleitet sei?

* Holtsmark, Johan Peter (1894–   ). Norwegian physicist. Professor of physics at the Technical University of Trondheim 1923–1942, at the University of Oslo 1942–1964.

Dass trotzdem Absorption der betreffenden Frequenz statthat, ist wohl ein ähnliches Problem, wie es in der Quantentheorie öfter vorkommt, dass die Elektronen (beziehsgw. das Atom) sozusagen ein besonderes Voraus-empfinden für Quanten haben.

Oder müsste das äussere elektrische Feld trotzdem die absorbierte Frequenz beeinflussen, auch wenn die beeinflussbare Endbahn nicht erreicht wird? Das möchte ich doch nicht glauben.

Verzeihen Sie, dass ich Sie noch einmal mit einer Frage belästige. Aber dies Problem beschäftigt mich schon lange, auch beabsichtige ich neue Versuche darüber, ich habe aber ausser jener Bemerkung hierüber bei Ihnen nichts weiter gefunden.

Mit freundlichen Grüssen
Ihr sehr ergebener
R. Ladenburg.

LETTER FROM NIELS BOHR TO R. LADENBURG (29 Sept. 20)

Kopenhagen, 29. Sept. 1920.

Lieber Professor Ladenburg,

In Antwort auf Ihren freundlichen Brief möchte ich zuerst bemerken, dass, obgleich ich wie in meinem Briefe des 16/7 1920 gesagt von der Absicht bin, dass wir jetzt noch nicht genügend über die Bewegungs- und Stabilitätsverhältnisse unterrichtet sind, um zu einer Entscheidung in der Frage nach der Elektronenkonfiguration kommen zu können, ich natürlich mit Ihrer Ansicht einverstanden bin, dass es ausserordentlich nützlich ist mit Analogien, die sich aus den chemischen Tatsachen darbieten, zu arbeiten zu versuchen, und dass ich in dieser Hinsicht die Arbeiten von Kossel und anderen sehr schätze; dass ich mich aber sehr zu Vorsichtigkeit gemahnt fühle von der Überzeugung aus, dass letzten Endes die ganze Frage nach der Erklärung der chemischen Stabilität nicht primär als eine geometrische aufgefasst werden darf, sondern vielmehr als ein dynamisches Problem angesehen werden muss.

Was Ihre Frage anbelangt über das Starkeffekt in Absorption, glaube ich erstens, dass man bestimmt darauf halten muss, wie man sich auch in Einzelheiten der Mechanismus der Absorption denken muss, dass das Absorptionsspektrum immer dem Emissionsspektrum entspricht in solcher Weise, dass eine Absorptionslinie durch die Wirkung äusserer Kräfte auf das Atom in genau derselben Weise wie die entsprechende Emissionslinie beeinflusst wird. Dies gilt aber nur für die regulären Wirkungen, die durch ein konstantes äusseres Kraftfeld hervorgebracht werden. Wenn aber die äusseren Störungen von einer irregulär variierenden, stossähnlicher Natur sind, möchte man sich denken, dass die Gesamtwirkung for Emission und Absorption unter Umständen ganz verschieden ausfiele, weil man sich schwer vorstellen kann, dass bei Absorption eine zufällig in dem "Endzustand" "lokal" auftretende Störung (Stoss) die grosse Wirkung auf die Emission der entsprechenden Linien haben würde genau dem Effekt auf die entsprechende Absorptionslinie entsprechen könnte.

(Hochachtungsvoll
N. Bohr).

LETTER FROM R. LADENBURG TO NIELS BOHR (12 Nov. 22)

Breslau, 12 Nov. 22.
Menzelstrasse 87

Sehr verehrter Herr Professor Bohr,

Mit aufrichtigem Dank bestätige ich die so freundliche Übersending Ihrer 3 Aufsätze über Spektren u. Atome, die gesammelt zu besitzen mir äusserst wertvoll ist. Zugleich drängt es mich, Ihnen meine aufrichtigen Glückwünsche zur Verleihung des Nobelpreises zu übersenden, von der ich heute mit ganz besonderer Freude gelesen habe; ich möchte die Gelegenheit zum Anlass nehmen Ihnen zu sagen, welch unvergleichliche Bedeutung Ihrer Untersuchungen über den Atombau für das ganze physikalische Denken und Empfinden unseres wissenschaftlichen Kreises ausüben: seit Jahren vergeht kaum ein Tag, an dem wir Kollegen nicht Fragen diskutieren, die durch Ihre Arbeiten aufgeworfen sind.

Schlieslich möchte ich mir noch erlauben Ihnen eine Frage vorzulegen nicht rein wissenschaftlicher, sondern mehr persönlicher Natur. Nachdem ich schon früher mehrfach über die neuen Vorstellungen vom Atombau und das periodische System Vorträge gehalten habe, musste ich letzthin im Berliner Colloquium auf besonderen Wunsch von Geheimrat Rubens, da er sich selbst nicht mehr kräftig genug fühlte, über Ihre letzte grosse Arbeit in der Zs. f. Physik und über die Aufklärungen referieren, die Sie uns darüber in Göttingen gegeben haben, und zwar in ausführlichster Weise, 4 Stunden lang. Daraufhin wurde ich mehrfach von Fachkollegen angesprochen, etwas hierüber drucken zu lassen, was auch für weiter entfernt Stehende geeignet wäre. Hiergegen habe ich mehrere Bedenken nicht unterdrücken können—ob es nicht verfrüht ist mit diesen Dingen an die breite Öffentlichkeit zu treten, da wir von einer abgeschlossenen Atomtheorie ja noch weit entfernt sind—und vor allem, ob es Ihnen recht sein würde, wenn ich eine solche Arbeit übernähme. Ich scheute mich immer Sie anzufragen, um Sie nicht zu behelligen, und habe nach längerem Zögern schliesslich doch angefangen eine kleine Broschüre über die "Quantentheorie des periodischen Systems" zu schreiben oder wenigstens zu entwerfen. Den Kern würden natürlich die Gedanken bilden, die Sie in den letzten Jahren publiziert haben, und ich bin mir ganz klar, dass mir eine solche Darstellung erst möglich geworden ist, seit ich in Göttingen Ihre Vorträge hören durfte. Ich wurde deshalb in der Einleitung natürlich besonders auf meine Teilnahme an den Göttinger "Bohrtagen" hinweisen. Aber es ist nun mein Wunsch, Sie von diesem Beginnen rechtzeitig zu unterrichten, um Ihnen Gelegenheit zu geben Bedenken, die Sie vielleicht gegen meinen Plan haben, mir freundlichst mitteilen zu wollen. Denn es liegt mir nichts ferner, als gerade in dieser Angelegenheit irgendwelche Wünsche Ihrerseits zu übergehen oder gar Ihnen in irgend einer Weise zuvorzukommen. Ich bitte Sie da wirklich ganz offen zu sprechen. Von Ihren in Göttingen mitgeteilten und anderweitig nicht publizierten Resultaten hatte ich nicht die Absicht Gebrauch zu machen—abgesehen höchstens von der Tafel, die die den verschiedenen Elementen zugehörigen Elektronenverteilungen enthält (wobei ich mir übrigens die Bemerkung erlauben möchte, dass wohl die Anordnung für die Atome von N, O, Fl u. ihre Homologe noch unsicher ist, da ja die Moleküle fast nur zweiatomig vorkommen, u. dass bei den seltenen Erden noch manche Rätsel liegen wie z.B. die in der Reihe schwankende Färbung und der wechselnde Magnetismus zeigen—so ist ja z.B. $CeCl_3$ farblos und paramagnetisch (Atommagnetismus $k \times 10^6 = +2$), dagegen 4 wertige Cerionen farbig; ebenso CeO_2, dessen Atommagnetismus aber nur 0.01 ist, etc.;—dass vielleicht 3 Elektronen in gleiche Bahnen mit $k = 1$ überhaupt nicht vorkommen können, aus denselben korrespondenzmässigen Gründen, die mehr als 4 Elektronen in gleichen Bahnen umzulaufen hindern, deshalb wird auch in Sc (Y, La) das neue Elektron nicht ebenso wie die 2 letzten Elektronen gebunden, und hiermit hängt vielleicht auch die Bildung des 3. L Niveaus, des 3. M Niveaus im Argon, des 3. N Niveaus im Kr etc. zusammen: wenn von 4 2_1 (bzw. 3_2, 4_2) Elektronen entfernt wird, bleibt eine "bekannte u. erprobte" Konfiguration zurück, wenn aber von 4 2_1 Elektronen (bzw. 3_1, 4_1 Elektronen) 1 fortgeht, bleibt eine unbekannte Konfiguration – u. nun sind 2 verschiedene Gruppierungen möglich. Ähnlich könnte man die andern überzähligen Niveaus deuten.)

Ferner würde ich ev. gern ein oder zwei Ihrer Tafeln der Spektralterme reproduzieren sowie schliesslich Ihre Berechnung bzw. Abschätzung der effektiven Quantenzahl des s-Term, aber nur wenn Sie mir hierzu die Erlaubnis geben würden, dass es Sie nicht stört.

Mit der Bitte zu entschuldigen, dass ich Ihre kostbare Zeit schon zu lange in Anspruch genommen habe

bin ich mit freundlichen Grüssen
Ihr stets sehr ergebener
R. Ladenburg

LETTER FROM NIELS BOHR TO R. LADENBURG (23 Dec. 22)

23. Dezember 22.

Sehr verehrter Herr Professor Ladenburg,

Ich danke Ihnen vielmals für Ihre Glückwünsche in Ihrem Briefe des 12. November, und bin sehr dankbar für die freundliche Gesinnung, der Sie darin Ausdruck gaben. Zugleich bitte ich Sie sehr zu entschuldigen, dass ich nicht früher auf Ihre Frage geantwortet habe. Wie Herr Pauli Ihnen sicher schon erzählt hat, ist meine Zeit aber in diesen Monaten so viel in Anspruch genommen gewesen, dass ich gar keine Zeit gehabt habe, die Sache in Ruhe zu überdenken. Ich war sehr interessiert zu hören, dass Sie daran gedacht haben, ein Büchlein über die Theorie des periodischen Systems zu schreiben, und es ist mir natürlich eine grosse Freude, dass Sie so viel Interesse an meinen Untersuchungen genommen haben. Was Ihr Bedenken gegen Ihren Plan betrifft, dass meine bisherigen Veröffentlichungen nur so unvollständig sind verglichen mit was ich in Göttingen vortrug, fühle ich gar keine Schwierigkeiten, und ausserdem wird im Laufe des Frühjahrs eine Reihe von Abhandlungen in der Zs. für Phys. erscheinen, die die verschiedenen Fragen in Einzelheiten behandelt. Die einzelne Schwierigkeit ist ja noch, dass die Arbeit in vielen wichtigen Punkten in Entwickelung begriffen ist und ich nicht immer weiss, ob ich in allen Einzelheiten genau noch dieselben Ansichten habe, wie in meinen formal ganz unvorbereiteten Göttinger Vorlesungen. Zum Beispiel wenn Sie um meine Zustimmung fragen, die damals gegebenen Tabellen zu publizieren, liegt es mir nur am Herzen, dass sie richtig sind. So halte ich z.B. die Quantenzahlen, die ich für die Spektren der Alkalimetalle und der alkalischen Erden gegeben habe, für sicher begründet; solche Spektren wie die der Silber und Zinkgruppe ist aber noch eine Unsicherheit in der Hauptquantenzahl der P-Terme vorhanden, denn die Untersuchung über das Starkeffekt des Quecksilbers, den ich in Göttingen erwähnte, hat gezeigt, dass sie für dieses Spektrum niedriger anzusetzen ist. Ein ähnlicher Punkt betrifft die Übersicht über die Elektronenbahnen im Normalzustand des Atoms, wo ich an einzelnen Punkten mich jetzt mit etwas grösserer Vorsicht ausdrucken mochte; eine Tabelle, die ich als verteidigbar ansehe, ist in einer gemeinsamen Arbeit mit Coster über Röntgenspektren und periodisches System, die in wenigen Wochen in Zs. für Phys. erscheint, eingefügt. Da werden Sie auch eine Stellungsnahme zu der Frage nach der Anzahl der Röntgenniveaus finden, eine Frage, die übrigens wohl noch nicht als geklärt betrachtet werden kann.

Was meine in Göttingen mitgeteilte Abschätzungsweise der Spektralterme betrifft, kann es mir natürlich auch nur eine Freude sein, wenn Sie darüber schreiben. Es ist ein Gesichtspunkt, der wesentlich ausarbeitungsfähig ist, wie ich es in meinen kommenden Arbeiten darzulegen beabsichtige. Natürlich ist die Abschätzung nur eine angenäherte aber keineswegs eine qualitative in dem Sinne wie z.B. Fuess in seiner eben erschienenen Arbeit es aufgefasst hat, denn eine quantitative Berechnung heisst beim jetzigen Standpunkt der Atomphysik wohl eben Spektralterme und andere Eigenschaften aus der Atomnummer und den Quantenzahlen der Elektronenbahnen im Atom ableiten. Im Gegensatz zu dem üblichen Verfahren mit fingierten Kraftfelders u.s.w. beabsichtigen meine Betrachtungen eben einen Schritt in die genannte Richtung zu machen.

Zusammenfassend kann ich Ihnen nur raten, Ihre Darstellung zu publizieren, und ich freue mich mit ihr bekannt zu machen. Wenn Sie es für wünschenswert halten, wird es mir eine Freude sein, die Korrekturbogen durchzublicken.

Mit freundlichen Grüssen,
Ihr sehr ergebener,

A. LANDÉ

Letter from Niels Bohr to A. Landé (24 Feb. 20)
Letter from A. Landé to Niels Bohr (25 July 20)
Letter from Niels Bohr to A. Landé (14 Feb. 21)
Letter from A. Landé to Niels Bohr (21 Feb. 21)
Letter from Kramers (for N. Bohr) to A. Landé (28 Feb. 21)

LETTER FROM NIELS BOHR TO A. LANDÉ (24 Feb. 20)

Gersonsvej 55, Hellerup. 24/2 1920.

Sehr geehrter Dr. Landé.

Ich danke Ihnen für die freundliche Zusendung der Korrektur von dem Auszug Ihrer Habilitationsschrift in der Physikalischen Zeitschrift. Es hat mich natürlich sehr interessiert damit Bekanntschaft zu machen, und ins besondere zu erfahren warum Ihre Resultate, wie Sie schon in Ihrem freundlichen Briefe vom 9′ August 1919 mir, ohne näheres über die Berechnungen zu erzählen, mitgeteilt haben, von den Resultaten der Arbeit von Kramers und mir abweichen. Bei durchlesen Ihrer Arbeit bin ich aber zu der Ansicht gekommen, dass diese Abweichung von der Verschiedenheit in dem Approximationsverfahren herrührt, und dass Ihre Berechnungsweise für den vorliegenen Zweck nicht besonders geeignet ist. Die Bahn des inneren Elektrons ist in erster Approximation in jedem Augenblick eine Elipse, und wie erwähnt in meinen Briefe vom 26′ Juni 1919 haben wir das Problem behandelt in solcher Weise, dass wir die sekulären Störungen, die das äussere Elektron in dieser Bahn hervorruft, berechnet haben mittels eines Verfahrens, das der Berechnung über perturbiert periodischen Systeme in dem 2. Teil meiner Abhandlung in der Kopenhagener Akademie ähnlich ist. In Ihrer Berechnung gehen Sie aber von einer "intermediären" Kreisbahn aus und berechnen direkt die "Störungen" in dieser Bewegung. Nun sind aber die Störungskräfte in Ihrem Falle (in Gegensatz zu unserem Falle) sehr klein in Verhaltnis zu der "Störungen" selbst. So ist die exzentrische Verschiebung der inneren Bahn nach Ihrer Ansicht proportional mit der 2/3 Potens des Verhaltnisses der Bahnradien; die Störungskräfte aber sind proportional mit der 2. Potens dieser Grösse. Um nur die erste Approximation erreichen zu hoffen sind Sie deshalb nach meiner Ansicht gezwungen alle Glieder, die vom 3. Grade in der Verschiebung sind, mitzunehmen; während Sie nur Glieder vom 2. Grade mitgenommen haben. Wenn man die fortgelassenen Glieder mitnimmt, bekommt man aber andere Resultate, und soweit ich sehen kann, gibt die etwas umständliche Rechnungen genau dieselben Resultate, die wir mit unserem Berechnungsverfahren direkt bekommen haben. Wie ich in dem erwähnten Briefe gesagt habe, bekommt man für die Exzentricitet der inneren Bahn eine Grösse, die einfact proportional mit dem Verhältnis der Winkelmomente des innerem und des äusserem Elektrons ist; und dieses gilt ganz allgemein, unabhängig von der Exzentricitet der äusseren Bahn. Unsere Veröffentlichung ist leider länger ausgeblieben, als ich damals dachte, weil Kramers' Erhohlung länger Zeit brauchte, als ich gehofft hatte, und weil ich in der Zwischenzeit mit anderer Arbeit mich zu beschäftigen genötig gewesen bin. Ich arbeite jezt aber wieder mit diesen Sachen, und ich hoffe, dass in kurzer Zeit, sowohl 3. Teil meiner groszer Abhandlung erscheinen wird, als auch die gemeinsame Arbeit von Kramers und mir über das Heliumspektrum, die jetzt so lange ungedrückt geliegen hat. Wegen der möglichst vollständigen Besprechung der Litteratur über die Spektralfragen, die ich in meiner Abhandlung zu geben versuche, werde ich sehr dankbar sein ein Eksemplar von Ihrer Habilitationsschrift zu bekommen, sobald es erschienen ist.

Mit freundlichen Grüssen

Ihr ergebener
N. Bohr.

LETTER FROM A. LANDÉ TO NIELS BOHR (25 July 20)

Institut F. Theoret. Physik
der Universität
Frankfurt a. M.
Robert Mayer Str. 2.

25. VII.20.

Sehr verehrter Herr Professor!

Beiliegend sende ich Ihnen die Ergebnisse einer in einigen Monaten erscheinenden Arbeit, welche meine Untersuchungen über Dynamik der räumlichen Atomstruktur zusammenfasst und eine Reihe von Folgerungen daraus zieht. Durch eine Modifikation meines ursprünglichen Modells haben Madelung und ich ein Bahnensystem herausbekommen, bei dem jedes der acht Elektronen eine volle Kreisbahn (ohne Knicke) beschreibt, und dabei ein besonders stabiles Würfelbahnensystem bildet, woraus sich dann eine Reihe von physikalischen und chemischen Daten allein mit Hilfe von universellen Konstanten. Atomnummern und Quantenzahlen berechnen lässt. Wenn es Sie interessiert, würde ich mich sehr freuen, Ihnen in Kopenhagen darüber persönlich berichten zu können. Sehr gespannt bin ich auf Ihre angekündigten Berechnungen über das Helium, besonders über die Art, wie Sie die Duplizität von He_{II} herausbekommen. Ich glaube, es wird nicht sehr schwer sein, mit Hilfe der Würfelatome auch die Alkalispektren zu berechnen; ich habe darüber schon einige vorläufige Resultate.

Mit den besten Grüssen

Ihr Ihnen sehr ergebener
A. Landé.

LETTER FROM NIELS BOHR TO A. LANDÉ (14 Feb. 21)

14. Februar 21.

Lieber Dr. Landé,

Vielen Dank für Ihren Brief und die Korrektur Ihrer Arbeit über die Kohäsionskraft im Diamanten, die mich natürlich sehr interessiert hat. Zur selben Zeit muss ich Sie um Verzeihung bitten, dass ich nicht früher auf Ihre freundliche Karte geantwortet habe. Auch wir in Kopenhagen waren sehr froh über Ihren Besuch und hoffen, dass es gelingen wird Sie zum Herbst wieder hierher zu bringen. Wie Sie wissen, war ich ja selbst besonders in Ihren Arbeiten interessiert, und in der Zwischenzeit habe ich mich viel mit den Problemen beschäftigt, über die wir gesprochen haben. Was die in Ihre Arbeit gemachten Annahmen anbelangt und Ihren freundlichen Hinweis auf einer Publikation von mir, muss ich jedoch gestehen, dass meine Ansichten über diesen Punkt sich in mehrfacher Weise geändert und, wie ich denke, geklärt haben. Ich denke jetzt, dass man nicht allein, wie ich Ihnen damals mitteilte, eine ungezwungene Erklärung für das Auftreten mehrkvantiger Bahnen im normalen Zustand des Atoms erreichen kann, sondern eine Weiterbildung desselben Argumentes hat mich in ungezwungener Weise zu der Annahme einer ganz bestimmten Anordnung dieser Bahnen im Atome geführt. Die Anordnung zeigt wohl eine ausgesprochene räumliche Symmetrie, gehört aber nicht zu der von Ihnen vorgeschlagenen Bewegungstype, sondern möchte am besten dadurch characterisiert werden, dass jedes Elektron im Atome sich im gewissen Masse unabhängig von den anderen bewegt. Ich habe diesen Gesichtspunkt zu einer Theorie des Atombaus entwickelt, wie ich in einem zur "Nature" eben gesandten Briefe, von dem ich Ihnen eingeschlossen eine Kopie des Manuskriptes schicke, angedeutet habe, und ich beabsichtige die näheren Resultate, die von ziemlich detailliertem Character sind, in einer ausführlichen Publikation baldigst mitzuteilen, die ich Ihnen schicken werde, so bald es fertig ist.

Mit freundlichsten Grüssen von uns allen hier und auch von Professor Franck, der sich hier gelegentlich aufhält um durch seine Ratschläge den Anfang der experimentellen Arbeiten im neuen Institut freundlichst zu unterstützen,

Ihr sehr ergebener,

LETTER FROM A. LANDÉ TO NIELS BOHR (21 Feb. 21)

Frankfurt 21. II. 21.

Sehr verehrter Herr Professor Bohr!

Herzlichen Dank für die Uebersendung Ihres Manuskripts, das ja für die künftige Atomtheorie von der allergrössten Bedeutung sein wird. Ueberhaupt scheint es mir, dass bis zum Erscheinen Ihrer ausführlichen Mitteilung es garkeinen Sinn hat, in der Atomtheorie weiter theoretisch zu arbeiten. Deshalb möchte ich Ihnen den Vorschlag machen, mir Ihren Brief an die "Nature" einige Zeit zu überlassen, damit ich ihn ins Deutsche übersetze und baldmöglichst an die Ztschr. für Physik gelangen lasse, sodass auch die deutschen Physiker möglichst schnell auf Ihre neuen Fortschritte hingewiesen werden. Ich erlaube mir also Ihr Manuskript bis auf weitere Nachricht noch hier zu behalten, und grüsse Sie herzlich.

Ihr Ihnen sehr ergebener
A. Landé.

Nicht mehr Odenwaldschule,
sondern *Frankfurt am Main*
Institut für theor. Physik, Robert Mayerstr. 2.

LETTER FROM KRAMERS (FOR BOHR) TO A. LANDÉ (28 Feb. 21)

28. Februar 21.

Sehr verehrter Dr. Landé,

Professor Bohr hat mich gebeten Ihnen einige Zeilen als Antwort auf Ihre letzten zwei Briefe an ihm zugehen zu lassen. Eben in diesen Tagen ist er ausserordentlich beschäftigt mit der Fertigstellung des neuen Instituts, das am 3. März eingeweiht werden soll und hat deshalb keine Gelegenheit selbst seine Korrespondenz zu besorgen.

Was Ihr freundliches Angebot betrifft den Brief an "Nature" auf deutsch zu übersetzen und zum Beispiel in die Zeitschrift für Physik zu publizieren, glaubt er, dass dieses nicht zweckmässig sein würde. Der Brief darf ja nicht anders aufgefasst werden als ein Einleg in der Diskussion über Atombau, die eben in dieser Zeit in "Nature" geführt wird, und würde, wenn er für sich allein an anderer Stelle publiziert würde, allzu fordrungsvoll erscheinen, zumal weil die Argumente für die in ihm enthaltenen neuen Anschauungen darin so gut wie gar nicht entwickelt werden sind. Ausserdem hofft Professor Bohr innerhalb weniger Wochen eine deutsche ausführliche Artikel über den betreffenden Gegenstand fertiggestellt zu haben, und da wird er Ihnen einen Abdruck des Manuskriptes sofort zugehen lassen.

Was Ihre Bemerkungen über den anomalen Zeemaneffekt betrifft, bittet er Ihnen zu sagen, dass er natürlich das grösste Interesse an diesen Sachen hat, aber dass er in diesen Tagen, wo er so sehr mit andern Dingen beschäftigt ist, ganz verhindert ist sich darin zu vertiefen und besonders zu prüfen, wie Ihre Resultate sich zu der theoretischen Anschauung verhalten. Auch über diese Fragen wird er aber nach der Fertigstellung des Instituts Ihnen eingehend schreiben.

Mit freundlichen Grüssen von Bohr und mirselbst,
Hochachtungsvoll

I. LANGMUIR

Letter from Niels Bohr to I. Langmuir (3 Dec. 20)

LETTER FROM NIELS BOHR TO I. LANGMUIR (3 Dec. 20)

3rd of December 20.

Dr. Irving Langmuir,
Research Laboratory. General Electric Co.
Schenectady. New York.

My dear Dr. Langmuir,

I thank you very much for your kind letter together with your most interesting papers on atomic structure, which I received a few weeks ago. I am naturally exceedingly interested in your work and in the tendencies, which are exhibited therein. In connection with Landé's work I have myself for some months been working on a paper, which I hope will appear in a few weeks, and which may interest you, because it bears especially upon a point, on which you express criticism in your letter. Hearing a lecture of Dr. Landé in Copenhagen in the beginning of October on structure of diamant crystall it struck me that it was possible on the quantum theory by means of a simple consideration to obtain a rational basis for Landé's assumptions, which at the same time offered an immediate explanation of the coupling between the different shells of the electrons in the atoms, which is exhibited so strigtenly by the chemical properties of the elements. In fact, the want of explanation of this coupling, which you mention in your letter, I have always felt to be the weakest point in my old preliminary work. The point of view, from which I then mainly considered the question, was the condition of the possibility of building an atom by successive binding of the single electrons; but due to the want of a sufficiently developed basis for the considerations this proved not always a quite reliable guidance. However I am still considering the mentioned point of view as a fundamental one, and for this reason I must confess doubt in the reality of the interesting models of hydrogen molecules and helium atoms, proposed in your recent papers. In difference to the model of a hydrogen molecule, proposed in my first papers, I cannot imagine any way, in which models of the type, proposed by you, might be formed by a natural process. For the rest I do not believe at all in my own models for the hydrogen molecule and the helium atom, but think that the motion of the electrons in these structures is of a more complex type, which do not at present allow of any simple mechanical treatment. As regards your question of the fixation of the stationary states of possible structures of the type proposed by you, I hold it for legitimate to fix these by application of the condition $\int p\mathrm{d}q = nh$, which at any rate do not seem to involve any additional difficulty as regards the question of the choice of coordinates, because for the type of motion under consideration the mentioned condition will lead to the same result for any choice of generaliced coordinates. The last point is considered in detail in my recent paper on the quantum theory of line spectra, to which you refer in your letter.

I regret very much that you have had difficulties in obtaining this paper, but due to the fact that the different parts of my paper are meant to form a single memoir, they were originally intended for simultaneous appearance in the publications of the Danish Academy, and till now I have only had a few private separate copies of the parts hitherto appeared, which long time ago have been dispersed. Due to the long delay in the publication of the two last parts of the paper the secretary of the society has however a few weeks age decided, that the different parts were to appear as separate editions as of course ought to have been the case from the beginning. Due to this decision I have just in these days received a number of reprints corresponding to the mentioned form of edition, and I take a pleasure with the same post to send you copies of the parts hitherto appeared together with a paper of Dr. Kramers on a connected subject.

In this letter I allow me to enclose a reprint of a lecture, which I gave in Berlin last April. It contains a general discussion of the problem of series spectra, but do only touch very lightly on the problem of the internal structure of the atom, which however will be dealt with in greater details in the new paper, of which I hope to send you a reprint very shortly.

Asking you to excuse the late answer, which is caused by my being away from Copenhagen the last weeks on a short holiday, and sending the kindest regards, I remain

Yours very truly

H. A. LORENTZ

Letter from H. A. Lorentz to Niels Bohr (10 Dec. 22)*

* [Complete letter given in Introduction, p. [29].]

J. C. McLENNAN

Letter from Niels Bohr to J. C. McLennan (14 Mar. 23)

LETTER FROM NIELS BOHR TO J. C. MCLENNAN (14 Mar. 23)

14th of March 23.

My dear McLennan,

I was very interested from your kind letter of March 1rst, which due to absence from Copenhagen I first received a few days ago, to hear that you are in Europe at present. It would be a very great pleasure for us, if you could come to Copenhagen. After June we are staying in a small country cottage, but if you come before that time, my wife and I should be delighted if you would be our guest during your stay.

I have just finished a paper containing a discussion of the spectra of the elements in connection with the ideas of atomic constitution, and I should look forward very much to discuss a number of problems with you as regards the interpretation of the spectroscopical material. Your blue prints containing data for all elements interested me very much. At the present state of the theory I have not ventured to publish definite proposals for the number of electrons in the sub-groups in the normal state of the atoms of all elements, although of course I have given much thought to this problem. With the same post I am sending you some reprints of recent papers from our institute. In the paper on X-rays spectra and periodic system" by Dr. Coster and myself you will find on page 344 a table giving a survey of the electronic configurations as far as I believe that the present state of the theory offers a valid foundation. Already in this table some uncertainties are marked for elements, where the spectroscopic evidence at hand is not sufficient to settle the number of electrons in the groups of highest quantum numbers. In view of the rapid progress in spectroscopy in the last years one may hope, however, in short time not only to settle these points, but also to reach definite conclusions regarding the elements omitted in the table.

As far as our means allow us, we are endeavouring here to contribute to the investigation of spectra. In this respect your advices will of course be of great value to us, if you can manage to visit Copenhagen. I intend to come to the British Association in Liverpool and possibly thereafter in the autumn to visit America, from where I have received invitations to lecture at various places.

With kind regards, and in the hope of seeing you soon here,

Yours sincerely

C. W. OSEEN

Letter from Niels Bohr to C. W. Oseen (30 Aug. 22)

LETTER FROM NIELS BOHR TO C. W. OSEEN (30 Aug. 22)

30. August 22.

Kære Ven,

Jeg vilde gerne takke Dig og Din Hustru for al Eders Venlighed og Gæstfrihed under det interessante og vellykkede Fysikermøde i Upsala. Jeg kan ikke sige, hvor stor en Glæde det er for mig, hver Gang vi træffes at føle hvor nær vi staar hinanden i vort almindelige Syn paa de principielle fysiske Problemer.

Jeg benytter Lejligheden til at sende en lille Regning, som Kramers har udført angaaende det af Wereide behandlede Spørgsmaal. Som vi talte om den sidste Aften, opfylder de af Wereide benyttede Kvantebetingelser ikke de Fordringer til Forbindelse mellem Atomets Periodicitetsegenskaber, som Korrespondensprincippet fordrer. Som det fremgaar af Kramers' Regning faar man da ogsaa noget ændrede Betingelser, naar man fører Bevægelsesligningerne tilbage til deres kanoniske Form, ved hvilken Fremgangsmaade man jo som bekendt sikrer sig, at alle Fordringer for Adiabatprincippets og Korrespondensprincippets Opfyldelse er tilfredsstillede. Jeg sender samtidig Kramers' Regning til Wereide.

Med de venligste Hilsener til Dig og Din Hustru fra os alle her, især fra min Hustru og

Din hengivne

Translation

30 August 1922.

Dear friend:

I should like to thank you and your wife for all your kindness and hospitality during the interesting and successful physics conference in Upsala. I cannot say how great a joy it is for me every time we meet to feel how close we are to each other in our general view of the fundamental problems of physics.

I take this opportunity to send you a little calculation carried out by Kramers on the question treated by Wereide. As we discussed the last evening, the quantum conditions used by Wereide do not satisfy the requirements of a connection with the periodicity properties of the atom demanded by the correspondence principle. In fact, as it follows from Kramers' calculation, one actually gets somewhat different conditions when the equations of motions are reduced to their canonical form, a procedure which, as is well known, insures that all requirements for the fulfilment of the adiabatic principle and the correspondence principle are satisfied. I am at the same time sending Kramers' calculation to Wereide.

With kindest regards to yourself and your wife from all of us here, especially from my wife and myself

Affectionately yours,
Niels Bohr.

A. RUBINOWICZ

LETTER FROM A. RUBINOWICZ TO NIELS BOHR (18 Apr. 21)

Laibach, 18. April 1921.

Hochgeehrter Herr Professor!

Vor allem möchte ich Ihnen für die gütige Uebersendung der Separata herzlichst danken, insbesondere für Ihren Nature-Brief, der mir so ausserordentlich viel Freude bereitet hat. Wenn auch der Nature-Brief uns die näheren Einzelheiten der Wunderwelt, die Sie uns nun neu erschlossen haben, nicht sehen sondern nur ahnen lässt, so viel glaube ich aber aus ihm doch entnehmen zu können, dass Sie uns durch Ihre Arbeit wieder auf eine Höhe führen von der aus wir nun wieder ein weites Feld als eine klare und schöne harmonische Einheit überblicken können. Wenn ich nicht wüsste wieviel Mühe Ihnen, wegen der allzupeinlichen Sorgfalt und Gewissenhaftigkeit, das Abfassen einer Abhandlung verursacht, würde ich so egoistisch sein, Sie zu bitten wenigstens einen etwas ausführlichen Bericht über Ihre neuen Gedanken zu veröffentlichen. So muss ich aber im allgemeinen Interesse hübsch bescheiden sein und die ausführliche Publikation still abwarten. Es wäre schön und in aller unser Interesse zu wünschen, wenn dieser Nature-Brief bei Ihnen wieder der Anfang einer Periode grösserer Publikationsgeschwindigkeit bedeuten würde.

Und nun kommt eine etwas weniger erquickliche Angelegenheit. Ich habe jetzt eine kleine Notiz über die Intensitätsdissymmetrie beim Stark-Effeckt an die Zeitschrift für Physik eingesendet in der ich darauf hinweise, dass (bis auf verschwindende Ausnahmen) im Feinstrukturbild der Balmerserie die eine Hälfte der Feinstrukturkomponenten Anfangsbahnen entspricht, die sich mehr in der Richtung des Feldes, die andere aber Anfangsbahnen die sich mehr in der dazu entgegengesetzten Richtung erstrecken. Die Intensitätsdissymmetrie ist dann so zu deuten, dass die durch die Kathode in das aufspaltende Feld eintretenden Atome zum grösseren Teil Anfangsbahnen haben, die sich vorzugsweise in der zur Flugrichtung der Kanalstrahlen entgegengesetzten Richtung erstrecken. Nachdem ich nun diese Notiz abgeschickt habe sind in mir Zweifel entstanden, ob Sie mich nicht etwa schon in Kopenhagen auf diese Tatsache aufmerksam gemacht haben (da Sie mir damals doch so viel über den Elektronenstoss erzählten) und mir nur die Angelegenheit deswegen nicht im Gedächtnisse haften geblieben ist, weil ich damals über den experimentellen Tatbestand bei der Intensitätsdissymmetrie so gut wie gar keine Vorstellungen hatte. Falls dies der Fall sein sollte, wäre ich Ihnen zum grössten Danke verpflichtet, wenn Sie mir dies telegraphieren wollten, damit ich den Abdruck der Note noch rechtzeitig verhindern kann. Die Korrektur habe ich bisher nicht bekommen, wenn sie kommt werde ich sie eine Zeitlang zurückhalten. Sie können sich wirklich garnicht denken, wie grässlich unangenehm mir der ganze Vorfall ist.

An die gnädige Frau bitte ich Sie meine ergebensten Handküsse ausrichten zu wollen.

Mit der vorzüglichsten Hochachtung
ergebenst
A. Rubinowicz.

LETTER FROM NIELS BOHR TO A. RUBINOWICZ (3 Jan. 22)

3. Januar 22.

Lieber Rubinowicz,

Ich war sehr froh zu hören, dass Sie glauben dass es möglich für Sie sein wird, hier in diesem Frühjahr zu kommen, und es wird eine grosse Hilfe für mich sein, wenn Sie wirklich daran denken können mir mit der Vorbereitung meiner Göttinger Vorträge behilflich zu sein. Ich kann Ihnen für Ihren Aufenthalt hier eine Summe von 1500 Kronen anbieten, über die ich für Hilfe in meiner Arbeit verfüge. Selbstverständlich soll ich Ihnen eine formale Einladung zwecks Ihres Urlaubsersuches in Laibach schicken. Ich möchte aber erst gern eben ein Wort von Ihnen haben, welche Zeit Sie denken kommen zu können. Wie ich Hevesy erzählt hatte, muss ich in März Vorträge in Cambridge halten, die voriges Jahr meiner Krankheit wegen aufgeschoben wurden; ich werde daher nur über April und Mai verfügen können für die Vorbereitung meiner Vorträge, die in Göttingen in Juni gehalten werden sollen. Sobald ich von Ihnen höre, ob dies Ihnen passen kann, werde ich die formale Einladung schicken.

Indem ich mich freue Sie hier wieder zu haben, sende ich die freundlichsten Grüsse für Sie und Ihre Frau von meiner Frau und mir selbst und die besten Wünschen für das neue Jahr.

Ihr sehr ergebener

LETTER FROM A. RUBINOWICZ TO NIELS BOHR (16 Nov. 22)

Wien, 16. November 1922.

Hochgeehrter Herr Professor!

Mit freudiger Genugtuung haben wir vernommen, dass Sie einer der höchsten Ehren, die die Wissenschaft auszuteilen hat, teilhaftig geworden sind. Eine Überraschung war dies für uns allerdings nicht, denn jeder der Ihre Arbeiten auch nur vom Hörensagen kennt, hätte den Nobelpreis Ihnen gerne schon längst verliehen. Wenn es, wie gesagt, eine der höchsten Ehren ist, so ist dies doch nur ein schwacher Ausdruck des grossen Dankes, den Ihnen die wissenschaftliche Welt schuldet, für die Fortschrittswelle, die Sie eingeleitet und in einzigartiger Weise fortgeführt haben. Wie stolz können wir nun sein, die wir den Vorzug haben, Sie persönlich zu kennen!

Damit will ich aber auch unsere allerergebensten und allerherzlichsten Glückwünsche beschliessen, denn ich fürchte, dass Sie mit dem Lesen der vielen Gratulationsbriefe sonst kaum bis zu Ihren nächsten Nobelpreis fertig werden. Also: Vivant sequentes!

Mit den allerergebensten Empfehlungen an Ihre verehrte Frau Gemahlin von meiner Frau und mir.

Ihr in vorzüglichster Hochachtung

ergebener
Rubinowicz

E. SCHRÖDINGER

LETTER FROM E. SCHRÖDINGER TO NIELS BOHR (7 Feb. 21)

Breslau, Tiergartenstrasse 7^{II}

Dear Mr. Bohr!

I have taken allowance to send you and Mr. Kramers each a separate copy of a note, concerning the model, which will perhaps be of some little interest to you, as it seems to me to fit very well with the new and most interesting views, you have expressed in your letter to "Nature" on the necessary interaction between the electrons of the different "shells". Of course, I know very well, that the simple way of calculating this interaction, followed in this my note, is merely a very rough approximation!–

May I add rather an impudent prayer? Your two important communications in the Danish Akademy from 1918 are so very difficult to procure for us in Germany. At Breslau we have but *one* copy, I know of, that is the one of Mr. Ladenburg. Would it be possible to you, to lett me have a copy? Beeing made Prof. Ch. Schäfers successor in the chair of theoretical physics at Breslau, I keep a college on the atom model just now, and so you may imagine, how thankfull I should be, to possess your important papers, especially because Sommerfelds beautifull work (which is of course our standard-work at present in Germany in all questions of the atom) differs, as you well know, in many respects widely from your own, most fascinating mode of viewing the problems.–

With my very best compliments to you and Mr. Kramers I stay

Yours sincerely
E. Schrödinger.

LETTER FROM NIELS BOHR TO E. SCHRÖDINGER (15 June 21)

15. Juni 21.

Sehr verehrter Herr Kollege,

In Antwort auf Ihren freundlichen Brief habe ich Ihnen die erwünschten Abhandlungen geschickt sowie die Dissertation von Kramers. Ihre Abhandlung in der Zeitschrift für Physik hat mich natürlich sehr interessiert. Ich habe übrigens selbst vor längerer Zeit genau dieselbe Überlegung angestellt und entsprechenden Berechnungen durchgeführt; zum Beispiel hatte ich in einem Vortrag in der hiesigen physikalischen Gesellschaft in Dezember voriges Jahr, wo ich meine Resultate hinsichtlich der Erklärung des periodischen Systems, die in der Note zu "Nature" angedeutet sind, vorlag, als Illustration der Wert mitgeteilt, den der Radius einer solchen idealisierten Kugelverteilung der Ladung der inneren Elektronen haben sollte um die Terme der scharfen Nebenserie des Lithiums genau wiedergeben zu können. Wegen Überanstrengung bin ich leider bisher nicht im Stande gewesen eine ausführliche Mitteilung über den Inhalt dieses Vortrages zu veröffentlichen. Ich versuche aber jetzt wieder mit der Arbeit anzufangen und hoffe in den nächsten Wochen eine solche Veröffentlichung fertig zu stellen.

Mit freundlichem Gruss, auch von Dr. Kramers,

Ihr ergebener

A. SOMMERFELD

LETTER FROM A. SOMMERFELD TO NIELS BOHR (7 Mar. 21)

[Postcard]

Lieber Bohr!

Herzlichen Dank für Ihren Brief an die "Nature". Er bedeutet offenbar den grössten Fortschritt im Atombau seit 1913. Ich habe, wie Sie 1916, die Consequenz daraus gezogen, einen Teil eines Ms., das ich an die Zeitschr. f. Phys. geschickt hatte, nachträglich zurückzuhalten. Ich hatte mir dort meine Gedanken gemacht, wie die im Atominnern systematisch anwachsenden Quantensummen an der Atomoberfläche abnehmen mögen. Aber das haben Sie nun fest begründet.

Ich bin in der unangenehmen Lage, schon wieder eine neue Auflage meines Buches schreiben zu müssen. Sie sind wohl damit einverstanden (wenn ich nichts höre nehme ich es an), dass mir, falls Sie etwas über Ihre neuen Resultate an die Ztschr. f. Phys. schicken, die Correktur hiervon zugeschickt wird. Jedenfalls werde ich Sie im Sommer in Göttingen treffen, um mir für mein Buch nähere Auskunft über Ihre letzten Entdeckungen zu holen. Herzl. Glückwunsch dazu!

Ihr

A. Sommerfeld.

LETTER FROM A. SOMMERFELD TO NIELS BOHR (25 Apr. 21)

München 25. IV. 21.

Lieber Bohr!

Vor zehn Jahren schrieb ich an Hilbert, als er sich nach Abschluss der Integralgleichungen gründlich überarbeitet hatte und ein Sanatorium aufsuchen musste: "Das mathematische Königreich, das Sie errichtet haben, ist schon eine Messe wert." Dasselbe möchte ich Ihnen sagen, da ich von Franck höre, dass Sie mit Ihrer Arbeitsfähigkeit streiken. Ihr mathematisch-physikalisches Königreich wird noch längeren Bestand haben und noch mehr Bürger zählen als das Hilbert'sche Integral-Gleichungs-Reich. Glauben Sie nur nicht, dass Ihre jetzige Arbeitspause etwas Ernstliches oder Besonderes ist. Ich finde es nur natürlich, dass Sie für Ihre neuesten Entdeckungen, die gewiss eine ungeheure Gedanken-Concentration erforderten, den Tribut der Menschlichkeit zahlen müssen. Und ich würde ihn, wenn nur mir so tiefe Einsichten beschert wären, *gerne* zahlen. Ich weiss übrigens von Ihnen weiter nichts, als dass Sie nicht nach Göttingen gehen werden und nicht in Brüssel gewesen sind. Vielleicht wäre es richtiger gewesen wenn Sie meine ärztliche Vorschrift befolgt hätten und mit uns zum Skilaufen gekommen wären.

In Göttingen hätte ich Sie besucht und hätte Sie über die Atombauten ausgefragt. So werde ich meine dritte Auflage schreiben müssen, ohne Genaueres über diese entscheidende Wendung der Elektronenbahnen zu wissen. In gewisser Weise wird dadurch das Geschäft vereinfacht und beschleunigt.

Ich bin zwar ganz davon durchdrungen, dass Ihr Weg der richtige ist—wenn Sie, wie es scheint, die Periodenzahlen 2, 8, 18, ... mathematisch nachkonstruiren können, so ist das ja die Erfüllung der kühnsten Hoffnungen der Atomphysik—. Ich bin aber ketzerisch genug zu denken, dass das eines Tages noch auf einem anderen weniger formalen, einheitlicheren Weg möglich sein wird. Allerdings erst dann, wenn man die kontinuirliche Elektrodynamik durch eine nadelförmige ersetzt haben wird.

Doch ich wollte Ihnen nichts Wissenschaftliches sondern nur Persönliches schreiben. Das Wissenschaftliche sollten Sie sich für die nächsten Wochen ganz vom Halse halten. Auch die Institutssorgen! Ich fürchte nur überhaupt, dass Sie sich mit Ihrer Doppelstellung als Institu⟨t⟩sdirektor und Direktor der Atomtheorie zu viel übernommen haben. Wenn ich Ihnen raten darf, so überlassen Sie nur die Institutsdirektion möglichst vollständig Hr. Collegen Hansen.

Baldige Erholung wünscht Ihnen

Ihr

A. Sommerfeld

LETTER FROM NIELS BOHR TO A. SOMMERFELD (16 Sept. 21)

16. September 21.

Lieber Sommerfeld,

Ich schäme mich sehr, dass Sie so lange nichts von mir gehört haben. Ich bin aber sehr dankbar gewesen für Ihre freundlichen Briefen und Teilnahme in meinem Unwohlsein. Ich fühle mich in den letzten Monaten viel besser und bin beschäftigt mit der Ausarbeitung eines ausführlichen Berichts über meine Anschauungen über Atombau. Es ist mit der Arbeit gut gegangen, und es scheint, dass man wirklich eine Menge von Tatsachen verstehen kann, nicht nur was den Serienspektren und den chemischen Eigenschaften anbelangt, im besonderen aber scheint es, dass man eine detaillierten Erklärung aller Einzelheiten der feineren Struktur der Röntgenspektren bekommt, die Ihre allgemeine Theorie derselben in schönster Weise bekräftigt. Während der Arbeit haben sich meine Anschauungen wesentlich geklärt, und manches erscheint noch einfacher, als ich hoffen dürfte. In diesem

Brief schliesse ich eine Kopie eines Briefes ein, den ich eben an "Nature" geschickt haben, und der, wie Sie sehen werden, in einem wesentlichen Punkt eine Berichtigung und Verbesserung meines vorigen Briefes darstellt. Ich hoffe bald mit dem Bericht fertig zu sein und werde Ihnen dieselbe natürlich dann so schnell wie möglich zukommen lassen.

Mit den besten Grüssen von meiner Frau und mir an Sie und alle gemeinsamen Freunde in München.

Ihr sehr ergebener,

INVENTORY OF MANUSCRIPTS IN THE NIELS BOHR ARCHIVE

DOCUMENTS RELATED TO THE APPLICATION OF THE CORRESPONDENCE PRINCIPLE AND TO THE INTERPRETATION OF THE PERIODIC SYSTEM

The manuscripts listed here were catalogued by Erik Rüdinger, with the assistance of H. K. E. Richter. This invèntory is excerpted from his cards: the complete catalogue forms part of the microfilms of the Bohr manuscripts and is on deposit in the Archives for the History of Quantum Physics in Berkeley and Philadelphia, as well as in Copenhagen. On the second line of each entry is given the number of the microfilm on which the manuscript is reproduced.

The short manuscript titles have been assigned by the cataloguers, as have the dates included in square brackets. These dates are tentative. In a couple of places they have been revised. Unbracketed dates are taken from the manuscripts. The following abbreviations are used: Da for Danish, En for English, Fr for French, Ge for German, and Mf for microfilm. The designation "Bohr MSS", which identifies this group of microfilms in the Archives for the History of Quantum Physics, has been omitted for brevity.

Numbers in the margin facing an item indicate the volume and pages on which this item is reproduced. Volumes 3 and 4 are indicated by a 3 or a 4, respectively. The numbers are followed by the letter E if only excerpts are given. Numbers referring to English translations are followed by the letter T.

1 *Atomteori* 18 Dec. 1917
Sheets, handwritten, 2 pp., Da, Mf 7

Apparently part of outline of a lecture.

2 *Principles of the quantum theory* [1917–1918]
Drafts, typewritten, handwritten, 50 pp., En, Mf 7

Probably various drafts to "On the Quantum Theory of Line Spectra". Contains important remarks concerning abandonment of conservation for individual quantum processes, and on the picturing of motions in different stationary states as points in Riemannian sheets.

3 *Quantum theory of line spectra. Part III* [1917–1918]
Manuscripts, handwritten and typewritten, 228 pp., En, Mf 7

Various drafts and manuscripts, one of which corresponds rather closely to the published version (Dan. Vid. Selsk. Skr. nat. mat. Afd. (8) IV, 1, 3 (1922)).

3 [185–200] 4 *Quantum theory of line spectra. Part IV* [1917–1918]
Sheets, handwritten, 36 pp., En, Mf 7

Draft of Part IV of "On the Quantum Theory of Line Spectra" (unpublished).

5 *Om grundstoffernes liniespektre* 14 Feb. 1918
Sheets, handwritten, 3 pp., Da, Mf 7

Outline of a lecture before the Physical Society of Copenhagen.

3 [54–64] 6 *Quantum theory of line spectra* Feb. 20, Apr. 30, 1918
Proof sheets with handwritten notes, 11 pp., En, Mf 7

Proof pages to Parts I and II of "On the Quantum Theory of Line Spectra", containing remarks later omitted from the published version.

7 *Grundlaget for kvanteteorien* 22 March 1918
Sheet, handwritten, 1 p., Da, Mf 7

Outline of lecture before the Royal Danish Academy of Sciences and Letters.

2 8 *Drawings to triatomic hydrogen molecule* [1918–1919]
Drawings, 6 pp., Mf 7

Drawings of four figures for the paper "On the Model of a Triatomic Hydrogen Molecule" (Medd. Kgl. Vetensk.-akad. Nobelinstitut **5**, no. 28 (1919)).

9 *Virkningen af sammenstød mellem atomsystemer og fri elektriske partikler* [about 1919]
Manuscript, calculations, 21 pp., Da, Mf 7

Outline and manuscript for a lecture.

10 *Mekaniske og termodynamiske grundlag for kvanteteorien* 13 and 27 Feb. 1919
Sheets, handwritten, 18 pp., Da, Mf 7

Outline of two lectures "On the mechanical and thermodynamical basis of the quantum theory" before the Physical Society of Copenhagen.

11 *Problems of the atom and the molecule I* 1919
Sheets, handwritten, 201 pp., En, Mf 8

Various drafts for the lecture in Leiden on 25 April 1919.

3 [202–216] 12 *Problems of the atom and the molecule II* 1919
Drafts, manuscripts, translations, handwritten and typewritten, 285 pp., En, Ge, Da, Mf 8

Drafts and manuscript for the lecture "On problems of the atom and the molecule" delivered in Leiden on 25 April 1919.

13 *Problems of the atom and the molecule III* 1919
Sheets, handwritten, 38 pp., En, Mf 8

Drafts of parts of the lecture delivered in Leiden on 25 April 1919.

14 *Trykkets indflydelse paa brintlinierne* 1919
Calculations, notes, handwritten, 15 pp., Da, Mf 8

Rough draft of treatment of the influence of pressure upon the appearance and sharpness of the hydrogen lines.

15 *Atomic physics and the problem of radiation* 10 Sept. 1919
Sheets, handwritten, 19 pp., En, Da, Mf 8

Outlines and drafts of a lecture given at the conference on atomic physics in Lund, Sweden, in September 1919.

3 [218–219] 16 *Rede an Sommerfeld* 22 Sept. 1919
Manuscript, handwritten, 2 pp., Ge, Mf 9

Speech of appreciation to Sommerfeld after his lecture before the Physical Society of Copenhagen.

3 [222–226] E, T 17 *Den nyere atomfysiks program* 2 Dec. 1919
Part of manuscript, typewritten, 13 pp., Da, Mf 8

Part of a lecture on the program of the contemporary atomic physics given before the Chemical Society of Copenhagen.

18 *Disentanglement of series spectra* [about 1920]
Notes and calculations, handwritten, 48 pp., Da, Ge, Mf 9

Notes concerning the disentanglement and interpretation of series spectra.

19 *Indflydelse af et elektrisk felt paa finstrukturen* [about 1920]
Tables, typewritten, 16 pp., Da, Mf 9

Calculations of the influence of an electric field on the fine structure of a hydrogen line.

3 [228–240] T 20 *Vekselvirkningen mellem lys og stof* 13 Feb. 1920
Manuscripts, typewritten, 27 pp., Da, Mf 9

Manuscript of a lecture on the interaction between light and matter delivered before the Royal Danish Academy of Sciences and Letters. Apparently intended for publication.

21 *Om grundstoffernes seriespektre* 1920
Draft, typewritten, handwritten, 37 pp., Da, Mf 9

Apparently an early version of the lecture delivered in Berlin on 27 April 1920. The corrections introduced in this draft are taken into account in the extended German version (Z. Phys. **2** (1920) 423).

3 [242–282] T 22 *Serienspektra der Elemente* 27 April 1920
Manuscripts and drafts, typewritten, handwritten, 190 pp., Ge, Da, Mf 9

Lecture delivered on 27 April 1920 before the German Physical Society in Berlin. The typed manuscript is practically identical with the version printed in Z. Phys. **2** (1920) 423. The larger part of the drafts deals with a more detailed discussion of the helium spectrum omitted from the actual lecture.

23 *Virkning af elektriske og magnetiske felter* 11 May 1920
Sheets, handwritten, 2 pp., Da, Mf 9

Outline of a lecture on the effect of electric and magnetic fields on spectral lines delivered in Uppsala, Sweden.

3 [48–52]E 24 *Generalization of correspondence principle* [1917?]
Sheets, handwritten, 13 pp., En, Da, Mf 9

In envelope labeled by Bohr: Første Antydning af Generalisation af Korrespondens-princippet (First indication of generalization of correspondence principle).

3 [325–337] T 25 *Fortale til tysk oversættelse* 1920
Manuscript, typewritten, handwritten, 45 pp., Da, Ge, Mf 9

Preface to the German translation of papers on atomic structure from 1913 to 1916 (Abhandlungen über Atombau aus den Jahren 1913–1916, Vieweg 1921). The "Geleitwort" follows the manuscript although several changes were introduced.

4 [44–69] T 26 *Atomernes bygning* 15 Dec. 1920
Draft, parts of manuscript, handwritten, typewritten, 73 pp., Da, Mf 9

Typed manuscripts with headings "Nogle Betragtninger over Atomernes Bygning" (Some considerations on the structure of atoms) and "Om vort Kendskab til Atomerne" (On our knowledge of the atoms). Handwritten draft with the former title to a lecture before the Physical Society of Copenhagen on 15 Dec. 1920.

27 *Forelæsning i Fysisk Forening* 15 Dec. 1920
Part of manuscript, handwritten 10 pp., Da, Mf 9

Apparently part of an elaborated version of the lecture on the interpretation of the periodic system given before the Physical Society of Copenhagen on 15 December 1920.

3 [293–301] T 28 *Indvielsen af Institutet* 3 March 1921
Manuscript, typewritten, 28 pp., Da, Mf 9

Speech on the occasion of the dedication of the Institute for Theoretical Physics of the University of Copenhagen.

3 [364–380] 29 *3rd Solvay Congress* April 1921
Manuscript, typewritten, 33 pp., En, Mf 9

On the application of the quantum theory to atomic problems. Presented at the Solvay Congress by P. Ehrenfest. Published in French translation in *Rapport et discussions du conseil de physique*, Paris 1923, p. 228. Also "Brief extract of contents of second half of Professor Bohr's report".

30 *Normaltilstanden af grundstoffernes atomer* Spring 1921
Sheets and part of manuscript, typewritten, handwritten, 29 pp., Da, Mf 9

"On the normal state of the atoms of the elements". "Note concerning German version of work on the atoms". Evidently planned for a German version of the lecture of 15 December 1920 before the Physical Society of Copenhagen.

31 *Perturbation theory* Spring 1921
Part of manuscript, typewritten, 13 pp., En, Mf. 9

Apparently intended as part of a paper by Bohr and Kramers.

3 [350–356] T 32 *Zur Frage der Polarisation* June 1921
Part of manuscript, typewritten, 4 pp., Da, Mf 9

Notes not used for the paper "On the question of the polarization of radiation" (Zur Frage der Polarisation der Strahlung in der Quantentheorie, Z. Phys. **6** (1921) 1).

33 *Elements within the first period* 1921
Part of manuscript, handwritten, 10 pp., En, Da, Mf 9

Notes concerning series spectra. Probably intended as part of Part B of Bohr's Solvay Report which was never submitted.

4 [100–121] 34 *Constitution of atoms* 1921
Part of manuscript, typewritten, handwritten, 27 pp., En, Mf 10

§1. Interpretation of the series spectra of the elements. General considerations of the process of binding of electrons by atoms. Intended originally as part of the second part of the Solvay Report. Later intended as paper to be published by the Royal Danish Academy.

4 [100–174] (T) 35 *Atomic theory and periodic system* 1921

Parts of manuscript, typewritten, handwritten, 155 pp., En, Da, Ge, Mf 10

Four envelopes labelled (1) § 1. Interpretation of the series spectra of the elements (2) § 2. Helium, (3) On the elements within the first period of the periodic system, and (4) Chapter V. After abandoning his plan to submit a second part of the Solvay report this material was intended for a paper to be published by the Royal Danish Academy.

3 [398–414] 36 *Application of the quantum theory to atomic problems in general* 1921

Manuscript, typewritten, handwritten, 57 pp., En, Mf 10

This manuscript contains such headings as Chapter 1: General principles, § 1 The fundamental postulates, § 2 The general properties of stationary states, § 3 The state relations of multiple periodic systems, and § 4 The radiation problem in the quantum theory. As follows from the introduction on carbon copies, this was intended as the first of a series of papers to be published by the Royal Danish Academy.

37 *Series spectra and atomic constitution* [1921–1922]

Part of manuscript, typewritten, 2 pp., En, Mf 10

Constitutes the beginning to, or the beginning to a section of, an unpublished manuscript.

38 *Bemerkungen über den Atombau* 1922

Manuscript, typewritten, 20 pp., Ge, Mf 10

This seems to be a paper promised in a footnote to "Der Bau der Atome und die physikalischen und chemischen Eigenschaften der Elemente" (Z. Phys. **9** (1921) 1).

39 *Lectures London and Cambridge* 6–22 March 1922

Outlines, handwritten, 9 pp., En, Da, Mf 10

3 [419–446] 40 *Guthrie lecture* 24 March 1922

Drafts and parts of manuscripts, handwritten, typewritten, 129 pp., En, Da, Mf 10

One manuscript is almost identical with the version published (The effect of electric and magnetic fields on spectral lines, Proc. Phys. Soc. London 1923, p. 275). Another manuscript has undergone greater alterations.

4 [341–419] T 41 *Die Theorie des Atombaus* 12–22 June 1922

Manuscript, typewritten, handwritten, 119 pp., 13 figs, Ge, Mf 10

Manuscripts and notes to seven lectures on the theory of atomic structure delivered in Göttingen in June 1922.

4 [421–424] T 42 *Forklaringen af det periodiske system* 24 Aug. 1922

Outline of lecture, part of manuscript, notes, 16 pp., Da, Mf 10

On the explanation of the periodic system, lecture delivered in Uppsala, Sweden, before the Second Nordic Physics Conference.

43 *Kvanteteoriens anvendelse* 1922

Part of manuscript, typewritten, 5 pp., Da, Ge, Mf 10

The first few pages of the paper "Über die Anwendung der Quantentheorie auf den Atombau. I. Die Grundpostulate der Quantentheorie". Z. Phys. **13** (1923) 117. (On the application of the quantum theory to atomic structure. I. The fundamental postulates of the quantum theory, Cambr. Univ. Press, 1924.)

44 *Ukomplet manuskript* 1922

Part of manuscript, handwritten, 16 pp., Da, Mf 10

Incomplete manuscript on quantum theory.

45 *Veröffentlichte Werke* 1922

Sheet, handwritten, 1 p., Ge, En, Mf 10

List of the most important of Bohr's papers through 1922.

4 [26–27] T 46 *Tale ved tildelingen af Nobelprisen* 30 December 1922

Manuscript, typewritten, 2 pp., Da, Mf 10

Speech delivered at banquet in Stockholm on the occasion of the award of the Nobel Prize.

47 *Forskellige optegnelser* 1923

Sheets, handwritten, 2 pp., Da, Mf 11

Various notes written by N. Bohr.

48 *Spectra of isotopes* 1923

Sheet, typewritten, 1 p., En, Mf 11

Refers to V. Thorsen, Naturwiss. **11** (1923) 78; "The difference between series spectra of isotopes", Nature **109** (1922) 746.

49 *Answer to Campbell* 1923

Draft, typewritten, 6 pp., En, Mf 11

Unpublished answer to an article by Norman Campbell "A static or dynamic atom", Nature **111** (1923) 569.

50 *Answer to Wood and Ellett* 1923

Draft, typewritten, 2 pp., En, Mf 11

Unpublished draft of an answer to an article by Wood and Ellett (Nature **111** (1923) 255).

51 *Siegbahn on Hafnium* 1923

Sheet, typewritten, 1 p., Swedish, Mf 11

From a speech by Siegbahn in Gothenburg referring to Dauvillier's claim concerning element 72.

52 *Atomteoretiske problemer* 1923

Part of manuscript, typewritten, 9 pp., Da, Mf 11

Brief discussion of certain problems of the theory of atoms.

53 §5. *Die Gruppeneinteilung der Elektronen im Atom* 1923

Part of drafts, typewritten, handwritten, 7 pp., Ge, Da, Mf 11

Consists of two parts and an outline of the arrangement of electrons in groups within the atoms.

54 *Komplexstruktur-Linien* 1923

Part of draft, handwritten, 8 pp., Ge, Mf 11

Deals with the Zeeman effect of line with complex structure.

55 *Table of doublet systems* [1923?]

Sheets, handwritten, 3 pp., En, Mf 11

Contains two tables of doublet systems of the metals in the three first groups and a table of singlet and triplet systems of metals in the second group.

56 *Atomic structure* (*Nature proof*)

Proof, 1 p., En, Mf 11

A discussion of Bohr's scheme of the quantum groups (outlined in Nature of July 7, 1923) by J. D. Main Smith, and an answer by Bohr. Neither part was ever published.

3 [532–558] T 57 *Anwendung der Quantentheorie auf den Atombau II* 1923
Manuscript, typewritten, 43 pp., Ge, Mf 11

On the application of the quantum theory to atomic structure II. Theory of series spectra. This was to be the second of a series of papers in Z. Phys. Because of the difficulty of explaining the anomalous Zeeman effect and the complex structure, it was never published.

58 *Toronto lectures* Oct. 1923
Outlines, handwritten, 5 pp., En, Mf 11

Outline of three lectures delivered at the University of Toronto 1–3 October 1923.

59 *Amherst lectures* Oct. 1923
Outlines, handwritten, 5 pp., En, Mf 11

Outline of five lectures delivered at Amherst University in October 1923.

3 [578–579] 60 *From the Liverpool meeting of the British Association* 22 Oct. 1923
Draft of manuscript, 5 pp., En, Mf 11

Letter to Prof. Rankine, enclosing an abstract of Bohr's remarks in the discussion of the origin of spectra. A published version is given in Nature **113** (1924) 223.

61 *Harvard lectures* 26 Oct. 1923
Outline, handwritten, 1 p., En, Mf 11

Outline of the second of two lectures delivered at Harvard University on October 26, 1923.

3 [45–46] 62 *Science Club lecture. On certain philosophical aspects of modern atomic theory* 29 Oct. 1923
Outline, handwritten, 1 p., En, Mf 11

63 *Columbia lecture* (slides) 19 Nov. 1923
Outline, handwritten, 3 pp., En, Mf 11

Outline of a single lecture at Columbia University. The slides are not extant.

64 *Yale lectures* Autumn 1923
Outlines, handwritten, 3 pp., En, Mf 11

Apparently outlines of nine lectures delivered at Yale University in October or November 1923. (Six of these lectures may have been the Silliman lectures.)

3 [582–601] 65 *Silliman lectures* [Autumn 1923]

Outlines, handwritten, typewritten, 44 pp., En, Mf 11

Outlines of six Silliman lectures delivered at Yale University in October or November 1923.

66 *On the selective reflection of X-rays* 1924

Manuscript, typewritten, 6 pp., En, Mf 11

Paper sent to P. Foote on February 6, 1924, for publication in Journ. of Opt. Soc. of Am. Cancelled May 6, because of results on the phenomenon in question obtained by W. Duane.

3 [458–500] 67 *On the application of the quantum theory to atomic structure* 1924

Manuscript, typewritten, 69 pp., En, Mf 11

This is a translation by L. F. Curtiss of Bohr's paper in Z. Phys. **13** (1923) 117. This translation was published in Proc. Camb. Phil. Soc., Suppl., p. 1 (1924).

INDEX*

* Frequently occurring topics, or commonly used terms, such as Planck's constant, Rydberg constant, and Stark effect, are generally not listed.